First International Conference on Construction, Infrastructure, and Materials (ICCIM 2019)

IOP Conference Series: Materials Science and Engineering
Volume 650

Jakarta, Indonesia
16 - 17 July 2019

Editors:

Joewono Prasetijo
Han Ay Lie
Tavio
Stefanus Adi Kristiawan

ISBN: 978-1-5108-9847-9
ISSN: 1757-8981

Printed from e-media with permission by:

Curran Associates, Inc.
57 Morehouse Lane
Red Hook, NY 12571

Some format issues inherent in the e-media version may also appear in this print version.

Institute of Physics
Dirac House, Temple Back
Bristol BS1 6BE UK

Phone: 44 1 17 929 7481
Fax: 44 1 17 920 0979

techtracking@iop.org

Additional copies of this publication are available from:

Curran Associates, Inc.
57 Morehouse Lane
Red Hook, NY 12571 USA
Phone: 845-758-0400
Fax: 845-758-2633
Email: curran@proceedings.com
Web: www.proceedings.com

Table of contents

Volume 650

First International Conference of Construction, Infrastructure, and Materials

16–17 July 2019, Jakarta, Indonesia

Accepted papers received: 16 September 2019
Published online: 29 October 2019

Preface

Papers

Construction Management & Infrastructure Development

Geotechnical Engineering

Structural Engineering & Construction Material

IOP Conf. Series: Materials Science and Engineering **650** (2019) 011001 doi:10.1088/1757-899X/650/1/011001

The First International Conference of Construction, Infrastructure, and Materials (ICCIM 2019)

Preface

The rapid advance of technology brings tremendous change and its impacts are upon every nation in the world. The fourth industrial industry will profoundly alter how things work and Civil Engineering too will advance from the current state. Stakeholders of Civil Engineering such as academicians, researchers, engineers, and policy makers must be ready to adapt and to progress in different mindsets of societies, different engineering method, different demand of infrastructures and different policies.

Therefore, Civil Engineering Undergraduate Program of Universitas Tarumanagara will hold The First International Conference of Construction, Infrastructure, and Materials (ICCIM) to provide the international platform for professionals of their expertise to share the information about the state of art of Civil Engineering. By having better understanding of the current state and by establishing firm collaboration with each other, Civil Engineering stakeholders will be more prepared to thrive in perpetually progressing civilization.

The theme of The First International Conference of Construction, Infrastructure, and Materials is Towards the Future of Civil Engineering Roles to Progressive Infrastructures Worldwide. The prospective participants are welcomed to submit papers with the following subjects :

1. Structural Engineering;
2. Construction Materials;
3. Geotechnical Engineering;
4. Transportation System and Engineering;
5. Construction Management;
6. Water Resources Engineering; and
7. Infrastructure Development.

We thank Prof. Dr. Leksmono Suryo Putranto of Universitas Tarumanagara, Indonesia; Prof. Dr. Bonaventura Hadikusumo of Asian Institute of Technology, Thailand; Dr. S. K. Ghosh of S. K. Ghosh Associates, USA; and Prof. Dr. Donguk Choi of Hankyong National University, South Korea for accepting our invitation to become the keynote speakers of this conference. We also thank all authors, reviewers, organizers, and sponsors for their great contribution.

On behalf of ICCIM 2019, we wish you all the best. Thank you.

Prof. Dr. Chaidir Anwar Makarim

IOP Conf. Series: Materials Science and Engineering **650** (2019) 011001 doi:10.1088/1757-899X/650/1/011001

ICCIM 2019 Logo

IOP Conf. Series: Materials Science and Engineering **650** (2019) 011001 doi:10.1088/1757-899X/650/1/011001

Sponsors Logo

IOP Conf. Series: Materials Science and Engineering **650** (2019) 011001 doi:10.1088/1757-899X/650/1/011001

List of Committees

Steering Committee

1. Prof. Dr. Edward G. Nawy (The State University of New Jersey, USA)
2. Dr. Widodo Kushartomo (Universitas Tarumanagara, Indonesia)
3. Prof. Dr. Tavio (Institut Teknologi Sepuluh Nopember, Indonesia)
4. Dr. Hugeng (Universitas Tarumanagara)

Editorial Board

1. Dr. Joewono Prasetijo (Universiti Tun Hussein Onn Malaysia, Malaysia)
2. Prof. Dr. Han Ay Lie (Universitas Diponegoro, Indonesia)
3. Prof. Dr. Tavio (Institut Teknologi Sepuluh Nopember, Indonesia)
4. Prof. Dr. Stefanus Adi Kristiawan (Universitas Sebelas Maret, Indonesia)

Technical Advisory Committee

1. Prof. Dr. Leksmono Suryo Putranto (Universitas Tarumanagara, Indonesia)
2. Prof. Dr. Fwa Tien Fang (Singapore National University, Singapore)
3. Prof. Dr. Yusak Susilo (KTH Royal Institute of Technology, Sweden)
4. Dr. Kardi Teknomo (Ateneo de Manila University, Philippines)
5. Dr. Sittha Jaensirisak (Ubon Ratchathani University, Thailand)
6. Dr. Ria Asih Aryani Soemitro (Institut Teknologi Sepuluh Nopember, Indonesia)
7. Dr. Priyantha Wedagama (Universitas Udayana, Indonesia)
8. Dr. Joewono Prasetijo (Universiti Tun Hussein Onn Malaysia, Malaysia)
9. Dr. Muslich Hartadi Sutanto (Universiti Teknologi Petronas, Malaysia).
10. Prof. Dr. I Nyoman Arya Thanaya (Universitas Udayana, Indonesia)
11. Dr. Muhammad Isran Ramli (Universitas Hassanuddin, Indonesia)
12. Dr. Bagus Hario Setiadji (Universitas Diponegoro, Indonesia)
13. Prof. Dr. Ludfi Djakfar (Universitas Brawijaya, Indonesia)
14. Prof. Dr. Sutanto Soehodho (Universitas Indonesia, Indonesia)
15. Dr. Yosritzal (Universitas Andalas, Indonesia)
16. Prof. Dr. Ahmad Munawar (Universitas Gadjah Mada, Indonesia)
17. Dr. Sony Sulaksono Wibowo (Institut Teknologi Bandung, Indonesia)
18. Dr. Imam Muthohar (Universitas Gadjah Mada, Indonesia)
19. Dr. Hera Widyastuti (Institut Teknologi Sepuluh Nopember, Indonesia)
20. Dr. Achmad Wicaksono (Universitas Brawijaya, Indonesia)
21. Prof. Dr. Ade Sjafruddin (Institut Teknologi Bandung, Indonesia)
22. Dr. Eng. M. Zudhy Irawan (Universitas Gadjah Mada, Indonesia)
23. Dr. Resdiansyah (Politeknik Enjinering Indorama, Indonesia)
24. Dr. Harun Al Rasjid Lubis (Institut Teknologi Bandung, Indonesia)
25. Dr. Sofyan M. Saleh (Universitas Syiah Kuala, Indonesia)
26. Dr. Medis Surbakti (Universitas Sumatera Utara, Indonesia)
27. Dr. Bayu Martanto Adji (Universitas Andalas, Indonesia)
28. Dr. Yossyafra (Universitas Andalas, Indonesia)
29. Dr. Caroline A. Sutandi (Universitas Parahyangan, Indonesia)

IOP Conf. Series: Materials Science and Engineering **650** (2019) 011001 doi:10.1088/1757-899X/650/1/011001

30. Dr. Bambang Sugeng Subagio (Institut Teknologi Bandung, Indonesia)
31. Dr. Hitapriya Suprayitno (Institut Teknologi Sepuluh Nopember, Indonesia)
32. Dr. Dimas Bayu Endrayana Dharmowijoyo (Universiti Teknologi Petronas, Malaysia)
33. Dr. Noor Mahmudah (Universitas Muhammadiyah Yogyakarta, Indonesia)
34. Dr. M. Asad Abdurrahman (Unversitas Hassanuddin, Indonesia)
35. Prof. Dr. Erika Bukhari (Universitas Sriwijaya, Indonesia)
36. Dr. Dewanti Marsoyo (Universitas Gadjah Mada, Indonesia)
37. Prof. Dr. Djwantoro Hardjito (Universitas Kristen Petra Surabaya, Indonesia)
38. Prof. Dr. Yoyong Arfiadi (Universitas Atma Jaya Yogyakarta, Indonesia)
39. Dr. Januarti Jaya Ekaputri, (Institut Teknologi Sepuluh Nopember, Indonesia)
40. Prof. Dr. Tavio (Institut Teknologi Sepuluh Nopember, Indonesia)
41. Prof. Dr. Han Ay Lie (Universitas Diponegoro, Indonesia)
42. Prof. Dr. Stefanus Adi Kristiawan (Universitas Sebelas Maret, Indonesia)
43. Prof. Dr. Roesdiman Soegiarso (Universitas Tarumanagara, Indonesia)
44. Prof. Dr. Antonius (Universitas Islam Sultan Agung, Indonesia)
45. Prof. Dr. Agus Setyo Muntohar (Universitas Muhammadiyah Yogyakarta, Indonesia)
46. Prof. Dr. Widjojo Adi Prakoso (Universitas Indonesia, Indonesia)
47. Prof. Dr. Krisna Mochtar (Institut Teknologi Indonesia, Indonesia)
48. Prof. Dr. Yusuf Latief (Universitas Indonesia, Indonesia)
49. Dr. Basuki Anondho (Universitas Tarumanagara, Indonesia)
50. Dr. Iwan B. Santoso (Universitas Tarumanagara, Indonesia)
51. Ayomi Dita Rarasati, S.T., M.T., Ph.D. (Universitas Indonesia, Indonesia)
52. Leni Sagita Riantini, S.T., M.T., Ph.D. (Universitas Indonesia, Indonesia)
53. Prof. Dr. Iwan Kridasantausa Hadihardaja (Institut Teknologi Bandung, Indonesia)
54. Prof. Dr. Indratmo Soekarno (Institut Teknologi Bandung, Indonesia)
55. Dr. Wati A Pranoto (Universitas Tarumanagara, Indonesia)

IOP Conf. Series: Materials Science and Engineering **650** (2019) 011001 doi:10.1088/1757-899X/650/1/011001

Photographs

IOP Conf. Series: Materials Science and Engineering **650** (2019) 011001 doi:10.1088/1757-899X/650/1/011001

ICCIM 2019
Jakarta, Indonesia

IOP Conf. Series: Materials Science and Engineering **650** (2019) 011002 doi:10.1088/1757-899X/650/1/011002

Peer review statement

All papers published in this volume of *IOP Conference Series: Materials Science and Engineering* have been peer reviewed through processes administered by the proceedings Editors. Reviews were conducted by expert referees to the professional and scientific standards expected of a proceedings journal published by IOP Publishing.

IOP Conf. Series: Materials Science and Engineering **650** (2019) 012001 doi:10.1088/1757-899X/650/1/012001

Indonesian contractor technological learning mechanism and its considerations

B Kusuma[1*], B W. Soemardi[1], K S Pribadi[1] and S Yuliar[1]

[1]Department of Civil Engineering, Institut Teknologi Bandung, Jln. Ganesha No. 10, Bandung, Indonesia

[*] bayukusuma@students.itb.ac.id

Abstract. The construction industry faces considerable challenges both in terms of rising construction capitalization and increasingly stringent consumer demand for project quality and costs. This challenge needs to be addressed appropriately by the contractor as an important actor who is directly in contact with the construction production process. Through mastery of technology, contractors are expected to overcome these challenges. The process of mastering technology is part of the contractor's effort in improving its technological capabilities through technological learning. This study aims to obtain the latest information related to the mechanism carried out by contractors in the process of mastering technology and the considerations taken. This research was conducted through a survey on contractors whom operating in Indonesia. Respondents were given questions to choose the mechanism they used from 32 mechanisms that had been identified by previous researchers, which were grouped into 3 groups, import, internal, and collaboration. The results showed a difference between the contractor groups (all contractor, large contractor, and medium contractor). The mechanism that is commonly used is by introducing a formal Quality Management system; documenting and analyzing all data and information related to project acquisition and project implementation; and participating in technology exhibitions. The results of this study will become worthwhile information for the government in formulating policy to improve the performance of contractors and as a further part in the effort to map the pattern of technological capability development carried out by Indonesian contractors.

1. Introduction

The Indonesian construction industry has not been able to provide reliable support from the national economy [1]. The dependence of the Indonesian construction industry on foreign steel and the limited availability of workers with higher education, [2]. The construction industry is opaque, fragmented, and full of inconsistent incentives so that it is often where the unproductive player can succeed [3]. The condition of the construction industry has resulted in operational failures within the company, including inefficient, inadequate design time to apply the latest thinking about project management and implementation, low-skilled workforce, and lack of investment in technology and digitalization that will help increase productivity [3].

Actors in the construction sector consist of at least service providers (consultants and contractors), service users (public and private), regulators, and communities with their respective supply chains [4].

The contractor, according to the level of complexity of the work, the amount of funds used, the amount of labor involved, and the multiplier effects for both the upstream and downstream industries, has a strategic position in the construction sector [5]. Contractors in the business in the construction sector require equipment with adequate types and quantities, experience in implementing high-performance projects, access to finance and foreign exchange [6].

As Indonesia's construction capitalization is increasing, the government's desire to accelerate infrastructure provision, there are at least two situations faced by construction in Indonesia, the construction products obtained must be timely, cost and quality; Indonesian contractors must be able to compete so that they become host in their own country. This situation indicates that the ability of Indonesian contractors needs to develop, especially the ability of technology to deal with existing dynamics. This shows that there is a need to explore deeply the conditions for developing national contractors' technological capabilities in Indonesia.

Knowledge and understanding of the various mechanism in contractor technological learning mechanism is needed to determine the focus of the necessary steps to improve construction business environment thereby increasing construction industry performance.

The purpose of this study is to identify what considerations and mechanisms are the main choices of contractors in developing their technological capabilities. The result will be useful information to improve Indonesia's construction productivity and competitiveness.

2. Contractor Technological Learning

One way to develop the construction industry is to improve the performance of construction companies, focusing on contractors and technology development, [7]. Therefore, it is important to know: 1) how the contractor's role in the dynamics of the construction industry is, 2) how big is the contribution of contractors in Indonesia to the dynamics of the construction industry in supporting national economic activities. As an illustration, Figure 1 is presented to illustrate the contractor's business pattern.

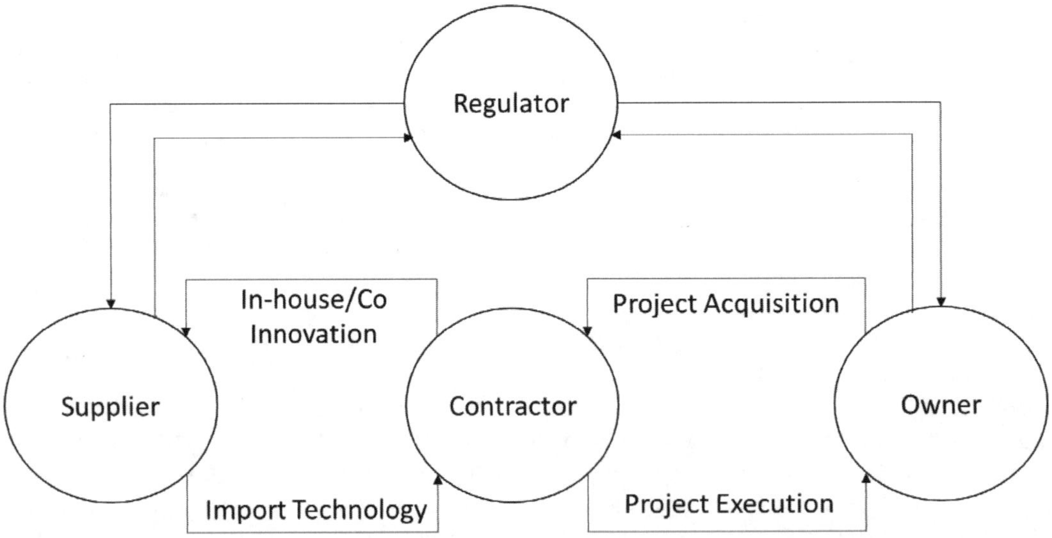

Figure 1. Contractor Business Pattern

The issue faced by contractors is competitiveness and productivity. Both issues greatly influenced the focus of the contractor in winning the competition to get the project and complete the project successfully. [8] suggests intensifying the use of technology to overcome productivity problems. [9] state that in the long run, competitiveness comes from the ability to build, at a lower cost and faster than competitors, the core competencies that spawn anticipated products. As revealed by [10] and [11] companies need to prepare conditions for internal capabilities before determining corrective measures to address changes in the business environment.

[12] state that technological capability is one of the factors that enables a country and its company to perform several functions (mainly innovation) that are important for economic development and international competitiveness. Efforts made by the Indonesian government to accelerate infrastructure provision are carried out through regulations on land acquisition, technological innovation, financing, leadership, and coordination between institutions [13].

To achieve a level of technological capability, a company can do so through the company's internal learning mechanism and by utilizing external learning mechanisms, for example through collaboration with government research laboratories or by networking with other companies [14]. Based on previous studies, mechanism that be used by contractor for technological learning have been obtained from works by [14], [15], [16], [17], [18], [19].

For construction actors in this case contractors, developing a learning mechanism system is generally seen as a technical process [20]. Technological learning strategies play a decisive role in guiding and monitoring the application of firm-level technological capability patterns effectively [16], [12]. [21] state that technological capabilities will affect the performance of companies, as well as [16] outline the relationship between learning technology, technological capabilities, and company performance.

Increased understanding of construction technology will help identify opportunities for improvement and help companies seek competitive advantage based on advanced technology [22]. The process of introducing new technology to the construction industry is quite slow when compared to other industries, especially in the field of field process automation using industrial robotics [23]. Companies can improve their technological knowledge and strengthen their technological capabilities through the process of searching and using external technology, [24].

[25] state that, industries in developing countries generally do not have the ability to compete with the latest technology. Important points in terms of technological capabilities, the absorption of higher technology in developing countries cannot be taken for granted. [26] describes at least three barriers experienced by latecomer companies in building technological capabilities namely, not having access to information and knowledge about the latest technological developments, lack of market demand for advanced technology, and research and education infrastructure often not providing companies with graduates trained, and information and knowledge about the latest technological developments.

[27] describes that there are enough people who have conducted studies on technology learning at the international level including the learning process, learning determinants, and modeling of technological learning. However, for Indonesia, studies on this subject are still limited and lead more to interactions in technology learning, technology learning models, and technological learning processes.

[28] said that contractors in carrying out construction projects use a combination of tacit and explicit technology including plant construction and equipment, project engineering, construction and management processes, and intuitive ideas that are included in project design, and managing processes construction. This shows that the contractor can conduct technology learning through the experience of implementing construction projects. [29] describe that through construction projects, individuals can learn through five approaches, namely learning through individual networking; learning through organizing; learning through experimenting; learning through reading; and learning through attending courses and seminars.

For contractors, individual learning needs to be facilitated so that it becomes corporate learning. To facilitate the process, as presented by [17], contractors can perform several mechanisms both at the locus of learning in the project and in the organization through pre-learning system processes, experience accumulation, knowledge articulation, codification / storage, dissemination / distribution.

Based on past research, in conducting technological learning, there are several aspects that need to be examined, namely:

- Factors that influence the contractor so that technological learning needs to be done.
- Factors that hinder contractors in doing technological learning.
- Considerations underlying the contractor in choosing a technological learning mechanism.
- Learning mechanism that is in accordance with the conditions of the contractor.
- Construction engineering elements that are expected to develop through technological learning.

The factors which were identified from research are used as a basis for preparing a questionnaire to investigate the considerations and mechanisms which are the main choices of contractors in developing their technological capabilities.

3. Research Method

3.1. Research methods

Data collection will be carried out by using questionnaire distributed to contractor that had qualification of medium and big in Indonesia. In this study, an ordinal measurement scale 1 to 5 was used to determine the effect level. Respondents were asked to rank factors related to how contractor deploy their technological learning according to the degree of importance.

In this study, the driving factors, obstacle factor, consideration factor, and learning mechanism have 5, 6, and 12 indicators, respectively. These indicators can be seen in Table 1. On the Driving factors, the 1-5 Likert scale indicating answer for "Very Low", "Low", "Average", "Strong", "Very Strong"; on Obstacle factors: "Yes" or "No"; on Consideration factors and on Learning Mechanism: "Never", "Rarely", "Occasionally", "Frequently", "Always"; and on construction engineering elements: "Not a Priority", "Low Priority", "Medium Priority", "High Priority", "Urgent Priority". For analyzing data by ordinal scale, a relative importance index (RII) was used by following equation (1):

$$\text{Relative Importance Index} = \frac{\sum_{i=1}^{5} WiXi}{\sum_{i=1}^{5} Xi} \quad (1)$$

Where:

W_i = the rating given to each factor by the respondent ranging from 1 to 5
X_i = the percentage of respondents scoring
i = the order number of respondents

The relative importance index (RII) for all factors was calculated. Meanwhile, the group index was calculated by taking the average of factors in each group [30]. The maximum value of the index is 5 when all respondent answered, "very high effect" and the minimum value of the index is 1 when all respondents answered, "affects with little degree". Since the results are obtained as decimal numbers instead of integer numbers, a specific scale should be established. Thus, 5 expressions are defined by the intervals of 0.80 to classify the effect level (see Figure 2).

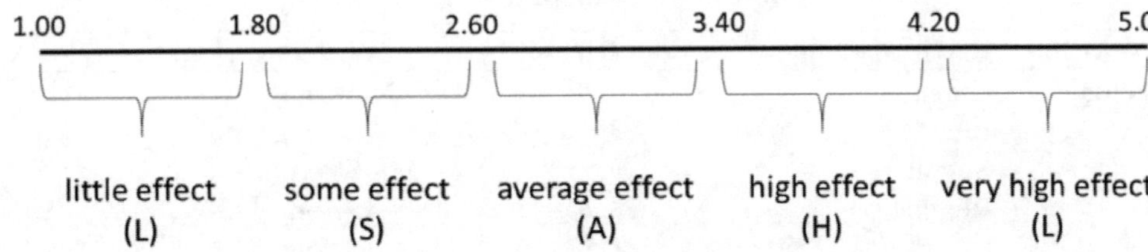

Figure 2. Evaluation scale.

3.2. Data analysis

As a result of surveying, mailing, and following up, a total of 100 questionnaire were returned and can be analyzed further from various district in Indonesia (see Table 1). A close personal contact with contractors is needed due to a lot of item in questionnaire to be filled. The respondents from large contractor (46%) and medium contractors (54%). The classification of large or small and medium contractors is based on valid regulations in Indonesia.

Table 1. Survey area and number of respondents.

No	Area of Survey	Number of respondents
1	West Java	44
2	Jakarta and surrounding area	37
3	Western Region	9
4	Eastern Region	10
		100

Meanwhile according to their working experience, the majority of respondents (70%) working experience more than 5 years, more over 39% have working experience more than 10 years. The experiences of the respondents include various projects from road ang bridges, water building and irrigation, and low-rise buildings to high rise buildings, respondents can choose more than one type of projects (see Table 2).

Table 2. Type of project.

Buildings < 8 Floors	Buildings > 8 Floors	Road & Bridges	Water Building & Irrigation	Industrial Building	Others	Number of respondents
31%	10%	33%	26%	7%	7%	100

4. Result and Discussion

Based on past researches, 32 learning mechanism have been identified and were groups into 3 groups according to their characteristics, namely: 1. Import (6 mechanism); 2. Internal (21 mechanism); 3. Collaboration (5 mechanism). The level of driving factor which influence contractor to deploy technological learning by each group of contractors is presented in Table 3. The first and second rank of the driving factor are same within all, large, and medium contractor, namely client and internal respectively. In all contractor the third rank of driving factor is regulation, in the large contractor is competitor, and in medium contractor is regulation. This is due to the fact that in the medium contractor most of the project acquisition processes do not compete with certain technological capabilities.

Result shows that the regulation is not the main factor that influence contractor to deploy technological learning in all group of contractors. This indicates that the prevailing laws and regulations in general have not been able to stimulate contractors to invest in developing their technological capabilities.

IOP Conf. Series: Materials Science and Engineering **650** (2019) 012001 doi:10.1088/1757-899X/650/1/012001

Table 3. The rank of driving factor.

Group Contractor	Driving Factor	RII	Effect	Rank
All contractor	Client	3.96	Strong	1
	Internal	3.89	Strong	2
	Supplier	3.53	Strong	5
	Regulation	3.70	Strong	3
	Competitor	3.58	Strong	4
Large	Client	4.09	Strong	1
	Internal	4.04	Strong	2
	Supplier	3.59	Strong	5
	Regulation	3.80	Strong	4
	Competitor	3.83	Strong	3
Medium	Client	3.85	Strong	1
	Internal	3.76	Strong	2
	Supplier	3.48	Strong	4
	Regulation	3.61	Strong	3
	Competitor	3.37	Average	5

The level of obstacle factor which hinder contractor to deploy technological learning by each group of contractors is presented in Table 4. In all contractor groups, the top three factors that hinder contractors from implementing technological learning are budget, lack of staff to support technology adoption, and lack of knowledge about what new technology is available. This result shows that most contractors have not been able to set aside income from the projects they have acquired for the development of the company, especially the development of their technological capabilities. For intermediate contractors, it seems they lack employees who are competent in conducting technological learning. In large contractors, it seems they lack knowledge of the latest technological developments that are in line with their business scope.

Table 4. The rank of obstacle factor.

Group Contractor	Obstacle Contactor	Number of Contractor	Rank
All contractor	Budget	59%	1
	Lack of staff to support technology adoption	50%	2
	Lack of knowledge about what new technology is available	47%	3
Large	Budget	43%	1
	Lack of staff to support technology adoption	30%	3
	Lack of knowledge about what new technology is available	39%	2
Medium	Budget	37%	1
	Lack of staff to support technology adoption	37%	1
	Lack of knowledge about what new technology is available	26%	3

The level of learning mechanism considerations used by contractors in determining what technological learning mechanisms are appropriate to the conditions of the company is presented in Table 5. The main consideration factor for all of the contractor groups is market/client/project needs.

IOP Conf. Series: Materials Science and Engineering **650** (2019) 012001 doi:10.1088/1757-899X/650/1/012001

Table 5. The rank of learning mechanism considerations.

Group Contractor	Consideration Factor	RII	Effect	Rank
All contractor	Market / client / project needs	3.82	Frequently	1
	Based on with the development of the contractor's current needs	3.78	Frequently	2
	Beneficial to both parties	3.77	Frequently	3
Large	Market / client / project needs	4.15	Frequently	1
	Based on with the development of the contractor's current needs	4.13	Frequently	2
	Beneficial to both parties	4.09	Frequently	3
Medium	Market / client / project needs	3.56	Frequently	1
	Based on with the development of the contractor's current needs	3.52	Frequently	2
	Beneficial to both parties	3.52	Frequently	2

The level of use by contractors to each group is presented in Table 6. In all contractor the level of use by all contractor in chronological order are: import, internal, and collaboration. While in medium contractor reflects the same order as in all contractor, the large contractor use internal as its main learning mechanism. The results indicate that medium contractors have not carried out continuous technological learning. The top three of learning mechanism that frequently used by contractor to develop their technological capability are introducing a formal Quality Management system; documenting and analyzing all data and information related to project acquisition and project implementation; and participating in technology exhibitions.

Table 6. The rank of group mechanism of all companies.

Group Contractor	Group mechanism	RII	Effect	Rank
All contractor	Import	3.43	Frequently	1
	Internal	3.29	Occasionally	2
	Collaboration	2.92	Occasionally	3
Large	Import	3.58	Frequently	2
	Internal	3.69	Frequently	1
	Collaboration	3.29	Occasionally	3
Medium	Import	3.31	Occasionally	1
	Internal	2.95	Occasionally	2
	Collaboration	2.61	Occasionally	3

The priority level of the construction engineering elements that are expected to develop from the implementation of technological learning is presented in Table 7. As expected, the main construction engineering element by which hopefully to be improve as an output from the deployment of technological learning is construction methods which is reflected in all group of contractors. While in large contractor the second priority is the development of construction equipment, in medium contractor the development of principles of construction engineering is the second priority.

Table 7. The rank of construction engineering element.

Group Contractor	Construction Engineering Element	RII	Effect	Rank
All contractor	Principles of Construction Engineering	3.66	High Priority	2
	Construction material	3.58	High Priority	4
	Construction Equipment	3.62	High Priority	3
	Construction Methods	3.92	High Priority	1
Large	Principles of Construction Engineering	3.83	High Priority	3
	Construction material	3.70	High Priority	4
	Construction Equipment	3.87	High Priority	2
	Construction Methods	4.13	High Priority	1
Medium	Principles of Construction Engineering	3.52	High Priority	2
	Construction material	3.48	High Priority	3
	Construction Equipment	3.41	High Priority	4
	Construction Methods	3.74	High Priority	1

5. Conclusion

Contractors carry out technological learning mainly because of client demands and internal needs. Constraints that generally apply to contractors in implementing this policy are related to budgets, employee capabilities, and proper knowledge of the latest technology. In addition to stimulation of influence factors, contractors also have certain considerations in determining what mechanisms will be used in implementing technology learning. The main consideration is what technology can meet market / user / project needs. This is in line with the findings of the influence factor. The technology learning mechanism used by contractors is generally more about learning from parties outside the contractor. Although all elements have a high priority level, especially construction methods, this shows that in general the contractor has not been able to distinctly determine what elements should be prioritized.

Acknowledgements

The authors would like to greatly appreciate the willingness of all respondent from various companies/institutions who took their valuable time to answer the survey that made this research possible.

References

[1] Budiwibowo A, Trigunarsyah B, Abidin IS and Soeparto HG 2009 Competitiveness of the Indonesian Construction Industry *Journal of Construction in Developing Countries* **14(1)** pp 51-68

[2] McKinsey Global Institute 2012 *The Archipelago Economy: Unleashing Indonesia's Potential McKinsey & Company*

[3] McKinsey Global Institute. Reinventing Construction 2017 *A Route to Higher Productivity McKinsey & Company*

[4] Osei V 2013 The Construction Industry And Its Linkages To The Ghanaian Economy-Polices To Improve The Sector's Performance International *Journal of Development and Economic Sustainability* **1(1)** pp 56-72

[5] Julison B, Wardani S P R and Wibowo M A 2017 Innovation Performance of Large Contractor in Indonesia: influencing factors and its impact on firm'sperformance *Procedia Engineering* **171** pp 370-378

[6] Danunagoro S 1976 Pembangunan Nasional Ditinjau Dari Industri Konstruksi Jakarta

[7] Ofori G 2015 Nature of the Construction Industry, Its Needs and Its Development: A Review of Four Decades of Research *Journal of Construction in Developing Countries* **20(2)** pp 115– 135

[8] The Boston Consulting Group 2015 *ECS Value Creators Report: Opportunities Amid Uncertainty Boston*

[9] Prahalad C K and Hamel G 1990 *The Core Competence of The Corporation Harvard Business Review* ***May-June*** pp 79-90

[10] The Boston Consulting Group 2016 *ECS Value Creators Report: Building Endurance* Boston

[11] KPMG 2016 *Building Technology Advantage*

[12] Sobanke V, Adegbite S, Ilori M and Egbetokun A 2014 Determinants of Technological Capability of Firms in a Developing Country *Procedia Engineering* **69 p**p 991 – 1000

[13] Hadimuljono M B 2017 *Terobosan Dalam Pembangunan Infrastruktur Untuk Mengejar Ketertinggalan* Bandung

[14] Kumar V, Kumar U and Persaud A 1999 Building Technological Capability Through Importing Technology: The Case of Indonesian Manufacturing Industry *Journal of Technology Transfer* **24 p**p 81-96

[15] Lall S 1992 Technological Capabilities and Industrialization *World Development* **20(2)** pp 165-186

[16] Liu J J, Qian J Y and Chen J 2006 Technological learning and firm-level technological capability building: analytical framework and evidence from Chinese manufacturing firms *Int. J. Technology Management* **36(1/2/3)** pp 190-208

[17] Mahdiputra S, Giddings B, Hogg K and Daws M 2005 Organisational Learning in construction companies in *Proc. the Fifth International Postgraduate Research Conference* **2** pp 696-711

[18] Marcelle G M 2004 Technological Learning: A Strategic Imperative for Firms in the Developing World *Massachusetts, USA: Edward Elgar Publishing*

[19] Ofori G 1992 Construction Technology Transfer: Issues and Options in *Canada: CIB 92 Conference Montreal*

[20] Sekarsari J 2005 Knowledge Management Employed by Firms: A Study of The Indonesian Construction Contractors In: *Proceedings of Seminar 25 Tahun MRK ITB.*

[21] Egmond E L C V and Smook R A F 2001 The Technology Mapping Methodology For Benchmarking The Cons0truction Performance *CIB World Building Congress* pp 1-12

[22] Tatum C B 1988 Classification System For Construction Technology *Journal of Construction Engineering and Management* **114(3)** pp 344-363

[23] Skibniewski M J and Chao L C 1992 Evaluation of Advanced Construction Technology with AHP Method *Journal of Construction Engineering and Management* **118(3)** pp 577-593

[24] Tsai K H and Wang J C 2008 External technology acquisition and firm performance: A longitudinal study *Journal of Business Venturing* **23** pp 91–112

[25] Putranto K, Stewart D, Moore G and Diatmoko R 2003 Implementing a technology strategy in developing countries: The experience of the Indonesian rolling stock industry *Technological Forecasting & Social Change* **70** pp 163-176

[26] Rousseva R 2008 Identifying Technological Capabilities with Different Degree of Coherence: The Challenge before Latecomer Companies to Enter the International Markets *Technological Forecasting and Social Change* **75(7)** pp 1007-1031

IOP Conf. Series: Materials Science and Engineering **650** (2019) 012001 doi:10.1088/1757-899X/650/1/012001

[27] Prihadyanti D 2015 Pembelajaran Teknologi di Perusahaan Manufaktur Berintensitas Teknologi Tinggi dan Menengah-Tinggi *Jurnal Manajemen Teknologi* **14(1)** pp 1-13

[28] Oti-Sarpong K and Leiringer R 2016 The Role Of Learning In Conctruction Technology Transfer: *A 'SCOT' Perspective Proceedings of the 32nd Annual ARCOM Conference*

[29] Wasif I, Josephson P E and Styhre A 2008 Individual learning in construction projects: professions and their approaches *The Australasian Journal of Construction Economics and Building* **8(2)** pp 50-60

[30] Soekiman A, Pribadi K S, Soemardi B W and Wirahadikusumah R D 2011 Factors Relating to Labor Productivity Affecting the Project Schedule Performance in Indonesia *Procedia Engineering* **14** pp 865-873

IOP Conf. Series: Materials Science and Engineering **650** (2019) 012002 doi:10.1088/1757-899X/650/1/012002

Strategic factor analysis of innovation adoption in Thai architects: A case study of Building Information Modeling

Grit Ngowtanasuwan

Construction Management Program, Faculty of Architecture, Urban Design, and Creative Arts, Mahasarakham University 44150 Thailand

grit_n@hotmail.com

Abstract. Inspiring competition and enhancing standards of living as the foundation of a progressive knowledge-based society, innovation has been acknowledged as having an indispensable influence on the economic development of nations for a long time. Currently being implemented in nations throughout the world for the design and application of design projects, building information modeling (BIM) involves computer technology in the structure and facility design industry. BIM's adaptive design technology is presently being distinguished and embraced by the Thai architectural design industry. Exploratory factor analysis (EFA) statistical approaches have been employed to examine, categorize and evaluate the factors associated with BIM adoption behavior in the industry. A total of 199 Thai architects with familiarity in the utilization of BIM were employed to accumulate the empirical data. Using EFA, factors were categorized into four major clusters including (1) adopter characteristics, (2) BIM characteristics, and (3) vendor characteristics, as well as (4) environmental characteristics. In order to articulate, exhibit, and deliberate an approach to develop BIM implementation by the Thai architectural design industry, the outcomes were applied.

1. Introduction

For a variety of research fields such as economics, business, engineering, science and sociology, the topic of innovation has gained increasing interest. In spite of this, similar words and concepts including change, invention, design and creativity have often been misunderstood to mean the same as 'innovation'. As precision by its very features restricts inventiveness, the characterization of innovation covers a wide range of conceptions. By completely altering the data input procedure from traditional coordination (x, y-axis) in computer-aided design (CAD) to 3-dimensional object orientation, resulting in significantly improved potential, recent enhancements in design technology have led to a new approach called 'Building Information Modelling' (BIM) [1]. For the execution of projects in numerous nations globally, BIM is computer technology being implemented in the building and facility design industry. The design technology of BIM is also being utilized by the Thai architectural design industry, though prevalent application throughout the nation remains somewhat limited by a lack of comprehension and awareness by architects. Therefore, the objective of this research is to study the factors related to BIM adoption in the Thai architects using exploratory factor analysis (EFA). This paper is an extension of from previous research [2], [3] that studied the factors related to BIM adoption behaviors of Thai architects and engineers (including civil, mechanical, electrical, sanitary engineers, etc.). In this research, Thai architects working in Thailand would be focused as samples. Researching and assessing the factors that detail the behaviour among Thai architects towards BIM adoption was the primary aim of this research study. Top managers, design managers, senior architects and architects should be able to utilize the results of this work to examine the present status of their planning. Also provided by the current work were strategies intended to reinforce the implementation of BIM by the Thai architectural design industry.

2. Research Background

As suggested by Troff [4], multiple factors relating to the effects of psychological or personal features, technology assessments, communication behaviour and socio-demographic qualities are concerned in the diffusion and adoption steps of innovation or technologies in civilization. Thus, consumer buyer behaviour comprises the study of how and why consumers buy certain goods and services. Diffusion is considered the process through which innovation is conveyed to participants of a social system using specific channels, as defined by Rogers [5]. A distinctive form of interaction involving the dissemination of messages distinguished as being new ideas can also be considered as diffusion, while any idea, practice, or object observed to be new by an entity or another unit of adoption is an innovation. Any form of innovation spreading into the Thai architectural industry is deemed to be BIM for the purposes of this study. Rogers [5] likewise offered that the diffusion of an innovation model may be characterized using five features that impact the rate of diffusion. The first feature is Relative Advantage, which is the level to which an innovation is thought to be superior to the idea that came before it by a certain group of users as determined by quantifiable measures including economic benefit, social standing, accessibility or gratification. The second feature is Compatibility, which is the level to which an innovation is thought to be in line with the standards, previous experiences and requirements of prospective users. An innovation that is well-suited with their values, norms or practices will be accepted and utilized more readily than one that is mismatched. The third feature is Complexity, which is the level to which an innovation is deemed to be problematic to figure out and use. The fourth feature is Trial Usability, which is the ease with which an innovation can be tested during a controlled period. Any innovation that is testable equates to decreased hesitation to a consumer who might be pondering it. The final feature is Observability. Consumers are more willing to adopt an innovation if it is easier for them to observe the results of its operation or function. Awareness means the state or ability to distinguish, feel or be aware of certain events, objects or sensory patterns, as defined by Wikipedia [6]. At this dimension of cognizance, sense information can be affirmed by an eyewitness without fundamentally suggesting comprehension. All the more comprehensively, mindfulness is the state or nature of monitoring something. In organic brain research, mindfulness is characterized as a human's or a creature's observation and intellectual response to a condition or occasion. The decision to fully employ an innovation as the best course of action available is the definition of adoption as found by Schiffman and Kanuk [7], while the decision not to adopt an innovation is defined as rejection. In the overall situation, adoption is viewed as an aspect of the overall procedure, whereby an entity or other decision-maker moves from the initial realization of an innovation to developing an opinion of that innovation and deciding whether or not adopt it, to the operation of the new innovation and verification of the decision. This is what is called the innovation-decision process. A conceptual framework specifying the factors manipulating the dissemination of new products was offered by Sorescu et al. [8]. They recognised four noteworthy gatherings of variables that influenced both the initial and subsequent purchases of a new item by consumers as adopter attributes, advancement qualities, firm qualities and condition qualities. Commonly employed to study the behavioural approval of technologies by people in society including smartphones and LED TVs, Davis [9] introduced a technology acceptance model (TAM) that became quite renowned by showing that the appearance of identified ease of use affected the recognised convenience of technology. The aspect of attitude towards using the technology is directly influenced by "perceived ease of use" as well as "perceived usefulness". Ultimately, the aspect of actual system use is influenced by the opinion towards using the technology. The factors concerning BIM adoption in design organisations by concentrating on architects' behavioural objectives in Korean firms were studied by Son et al. [10], which identified that management sponsorship, particular standards, compatibility and computer self- proficiency were significant factors influencing the architects' behavioural intentions to implement BIM. BIM design technology has been adopted by Thailand's architectural design industry. Still, a logjam for BIM adoption has been created by the present shortage of administrators with suitable knowledge and awareness. BIM execution and implementation has been examined by numerous academics [11]. A significant degree of acceptance into current system practices with obvious concentration and norms aimed at satisfying the requirements of the project as well as clients is needed for the implementation of BIM [12-13]. BIM acceptance by architectural firms in other nations such as India was slow and affected by expertise, trialability and management support, as identified by Ahuja et al. [14]. The primary obstacles to BIM use by engineering consulting firms in Malaysia included a lack of proficient personnel, supervision and governmental support [15], while the critical factors for BIM implementation by architects in China comprised enthusiasm, technical flaws in BIM and BIM competence [16]. Currently, Thai architectural design firms are utilizing BIM to execute design projects.

3. Research Method

3.1. Factor Identifications

For the purpose of gaining the unification of belief regarding actual knowledge lobbied from experts within particular fields, the Delphi technique is a commonly utilized and acknowledged approach [17]. Designed to carry out comprehensive examinations and discussions of a specific issue, this technique is intended as a group communication process among experts. Until such a time as agreement has been deemed to have been accomplished, the process can be repeated over and over again. Through identification of the factors, related theories and literature available including textbooks and research articles, this approach was employed to verify the aspects as well as the theoretical research model. Experts were employed to run through, inventory and substantiate all factors. Based on their particular experience, seven experts in correlated fields were selected as follows:

- Four experts from the Thai architectural design industry (senior architects) possessing more than 10 years of experience,
- Two BIM specialists from academic institutes with expertise in other enquiry procedures, and
- A marketing field professional from a private firm.

Through agreement of estimation among the seven specialists directly as well as independently beginning from specialists 1 to 7 and back to specialist 1, the listed factors were verified. Exclusive of having any contact with each other a minimum of three times for each expert until agreement was overwhelming, all specialists considered the listed factors cautiously and discretely. The listed factors were composed and prepared for analysis after they were corroborated and catalogued by the seven specialists. Table 1 provides a list of the preliminary factors and details.

Table 1. Initial factors and items.

Factor	Item
Adopter characteristics :AC	AC1: experience AC2: liking the new technology AC3: liking in using tools, devices, etc.
BIM characteristics :BC	BC1: relative advantage BC2: quality BC3: ease of use BC4: compatibility
Vendor characteristics :VC	VC1: vendor reputation VC2: vendor service VC3: trialability
Environment characteristics :EC	VC4: vendor marketing effort EC1: BIM adoption by colleagues EC2: BIM adoption by general architects and others EC3: pressure by boss or client EC4: BIM training

3.2. Questionnaire development

Data collection was carried out using a questionnaire survey in order to determine the principal causal factors leading to Thai architects adopting BIM. The questionnaire was subdivided into five factor categories with the items selected as listed in Table 1. The initial section covered demographic data including professional and organizational background of the respondents (3 items). The second section covered the typical characteristics of adopters (3 items). The third section covered the characteristics of BIM (4 items), while the fifth section covered characteristics of the vendors (4 items). The final section covered environmental characteristics (4 items). The demographic data were presented by frequency percentage, and the other sections asked respondents to answer using a 5-point Likert scale from 'strongly disagreement' to 'strongly agreement'.

3.3. Tests of reliability and validity

Validity and reliability tests were employed to ensure that the questionnaire items were suitable for their purpose. The validity testing involved interviews with seven experts to discuss the factors which were identified from Section 3.1. The expert reviewers confirmed that the items selected were appropriate for the measurement of the model, while also making further suggestions for items which might improve the model in this particular context. This is an import step in ensuring that the items are valid while avoiding confusion and ambiguity. To analyze the reliability of the items, Cronbach's alpha was used in assessing the responses in a pilot study which involved 30 participants, who were Thai architects, in order to match the study population. The fifteen questionnaire items measured by Likert scale (AC [3 items], BC [4 items], VC [4 items], and EC [4 items]) underwent analysis using statistical software. The overall Cronbach's alpha coefficient for the 15 items was 0.912, and it was observed that the result would be 0.935 without the presence of item AC1. Accordingly, AC1 (experience) was removed from the model. For each of AC, BC, VC, and EC categories, the Cronbach's alpha coefficients were 0.734, 0.872, 0.885, and 0.831 respectively. Since none of the coefficients fell below 0.7, it can be concluded that the questionnaire was reliable [18].

3.4. Data collection
Having completed the questionnaire designed, the next stage involved the selection of the sample group of BIM-experienced Thai architects using a convenience non-probability sampling approach. The respondents were predominantly based in Bangkok, with a minority coming from other large urban centers nearby. The survey was carried out over a three-month period. Interviews were first conducted face-to-face in order to ensure that all the respondents fully understood the purposes of the research. A total of 250 questionnaires were distributed, while 199 were accepted for the study and thus underwent further analysis.

4. Results

4.1. Descriptive results
The profiles of the 199 respondents are given in the form of the results shown in Table 2.

Table 2. Descriptive results.

Description	Frequency	Percentage
Major profession		
- Building architecture	178	89.5
- Urban design	7	3.5
- Interior	8	4.0
- Landscape	6	3.0
Organizational background		
- Private sector	135	67.8
- Government sector	30	15.1
- Own business/freelance	34	17.1

4.2. Exploratory factor analysis
Exploratory factor analysis (EFA) is a statistical technique which can be applied in order to determine the structure which underpins large sets of variables [19]. EFA can take these variables and eliminate those which are irrelevant, leaving only the summary factors. This study used EFA to establish the set of factors which influence the adoption of BIM among members of the Thai architectural community. The implementation of EFA was performed with varimax rotation which required statistical software from which the output presented the Kaiser-Meyer-Olkin (KMO) measure of sampling adequacy = .824 (KMO>0.7) [20]. Meanwhile, the value reported for Bartlett's test of sphericity was significant at = .000 (less than .05) with and approximate chi-square value = 1314.32, and df = 91. Any factor loading values below 0.5 were discarded. The extraction of EFA was conducted on the basis of initial eigenvalues in exceeding 1 [21] as shown in Table 3, while the output analysis displayed four components as the number of factors for this particular EFA. The percentage value for explained variance was found to be 69.9%.

Table 3. Total variance explained.

Component	Initial Eigenvalues			Extraction Sums of Squared Loadings			Rotation Sums of Squared Loadings		
	Total	% of Variance	Cumulative %	Total	% of Variance	Cumulative %	Total	% of Variance	Cumulative %
1	5.487	39.192	39.192	5.487	39.192	39.192	3.010	21.503	21.503
2	1.753	12.520	51.712	1.753	12.520	51.712	2.783	19.878	41.381
3	1.544	11.029	62.741	1.544	11.029	62.741	2.214	15.811	57.192
4	1.002	7.155	69.896	1.002	7.155	69.896	1.779	12.704	69.896
5	.788	5.631	75.527						
6	.621	4.435	79.962						
7	.597	4.263	84.225						
8	.469	3.352	87.577						
9	.360	2.572	90.149						
10	.344	2.458	92.607						
11	.318	2.270	94.877						
12	.270	1.931	96.808						
13	.228	1.632	98.440						
14	.218	1.560	100.000						

The output from EFA indicated that the 14 remaining items could be categorized into four groups, which were: factor 1 (BC1, BC2, BC3, BC4, and VC3); factor 2 (VC1, VC2, VC4, and EC4); factor 3 (EC1, EC2, and EC3), and factor 4 (AC2, and AC3). For these factors, the average values for Cronbach's alpha were 0.765, 0.792, 0.814, and 0.839 respectively, and given that Nunally [18] reports that a value greater than 0.7 can be deemed acceptable, these factors are acceptable, with the overall value for the four factors being 0.869.

Table 4. Factor loading and Cronbach's alpha.

Item	Factor 1	Factor 2	Factor 3	Factor 4	Cronbach's alpha	
BC2	.763	.146	.092	.291	0.839	0.869
BC1	.742	.266	.038	.066		
BC3	.730	.055	.152	.226		
VC3	.655	.480	.153	.123		
BC4	.578	.300	.345	.217		
VC2	.068	.850	.242	.118	0.814	
VC1	.222	.781	.047	.106		
VC4	.292	.775	.086	.182		
EC4	.478	.552	.070	-.268		
EC2	.054	.067	.862	.125	0.765	
EC1	.144	.029	.862	.071		
EC3	.181	.304	.686	-.125		
AC2	.160	.154	.094	.864	0.792	
AC3	.363	.084	.001	.814		

Table 4 presents the results of EFA for the causal factors identified for the adoption of BIM by Thai architects, separated into four principal factor groups. The four groups were names as Factor 1 (BIM characteristics); Factor 2 (vendor characteristics); Factor 3 (environment characteristics), and Factor 4 (adopter characteristics).

5. Conclusions

This study conducted factor analysis in the context of BIM adoption among Thai architects, using a total of 199 sample questionnaires completed by respondents who were professionals in the Thai architectural industry and had experience in BIM use. The data were then analyzed in order to confirm the relevant factors. The demographic profiles of the participants are shown in Table 2. EFA was used to identify and confirm the relevant factors which influence the adoption of BIM. The four factors identified by EFA were: Factor 1 (characteristics of BIM including ease of use, quality, trialability, compatibility, and relative advantage); Factor 2 (characteristics of vendors including service, reputation, marketing, and BIM training); Factor 3 (characteristics of the environment including the BIM adoption by other architects, the adoption by colleagues, and the influence from management or customers), and Factor 4 (characteristics of typical adopters, including a liking for new technology and willingness to try new devices or tools). In order to make recommendations concerning the adoption of BIM in Thailand's architectural sector, Factors 1 and 2 from the factor loading values shown in Table 4 were chosen to develop the guidelines influencing each of the factors. The factors for BIM adoption by Thai architects are presented in Table 5.

Table 5. Factors of BIM adoption by Thai architects.

BIM Adoption	BIM characteristics	-	Quality
		-	Relative advantage
	Vendor characteristics	-	Vendor service
		-	Vendor reputation
	Environment characteristics	-	BIM adoption by general architects & others
		-	BIM adoption by colleagues
	Adopter characteristics	-	Liking the new technology
		-	Liking using tools, devices, etc.

The factors and items which are shown in Table 5 might be employed to formulate strategies promoting the adoption of BIM among Thailand's architects in order to create a competitive advantage. These strategies might be described briefly as follows: Strategy One: BIM should be promoted through an emphasis upon the advantages which are offered by the attributes of BIM. The ideal candidates to deliver such a strategy would be academics and experts in institutes of higher learning. Other professional organizations also have a role to play, such as the Association of Siamese Architects (ASA) and the Architect Council of Thailand (ACT). This policy would serve to encourage Thai design companies and their personnel to better understand and implement BIM in their professional practice. Strategy Two: BIM must be presented appropriately to interested architects through displaying a professional image, positive reputation, strong company profile and excellent services. This is an approach which must be implemented by BIM vendors who have a vital role to play in marketing. Strategy Three: Conduct observations and surveys to examine the extent to which BIM is adopted and applied in the general field of architecture and design. The findings can then be used to further support the effective use of BIM within architectural design firms. Design managers or senior architects should employ this strategy in order to keep pace with the latest advances in their industry and to maintain competitiveness. Strategy Four: BIM vendors should focus strongly upon those architects who express a liking for new technology and are willing to utilize the latest tools, such as BIM. This strategy will lead vendors to potential customers.

Acknowledgement

This research was funded by a grant provided by the National Research Council of Thailand through Mahasarakham University, Thailand (No. 5805038/2558).

References

[1] Barlish K and Sullivan K 2012 How to measure the benefits of BIM - A case study approach *Automation in Construction* Vol **24** pp 149-159

[2] Ngowtanasuwan G 2017 A causal model of BIM adoption in the Thai architectural and engineering design industry *Journal of Procedia Engineering* Vol **180** pp 966-974

[3] Ngowtanasuwan G 2017 Factor analysis of BIM adoption in the Thai architectural and engineering design in dustry Suranaree Journal of Science and Technology **24 (4)** pp 415-422

[4] Trott P 2008 *Innovation Management and New Product Development* 4th edition (Edinburgh: Pearson Education Limited)

[5] Roger E M 2003 *Diffusion of Innovations* Fifth Edition (New York: Free Press Ltd)

[6] Wikipedia the free encyclopedia 2019 *Awareness* [cited 2019 Jan 1] Avaiable from https://en.wikipedia.org/wiki/Awareness

[7] Schiffman L G and Kanuk L L 2004 *Consumer Behavior* 8th edition (New Jersy: Pearson Practice Mall) pp 155-6 and pp 258-9

[8] Sorescu A B, Chandy R K and Prabhu J C 2003 Sources and financial consequences of radical innovation: insights from pharmaceuticals *Journal of Marketing* Vol **67** (4) pp 82–102

[9] Davis F 1985 *A technology acceptance model for empirically testing new end-user information systems: theory and results* Ph.D dissertation (Massachusetts: MIT Sloan School of Management Cambridge)

[10] Son H, Lee S and Kim C 2015 What drives the adoption of building information modeling in design organizations? An empirical investigation of antecedents affecting architects' behavioral intentions *Automation in Construction* Vol **49** pp 92-99

[11] Succar B and Kassem M 2015 Macro-BIM adoption: Conceptual structure *Automation in Construction* Vol **57** pp 64-79

[12] Rowlinson S 2017 Building information modelling, integrated project delivery and all that *Construction Innovation* Vol **17** No 1 pp 45-49

[13] Yalcinkaya M and Singh V 2015 Patterns and trends in Building Information Modeling (BIM) research: A Latent Semantic Analysis *Automation in Construction* Vol **59** pp 68-80

[14] Ahuja R, Jain M, Sawhney A and Arif M 2016 Adoption of BIM by architectural firms in India: technology-organization-environment perspective *Architectural Engineering and Design Management* Vol **12** No 4 pp 311-330

[15] Rogers J, Chong H Y and Preece C 2015 Adoption of Building Information Modelling technology (BIM): Perspectives from Malaysian engineering consulting services firms *Engineering Construction & Architectural Management* Vol **22** pp 424-445

[16] Ding Z, Zuo J, Wu J and Wang J Y 2015 Key factors for the BIM adoption by architects: a China study *Engineering Construction & Architectural Management* Vol **22** No 6 pp 732-748

[17] Hsu C C and Sandford B A 2007 The Delphi Technique: Making Sense of Consensus *Practical Assessment Research & Evaluation* Vol **12** (10)

[18] Nunnally J C 1978 *Psychometric Theory* 2nd edition (New York: McGraw)

[19] Norris M and Lecavalier L 2010 Evaluating the Use of Exploratory Factor Analysis in Developmental Disability Psychological Research *Journal of Autism and Developmental Disorders* Vol **40** (1) pp 8-20

[20] Tabachnik B G and Fidel L S 2001 Using multivariate statistics. 4th edition (Massachusetts: Allyn & Bacon Needharm Hights)

[21] Kaiser K F 1958 The varimax criterion for analytic rotation in factor analysis *Psychometrika* Vol **23** (3) pp 187-200

IOP Conf. Series: Materials Science and Engineering **650** (2019) 012003 doi:10.1088/1757-899X/650/1/012003

A conceptual framework of cost of safety model in infrastructure project based on work breakdown structure (wbs) to improve safety policy in indonesia

R Fitriani[1] and Y Latief[1]

[1] Universitas Indonesia, Depok, Indonesia

ratih.sudwiyanto@gmail.com; latief73@eng.ui.ac.id

Abstract. The high accidents rate in infrastructure projects in Indonesia in the last 1 (one) year due to lack implementation of the Occupational Health and Safety Management System in its entire project. One of the key indicators that influence it is the safety cost allocation. Currently, there is no structured guidelines to prepare cost of safety that is in accordance with the conditions of infrastructure projects in Indonesia. The purpose of this study is to address this gap by conceptualizing cost of construction safety as a more structured model for construction safety improvement. A conceptual framework which is not only integrate construction project management theory and regulation in Indonesia, but also consider project scope, project duration, project location, hazard analysis, and safety program, is proposed. This conceptual paper contributes to the emerging literature on safety cost and produce a systematic model of safety costs by involving appropriate risk control, and considering applicable regulations.

1. Introduction

The Indonesian government for the 2015-2019 period focused on infrastructure development, i.e. road, toll road, port, and airport constructions in necessary areas. This is required, considering that the condition of infrastructure in Indonesia is still far from ideal, and has even been depreciating in the last two decades. Study from Bloomberg Mckinsey in 2013, in kemenkeu.go.id [1] stated that compared to other countries in the world which had an average of around 70% of GDP invested in infrastructure stock, that of Indonesia was only 38% which decreased from 49% in 1995. According to the 2015-2016 Global Competitiveness Report compiled by the World Economic Forum (WEF)[2], Indonesia ranks 80th out of 137 countries in terms of infrastructure quality especially roads, which is a down grade from 72nd in the previous year [3]. President Joko Widodo stated that the acceleration of infrastructure development is needed to create higher economic growth, both in the medium and long term.

However, the infrastructure development program was not implemented smoothly, and as a result, there were recurring construction accidents on national scale construction projects from the end of 2017 to the beginning of 2018. Recorded from August 2017 to December 2018, there were thirty-five accidents on construction projects and still increasing in number. This led to the Ministry of Public Works and Settlements Indonesia (MPW) to suspend construction work temporarily in February 2018, especially on projects involving elevated construction. Consequentially, the suspension led to the loss of as much as billions of rupiahs per day. Besides economic losses, construction accidents also cause casualties such as death. Also, what we least realize is that recurring construction accidents can result in public anxiety in both the construction workers and the public. When communities around these projects are anxious, they can reject development, and as a result hampering infrastructure development of the country as a whole which is a priority at this time.

The Construction Safety Committee identified general findings that contractors do not prioritize construction safety, namely the implementation of a construction safety and health system has not been optimally implemented. This is also supported by the evaluation results related to the Orderly Implementation of Construction Services conducted by the Directorate General of Construction of the MPWS which stated that the implementation of construction SHS in MPW was only 27.43% in 2016 and 34% in 2017. One of the key factors that emerged in the evaluation as causes of this low implementation was related to the cost of Occupational Safety and Health (OSH).

The procedure for calculating the cost of implementing construction safety is not strictly regulated in the laws and regulations. In PW Minister Regulation No. 05 of 2014 about the Occupational Safety and Health Management System in the construction sector, it is stated that OHS cost is part of the total cost which includes: safety plan preparation; safety socialization and promotion; work protective equipment; personal protective equipment; insurance and licensing; safety personnel; health facilities; signs; and other things related to safety risk control.

However, in Minister Regulation No. 31 of 2015, Amendments to the Third Minister of PW Regulation Number 07/PRT/M/2011 about Standards and Guidelines for the Procurement of Construction Works and Consultancy Services, safety costs are included in the overhead costs. The two regulations only regulate OSH costs administratively in the preparation of government budgets, but not for the procedures concerning the preparation of OHS costs, starting from hazard control. Against this backdrop, the purpose of this paper is to conceptualise the cost of construction safety as a more detailed and structured model, that results in accurate calculations.

2. Cost of safety

The construction sector has the highest risk of accidents compared to others [4]. According to the MPW, the number of construction accidents in the past year in Indonesia has increased. In Malaysia, accidents in the construction industry reached 3 times greater than other industries [5]. One of the key factors of an increase in the number of these accidents in Indonesia is the cost of safety (COS). Some researchers have divided the COS into several parts, and they include [6] [7] (fig.1):

- An insurance cost is that which must be paid by the employer to the insurance company to bear the costs incurred in the event of an accident or illness due to work on workers. This amount depends on the job category of workers.
- Prevention costs are costs calculated by the employer to meet the applicable safety regulations during the course of completing their work.
- Accident costs are those required when a work accident occurs, with the amount depending on the severity. Elias [7] divided the cost of accident into 2 (two) direct and indirect costs of accident.
- Recovery cost is that which must be incurred by the employer to workers affected by work-related accidents, after the second day of treatment to recover and be able to resume work. This is regulated by the Spanish social security system.

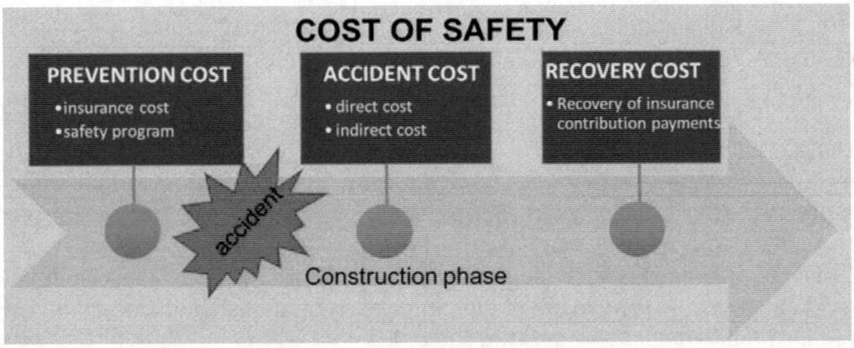

Figure 1. Cost of Safety [6] [7]

There have been several studies conducted on safety costs, and most have focused on the costs of preventing construction accidents. COS is influenced by several factors including: Project scope [4], [5]; Project duration [4]; Man-hour values [5]; Recovery cost [6]; Project budget [8]; Project phase [8], [9]; Project progress [9]; Material operating budget [9]; Number of accidents [4], [9]; Number of workers [9]; Number of subcons [9]; Type of labour contract [9]; Prevention cost [4], [5], [6], [8], [9], [10], [11]; Accident cost [6], [9], [10]; Safety performance [10].

One of the fundamental problems that led to the low implementation of construction safety management systems in Indonesia was the absence of safety cost calculation standards. According to Lopez [9], the COS in construction projects was influenced by several factors, namely: the scope and duration of the project, the number of work accidents, the safety component, and the costs when an accident occurs. The optimal amount of OHS costs is 1% of the total project [4], or equal to 1.92% of the total project cost, according to Gurcanli [5] by considering the work component and the hours of labour.

Table 1. Variation Cost of Prevention

Research	Value of prevention cost to project cost (%)
Shohet, 2018 [4]	1.00
Gurcanli, 2015 [5]	1.92
Pellicer, 2014 [6]	1.54
Son, 2000 [12]	1.2 to 1.3
Abdul Hamid, 2014 [13]	0.15 to 1.08

With the mathematical formulation method, according to Pellicer [6], the prevention costs for construction projects add up to around 1.54% of the total project costs. Abdul Hamid stated that the cost of compliance with health and safety management system among contractor in construction industry varies from minimum 0.15% to 1.08% with average of 0.41%, while Son found that efficient safety control cost should be 1.2 - 1.3% of total contract costs (table I).

Pellicer's calculation also states that safety costs add up to 5% of the project's value by calculating insurance costs and costs when a construction accident occurs. According to Toutounchian [8], in studies conducted on oil and gas projects, the estimated cost of safety was different at every stage of the project. The influencing variables are: the total budget, project stages and prevention costs (first aid kit, etc.). Lu [10] stated that, a low level of construction safety would result in a high accident rate in the projects. The influential factors are prevention and accident costs, as well as safety performance assessment. Lopez [9] concluded that the number of occurred accidents was directly proportional to the number of workers, subcons and the amount of safety costs, but inversely proportional to the cost of prevention.

It was stated in PMBOK construction extension that the most commonly used construction

IOP Conf. Series: Materials Science and Engineering **650** (2019) 012003 doi:10.1088/1757-899X/650/1/012003

techniques for estimation are analogous, parametric, bottom-up, three-point estimation, and Monte Carlo simulation. The more accurate the estimates means the more detailed information about the project, and the more consuming time and resources to develop.

3. Work Breakdown Structure

Making a WBS is a process that describes work and project results of individual components in the form of a top-down list. It also means hierarchically describing the components to be built, as well as the work associated with them. Each WBS represents a more detailed work item. It is a system that divides a project into manageable work packages, components, or WBS elements in order to provide a common framework for scheduling scope, costs, allocation of responsibilities, communication, risk assessment, monitoring and control.

According to PMBOK construction extension [14], each construction project, regardless of size and complexity, continues to face a variety of uncertain situations due to the common factors within the construction industry. Hazard risk can be identified using several approaches, one of which is by following a work breakdown structure.

4. Public Works and Human Settlements Infrastructures

The Ministry of PW has duties and functions as regulators as well as owners in the national infrastructure development of Indonesia. According to Minister of Public Works Regulation No. 8 of 2018 about the 2015-2019 MPW Strategic Plan [15], there are various types of infrastructures built by the MPW in 3 (three) organizational units with the largest budget as the object of this study, there are dam, bridge, road, and flat.

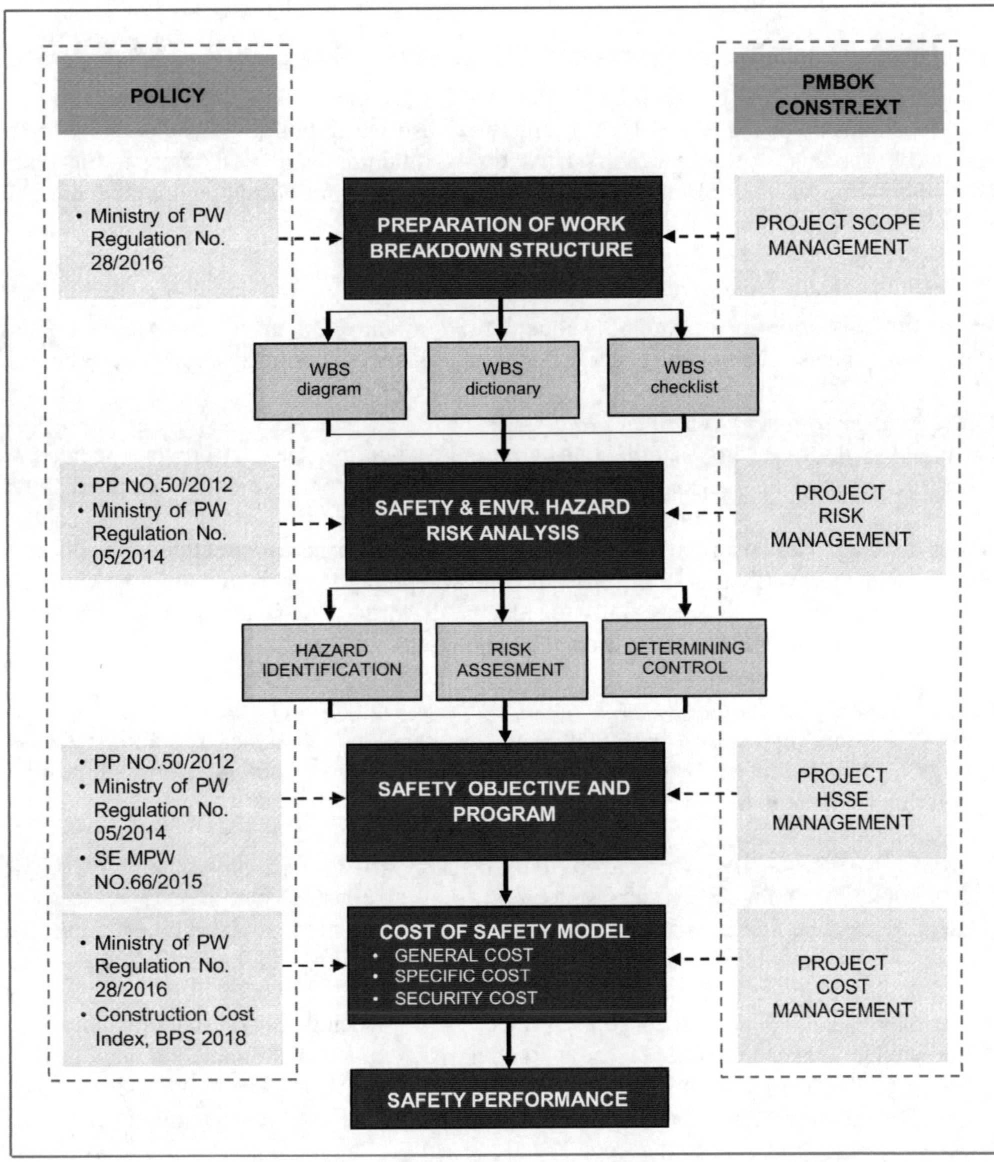

Figure 2. Conceptual Framework of Cost of Construction Safety

5. Conceptual Framework Discussion

The proposed conceptual framework of the cost of construction safety in this research begins with determining the type of MPW infrastructure that will be examined as a study boundary.

The types of MPW infrastructure that will be used as objects of this research are roads, bridges, dams and flats. In general, this research is divided into 4 major stages which are carried out sequentially (figure 2), and they include:

5.1. Preparation of work components / Work Breakdown Structure (WBS).

After determining the type of infrastructure to be studied, the next step is to develop WBS on buildings, roads, bridges and dams which include the preparation of WBS diagrams, dictionaries and checklists. This arrangement is carried out by referring to:

5.1.1. Ministry of Public Works regulation *No 28 Year 2016 about analysis work unit price.*

In appendix part 4, Analysis of Work Unit Prices (AHSP) in the field of Cipta Karya, it is stated that compiling an analysis of the price of work units in constructing buildings begins by describing the scope of work, and this is called the Work Breakdown Structure (WBS). Likewise in this research, in accordance with the Ministerial regulation, the preparation of safety costs begins with compiling the WBS.

5.1.2. PMBOK Construction Extension

In construction projects, scope planning, scope definition and WBS are the first steps in determining the framework. This is stated in one body of knowledge known as project scope management.

5.2. Safety and environment hazard risk analysis

Furthermore, to discover the hazard control measures put in place to prevent accidents during the construction process, an analysis of safety and environmental risk hazards have been prepared. The results of this analysis include risk control, which begins with hazard identification, risk assessment, and establishing risk control for hazards related to work activities in accordance with the WBS that had been previously prepared. This analysis is carried out by considering the project location especially to identify hazards with projects in certain locations. In conducting this analysis, there are references, among others:

5.2.1. *Government Regulation No. 50 Year 2012 about Safety System.* In appendix II points 2.1.1 and 2.1.3 this regulation states that identification of potential hazards and risk assessment must be considered when formulating a safety plan. If risk control efforts are required, it is determined through the level of control.

5.2.2. *Ministry of Public Works regulation No. 05 Year 2014 about construction safety system.* In article 16, in this regulation, it is stated that in every construction project, service users must identify and determine potential hazards. potential hazards. On the other hand, article 19 point b states that the service provider must submit a safety plan in the bidding document.

5.2.3. *PMBOK Construction Extension.* Project Risk Management, one of the bodies of knowledge, states that one of the potential risks in a construction project is safety and security. To identify hazard risk means to perform Hazard Identification, Risk Assessment, and Determining Control (HIRADC). The four strategies of mitigating negative risks are avoid, transfer, mitigate, and accept.

5.3. Safety target and program

After hazard controls are set, the safety targets and programs will be prepared to ensure the safety of the project. This safety program will then calculate its resource requirements. The preparation of objectives and safety programs refers to:

5.3.1. *Government Regulation No. 50 Year 2012 about Safety System.* In appendix II point 2.1.4 it is stated that measurable goals and objectives are required in developing a safety strategic plan in controlling risk.

5.3.2. *Ministry of Public Works regulation No. 05 Year 2014 about construction safety system.* Section 20 stated that the cost of OHSMS Construction implementation is allocated with the general cost.

5.3.3. *Ministry of Public Works regulation No 66 Year 2015 about cost of construction safety system.* In point E, the details of the implementation of construction OHSMS are expressed in the preparation of the analysis of the work unit price (AHSP) in the MPW field, in accordance with item G.

5.3.4. *PMBOK Construction Extension.* One of the bodies of knowledge, Health Project Safety Security and Environment Management, states that there are several methods of maintaining workplace health, to ensure safety at work, to guarantee security within the workplace, and to preserve the work environment.

5.4. Last step is to develop cost of safety model

The preparation of these costs is based on existing activities in the safety program using the bottom-Up (detailed) estimating method, namely, the method of estimating costs that are influenced by the size and complexity of individual activities, work packages, and work components.

According to Popescu in Zhang [16], one of the aspects that influence the cost of a construction project is its location. Some locations have different price indexes which affect the cost of materials, labour, and equipment. This was also stated by Pietlock in Zhang, saying that there is a location factor used in calculating the price of the same project at a different location. In Indonesia, there was a construction cost index (IKK) issued by Badan Pusat Statistik (BPS) since 2003. According to BPS [17], "IKK is used as a proxy of geographical adversity of a region, where the more difficult the geographical location, the higher price level in the area." Five provinces in Indonesia with the highest construction cost indexes are Papua, West Papua, Riau Islands, Maluku, and then Bali.

In preparing the construction safety cost model, the location factors are also considered in accordance with the construction cost index issued by BPS. In addition, this model also considers:

5.4.1. *Ministry of Public Works regulation No. 28 Year 2016 about analysis work unit price on section public work.* In the appendix, it is stated that in the structure of the work unit price analysis, safety costs are included in the general costs, to be exact, in indirect costs. In appendix table 16, safety cost items include general safety, special safety, and unit price safety.

5.4.2. *PMBOK Construction Extension.* A body of knowledge, project cost management states that the prerequisites for bottom-up estimation are clear and detailed scopes including documents such as WBS, which are issued for drawings and construction specifications. Detailed estimation techniques produce transparent estimates and structures for more accurate and reliable projects.

6. Conclusion

This four step detailed and structured conceptual framework results cost of safety in accurate calculations and improve safety performance in construction projects. By integrating construction project management theory and regulation in Indonesia, and also considers project scope, duration, and location, as well as hazard analysis, and safety programs, this conceptual framework can be used as structured guidelines in place to prepare the cost of safety in accordance with the present conditions of infrastructure projects in Indonesia.

Acknowledgments

The Authors would like to thank the financial support provided by Universitas Indonesia through PIT 9 funding scheme under Grant number NKB-0087/UN2.R3.1/HKP.05.00/2019 managed by the Directorate for Research and Public Services (DRPM) Universitas Indonesia.

Reference

[1] Kemenkeu 2017, *Presiden Jokowi: Infrastruktur untuk Pertumbuhan Ekonomi Lebih Tinggi.*

IOP Conf. Series: Materials Science and Engineering **650** (2019) 012003 doi:10.1088/1757-899X/650/1/012003

[2] Schwab K, The Global Competitiveness Report 2015-2016. 2015, World Economic Forum.

[3] Schwab K, The Global Competitiveness Report 2014-2015. 2014, World Economic Forum.

[4] Shohet I M, Luzi M, and Tarshis M 2018 *Safety Sci.* **104** pp 231-238

[5] Gurcanli G E, Bilir S, and Sevim M 2015 *Safety Sci.* **80** pp 1-12

[6] Pellicer E, Carvajal G I, Rubio M C and Catala J 2015 KSCE *Jour. of Civ. Eng.* **18**(7) pp 1955-1965

[7] Ikpe E, Hammon F, and Oloke D 2012 *Jour. of Constr. Eng. and Mangmt.* **138**(8) pp 991-998

[8] Toutounchian S, Abbaspour M, Dana T and Abedi Z 2018 *Safety Sci.* **106** pp 35-46

[9] Alonso M L, Davila M P I, Gamez M C R and Munoz T G 2013 *Safety Sci* **60** pp 151-159

10] Lu M, Cheung C M, Li H and Hsu S C 2016. *Acc. Analy. and Prev.* **94** pp 8-17

[11] Ikpe E, Hammond F, Proverbs D and Oloke D 2011 *Int. Jour. Saf. and Sec. Eng.* **1**(3) pp 298-311

[12] Son K S, Melchers R E and Kal W M 2000 *Rel. Eng. & Sys. Safe.* **68** pp 187-194

[13] Abdul Hamid A R, Bachan S and Salleh A S M 2014 *Nat. Sem. on Civ. Eng. Res.* (UTM Skudai)

[14] Project M I 2016 *Construction Extension to the PMBOK Guide*. (Pennsylvania)

[15] Ministry of Public Works Regulation No. 08/PRT/M/2018, Indonesia.

[16] Zhang S, Migliaccio G C, Zandbergen P A and Guindani M *Jour. of Constr. Eng. and Man.* **140**(6): 4014015.

[17] Badan Pusat Statistik, *Construction Cost Index 2018.*

IOP Conf. Series: Materials Science and Engineering **650** (2019) 012004 doi:10.1088/1757-899X/650/1/012004

Purchasing Strategy of Small-sized Contractors for Building Projects in The Greater Bandung Areas

R Irfanto[1], F S C S Maisarah[2], M Abduh[3]

[1]Postgraduate student in Civil Engineering Masters Program, Faculty of Civil and Environmental Engineering, Institut Teknologi Bandung, Bandung.

[2]Doctoral student in Civil Engineering Doctoral Program, Faculty of Civil and Environmental Engineering, Institut Teknologi Bandung, Bandung.

[3]Lecturer in Civil Engineering, Faculty of Civil and Environmental Engineering, Institut Teknologi Bandung, Bandung.

ryobiirfanto@gmail.com; fauziah.scs.maisarah@gmail.com; abduh@si.itb.ac.id

Abstract. The purchasing strategy of commodities in a construction project is one of the things that determine the success of a project. Commodities in construction projects are very diverse, thus the commodity purchasing strategy will be focused on strategic commodities. Important strategic commodities are determined by contractors in managing logistics to support the implementation of the projects they are working on. This paper discusses the purchasing strategy for small-sized contractors who are usually working on building projects in Greater Bandung areas. The methodology used for this study were identifying strategic commodities by conducting a survey of 10 small-sized contractors in Greater Bandung areas, getting an assessment of strategic commodities, and to trace the historical data of the project, confirming the value of purchasing strategic commodities, then analyze the appropriate purchasing strategy by doing literature review while considering the nature of strategic commodities. The results of the study show that strategic commodities in buildings are sand, cement, concrete, reinforcing steel, brick, and glass.

Keyword: purchasing strategy, small-sized contractor, building projects, strategic commodities

1. Introduction

Commodity purchasing in construction projects is in the range of 60-80% of the total value of construction [1]. The great value of purchasing construction commodities makes it need to be considered in making commodity purchasing. Project success is determined largely by making the most appropriate purchasing decisions and having the ability to coordinate different suppliers. The right purchasing decision and coordination are part of the company's purchasing strategy [2].

Small-sized contractors are vulnerable to changes in the company's cash flow. Some of the problems experienced by small-sized contractors are the problem of commodity cost fluctuations [3]. One of the factors to reduce commodity cost fluctuations is a good purchasing strategy

[4]. The consideration of choosing suppliers for purchasing commodities prioritized by small-sized contractors is location, price, speed, and quality of suppliers.

The type of dominant work carried out by small-sized contractors is the construction and renovation of simple buildings, rehabilitation of road infrastructure, and specific occupational specialties [5]. The purchasing strategy cannot be done on all types of building construction commodities so that the purchasing strategy will be focused on strategic commodities, which are large-value commodities and in the long term the availability of commodities will affect the construction process [6].

This paper discusses the study of suitable purchasing strategies for small-sized building contractors in the Greater Bandung areas. The expected results of this study are in the form of recommendations on strategic commodity purchasing strategies for small-sized contractors in Greater Bandung areas and are expected to provide benefits to small-sized contractors, as input to improve commodity purchasing systems and take strategic steps on commodity needs.

2. Literature Study

2.1 Kraljic Matrix

The definition of strategic commodities is a commodity that has a large value and in the long term, the availability of commodities influences the production process, in this case, the process of implementing construction [6]. Strategic commodities are closely related to existing purchasing practices in companies [8]. One of the main steps in the purchasing process involves identifying and selecting suppliers whose costs, quality, technology, timeliness, dependency, and best services meet the needs of the organization. Strategic commodity purchasing takes a strategic approach to supplier selection - an approach that is more aligned with organizational strategy. The strategic purchasing for commodities is not only applied at one time but needs sustainability to ensure that supply remains despite changes in the company's internal and external environment.

Kraljic introduced the first portfolio that explained purchasing and supply management. Kraljic [6] compiled a portfolio matrix that classifies products based on 2 dimensions, namely value and supply risk (low and high). Kraljic matrix has been widely used in different industries as an efficient tool for developing different purchasing strategies. Until now the matrix in the first portfolio of Kraljic is the best practice to determine strategic commodities by many industrial companies. Strategic commodities are important to be applied to contractors because they do not allow contractors to manage all types of construction commodities, and are not effective when contractors focus on commodities that have small value or their existence in construction is not too important.

Commodity classification is divided into two vectors, namely risk and value. The following is a description of the classification based on vectors:

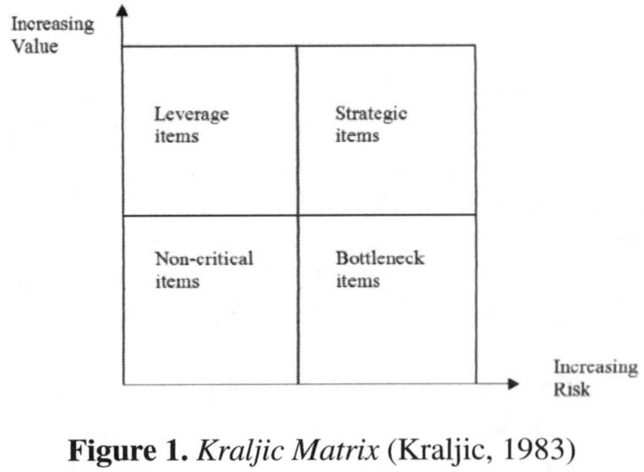

Figure 1. *Kraljic Matrix* (Kraljic, 1983)

IOP Conf. Series: Materials Science and Engineering **650** (2019) 012004 doi:10.1088/1757-899X/650/1/012004

Leverage: Low risk with high value. In this condition, many alternative suppliers and commodities are easily found on the market, but goods cannot be replaced with alternatives (high-value commodities). At this level, the commodity can be considered safe because it has many suppliers.

Strategic: High risk with high values. In this condition, suppliers are not many, commodities are rather difficult to find on the market, and commodities are difficult to replace with other alternative commodities. At this level the commodity needs more attention because the difficulty of finding suppliers can cause disruption to construction projects.

Bottleneck: High risk with low value. In this condition, suppliers are not many and commodities are rather difficult to find on the market, but commodities can be replaced with commodity alternatives when allowed by the project owner. At this level, the commodities need to be considered because the commodity market can be said to be monopolistic with a few of suppliers.

Non-critical: Low risk with low value too. In this condition, there are many alternative suppliers, commodities are easily found on the market, and commodities can be replaced with commodity alternatives when allowed by the project owners. At this level, the commodity does not need to be considered because the commodity market is very free and it is easy to find alternatives.

2.2 Purchasing Strategies

The contractors need to have purchasing or purchasing strategy to improve efficiency in commodity management. The purchasing strategy for each commodity varies, but one way that can be done is to identify commodities that are strategic then adjust their purchasing strategies [9]. There are several ways to manage purchasing strategies based on the quadrant in the Kraljic matrix as follows:

1. *Strategic*
   a. Maintaining strategic partnerships: To offset supply risk, the company builds partnerships with its suppliers. Mutual trust and commitment related to intensive relationships tend to reduce the risk of supply to a minimum. Close collaboration with suppliers will lead to improved product quality, delivery reliability, a grace period, product development, product design, and reduced costs. This situation can be characterized as one with balanced strength between buyers and suppliers.
   b. Accepting locked partnerships: This strategy often occurs when buyers are subject to unfavorable supplier conditions and cannot withdraw from the situation. The locked position may be due to the fact that the supplier holds a patent for a particular product and therefore has monopoly power to a certain extent. This situation can be categorized as a situation dominated by suppliers.
   c. Stopping partnerships: This strategy is used when supplier performance becomes unacceptable and irreparable. Buyers will try to reduce their dependence on suppliers. One way to achieve this is to find alternative suppliers. In this situation the buyer still depends on the supplier.
2. *Bottleneck*
   a. Accepting dependency, reduce negative consequences: The main focus of this strategy is to ensure supply even with additional costs if necessary. By conducting risk analysis, companies can identify the most important bottleneck products and consider their implications.
   b. Reducing dependency and risk, find other solutions: This strategy is directed at reducing dependence on suppliers. The most common way to achieve this is to expand product specifications or find new suppliers, and replace with other alternative commodities.
3. *Leverage*
   a. Exploiting purchasing power: In this strategy the company pursues competitive offers.

Because suppliers and products can be exchanged, there is no need for long-term supply contracts. Ordering commodities is placed as an administrative formality. The ability to choose high suppliers is then used to get a better deal with suppliers. Therefore, this scenario is characterized by buyer dominance.

b. Developing strategic partnerships: In some cases, one way of development is to create strategic partnerships with suppliers. In this scenario, a balanced power position is expected between buyers and suppliers. The development of close collaboration with suppliers will lead to improved product quality, delivery reliability, grace time, product development, product design, and will result in cost reduction as in the strategy of maintaining strategic partnerships (strategic quadrant).

4. *Non-critical Quadrant*
   a. Bulk Order Systems: Handling non-critical products requires a purchasing strategy that aims to reduce the complexity of logistics and administration. Contract systems are generally recommended as a way of doing business with routine product suppliers.
   b. Individual ordering, efficient processing: Whenever it is not possible to unite purchasing terms, professional buyers adopt a kind of individual order, for example by using a purchasing card. This strategy aims to reduce indirect purchasing costs associated with administrative activities, such as ordering and billing.

3. Methodology

The conceptual framework for the study of purchasing strategies for small-sized contractors is to test the strategic level of the commodity based on questionnaires and contract documents, from the strategic level testing to obtain estimates of strategic commodities, then contracts are used to confirm strategic commodities [10]. After obtaining strategic commodities, then the strategy of purchasing commodities that are suitable are analysed by conducting literature studies so that the conceptual framework can be described as shown in the figure below.

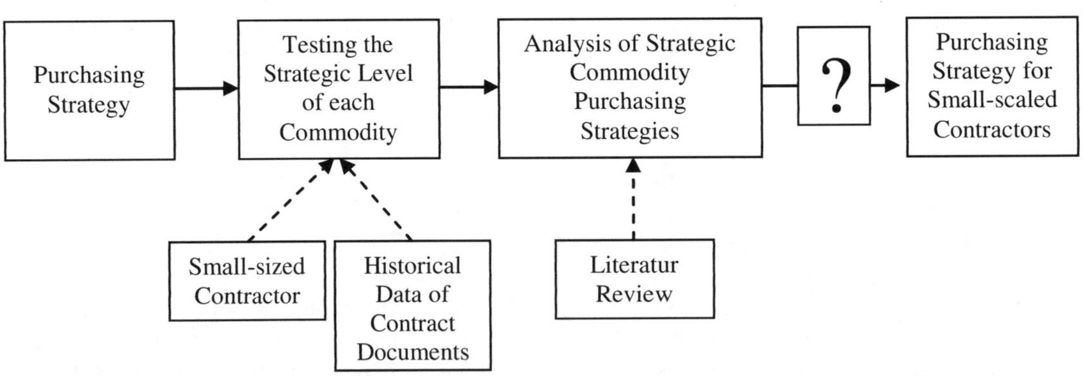

Figure 2. Conceptual Framework

From the conceptual framework, the things that could be done are:

1. Strategic Commodity Testing: From this test, we will find out the current strategic level of commodity and also grouping the dominant commodity needs in building works.
2. Strategic Commodity Purchasing Strategy: From the Kraljic Matrix, which has been confirmed by contract documents, a good system strategy for commodities that will be better managed will be analyzed. The purchasing of these strategic commodities will have an impact on the value and risk of commodities in the future.

4. Analysed Data and Results

4.1 Kraljic Matrix Strategic Commodities and Contract Documents

Strategic commodity assessments were obtained by conducting a survey of 10 small-sized contractors in Greater Bandung areas. The indicators on the questionnaire were made referring to the explanation of Irfanto et al. [10]. Based on the strategic commodity test questionnaire the Kraljic matrix results are obtained as shown below.

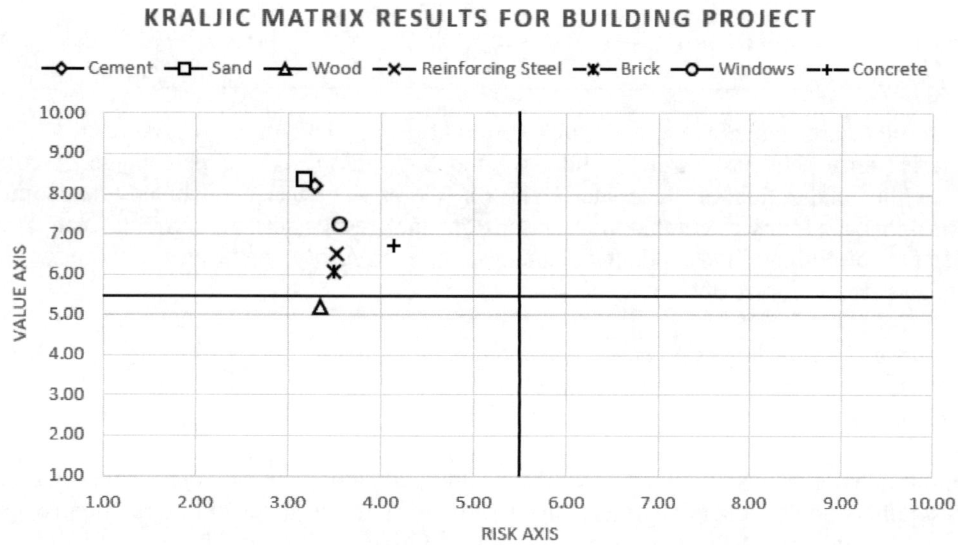

Figure 3. Kraljic matrix Result for Building Construction

From the matrix results, it can be seen that the construction commodities tested turned out to be in the leverage and non-critical quadrants. In the leverage quadrant, there are commodities such as sand, cement, glass, reinforced steel, concrete, and brick, while timber commodities are in the non-critical quadrant. Since the value of the risk attribute is not too different, the determination of strategic commodities can be done by focusing on the value attribute. The axis value is as follows:

Table 1. The Value Attribute of Kraljic matrix

Commodities	Value Attribute
Sand	8.38
Cement	8.2
Windows	7.25
Concrete	6.73
Reinforcing Steel	6.5
Brick	6.08
Wood	5.2

Based on the results of attribute values in table 1, it can be seen that timber commodities are the lowest and are in the non-critical quadrant so that the commodities cannot be strategic commodities. It was compared with contract documents, namely recapitulation from Budget Planning (RAB), Bill of Quantity (BoQ), and Analysis of Work Unit Prices (AHSP) to support the statement. Based on the average recapitulation of the contract documents that have been studied, the results obtained are as shown in the table below.

Table 2. Percentage of Commodity Costs

Commodities	Percentage of Commodity Costs from Total Building Prices
Reinforcing Steel	14.04%
Concrete	8.17%
Cement	6.77%
Brick	6.57%
Sand	4.42%
Windows	2.87%
Wood	0.94%

The results from table 2 are not much different from the Kraljic matrix results. The wood commodities are in the last position and have the last ranking or the percentage of commodity values to the total value of the material is very low, thus it is concluded that commodities wood is not a strategic commodity. Based on the results of the Kraljic matrix and the recapitulation of commodity contract documents, the strategic commodities are reinforcing steel, cement, glass, sand, brick and concrete.

4.2 Strategies for Purchasing Strategic Commodities

Based on the results of the Kraljic matrix, it is found that at present the commodity is at the level of leverage. There are 2 choices of strategies for managing commodities that are in the leverage quadrant [9]:

1. Exploiting purchasing power: In this strategy, the company pursues competitive offers. Because of suppliers and products can be exchanged, there is no need for long-term supply contracts. Ordering commodities is placed as an administrative formality. The ability to choose high suppliers is then used to get a better deal with suppliers. Therefore, this scenario is characterized by the dominance of buyers.
2. Developing a strategic partnership: In some cases, one way of development is to create strategic partnerships with suppliers. In this scenario, it is expected to have a balanced power position between buyers and suppliers. The development of close collaboration with suppliers will lead to improved product quality, delivery reliability, a grace period, product development, product design, and it will result in cost reduction as in the strategy of maintaining strategic partnerships (strategic quadrant).

Based on the results of the Kraljic matrix, the risk attributes are at a low value, which means that availability in the market, supplier performance, and commodity flexibility is still good. One consideration in determining the purchasing strategy of each commodity is that cement, glass, brick and reinforcing steel commodities do not have significant price differences between suppliers because the commodity is a factory manufacturing commodity that has certain standards. The concrete commodities are commodities that can not be stored for a long time, thus it must be used immediately. The sand commodity is a commodity that does not have a fixed standard with diverse quality and diverse prices. Reinforcing steel commodities have a problem which is supply of commodities at certain times [11].

Table 3. Commodity Purchasing Strategy

Commodities	Special properties	Purchasing Strategy	Detail
Cement	Manufacture	Exploiting Purchasing Power	This type of commodity is always produced without waiting for orders
Sand	Bulk	Develop a strategic partnership	To increase the efficiency of purchase value
Brick	Manufacture	Exploiting Purchasing Power	This type of commodity is always produced without waiting for orders
Windows	Manufacture	Exploiting Purchasing Power	This type of commodity is always produced without waiting for orders
Concrete	*Just-In-Time*	Develop a strategic partnership	Commodities must directly used, and shipping based on orders
Reinforcing Steel	Manufacture	Develop a strategic partnership	Ensure availability of commodities when needed

From the results of the considerations and purchasing strategies, cement, brick, and glass for buildings can be used as purchasing power exploitation strategies. This is because cement commodities are manufacturing commodities, thus it is always available. Sand commodities in buildings have varied prices without definite standards so that it is more possible to create strategic partnerships to increase the efficiency of the value of purchases in buildings. Concrete commodities in buildings have a high value, but concrete commodities are commodities that must be directly used when arriving in the field so that a suitable purchasing strategy is to create a strategic partnership. Rebar steel commodities in buildings have a value that is not too high so that to ensure that the commodity can be available is to make a strategic partnership.

5. Conclusion

The results obtained from this paper are strategic commodities according to small-sized contractors are cement, sand, concrete, reinforcing steel, brick, and glass with a strategy that is suitable for the current purchasing of each commodity for cement, glass, brick carried out exploiting purchasing power, while the commodities of sand, concrete, and reinforcing steel can be made to create strategic partnerships with suppliers to ensure availability and quality and improve the efficiency of time and costs of purchasing these commodities. This purchasing strategy is expected to help small-sized contractors in managing commodity purchases so that small-sized contractors can increase the efficiency of purchasing construction commodities.

References

[1] Eksekär A and Rudberg M 2015 Third-party Logistics in Construction: Perspective from Supplier and Transport Providers. *22nd International Annual EurOMA Conf.* (Switzerland, July 2015)

[2] Bildsten L 2016 *Purchasing in Construction Companies* (Netherland: University of Twente)

[3] Raghavan V S dan Kumar K 2015 Problem Faced by Small Scale Contractors in India *International Research Journal of Engineering and Technology (IRJET)* ***02****(02)* pp 105-107.

[4] Zhou H and Jianhua Z 2013 Analysis on Factor to Cause the Price Changed of Building Material. *Advanced Material Research* ***683*** pp 668-672

[5] Johannessen B 2009 Training Guidelines: *Contracting Local Infrastructure Works* ILO (International Labour Organization).

IOP Conf. Series: Materials Science and Engineering **650** (2019) 012004 doi:10.1088/1757-899X/650/1/012004

[6] Kraljic P 1983 Purchasing Must Become Supply Management *Harvard Business Review*

[7] Bildsten´ L 2010 Applying The Kraljic Model To The Construction Sector: The Case Of A Prefab Housing Factory *Procs 26th Annual ARCOM Conf.* (United Kingdom: Association of Researchers in Construction Management)

[8] Rendon and Rene G 2005 *Comodity Sourcing Strategies: Supply Management Action* Naval Postgraduate School (Montery, California)

[9] Caniels M C J 2005 Purchasing strategies in the Kraljic matrix—A power and dependence perspective *Journals of Purchasing and Supply Management* pp 141-155

[10] Irfanto R, Maisarah F S C S, Abduh M 2018 Metoda Identifikasi Komoditas Strategis untuk Kontraktor Kecil di Bandung Raya *Prosiding Konferensi Nasional Pascasarjana Teknik Sipil (KNPTS) 2018* pp 19-26

[11] Kementerian Pekerjaan Umum 2012 *Kajian Rantai Pasok Baja Konstruksi untuk Mendukung Investasi Infrastruktur* (Badan Pembinaan Konstruksi: Pusat Pembinaan Sumberdaya Investasi)

[12] Fereira L 2012 Application of Kraljic's Purchasing Portofolio Matrix In Construction Industry - A Case Study *Internasional Conference on Industrial Engineering and Operation Management*

Financial and institutional scheme of aerotropolis conceptual design

M A Berawi[1], Z N Affiandi[2], P Miraj[2,3] and Gunawan[2]

[1]Faculty of Engineering, Universitas Indonesia, Kampus Baru UI Depok 16424, Indonesia

[2]Center for Sustainable Infrastructure Development, Universitas Indonesia, Kampus Baru UI Depok 16424, Indonesia

[3]Faculty of Engineering, Universitas Pancasila, Srengseng Sawah Jakarta Selatan 12640, Indonesia

*maberawi@eng.ui.ac.id

Abstract. The concept of aerotropolis expected to create an integrated transport, effective and efficient airport. It facilitates higher mobility and generates new economic center for regional development. Aerotropolis concept in Indonesia is facing some obstacles such as complex bureaucracy, inadequate infrastructure, and limited funding. This research aims to produce an alternative concept of aerotropolis by taking into account the project feasibility. The research will use life-cycle cost method and involved initial cost, operational, maintenance cost and revenue for project evaluation. The aerotropolis in Indonesia shall consist of the airport, industrial development, mixed-use and supporting infrastructure. The result shows that 15.81% of the internal rate of return might be obtained from this conceptual design. The study also formulates a suitable institutional scheme by considering the cost-sharing between public and private in the project life cycle.

1. Introduction

The role of transportation is significant as the core of economic, social, political, cultural, defense, and security of a nation [1, 2]. The government of Indonesia attempt to create an effective and efficient transportation system for national competitiveness. The transport system should improve human mobility, goods, and services and translate by land, sea, and air transportation systems. Each mode of transportation has its advantages and characteristics. Compared with other modes of transportation, air transport mode offers faster and more efficient system, and capability to reach remote areas [3, 4].

Most airports in Indonesia are located far from residential areas, or not available in specific locations and regions. In other countries, the role of the airport has grown not only act to transport people and goods from an area to the others but also becomes business center and economic hub. This new role of the airport will attract business, commercial and industrial activities. In the longer term, the airport may change the landscape of economic activities and regional development [5, 6].

The concept of aerotropolis attempt to integrate people mobility with economic activities by taking an airport as the center. It may create a new form of areas within site from offices, hotels, entertainment centers, and others. The concept expected propose higher mobility that exists around the airport and multiplies it through supporting infrastructure. Many airports have applied this concept, for instance, Changi Airport in Singapore built its own cinema, Beijing Capital Airport in China provide a bank facility, Hong Kong International Airport offer a large shopping center facility, Las Vegas has museum facilities, Frankfurt Airport in Germany owns hospital, Schiphol Airport in Netherland built art museum, and Stockholm Arlanda in Sweden has a multipurpose church that has been used for more than 400 marriages. [7, 8].

Aerotropolis concept in Indonesia facing some obstacles such as complex bureaucracy, inadequate infrastructure, and limited funding. In Indonesia, the development of the aerotropolis concept is still in the stages of Aerospace Park. It is an industrial area focusing on the maintenance of aircraft. This condition arises due to the increasing fleet ownership from the aircraft company in Southeast Asia. The challenge for airport maintenance is high and thus providing complete facilities such as maintenance, repair, and overhaul are significant to be champion in this competition.

This research attempts to evaluate the concept of Aerotropolis by taking into account Lampung as the case study. Despite lower population and movement of aircraft, the airport has vast land availability about 9,650 square meters for development. The airport (Radin Inten II airport) location also strategically located with other Asian countries and the Middle East region. The country also has a similar airport (middle class) type; thus the result may be replicated into other airports to generate economic growth.

2. Methodology

This research combines qualitative and quantitative approaches to generate the targeted result. Firstly, desk study was used to evaluate the literature study about the aerotropolis concept and benchmark the assumption for analysis. It comprises of inflation, discount rate, weighted average cost of capital (WACC), and other financial aspects. The concept design then investigated by using life cycle cost by taking into account the initial cost, operation, and maintenance cost and revenue to generate the internal rate of return (IRR). It will confirm the capability of additional functionality to reach the expected IRR [9, 10].

When the IRR failed to meet particular value, further sharing scheme will be applied. It consists of four scenarios; initial cost sharing, operation, and maintenance cost sharing, initial cost+operation, and maintenance cost sharing and initial cost+operation and maintenance cost+revenue sharing. The scenario will govern the responsibility of public and private in the project and propose the optimum result in term of financial feasibility. This scenario will be used to formulate the institutional scheme on the project. The in-depth interview will be conducted with three experts on infrastructure policy with more than ten years of experience. The instruments use a semi-structured questionnaire and take 15 – 30 minutes to collect sufficient data to be analyzed. The result from the in-depth interview and financial analysis will support the decision to develop robust financial scenario [11].

3. Result and Discussion

3.1. Design and Development

The design and components were using previous research development and published by the research team [8]. The aerotropolis concept consists of four components such as airport terminal, industrial zone, mixed-use area and supporting infrastructure. The concept visualizes as follows.

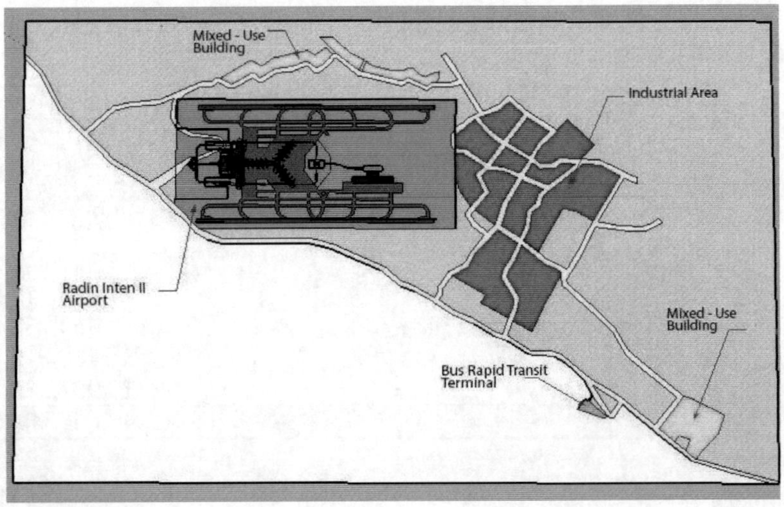

Figure 1. The concept of Aerotropolis in Lampung Province
Source: [8]

3.2. Initial Cost

The initial cost was divided based on the components of aerotropolis. The airport is an international airport class occupying the site for about 770 Ha. The size of the terminal estimated around 62.5 Ha and the runway stretch about 3,500 m. In term of an airport terminal, the analysis considered the land acquisition, landside area, airside area, and contingency cost. The investment cost also involved the cost construction index of Lampung area (0.88) compared with the national construction cost index (0.40). The result shows that the total cost estimated about 20,524,093,076,791 rupiah or equal to US$ 1,425,284,241.

Industrial development cost considered six industries in aerotropolis ranging from cassava, corn, sugar, pineapple, palm oil, and coffee. Baseline cost of cassava is generated from related cassava industry in Malaysia and converted into Indonesian price. On the other hand, corn processing industry used the price component from Iowa State University in the United States and converted into the standard price from Bogor Agricultural Institute. The rest of the industries was using the baseline cost from companies or reports from Indonesia and consider construction cost index for the analysis. The total cost for the industrial area development about 5,365,303,968,008 rupiah or equal to US$ 372,590,553.

The mixed-use area consists of two locations; one area has land parcel about 445,786 square meters while others have 570,319 square meters. Both of them has commercials, residential, office, hotels, road access, and open space. The different is about the size of each function. The smaller size of a mixed-use area requires a 25,699,873,591,185 rupiah of investment cost, while the larger mixed-use needs a 33,305,339,412,010 rupiah. The total of both mixed-use is about 59,005,213,003,195 rupiah or equal to US$ 4,097,584,236.

Supporting infrastructure in the aerotropolis concept design is divided into four components from water processing, waste processing, geothermal power plant, road access and bus rapid transit (BRT). The cost analysis the supporting infrastructure has been considered land acquisition and cost of construction for each component. Overall, the investment cost of aerotropolis conceptual design is summarized as follows.

Table 1. Cost of Investment for Aerotropolis Conceptual Design in Lampung Province

No	Component	Cost of Investment (US$)
1	Airport	1,425,284,241.44
2	Industrial development	372,590,553.33
3	Mixed-use	4,097,584,236.33
4	Supporting Infrastructure	
a	Water processing	17,341,446.86
b	Waste processing	7,025,716.30
c	Geothermal power plant	178,433,778.14
d	BRT	31,480,078.76
e	Road access	17,921,533.11
	TOTAL	6,147,661,584.27

3.3. Operation and Maintenance

Operation and maintenance (OM) cost in the study considered the functions that consist of the airport, industrial development, mixed-use and supporting infrastructure. In term of operation and maintenance cost of the airport, the estimation considered three development phase. In the first phase, the airport which operated 9,000 square meters will be estimated its OM cost up to 2025. The second phase, the airport will be developed into 416,000 square meters and OM cost will be forecasted to 2030. The last phase, the airport expanded into 625,000 square meters and the OM cost estimated to 2045.

Each industry in industrial area function has different OM cost depends on its characteristics. The coffee industry is the highest contributor to OM cost among others as about US$ 662,395,978.75. It followed by pineapple can industry that requires US$ 243,552,573.44 and cassava processing as about US$ 177,694,780.85 for operation and maintenance. Other industries generate lower OM cost below US$ 100,000,000 such as corn (US$ 84,770,444.25), sugar (US$ 7,888,446.60) and palm oil (US$ 9,081,728.85).

Each property in the mixed-use area has different OM cost depends on the operational year and the size. Office building contributes as the highest OM cost due to higher floor area ratio (FAR) compared to other types of buildings (residential, commercials, and hotels). The mixed-use area is developed into four phases; current to 2030, 2030-2035, 2035-2040 and 2040-2045. Each phase will have different size of development and construction, subsequently, affect the operation and maintenance.

Three infrastructure from five supporting infrastructures such as water treatment plant, wastewater treatment plant, and geothermal power plant is analyzed by considering the production unit. In term of the water treatment plant, 11,537 rupiahs per cubic meter is required for operation and maintenance. On the other hand, wastewater treatment plant needs a Rp 2,395 for production. Last, geothermal power plant requires Rp 508.1 billion to produce a 394,200 MWh electricity to electrify the overall industrial estate.

OM cost of BRT was analyzed by multiplying the operating cost per bus per km by taking into account a bus distance estimated around 412 km per day, the amount of bus and daily passengers capacity. In 2030, the capacity is about 61,753 passenger/day and covered by 36 unit of buses. It increases into 85,270 passenger/day in 2035 and covered by 48 buses. In 2045, the daily passengers have reached 159,313 and shall be accommodated by 91 buses. The OM cost will then increases every five years to cover estimated BRT demand.

Last, road maintenance uses a similar price in the capital city of Indonesia as about 75,000 rupiah or equal to US$ 5.21 per square meter. It then converted into the airport location in Lampung province by considered the construction cost index. The OM cost of each function is summarized in the following table.

IOP Conf. Series: Materials Science and Engineering **650** (2019) 012005 doi:10.1088/1757-899X/650/1/012005

Table 2. Operation and Maintenance Cost of Aerotropolis Conceptual Design in Lampung Province

No	Component	OM Cost (US$)
1	Airport	428,064,339.68
2	Industrial development	1,185,383,952.74
3	Mixed-use	465,442,032.11
4	Supporting Infrastructure	
a	Water processing	
b	Waste processing	40,648,601.16
c	Geothermal power plant	
d	BRT	2,346,431.18
e	Road access	2,543,021.26
	TOTAL	2,124,428,378.13

3.4. Revenue

Each function in the aerotropolis conceptual design generates different revenues. Source of revenue from airport divided into aeronautical revenue and non-aeronautical revenue. The aeronautical revenue is generated from services or facilities related to the operation of aircraft and cargo and its supporting facilities. On the other hand, the non-aeronautics revenue is obtained from the facilities of passengers available at airports outside the aeronautical sector.

Revenues from the aeronautical sector are generated from landing and aircraft storage, air passenger service, aircraft service, aviobridge supply, counter usage and conveyor by airline, and baggage handling system. As for the revenue generated from the cargo, the sector is generated from the tariff of cargo service and cargo loading by calculating the weight of cargo. This research assumed the workload unit from similar tariff in Schipol Airport, Netherlands (9.93 Euro for passenger and 100 kg of cargo) and converted into Indonesian context. The result shows that passenger will be charge 234,514 rupiah or equal to US$ 16.29 and 2,345,140 rupiah or equal to US$ 162,86 per ton cargo.

Revenue from the non-aeronautical sector is generated from two sectors, namely the merchandise sales and commercial land leasing at the airport. The assumption uses a similar case from the Annual Report of Incheon Airport in South Korea in 2014. The merchandise sales estimated about 33,355 rupiah or equal to US$ 2.32 per pax and 15,282,322 rupiahs or US$ 1,061.27 for commercial land leasing.

Revenue from industry consist of two types; revenue from the processing industry and its supporting infrastructure. The processing industry will be considered the capacity production, required land for development, multiplier factor and commodity price. Moreover, by taking into account the inflation, the selling price for each commodity will be adjusted. The summary of revenue can be seen in Table 3.

Table 3. Revenue Component from Each Industry

No	Component	Land area (Ha)	Multiplier factor	Capacity Production	Price (US$)
1	Cassava	190	95	684,000 ton/year	0.28
2	Corn	44	11.52	4,360,158 m^3/year	0.11
3	Sugar	20	13.33	213,333 ton/year	0.87
4	Pineapple	18.72	187.10	673,560 ton/year	0.76
5	Palm Oil	18.71	3.12	1,403,250 ton/year	0.76
6	Coffee	10	250	15,000 ton/year	3.13

Supporting industry such as water treatment plant, wastewater treatment plant, and geothermal power plant also generate significant revenue for the aerotropolis development. The analysis will consider the capacity, production, land availability, needs of the area and tariff for each service. Supply of clean water in the area will be charged as about 15,046 rupiahs or US$ 1.04 per cubic meter. On the other hand, wastewater processing will be charged as about 21,065 rupiahs or US$ 1.46 per cubic meter. Last, geothermal power plant involved industrial area needs and generated the capacity for about 50 MW.

In term of mixed-use area, the analysis follows property price from each building to generate revenue. Office, commercial land and hotel use renting price, while apartment will be sold for users. Each of them has a different unit price and occupancy rate, therefore, generates different revenue. Office contributes as the highest contributor to revenue, follows by the hotel, commercial and apartment.

Bus Rapid Transit (BRT) will also produce income for the aerotropolis. It will be operated on the aerotropolis area to accommodate passenger movement from the residential or other areas into the center of economic activities. The route spans for about 39.82 km and a total of 40 bus stops. It will serve 16,178,517 passengers/year and revenue for about US$ 7,752,205.95.

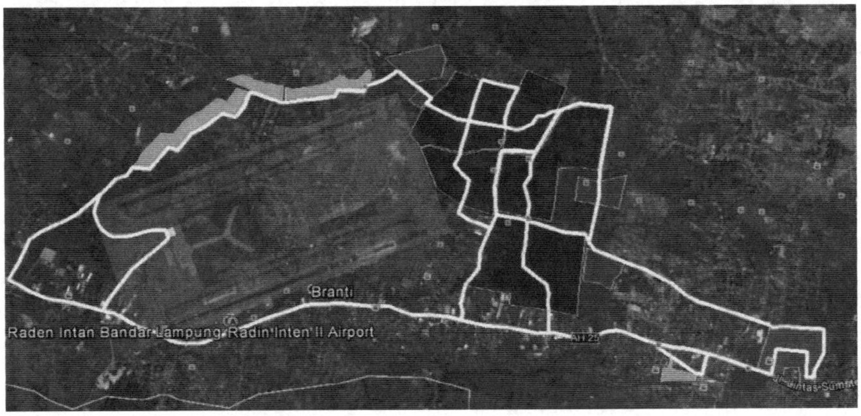

Figure 2. BRT Route in Aerotropolis Conceptual Design

Based on the initial cost, operation and maintenance cost and revenue from the above analysis, the IRR of the aerotropolis concept is 8.80%. Despite producing a positive value, this result is argued below expected WACC of infrastructure development in Indonesia as about 11.00%. The financial scenario in the following sub-section will formulate the scheme to achieve the targeted result of IRR.

3.5. Financial Analysis

In the financial scheme, this research generates 42 sharing scenarios which divided based on project life cycle. There are three initial cost-sharing scenarios, three OM cost-sharing scenario, nine initial cost
plus the OM cost-sharing scenario, and 27 initial cost + OM cost + revenue sharing scenarios. These scenarios will generate optimum IRR for the project. When analyzing the initial sharing scenario, scenario 3 produced an optimum result among others. The scenario offers government sunk cost into the project about 40% from the investment cost and the rest will be handled by the private investors. They will also be responsible for operation and maintenance as well as receiving income from carrying the services. The scenario generates 14.16% of IRR with positive NPV.

In the scenario where operation and maintenance cost is shared between government and private investors, scenario five is selected as the best contributor in term of IRR. This scheme offers private investors responsible for initial cost, but the operation and maintenance will be divided equally between parties. Revenue from this project will be claimed by the private investors. This scenario generates an IRR of about 17.18% and a positive NPV.

This research also simulates where the sharing is not only in initial cost phase but also in the operation and maintenance phase. The analysis shows that scenario 11 proposes as the best result by taking into account equal responsibility in initial cost, OM cost and private investors gains all the revenue. IRR on this scenario generates high return compared to the previous scenario as about 26.46% with a positive NPV due to the minimum cost is borne by the private investors but gained maximum income.

Last, all phases in the life cycle interfered with sharing between the government and private sectors. Scenario 39 is selected as the best result by considering 40% of the initial cost should be handled by the government, and they will also be involved equal distribution in operation and maintenance. However, unlike other previous scenarios, in this one, the government will obtain 20% of the revenue, and the rest will belong to the private investors. The result shows that 18.77% of IRR was generated with a positive NPV. The summary of this analysis is shown in the following table.

Table 4. Revenue Component from Each Industry From Private Investor Perspectives

Scenario	Initial Cost Sharing	OM Cost Sharing	Revenue Sharing	IRR	NPV
3	60%	x	x	14.16%	Positive
5	x	50%	x	17.18%	Positive
11	50%	50%	x	26.46%	Positive
39	60%	50%	80%	18.77%	Positive

Based on the above analysis, each function (airport, industrial area, mixed-use, and others) in the aerotropolis are divided into selected parties responsibility. The result shows that private investors contributed 60.41% to the initial cost, 62.37% to the operation and maintenance cost and gained 78.13% of revenue. The result offers a sharing between the government and private investors in every life cycle stages without compromise the IRR and NPV. The IRR argued adequately to attract private investors with 15.81% to this project. Furthermore, in this scenario, the government is not only involved in funding the project but also has the opportunity of gaining income. This revenue might significantly contribute to helping other projects that have financial issues.

3.6. Institutional Scheme

The institution scheme follows the life cycle stage of project construction. In the development stage, planning consultant that consist of an airport consultant collaborates with industry consultant conduct the initial and planning from regional development to the investment scheme. The proposal submitted to the Ministry of Planning Agency for tendering and evaluation of its feasibility. The project will be supported by a national guarantee fund when feasibility is reached.

A special purpose vehicle formed by the government and private investors will be responsible manage the project. The government will support the initial cost in term of the airport, supporting infrastructure as well as cassava and corn processing industries. The private party will cover the initial cost of mixed-use and processing industries from sugar, pineapple can, palm oil and coffee. Each of the functions will be operated by respective parties. State-owned enterprises operate functions from the government side and private corporation for private investors side. The research produces an institutional scheme that can be seen as the following figure.

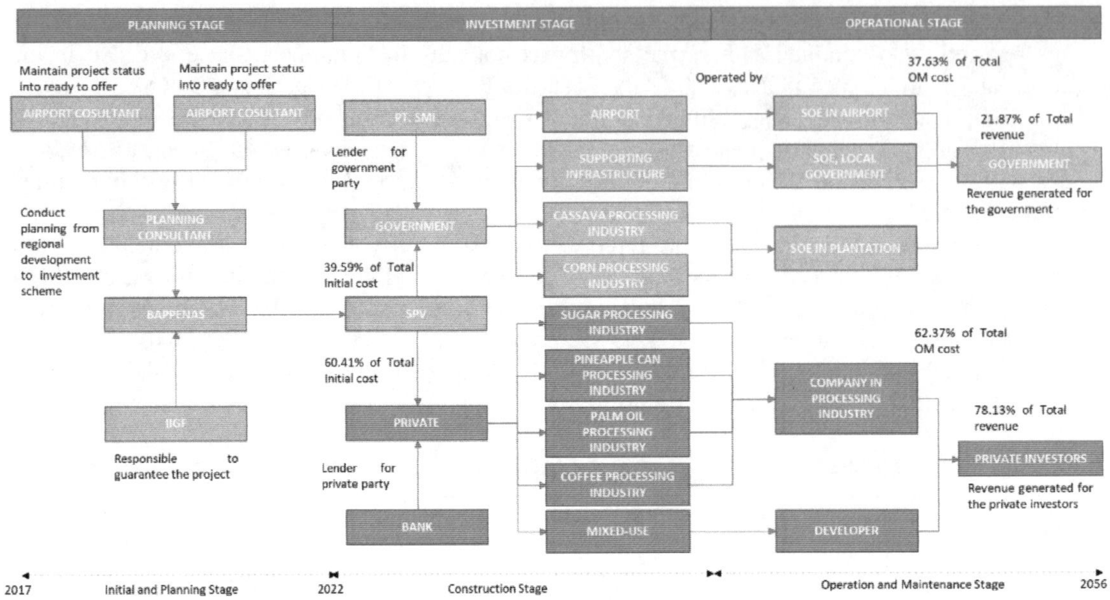

Figure 3. Institutional Scheme in Aerotropolis Conceptual Design

4. Conclusion

Aerotropolis propose greater connectivity and increase the economic growth of a city, local governments, and a nation. This research attempt to generate alternative conceptual design by integrating transportation, industrial development, mixed-use and supporting infrastructure. Based on the financial analysis, these integrated functions produce 8.80% of the internal rate of return (IRR). Further analysis through scenario sharing such as initial cost sharing, operation, and maintenance cost sharing, initial cost+operation and maintenance cost sharing and initial cost+operation and maintenance cost+revenue sharing have established optimum scenario both for public and private interest.

The scenario that considers 40% of initial cost and 50% operation and maintenance cost tackled by the government arise as the best alternative among others. Despite government sunk cost in each stage, the public will gain 20% of the revenue which may be used to support other infrastructure projects in the nation. This research also proposes an alternative institution scheme that may be used by the government institutions in handling such projects in Indonesia. The result of this research also can be used by the private sector as an alternative to collaborate with the government entity such as local government, SOE, and many others to develop aerotropolis concept in the country.

Acknowledgement

This research was fully supported by grant from Universitas Indonesia PITTA 2018.

References

[1] Démurger S 2001 *J. Comp. Econ.* **29** p 95
[2] Button K and Taylor S 2000 *J. Air Transp Manag.* **6** p 209
[3] Fu X, Oum T H and Zhang A 2010 *Transportation Journal.* **24**
[4] González-Savignat M 2004 *JTEP.* **38** pp 77-107
[5] Vickerman R 1995 *Transport Policy.* **2** p 225
[6] Démurger S *et al* 2002 *Asian Economic Papers.* **1** p 146
[7] Peoples J 2014 The economics of international airline transport. Emerald Group Publishing.

IOP Conf. Series: Materials Science and Engineering **650** (2019) 012005 doi:10.1088/1757-899X/650/1/012005

[8] Berawi M A, Miraj P, Adhityo A D and Sakti GR 2017 *IOP Conf Ser Earth Environ Sci.* **109** 012009
[9] Berawi M A, Ibrahim B E and Miraj P 2019 *JDBE* **19** p 40
[10] Karim S B A, Abdul-Rahman H, Berawi M A and Jaapar A 2017 A review on the issues and strategies of stakeholder management in the construction industry. *In Meeting and Conference on Management in Construction and Researchers Association (MICRA)*
[11] Rahman H Z *et al* 2018. *IJTech.* **3** p 549

Top Ten Similarity Ranking for Project Delay Factors in Construction Industry

Ade Asmi[1], Aurino Djamaris[2], Mohammad Ihsan[1]

[1]Civil Engineering Dept, Universitas Bakrie, Jl HR Rasuna Said Kav 22 Kuningan, Jakarta Selatan 12920

[2]Management Dept, Universitas Bakrie, Jl HR Rasuna Said Kav 22 Kuningan, Jakarta Selatan 12920

ade.asmi@bakrie.ac.id; aurino@bakrie.ac.id; mohammad.ihsan@bakrie.ac.id

Abstract. Delays that occur on a construction projects can cause project performance inefficiencies and can even add to the project cost. Factors due to project delays consist of many things that can cause project delays. This study aims to find a top ten similarity factor that causes delays in a construction projects based on the results of research in Jakarta construction projects against previous studies in other countries. From a total of 100 distributed questionnaires, 88 (88%) were filled in correctly and returned as data analysis material. The Data from the survey results were analyzed using Statistical Software Package (SPSS) to produce a ranking for Project Delay Factor. The ranking of the study was then re- analyzed based on the similarity of the previous research so that the similar Project Delay Factor Ranking was obtained.

Keywords: Project Delay Factor

1. Introduction

Delay factors in construction projects can be caused by many factors that influence one another. Factors that can cause delays in this project are very crucial and need to be identified to improve and anticipate problems with delays in construction projects. For this reason, a review of the literature on comprehensive project delays was carried out for identifying the anticipation of the risk of factors causing delays in construction projects.

Jackson and Steven (2001) conducted an analysis of the causes of delays in construction projects in Ilorin using a questionnaire survey and found that the main factor of delay was "price fluctuations in material or labor"; "irregular orders", "late employee salaries", "lack of proper analysis in the process of project bidding", "selection of contractors who are less competent", "bidding on improper project prices and ever-changing requests from clients".

Jackson (2002) investigated the causes of late construction projects in the United Kingdom through survey questionnaires and found results of research that delays were caused by "additional costs at the time of design changes", "design change factors", "information availability", "estimation methods", "project management and design team performance".

Chang (2002) wants to research the causes of construction project delays through four project cases to calculate their contribution to engineering design projects in America. The results of this

IOP Conf. Series: Materials Science and Engineering **650** (2019) 012006 doi:10.1088/1757-899X/650/1/012006

study found that the main factor of the delay in construction projects was "the request of the owner to increase the scope of work and additional work".

Frimpong et al. (2003) conducted a research survey using a questionnaire consisting of 26 factors to obtain the main contributor to the delay in the Groundwater Drilling project in Ghana. Base on the 26 factors tested, 10 main factors were "payment difficulties"; "poor contract management"; "material procurement"; " inflation"; "financial difficulties from contractors"; and "escalation of material prices"; "payments during construction, planning and poor scheduling"; "bad weather and preparation of incorrect project cost estimates".

Koushki et al. (2005) conducted a study of the delay in housing projects funded by the private sector in Kuwait. There were 3 main factors that caused delays, including problems related to the contractor, material and financial difficulties of the owner.

Omoregie and Radford (2006) found that the main factor in delays in infrastructure projects in Nigeria was "price fluctuations"; "financing and payments from completed work"; "poor contract management"; "schedule delays"; "changes in conditions on the ground"; "inaccurate estimates"; "material shortages"; "additional work"; "design changes"; "subcontractors and supplier nominations"; "weather", "non-compliance with the conditions of the work contract"; "errors and differences in the conditions of the employment contract".

Azhar et al. (2008) conducted an investigation into the delay in construction projects in Pakistan. Research studies were carried out using a questionnaire consisting of 42 factors showing that the 10 main factors in delay were fluctuations in "material prices"; "high costs for engine maintenance"; "too low supply"; "procedures for procurement of goods and materials", "procurement phase", "methods of improper cost estimates", "additional work", "improper planning", "inadequate government policies".

Le-Hoai et al. (2008) investigated the causes of delays in construction projects in Vietnam using questionnaire surveys. Investigations involved 21 factors and 5 general factors including "field management and poor supervision"; "poor project management"; "financial difficulties from the owner"; "financial difficulties from contractors and design changes".

Enshassi et al. (2009) conducted a questionnaire study survey consisting of 42 factors to investigate the main causes of delays in construction projects in Gaza among contractors, consultants and clients. The results of the study show that the 10 main factors that cause delays received by 3 parties (contractors, consultants and clients) are increases in material prices, late construction, material and equipment supplies by contractors, fluctuations in construction material prices, exchange rates against the US dollar, monopoly of project material by several material providers, source constraints: funding and lack of preparation from related parties, lack of funds in planning or supervision during the contract stages before and after, improvements to drawings during the construction phase, design changes and lack of accuracy in taking quantities material.

Kaliba et al. (2009) conducted a study to determine the contributors to the late construction of flat road construction projects in Zambia. The results of the study show that the main cause of delays is the change in construction costs caused by weather caused by heavy rains and floods, changes in the scope of work, protection of the environment and mitigation of costs, schedule delays, strikes from labor, technical challenges, inflation and local government pressure.

Ameh et al. (2010) investigated the causes of construction delays in 53 telecommunication projects in Nigeria through questionnaire surveys. The survey results show that 7 main factors are "inexperienced contractors"; "material prices"; "fluctuations in material prices"; "repeated design changes"; "economic stability" and "high interest rates on loans imposed on contractors"; "types of payments"; "loans and payments".

Ade Asmi et al. (2016) investigate the factors that have been identified in the literature review study where these factors cause delays in the construction industry. A comprehensive review of 46 articles produced 78 general factors from construction delays that were used in the investigation of factors that caused construction delays in the construction industry in Indonesia.

2. Analysis

Based on previous research (Ade Asmi et al 2016), this study identifies general factors that cause delays; and to assess in a hierarchical manner the causes of delays in construction projects in Jakarta. This research is based on two approaches to research methodology including mapping literature reviews and questionnaires. Both of these methods act as supplements to each other which make data collection more comprehensive. From the factors that have been identified, through the literature review literature from previous research. Then a structured questionnaire was drafted and randomly designed and distributed to obtain the ranking of the study results. The top ten results of the study are as follows:

Table 1. Top 10 Project Delay Factors (Ade Asmi et al 2016)

Project Delay Factor	Mean Rank Value	Ranking
Frequent design changes	39.05	1
Financial difficulties of owner	37.14	2
Delay in progress payment by owner	36.92	3
Schedule delay	36.16	4
Incompetent subcontractors	36.08	5
Late delivery of materials and equipment	34.98	6
Poor design and delays in design	34.34	7
Delay in material	34.19	8
Labour productivity	34.16	9
Delay preparation and approval of drawings	34.12	10

From the results of **Table 1**, project delay factors will be analyzed based on the similarity ranking of the results of the project delay factors research from other countries as shown in **Table 3** and **Table 3** as follows.

Table 2. Delay factors of Project by Country (a)

No.	Author	Project Delay by Country	Delay Factors
1	Ogunlana S.O et al.,(1996)	High rise building construction project: Bangkok, Thailand"	• "Shortages of construction materials"; "Shortage of site workers"; "Frequent changes by owners"
2	Chan and Kumaraswamy (1997)	Time overrun in construction projects, Hong Kong	• "Poor site management" ; "Unforeseen ground condition" ; "Low speed of decision making"; "Client-initiated variation"; "Necessary variations of works"
3	Al-Khalil M and Al- Ghafly (1999)	Public utility projects, Saudi Arabia	• "Cash flow and financial difficulties"; "Difficulties in getting work permit"; "Practice of assigning contract to lowest bidder"; "Underestimate project duration"; "Effect of sub- surface condition"; "Changes in scope of projects"; "Ineffective planning and scheduling by contractors"; "Shortage of manpower".

Table 2. Delay factors of Project by Country (b)

No.	Author	Project Delay by Country	Delay Factors
4	Odeh A.M and Battaineh (2002)	Traditional contracts of Large construction industry, Jordan	• "Inadequate contractor experience"; "Owner interference" ; "Finance and payments of completed work"; "Slow decision making by owners"; "Labor productivity"
5	Frimpong Y, Oluwoye J , Crawford (2003)	Groundwater construction Projects, Ghana	• "Monthly payment difficulties"; "Poor contractor management"; "Material procurement";" Poor technical performances"; "Escalation of material prices"
6	Long N.D, Ogunlana S , Quang T and Lam K.C(2004)	Large construction project, Vietnam	• "Incompetent designers and contractors"; "Poor estimation and change management".; "Social and technological issues"; "Site related issues"; "Improper techniques and tools".
7	Assaf S.A, Al-Hejji S (2006)	Large construction projects, Saudi Arabia	• "Change order"; "Delay in progress payment"; "Ineffective planning and scheduling of project by contractor"; "Poor site management and supervision by contractor"; "Shortage of labors"; "Difficulties in financing project by contractor"
8	Fong N.K, Wong L.Y and Wong L.T (2006)	Fire services installation related contributors of construction delays in building construction work, Hongkong	• Improper site co-ordination management of the electrical and mechanical installation.; Lack of timely decision making of the client; Defects identified during the fire services Inspection.
9	Sambasivan M, Soon Y.W (2007)	Causes and effects of delays in Construction industry, Malaysia	• "Contractor's improper planning"; "Contractor's poor site management"; "Inadequate contractor experience"; "Inadequate client's finance and payments for completed work"; "Problems with subcontractors"; "Shortage in materials"; "Labor supply"; "Equipment availability and Failure"; "Lack of communication between parties"; "Mistakes during the construction stage"
10	Sweis, G., Sweis, R., Hammad, A.A., and Shboul, A. (2007).	Delays in construction projects, Jordan	• "Financial difficulties faced by contractor"; "Too many change orders by owner"; "Poor planning and scheduling of the project by contractor".
11	Muhammad Saiful Islam, and Bambang Trigunarsyah (2017)	Construction Delays in Developing Countries	• "Improper planning and scheduling"; "Lack of experience"; "Incompetent project team"; "Improper progress monitoring and cost control"; "Inadequate site inspection"; "Inaccurate time and cost estimation"
12	Mohd Razaki Abdullah, Ismail Abdul Rahman, Ade Asmi Abdul Azis (2009)	Large Mara Construction Projects Based on Project Management Consultant Perspective, Malaysia	• "Cash flow and financial difficulties faced by contractors"; "Contractor's poor site management"; "Inadequate contractor experience"; "Shortage of site workers"; "Ineffective planning and scheduling by contractors"; "Escalation of material prices"; "Practice of assigning contract to lowest bidder"; "Problems with subcontractors"; "Lack of communication between parties"; "Poor estimation and change management"; "Material procurement"; "Incompetent designers and contractors"; "Underestimate project duration"; "Low speed of decision making"; "Unforeseen ground condition"; "Changes in scope of projects"; "Frequent changes by owners"; "Owner interference"

Table 3. Similarity Ranking

No	Author	(1) Frequent design changes	(2) Financial difficulties of owner	(3) Delay in progress payment by owner	(4) Schedule delay	(5) Incompetent subcontractors	(6) Late delivery of equipment	(7) Poor design and delays in design	(8) Delay in material	(9) Labour productivity	(10) Delay preparation and approval of drawings
1	Ogunlana S.O et al.,(1996)	✓	-	-	-	-	-	✓	✓	✓	✓
2	Chan and Kumaraswamy (1997)	✓	-	-	-	-	-	✓	-	-	-
3	Al-Khalil M and Al-Ghafly (1999)	✓	✓	✓	✓	-	-	-	-	✓	-
4	Odeh A.M and Battaineh (2002)	✓	✓	✓	✓	✓	-	-	-	✓	✓
5	Frimpong Y, Oluwoye J , Crawford (2003)	-	✓	✓	✓	✓	-	-	✓	-	-
6	Long N.D, Ogunlana S , Quang T and Lam K.C(2004)	-	-	-	✓	✓	-	✓		-	✓
7	Assaf S.A, Al- Hejji S (2006)	✓	✓	✓	✓	✓	-	-		✓	-
8	Fong N.K, Wong L.Y and Wong L.T (2006)	✓	-	-	✓	-	-	-	-	-	-
9	Sambasivan M, Soon Y.W (2007)	-	✓	✓	✓	✓	✓	-	✓	✓	-
10	Sweis, G., Sweis, R., Hammad, A.A., and Shboul, A. (2007).	✓	✓		✓	-	-	-	-	-	✓
11	*Muhammad Saiful Islam, and Bambang Trigunarsyah (2017)*	-	✓	-	✓	✓	✓	-	-	-	-
12	Mohd Razaki Abdullah, Ismail Abdul Rahman, Ade Asmi Abdul Azis (2009)	✓	-	-	✓	✓	✓	✓	✓	-	✓
	Frequency	**8**	**7**	**5**	**10**	**6**	**3**	**4**	**4**	**5**	**5**

Based on analysis of **Table 3** similarity ranking, the Top ten similarity ranking can be obtained as follows:

Table 4. Top ten Similarity Ranking

Project Delay Factor	Frequency	Ranking
Schedule delay	10	1
Frequent design changes	8	2
Financial difficulties of owner	7	3
Incompetent subcontractors	6	4
Delay in progress payment by owner	5	5
Labour productivity	5	6
Delay preparation and approval of drawings	5	7
Poor design and delays in design	4	8
Delay in material	4	9
Late delivery of equipment	3	10

Based on **Table 4** above, "Schedule delay is ranked first. Factors that have the same frequency similarity ranking will be determined based on the priority of the research ranking (Ade Asmi et al 2016). For example, such as the delay factor "Delay in progress payment by owner" is taken as the number 5 because in the ranking study (Ade Asmi et al. 2016) the ranking is higher than the "Labour productivity" which is ranked sixth. Then for the "Labor productivity" factor, it has a higher rating than the "Delay preparation and approval of drawings" factor so that the "Delay preparation and approval of drawings" factor is ranked seventh. Likewise for the factor "Poor design and delays in design" has a higher rating than the factor "Delay in material" which is ranked eighth.

3. Conclusion

From the results of **Table 4**, it is clear that the "Schedule delay" factor is a factor that must be considered in project management to avoid delays. So that the absolute ability to manage, monitor and supervise project schedules is very important especially in maintaining critical path activities. For the factor "Frequent design changes" is a note as a factor that must be considered in running the project to avoid delays. Therefore, it is necessary to minimize the risk of events using the Design and Build scheme to avoid the many design changes in project implementation.

For the "Financial difficulties of owner" factor where the owner faces financial difficulties while the "Delay in progress payment by owner" factor is a condition where the payment is delayed by the owner but the owner does not experience financial difficulties significantly and can still make payments even though it's too late. For the "Financial difficulties of owner" factor and the "Delay in progress payment by owner" factor, we need to watch out for construction practitioners in terms of choosing projects, especially for the project owner itself to avoid this. For example, by checking the track record of the owner, the selection of projects originating from the government or the private sector also needs attention. Factor "Incompetent subcontractors" becomes a matter that needs to be considered to support the main contractor in implementing the project. So that the initial process of selecting a subcontractor is very important. For the "Labor productivity" factor related to resource leveling or management of Human Resource management is very important because the project costs will increase if construction practitioners cannot control "Labor productivity" as expected.

The "Delay preparation and approval of drawings" factor is required for the ability to coordinate preparation and approval of drawings within the project. For Factors "Poor design and delays in design" The need for vigilance and anticipation of design changes causes delays or changes in design by owner. For the "Delay in material" factor Material management capability is needed to anticipate material availability. For the "Late delivery of equipment" factor, it is necessary to manage the equipment settings in the project, especially heavy equipment.

References

[1] Jackson O and Steven O 2001 Management of cost overrun in selected building construction project in Ilorin *Review of Business and Finance.* **3**(1)

[2] Jackson S 2002 *Project cost overruns and risk management.* Paper presented at the Greenwood, D. (Ed.) Proceedings of Association of Researchers in Construction Management 18th Annual ARCOM Conference, Newcastle.

[3] Chang A S T 2002 Reasons for cost and schedule increase for engineering design projects *Journal of Management in Engineering.* **18**(1) pp 29-36

[4] Frimpong Y, Oluwoye J and Crawford L 2003 Causes of delay and cost overruns in construction of groundwater projects in a developing countries; Ghana as a case study *International Journal of Project Management.* **21** pp 321-326

[5] Koushki P A, Al-Rashid K and Kartam N 2005 Delays and cost increases in the construction of private residential projects in Kuwait *Construction Management and Economics.* **23**(3) pp 285-294

[6] Omoregie A and Radford D 2006 Infrastructure delays dan cost escalation: Causes dan effects in Nigeria. papers presented at the proceeding of sixth international postgraduate research conference

[7] Azhar N, Farooqui R U and Ahmed S M 2008 Cost overrun factors in construction industry of Pakistan. Paper presented at the first international conference on construction in developing countries (ICCIDC-I) "Advancing and Integrating Construction Education, Research and Practice

[8] Le-Hoai L, Lee Y D and Lee J Y 2008 Delay and cost overruns in Vietnam large construction projects: A comparison with other selected countries KSC *Journal of Civil Engineering.*

[9] Enshassi A, Lisk R, Sawalhi I and Radwan I 2003 Contributors to construction delays in Palestine. *The Journal of American Institute of Constructors.* **27**(2) pp 54-53

[10] Kaliba C, Muya M and Mumba K 2009 Cost escalation and schedule delays in road construction projects in Zambia *International Journal of Project Management.* **27** pp 522-531

[11] Ameh O J, Soyingbe A A and Odusami K T 2010. Significant factors causing cost overruns in telecommunication projects in Nigeria *Journal of Construction in Developing Countries.* **15**

[12] Asmi A, Pratama J C and Safrilah S 2016 Identifikasi Faktor-Faktor Keterlambatan Dalam Proyek Konstruksi di Jakarta. (Jakarta: Muhammadiyah University)

[13] Ogunlana S O and Promkuntong K 1996 Construction delays in a fast-growing economy: Comparing thailand with other economies *International Journal of Project Management.* **14**(1) pp 37-45

[14] Chan D W M and Kumaraswamy M M 1997 A comparative study of causes of time overrun in Hongkong construction projects *International Journal of Project Management.* **15**(1) pp 55-63

[15] Al-Khalil M and Al-Ghafly M 1999 Important causes of delay in public utility projects in Saudi Arabia *Const. Management Econ.* **17** (5) pp 647–655

[16] Odeh A M and Battaineh H T 2002 Causes of construction delay: traditional contracts *International Journal of Project Management.* **20**(1) pp 67-73

[17] Long N D, Ogunlana S, Quang T and Lam K C 2004. Large construction projects in developing countries: a case study from Vietnam *International Journal of Project Management.* **22** pp 553-561

[18] Assaf S and Al-Hejji S 2006 Causes of delay in large construction projects *International Journal of Project Management* **24**(4) pp 349–357

[19] Fong N K, Wong L Y and Wong L T 2006 Fire services installation related contributors of construction delays *Building and Environment.* **41** pp 211-222

[20] Sambasivan M and Soon Y W 2007 Causes and effects of delays in Malaysian construction industry *International Journal of Project Management* **25** pp 517-526

[21] Sweis G, Sweis R, Hammad A A and Shboul A 2007 Delays in construction projects: The case of Jordan *Int. J. Project Management*, Article in Press.

[22] Islam M S and Trigunarsyah B 2017 Construction Delays in Developing Countries: A Review *KICEM Journal of Construction Engineering and Project Management Online* ISSN 2233-9582.

[23] Abdullah M R, Rahman I A and Azis A A A 2009 Delay in Large Mara Construction Projects Based on Project Management Consultant Perspective. *Proceedings of MUCEET 2009 Malaysian Technical Universities Conference on Engineering and Technology* (Pahang, Malaysia:MS Garden,Kuantan, June 2009)

IOP Conf. Series: Materials Science and Engineering **650** (2019) 012007 doi:10.1088/1757-899X/650/1/012007

Proposed SDI equations to improve the effectiveness in evaluating crack damage on the road pavement

B H Setiadji*

Department of Civil Engineering, Faculty of Engineering, Diponegoro University, Jl. Prof. Soedarto, SH., Tembalang, Semarang, Indonesia, 50275

*bhsetiadji@ft.undip.ac.id

Abstract. Road pavement has a life-cycle, where the age of the road pavement starts from design to the end of service life, so that maintenance works are one of the most important stages. To enable carrying out maintenance works, a process of assessment of road conditions, both from structural and functional, is required. At present, one of the evaluations of the functional condition of the road is using surface distress index/SDI. The SDI method used today has a simple procedure and is easy to use. However, in the term of accuracy in estimating the functional condition, it will be seen that the SDI parameter is far from satisfying. In this study, an effort to increase the effectiveness of SDI parameters especially for crack damage was carried out. The improvement of SDI parameter was conducted by developing a non-linear equation for each type of crack damage based on the deduct value curve of the PCI method, with a coefficient of determination R^2 of at least 0.99. The proposed SDI equations have been calibrated and produced an error less than 6.2%. A comparison between the value of the proposed and the existing SDI also was presented and the results showed that the similarities between the two SDI values were only 19.86%. The rest of the existing SDI value was under- estimate proposed SDI, but there was also an existing SDI value over-estimated proposed SDI, especially in the road segment which was only found one type of crack damage with low/medium severity and low density.

1. Introduction

Road pavement has a repetitive life cycle, starting from planning, construction, operational, maintenance/rehabilitation, reconstruction and then this process returns to the operational stage and so on. Of all these processes, maintenance/rehabilitation (M & R) is a stage that has the longest period compared to the other stages, and together with operational and reconstruction stages, M & R will last until the end of the service life of the road pavement.

To determine the type of maintenance work required, the road agency will conduct a process of evaluating road conditions, to ensure that the proposed maintenance work is appropriate. This is important due to the limit amount of the available budget so that only the most effective and efficient one should be processed. This underlines how important the process of road condition evaluation is. Two kinds of condition evaluation that are usually conducted namely functional and structural condition evaluations.

Functional condition evaluation is performed to find out whether the road pavement still has an adequate level of performance to be able to carry out its functions. This evaluation on a flexible pavement usually consists of 3 types, i.e. evaluation of road deterioration, roughness, and skid

IOP Conf. Series: Materials Science and Engineering **650** (2019) 012007 doi:10.1088/1757-899X/650/1/012007

resistance. In Indonesia, one parameter that is commonly used to evaluate the functional condition of the road is called surface distress index (SDI) [1]. The SDI parameter is very simple and easy to use because it only evaluates 3 types of road damage (i.e. cracking, potholes and ruts), compared to 19 types of road damage identified in PCI (Pavement Condition Index) method [2,3], another parameter that is used to evaluate the road functional condition.

One type of damage in SDI, which is crack damage, is only represented by one type of crack damage compared to seven types of crack damage in the PCI method. With all the features, the PCI method is recognized by AASHTO as one of the most comprehensive methods in estimating the functional condition of the road [4,5,6]. However, with its comprehensive level, the measurement and analysis procedure of the PCI method becomes very complex. In this point of view, the advantages and disadvantages of SDI actually lie. The advantages of the SDI are in terms of simplicity and ease of measurement of the functional condition of the road in practice by surveyors with low to medium education levels, while the weakness of the SDI is the simplification can lead to inaccurate estimation of road functional condition, which in turn will affect the inaccuracy in the selection of road handling types.

This research is an initial part of a series of studies that aims to optimize the use of SDI parameters for the purpose of evaluating the functional condition of the road. As an initial stage, this research is intended to propose an increase in the effectiveness of SDI parameters especially for evaluating the crack damage on the road pavement.

2. Surface Distress Index (SDI)

Surface distress index (SDI) is a scale of road functional condition obtained from visual observations (road condition survey/RCS) and detailed measurements on road damage that occurs in the field. In observing road damage in the field, the road evaluated has to be divided into 100 m long segments based on the Road Condition Survey (RCS) Guide No. SMD-03/RCS [1]. According to this guide, there are 4 criteria for determining the SDI index, namely the total area of the crack, the average crack width, the number of holes per km of road length, and the average rut depth. Especially for crack damage, the SDI equations for this kind of damage are as follows.

- SDI_1 as a function of total area of crack

$SDI_1 = 5$	if total area of crack is less than 10%	(1)
$SDI_1 = 20$	if total area of crack is in between 10% and 30%	
$SDI_1 = 40$	if total area of crack is more than 30%	

- SDI_2 as a function of average crack width and SDI_1

$SDI_2 = SDI_1$	if average crack width is less than 3 mm	(2)
$SDI_2 = SDI_1 * 2$	if average crack width is more than 3 mm	

According to the equations above, the maximum SDI value for the criteria of total area of crack and average crack width equals to 80. SDI parameter has main advantages in the form of simplicity and ease of use, but this parameter has several fundamental weaknesses, namely:

a. There is only one type of crack damage, although in practice crack damage has more than one type with various characteristics;
b. All the results of SDI parameter analysis are discrete (not continuous) so the results of different damage measurements can get the same SDI value

3. Research Methodology

This research consisted of three main stages as follows.

- Collection of data related to damage. The types of data needed in this study include the type of crack damage, dimensions of crack damage (expressed as the ratio between the extent of damage to the area evaluated or named as density in %), and the severity of the crack damage.

Variations of the data refer to the PCI method [2, 3], i.e. there are 7 types of crack damage and 3 damage severity levels. The data is collected using road condition survey (RCS) method developed by Setiadji et al. [7]. There are 9 RCS locations, which is consisting of urban, provincial and national roads. Seven of nine RCS locations around Semarang is shown in figure 1, while two other locations, that is, Bawen - Salatiga road and the Weleri - Temanggung road, are at a considerable distance from Semarang city.

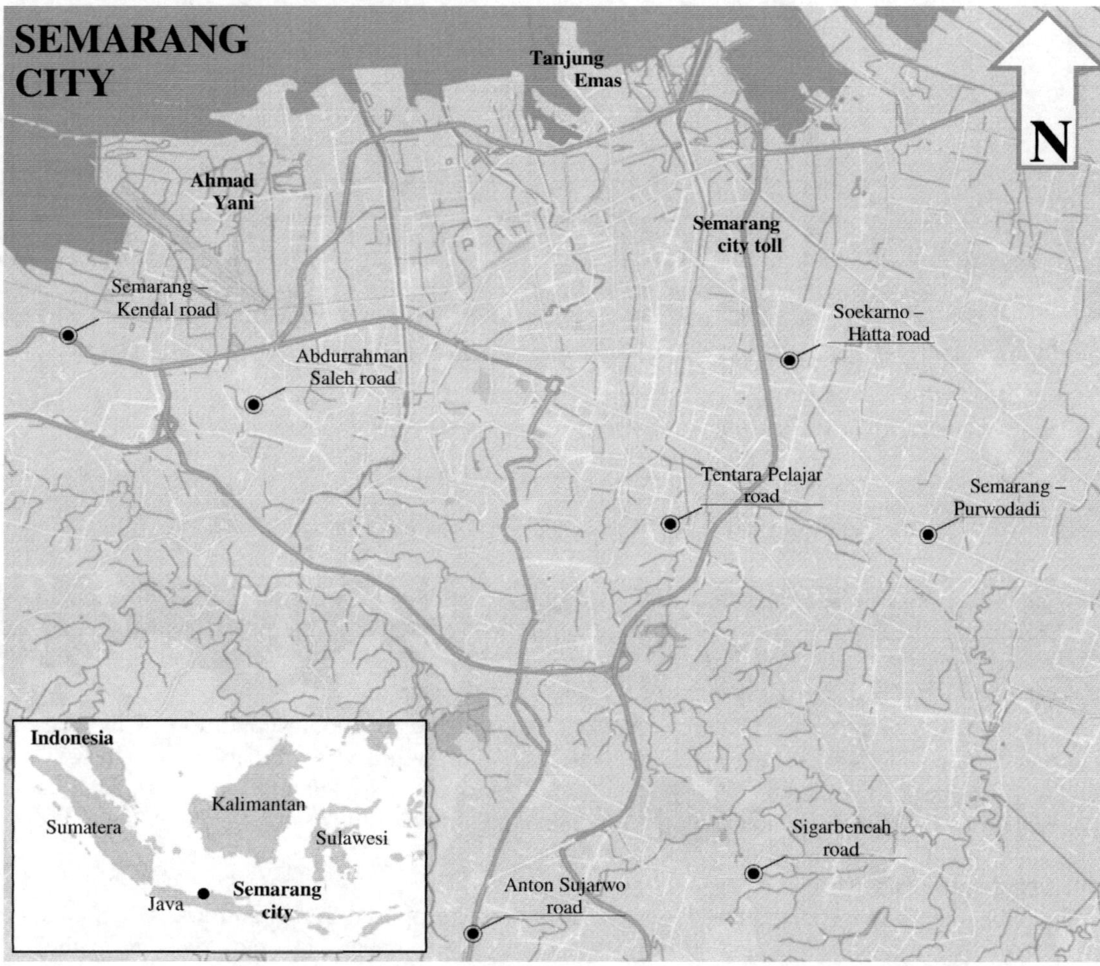

Figure 1. Locations of road condition survey around Semarang city

- Development of the proposed SDI equations. As a reference for developing the equations, PCI method were used. In PCI method, there are 2 steps of determining PCI values: (i) to score on the negative effects generated by each type of distress based on distress density and level of severity, called as deduct value; (2) to calculate PCI value where the summation of deduct values produced by different distresses will be used to subtract from possible maximum PCI, i.e. 100 [8]. In step (2), it is possible to make adjustments to the total deduct value so that the calculated PCI is not negative. In this study, only step (1) of the PCI method was adopted for the development of proposed SDI value. Step (1) of the PCI method is similar to the determination of the SDI value, where distress density and level of severity variables (see Fig. 2), are similar to the SDI_1 and SDI_2 variables (equations 1 and 2) in the SDI method. In this part, the SDI_1 variable, which has a discrete value, needs to be improved by adopting the density variable from the PCI method which has a continuous characteristic. According to this similarity, the proposed SDI values can be assumed as a function of the deduct value curves. However, the total score of deduct value in the PCI method is different from the total score of the damage of the SDI method, which is a maximum of 100 and 80 respectively. Therefore, it is necessary to adjust the total score of deduct value so that the total score can be used in the SDI method.

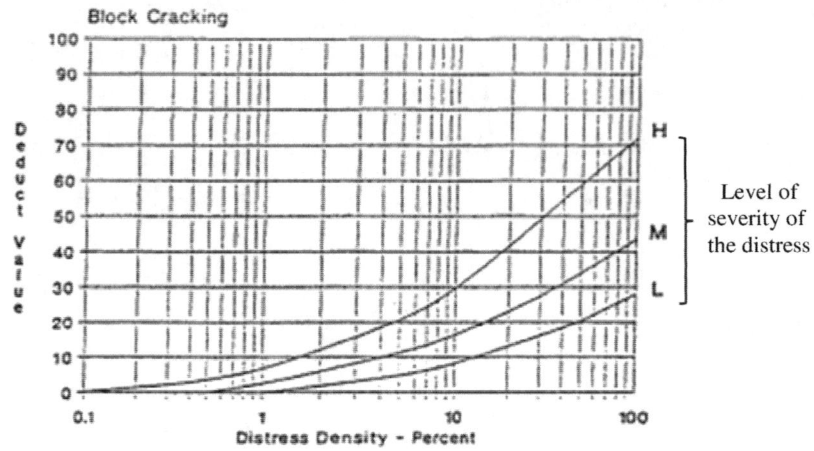

Figure 2. Example of deduct value curve in PCI method [2, 3]

In order to allow the proposed SDI value is still usable in RCS Guide [1], the maximum existing SDI value was adopted. Adjustment of the proposed maximum SDI value for the crack damage to the total proposed maximum SDI value (taking into account pothole and rut damages) will be carried out at the next stage of this research.

- The use of the proposed SDI equations to evaluate the functional condition of the road by using data collected in stage 1. Subsequent analysis was carried out by comparing the existing and proposed SDI values from the equations developed to see the effectiveness of the use of the proposed SDI values to assess the functional condition of the road.

4. Results and Discussion

The most important step in this study was to develop the proposed SDI equations. As explained in section 3 previously, the deduct value (DV) curve of seven types of crack damage in the PCI method was used to develop the proposed SDI equations. Each DV curve has 3 levels of severity, i.e. high, medium and low, and each level of severity has a different density range. A density range of up to 100% can be found on the DV curve of alligator cracking, block cracking and slippage cracking. As for other types of damage, the density range varies between 0 - 30%.

Using hypothetical data from the density parameters, DV values will be obtained for each severity level, where the DV values obtained will form the following sequence: $DV_{\text{low severity}} < DV_{\text{medium severity}} < DV_{\text{high severity}}$. Based on the DV value of each density, the SDI value can be calculated using the following equation.

$$SDI = 80 * DV / DV_{max} \quad (3)$$

In which: 80 is the maximum *SDI* value for crack damage, *DV* is the deduct value for each density, DV_{max} is the maximum *DV* value for crack damage, i.e., in this case, DV, it is equal to 91

IOP Conf. Series: Materials Science and Engineering **650** (2019) 012007 doi:10.1088/1757-899X/650/1/012007

in alligator cracking or slippage cracking damage. The proposed SDI equations developed using equation (3) is as presented in table 1. The SDI equations in table 1 are the improvement of the SDI curves that have been developed by Setiadji et al. [7].

From table 1, except for edge cracking, the SDI equations for all types of crack damage is divided into two, namely the SDI equation for density between 0% - 10% and the SDI equation for the density of more than 10%. This is because it is difficult to model SDI in one equation for all range of density values. By making the SDI equation based on two ranges of density values, all equations in table 1 can have a minimum coefficient of determination (R2) equals to 0.99. Due to a large number of SDI equations produced, to make it easier for ones to use the SDI equation, development of an application based on the proposed SDI equations, called e-SDI, will be initiated at the next stage of research.

Table 1. Proposed SDI equations

Crack types[a] - level of severity[b]	SDI equations for $0 < x < 10$[c]	SDI equations for $x \geq 10\%$
AC – L	$y = 0{,}0675x^3 - 1{,}3606x^2 + 9{,}6465x + 1{,}0849$	$y = 5E\text{-}05x^3 - 0{,}0118x^2 + 1{,}021x + 19{,}8$
AC – M	$y = 0{,}1819x^3 - 3{,}1301x^2 + 16{,}706x + 5{,}3688$	$y = 5E\text{-}05x^3 - 0{,}0118x^2 + 1{,}0225x + 32{,}465$
AC – H	$y = 0{,}2086x^3 - 3{,}6735x^2 + 20{,}365x + 9{,}6453$	$y = 11{,}154\ln(x) + 28{,}859$
BC – L	$y = -0{,}0186x^3 + 0{,}2614x^2 + 0{,}0422x - 0{,}0212$	$y = 2E\text{-}05x^3 - 0{,}005x^2 + 0{,}4878x + 3{,}6144$
BC – M	$y = -0{,}0171x^3 + 0{,}1596x^2 + 1{,}6502x - 0{,}4228$	$y = 1E\text{-}05x^3 - 0{,}0043x^2 + 0{,}5773x + 9{,}8729$
BC – H	$y = 0{,}0622x^3 - 1{,}1344x^2 + 7{,}8845x - 1{,}2327$	$y = 5E\text{-}05x^3 - 0{,}0126x^2 + 1{,}2512x + 15{,}333$
LC/TC – L	$y = -0{,}0051x^3 - 0{,}0052x^2 + 2{,}1948x - 0{,}5303$	$y = -0{,}0088x^2 + 0{,}8352x + 8{,}3516$
LC/TC – M	$y = -0{,}2536x^2 + 5{,}1317x + 1{,}0612$	$y = -0{,}0176x^2 + 1{,}3187x + 15{,}824$
LC/TC - H	$y = -0{,}4332x^2 + 9{,}1908x + 5{,}652$	$y = -0{,}044x^2 + 2{,}9011x + 29{,}89$
EC – L	y = 0,0062x3 - 0,2136x2 + 2,4256x + 0,5135	
EC – M	$y = 0{,}0072x^3 - 0{,}2749x^2 + 3{,}6453x + 3{,}9194$	
EC – H	$y = 0{,}009x^3 - 0{,}3785x^2 + 5{,}5896x + 7{,}1162$	
JRC – L	$y = 0{,}0251x^3 - 0{,}4627x^2 + 3{,}5049x - 0{,}7364$	$y = -0{,}0044x^2 + 0{,}6593x + 7{,}033$
JRC – M	$y = 0{,}0674x^3 - 1{,}188x^2 + 7{,}9693x + 0{,}7459$	$y = -0{,}0308x^2 + 1{,}5824x + 16{,}264$
JRC – H	$y = 0{,}1135x^3 - 2{,}0996x^2 + 14{,}797x + 2{,}073$	$y = -0{,}0352x^2 + 1{,}8901x + 38{,}242$
SC – L	$y = 0{,}0333x^3 - 0{,}646x^2 + 5{,}6258x - 0{,}3221$	$y = 3E\text{-}05x^3 - 0{,}0092x^2 + 0{,}8766x + 16{,}346$
SC – M	$y = 0{,}0596x^3 - 1{,}2003x^2 + 9{,}6815x + 1{,}3868$	$y = 7E\text{-}05x^3 - 0{,}014x^2 + 1{,}073x + 29{,}157$
SC – H	$y = 0{,}1087x^3 - 2{,}1719x^2 + 16{,}55x + 2{,}7742$	$y = 7E\text{-}05x^3 - 0{,}014x^2 + 1{,}0328x + 51{,}234$

[a] AC = alligator cracking, BC = block cracking, LC = longitudinal cracking, TC = transversal cracking, EC = edge cracking, JRC = joint reflection cracking, SC = slippage cracking
[b] L = low severity, M = medium severity, H = high severity
[c] x = density (%), y = SDI

Furthermore, a calibration process was necessary to be performed to ensure that the accuracy of the equations before the equations are used in practice. The calibration process was done by comparing two SDI values, i.e. the SDI value obtained from the conversion of DV and SDI values generated by the SDI equations in table 1. The results of the SDI equation calibration process are presented in table 2.

IOP Conf. Series: Materials Science and Engineering **650** (2019) 012007 doi:10.1088/1757-899X/650/1/012007

From table 1, it can be seen that all equations can predict the SDI values with a difference of less than 5 points (6.2%). This shows that the proposed SDI equations can estimate all SDI values quite accurately. In addition, the following guidelines are required for the use of the equations in practice.

a) Determine the density value (i.e. the ratio of the area of damage to the area evaluated) for each type of crack damage that is obtained from the results of road condition survey (RCS);
b) The maximum density value of edge crack, joint reflection crack, and longitudinal/transversal cracks are 20%, 30%, and 30%., respectively. For the other types of cracks, the maximum density is 100%;
c) Each type of crack damage with different severity will produce individual SDI values. The individual SDI values of each crack damage and severity level will be accumulated as the total SDI value for the road segment reviewed;
d) If the total SDI value is greater than 80, then the value used is 80. This is because, at this time, the maximum SDI value for crack damage is 80. However, if the total of SDI value is less from 80, then that value is expressed as the total SDI value for the segment.

Table 2. The calibration results for SDI equations

Crack types – level of severity	Average deviation of SDI	Largest deviation of SDI
AC – L	0.67	2.12
AC – M	0.63	2.74
AC – H	0.68	1.72
BC – L	0.27	1.34
BC – M	0.88	3.08
BC – H	0.51	2.04
LC/TC – L	0.12	0.41
LC/TC – M	0.75	1.97
LC/TC – H	0.71	2.29
EC – L	0.23	0.75
EC – M	0.24	0.62
EC – H	0.25	0.85
JRC – L	0.11	0.39
JRC – M	0.34	0.83
JRC – H	0.41	1.77
SC – L	0.71	4.59
SC – M	0.88	4.92
SC – H	0.85	4.51

In the proposed SDI method, each road segment can produce SDI values continuously with more accuracy (not discrete, as indicated by the existing SDI method), where this will be very helpful in doing ranking based on the condition value of the road segment that must be handled. However, because in the SDI method, the total SDI value is a maximum of 80, so that for the road with a value of SDI equals to 80, it indicates that the road is in a condition of severe damage.

To help provide further information on what type of crack damage is dominant, which will be useful in determining the type of handling quickly, it is proposed to provide an abbreviation of the type of crack after the SDI value. For example, if the value of the SDI is 80 AC, it means that the crack damage in that segment is quite significant with the dominant type of crack damage being AC (alligator cracking).

Furthermore, the proposed SDI equations need to be implemented using field data to see the accuracy of this method in predicting SDI values (see tables 3 and 4). In addition, the proposed SDI method will be compared with the existing SDI method to assess effectiveness in delivering information on functional conditions of a road. Table 3 is a comparison of the SDI values produced by the proposed method and the existing method in an urban road, while table 4 presents the same comparison with the road damage data taken on a national road.

Table 3. Comparison between the existing and proposed SDI based on road condition survey on Abdurrahman Saleh Road

Sta.	Crack types[a] & level of severity[b]	Density (%)	Average crack width (mm)	Average crack width per segment (mm)	Total Density (%)	Existing SDI	Proposed SDI per type of crack	Total proposed SDI	Applied proposed SDI
0+100	AC – H	0.96	40	28.33	5.31	10	25.99	63.44	63.44 AC
	AC – M	0.74	30				16.09		
	AC – L	3.61	15				21.35		
0+200	AC – L	1.02	10	22.5	2.83	10	9.58	36.01	36.01 AC
	AC – M	1.81	35				26.43		
0+300	AC – H	0.72	40	40	10.26	40	22.48	98.84	80.00 AC
	AC – M	6.22	30				31.95		
	AC – H	3.32	50				44.40		
0+400	AC – M	3.24	30	17	13.06	40	32.82	94.24	80.00 AC
	AC – L	0.65	10				6.80		
	AC – L	1.62	10				13.43		
	AC – M	3.11	25				32.52		
	LC – L	4.44	10				8.67		
0+500	LC – L	1.67	10	21.67	8.74	10	3.10	34.59	34.59 AC
	BC – L	0.33	35				0.02		
	AC – M	6.74	20				31.47		
0+600	LC – L	0.92	5	5	0.92	10	1.48	1.48	1.48 LC
0+800	AC – L	0.08	10	10	0.63	10	1.85	7.84	7.84 AC
	AC – L	0.55	10				5.99		

[a] AC = alligator cracking, LC = longitudinal cracking
[b] L = low severity, M = medium severity, H = high severity

Table 4. Comparison between the existing and proposed SDI based on road condition survey on Semarang – Purwodadi road (a)

Sta.	Crack types[a] & level of severity[b]	Density (%)	Average crack width (mm)	Average crack width per segment (mm)	Total Density (%)	Existing SDI	Proposed SDI per type of crack	Total proposed SDI	Applied proposed SDI
1+100	LC - L	0.8	10	10	0.8	10	1.22	1.22	1.22 LC
1+200	AC - M	0.61	20	20	0.61	10	14.44	14.44	14.44 AC
1+400	AC - H	1.72	40	40	1.72	10	34.87	34.87	34.87 AC
1+500	AC - M	0.55	20	15	2.52	10	13.64	17.37	17.37 AC
	LC - L	1.97	10				3.73		
1+600	AC - M	0.75	20	20	0.75	10	16.21	16.21	16.21 AC
1+700	AC - M	0.34	15	15	0.34	10	10.69	10.69	10.69 AC
1+900	AC - M	0.43	10	20	1.49	10	11.99	31.76	31.76 AC
	AC - M	1.06	30				19.78		
4+100	AC - M	0.22	20	20	0.22	10	8.89	8.89	8.89 AC
4+400	LC - L	1.41	10	10	1.41	10	2.54	2.54	2.54 LC

Table 4. Comparison between the existing and proposed SDI based on road condition survey on Semarang – Purwodadi road (b)

Sta.	Crack types[a] & level of severity[b]	Density (%)	Average crack width (mm)	Average crack width per segment (mm)	Total Density (%)	Existing SDI	Proposed SDI per type of crack	Total proposed SDI	Applied proposed SDI
4+600	AC - L	0.25	5	5	0.25	10	3.41	3.41	3.41 AC
4+900	AC - L	0.48	10	10	3.31	10	5.41	10.93	10.93 AC/LC

a AC = alligator cracking, EC = edge cracking
b L = low severity, M = medium severity, H = high severity

From the results of the determination of the proposed SDI value of the nine evaluated road segments (where two road segments are shown in tables 3 and 4), it was found that generally, the proposed SDI value was higher than the existing one. Of the 141 evaluated road segments, only 28 road segments, or around 19.86%, have the proposed SDI value similar to the existing SDI (in this case, the similarity is expressed by the difference between the two by maximum values of 5 points, as shown in tables 3 and 4 parts marked with shading). The rest of the existing SDI value was under- estimate the proposed one, but there was also the existing SDI value over-estimate the proposed one, especially in a road segment that is only found one type of crack damage with low/medium severity and low density.

5. Conclusions

This paper presented a proposed surface distress index (SDI) equations to improve the effectiveness of assessing the functional condition of the road, especially for crack damage. The proposed SDI equations developed were functions of maximum SDI value for crack damage, PCI's deduct value for each density, and PCI's maximum deduct value for crack damage. It resulted in: (i) the proposed SDI equations can predict the SDI values with a difference of less than 6.2%. This shows that the proposed SDI equations can estimate SDI values quite accurately, (ii) the similarities between the proposed and existing SDI were only 19.86%. Most of the existing SDI value was under-estimate proposed one, but there was also an existing SDI value over-estimated proposed one, especially in the road segment which was only found one type of crack damage with low/medium severity and low density

References

[1] Directorate General of Highway 2011 *Road Condition Survey Guide No. SMC-03/RCS*
[2] Shahin M Y 2005 *Pavement Management for Airports, Roads, and Parking Lots*, 2nd Edition (New York: Springer)
[3] ASTM 2007 *Standard Practice for Roads and Parking Lots Pavement Condition Index Surveys*, ASTM D6433-07 (West Conshohocken: ASTM)
[4] AASHTO 2012 *Pavement Management Guide*, 2nd Edition p 200
[5] Karim F M A, Rubasi K A H and Saleh AA 2016 The road pavement condition index (PCI) evaluation and maintenance: a case study of Yemen *Organization, Technology, and Management in Construction* (Zagreb: Sciendo) **8** pp 1446–1455
[6] Setyawan A, Nainggolan J and Budiarto A 2015 Predicting the remaining service life of road using pavement condition index *Procedia Engineering* **125** pp 417– 423
[7] Setiadji B H, Supriyono and Purwanto D 2018 Surface Distress Index Updates to Improve Crack Damage Evaluation, *Proc. the 11th Asia Pacific Conference on Transportation and the Environment* Malang Indonesia (Malang: Brawijaya University)
[8] Shahin M Y and Kohn S D 1979 *Development of Pavement Condition Rating Procedure for Roads, Streets and Parking Lots, Volume 1: Condition Rating Procedure*, Technical Report CERL-TR-M-268 p 87

IOP Conf. Series: Materials Science and Engineering **650** (2019) 012008 doi:10.1088/1757-899X/650/1/012008

Benefits and implications of the different types of quality management in the Malaysian construction industry

Mazlina Zaira Mohammad[1*], Marina Viana[2] and Ahmed Talal Raghib[1]

[1]Universiti Teknologi MARA, Shah Alam, Selangor, Malaysia
[2]Asian Institute of Technology, Pathumthani, Bangkok, Thailand

*mazlinazaira@uitm.edu.my

Abstract. The quality management system (QMS) is defined in terms of quality assurance, quality control, and quality planning. It is necessary for the construction field to confirm the projects will to be finished with success and without any obstacles for the best quality, explicit amount and at a minimum attainable value. The aim of this article is to produce a crucial and constructive analysis through a revision of five revealed works of literature that relate to QMS in the construction field based on Malaysian research setting. Relevant articles associated with the QMS in Malaysia has been studied. The main findings at those researches indicated that the satisfaction of the owner or the client, improved management and work capacity of the company, and facilitate the project management team to expeditiously manage the development activities for the project life cycle. More studies can use constant performance indicators at the project to check the efficacy of QMS in many samples to enhance the findings.

1. Introduction

The issue of quality has become a heavy concern inside the development of the construction industry. There are several advantages that can be gained by International Organization for Standardization (ISO). ISO is an international organization that promotes a standard for various type of companies and products. However, this does not mean that if a company has succeeded in obtaining certification, they will be ready to expertise all the useful results. Kiev [1] stated in his study that the benefits obtained by construction companies as a result of the ISO implementation were small and indicated that the level of improvement was in line with their original expectations. That is why most construction companies are reluctant to implement the system because they are uncertain whether or not the enforced system is worth it particularly once an enormous outlay was invested with in implementing the system [1].

According to researchers, the Quality Assessment System in Construction (QLASSIC), the ISO 9001 quality management system (QMS), the quality assurance (QA) system, and the quality control (QC) system are among the many types of standards which provide a number of indicators on the way to establishing a high-quality process management system to manage the processes that affect their products or services. It was initially developed in 1987 to improve the service and merchandise standards provided by completely different business sectors [2]. QLASSIC is an important tool used to assess the degree of quality of a building work supported by the quality standards. QLASSIC score is considered One of QLASSIC's key outcomes is that it denotes the quality extent of the construction projects being evaluated [3]. In the field of construction, the quality management system had been enforced and widely adopted, especially those firms are able to handle mega projects. Although numerous research was carried out to review and analyze the relationship between the quality management system and various industries (e.g., manufacturing, food, service, etc.). There is a shortage of relevant construction industry studies. As a result, the importance of project works and cost accounting in the construction industry as compared to the quality management system is given by these researchers [4].

The main goals of this article are: To spot the various varieties of quality management system within the Malaysian construction field. To spot the advantages of the application of a quality management system in the construction companies in Malaysia. To spot the issues encountered by the firms of construction in the application of the system of quality management. To spot the methods adopted by construction firms in overcoming or mitigating the implementation dilemmas.

2. Major findings

Benefits gained from implementing quality management, there are several advantages which will be achieved through quality management. Keng and Kamil [2] done a research by surveying construction companies, it was found that the benefits gained by contractors are: a common language for communicating quality assurance. Major general guidance to a top quality management system for a company or establishment in any business. Facilitates and promotes auditing and certification by third parties. Rise the confidence of the owner. Raise the competition for higher quality merchandise and services. Raise the satisfaction of the client on quality and reduce the prices to make the correction for the problematic work.

Keng and Kamil [2] has categorized the advantages that are fulfilled by construction firms into two classes that internal and external benefits. Internal benefits: increased company communication, improve the documentation, improved the technique of operating, improved quality of labor done, bigger owner or client focus, improved worker morale, improved performance appraisal, and increased potency and productivity. External benefits: access to domestic market, higher competitive edge, higher perceived quality of labor done, improved profitableness, access to overseas market, having a valuable selling tool, improved consumer satisfaction, improved the relations with provider or the supplier.

The researcher has classified his discovery of the advantages seasoned in the ISO 9000 registered corporations into three classes that are major advantages to employees, a benefit to processes an advantage to a business. As an advantage to staff: improved shared aims, clear operating procedure, scale back employees conflicts, lower employee turnover, an additional suggestion from employees. As an advantage to operation: reduced the waste of materials, enhanced potency, raised product quality or service quality, higher management of subcontractors which resulted in lower operating prices, and enhanced amount of production. As an advantage to the business: enhanced sales to the current clients, additional new native clients, decreased the complaints, raised the profitability, more and more new clients abroad. The researcher also admitted that ISO 9000 certification's most strategic profit grant associate improved the relationship with clients the led to associate enhanced market share. The most operational undertaking blessings is that the operational process documentation that raises worker's abilities and permits the corporate to reply rapidly to cause order changes of the market. Additionally, to it, the ISO 9000 QMS additionally provides positive implications for shorter completion of the project, better control, and management of the risk, and decreased issues in errors liability amount.

According to Zakuan [4], the advantages achieved by corporations by adopting QMS, is divided into two classes that are considered as internal advantages and external advantages. Internal advantages match with the benefits which will be gathered from the establishment that are helpful for the workers and the internal functioning of the establishment. As an example of advantages gained from an inside perspective are increased company communication, the documentation was enhanced, enhancement for the technique of operating, enhancement for the quality of labor done, enhancement for the morale of the worker, accrued potency and productivity, and less waste of the material. The external advantages match with advantages which will be noticed from the perspective of the client that is beneficial for the company itself. As an example of advantages obtained from an external viewpoint are increased firm's place in the market, accrued gain, accrued sales with the current client, enhanced the relationships with provider or supplier, a lot of new native clients, decreased the amount of errors, completion of project in the right time as planned, a lot of Contracts or agreements and facilitate the reach to the market.

Again research by Zakuan [4], the investigator explicit that that application of the quality management system will enhance the dilemma of communication between the workers; decrease mistakes, rework, and the big amount of the waste of the materials; have higher management and subcontractor and supplier control. Therefore, the gain of better productivity, and usually market sharing is accrued and conjointly alter contractors to fulfill the consumer necessities. Quality management could be management operate that ensures Providing products, services, operations or processes that meet client's demand and conjointly needs a repetitive enhancement. The quality process results in the aggressiveness and property of the establishment [5]. Clear advantages incontestable via the implementation of QMS. Effects of client satisfaction are being declared by Kiew and Ismail [1] as "for the primary time a serious study has shown that implementation of TQM is completely related to home-buyer satisfaction".

QMS is employed as follow: To enhance the company's quality image within the construction industry. To raise the company's potency and management. To resolve the issues with poor quality arising construction processes. to cut back the failure prices and liability risks. To satisfy the mandate from the government. To satisfy the needs of the non-public owners/ clients. To be a benchmark to start implementing total quality management.

Ali [3] indicated that quality provides a crucial climate for creativity and advancement of construction creativity. By adopting correct QMS the company is anticipated to control consistently the time, the scope, the value, and the resources to fulfill the goals of the project. Consequently, it encourages to deal with a number of modern problems encountered in the construction field i.e. the errors of the work, the extra costs and project's delaying. Usually, most companies employed by ISO 9001 certified contractors point out their quality performance range of the complaints of the customer, rating from the feedback forms of the client, waste of materials, variety of work errors, failing in testing results and the like. Apart from those metrics, a score of QLASSIC can be a beneficial tool to evaluate the projects' performance.

QLASSIC was created and enhanced to fulfill the subsequent goals: to create a benchmark for the quality of the workmanship. To assess the quality of accomplishment of the residential projects. To assess the quality of accomplishment of a residential project supported on the approved standards. To evaluate the interpretation of subcontractors and main contractor supported on quality of accomplishment. To assemble information and data for applied mathematics analysis.

According to the reviewing of the literature, several kinds of research have shown the benefits in the application of QMS in the execution corporations. Ali [3] had abstracted the category of benefits as follows: Raising the communications and the relationships between all parties. Mitigate process and decrease reworking. Cost and time are going to be reduced. Raise the effectiveness of the work.

3. Issues of quality management implementation

The road to obtaining the certification of ISO 9001 is rarely interesting and sleek. Many types of research on the system of quality management point out this matter which obviously clarifies that the adopting of an efficient quality program isn't a straightforward duty. Between the dilemmas of application that had been known by the primary researchers are combating to alter is one of the intense dilemmas mentioned by Keng and Kamil [2]. As the researcher expressed that workers' combating is the most crucial issue that becomes an obstruction for the application of the ISO quality.

The main issues for the combating are as follow: modification is a danger for them. It makes doubt and moves one from a foothold of the rest as one's information and ability isn't any longer enough. Recent information and skills gaining have become compulsory. Modification needs further work. The procedures of the application need additional works like documentation and coaching the workers. The benefits of the modification may not outweigh the prices in question. Workers may be concerned that the additional effort and spending the time on creating and implicating ISO 9001 QMS will not fulfill or meet the company advantages. They are much busy with the work and have no much time to carry out a replacement effort like ISO 9001 QMS. In fact, some establishment ought to set up and perform ISO while not further employees. Shortage of commitment from prime management.

Management shortage is considered one of the most important issues faced during the application mentioned by Keng and Kamil [2]. Expressed the rationale of the matter was related of miss understanding of the advantages of the system of quality. He also mentioned that many companies ' prime management do not place quality because of the priority versus the issue of time and price. Interpreting the quality and needs of the quality system is difficult: Zakuan and Saman [4] expressed that the quality and demand of the standard system is difficult to know. He pointed out that transforming the standards into the appropriate action requires a full understanding of the thinking and philosophies of the standards of the ISO. While not fully understanding, it would be complicated to realize the specified commonplace of quality. Keng and Kamil [2] the same thing that some difficulties appeared when they tried to interpret the standards ' requirements within the construction context. However, due to a shortage of data and resources, it is found that it is difficult to identify the needs of ISO 9001. Perceived in the work to be redoubled. Usman, n., Said, i. (2013) expressed that the ISO 9001 application suggests that additional documentation work should be handled throughout the application phase. He further stated that cognitive content or the shortage of a obvious understanding of the basic requirements of the documentation is the main reasons of the creation of a large amount of hard-to-use work. However, this may be the first defect of most organizations in some state of affairs that they will not have unbroken enough records or important data for observing, assessing and to review the performance of the establishments. As coaching is vital to give the idea of QMS understanding and familiarity.

This disadvantage may be due to management's reluctance to commit time and expenditures to train the employees. He mentioned jointly that some organization felt that QMS was all about simply documenting their quality system and not listening to the importance of training the staff. Moreover, Ali [3] argued that lack of assurance even supposing they need exerted heaps of effort coaching between workers They made a heavy bank on the quality manager. The QMS idea is clearly understood by the UN agency. Consequently, they will simply hear the instruction given and then execute it in accordance while not really knowing why they do it. This created unhealthy dependence on the associated degree on the standard manager, So the standard idea gave the impression of becoming a' one man ' show. A high content of foreign workers and lack of skilled labor. Ali, m. (2014) found that a high labor shortage caused the majority of local subcontractors, such as brick builder, steel fixers, plaster' makers and tile workers, to engage in foreign labor holding the majority of the work with 1 or 2 expert staff. It is therefore exhausting to expect these subcontractors to provide products and services of sensitive quality below the standard idea of assurance.

Lack of management or collaboration of appointive sub-contractor. Since the consumer presented the appointing subcontractor, the degree of management of the most contractor was restricted over the appointing of the subcontractor. This drawback becomes additional crucial once non- conformance is detected and correction got to be taken. Co-operation from appointive sub-contractors will occasionally be tough [5]. High implementation and maintenance cost: [5] United that a system of high quality requires a high value for application. As a result, several establishments are reluctant to implement QMS because they are uncertain whether or not the advantages obtain can cowl the price concerned. However, several establishments misconstrue the value of the quality because they usually understand that the application of it is an additional costing. In fact, prices are not the standard. Non- compliance with quality, however, is more dear. This may be related to the cost of errors" correction, reworking, and customer grievance reaction [5]. Meagre the time: many studies jointly reported that meager time has become a barrier to the successful application of a good system of quality. Kiew and Ismail [1] mentioned that this drawback was because of poor designing so inflicting delay since the corporate did not meet the proposed point in time. He was conjointly explicit that the meager time to perform is because of the shortage of human resources and the capital.

Shortage of communication: Kiew and Ismail [1] declared that shortage of the external and the internal communication conjointly are the obstructions for the application of a good system of quality. He conjointly declared that little communication within the organization was as a result of restricted resources and data, shortage of expertise plus coaching. Kiew and Ismail [1] meanwhile, it means that there is a downside in communication between staff attributable to language differences and cultural differences between the workforce. Shortage of consultant participation: despite the fact that the contractor has performed a high-quality management system, the contractor's performance usually suffers from consultants who have not performed quality assurance despite the fact that they need tons of attempts to enhance quality. For example, discrepancies in drawings and specifications on an ongoing basis have led to delays in construction, thus reducing the efficacy of quality assurance [4]. A shortage of participation from owner or client. There is a shortage of customer involvement throughout the application of the process. If the owner support exists, the application will be formed solely to meet the needs of the owner. The establishments are oblique to carry out QMS as it becomes a precedent for its owner or customer services. Kiew and Ismail [1] else stated that clients need to have ISO certification from the consultant and the contractor, but they do not have such quality data themselves. So much weak system design: the organization may fail to recognize or understand the basic needs of quality management. Therefore, it'll result in uneconomical, government officials and high paper generating system.

Kiew and Ismail [1] conjointly declared that between the dilemmas throughout the implementation method is no well-structured system of quality and steps. A shortage of technical experience and cleverness: it is very crucial to appoint a quality consultant. While their experience or inappropriate participation of advisors will not hinder carrying out of the quality. Zakuan and Saman [4] conjointly stated that failing in appointing adviser of quality may result in excess delay, initial dismissal by external registrars and discontent of workers. Lack of resources. Keng and Kamil [2] stated that the lack of resources to carry out and retain the quality system is due to the poor quality of the culture of the work, a shortage of data and a low level of learning. She jointly argued that the shortage of quality management practices of resources in terms of the human resources and the capital has resulted in delays in obtaining the certification of ISO 9001 QMS.

Other issues were expressed by Keng and Kamil [2]. He expressed that among the assorted issues known the contractors who discourage carrying out of the ISO 9000 standards in their companies are losing staff productivity due to the attempts of learning and carrying out the modern system. Personnel financial obligation, remote job site creating its commitment, weak communication, shortage of coaching for the managers and employees, combating to alteration, and therefore lack of customer and alternative authority involvement. Technical problems: Problems are caused by an external issue during this class that directly laborious to regulate and track quality system implementation altogether sites. He conjointly mentioned that almost all establishments seek to keep the certification of ISO 9001 as a ' work permit '. However not ordering an additional enhancement in quality. It is stated jointly that it was a drag that employees were employed to figure out superintendence instead of following procedural manuals and that QMS was not applied to the subcontracting system.

According to the review of the literature, the issues faced throughout the method of carrying out by ISO-certified firms will be summarized into two categories of people-related issues and technical issues. Issues relating to people: problems during this class are caused by individuals concerned within the implementation method. It's going to occur throughout performing of the construction and doing the management on the positioning additionally as the steps of the work at the positioning. samples of the dilemma make up this class are the shortage of prime management and employees aren't caused by the behavior of people. Such problems could reflect the disadvantages of the quality system itself, making it difficult to carry them. Samples of problems that make up this classification are high carrying out and maintenance value, lack of resources, difficult interpretation of the quality and demand of the quality system, hyperbolic in the workplace and remote work, making it difficult to regulate and monitor the quality system carrying out.

4. Strength and weakness

It was noted that the five journals have achieved their main objective. The theories and analysis supported each other which made their finding are strong in comparison. However, there were some limitations of the research is the results can not be generalized statistically. It also focused on the supporting role of quality practices in managing strategically crucial processes of creativity. In the research, time was the main limitation in collecting enough data and carrying out interviews [5].

5. Technical evaluation

The researchers conducted researches through quantitative and quantitative analysis. Case studies were used as the researchers intend to support their arguments by associate in-depth analysis of someone, a gaggle of staff, a company or a specific project. The case study approach was adopted for the analysis. The tools used for information assortment was interviewing. It was mentioned that interview permits the greatest chance to get what somebody thinks or feels, and the way they react to huge problems. The shape of the interview used was semi-structured that uses a mixture of 'open' and 'close-ended' queries. This method is employed so as to allow flexibility to the scientist and therefore the responder throughout the interview session [2].

The reasons for conducting a case study for this analysis are to appear into details the advantages achieved, issues encountered and therefore the ways utilized by ISO certified contractors to combat the implementation dilemmas. Case study approach enables scientist to answer the 'why' and 'how' queries once the respondents are needed to allow justifications on the chosen issue Keng and Kamil [2]. Said and Usman [5] performed the analysis in Malaysian construction corporations. The technique which was used within the study as a result of direct interview survey would facilitate the queries, remove uncertainties, and certify the queries are totally well known by the interviewee before they give answers to the questionnaire. In addition to that, this technique can Scale back the "nonresponse" rate and gain more information throughout the interview session on QMS maintenance. The following sections provide a close description of the survey methodology [4].

IOP Conf. Series: Materials Science and Engineering **650** (2019) 012008 doi:10.1088/1757-899X/650/1/012008

6. Future research

As mentioned in some researches, the restrictions of the researches were that the results can not be generalized statistically. It jointly focused on the supplementary role of quality management processes in managing strategically necessary processes of creation. During the researches, the time factor was a significant limit to gather decent knowledge and perform interviews. Also, there were no clear examples for a few forms of quality like internal control and quality assurance. As a result of that, the reader still confused between them. Therefore, for the longer term researches, it's counseled to elucidate with real examples from the sensible life every sort and the way it may be enforced to clarify every sort rather than giving solely definitions. conjointly for doing such researches, the researchers should take enough time to gather adequate knowledge.

7. Conclusion

This article was launched to look at 5 journals associated with quality management in the construction field in Malaysia and build an outline for them. The results show that owner or customer satisfaction and the time reducing have a positive and important related to QMS. During this section, all the findings are summarized. This section can discuss the common advantages, issues, and techniques known within the findings from the previous section.

Benefits: Improved management and work efficiency. Improved communication. higher documentation management. Increase the owner or client confidence and satisfaction. Also adopting proper quality management raised the company's company image. Raised range of project undertaken. Improved quality of labor done on the site. A quality set up by strictly following these documents, contractors are able to forestall any errors like a rejection of sub-standard materials. Therefore, this leads to value maintaining through avoidance of errors. Secondly, all construction activities were strictly supervised by the positioning supervisor, Site officer likewise as a project manager. this can be to make sure that construction employees meted out the work following the proper method, to the standard of workmanship needed whereas yielding with the protection necessities. The standard of labor was therefore secured with fewer defects and material wastage and accidents on the site. On-time completion, corporations argued that systematic management of labor beneath the ISO standards did advantages them by enabling the current projects to be delivered on time. However, in some project, there are surprising events occur that hinder the comes to be completed in the desired time albeit they meted out the work following the needs.

Problems and strategies: Awareness and employee's commitment: so as to combat this quandary, all respondents mentioned that their company sent all staff for coaching. Usually, there are two main coaching given which are the notice or awareness course and auditing coaching. Either the ISO authority employed by the corporate, CIDB or alternative external bodies conducts these coaching. Except that, the team working towards the task could be awakened strategy so that staff can share their data, interests, and experiences to unravel issues and this can be effective to extend their commitment towards the duty. Lack of cooperation from consumer, authority and/or sub-contractor: corporations encounter issues with their consumer and consultants. Firms face problems with the consultants who didn't provide full cooperation towards them even supposing they grasp that the contractors are implementing ISO. A number of the consumer/clients and consultants that they're operating with tend to try and do the works their own manner.

In distinction, one company mentioned that they get full cooperation from their consumer and authority. However, they had some problems with their sub-contractor who were unsuccessful in meeting the ISO needs as tough to monitor quality system on the remote website: Some corporations in agreement that it's tough to observe the standard quality system at the remote site. In overcoming this drawback, company b allocates one in every of their quality unit employees to observe all activities there. They applied cross-auditing between the site, which ends in time, and the company's price savings. Meanwhile, for any company, if the issues are unable to be sorted out by the project team on site, home base ought to be told as more action are often applied. Lack of resources: Some corporations intimate scant of human resources in implementing ISO particularly throughout the document preparation. Since this company is assessed as a little and medium-sized company, they have to mobilize all of their staff. External audit and internal audit so as to make sure that the systems applied are operating properly. Corrective action is created where there are errors in carrying out of the system. It can jointly facilitate the spotting of the contractor if any resources are needed to carry out the system.

References

[1] Kiew P N and Syuhaida I 2016 Integration of quality management system in the malaysian construction industry *J. Organ. Manag. Stud.* pp 1–9

[2] Keng C T and Kamil S Z 2016 Implementation of iso quality management system in construction companies of malaysia *J. Tech. Manag. Bus.* **3** pp 1–23

[3] Ali C 2014 Exploring the potential of integration quality assessment system in construction (QLASSIC) with ISO 9001 quality management system (QMS) *Int. J. Qual. Res.* **8** pp 73– 86

[4] Zakuan N and Saman M Z 2014 Using project performance to measure effectiveness of quality management system maintenance and practices in construction industry *Sci. World J.* pp 1-9

[5] Said I and Usman N 2013 Evaluation of innovative quality management practices: a case study *J. Ent. Bus.* pp 11-20

Developers' perception of cost drivers influencing project cost performance on Surabaya high-rise residential, Indonesia

D Chicoca[1] and C Utomo[1]

[1] Institut Teknologi Sepuluh Nopember, Kota SBY, Jawa Timur 60111, Indonesia

dchicoca@hotmail.com; christiono@ce.its.ac.id

Abstract. Good cost performance is one of the fundamental criteria of a successful project. The major challenges developers have in this process is to balance the budget and project cost performance. Most of the scholars agreed that the construction indu stry in any country is complex because it involves a broad range of stakeholders and has wide-ranging linkages with other areas. Some of these factors are also unknown at the project level, which create difficulty for project managers to assign a favourable contingency cost. Therefore, this work came to establish factors influencing project cost performance of high-rise residential development through macro indicators based on developer's perspective. It sought to identify these factors in an effort to shed more light on how cost drivers affect project cost performance. From the study findings, there are seven cost drivers. These drivers are analysed based on significance and impact using multiple regression analysis. The top most in terms of significance are property market and construction industry, project characteristic factor, socio economic factor and project component costs factor. The least one out of the seven factors is the statuary regulation factor. In relation to impact, project stakeholder factor has the highest impact, followed by the property market and construction industry, project characteristic factor and project component costs factor. The least one is socio economic factors.

1. Introduction

One of the major challenges facing developers is balancing the high-rise apartment budget and project management performance. Cost is arguably the most important concern in any business endeavor, not least in the construction industry [1]. Poor cost performance in construction projects has become a major concern for both contractors and clients [2, 3]. However, because the business model for high-rise residential is being sold out based on cost per square meters, the developers also need cost efficiency since project efficiency impacts customer and business success [4].

The identification of cost drivers for building development is crucial to improve cost performance. A study conducted in the Singapore construction industry [5], investigated factors contributing to project costs. The findings identified three main factors, which were contractor, owner and financial management. Additionally,[3, 5, 6] identified the factors that significantly influence construction costs in different countries. The factors are directly related to the construction organizations' responsibilities for managing the cost, socio-cultural, economic and external conditions as well as the technological and political environments within which they operate. Despite the varied literature on key cost drivers for building development, none of the presented research illustrates the significance of these factors on project cost performance for high-rise residential, particular in Surabaya construction market industry. The prior research focused on analyzing the influence of stakeholder on project success [7], risks factors affecting project success [8] and exploring the relationship between building development and significant factors based on New Zealand context industry [9]. The construction industry still suffers from poor project performance because of its nature, which is fragmented by the work of different stakeholders and sub-processes behind it [10, 11]. Apart from that, [12] investigated risks to predict project cost performance level by assessing risk factors of building construction in South Korea. The findings indicate that predicting cost performance is not simple. It is dependent on numerous factors, not only factors innate to the project but also project-independent circumstances. Clearly, it can be seen that cost driver factors for high-rise building development was not explicitly investigated in these studies. Previous research assessed some of the core components of projects through a micro perspective. Each of these studies provided valuable information but still, there is a need to identify and clarify key costs drivers for high-rise apartments from Surabaya's construction industry.

The micro perspective, often tend to be opposed to stepping back in not looking at the big picture. Meanwhile, for developers, the macro perspective would be the most suitable step because it exploits the profit margin over the past years and the current position of the business industry. In addition, a few research presented the relationship between categorized high-rise residential development factors and project cost performance (cost escalation and profit) that is essential for successfully managing construction projects. Keeping construction projects within estimated costs and schedules require good practices and careful judgment. Therefore, apart from the issue of flotation, uncertainty, and contingency in the present high-rise building development there are other cost drivers that influence the performance of project costs that need to be considered. This paper came to establish these factors through macro indicators, by showing the impact of cost drivers on project cost performance in Surabaya.

2. Literature Review

2.1 The concept of cost driver factors

Based on the interests of this research, cost driver factors on this work are being presented into the accounting system. The strategic purpose of a cost driver means the factor that causes a change at the cost level or total cost [13]. Although construction accounting differs from normal accounting, the basic principles of accountancy are applicable for construction projects. Any firm incurs a cost when it uses a resource or performs a specific activity for some purpose. Therefore, for a company or developers to compete on the basis of cost leadership, management of key cost drivers is essential in order to avoid unnecessary losses and maximize the profits. Cost drivers should always be taken into construction accounting because it helps to identify opportunities to improve project cost performance within a construction development and disposal of the facilities.

2.2 Project cost performance.

Project cost performance is a measure of the value of the work completed compared to the actual cost or progress made on the project and it is one of the fundamental criteria for the success of any project. Several studies concluded that the construction industry has been facing poor cost performance (cost overruns and time extension). Most construction managers have been addressing this issue just by considering a contingency cost, without having the full picture of unseen events at neither project nor the corporate level. Developers understand that besides cost escalation, profit (total cost/revenue) should also be included into account for cost performance index. For project managers, their primary project planning is to identify costs deviations while developers suggest a possible way for cost savings. The purpose of project management and control is not sufficient to consider only the past record of costs but also the revenues incurred in a building project. Therefore, good managers should focus upon future revenues, future costs and project technical problems [14].

2.3 Cost driver factors on building development

Based on the literature review, various studies had been conducted towards building development. Most of them focused on finding variables for keeping construction projects within estimated costs and predicting cost performance [3, 5, 6, 9, 12, 15–17]. All the studies were important but the work of [9] took this subject to another level. The study identified 45 indicators that influence building development costs and sorts them into seven categories. The seven categories are project component cost factor, project characteristics factor, project stakeholders' influence factor, property market and construction industry factor, statutory and regulatory factor, national and global dynamics, and the socio-economic factor. This research discovered that property market and the construction industry factor, national and global dynamics and socio economic factors are having significant positive effects on building development costs. While, statutory and regulatory factors and others costs are affecting building development negatively in New Zealand.

3. Methodology

3.1. Data collection procedure

The study employed exploratory, survey and descriptive research design. The exploratory design was partly utilized in the introductory part of the proposal and literature review. Towards this research, a data method was used to collect information about the entire population. The sampling method was used because does not demand a high budget and brings the population to a manageable number. The instrument was formulated based on the objectives and research questions. The questionnaire was divided into three parts.

In part A, general information about respondents was asked so that their characters could be identified. In part B, the question was asked about the extent of the highlighted cost drivers on Surabaya high-rise apartment development. Finally, in part C, the question investigated the significance of cost drivers on project cost performance. This design was used to clarify how far scholars and construction stakeholders in Surabaya are concerned about this subject. The data were collected with a time frame and developers' were included in the study as they are the ones who know the company's revenue. The proposed conceptual framework (Figure 1) shows the relationships between cost drivers and Project cost performance. The model was an extension developed based on the work of Zhao that presented the categorical factors on building development costs [9].

The involved exploratory design was crucial in reviewing the existent literature as well as discussions with experts in the subject matter. The findings provided useful insights for identifying important cost drivers affecting project cost performance in Surabaya. Furthermore, statistical analysis was carried out to verify the fitness of the proposed model.

3.2. Data analysis technique

According to [18] the process of data analysis involves several stages. For instance, in this study, the primary stage focused on structuring the questionnaire. The questionnaire was checked several times for completeness and consistency. Afterward, variables were coded. Data were analysed using descriptive analysis such as mean scores, standard deviations, frequencies distributions, and percentages. Furthermore, an analysis of variance (ANOVA) test was done to verify either the overall model is significant for project cost performance or not, which is a necessary step before proceeding with causal analysis. Moreover, multiple regressions were also performed in order to establish the reliability and validity of the proposed model. The study used both descriptive and inferential statistics to analysed data from the questionnaires. Multiple linear regression (MLR) analysis was used to establish the nature and magnitude of the relationship between variables. Below is the MLR equation:

$$Y=\beta0+\beta1X1+\beta2X2+\beta3X3+\beta4X4+e \quad (1)$$

Where;

Y = Project Cost Performance

(PCP) $\beta0$ = Constant Term

β1, β2, β3, β4 = Beta coefficients
X1 = Project Characteristics Factor (PCF)
X2 = Project Component Cost Factor (PCCF)
X3 = Project Stakeholders' Influence Factor (PSIF)
X4 = Statutory and Regulation Factor (SRF)
X5 = Socio-Economic Factor (SEF)
X6 = Property Market and Construction Industry Factor (PMCIF)
X7 = National and Global Dynamic Factor (NGDF)
e = Error Term

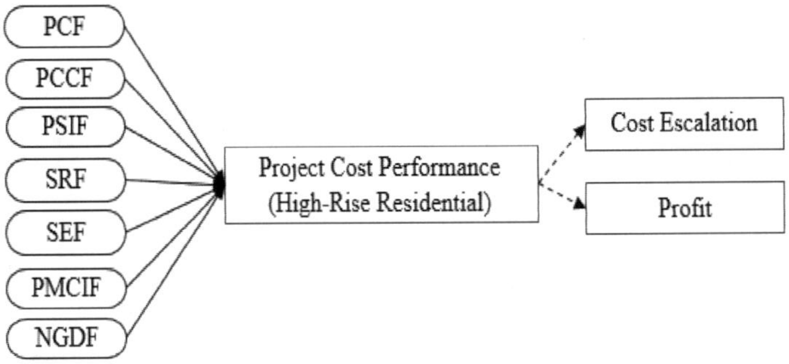

Figure 1. Research framework Model.

4. Results and Discussion

4.1. Developers' characteristic

The profile of developers who were involved in the study is those who have been active in Surabaya construction industry. Based on literature and field survey were identified 35 majors high-rise apartment developers in Surabaya. 35 self- administered questionnaires were distributed. 28 questionnaires were duly returned, representing a 70% response rate, which is arguably sufficient to proceed with the data analysis. Further analysis showed that out of 28 respondents who participated in the study, 25 (91.7%) respondents were male while 3 (8.3%) were female. The findings also indicate that respondents have diploma classifications at the bare minimum as presented in Table 1. In addition, respondents were also asked to indicate their experience background on this field and findings show that 54% (Above 25 years of experience), 35% (16-25 years), 8% (11-15 years) and 3% (5-10 years).

Table 1. Education level of respondents.

Level of education	Frequency	Percent
Certificate	2	7.1
Diploma	4	14.3
Undergraduate Degree	6	21.4
Postgraduate Degree	16	57.2
Total	28	100%

4.2. Cost drivers that influence project cost performance of high-rise residential development

The finding shows that there are seven significant categories of variables affecting high-rise residential development. In order to determine the factors that influence project cost performance, these categories of variables were analysed based on the cumulative Mean as presented the Table 2. The Table presents the results of descriptive statistical analysis for project cost performance on high-rise residential.

Table 2. Significant factors on project cost performance.

Cost drivers Categories on Project Cost performance	Mean	Std. Deviation	N
Property Market and Construction Industry Factor (PMCIF)	5.87	0.488	28
Project Characteritic Factor (PCF)	5.54	0.358	28
Socion Economic Factor (SEF)	5.36	0.430	28
Project Component Costs Factor (PCCF)	5.20	0.715	28
National and Global Dynamics (NGD)	5.03	0.538	28
Project Stakeholder Factor (PSIF)	4.46	0.700	28
Statutory Regulation Factor (SRF)	4.15	0.729	28
Mean	5.09		

Table 2, presents the Mean and standard deviation analysis results. Means and standard deviations were computed based on a Likert scale of 6 points. Property market and construction industry factors (PMCIF) was supported with a mean of 5.87, project characteristics factor (PCF) was supported with a mean of 5.54, and socio economic factor (SEF) was supported with a mean of 5.36. These are the categories that most influence project cost performance for high-rise residential in Surabaya and the least significant factor or category is statutory regulation factor (SRF) which was supported with a mean of 4.15, and overall the mean of mean project cost performance is 5.09. Even though the finding shows a different degree of acceptance for each factor, all were confirmed as relevant for high-rise apartment development.

4.3. The significance of all the variables in the presented model for project cost performance

In order to test for sampling adequacy and correlation, the Pearson test was computed and the value was below 0.80, which means there is no multicollinearity in the proposed model. Besides, a variance inflation factor test (VIF) was also performed and its value was less than 10 (See Table 5), hence we do conclude that there is no multicollinearity between the independent variables. The assumption of having a normal data distribution has been satisfied and it led to continuing with F-test analysis. These tests indicate that the analysed factors were appropriate for the set of data and the F-test was computed to measure association among variables. F-test shows that the overall model is significant for project cost performance with *P-value* (0.000), which indicates that the correlation matrix was less than 0.05 and implied that there was a strong association among variables.

For the measurement to what extent these factors contributed to project cost performance, Multiple Linear Regression (MLR) analysis was performed and this method was adopted because the prior assumptions are satisfied. A normality test was performed to measure whether the model with seven predictors predicts the overall changes in cost performance. The test indicates that the model that comprises of PMCIF, PCF, SEF, PCCF, NGD, PSIF and SRF does better than having singular factors. Table 3 shows the correlation coefficient (R) is 0.989. This means the relation between cost driver's categories and project cost performance is strong. The value of the determination coefficient (R^2) is 0.979. It means 98% of PCP is affected by PMCIF, SEF, SRF, PCCF, NGD, PCF, PSIF and the remains 2% implies other variables that cannot be explained by the model. Table 3 presents the significance and the person correlation of the model.

The correlation analysis was taken in order to verify the possibility of any multi-collinearity between the independent variables. If the Pearson Correlation is greater than 0.80 then there is multicollinearity and compromise the assumption of MLR. In this study, the correlation between each variable is less than 0.80. This shows that the variables are not highly correlated hence the assumption of MLR is satisfied. In addition, Table 4 shows the ANOVA, with a *p-value* (0.000) which is less than 0.05. Hence, it is concluded that the model is significant for Surabaya construction market (project cost performance). Altogether, since all the overall model and independent variables were significant then the MLR equation could be generated. The variables of logit equation, coefficients, and inflation factor test (VIF) values are presented in Table 5.

Table 3. Correlation coefficient (R) and Person

Model		R	R Square	Adjusted RS	Std. Error
1	Regression	.989[a]	.979	.971	.06021

		PCP	PCF	PCCF	PSIF	SRF	NGDF	SEF	PMCIF
Pearson	PCP	1.000	.135	-.376	-.264	.276	.307	.467	.363
Correlation	PCF	.135	1.000	.577	.370	-.405	.164	.013	.484
	PCCF	-.376	.577	1.000	.695	-.617	-.074	-.551	.069
	PSIF	-.264	.370	.695	1.000	-.276	.096	-.424	-.124
	SRF	.276	-.405	-.617	-.276	1.000	.593	.348	-.279
	NGDF	.307	.164	-.074	.096	.593	1.000	.185	.015
	SEF	.467	.013	-.551	-.424	.348	.185	1.000	.579
	PMCIF	.363	.484	.069	-.124	-.279	.015	.579	1.000

[a] Dependent Variable: Project Cost Performance for high-rise residential
[b] Predictors: (Constant), PMCIF, SEF, SRF, PCCF, NGD, PCF and PSIF

Table 4. The overall model (ANOVA[a])

Model		Sum of Squares	df	Mean Square	F	Sig.
1	Regression	3.356	7	.479	132.3	.000[b]
	Residual	.073	20	.004		
	Total	3.429	27			

Table 5 Variables of logit equation and coefficient

Model		Unstandardized Coefficients		Standardized Coefficients	t	Sig.	
		B	Std. Error	Beta			VIF
1	(Constant)	2.497	.116		21.449	.000	
	PCF	.016	.005	.196	2.982	.007	4.079
	PCCF	.014	.006	.129	2.520	.020	2.478
	PSIF	.019	.005	.233	3.520	.002	4.157
	SRF	.010	.004	.108	2.657	.015	1.569
	NGD	.013	.005	.148	2.581	.018	3.127
	SEF	.005	.002	.122	2.698	.014	1.930
	PMCIF	.017	.004	.263	4.282	.000	3.569

Under the coefficient column B, it can be seen that all seven predictors are positively related to logit or standard error of Project cost performance coefficients. PSIF has a high influence compared to the

IOP Conf. Series: Materials Science and Engineering **650** (2019) 012009 doi:10.1088/1757-899X/650/1/012009

rest of the predictors as the unit increases. As the unit increases, it shows the logit of cost performance increase by 0.19. Developers see project stakeholder factors as one of the least significant cost drivers. However, it has the highest impact on project cost performance. By looking at the odd i.e. Exp (B), an increase of one unit on PMCIF increases the odds of cost performance by 0.263 while controlling other predictors. Next, the high odds increase is on PSIF (0.233) while controlling other variables. Odds for SEF, SRF, PCCF, NGD, PCF, and PSIF are also respectively presented. From Table 5, we can find the regression model for Project Cost Performance in Surabaya as follows:

$$PCP = 2.497 + 0.019PSIF + 0.017PMCIF + 0.016PCF + 0.014PCCF + 0.013NGD + 0.010SRF + 0.005SEF \quad (2)$$

This equation model represents the influence of cost drivers on cost performance of high-rise residential in Surabaya, by showing that one unit change in one of the categories would improve the project cost performance significantly. For instance, a constant of 2.497 means that if the project characteristics factor (PCF),project component cost(PCCF),project stakeholders factors (PSIF), socio economic factor (SEF), statutory regulation factors (SRF), national and global dynamics (NGD), property market and construction industry factors (PMCIF) all be rated as zero or without any positive changes, then project cost performance for high-rise residential would have an average of 2.497. This value is not absolute; still, there is a possibility that the project would be under budget if nothing happens.

4.3.1 Property Market and Construction Industry Factor (PMCIF). This factor could be impacting project cost performance as a matter of fact that the single largest input in construction, which is the building materials, it is part of this category. Typically it accounts for over 50 percent of the total construction costs. The cost of construction is usually high because of the high cost of materials, labor and equipment. Thus, it makes sense for developers to be highly concerned about the availability of supply and demand. It is their duty to ensure having enough budgets for the project's needs. The magnitude of unfulfilled demand for construction is related to the limited supply capacity of the construction industry considering the fact it is a vital cost driver for successful planning. [19] States when demand for the apartment is low, the priority of developers' lies in sustaining their properties' sales momentum, which is increasingly being challenged by the currently tepid market. Therefore, attractive product offerings with strong value propositions, such as location, concept and a clear target market are critical to driving sales.

4.3.2. Project component costs factor (PCCF). In general, cost is among the major consideration throughout the project management life cycle. The findings show that developers are aware that projects are subjected to cost escalation, which is one of the main problems in construction industry. The issue may be found both in developing and developed countries. Proper cost control is important as it is the general trend towards greater cost-effectiveness and ensures construction costs not solely in the context of initial costs, but in terms of life-cycle costs or total cost appraisal. Thus, it is highly recommended to oversee the construction costs both from a project and cooperate level.

4.3.3. Socio Economic Factor (SEF). Socio economic condition has an influence on high-rise apartment development. When the economy is booming, it boosts up the investor confidence. Cities with an appealing investment confidence index may attract more investors, which is a crucial element for the property market. Developer are truly concerned about this because taking a look at high-rise apartment market, it has only two types of costumers which are Investors and end-users. Therefore, a significant positive change in the socio economic index tends to attract more projects. To some extents the confidence level for Surabaya economy is satisfactory and it is expected to be even better with development of the country's economy.

4.3.4. Global and national dynamics (NGD). This driver might be included as a matter of fact that construction industry is integrated with national and global dynamics. Developers are willing to ensure that the investments are aligned with the local market and its surroundings. The finding is in agreement with the 2018 Surabaya market report, stating that the market sentiment index in Surabaya remains soft particularly in the high-interest rate environment [19, 20].

4.3.5. Project characteristic factors (PCF). Based on the prior market sentiment year it was expected that developers would be concerned with project characteristic factors. The construction process may be considered the most complex undertaking in any industry and adversely affects building project costs. Surabaya developers are considering this factor by taking a look at tech innovation. This finding is supported by [9], good procedures within project developments make a huge positive impact, by adding more value to the investment. The misunderstanding of the procedures and methods may result in a delay on the construction site and drive up building costs.

4.3.6. Project stakeholder Factors (PSIF). Developers are highly concerned with stakeholder engagement, especially when it comes to contractors about their roles on the project development. Contractors are responsible to deliver the project as requested. Contractor's performance is critical to the success of any construction building because it is their duty to convert designs into practical reality. To ensure successful project execution there must be adequate financial capacity. [21] Supports this implication by saying that monetary strength of contractors and sufficient cashflow is critical in keeping construction progress as planned.

4.3.7. Statuary Regulatory Factor (SRF). Regulatory status is a crucial fundamental factor for project development. Surabaya had successfully improved the issues of building permits as suggested in the prior study [22], that changes in the local government building would unblock construction project development in Surabaya. Therefore, developer seems to be satisfied with the current situation, but still there is a need to facilitate the building compliance process in order to have a well-planned construction and avoid unnecessary costs.

5. Conclusion

The study established that there are positive correlations between cost drivers and project cost performance in Surabaya. Seven cost drivers were identified and analysed based on significance and impact. The top most in terms of significance are property market and construction industry, project characteristic factor, socio economic factor and project component costs factor. The least one out of the seven factors is the statuary regulation factor. In relation to impact, project stakeholder factor has the highest impact, followed by the property market and construction industry, project characteristic factor and project component costs factor. The least one is socio economic factors. In order to predict project cost performance of high-rise apartment development favourably, suggestions are made for considering closely the identified macro factors. Additionally, further research is required to assess the other variables that cannot be explained by the model.

Acknowledgments
This research was supported by the Ministry of Research, Technology and Higher Education, by Republic of Indonesia and Angola. Authors also wishing to acknowledge the cooperation of high-rise residential' developers in Surabaya.

References

[1] Bari N A A,Yusuff R, Ismail N, Jaapar A and Ahmad R 2012 *Procedia - Soc. Behav. Sci.* **35** pp 689–696
[2] Xiao H and Proverbs D 2002 *Int. J. Qual. Reliab. Manag.* **19** pp 672–687
[3] Graham M W 2010 *Managing Construction Projects 2nd ed* Wiley-Blackwell (UK)
[4] Aaron S and Dov D 1997 *Proj. Manag. J.* **28** pp 25–13
[5] Lean S and Park M 2007 Constr Manag Econ (March 2005) **23** pp 295–304
[6] Elhag T M S 2005 *Int. J. Proj. Manag.* **23** pp 538–545

IOP Conf. Series: Materials Science and Engineering **650** (2019) 012009 doi:10.1088/1757-899X/650/1/012009

[7] Chandra H P, Wiguna I PA and Kaming P F 2012 *Int. J. Acad. Res.* **4** pp 41–46
[8] Chandra H P 2015 *In: Procedia Eng. Elsevier B.V.,* pp 53–59
[9] Zhao L, Mbachu J and Domingo N 2017 *Buildings article* **7** p 57
[10] Rahman M M and Kumaraswamy M M 2004 *J. Manag. Eng.* **20** pp 147 –161
[11] Harper C M 2014 *Measuring project integration using relational contract theory* Vol. 3635836 University of Colorado, Boulder CU Scholar pp 38–184
[12] Cha H S and Shin K Y 2011 *J. Asian Archit. Build. Eng.* **10** pp 437– 444
[13] Blocher E J, Stout D E and Cokins G 2010 *Cost Management: A Strategic Emphasis* 5th Ed (New York: McGraw-Hill/Irwin) pp 17-55
[14] Hendrickson C 1989 *Project Management for Construction Fundamental Concepts for Owners, Engineers, Architects and Builders* Vol 2 Ed Prentice Hall (Pittsburgh) pp 5 –314
[15] Iyer K C and Jha K N 2005 **23** pp 283–295
[16] Griffis F H B, Asce F and Choi H 2013 pp 2–9
[17] Akanni P O, Oke A E and Akpomiemie O A 2015 *HBRC J.* **11** pp 91–97
[18] Glasgow P A 2005 Cent. Eval. Res. **55** pp 311–327
[19] Colliers International 2017 *Surabaya Property Market Report*
[20] Colliers International 2018 *Surabaya Property Market Report*
[21] Ameh O and Osegbo E 2014 *Int. J. Constr. Supply Chain Manag.* **1** pp 56–67
[22] Susilawati C and Sutoto Y 2010 *Proc. Int. Conf. Constr. Real Estate Manag.* **1-3** pp 397–402

IOP Conf. Series: Materials Science and Engineering **650** (2019) 012010 doi:10.1088/1757-899X/650/1/012010

Implementing Three-Variance Approach for Project Time and Cost Control in a Building Construction Project (Case Study: A Project in West Java)

L Friescilia[1] and O Gondokusumo[2]

[1] Tarumanagara University, Department of Civil Engineering Jakarta 11440, Indonesia
[2] Tarumanagara University, Department of Civil Engineering Jakarta 11440, Indonesia

lilyanafd@gmail.com; onnygk@gmail.com

Abstract. Earned Value Method (EVM) is one of project controlling methods. This method has been widely used in many projects since long time ago. However, this traditional EVM still has some shortcomings that related to time aspect, such as the use of monetary units, also the time aspect and work aspect are not measured separately. To overcome these shortcomings, several researches have been carried out and one of them is Three-Variance Approach (TVA). This paper proposed the use of EVM and TVA and compared the result from those methods. This paper also aimed to identify the strengths and problems that may arise when using TVA in a real project. The results are visualized with graphs, then be gathered into three categories; project status, forecasting, and trend. Project status determines the condition of the project in each period. Forecasting enables one to estimate the final condition of a project. The improvement of project condition from time to time can be seen in trend.

1. Introduction

Good planning and controlling need to be done to enhance the success of a project. Project performance can be measured with these indicators: cost, quality, time and safety [1]. There are a variety of methods that can be used for project control, one of them is Earned Value Method (EVM) [2]. This method has long been used as a tool for project control [3]. Problems and project status can be identified earlier and accurately with EVM, so that it can be used as a basis for improvement of the project condition [4]. However, there have been several issues related to the time aspect on EVM, such as the use of monetary units, and the misinterpretation of schedule variance [5] [6]. To overcome these issues, there have been various studies for development of EVM, and Three-Variance Approach (TVA) is one of the developments made [7] [8] [9] [10]. This paper aims to implement TVA in a real project, to compare the result from EVM and TVA, and to identify the strengths and problems that may arise when implementing TVA in a real project.

2. Literature Review

2.1. Earned Value Method (EVM)

EVM as one of project controlling methods is based on the concept of measuring the progress of the work that has been completed at a time, and evaluated by the help of the amount of budget provided for the work [11]. EVM has been used and studied in various countries such as Italy [12], India [13] and Indonesia [14]. EVM focuses on two aspects: cost and schedule. Calculations with EVM can be briefly seen in table 1.

Table 1. Summary of Earned Value Method[a].

Description of project performance	Cost	Schedule
Variance	CV = EV - AC	SV = EV - PV
	CV < 0, over budget	SV < 0, behind schedule
	CV = 0, on budget	SV = 0, on schedule
	CV > 0, under budget	SV > 0, ahead of schedule
Performance Index	CPI = EV/AC	SPI = EV/PV
	CPI < 1, over budget	SPI < 1, behind schedule
	CPI = 1, on budget	SPI = 1, on schedule
	CPI > 1, under budget	SPI > 1, ahead of schedule
Critical Ratio	CPI x SPI	
Forecast	ETC = EAC - AC	-
	(1) EAC = BAC/CPI	-
	(2) EAC = AC + BAC - EV	-
	(3) $EAC = AC + \frac{BAC - EV}{CPI \times SPI}$	-

[a] All the abbreviations used in this table can be seen in section 6 (Notation).

EVM result can be visualized with graphs as seen in [7]. With the help of performance index, project performance can be evaluated (known as project trend) in two-dimensional graph [15]. EVM is useful and can form a basis to consider the improvement of the project [4]. However, in addition to the benefits gained, the use of EVM is also constrained by some issues, one of them is on time aspect as stated in [5] and [6]. The shortcomings are:

- SV is measured with monetary units, and it can be difficult to understand since 'schedule' is a term that is related to time units.
- SV = 0 is used in showing that the project has been completed, but when the project is on schedule, the value of SV is equal to zero, which leads to confusion.
- SV always converges to zero when it approaches the end of the project. When finished, SV = 0 can cast doubts whether the project is on schedule and overdue.

2.2. Three-Variance Approach

The project observations with TVA are reviewed in three aspects: cost, time, and scope (range of work). TVA also uses the Earned Time Method concept to determine the work rate. Work rate (WR) is the amount of work completed per unit of time. There are three types of work rate used in the observation process with TVA. These three work-rate are listed below:

- WR_S, is scheduled average work rate between two observation points, such that the corresponding change in PV can be observed. WR_S can be used to estimate ET with respect to EV at an observation time t_1, within a given planned project duration.
- WR_P, is planned average work rate from the beginning of the project ($t_0 = 0$ dan $PV_0 = 0$) to the observation point t_1. WR_P can also be used when the observation point t_1 is within the planned project duration.
- WR_M, is mean work rate of the project from beginning of the project ($t_0 = 0$ dan $PV_0 = 0$) to the end ($\tilde{t}$ = PD and PV = BAC). WR_M can be used to estimate ET at an observation point t_1 once the planned project duration has been exceeded.

When the project is overdue, two artificial measures are used to analyze the data, called Artificial Planned Work (APW) and Artificial Planned Value (PV′). Table 2 shows the formulas used in TVA.

IOP Conf. Series: Materials Science and Engineering **650** (2019) 012010 doi:10.1088/1757-899X/650/1/012010

Table 2. Summary of Three-Variance Approach (Chang and Yu, 2018)[a].

Description of project performance	Formulas of each EVM factor type		
	Work	Time	Cost
Variance	WV = EV - PV	TV = ET - AT	CV = EV - AC
	WV < 0, less work	TV < 0, behind schedule	CV < 0, over budget
	WV = 0, same work	TV = 0, on schedule	CV = 0, on budget
	WV > 0, more work	TV > 0, ahead of schedule	CV > 0, under budget
	-	$ET = EV/WR_S$,	-
	-	or EV/WR_P	-
	AT > PD	AT > PD	AT > PD
	PV' = BAC + APW	$ET = EV/WR_M$	Same as above
	$APW = WR_M \times ALT$	ALT = AT - PD	-
	WV' = EV - PV'	-	-
Performance Index	WPI = EV /PV	TPI = ET/AT	CPI = EV/AC
	WPI < 1, less work	TPI < 1, behind schedule	CPI < 1, over budget
	WPI = 1, same work	TPI = 1, on schedule	CPI = 1, on budget
	WPI > 1, more work	TPI > 1, ahead of schedule	CPI > 1, under budget
	AT > PD	AT > PD	AT > PD
	WPI' = EV/PV'	Same as above	Same as above
Forecast	ROW = BAC - EV	ETTC = (PD - ET)/TPI	ETC = (BAC - EV)/CPI
	EWAC = BAC	ETAC = PD/TPI	EAC = BAC/CPI

[a] All the abbreviations used in this table can be seen in section 6 (Notation).

3. Research Methodology

This research takes a real construction project in West Java as an example. The analysis was done in steps as listed below:

- Literature study and project data collection
  Project data that are needed to perform the analysis are bill of quantity, master schedule, project reports (daily report, weekly report, progress report), project cost data, and variation order data.
- Data processing
  All the data that has been collected further processed into Planned Value (PV), Earned Value (EV), and Actual Cost (AC) in accordance with the periods of observation used.
- Analysis and result interpretation
  Analysis with EVM and TVA are carried out using the PV data, EV, and AC from every period of observation. Figure 1 can help show the steps of calculation using TVA.

Arrow that points to each box indicating the prerequisite data needed for each calculation. The results of calculation with EVM is then compared with results from TVA, and then visualized with graphs and tables.

IOP Conf. Series: Materials Science and Engineering **650** (2019) 012010 doi:10.1088/1757-899X/650/1/012010

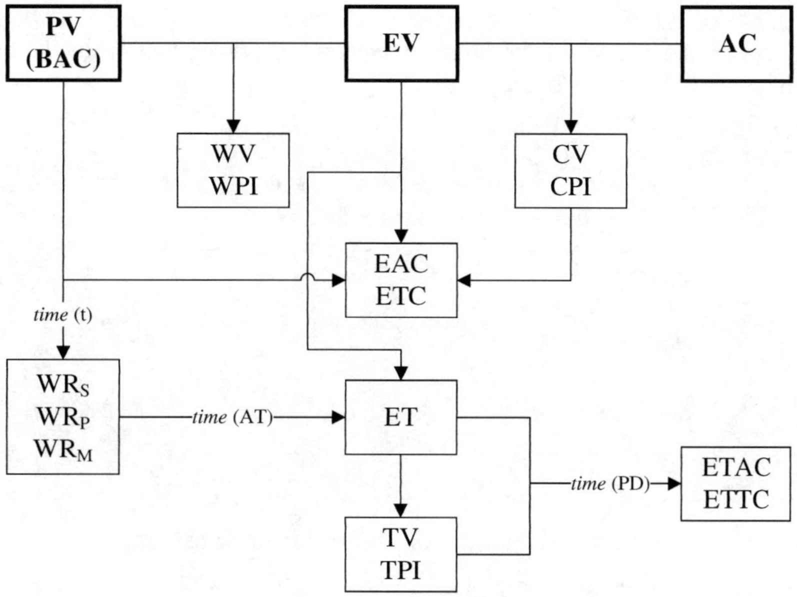

Figure 1. Step of analysis with Three-Variance Approach.

4. Result and Discussion

Scope of work reviewed in this research is Mechanical Electrical works. The observation is performed periodically every three months. Table 3 shows the result of analysis with EVM and table 4 shows the results of analysis with TVA. The results of the calculations are then classified into three main sections. They are project status, forecasting, and trend. The classification is done to make the result more systematic. To simplify monitoring the classification, factors in the table are given colors. Red denotes project status, and green denotes forecasting. The trend is described separately with a graph that shows relationship of both cost and schedule performance index in two-dimensional graph (figure 2) and cost, time, and work performance index in three-dimensional graph (figure 7; figure 8).

Table 3. The result of analysis with Earned Value Method (1st period to 9th period)[a].

Factors	Dimension	Period								
		1	2	3	4	5	6	7	8	9
PV	IDR (mil.)	154,43	3.189,79	14.590,00	14.590,00	14.590,00	14.590,00	14.590,00	14.590,00	14.590,00
EV	IDR (mil.)	17,63	453,16	2.917,58	7.440,93	11.195,72	12.684,60	13.493,74	14.158,49	14.590,00
AC	IDR (mil.)	233,38	799,85	2.812,76	6.449,75	9.541,58	10.987,22	11.975,02	12.700,73	13.243,00
CV	IDR (mil.)	-215,75	-346,68	104,81	991,18	1.654,13	1.697,37	1.518,72	1.457,76	1.347,00
CPI	-	0,08	0,57	1,04	1,15	1,17	1,15	1,13	1,11	1,10
SV	IDR (mil.)	-136,80	-2.736,63	-11.672,42	-7.149,07	-3.394,28	-1.905,40	-1.096,26	-431,51	0,00
SPI	-	0,11	0,14	0,20	0,51	0,77	0,87	0,92	0,97	1,00
EAC (1)	IDR (mil.)	193.191,64	25.751,79	14.065,86	12.646,53	12.434,37	12.637,66	12.947,90	13.087,81	13.243,00
EAC (2)	IDR (mil.)	14.805,75	14.936,68	14.485,19	13.598,82	12.935,87	12.892,63	13.071,28	13.132,24	13.243,00
EAC (3)	IDR (mil.)	1.690.915,25	176.434,63	59.086,44	18.600,25	13.311,40	12.885,57	13.026,94	13.099,60	13.243,00
ETC (1)	IDR (mil.)	192.958,26	24.951,94	11.253,10	6.196,77	2.892,79	1.650,43	972,88	387,08	0,00
ETC (2)	IDR (mil.)	14.572,37	14.136,84	11.672,42	7.149,07	3.394,28	1.905,40	1.096,26	431,51	0,00
ETC (3)	IDR (mil.)	1.690.681,87	175.634,78	56.273,67	12.150,50	3.769,81	1.898,35	1.051,92	398,88	0,00
Crit. Ratio	-	0,01	0,08	0,21	0,59	0,90	1,00	1,04	1,08	1,10

[a] All the abbreviations used in this table can be seen in section 6 (Notation).

The EVM calculation results on table 3 gives the basic information that was provided by EVM. With only two aspects used in EVM (cost and schedule), the information given is less than TVA.

EVM can only show the estimation in cost aspect (EAC and ETC). In Figure 2, the project trend in two-dimensional graph can show the direction of the project movement from the previous period, which is, still, limited to two aspects only.

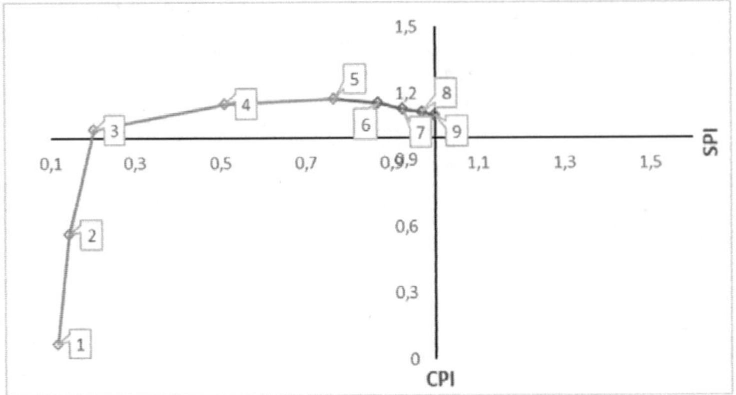

Figure 2. Trend (EVM) in two-dimensional graph.

The movement of project conditions is given different colors. While the color green indicates the project trend improves on both aspects (cost and schedule), blue indicates only one aspect improves (either cost or schedule). In contrast, the color red indicates the trend deteriorates in both aspects. The TVA calculation results can be seen in table 4.

Table 4. The result of analysis with Three-Variance Approach (1st period to 9th period)[a].

Factors	Dimension	Period								
		1	2	3	4	5	6	7	8	9
PV	IDR (mil.)	154,43	3.189,79	14.590,00	-	-	-	-	-	-
ALT	month	-	-	-	3,00	6,00	9,00	12,00	15,00	17,00
PV'	IDR (mil.)	-	-	-	19.453,33	24.316,67	29.180,00	34.043,33	38.906,67	42.148,89
EV	IDR (mil.)	17,63	453,16	2.917,58	7.440,93	11.195,72	12.684,60	13.493,74	14.158,49	14.590,00
AC	IDR (mil.)	233,38	799,85	2.812,76	6.449,75	9.541,58	10.987,22	11.975,02	12.700,73	13.243,00
WV	IDR (mil.)	-136,80	-2.736,63	-11.672,42	-7.149,07	-3.394,28	-1.905,40	-1.096,26	-431,51	0,00
WPI	-	0,11	0,14	0,20	0,51	0,77	0,87	0,92	0,97	1,00
ROW	IDR (mil.)	14.572,37	14.136,84	11.672,42	7.149,07	3.394,28	1.905,40	1.096,26	431,51	0,00
PCR	-	0,01	0,22	1,00	1,00	1,00	1,00	1,00	1,00	1,00
WCR	-	0,00	0,03	0,20	0,51	0,77	0,87	0,92	0,97	1,00
WRs	IDR (mil.)/mo.	51,48	1.011,79	3.800,07	1.621,11	1.621,11	1.621,11	1.621,11	1.621,11	1.621,11
ETs	month	0,34	3,30	5,93	4,59	6,91	7,82	8,32	8,73	9,00
TVs	month	-2,66	-2,70	-3,07	-7,41	-8,09	-10,18	-12,68	-15,27	-17,00
TPIs	-	0,11	0,55	0,66	0,38	0,46	0,43	0,40	0,36	0,35
ETACs	month	78,86	16,39	13,66	23,53	19,55	20,70	22,71	24,73	26,00
ETTCs	month	75,86	10,39	4,66	11,53	4,55	2,70	1,71	0,73	0,00
WRp	IDR (mil.)/mo.	51,48	531,63	1.621,11	1.621,11	1.621,11	1.621,11	1.621,11	1.621,11	1.621,11
ETp	month	0,34	0,85	1,80	4,59	6,91	7,82	8,32	8,73	9,00
TVp	month	-2,66	-5,15	-7,20	-7,41	-8,09	-10,18	-12,68	-15,27	-17,00
TPIp	-	0,11	0,14	0,20	0,38	0,46	0,43	0,40	0,36	0,35
ETACp	month	78,86	63,35	45,01	23,53	19,55	20,70	22,71	24,73	26,00
ETTCp	month	75,86	57,35	36,01	11,53	4,55	2,70	1,71	0,73	0,00
CV	IDR (mil.)	-215,75	-346,68	104,81	991,18	1.654,13	1.697,37	1.518,72	1.457,76	1.347,00
CPI	-	0,08	0,57	1,04	1,15	1,17	1,15	1,13	1,11	1,10
ETC	IDR (mil.)	192.958,26	24.951,94	11.253,10	6.196,77	2.892,79	1.650,43	972,88	387,08	0,00
EAC	IDR (mil.)	193.191,64	25.751,79	14.065,86	12.646,53	12.434,37	12.637,66	12.947,90	13.087,81	13.243,00

[a] All the abbreviations used in this table can be seen in section 6 (Notation).

From the TVA results in table 4, it can be seen that TVA provides more information than traditional EVM. TVA gives its users information in three aspects, namely cost, time, and work. Information about the work that has been completed at one particular time can be measured in the work aspect, while time aspect only focuses on monitoring the completion time of works in the project. The information provided on the cost aspect is in no way different from the traditional EVM. The use of work rate concept (WR_P, WR_S, and WR_M) can deliver information about the value of work on each period based on the PV data. In TVA, there are cost estimation (EAC and ETC) and also time estimation: ETAC and ETTC. Figure 3 to Figure 6 show the result of TVA if displayed in a cartesian graph.

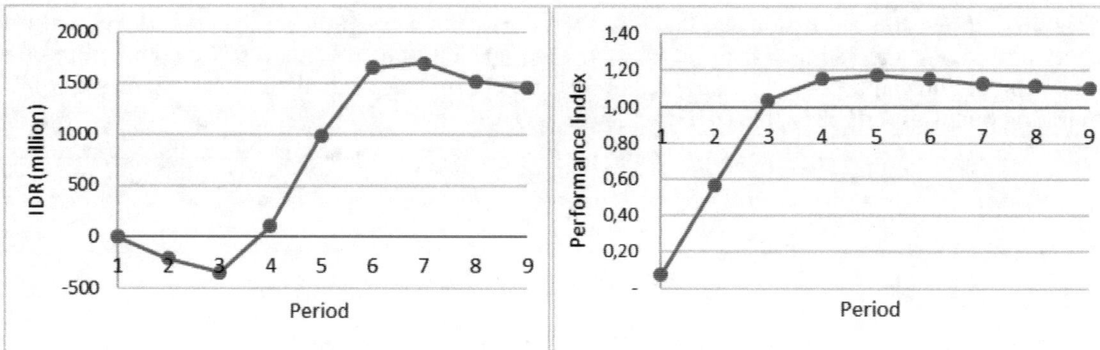

(a) (b)

Figure 3. Result from Three-Variance Approach (a) Cost Variance (b) Cost Performance Index

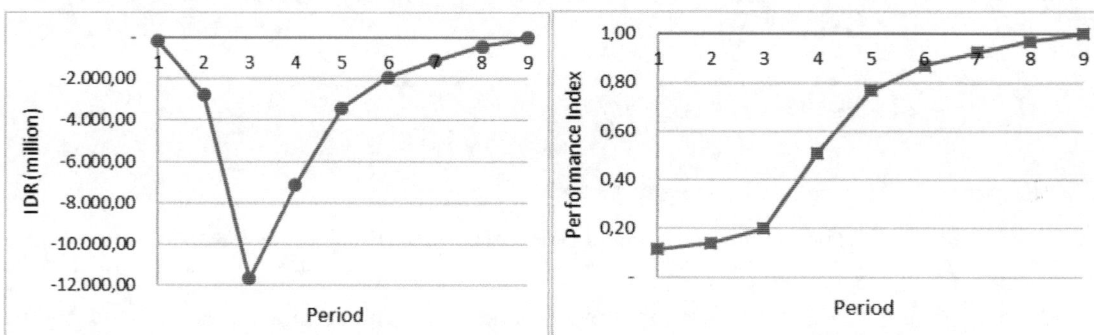

(a) (b)

Figure 4. Result from Three-Variance Approach (a) Work Variance (b) Work Performance Index

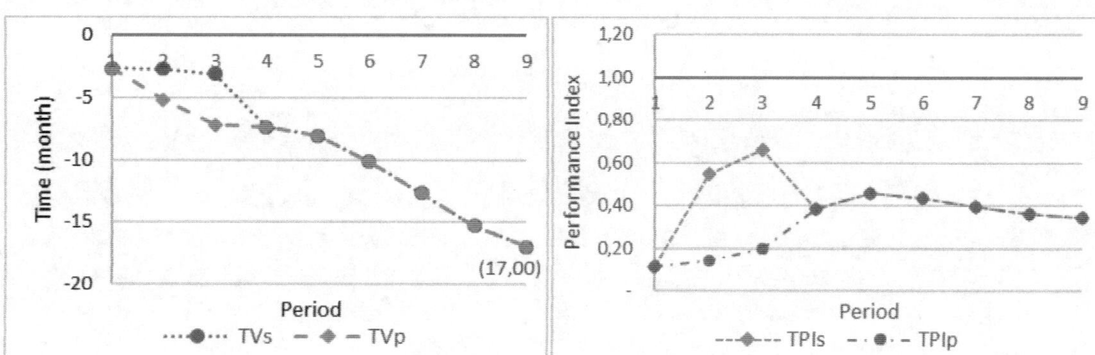

(a) (b)

Figure 5. Result from Three-Variance Approach (a) Time Variance (b) Time Performance Index

In Figure 3 to Figure 5, if the calculation results are above the red line, then the project status in the observed aspect is in good condition. On the contrary, if the calculation results are below the line, then the aspect is monitored in bad condition. By using TVA, the separation between the work and time aspects can be clearly seen at the end of the project. In Figure 4, when the project is finished, the variance line moves towards 0, and the performance index moves towards 1, indicating that the work is completed (regardless of the project on schedule or behind schedule). At the end of the project as in figure 5, the graph deviates from the center of the X axis. The deviation of the graph shows whether the project finished ahead or behind the schedule. The line on the graph of the time aspect in Figure 5 will only be at the center of the X axis if the project is right on schedule.

Figure 6 shows the estimation result using TVA, which is estimation on the cost aspect and time aspect. With TVA, the estimation of how much time needed until the project is completed in every observation period can be known. In the last observation period, the calculation result shows a number of duration that is exactly the same as the actual project duration.

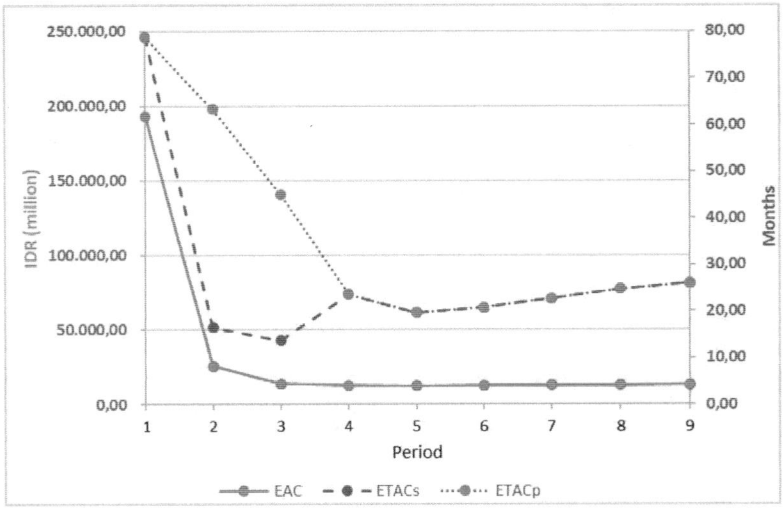

Figure 6. The graphic of time and cost forecasting from Three-Variance Approach

The project trend in TVA can be seen in three-dimensional graph as seen in Figure 7 and Figure 8. There are two different graphs, because there are two types of TPI used. The lines on the graph are given different colors to indicate the condition of the project in each period, which is improved or deteriorated, either in one aspect or all the three aspects (same as the colors used in figure 2).

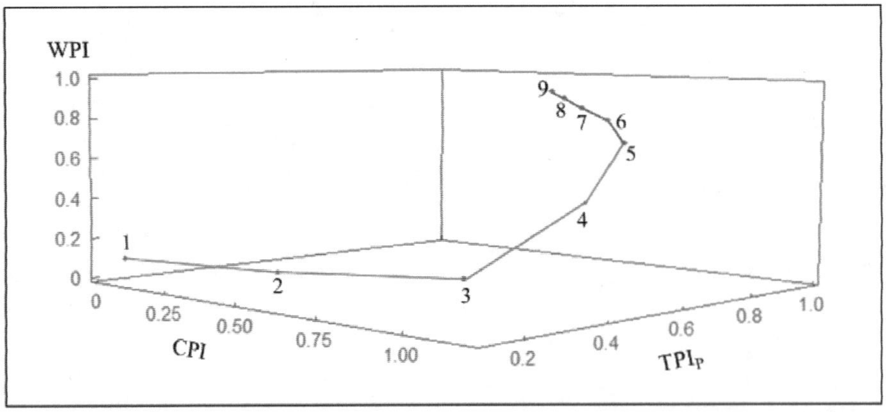

Figure 7. Trend from Three-Variance Approach with TPI_P

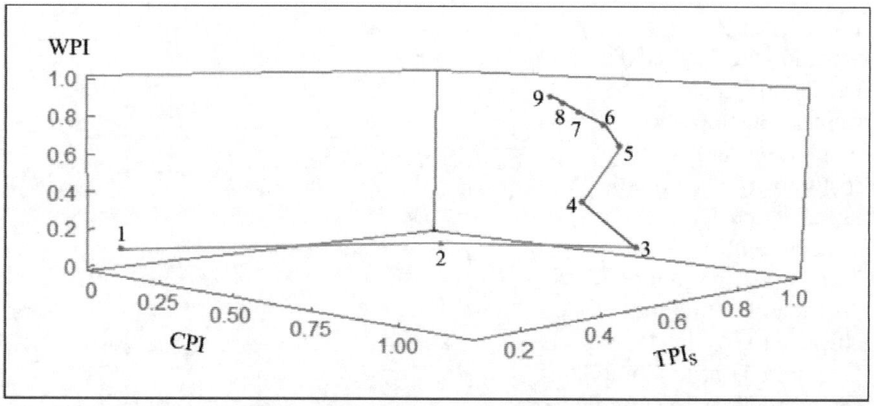

Figure 8. Trend from Three-Variance Approach with TPI_S

It is difficult to see the 3D graph especially when the graphs are printed. Therefore, the displacement vector is calculated to help interpret the 3D graph. Table 5 shows the result of displacement vector (The example on table 5 uses TPI_P).

Table 5. Vector analysis in three-dimension trend

Period	x	y	z	Vector			Description		
	CPI	TPI_P	WPI	x_2-x_1	y_2-y_1	z_2-z_1	Cost	Time	Work
1	0,08	0,11	0,11						
2	0,57	0,14	0,14	0,49	0,03	0,03	better	better	better
3	1,04	0,20	0,20	0,47	0,06	0,06	better	better	better
4	1,15	0,38	0,51	0,12	0,18	0,31	better	better	better
5	1,17	0,46	0,77	0,02	0,08	0,26	better	better	better
6	1,15	0,43	0,87	-0,02	-0,03	0,10	worse	worse	better
7	1,13	0,40	0,92	-0,03	-0,04	0,06	worse	worse	better
8	1,11	0,36	0,97	-0,01	-0,03	0,05	worse	worse	better
9	1,10	0,35	1,00	-0,01	-0,02	0,03	worse	worse	better

With the concept of vector, all the displacement vectors in each period can be found. If previous period is represented by $\bar{\mathbf{a}}$ (x_1, y_1, z_1) and current period is represented by $\bar{\mathbf{b}}$ (x_2, y_2, z_2), then the displacement vector, showing the project trend, is $\bar{\mathbf{c}}$ (x_2-x_1, y_2-y_1, z_2-z_1). If $\bar{\mathbf{c}}$ is positive, the project trend is getting better, otherwise if $\bar{\mathbf{c}}$ is negative, the project trend is getting worse.

5. Conclusion

Based on the research conducted, Three-Variance Approach (TVA) as a development from Earned Value Method (EVM) can be used to control project time and project cost. Time and work that measured separately in TVA can give more details on every aspect. TVA analysis results are classified into project status, forecasting, and trend, so that the result can be more systematic. Project status and forecasting can be visualized in two-dimensional graph to make the project reports more understandable. Trend can be seen in three-dimensional graph. Three-dimensional graph readings are assisted with the concept of displacement vectors. With the help of vector concept, it is easier to read the three-dimensional graphs when the graphs are printed. It is recommended for further studies to develop a software to conduct TVA analysis, thus simplifying users in working on this new method.

6. Notation

The following symbols are used in this paper:

AC = Actual Cost

AT	= Actual Time
ALT	= Accumulated Lagged Time
APW	= Artificial Planned Work
BAC	= Budget at Completion
CV	= Cost Variance
CPI	= Cost Performance Index
ET	= Earned Time
EV	= Earned Value
EAC	= Estimate at Completion
EVM	= Earned Value Method
ETC	= Estimate to Complete
ETAC	= Estimated Time at Completion
ETTC	= Estimated Time to Complete
EWAC	= Estimated Work at Completion
PD	= Planned Duration
PV	= Planned Value
PV'	= Artificial Planned Value
PCR	= Planned Completion Rate
ROW	= Rest of Work
SV	= Schedule Variance
SPI	= Schedule Performance Index
TV	= Time Variance
TPI	= Time Performance Index
TVA	= Three-Variance Approach
WR	= Work Rate
WV	= Work Variance
WCR	= Work Completion Rate
WPI	= Work Performance Index

References

[1] Husen A 2011 *Manajemen Proyek Perencanaan, Penjadwalan, dan Pengendalian Proyek (Edisi Revisi)* (Yogyakarta: Andi)
[2] Project Management Institute 2013 *A Guide to the Project Management Body of Knowledge 5th Edition* (Pennsylvania: Newtown Square)
[3] Fleming Q W and Koppelman J M 1997 Earned value project management *Cost Eng.* **39(2)** 13–15
[4] Kerzner H 2009 *Project Management: A System Approach to Planning, Scheduling, and Controlling 10th Edition* (New Jersey: Wiley)
[5] Lipke W 2003 Schedule is different *Measurable News* **31(1)** 1-10
[6] Vanhoucke M and Vandevoorde S 2007 A simulation and evaluation of earned value metrics to forecast the project duration *J. Oper. Res. Soc.* **58(10)** 1361-74
[7] Anbari F T 2003 Earned value project management method and extensions *Proj. Manage. J.* **34(4)** 12-23
[8] Jacob D 2003 Forecasting project schedule completion with earned value metrics *Measurable News* **1(11)** 7-9
[9] Khamooshi H and Golafshani H 2013 EDM: Earned Duration Management, a new approach to schedule performance management and measurement" *Int. J. Proj. Manag.* http://dx.doi.org/1016/j.ijproman.2013.11.002
[10] Chang C J and Yu S W 2018 Three-variance approach for updating earned value management *J. of Const. Eng. Manage.* **144(6)** 04018045 1-10
[11] Widiasanti I and Lenggogeni 2013 *Manajemen Konstruksi* (Bandung: Rosda)

[12] De Marco A and Narbaev T 2013 Earned value-based performance monitoring of facility construction projects *J. of Facilities Manage.* **11(1)** 69-80

[13] Nagaraj B, Brijbhushan S, and Maneeth P D 2017 Scheduling, monitoring and cost analysis by earned value method of phase-2 (reach-2a from pier No. 443 to pier No. 470) using Primavera P6 *Int. J. of Eng. Development and Research* **5(3)** 625-32

[14] Sediyanto and Hidayat A 2017 Analisa kinerja biaya dan waktu pada pelaksanaan proyek konstruksi dengan metode earned value (studi kasus proyek konstruksi mall dan hotel X di Pekanbaru) *Jurnal Ilmu Teknik dan Komputer* **1(1)** 36-51

[15] Oberlender G D 2000 *Project Management for Engineering and Construction 2nd Edition* (USA: McGraw Hill)

IOP Conf. Series: Materials Science and Engineering **650** (2019) 012011 doi:10.1088/1757-899X/650/1/012011

Risk analysis of traffic congestion due to problem in heavy vehicles: a concept

P A Kesuma*, M A Rohman* and C A Prastyanto

Department of Civil Engineering, Institut Teknologi Sepuluh Nopember (ITS), Surabaya, Indonesia

*auliakesumaputra@gmail.com; arif.its@gmail.com

Abstract. One of the most common problems on roads is traffic congestion caused by problems with heavy vehicles (e.g. trucks, trailers). The consequences of this problem often harm the wider community, both in material and non-material forms such as fuel waste and loss of time. Several previous studies have tried to look for factors that cause traffic congestion, but it has not been able to provide a complete understanding of the root causes and who are the stakeholders who must have an important role in solving this problem. This study aims to develop a research framework to get more better understanding about congestion caused by problems with heavy vehicles, which can be categorized as a type of non-recurrent congestion (NRC) that is dynamic and unpredictable. Risk management prioritizes a proactive approach before an event occurs, so the impact of risk can be minimized. Probability and Impact Matrix techniques are used to determine risk priority. Furthermore, to find the root causes of traffic congestion problems, the Fault Tree Analysis (FTA) method and Social Network Analysis (SNA) are considered to be quite potential to be used to describe complex patterns of relationships among stakeholders. The results of this study are expected to provide more complete information to understand the problem of congestion due to the problem of heavy vehicles, so that a more comprehensive solution can be formulated.

1. Introduction

The highway is one of the infrastructure facilities in the transportation system that has important functions to connect between the region to support accessibility, mobility, and provide many benefits in life [1]. However, the development of the construction industry sector also increased demand for highways, one of which can be seen from an increase in the number of heavy vehicles carrying logistics projects. This condition triggers a variety of serious problems on the highways, including traffic congestion.

Some conditions that can trigger and aggravate traffic congestion, i.e. population growth, economic growth, increased vehicle count, low road capacity, poor city planning and supervision, accidents, and natural factors [2]. Traffic congestion is a global phenomenon predicted to deteriorate in the future [3]. The impact of congestion can be material losses such as energy inefficiencies and pollution on non- material impacts such as time and opportunity wasted due to the psychological impact of increased pressure on human life [4][5]. If ignored, the traffic congestion problem will not only impact the productivity of a project, but it can also reduce the competitiveness of a country.

Although there are many factors that cause traffic congestion, in generally divided into two categories, that is recurrent traffic congestion (RC) and non-recurrent (NRC). RC is a type of traffic congestion that occurs repeatedly or routinely [1]. An example is congestion caused by periods of busy times such as morning and evening [6]. Whereas NRC, is a type of congestion that occurs due to unpredictable events, i.e. traffic congestions that occur due to vehicle damage, accidents, weather conditions, natural disasters or human factors [7]. NRC traffic congestion problem is a dynamic and complex event.

Traffic congestion is a complex problem because of the many factors and stakeholders involved there. Stakeholders involved such as the government, environmental services, transportation departments, highway departments, traffic police, road users, communities and others [8]. The complexity of the congestion problem above is actually quite right when viewed from a risk management perspective. The risk management approach offers a fairly appropriate concept to see the dynamics and complexity of problems, especially the type of NRC traffic congestion. In addition, risk management is a more proactive approach and does not rely solely on reactive responses to risk issues. With this proactive perspective, it is expected that the negative impacts can be minimized or even eliminated. In addition, the risk management concept provides a way to respond to risk and make risk allocation to the most appropriate parties to better manage risk. The concept of risk management aims to increase the probability of a positive impact or opportunity and reduce the risk probability of a negative or threatening impact [9]. With the concept of risk management, congestion problems must be seen from a more comprehensive understanding by looking at all types of risks and what stakeholders influence efforts to solve problems while making the right risk allocation concept.

Efforts to understand NRC traffic congestion problem more comprehensively, the concept of risk management has several general stages such as risk identification, risk analysis, and risk response planning [10]. Fault tree analysis (FTA) is one technique that can be used in risk analysis to find the core of the problem. According to [11] FTA is an analytical technique carried out to find out the root causes and the possibility of an undesirable event. Social network analysis (SNA) is one approach to understanding the complex relationships between problems in a structured network system with mathematical foundations of graph theory [12]. Through the SNA the relationship between risk factors for congestion and its relationship with stakeholders can be well represented. Through the SNA the relationship between risk factors for congestion and its relationship with stakeholders can be well represented. SNA has been widely used in various construction industry areas such as construction management, transportation planning, and construction safety [13].

Based on the description of these conditions, this study proposes the concept of risk management in solving this problem by applying FTA to find the root cause of the problem. Then the SNA is used to illustrate the role of stakeholders in road traffic congestion issues, especially for NRC congestion due to problems with heavy vehicles. It is hoped that a complete understanding of the NRC congestion problems can be the basis for solving the problems that occur and allocating risks to those stakeholders who have the most ability to manage these risks.

2. Literature Review

The theory that underlies this research includes road traffic congestion problems and the concept of risk management as an alternative solution to this problem.

2.1 Highway traffic congestion

The traffic congestion often occurs especially on highways are the main access between regions and are a very serious problem experienced by many big cities in developing countries including Indonesia. Traffic congestion is a condition of highway traffic flow that stops or moves very slowly [8]. The highway in question is a form of paths on the surface of the earth made by humans with the size and type of construction so that it can be used to channel traffic for people, animals and vehicles that transport goods from one place to another easily and quickly [14]. Two circulation domains on the road, such as passengers and goods that use the same road infrastructure, result in road infrastructure use that exceeds capacity [2]. This is reflected in the decline in speed, increase in travel time, and queue of vehicles. [3] Argues that are two perspectives of traffic congestion problems, first perspective is that congestion can be considered as an indicator of a region's economic growth and the second perspective is an indicator of the decline or shortcomings of a region.

Traffic congestion problems that occur generally are caused by an imbalance between supply (available transportation facilities) and demand (transportation demand). Availability of transportation facilities influenced by history and geography; Transport and operational management; And the level of investment on highways, while demand is affected by the level of travel concentrations in space and time [1].

According to [2] which mention some conditions that can cause and worsen the condition of road congestion, such as population growth, economic growth, increased employment opportunities, lack of investment in road infrastructure, increased number of vehicles, low road capacity, road layout, poor traffic management, violations of road rules, lack of parking lots, poor city planning and supervision, inadequate public transportation, increased use of private vehicles, accidents, an event, road works, bad weather.

Traffic congestion problems in developing countries are caused by several reasons such as; unplanned cities, poor road user discipline, alternative traffic, ancient traffic management systems, improper track management [1]. The results of other studies conducted by [15] find several lengths of traffic congestion affected by conditions such as; pedestrian crossing by 0.4%, road works by 4.5%, bottlenecks at 9.3%, traffic accidents by 12.9%, crossroads at 72.9%, besides, factors the weather is also the cause where traffic conditions increase by 20% and the snow conditions increase by 45%. Overall [15] also concluded that traffic congestion that often occurs is due to time and work schedules reaching 82.6% of total traffic congestion. While congestion caused by unpredictable things such as accidents and road works is only 17.4% of total traffic congestion.

In general it according to [1] dividing highway congestion into two categories: (1) Recurrent traffic congestion (RC), are repetitions that occur repeatedly because of busy periods such as morning and evening. (2) Non-recurrent traffic congestion (NRC), which are congestion that occur due to unexpected events such as accidents; damage to the vehicle; road repair; bad weather; an event that causes a sudden surge in demand; and natural disasters or human errors. NRC traffic congestion can trigger new bottlenecks if they occur within a period of time that is not busy, but it is also possible to aggravate congestion conditions if they occur during periods of busy time.

2.2 *Traffic congestion review of the risk management concept*

The NRC traffic congestion problem is quite appropriate for review from the risk management perspective. This type of congestion is caused by events that are dynamic or contain uncertainty. Risk in the scope of the project is an event that can occur and will have an influence on the project objectives, which in this risk has an impact on loss or profit [16]. Risk in the operational scope refers to things that have a negative impact on the project that is already operating [17]. Traffic congestion affects travel costs, travel time, mobility, accessibility, productivity, and also impacts on the environment such as air pollution and global warming [1].

More about NRC traffic congestion issues, especially on roads that are passed by various types of vehicles such as light vehicles, heavy vehicles, and motorbikes. This often raises several unpredictable events such as damage to vehicles to accidents that have an impact on the obstacles of a road. Some of the problems that often occur in vehicles, especially in the heavy vehicle class such as over-loading that can trigger damage on tire, damage on vehicle braking system, damage on machine system [18][19][20].

Furthermore, some variables on NRC traffic congestion risk related to problems in heavy vehicles were identified based on some of the literature presented in Table 1.

Table 1. Risk variable of NRC traffic congestion

Risk Code	Risk event	Source
R1	Failure of tire parts such as; Flat tire or blow-out	[18][19]
R2	Engine part failure such as; the engine is off and cannot be turned on	[20][18][19]
R3	Damage on the axle	[18]
R4	Vehicle braking systems failure	[19][21]
R5	Vehicle lighting systems failure	[19]
R6	Disruption of vehicle electrical systems such as; battery doesn't work	[18]
R7	Damage on the bearing part of the wheel	[18]
R8	Involved in a crashes with another vehicle	[18]
R9	The vehicle is overturned or crashes into the road divider	[18]
R10	The vehicle spills dangerous material on the highway	[17]

In addition, it is also necessary to identify stakeholders involved in the highway as presented in Table 2. This aims to explore information and find roles or ownership of the risks that occur on the highways.

Table 2. Highways stakeholders

Category	Stakeholder	Source
Goverment	National development planning agency (BAPPENAS)	[8]
	Regional planning and development agency (BAPPEDA)	[8]
	Directorate general of regional development (BANGDA)	[8]
	Local government (PEMDA)	[8]
	Department of transportation (Dishub)	[8]
	Department of highways (Bina Marga)	[8]
	Road traffic and transport service (DLLAJ)	[8]
	Traffic police	[8]
	Transport organization (Organda)	[8]
Road User	Vehicle driver	[15]
	Bicyclist	[15]
	Pedestrians	[15][2]
	Passengers	[15]
	Community	[15][8]

3. Research methodology

The concept of the proposed framework consists of three main steps, that is measuring the magnitude of risk, root cause analysis, and analyzing the role of stakeholders as presented in Figure 1. The application of this methodology is expected to produce a more comprehensive understanding of NRC traffic congestion problems mainly due to heavy vehicle problems.

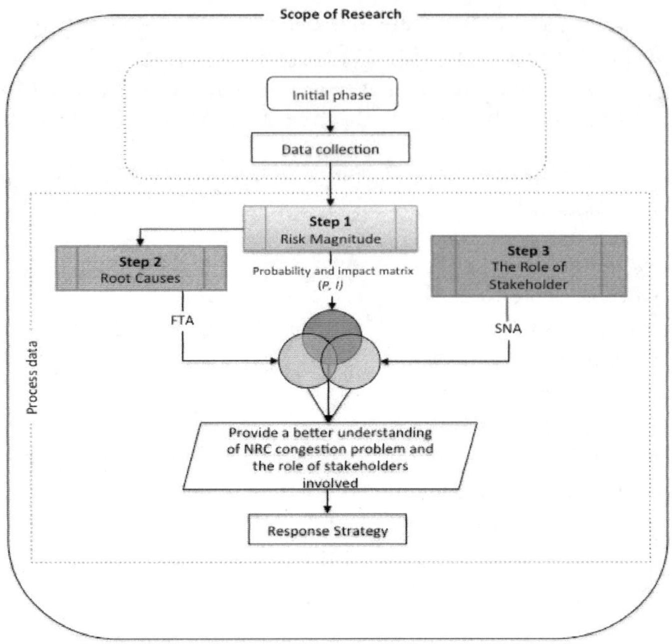

Figure 1. Framework development

3.1 Risk magnitude (step 1)

Risk measuring is an important part in risk management, Figures 2 showing the flowchart models proposed to analyze potential risks.

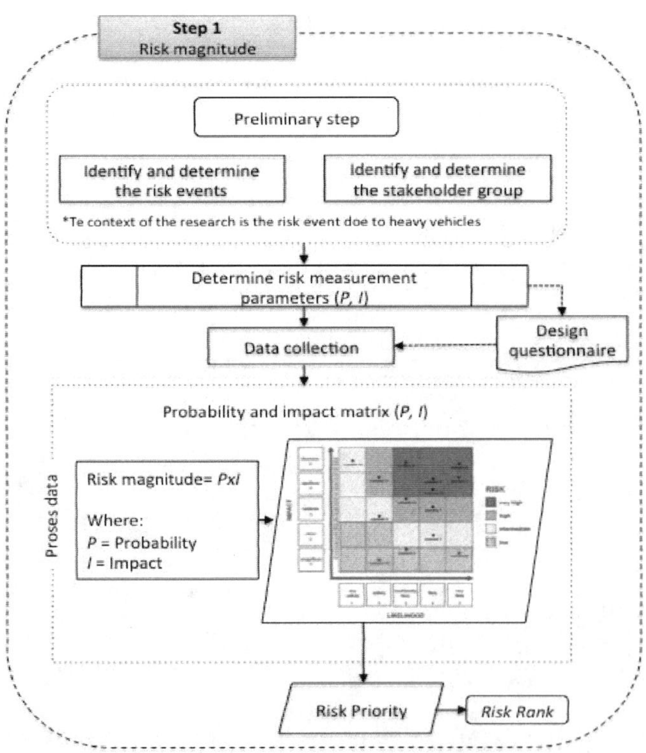

Figure 2. Risk magnitude model flowchart

IOP Conf. Series: Materials Science and Engineering **650** (2019) 012011 doi:10.1088/1757-899X/650/1/012011

According to [10] who argues that risk has 2 (two) main components in each risk event, namely the probability of occurrence of a risk event, and the consequences or impacts of that risk. From these 2 (two) components, can be defined through Equation 1. as follows:

$$Risk = f(P,I) \tag{1}$$

where;
P = Probability,
I = Impact.

3.2 Root causes analysis (step 2)
FTA is an analytical technique that is carried out to find out the root cause and the possibility of unwanted events [9]. The FTA diagram presents a logical graphical relationship between the problem and its causes. [22] It states that FTA is deductive, where analysis begins with system failure or peak events and then works backwards from top to bottom as the main cause. The steps proposed to analyze the root of the problem using Fault Tree analysis (FTA) are presented in Figure 3.

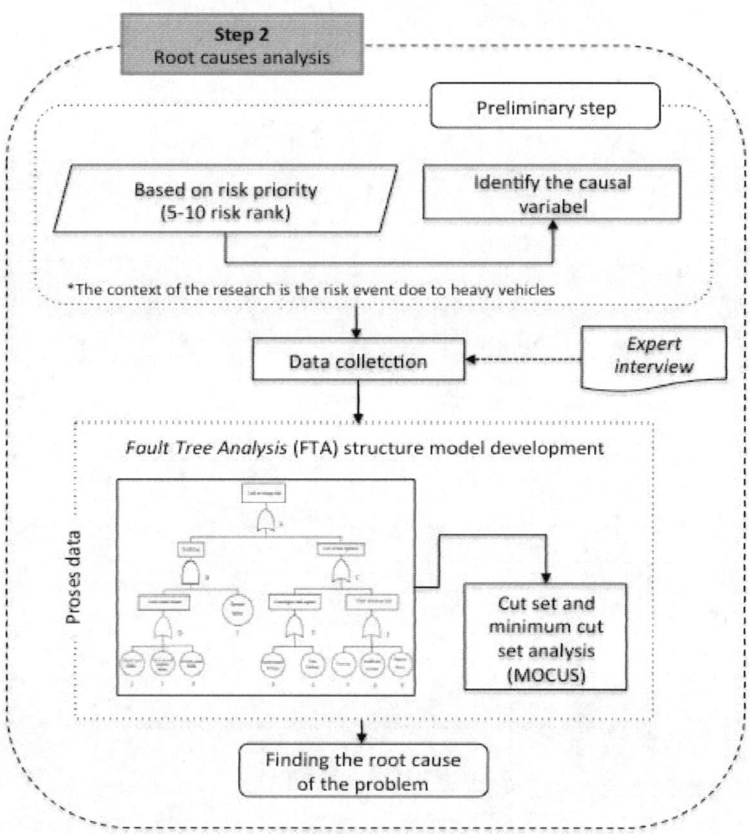

Figure 3. Root causes analysis model flowchart

3.3 Stakeholder analysis (step 3)
This stage applies the method of analyzing social networking (SNA) to provide information about the role of stakeholders involved in the problem of NRC traffic congestion, as

presented in Figure 4. SNA is a social network structure consisting of a set of connected social units. Two basic components are defined in social network: sets of actors (nodes) and sets of ties depicting the interconnections [23]. The relationship between sets of components produces a social network structure. The social network can be established with investigating the ties and the strength of the ties between actors base on interviewing stakeholders involved [23].

The key concept of SNA is centrality, can be measured by the degree of centrality (DC). DC is the number of connections that nodes have and describes the level of connection with other actors [23]. DC can be shown in Equation 2. as follows:

$$C_D\,(N_i) = \sum_{j=1}^{8} Xij \tag{2}$$

where:

$C_D\,(N_i)$ = Denotes the degree centrality for node i,

$\sum_{j=1}^{8} Xij$ = Sum of the intensity of direct ties that node i to the $g-1$ but excluding j (exclude node itself).

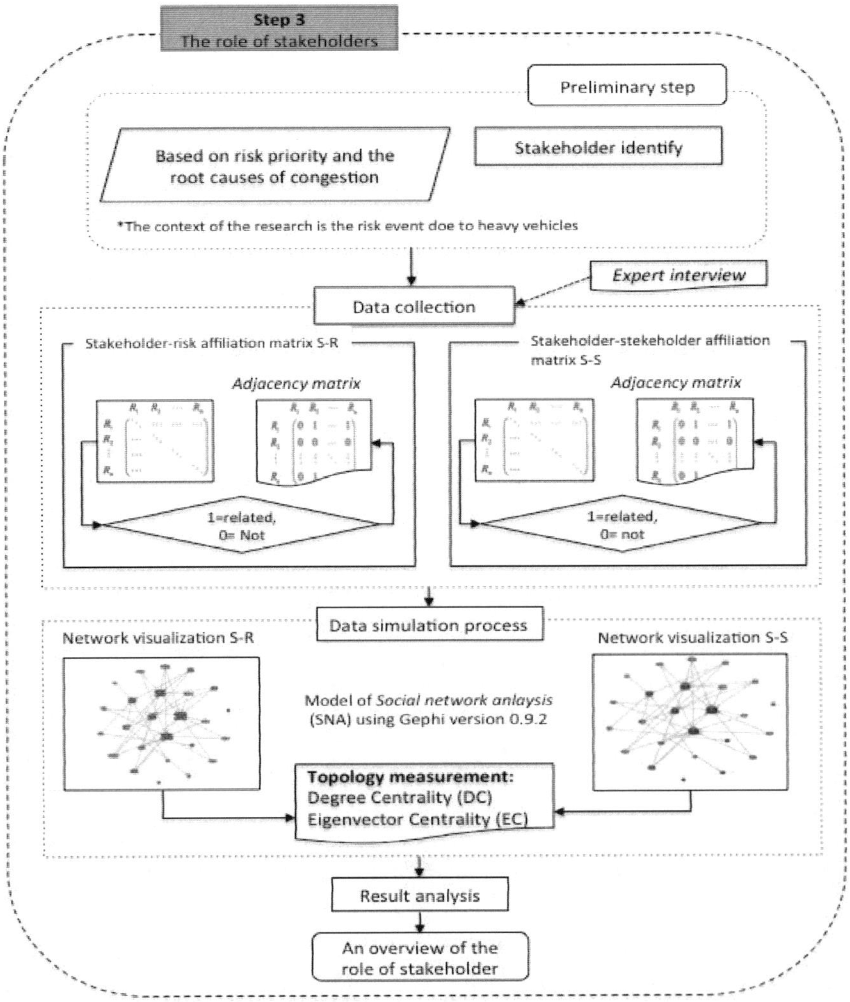

Figure 4. Stakeholders analysis model flowchart

4. Conclusion and recommendation

Based on the background of the problem and the literature review conducted, there is no study that comprehensively analyzes the problem of traffic congestion especially for NRC traffic congestion due to problems in heavy vehicles. Therefore, this study proposes the concept of risk management in solving this problem by applying FTA to find the root of the problem. Then SNA is used to visualize and map NRC traffic congestion problems because of problems with heavy vehicles. The hope is that this research provides a complete understanding of the problem of traffic congestion NRC can be a basis for addressing the problems that occur and allocating risks to stakeholders who have the greatest ability to manage this risk.

Acknowledgements

The authors thanks the Ministry of Research, Technology, and Higher Education for providing grants to perform this research and funded through ITS 2019 postgraduate research. We also appreciate the experts involved in the preliminary survey in this study.

References

[1] Falcochio J C and Levinson H S 2015 *Road Traffic Congestion: A Concise Guide* (London: Springer International Publishing Switzerland)
[2] Agyapong F and Ojo T K 2018 Managing traffic congestion in the accra central market, Ghana *Journal of Urban Management.* **7** pp 85-96
[3] Kiunsi R B 2013 A review of traffic congestion in Dar es Salam City from the physical planning perspective *Journal of Sustainable Development.* **6** pp 94-103
[4] Bharadwaj S, Ballare S, Rohit and Chandel M K 2017 Impact of congestion on greenhouse gas emission for road transport in Mumbai metropolitan region *Transportation Reaserch Procedia.* **25** pp 3538-3551
[5] Hopkins J L and McKay J 2018 Investigating 'anywhere working' as a mechanism for alleviating traffic congestion in smart cities *Technological Forecasting & Social Change*
[6] Anbaroglu B, Heydecker B and Cheng T 2014 Spatio-temporal clustering for non-recurrent traffic congestion detection on urban road network *Trans. Reasearch Part C* **48** pp 47-65
[7] Chung Y 2011 Asessment of non-recurrent traffic congestion caused by freeway work zones and its statistical analysis with unobseved heterogenety *Transportation Policy* **18** pp 587-594
[8] Tamin O Z 2000 *Perencanaan dan Pemodelan Sistem Transportasi* (Bandung: Penerbit ITB)
[9] PMI 2013 *A Guide to the Project Management Body of Knowledge, Fifth Edition* (Pennsylvania: Project Management Institute, Inc.)
[10] Kerzner H 2009 *Project Management Tenth Edition (*New Jersey: John Wiley & Sons, Inc.)
[11] Ericson C A 2005 *Hazard Analysis Techniques for System Safety* (New Jersey: John Wiley & Sons, Inc.)
[12] Otte E and Rousseau R 2002 Social network analysis: a powerful strategy, also the information sciences *J. Inf. Sci.* **28** pp 441-453
[13] Eteifa O S and El-adaway H I 2017 Using social network analysis to model the interaction between root causes of fatalities in construction industry *J. Management in Engineering* **34**
[14] Clarkson H O 1999 *Teknik Jalan Raya Jilid 1* (Jakarta: Gramedia)
[15] Stepanchuk O, Bieliatynskyi A, Pylypenko O and Stepanchuk S 2017 Surveying on traffic congestion on arterial roads of Kyif City *Procedia Engineering* **187** pp 14-21
[16] Cooper D F, Grey S, Raymond G and Walker P 2005 *Project Risk Management Guidelines: Managing Risk in Large Projects and Complex Procurements* (West Sussex: John Wiley & Sons Ltd.)

[17] Ghazali F E M 2009 Operational risks for highway projects in Malaysia *Proc. of World Academy of Science Engineering and Technoogy* **41**

[18] Jacob B and Beaumelle V F L 2010 Improving truck safety: potential of weigh-inmotion technology *International Association of Traffic and Safety Sciences* **34** pp 9-15

[19] Teoh E R, Carter D L, Smith S and McCartt A T 2017 Crash risk factors for interstate large trucks in North Carolina *Journal of Safety Research* **62** pp 13-21

[20] Blazquez C A, Picate B, Calderon J F and Losada F 2018 Spatial autocorellation analysis of cargo truck on highway crashes in Chile *Accident Analysis and Prevention* **120** pp 195-210

[21] Zhang M, Kecojevic V and Komljenovic D 2014 Invetigation of haul truck-related fatal accidents in surface mining using faul tree analysis *Safety Science* **65** pp 106-117

[22] Kabir S 2017 An overview of fault tree analysis and its application in model based dependability analysis *Expert System With Applications* **77** pp 114-135

[23] Doloi H 2012 Assessing stakeholders' influence on social performance of infrastructure projects *Facilities* **30** pp 531-550

Proposed maintenance system of Gelora Sriwijaya Jakabaring Stadium (case study: lighting utilities)

Muhammad Syazari Yazuar[1], Senot Sangadji[2], Stefanus Adi Kristiawan[2]

[1]Master Program in Infrastructure Maintenance and Rehabilitation Management, Civil Engineering Department, Faculty of Engineering, Sebelas Maret University, Surakarta, Indonesia

[2]Civil Engineering Department, Faculty of Engineering, Sebelas Maret University, Surakarta, Indonesia

m_syazari_y@yahoo.co.id; s.sangadji@ft.uns.ac.id; s.a.kristiawan@ft.uns.ac.id

Abstract. Maintenance of the stadium has been done for the 18th Asian Games 2018. However, the previous maintenances were not comprehensive and held in the run up to the big events only. It affects the performance of its components utilities. A new local company was established by the Government of South Sumatra Province, in order to make a good maintenance of the stadium. However, the planned maintenance has not been actualized by the company yet. Many obstacles encountered in stadium maintenance, including regulatory, complexity of utilities, and stakeholder's influences. This paper aims to develop a maintenance system that is in accordance with existing conditions and applicable regulations. This study is devoted to electrical lighting systems because almost every week the stadium is also used for football matches. The methodology was divided into three parts, i.e.: rating for each parts of the utility, descriptive qualitative for regulatory identification, and stakeholders analysis, and they are developed to create a maintenance system. The result is a system that is proposed to find out the most important stakeholders in the maintenance of the highest priority of lighting utility based on official assessment.

Keywords: stadium, maintenance, regulation, lighting utilities, stakeholders

1. Introduction

Since it was built in 2002, Gelora Sriwijaya Jakabaring (GSJ) Stadium was functioned for the first time in the National Sports Week (PON) in 2004. In the meantime, there was no planned maintenance system. Maintenance was only did when there were great-scale sports events such as Asian Football Cup (2007), SEA Games (2011), Islamic Solidarity Games (2013), ASEAN University Games (2014), and Asian Games (2018). Maintenance was only carried out during the implementation of project construction, and it ended when the executive contractor conducted the second handover.

The maintenance of project construction was not conducted by maintenance-specialist contractors. Every time the maintenance would be performed, the contractor qualification auctioned was general contractor qualification. Moreover, the maintenance merely fixed major damages on existing installation network and seemingly easy installations, for instance, arena lighting.

Aragones [1] say that the most influential stakeholders are the contractor and the signaling systems provider with almost 40% of influence on the project, because these two stakeholders have problems (delays, overruns or lack of specifications), the project will be severely affected.

Before 2018, the budgeting process involved many parties. Afterward, the Provincial Government established a Municipally Owned Corporation named PT. Jakabaring Sport City (PT. JSC). This corporation was in charge of managing the area of Jakabaring Sport City, including maintenance of the whole venues. As a newly established corporation, the maintenance process done by PT. JSC, however, was not sufficient due to several reasons as follows: a small number of personnel (organization structure compared to the number of venues), great coverage of task but no capable personnel who was responsible for each venue. The applicable regulations had not been fully implemented by the company.

Based on Chanter [2], in a nutshell, maintenance can be divided into two types: Planned Maintenance and Unplanned Maintenance. Planned maintenance is divided into preventive maintenance and corrective maintenance. For preventive maintenance, there are two types: scheduled maintenance and condition/situational maintenance. According to Chen [3], preventive maintenance (PM) is required for good equipment management and protection of company assets. Minimum standard of PM company must be applied. Special maintenance procedures must be available from the equipment department on the most major equipment. If certain standards are not available, the recommendations of manufacturer's minimum maintenance must be used. Stakeholders are people, groups, or institutions having interests in policies, programs or projects [8].

According to AFC regulation, the lighting system has a significant assessment score. The stadium lighting system is the most numerous system and the most electricity-consuming system. Arena lighting is an absolute qualification to fulfill the Olympic Organization and international broadcasting television standard. The complexity of the requirements of each standard influences the final assessment of the feasibility of the stadium.

Maintenance of stadium lighting systems was influenced by the involvement of various stakeholders, both internal and external. Each of them had an influence on each maintenance process. Based on these influences, they were divided into two categories: internal and external. The involvement of various parties in stadium maintenance was also based on applicable regulations in Indonesia and international standard.

As stated early, maintenance of the stadium could be categorized as incidental maintenance. It only held as the requirements of certain sports events. The Provincial Government has established a regional company to maintain the stadium. However, the effort was constrained by: many rules, complicated technical aspects and the involvement of stakeholders. In order to solve these problems, this study proposed a planned maintenance system of the lighting utility network of GSJ Stadium specifically.

2. Steps to be considered in the development of planned maintenance

In developing a planned maintenance system of stadium, holistic identification was needed. Technical aspects in maintenance were important but it still need to aware on another aspects that affected such as applicable regulations and stakeholders. Therefore, this paper tried to build a maintenance system that was composed of various influenced aspects.

2.1. Component, Installation and Network Rating

Based on installed parts, the GSJ stadium lighting utility could be divided into three categories, namely: network, installation and components. The component referred to an electrical object that could be functioned when strung together with another electrical objects. An installation is a series of components arranged for an electrical function. A network is a group of several similar installations.

At this stage, the scoring method is used for each component, installation and network. The network is a combination of all similar installations. Installation is an arrangement of components that

IOP Conf. Series: Materials Science and Engineering **650** (2019) 012012 doi:10.1088/1757-899X/650/1/012012

are arranged into one part and function properly. The scoring system used is the division of levels into 10 parts or a scale of 1 to 10. This division is based on the assessment in accordance to the building regulations of the Ministry of Public Works. In the regulation, it was stated that the value for minor damage worth <10%, so the building could be considered in good condition [5],[6].

2.2. Identification of Regulation

Regarding building maintenance, each country has its own regulations. These regulations were applied to all buildings, especially government-owned buildings that are used for public purposes.

Thus, it was necessary to collect data on all regulations that apply in Indonesia related to buildings, electrical installations and stakeholders. All collected regulations were grouped according to the category and issuing agency. With a qualitative descriptive method, the next step was to identify the regulations that contain the most explanation regarding the maintenance of the utility of the lighting system in special building buildings. With comparative study techniques, identification of regulations was carried out to see the relationship between the regulations which became the main reference with the regulations that become supporters' references.

2.3. Stakeholder Analysis

As a public building, the stadium must be well maintained. With the complexity of the installed utilities, its maintenance involved many parties. Stakeholders were involved in the aspects of implementation of maintenance and verification. Therefore, beside the technical aspects, stadium maintenance must consider the level of involvement of each stakeholder.

Based on PMBOK in 5th Edition [9], the steps in conducting stakeholder analysis are as follows:

- Identifying all stakeholders and related information such as roles, departments, interests, knowledge, expectations, and level of influence.
- Analyzing the potential impacts or support for each stakeholder that can be developed and classified for strategy development. In a large stakeholder community, it is necessary to prioritize stakeholders to ensure efficient efforts to communicate and manage their expectations.
- Assessing how key stakeholders react or respond to various situations to plan how to influence them in order to increase their support and anticipate the negative impacts that might arise.

In conducting this analysis, there are various classification models, namely:

- Power/interest grid, based on the level of power and interests of the stakeholders based on project results/outcomes (see Figure 1).
- Power/influence grid, based on the level of power and their influence/involvement in the project.
- Influence/impact grid, based on the level of influence/involvement in the project and their ability to provide the impact towards the project (plan - implementation).
- Salience model, which describes stakeholder classes based on their power, level of importance, and legitimacy.

The rating scale is in the form of a range of 1-5 with the following conditions: Power level. Level (1) means very low - Level (5) means very high

Interest level. Level (1) means very small - Level (5) means very large

Support level. Level (1) means very inhibiting - Level (5) means very helpful

Influence level. Level (1) means very small - Level (5) means very high.

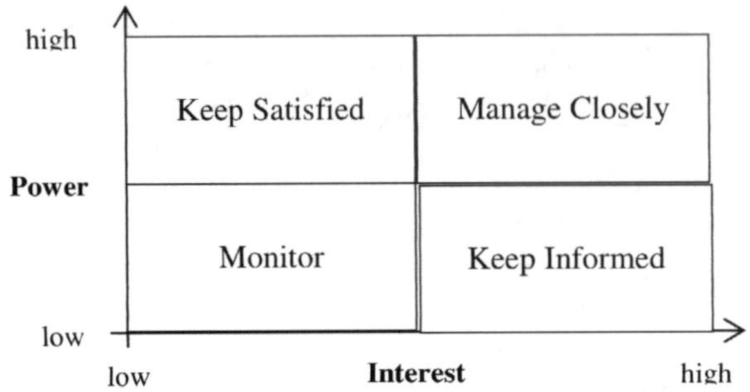

Figure 1. Grid Power/interest of Stakeholders

3. Proposed Maintenance System of the Stadium

As a special building that has high utility complexity and certain characteristics, the stadium requires a maintenance system that is prepared according to the applicable regulations and the existing conditions of the building.

Based on Minister of Public Works Regulation No.25/2007 [6], GSJ Stadium was included in the special building category. Until now there are no guidelines for the maintenance of special buildings in Indonesia. Maintenance guidelines for public building buildings are Minister of Public Works Regulation No. 16 of 2010 [4]. This regulation is the main reference for assessing the extent of damage to the utility of electrical lighting systems for buildings in general.

As a sports building, Stadium had a lot of constituent components that must be installed according to the standards of various organizations. Each stadium has a unique structure, architecture and MEP that is different from other stadiums. In order to organize certain events, the stadium must fulfil the organizer's qualifications. In addition, the utility of lighting as an electrical system is also required to be installed according to Indonesia's national standards in Electrical Installation (SNI 0225: 2011) [6].

3.1. Component, Installation and Network Rating

The first step in the preparation of a maintenance system for electrical installations for the lighting of stadium buildings was to compile the rate of each network listed in Table 1. The type of the lighting network was sorted from the highest score to the lowest one. Assessment scores were calculated based on location, power, difficulty in loading and unloading, duration of use, and the number of points. Each type of light network was also weighted by rating again for each point, based on location, price, and difficulty in loading and unloading. At each point of the light, an assessment is also carried out to determine the Rating of each constituent component, based on location, price, and difficulty in unloading.

Table 1. Type of Lighting Network Ratings

Lighting Network	Place	Power (watt)	Price (Rupiah)	Duration (per year)	Installation	Total Score
Area						
Mini 300						
Roof Facade						
Tango						
Songket Area						
Under the Tribune						
VVIP						
Downlight						
TL						
Ordinary Fittungs						

IOP Conf. Series: Materials Science and Engineering **650** (2019) 012012 doi:10.1088/1757-899X/650/1/012012

Table 2. Example ratings of difficulty level of repair

Category	score	Salary Convertion				
		SNI coefficient	interpolation	Salary *) (Rupiah)	Value (Rupiah)	
		(a1) helper (a2) worker	(b1) helper (b2) worker	(c1) helper (c2) worker	= [(a1)*(b1)*(c1)+ [(a2)*(b2)*(c2)	Amount
hardest	10	0.01	1.5	43,000	645	6,292.5
		0.06	1.5	62,750	5,647.5	
	9	0.01	1.4	43,000	602	5,873
		0.06	1.4	62,750	5,271	
	8	0.01	1.3	43,000	559	5,453.5
		0.06	1.3	62,750	4,894.5	
	7	0.01	1.2	43,000	516	5,034
		0.06	1.2	62,750	4,518	
	6	0.01	1.1	43,000	473	4,614.5
		0.06	1.1	62,750	4,141.5	
	5	0.01	1.0	43,000	430	4,195
		0.06	1.0	62,750	3,765	
	4	0.01	0.9	43,000	387	3,775.5
		0.06	0.9	62,750	3,388.5	
	3	0.01	0.8	43,000	344	3,356
		0.06	0.8	62,750	3,012	
	2	0.01	0.7	43,000	301	2,936.5
		0.06	0.7	62,750	2,635.5	
easiest	1	0.01	0.6	43,000	258	2,517
		0.06	0.6	62,750	2,259	

*) note: salary follow the current inflation

The coefficient above comes from the unit price standard issued by the Ministry of Public Works and Housing [6]. Unit price is the price issued by the local Regency/City Government and experiences adjustments every year. For interpolation, it is standardized that the easy category has the same value as the coefficient multiplied by the unit price that applies to the count per light point, without taking into account the different comparisons of the length of cable used for similar networks. Cable length affects the calculation of the price column. For the power column, the score is calculated based on the wattage capacity used. As it is presented in the table below:

IOP Conf. Series: Materials Science and Engineering **650** (2019) 012012 doi:10.1088/1757-899X/650/1/012012

Table 3. Scoring based on power requirements

Category	Scale	Percentage
lowest	1	< 10 %
	2	< 20 %
	3	< 30 %
	4	< 40 %
	5	< 50 %
	6	< 60 %
	7	< 70 %
	8	< 80 %
	9	< 90 %
highest	10	< 100 %

Calculation of price scores; the duration and calculation of the next lighting ratings also follow the pattern above. Compilation of network ratings is done by considering the real conditions in the field. Because the utilities that will be prioritized on each network, have their own characteristics.

3.2. Identification of Regulation

After determining the network hierarchy, the next step is to identify the relevant valid rules from various agencies. The total number of related rules are 65 items, consist of: 21 of (high state institutions); 29 (Ministry of Public Works and Housing); 16 (Ministry of Energy and Mineral Resources); 1 (Ministry of Youth and Sports); 1 (Ministry of Industry); 1 (PT. PLN) and 6 SNI (National Standardization Agency).

The results of identification of regulations with comparative methods, comparing between rules. The fact is that in general there are fundamental technical differences between regulations originating from technical ministries and applicable SNI. these differences are not contradictory, but are used to complement each other.

Technically, the Ministry of Public Works regulation was based on visual observations, while SNI is more likely to do a lot of testing with sufficient time long. However, the implementation of maintenance measures based on visual observations must still meet SNI requirements in general, so that the main reference to be used in this paper is Minister of Public Works Regulation No. 16 of 2010 [4] and National Electrical Standard of Indonesia 2011 [7].

Visual evaluation for maintenance used the forms according to the Minister of Public Works Regulation No. 16 of 2010. The form is solely taken in the sections which contain utilities in general and electrical installations in particular as shown in the figure below:

Assessment Form (based on Regulation Ministry of Public Works No. 16/2010 page 136)

Lamp Type: **TL 2x36**

1. Location/Code : East / 4.a.T.3.30
2. Chamber : 3rd Floor (under tribune)
3. Chamber type : [ ] class [ ] library [ ] meeting/conference [ ] lobby [√] gym/sport [ ] office [ ] toilet [ ] building connector [ ] others: ....
4. feature type : [ ] inside [ ] hung [√] on surface [ ] others:............
5. diffuser type : [√] metal [ ] plastic [ ] none [ ] parabolic [ ] others:............
6. Lamp type : TL (Tube Luminescent)
7. ballast type : [√] electronic [ ] magnetic
8. feature amount : 2
9. lux of lighting : 30
10. controller type : [ ] automatic [√] manual (on/off) [ ] dimmer [ ] Others: ....
11. age : 15 year
12. damage level :

	none 0%	bad < 1 %	worse 1 - 10 %	worst > 10 %
> decreased quality	[ ]	[ ]	[√]	[ ]
> faded color	[ ]	[ ]	[√]	[ ]
> blink rate	[√]	[ ]	[ ]	[ ]
> glowing level	[ ]	[√]	[ ]	[ ]
> less control	[√]	[ ]	[ ]	[ ]
> repair need	[ ]	[√]	[ ]	[ ]
> buzzing level	[√]	[ ]	[ ]	[ ]
> instability of light	[√]	[ ]	[ ]	[ ]
> light emitting level	[ ]	[ ]	[√]	[ ]

13 general condition : [ ] bad [√] worse [ ] good [ ] best

14 remaining service life : 1 year

15 comments : ....
....

16 Supervisor : .... date ....

Figure 2. Assessment form

IOP Conf. Series: Materials Science and Engineering **650** (2019) 012012 doi:10.1088/1757-899X/650/1/012012

3.3. Stage Stakeholder Analysis

The next step is to conduct stakeholder analysis. The first step of stakeholder analysis is to identify all relevant parties and carry out registered stakeholders. There are two types of stakeholders, namely internal and external.

Internal stakeholders consist of: Government of South Sumatra Province: PT. JSC (Directors, ME Division; and Hygiene Task Force); Housing and Environment Agency (Building Management Division); Financial and Asset Agency (Asset Division); PSSI (Football Association of Indonesia) of South Sumatra; KONI (National Sports Committee of Indonesia) of South Sumatra, and Regional House of Representatives of South Sumatra Province: Commission IV of Infrastructure.

External stakeholders consist of: 1) Sports Affairs, e.g: PSSI (Football Association of Indonesia) Center, KONI (National Sports Committee of Indonesia), AFC/FIFA/OCA, Sriwijaya FC; 2) Electricity Affairs: PT. PLN (National Electricity Company), AKLI (Association of Electrical Contractors and Indonesian Mechanics, ME Outsourcing Consultant of PT. JSC; 3) Job Affairs: Contractor, Sub Contractor, Mechanical Electrical Supervisor Consultant.

Table 4. Stakeholders identification

Stakeholders	internal / external	scale 1-5 (weakest - strongest)			
		interest level	Power level	influence level	Support level
Director of PT. JSC	internal	5	5	4	4
ME Division of PT. JSC	internal	5	5	5	5
Cleaning Service of PT. JSC	internal	4	4	5	5
Building Management Division of Housing and Environment Agency	internal	2	3	5	5
Asset Division of Financial and Asset Agency	internal	5	4	2	1
Regional House of Representative Assembly of South Sumatera Province	internal	2	5	4	1
AKLI (Electrical Association)	external	1	1	4	5
PT. PLN (Electrical Company)	external	2	1	2	1
Outsourcing consultant of PT. JSC	external	2	3	3	4
Independent Mechanical Electrical Supervisor Consultant	external	1	2	4	4
Main Contractor	external	5	5	5	5
Sub Contractor	external	5	3	5	5
PSSI (National Football Association)	external	4	2	5	3
AFC (Asian Football Confenderation)	external	1	2	1	2
KONI (National Olympic Commitee)	external	1	2	1	2
OCA (Olympic Committee of Asia)	external	1	2	1	2
PSSI Province Level	internal	5	2	4	4
KONI Province Level	internal	4	1	4	3
Team and Management of Sriwijaya FC	external	1	1	2	1
Supporter of Sriwijaya FC	external	4	1	5	1

Based on the assessment score in Table 4, it was recognized that the highest urgence level with the highest influence for the maintenance of lighting system utilities is the Director of PT. JSC and ME Division of PT. JSC. See the figure below:

IOP Conf. Series: Materials Science and Engineering **650** (2019) 012012 doi:10.1088/1757-899X/650/1/012012

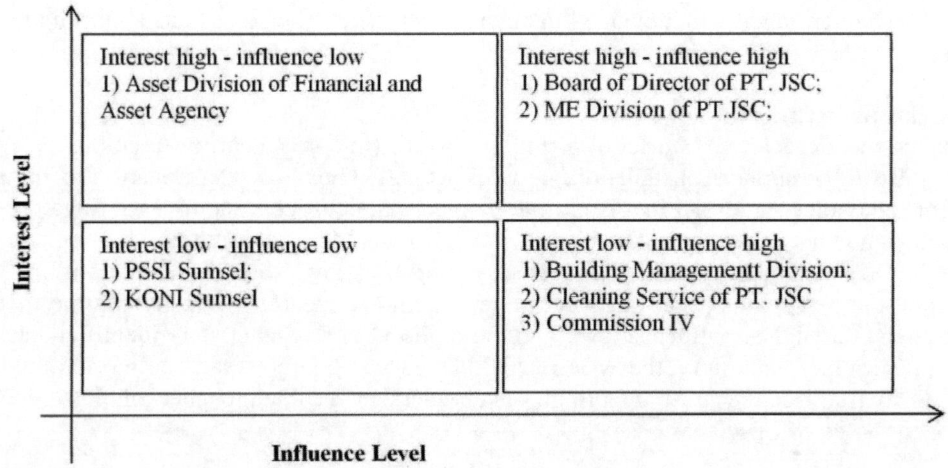

Figure 3. Analysis model stakeholder internal

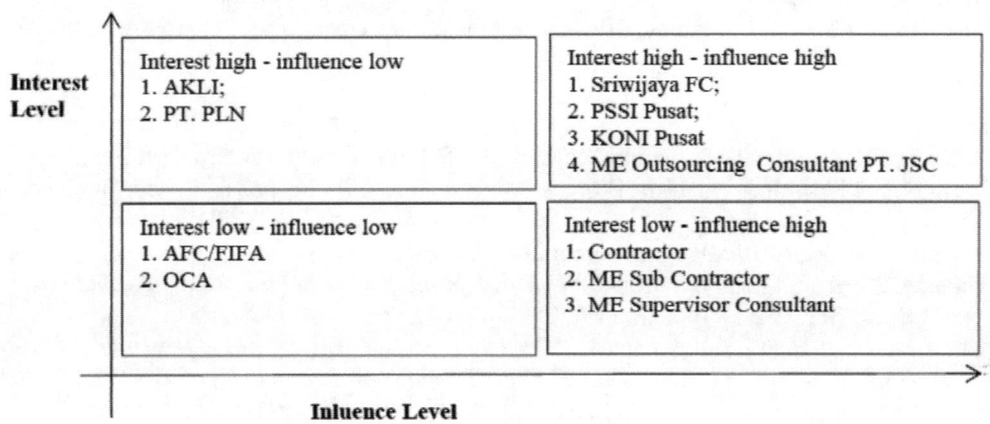

Figure 4. Analysis model stakeholder external

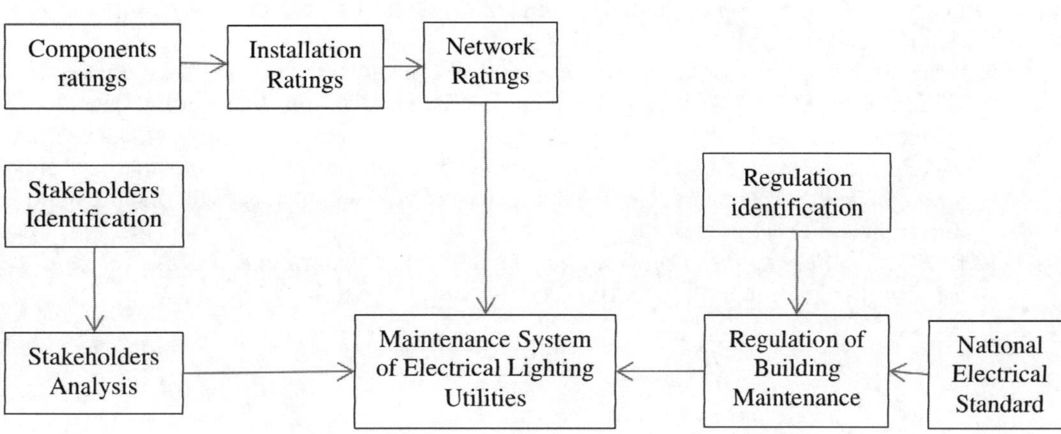

Figure 5. Flowchart of maintenance system

IOP Conf. Series: Materials Science and Engineering **650** (2019) 012012 doi:10.1088/1757-899X/650/1/012012

All parts are arranged into a planned maintenance system scheme in the stadium lighting electrical system.

4. Concluding Remark

This paper was expected to be useful to find out 'what' (in component, installation, network ratings); 'where' (also in component, installation, network ratings); 'how' (in identification of regulation); and 'who' (in stakeholder analysis) involved in a planned maintenance system. By ratings the component, installations and networks, the most prioritized of them could be determined. Thorough the ratings, the place of each component, installation, network could be known directly. By identifying the selected regulation, the best maintenance steps could be determined clearly. By analyzing the involvement of stakeholders, the highest influence level and the highest interest level were identified easily. Thus, in order to determine 'when/time', the system would be easy to be implemented. In brief, this System was proposed to find out the most important stakeholders in the maintenance of the highest priorities component based on official assessment.

Acknowledgements

The authors are grateful to the Ministry of Public Works and Public Housing of Republic of Indonesia for funding this research as part of scholarship master program in collaboration with Sebelas Maret University. The authors also expressed their gratitude to the Government of South Sumatra Province for providing technical support.

Reference

[1] Aragones P Beltran, M Garcia-Melon, and Montesinos J Valera 2017 How to Assess Stakeholders' Influence in Project Management? A Proposal Based on The Analytic Network Process *Int. J. Proj.Manag.* **35**:451-462. http://dx.doi.org/10.1016/j.ijproman.2017.01.001

[2] Chanter Barrie and Peter Swallow 2007 *Building maintenance management – 2nd ed* (Oxford: Blackwell Publishing Ltd)

[3] Chen Wai-Fah,Liew, J Y Richard 2003 *Civil engineering--Handbooks, manuals, etc. (New directions in civil engineering)* CRC Press LLC (Washington D.C.) ISBN 0-8493-0958-1

[4] Ministry of Public Works and Housing 2010 *Periodic Inspection of Building Technical Guidelines Regulation Number 16* (Jakarta)

[5] Ministry of Public Works and Housing 2018 *Construction of State Buildings Regulation Number 22* (Jakarta)

[6] Ministry of Public Works and Housing 2007 *Building Feasibility Certificate Regulation Number 25* (Jakarta)

[7] National Standard of Indonesia 2011 *General Requirements for Electrical Installation 0225:2011, ICS 91.140.50* (Jakarta: National Standardization Board of Indonesia)

[8] Partridge K, Jackson P, Wheeler, and D Zohar A 2005 *From Words to Action The Stakeholder Engagement Manual Volume 1: The Guide to Practitioners' , Perspectives on Stakeholder Engagement Stakeholder* (New York: Research Associates Canada Inc., United Nations Environment Programme)

[9] PMI 2013 *A guide to the project management body of knowledge (PMBOK® guide) -- Fifth edition* (Pennsylvania: Project Management Institute, Inc.)

IOP Conf. Series: Materials Science and Engineering **650** (2019) 012013 doi:10.1088/1757-899X/650/1/012013

Developing affordable housing for low-income households in developing countries using participative payment

M A Berawi[1*], R Aulidtyo[2], P Miraj[2,3] and Gunawan[2]

[1]Faculty of Engineering, Universitas Indonesia, Kampus Baru UI Depok 16424, Indonesia
[2]Center for Sustainable Infrastructure Development, Universitas Indonesia, Kampus Baru UI Depok 16424, Indonesia
[3]Faculty of Engineering, Universitas Pancasila, Srengseng Sawah Jakarta Selatan 12640, Indonesia

E-mail: *maberawi@eng.ui.ac.id

Abstract. The government of Indonesia attempt to increase the standard living of the low-income households by providing affordable housing. The concept is implemented in Jakarta-the capital city of Indonesia as the pilot project in the past years. However, the operation faces several obstacles, and one of them is the debt of rental fees from the households from 2013 to 2017 about US$ 2,013,888.89 for all affordable housing exist in Jakarta. This research attempt to evaluate the potential business that can be implemented on the site to improve the payment capability from users. Marunda Affordable Housing in North Jakarta and Cipinang Besar Selatan in East Jakarta will be used as the case studies. Pairwise comparison and life-cycle cost approach are adopted to achieve the targeted objectives. The result shows a different type of business that can be applied to both case studies. One is related to marine processing industry, while the other is suitable with the manufacturing industry. Both business proposals generate a significant rate of return above minimum attractiveness rate of return in the related industry.

1. Introduction

The government of Indonesia aims to accelerate the country's competitiveness through strategic infrastructure project [1, 2]. It was proposed through administrative regulation No 2 in 2016 about the acceleration of national infrastructure project and one of them about housing. The government attempt to build one million affordable housing across the country which 70% of them allocated for low-income community and the rest will serve for commercial housing. This program is implemented in all provinces in Indonesia and divided into three stages, with the primary target location at the first stage is the capital city of Indonesia-DKI Jakarta.

According to the report [3], about 48.91% or equal to 4.89 million citizen of Jakarta live in slum area or improper house due to the exponential price of residential and high demand of houses every year. As vacant land for development is limited, vertical houses are the solution to provide an appropriate house for the low-income community. In the long term, it is expected that through a better place for a living may increase their ability in finding jobs and reduce the poverty level of the society [4, 5]. However, the operation of affordable housing in Jakarta face several obstacles, and one of them is the payment for rental fees by the user is low. Based on the report of the government of Jakarta, the users have the debt of about 29 billion rupiah or equal to US$ 2,013,888.89 from 2013 to 2017 periods for all the affordable housing in Jakarta. Most of this low-income user had income below US$ 100 and was used for daily expenses such as transport, meals or sometimes children's tuition fees. Their earnings almost equal to the expenditure and therefore, the budget for rental fees is not available.

This research attempt to generate an alternative solution for the minimum capability of the households in paying the rental fees by taking into account a participative payment. The research will investigate the potential business that can be applied in the area considering the location, potential revenue that can be generated, employees needed for running the business and other related factors. The research output expected can be used to mitigate the increasing number of debt by the households and proposed a solution for related stakeholders in developing policy and regulations about affordable housing for low-income households in the near future.

2. Methodology

This research use a three stages approach to achieve the research objectives. In the first stage, a desk study from the literature review and benchmarking was used to generate variables and similar condition from other countries [6]. This stage will produce variables to assess the optimum business that potentially applied in the affordable housing. In the second stage, this research adopts a pairwise comparison for the analysis to determine the best selection for the business. It is a method to evaluate decision making by weighting the available criteria and proposed the alternative solution [7, 8].

In this research, four indicators were identified and will be compared with each other by using four-scale of importance level. The indicator composes Compound Annual Growth Rate (CAGR), net revenue, human resources, and market entry factor. Zero for less important, one for slightly important, two for moderate important and three for high important. Each score on each indicator will be sum up and the highest score selected as the best alternative. Pairwise comparison tabulation sample summarized in the following table.

Table 1. A sample of Pairwise Comparison Tabulation

	B		C		D		E		Score	%	Key Letter
A	B	3	A	2	D	1	A	2	4	22%	A
	B		C	1	B	2	B	2	7	39%	B
			C		D	3	E	1	1	6%	C
					D		D	2	6	33%	D
							Total		18	100%	

The weighted score of each indicator will be processed by adding an alternative solution to increase the business activity of an affordable house which may support the lack of compliance in paying the house rent to the government. The score from indicator will be multiplied by the score from the alternative solution to produce the total score and the highest ranking. In the last stage, an evaluation of the selected solution will be performed by taking into account the investment cost, operation, and maintenance cost and revenue. The analysis will produce the internal rate of return (IRR), and the value has to meet minimum discount rate and minimum attractive rate of return (MARR) of investors. MARR shall determine the project may attract private investors or shall remain government side and further subsidy required [9, 10].

This research uses a case study in two locations (Marunda and Cipinang Besar Selatan) of affordable housing managed by the government. Moreover, an in-depth interview with two experts related to affordable housing is conducted to evaluate and validate the research findings. One expert is a deputy of human resources in Ministry of Micro, Small and Medium Enterprises, Republic of Indonesia and the other is DKI Jakarta Expert Staff. The interview uses a semi-structured instrument and lasts for about 15-30 minutes. The result will be cross-checked the research findings, and a recommendation will be provided for the final report [11, 12].

3. Result and Discussion

Firstly, a pricing analysis was evaluated to determine the baseline cost of current retribution in the affordable housing. According to local regulation of the Jakarta capital city, No 3 in 2012 shows that monthly payment rental was different based on its location and characteristics. Those who lived in higher level and follows government program pay less rental fees. For instance, a user within the government program has to submit about 234,000 rupiah or equal to US$ 16.25 for a 30 square meter located on the first floor. On the other hand, the general user has to pay more than twice around US$ 35.28 for the same lot. A 36 square meter lot also available with higher price about 281,000 rupiah or equal to US$ 19.51 in the similar floor. Based on data from [3] and a decent living needs survey in the capital city in 2017 shows minimum wage province is US$ 233. The survey estimates 41.89% or 1,405,723 rupiahs (US$ 97.62) from the wage allocated for the housing. This figure will be used to determine which business that available to be implemented on affordable housing to support this needs.

3.1. Potential Business in the Two Locations

Marunda affordable housing is located in north Jakarta, close to the coastal area, therefore has enormous potential in fishery business. National and regional gross domestic product also support the argument that fishery sector increased significantly since 2014, especially in large and medium seafood processing industry, salt extraction, and tourism.

According to data from the Ministry of Marine Affairs and Fisheries of the Republic of Indonesia until 2017, the total national fishery production reached 23.26 million tons contributed from 6.04 million tons of captured fishery and 17.22 tons of fish farming. However, from those potential, only several businesses in the fishery that can be applied on a small and medium scale. One of them in aquaculture (shrimp processing industry) and small-medium scale of fish processing (catfish fillet processing industry). The reports from similar source had shown that the industries ranked first and sixth respectively from the overall fishery sector.

On the other hand, the analysis also investigates commodities from except the oil and gas sector and found out ten potential commodities for export one of them is crafting. This business contributes US$ 71.3 million per month all over the nation. The demand also increases along the US and Europe from Indonesia such as craft knitting.

Compared to Marunda affordable housing that has natural resources from the fishery, Cipinang Besar Selatan affordable housing has potential due to its location near the trading market such as Rice Central Market and Toys Center. The commodity that can be delivered from this area shall relate to manufacturing processing industry. The research proposes education toys as part of the industry considering the location and source of material such as plastics and woods are nearby the location. Handicraft from scraps or secondhand also available for development. Homemakers will be empowered to process how to decorate and to form the craft for selling purpose. The local government and related ministry may be invited to conduct workshop or entrepreneur training. Last, the user also invited to collaborate in processing brown rice. Currently, many people have been considered healthy life and consuming brown rice has become part of people daily life. This potential should be captured and not to mention the location of this affordable housing near the rice center, so the marketing process no longer a problem.

3.2. Pairwise Comparison Analysis

The indicators were delivered to experts in the industry and produce the result for further analysis. Overall, market entry is the highest indicator among others as about 33.33%, followed by net revenue for about 26.67%. CAGR and human resources share the similar result with 20.00%. The pairwise analysis from experts answer can be seen in the table 2.

The pairwise comparison result is then processed to each affordable housing in Marunda and Cipinang Besar Selatan. Decision matrix on Marunda affordable housing shows close result between catfish fillet processing industry and knitting business with 4.12% differences.

Table 2. The Result of Pairwise Comparison

	B		C		D		Score	%	Key Letter
A	A	3	C	3	D	2	3	20.00%	A = CAGR
	B		B	2	B	2	4	26.67%	B = Nett Revenue
			C		D	3	3	20.00%	C = Human Resources
					D		5	33.33%	D = Market Entry Factors
			Total				15	100.00%	

Catfish fillet processing industry is ahead on two indicators consist of CAGR score and human resource score with 66.67% and 83.33% respectively. On the other hand, knitting business surpasses other business alternatives in revenue and entry market with 71.43% and 66.67% in sequence. Despite leading equally on two indicators, knitting business has a lower score in CAGR and human resource indicators, thus ranked below fish filet processing industry. The detail of the decision matrix score shown in the table 3.

Table 3. Decision Matrix on Marunda Affordable Housing

Indicator	**CAGR**	***Nett Revenue***	**Human Resources**	**Market Entry Factor** 33,33%	**Total Score**	**Rank**
Weighted Score	20,00%	26,67%	20,00%			
		Alternative Recommendation				
Shrimp Processing Industry	33,33%	0,00%	0,00%	0,00%	6,67%	3
Score	6,67%	0,00%	0,00%	0,00%		
Catfish Fillet Processing Industry	66,67%	28.57%	83,33%	33,33%	48,73%	1
Score	13,33%	7,62%	16,67%	11,11%		
Knitting Industry	0,00%	71,43%	16,67%	66,67%	44,61%	2
Score	0,00%	19,05%	3,33%	22,22%		

Decision matrix on Cipinang Besar Selatan affordable housing shows toy processing industry as the recommendation for business in the site. Despite low score on market entry, toy processing industry score high on three indicators consist of CAGR, net revenue and human resource with 62.50%, 80.00 and 71.43% respectively. On the other hand, brown rice processing industry placed second only score high on market entry and not followed by other indicators. The detail of the decision matrix score shown in the table 4.

Table 4. Decision Matrix on Cipinang Besar Selatan Affordable Housing

Indicator	**CAGR**	**Nett Revenue**	**Human Resources**	**Market Entry Factor**	**Total Score**	**Rank**
Weighted Score	20,00%	26,67%	20,00%	33,33%		
Alternative Recommendation						
Brown Rice Processing Industry	0,00%	20,00%	28,57%	83,33%	38,82%	2
Score	0,00%	5,33%	5,71%	27,77%		
Toy Processing Industry	62,5%	80,00%	71,43%	0,00%	48,12%	1
Score	12,50%	21,34%	14,29%	0,00%		
Handicraft Industry	37,50%	0,00%	0,00%	16,67%	13,06%	3
Score	7,50%	0,00%	0,00%	5,56%		

3.3. Marunda Affordable Housing Financial Analysis
The financial analysis considers catfish fillet processing industry for Marunda Affordable Housing. In term of catfish fillet processing industry, considerations have been investigated. The minimum size for this industry is 1.34 hectare, and Marunda has adequate land for development of more than five hectares. For buildings, a similar corporation running similar business was used. A 0.3 Ha of factory building combined with a 0.2 Ha of the warehouse will be constructed to support the industry. A 500 household that failed to submit rental fees expected as the employee and it may be adjusted based on the needs or technical requirement. Investment cost also considers related equipment and gears as well as training/workshop for development.

Operation and maintenance cost of this industry consist of building operation, depreciation, raw materials, electricity and water, marketing, employee salary, rental fees of affordable housing. Rental fees are added to cover the failure payment by the user which about 560 unit. Inflation was also taking into account based on Bank of Indonesia about 3.8% per year. The operation and maintenance and investment cost shall be covered by the manufacturing income. The research assumes that the production is about 60% and the product is 80% will be sold. It will produce about US$ 600,967.95. This figure generates an internal rate of return for about 20% or higher than the industry's WACC around 13.51%.

3.4. Cipinang Besar Selatan Affordable Housing Financial Analysis
The financial analysis considers toy processing industry for Cipinang Besar Affordable Housing. The housing has vacant land about 2,000 square meter and such dimension capable of accommodating the toy processing industry. The toy factory itself require an 800 square meter with a warehouse about 100 square meters. Other equipment and gear for running the industry will be considered for investment cost calculation.

The investment for this industry will increase in the year fourth and seventh due to production equipment only last about approximately three years; therefore new purchasing is required while monthly maintenance is performed. Overall the investment cost is US$ 516,894.24 includes tax. This figure is much lower about almost a quarter compared to investment cost required for catfish fillet processing industry in Marunda affordable housing. With such investment, this type of industry categorized as a medium-sized industry.

The operation and maintenance and investment cost shall be covered by the manufacturing income. The research assumes that 6,000 unit per month were produced and recruit 40 employees from the housing. The product estimated to be sold around 80% with US$ 9.03 per toy. The result shows a potential rate of return of about 15% or higher than industry’s WACC around 13.51%.

4. Conclusion
Affordable housing, particularly in the capital city of Indonesia, provides a housing solution for those who incapable to afford the increasing landed house price in Jakarta. Despite, low rental fees charged

by the government, many of them fail to pay for months. This research uses two case studies located in Marunda Affordable Housing in North Jakarta and Cipinang Besar Selatan in East Jakarta. The result shows that potential business that can be applied in each affordable housing by inviting those who yet to pay the rental fees. A catfish fillet processing industry was proposed as a solution for Marunda Affordable Housing, while a toy processing industry was selected as the optimum industry for Cipinang Besar Selatan.

The life cycle cost analysis shows that both industries might contribute not only pay the rental fees for the user but also provide income for their daily activities. The industry in Marunda Affordable Housing generates a significant rate of return about 20%. Although Cipinang Besar Selatan shows a lower rate of return compared from the other affordable housing about 15%, both of them surpass the minimum attractive rate of return from the investor perspective. Further research is recommended to be conducted by taking into account the division of responsibility between the government and private sector. Its attempt to not only release rental fees from the user but in the longer term to provide sustainability of social infrastructure. The government also suggested formulating policy and regulation to facilitate cost sharing and revenue sharing among stakeholders involved. Regular supervision also required to mitigate corruption or other means of bribery to the project and maintain its sustainability.

Acknowledgments
This research was fully supported by a grant from Universitas Indonesia PITTA 2018.

References
[1] Berawi M A, Susantono B, Miraj P, and Nurmadinah F 2018 *Aviation* **22** 115-128
[2] Rahman H Z et al 2018 *IJTech* **3** 549-557
[3] Jakarta Central Bureau of Statistics 2017 *Poverty Level in DKI Jakarta* https://jakarta.bps.go.id/
[4] Rohe W M and Stegman M A 1994 *J Am Plann Assoc* **60** 173-184
[5] Susin S 2002 *J Public Econ* **83** 109-152
[6] Berawi M A, Ibrahim B E, and Miraj P 2019 *JDBE* **19** 40-48
[7] Negahban S, Oh S, and Shah D 2016 *Operations Research* **65** 266-287
[8] Abdul-Rahman H and Berawi M A *Quality Assurance* **9** 5-30
[9] Hartman J C and Schafrick I C 2004 *The Engineering Economist* **49** 139-158
[10] Bas 2013 *Appl Math Comput* **219** 5901-5908
[11] Longhurst R 2003 *Key methods in geography* 143-156
[12] Karim S B A, Abdul-Rahman H, Berawi M A, and Jaapar A 2017 A review on the issues and strategies of stakeholder management in the construction industry *In Meeting and Conference on Management in Construction and Researchers Association (MICRA)*

IOP Conf. Series: Materials Science and Engineering **650** (2019) 012014 doi:10.1088/1757-899X/650/1/012014

Finance-based scheduling using integer programming method for building construction project in Jakarta

Ervina[1] and O Gondokusumo[2]

[1] Master Student of Civil Engineering Department, Tarumanagara University, Jakarta, Indonesia
[2] Senior Lecturer of Civil Engineering Department, Tarumanagara University, Jakarta, Indonesia

E-mail: ervina.elizabeth@gmail.com, onnygk@gmail.com

Abstract. Construction project management consists of planning, scheduling, and controlling. Planning is important because it's set down the target of project to achieve successfulness. Successfulness of a project consists of several aspects, including: cost, quality, time, and HSE (Health, Safety, and Environment). In many cases, scheduling is optimized with minimum costs for achieving a project's goal. This doesn't necessarily guarantee that contactors can provide the budget according to the scheduling. This paper will focus on optimizing scheduling by considering financial capabilities. The method used to optimize scheduling by considering financial capabilities is called Finance-based Scheduling Method. The stages to do finance-based scheduling method, are: arranging initial scheme, arranging extension scheme, and formulating an integer programming model. This study have purpose to obtain optimal scheduling by considering contractor's financial capabilities. Analysis of this study uses integer programming to find the most optimal solution. However, integer programming requires different modeling for each case and result from it must be simplifying become a bar chart so it's make easier to understand.

1. Introduction

Construction project is a series of activities which carried out in achieving certain goal. However, the implementation is limited by time and resources needed. In order to achieve this goal, project planning is one of the important stages. It is important to determine the stage of activities and specified criteria.

Scheduling is one of the implementation of project planning, which is ideally carried out before construction begin. This process is important to do considering, because we need to determine relationship between the activities to achieve project's goal. The relationship between activities can be described in a network diagram which is can used to calculate duration of delivering project until finish and determine critical path. Determination of critical paths is also important to do because the activities located on this path will receive more attention during construction. If there is activity on critical path that is late for completion, it will affect overall duration and cost of the project. There are several ways to manage project scheduling, including: s-curves, bar charts, linear programming, and integer programming.

After completing scheduling, cash flow analysis can be done to get an estimated cost. Cashflow analysis is important for construction project, because we can calculate the maximum costs needed in each period. Maximum cost need to be adjusted to the budget that company can provide for the project. Furthermore, the budget that company provide called by credit limits.

In most cases, scheduling only focus on optimizing duration with minimum costs, without considering credit limits (financial capabilities). In fact, it is important to adjust the maximum cost needed each period with credit limits. This research present a scheduling method that considers financial capabilities, referred as finance-based scheduling method.

2. Literature review

2.1. Project scheduling

Project scheduling can be seen as a science compared with art of planning, it defines as logic, sequence and the calculation of critical path, float, start and finish dates of individual activities, and thus determines the feasibility of delivering the project within desired completion dates and budget [1]. Scheduling on construction projects consists of time, labor, equipment, materials, and finance. Accuracy of scheduling in project implementation greatly influences the avoidance of many losses, such as swelling construction costs, late project submission, and disputes or claims [2].

Some of the most commonly used scheduling methods are as follows: gantt or bar charts, milestone charts, line of balance, networks, program evaluation and review techniques, arrow diagram network, precedence diagram network, graphical evaluation and review technique [3]. PERT was originally developed in 1958 and 1959 to meet the needs of "age of massive engineering" where the techniques of Taylor and Gantt were inapplicable. Around the same time a company introduced a similar method known as the Critical Path Method (CPM) which also spread widely and was mainly concentrated in the construction industry [4]. The critical path is the longest path through the network and determines the duration of the project, as well as the shortest amount of time needed to complete the project.

2.2. Project financial management

Construction finance management is a high risk business with historically low profit margins. Control over costs, cash flow, and adequate project funding is very important for the success of each construction business [5]. An early estimate is also important to the project team because it becomes one of the key project parameters. It helps formulate execution strategies and provides a basis plan engineering and construction. [6]

Contractor's cash inflows and outflows are not always same each period throughout duration of project are generally caused by retention. Retention is a number of payments held by the project owner as a guarantee that the building built by the contractor is performing properly. Therefore, the contractor will experience a deficit and can experience financial problems. Contractors need to use their capital or borrow a certain amount of money to finance the implementation of the project. However, there are additional costs caused by borrowed need to be prepared which must be taken into account in the preparation stage for bidding [7].

Scheduling process is useful for determining the time starting from activities to complete the project in minimum time. For some project schedules, there are cash requirements specified. The schedule must be designed in such a way that the project's cash requirements for each period do not exceed the allocated credit limits. Small credit limits lead to an extension of the duration of the project and result in excess indirect costs [8].

Many of scheduling method focuses on finding the optimum duration with minimum costs, or what is known as a time-cost trade-off. Many research has been done to develop mathematical modeling to solve this time-cost trade-off problem, among others: the use of mixed-integer programming, linear/integer programming hybrid method [9], genetic algorithm [10]. However, scheduling with the time-cost trade-off method does not consider scheduling aspects that have budget constraints (credit limits).

IOP Conf. Series: Materials Science and Engineering **650** (2019) 012014 doi:10.1088/1757-899X/650/1/012014

2.3. Finance-based scheduling Method

Finance-Based Scheduling is used to carry out projects combines with financing, project duration, and budget (credit) needed [11]. To achieve profitable businesses, contractors must focus on minimizing three things. First, reducing financing definitely increases profits, but requires an increase in budgets that need to be prepared to shorten the duration. Second, minimizing budget to increase likelihood of being approved by a third party (bank) and giving effect being able to negotiate better interest rates. It can also be used to determine time of loan repayment, but it certainly results in an increase in duration and interest of the loan. Third, shortening the duration increases profits by reducing overhead costs and borrowing costs but requires a high amount of budget. Therefore, the cost of loan, duration of project, and the amount of budget needed are a set of contractor problems.

Elazouni and Gab-Allah in their study proposed a method for optimized scheduling with a critical path method that could be financially adequate (credit limits fulfilled). So that the schedule generated by critical path method can be executed financially [12].

3. Methodology

In this research, the main problem is to arrange scheduling of multi-storey building project in Jakarta using the finance-based scheduling method. By using this method, it allows scheduling to adjusted the financial capabilities of contractor when delivering a construction project. The stages in this study are summarized in the form of a diagram that can be seen in Figure 1.

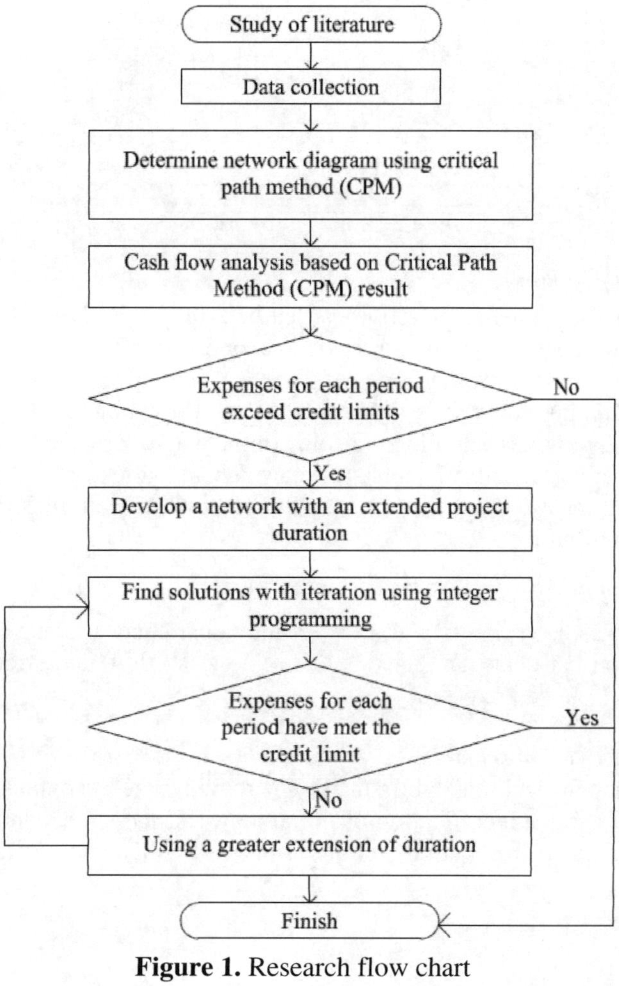

Figure 1. Research flow chart

IOP Conf. Series: Materials Science and Engineering **650** (2019) 012014 doi:10.1088/1757-899X/650/1/012014

3.1. Determining the network

Initial network that has been arrange using critical path method (CPM) is presented in form of bar chart, this diagram describes time information of each activities during project as in Figure 2. The Critical Path Method is carried out analysis of advanced calculations and countdown calculations, which are useful to find out the total float (TF) and free float (FF) of each activities. Total float represents maximum shift in an activity without affecting duration of the project (T). Critical activity is sign with TF = FF = 0, by using Critical Path Method, it can be seen the fastest duration to complete the project's activities.

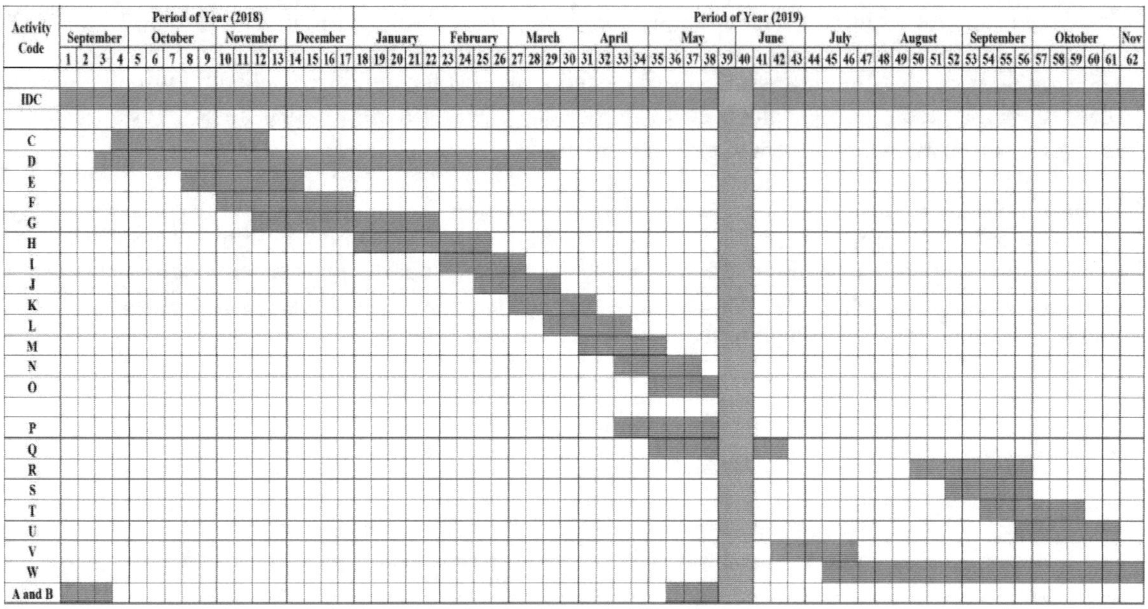

Figure 2. Initial network bar chart

3.2. Critical Path Method (CPM)

Critical path method is the sequence of activities that determines duration of the project. Critical path method offers countless benefits such as reduce delays, optimize efficiency, float calculation, visualize dependencies, and improve organization.

The purpose of analyzing network is to predict the total duration needed until the construction is completed. This method provides a graphical display of the network (activity on node network) through precedence diagram network. Critical path shows which activities are important to received more attention in maintaining the project completion on time and which activities that can still shift without affecting total duration of project.

3.3. Cash flow analysis

Network diagram that has been created will convert into a bar chart to make it easier for other users to read schedule. Based on bar chart, an analysis of project cash flow is carried out. So, the amount of expenses in each period can be predicted.

3.4. Project duration extension scheme

Basic objective of finance-based scheduling method is to minimize the extension of duration. Network extension such as Figure 3 is a modification of initial network that allows an increase the extension of duration (M). The results is increasing project duration by (T + M), and increasing total float value for each activities by Jk.

IOP Conf. Series: Materials Science and Engineering **650** (2019) 012014 doi:10.1088/1757-899X/650/1/012014

Total float that adjusts to duration extension is the time space for an activity can shift without affecting the duration of the project that has been extended. Total float adjustment can be seen in Equation 1.

$$J_k = TF_k + M \tag{1}$$

Cash flow analysis is carried out from point of view of the construction company, starting at the beginning of project and running over time, modeling cash flow sequentially adds income and expense [5]. The following are the equations used for cash flow analysis in the period can be seen in Equations 2 to 4.

$$Et = \sum_{i=1}^{m} Y_t \tag{2}$$

$$P_t = KE_t \tag{3}$$

$$F_t = N_{t-1} + E_t \tag{4}$$

Information:

m = extension of duration (in units of time)
K = multiplier to determine the amount of payment
F_t = cumulative cash flow at the end od the period t
P_t = value of receipt at the end of period t
E_t = expense in period t

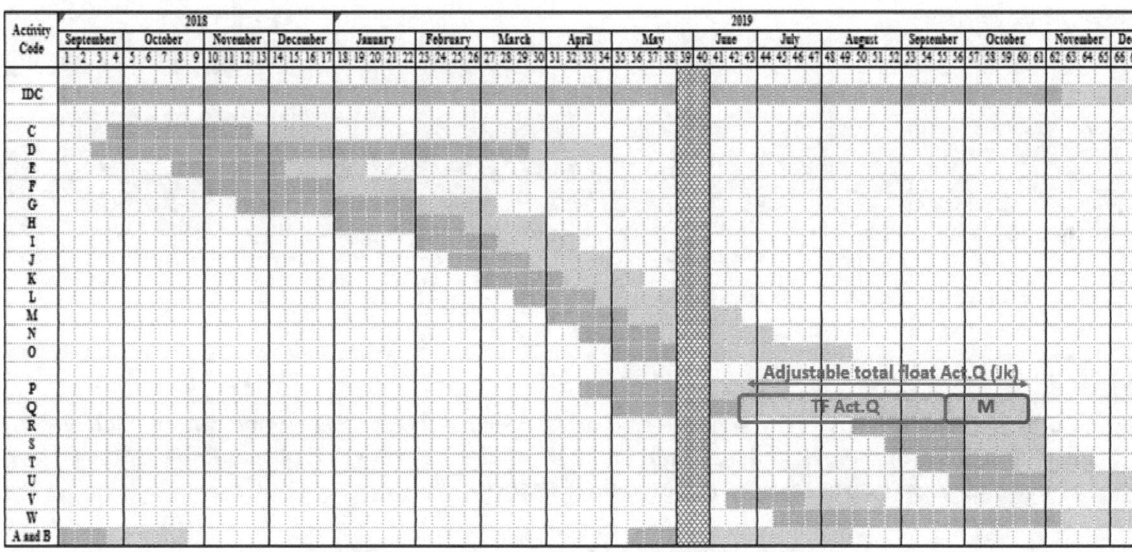

Figure 3. Network bar chart with extension

3.5. Integer programming modelling

In this study, integer programming use to find optimal solution from existing case. By using integer programming, we have to compose a mathematical models according to the case of research objects to be input in linear programming. Of course, the solution will be produced is very dependent on the inputted model.

IOP Conf. Series: Materials Science and Engineering **650** (2019) 012014 doi:10.1088/1757-899X/650/1/012014

3.6. Integer programming analysis

Integer programming uses mathematical models to describe problems as a constraint. The word programming does not refer to computer programming, but is a basic naming for planning so that this tool can be used in planning to get optimal results. All equations are input as linear mathematical functions, and the variables are given integer commands. Giving this command is used to ensure that the solution provided by integer programming is an integer value. The models that must be fulfilled when using integer programming, among others.

3.6.1. Objective function.

As mentioned above, the purpose of this model is to minimize the total extension of schedule through minimizing the shift from the last critical activity n. Thus the objective function can be formulated as can be seen in Equation 5 below.

$$\text{minimize } z = xn \tag{5}$$

Information:
x = shift in an activity
z = destination function

3.6.2. Activity shifting constraints.

This constraint ensures that shifts in activities (x_k) with k = 1,2, ..., n. N is the sum of all activities, in the form of integers in each extended total float. Constraint of shifting activities can be represented by Equation 6 and Equation 7 below. Through this equation also ensures that it only produces one value at the same time.

$$X_k = \sum_{j=1}^{Jk} jS_{kj} \tag{6}$$

$$\sum_{j=1}^{Jk} jS_{kj} \leq 1 \tag{7}$$

Information:
J_k = *total float adjustment in activities k*
$\boldsymbol{S_{kj}}$ = integer variable $\boldsymbol{S_{kj}}$ € {0,1}
k = 1, 2, ..., n

3.6.3. Activity sequance constraints.

This shifting constraint between activities are needed because the earliest completion of each activity (Q) must be equal to or more than the completion of the earliest activity (K) plus duration of the activity (Q), which can be represented mathematically like Equation 8 below.

$$EFq \geq (EFk + Dq) \tag{8}$$

Information:
EFq = the earliest completion time of activity Q
EFk = the earliest completion time of activity K
Dq = duration of activity Q
k = 1, 2, ..., n for all q ϵ Q_k

3.6.4. Specify credit limits constraints.

Negative cash flows for each period (t) including accumulated interest on loans may not exceed the specified credit limits. These constraint can be formulated like Equation 9 below.

$$\widehat{Ft} \leq W \tag{9}$$

Information:
W = specified credit limits value

The term $\widehat{Ft}$ can be formulated for each project period in the search for expenses for each activity. Furthermore, integer programming modeling for expense formulations can be seen in Equations 10 through equation 12.

$$Y_{ki} = (1 - \sum_{j=1}^{Jk} Skj)\, R_k;\ ES_k \leq i \leq EF_k \tag{10}$$

$$Y_{ki} = S_{kj} \cdot R_k;\ ES_k + j \leq i \leq EF_k + j;\ j = 1, 2, ..., J_k \tag{11}$$

$$Y_{ki} = 0 \text{ otherwise} \tag{12}$$

Expenses for K activities (R_k) will be considered from the start time until the completion of activity K. Based on integer programming modeling, the result will be an integer variable that shows a shift, for example: $S_{k3} = 1$ shows a shift of K activities for 3 days.

4. Analysis and Result

In determining the duration of each job, it takes different productivity of workers and tools for each company. Table 1 shows the activity's code, activity name, duration and cost for each activity and relationship for each activity.

Table 1. Project data

Code	Scope of work	Duration	Predecessors	Cost
A	Scattered	21 d		-Rp143,362,832
C	Foundation	58 d	Afs+2 d	Rp891,623,788
D	Dewatering	185 d	aSS+15 d, jFF	Rp225,710,442
E	Basement (Digging)	45 d	cSS+27 d	Rp990,200,696
F	Cutting pile	45 d	eFF+19 d	Rp64,222,088
G	Structure (Basement)	65 d	fSS+14 d	Rp4,769,600,943
H	Structure (GF)	52 d	gFF+26 d	Rp2,691,836,066
I	Structure (1st floor)	30 d	hFF+14 d	Rp3,435,104,148
J	Structure (2nd floor)	30 d	iFF+14 d	Rp3,703,451,280
K	Structure (3rd floor)	30 d	jFF+14 d	Rp3,018,319,139
L	Structure (4th floor)	30 d	kFF+14 d	Rp2,801,573,585
M	Structure (Roof)	30 d	lFF+ 14 d;pSS-12d	Rp2,457,464,922
N	Structure (LMR)	25 d	mSS+15 d	Rp23,003,994
O	Structure (Steel roof)	20 d	m; Nff+10 d	Rp152,178,576
P	Architecture (Basement)	35 d	jSS+54 d	Rp2,314,297,885
Q	Architecture (GF)	35 d	kSS+54 d	Rp4,041,810,292
R	Architecture (1st floor)	35 d	lSS+130 d	Rp5,069,234,120
S	Architecture (2nd floor)	34 d	rSS+14 d	Rp5,069,234,120
T	Architecture (3rd floor)	34 d	sSS+14 d	Rp5,069,234,120
U	Architecture (4th floor)	36 d	tSS+14 d; nSS+149d	Rp5,069,234,120
V	Exterior façade & canopy	37 d	sSS+22 d; tFF+6 d	Rp232,390,610
B	Scattered (Ex – Building)	15 d	pSS+24 d	-Rp143,362,832
W	External works	120 d	vFF+14 d; l; bFS-1d	Rp3,787,025,815

Based on the results of the analysis on some values of credit limits, there is a relationship curve between the extension duration of project and credit limits, as can be seen in Figure 4. On the curve it can be seen that graph tends to descending to right. This indicates that smaller credit limits value, cause longer the overall duration of project. However, the smaller credit value does not always cause the overall duration of the project extended, because each credit limits has its own extension duration range.

Maximum credit value is Rp. 10,653.98 (in millions), but based on Figure 4, up to a credit limits of Rp. 10,000.00 (in millions) has not caused an increase in overall project duration. This is because integer programming only shifts the start time of activities that are not included in the critical path, so it does not cause an extension in overall project duration.

Maximum value of duration extension which input on integer programming modelling is quite important. Therefore, this is related to limitations value of credit limits that can be analyzed. In this study, the maximum extension of the duration project used in this research is 5 weeks, thus causing the credit limits that could be analyzed only up to Rp. 3,800 (in millions). Based on the results of the analysis, minimum value of funds needed for each extension of duration range is obtained. This can help contractor to provide sufficient funds for execute project.

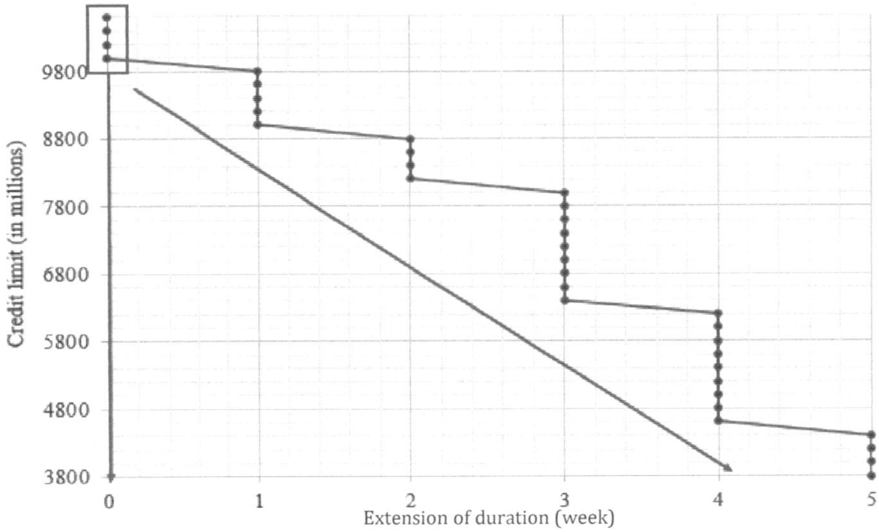

Figure 4. The relationship curve between extension of duration and credit limits

Finance-based scheduling method is not user-friendly because it needs to make a model that is suitable for each problem. If the limit is changed, a new model must be made before input to program. The results of integer programming analysis also need to convert into bar chart to make it easier for other users to read the results of scheduling using the finance-based scheduling method, as can be seen in Figure 5.

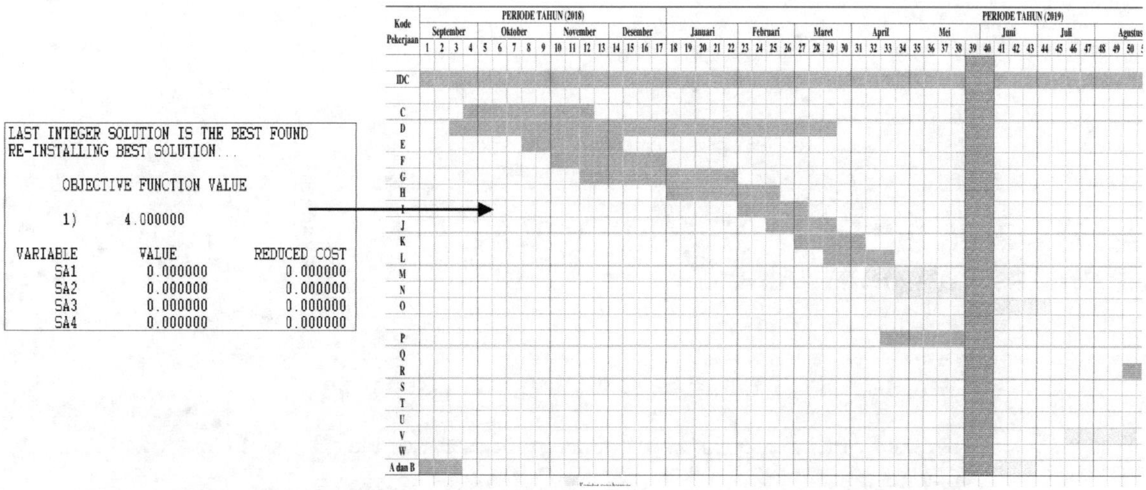

Figure 5. Simplification of integer programming results into a bar chart

5. Conclusion

The result of the study shows that the lower credit limits, overall project duration will become longer. However, lowering credit limits does not always lead to an extend the duration of the project, because for each value of duration has credit limits range. Optimizing the scheduling of construction projects with finance-based scheduling method using integer programming is also useful in tender process and during construction, because it produces a duration for project that fulfilled the financial capabilities of the contractor. However, it is necessary to make a mathematical model that fits each problem to find the optimum solution using a linear programming software which are available in the market. The generated optimum scheduling as the solution needs to be convert into a bar chart for make it easy to understand.

References

[1] Association for Projcet Management 2015 *Planning, scheduling, monitoring, and control* (Ibis House, Regent Park) Buckinghamshire

[2] Widiasanti I and Lenggogeni 2013 *Manajemen konstruksi* (Indonesia: PT Remaja Rosdakarya)

[3] O'Brien J J and Plotnick F L 2009 *CPM in construction management sixth edition* (New York: McGraw-Hill)

[4] Harold K 2006 *Project management: a system approach to planning, scheduling, and controlling* eight edition

[5] Levy S M 2006 *Project managemen in construction* (New York: McGraw-Hill), chapter 5 pp 10

[6] Garold D O 2006 *Project management for engineering and construction* (Oklahoma State Engineering)

[7] Alavipour S and Arditi D 2018 *Optimizing financing cost in construction projects with fixed project duration* J. Constr. Eng. Manage 144(4): 04018012

[8] Elazouni A and Metwally F 2005 *Finance-based scheduling : tool to maximize project profit using improved genetic algorithms* J. Constr. Eng. Manage 131 (4): 400-412

[9] Liu L, Burns, S A and Feng C W 1995 *Construction time-cost trade-off analysis using LP/IP hybrid method* J. Constr. Eng. Manage 121(4):446–454

[10] Li H and Love P 1997 *Using improved genetic algorithms to facilitate time-cost optimization* J. Constr. Eng. Manage 123(3): 233–237

[11] Abido M and Elazouni A 2011 *Multiobjective evolutionary finance-based scheduling: entire projects' portfolio* J. Comput. Civ. Eng 25(1): 85-97

[12] Elazouni A and Gab-Allah A 2004 *Finance-based scheduling of construction projects using integer programming* J. Constr. Eng. Manage 130(1): 15-24

Developing port-city conceptual design to improve regional industry competitiveness

M A Berawi[1*], T E Muljono[2], P Miraj[2,3] and Gunawan[2]

[1]Faculty of Engineering, Universitas Indonesia, Kampus Baru UI Depok 16424, Indonesia
[2]Center for Sustainable Infrastructure Development, Universitas Indonesia, Kampus Baru UI Depok 16424, Indonesia
[3]Faculty of Engineering, Universitas Pancasila, Srengseng Sawah Jakarta Selatan 12640, Indonesia

E-mail: *maberawi@eng.ui.ac.id

Abstract. As an archipelago country, Indonesia continues to face significant problems with port infrastructures and coastal area including financial shortage, complex bureaucracy not to mention development priority. Port-city is one of a city concept aims to facilitate the regional development by taking port as the hub for economic activities and mobility. This research use Dumai port in Riau Province, the Republic of Indonesia as the case study to evaluate the potential implementation of the port-city concept. Two methods are being used to evaluate the project feasibility namely life-cycle cost approach and public-private partnership scheme. It will regulate the division of responsibility between identified parties. The result shows three components to develop the concept including the port, industrial area and supporting infrastructure. This concept may generate 10.37% of the internal rate of return. This research also suggest public-private partnership scheme to generate optimum division of responsibility between each parties in the project.

1. Introduction

As an archipelago country, Indonesia has a high dependency on the marine transportation. In 2009, Indonesia's ports handled an estimate of 968 million tons of cargo [1]. The contribution of domestic and international cargo in Indonesia almost similar, with a ratio of 44% to 56% respectively. With the increasing population and economic growth, this figure will continue to increase soon.

The government attempt to respond to this need by issue the regulations of the Republic of Indonesia Number 16, 2017 about Marine policy in Indonesia. In this regulation, the government aims to improve the welfare of coastal communities and islands by developing the port and its surrounding area. One of the targeted locations in this policy is the city of Dumai, located in the Sumatera Island, and Riau Province of Indonesia. The location can be seen in Figure 1. Despite its potential to the economic growth and city expansion, the Dumai Port has no plan for expansion or terminal extended shortly.

Port cities can be effective when three factors are involved, namely a competitive port, economic growth of a city generated from port activities, and by negative impacts due to the presence of the seaport is mitigated [2, 3]. To achieve the aforementioned factors, it is essential for the city to plan

IOP Conf. Series: Materials Science and Engineering **650** (2019) 012015 doi:10.1088/1757-899X/650/1/012015

comprehensively the surrounding regions based on three essential pillars such as maritime cluster, industrial area, and urban waterfront [4, 5, 6].

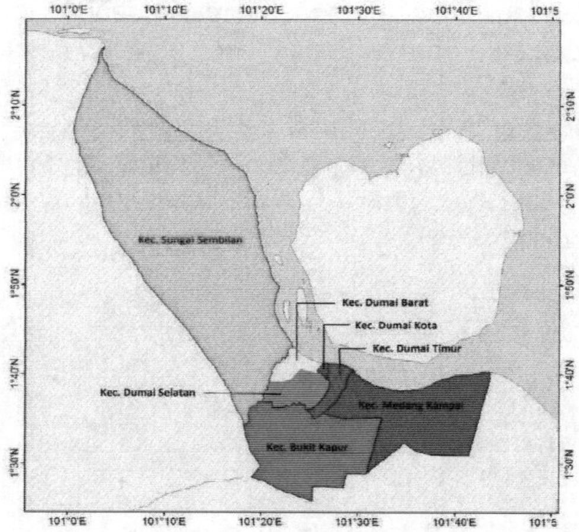

Figure 1. The City Area of Dumai in Riau Province

Port-cities concept is vital to support government vision as the maritime powerhouse [7, 8]. Dumai port is one area that potentially improves the economic competitiveness and generates revenue for local government. This research aims to develop Dumai port and its surroundings by taking into account coastal city development and cost-sharing through a public-private partnership.

2. Literature Study

The most accepted development model of how the port developed into a port city was the Anyport model by Bird, which explained the expansion of the port. This model shows that port development arises due to changing maritime technology. The Anyport model will show a change from spatial relations between ports and cities. As time goes by, the port will be built away from the business center of the city. In this model, three stages occur in the development of the port and the surrounding area; setting, expansion, and specialization. (Bird, 1980). These three stages can be explained as follows:

- Setting is the first stage of the construction of the port. In general, the first consideration of port construction is due to geographical factors. Port usually appears as a result of marine activities, such as fisheries, and shipbuilding. Before the industrial revolution, the port had existed and developed, but it was still pure, without significant changes in the shape and function of the port and the surrounding area. Most of the port activities are related to warehousing and wholesale trade located in the surrounding area.
- Expansion is the stage that occurs due to the industrial revolution. The industrial revolution made industry and transportation significantly increase. The area around the port was developed to accommodate the increasing number of passenger and freight boats. Increasing the size and number of ships encourages the construction of new terminals. Also, the integration of rail lines with ports has made access to hinterland easier. Activities related to ports are now developing to cover industrial activities.
- Specialization is the port development stage to achieve special functions to handle goods ships such as containers, agricultural products, oil, and coal. This phase causes warehousing needs to increase dramatically. Larger capacity vessels also require the construction of deep sea ports which often lead to the construction of new ports and facilities.

Although the Anyport model explains, in general, the development of the port and the development model that can be carried out, there are still some weaknesses in Bird's view According to (Notteboom & Rodrigue, 2005), this model does not explain the increasing port that acts as transhipment hub in the maritime network. With the emerged of a very long shipping lane, a port with very remote land access arises because of its function as a transit point.

Also, the Bird model does not take into account the land area as a driver in the dynamics of development. Regionalization means that development is more integrated by the context of the existing land. Regionalization means that development is more integrated by the context of the existing land. This concept can be illustrated as follows:

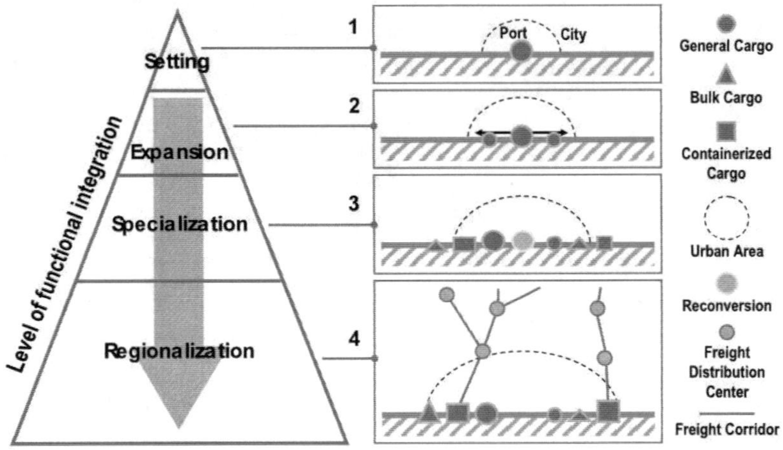

Figure 2. Illustration of port evolution in the concept of regionalization

3. Methodology

The research follows four stages to reach the targeted result. Firstly, the study determines the components required for Dumai port development. This stage starts by identifying the needs of the port infrastructure through desk study and secondary data from similar port cities across the globe. Secondly, research methodology is determined by considering case study and conceptual design of port-city from benchmarking process. Thirdly, a life-cycle cost analysis is used to evaluate the concept design by involving initial cost, operation & maintenance, and revenue [9, 10]. The result will then processed using a public-private partnership to generate division of responsibility between both parties. Last, an in-depth interview with three Indonesian infrastructure experts will be performed to confirm the result of the life cycle cost [11].

IOP Conf. Series: Materials Science and Engineering **650** (2019) 012015 doi:10.1088/1757-899X/650/1/012015

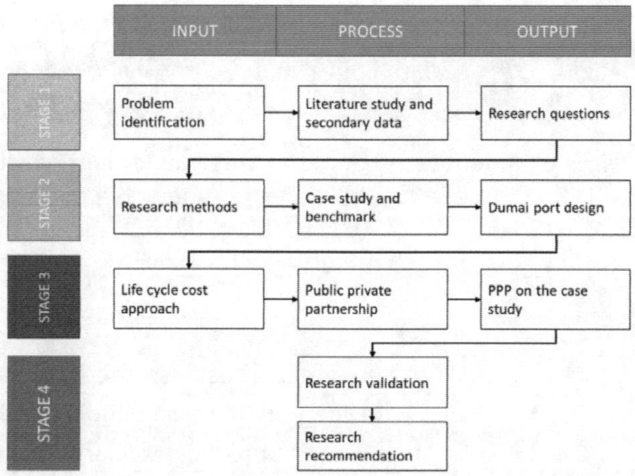

Figure 3. Research Framework

4. Result and Discussion

4.1. Design and Development

The concept design of Dumai port-city follows the coastal development standards and related regulations. The concept design represents by zoning design on Dumai port map. Designing this zoning attempt to minimize adverse impacts such as pollution, noise, and others as well as to increase land use compatibility.

Figure 4. The concept of Dumai's Port City in Riau Province

The zoning considers the negative impact of port and industrial area to other types of buildings and surroundings. The industrial area needs a specific space as an overall buffer as about 150 meters and 100 meters for primary buffer area. It is defined as a neutral area in between different type of buildings to minimize disturbance from one area and vice versa. In this research, 200 meters was implemented for buffer zone area and will be used for logistics area and retail. The zoning area can be seen as figure 4.

The zoning regulates the area into several districts. Port and industrial area located in the coastal area represented by white color in the above figure. The port consists of three types of the terminal; liquid bulk terminal, dry bulk terminal, and container terminal. On the other hand, the industrial area comprisesof marmalade jam processing, pineapple processing, modified cassava flour (MoCaF), and crude palm oil (CPO). Detail of the other nine districts can be seen in Table 1.

IOP Conf. Series: Materials Science and Engineering **650** (2019) 012015 doi:10.1088/1757-899X/650/1/012015

Table 1. The design concept in the area of Dumai Port-City

Zone	Color	Area (Ha)	Information
Port & Industrial Area		390	Supported by water treatment plant, wastewater treatment plant and renewable power plant
District 1		29	Commercials consist of retail and warehouse
District 2		35	Warehouse area
District 3		46	
District 4		43	Retail, park, business, modern market and resort
District 5		22	Residential (low rise building and landed house)
District 6		82	School, hospital, and commercials
District 7		34	Residential (villa and landed house)
District 8		40	Landed house and allocated for high-income
District 9		47	Landed house and allocated for middle to low income

4.2. Initial Cost

The initial cost of the port consists of a liquid bulk terminal, dry bulk terminal, and container terminal. Liquid bulk terminal divided by dock, mooring, and storage. The dry bulk terminal has larger components than the liquid terminal by adding types of equipment, warehouse and stacking field. The industrial area that consists of several industries has its particular components and related types of equipment. Initial cost also considers land acquisition and reclamation as well as required utilities such as power plant, water treatment plant and wastewater treatment plant. The assumptions in determining the cost of components are based on various standards such as ARCADIS and standard from Ministry of Public Works. The summary of the initial cost of the project can be seen in the following table.

Table 2 (a). Breakdown Cost of Port-City Concept in Dumai

No	Component	Cost of Investment (US$)
1	Port Terminal	
a	Liquid terminal	18,356,421.48
b	Dry bulk terminal	32,950,330.21
c	Container terminal	211,305,552.92
2	Industrial area	
a	Marmalade jam	1,286,267.33
b	Pinapple processing	4,097,759.94
c	MoCaF	61,666.67
d	CPO	43,457,283.41
3	Land	
	Land acquisition	1,609,057,514.40
	Reclamation	26,337,067.67

IOP Conf. Series: Materials Science and Engineering **650** (2019) 012015 doi:10.1088/1757-899X/650/1/012015

Table 2 (b). Breakdown Cost of Port-City Concept in Dumai

No	Component	Cost of Investment (US$)
4	Utilities	
a	Renewable Power Plant	63,729.89
b	WTP	6,238,622.7
c	WWTP	20,857,837.1
	TOTAL	1,974,070,053.8

4.3. Operation and Maintenance

Operation and maintenance (OM) cost in the study considered the functions that consist of the port, industrial development and supporting infrastructure. The OM cost considered the operational stages of each function and divided into three; 2023, 2025 and 2027. The assumption to generate the OM baseline cost derived from various reports and standards such as the Ministry of Public Works, local government standard and literature study. These costs are increasing by follows inflation of about 5.12%. The OM cost of each function is summarized in the following table.

Table 3. Operation and Maintenance Cost of Dumai's Port City

No	Component	2023	2025	2027
1	Port	3,458,344.12	3,845,575.78	4,276,165.86
2	Industrial development	9,898,459.11	19,261,886.09	30,598,063.27
3	Supporting Infrastructure	5,531,978.39	10,764,941.94	17,100,421.69
	TOTAL	18,888,781.62	33,872,403.81	51,974,650.82

4.4. Revenue

Revenue is generated from particular functions in the conceptual design of Dumai Port City. It consists of fourteen sources of income such as retail, warehouse, resort, business area, modern market, low-income apartment, landed house, schools (elementary, junior high, senior high), hospital, hotel, villa, and malls. The size of each source will be divided which one being sold, rent, charged retribution or other means of income collection.

Table 4. Source of Income for Project Revenue

Sources of Income	Category	Charge	Unit
Retail	Sell	109,375.00	Per unit
	Rent	7,291.67	
	Retribution	3.33	Per m^2
Warehousing	Sell	68,333.62	Per unit
	Rent	6,212.15	
Resort	Rent	885.42	Per unit
Business Area	Sell	445,601.85	Per unit
	Rent	40,509.26	
Modern Market	Rent	340.28	Per unit
Low-income apartment	Sell	10,264.40	Per unit
	Rent	598.76	
	Retribution	3.33	Per m^2
Landed House	Sell	128,685.90	Per unit
	Rent	4,289.53	
	Retribution	3.33	Per m^2
Elementary School	Building maintenance	191.67	Per student
	Tuition	327.75	
Junior High School	Building maintenance	263.54	Per student
	Tuition	396.75	
Senior High School	Building maintenance	263.54	Per student
	Tuition	396.75	
Hospital	Rent	8,450.87	Per patient
Hotel	Rent	18,325.00	Per room
Villa	Sell	58,603.74	Per unit
	Rent	2,930.19	
	Retribution	5.83	Per m^2
Mall	Rent	1,763.24	Per room

Those sources of income that is sold to the users such as retail, warehouse, business area, low-income apartment, landed house and villa estimated to complete in five years after the construction stage. On the other hand, hospital revenue is based on 60% of patient occupancy rate. Further, income from schools is from building maintenance and tuition fees.

From the initial cost, operation and maintenance cost and revenue from each function in the project, the result shows an internal rate of return for about 10.37%. An industrial area, mainly palm oil processing, contributes a more considerable revenue, thus affect positively to the overall revenue of the project. However, this rate of return is slightly below the expected weighted average cost of capital (WACC) from construction company/investors perspective as about 11%. Thus, a public-private partnership scheme will be used to analyze further the project to reach an expected rate of return.

4.5. Financial Analysis

In conducting the financial analysis, the research has to reach a specific value of WACC by the companies. It is a capital calculation from the corporation by taking into account the equity, debt and debt ratio from the internal company [12]. The value varied among others, for instance, the industrial sector has 10.51% of WACC. While the chemical sector shows 11.51%, real estate around 11.10% and health sector about 10.31%. Chemical sector is the highest WACC related to the project; therefore this value is set for the baseline for financial analysis. The value must be surpassed in order the project declared feasible in term of technical and financial consideration.

The research analyzes 42 scenarios to generate an optimum result of sharing between the public and private investors. The division of responsibility between two parties conducted in four models; initial

IOP Conf. Series: Materials Science and Engineering **650** (2019) 012015 doi:10.1088/1757-899X/650/1/012015

cost, operation, and maintenance cost, initial cost and operation, and maintenance cost as well as initial cost, operation and maintenance and revenue. In initial cost, three scenarios were delivered, a 60% government sharing with 40% private, equal distribution and 40% government and 60% private. Moreover, the result shows that 40% of share from the government reach a rate of return of about 15.88%. It is a typical scheme that has been applied in many other nations through build-operate-transfer, where the government support the infrastructure and private gain the revenue and maintain the infrastructure. Surely the public may access the infrastructure, but sunk cost from government relatively large and sometimes may affect the state budget.

The second model is the sharing conducted in operation and maintenance. The scheme generates three scenarios with a similar division of responsibility from initial cost. Limited implementation found using this type of model around the world, and the result shows insignificant IRR. Scenario 4 as the government contributes about 60% of operation and maintenance cost produce 12.02% of IRR. While the lowest, produce 11.49% of IRR.

The third model is sharing concept in initial cost and operation and maintenance cost. Nine scenarios were evaluated, and the lowest IRR about 17.49% and the highest is 25.93%. This scheme offers a higher rate of return for the private investors due to government involvement in the initial stage and operation and maintenance stage. In contrary, the government has to bear a higher cost to accommodate not only initial cost but also operation and maintenance cost.

The fourth model proposes a cost-sharing in every project life cycle including the revenue. A 27 scenario was elaborated by this research, and the result shows a range of rate of return started by 14.04% to 20.88%. The maximum IRR is used as the baseline for further evaluation where the private contributed 40% in initial cost and operation and maintenance cost and gained 80% of the profit from the project.

The evaluation was conducted by dividing each function in the project to either belong to private or public responsibility. For instance, private will contribute for industrial construction (70% of palm oil, marmalade jam industry, MoCF processing industry and pineapple processing industry), 40% of reclamation and mixed-use area (road, greenery, mosque, retail, warehouse, business area, resort, landed house, schools, hotel, villa, and mall). The government contributes in industrial construction (30% of palm oil), 60% of reclamation, land acquisition, utilities (power plant, WTP, and WWTP), port (dry bulk terminal, liquid terminal, and container terminal) and mixed-use area (modern market, low-income apartment, and hospital). The detail of the division of responsibility between private investors and government summarized in figure 5.

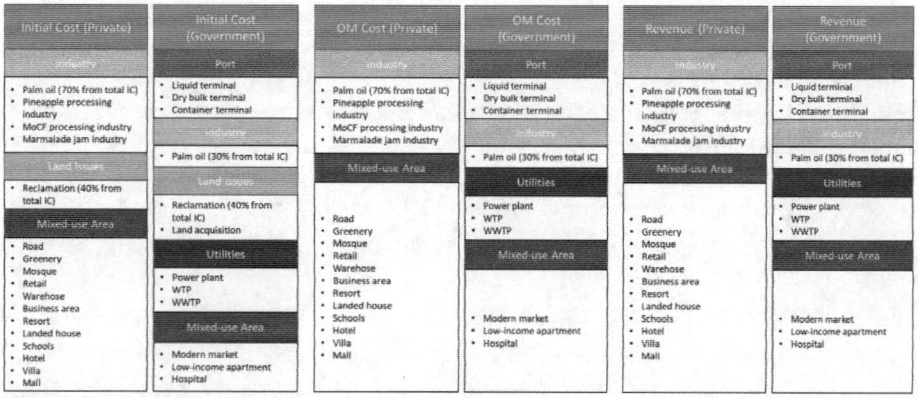

Figure 5. Divison of Responsibility between the Government and Private Investors

Based on the above division of responsibility, the final scene is produced. The government should support the initial cost about 59.38% and invest in operation and maintenance around 31.74% but obtained revenue of 33.96%. This scheme generates optimum IRR about 15.41%. It is argued that the value is adequate to attract private investors to the project and the support from the government gained maximum benefits for the people.

5. Conclusion

The port-city concept aims to support government vision as a global player in maritime transportation. Its existence also attempts to improve regional competitiveness by creating new economic activities and increase the connectivity and mobility of people and goods. This research use a case study of Dumai Port in Riau Province, Republic of Indonesia. The result shows that the port city may consist of port, industrial area, and supporting infrastructure. Port divided into the liquid terminal, dry bulk terminal, and container terminal. The industrial area comprises of marmalade jam processing, pineapple processing, modified cassava flour and palm oil. The last type of industry is the highest contributor among others in term of revenue. Last, utilities consist of a renewable power plant, water treatment, and wastewater treatment plant.

The life cycle cost analysis illustrated the concept of the port city in this research generates 10.37% of the internal rate of return. An evaluation through sharing concept in every construction project stages proposes a significant increase of IRR to 15.41%. The scenario suggests the private contributes 40.62% of the cost in the initial stage and 68.87% of the cost in the operation and maintenance stage. From that share, they will receive 66.04% of the total project revenue.

Acknowledgement

This research was fully supported by grants from Ministry of Higher Research and Education, Republic of Indonesia and Universitas Indonesia PITTA 2018.

References

[1] Nathan Associates 2011 *Technical Inputs for NPMP Revision Final Report. Jakarta: Australia-Indonesia Partnership Indonesia Infrastructure Initiative*
[2] Merk O *2014 The effectiveness of port-city governance. Alix, Y. Delsalle & B. Comtois, C.(eds) Port-City governance. Editions EMS* **233** 245
[3] Berawi M A, Miraj P, Gunawan, and Islamiah E R *2018 IOP Conf Ser Earth Environ Sci* **1977** 040001
[4] Hesse M and Rodrigue J P 2006 *Growth and Change* **37** 499
[5] Jung B M 2011 *AJSL* **27** 1
[6] Hall P and Clark A 2004 *Transforming urban waterfronts: fixity and flow* **17**
[7] Gipouloux F 2011 *The Asian Mediterranean: port cities and trading networks in China, Japan and South Asia* (Edward Elgar Publishing)
[8] Hoyle B S 1989 *Geoforum* **20** 429
[9] Berawi M A, Ibrahim B E, and Miraj P 2019 *JDBE* **19** 40
[10] Rahman H Z et al 2018 *IJTech* **3** 549
[11] Berawi M A, Susantono B, Miraj P, and Nurmadinah F 2018 *Aviation* **22** 115
[12] Fernandez P 2010 *Business Valuation Review* **29** 4

IOP Conf. Series: Materials Science and Engineering **650** (2019) 012016 doi:10.1088/1757-899X/650/1/012016

Near zero energy house design parameter implementation in cluster housing – a case study

Y Latief[1], M A Berawi[1], L S R Supriadi[1*], A B Koesalamwardi[1] and A Herzanita[2]

[1] Civil Engineering Department, Faculty of Engineering, Universitas Indonesia, Indonsesia

[2] Civil Engineering Department, Faculty of Engineering, Universitas Pancasila, Indonesia

E-mail: *lsagita@eng.ui.ac.id

Abstract. The sustainable concept of housing supports a house in reducing the dependence on fossil-based energy. The environmental impact of a housing building, particularly CO^2 emissions, depends on the house overall design, including site planning, configuration, materials, construction, and operation where every aspect affected the house environmental performance. Nature has provide sufficient energy to tropical climate environment like Indonesia. Nevertheless, economically viable cluster housing is the other issue that encounter high costs housing from energy efficient building such as the Near Zero Energy House (nZEH) due to high price of materials and equipment used. The objective of the research is to understand how nZEH design parameter has been implemented and give impacts on architectural form of housing design. Methods used in this study is qualitative approach using research instruments among others: checklist, interview form and observation log book. From the case study conducted in this research, it was found that the object, the cluster design housing, are poorly incorporate the nZEH design parameter due to costly reasoning that impact to initial cost of the housing building. Among the design parameter which has been suggested are limited to passive design, fenestration, HVAC, lighting, and minor consideration to operation and maintenance.

Keywords: Design Parameter, Building Specification, Construction, Cluster, Housing

1. Introduction

As a developing country, Indonesia's economic development depends on its energy sustainability that requires attention to environmental aspects. In a smaller scope such as housing development, shifting forward to a sustainable concept of housing entitle a house to play a significant role by reducing the dependence on fossil based energy appliances among others. Thereby, environmental impact of a housing building, particularly CO_2 emissions, depends on the house overall design, including site planning, configuration, materials, construction, and operation where every aspect affected the house environmental performance. To improve the energy performance of a house, nZEH is proven as a cost-effective solution to overcome climate change and increase energy security. Furthermore, it can support energy conservation which could address issues due to climate change such as the increase of global temperature and the reduction of fossil fuel energy. In adoption of nZEH concept, using the energy collected from the sun, it enables the building to generate its own saved energy which minimize the utilization of energy from sources outside the building [1].

Indonesia is an ideal location for development of renewable energy i.e. solar energy and its implementation and application in nZEH technology because it is located in the equator and illuminated by the sun throughout the year. Nonetheless, this potential has not been fully explored due to the high cost of solar power generation systems technology. Currently, the use of solar panels on the building technology in Indonesia, is still very rare. The practical uses of solar panel technology is limited to a street lamp, or a building located in remote places far from the PLN (*Perusahaan Listrik Negara* – a state owned electric company) grid [2] [3] [4].

This issue also causes a contradiction since design of energy efficient buildings typically uses expensive materials and technology, which affects the overall cost of construction. It needs a balance and optimization of the use of "green" technology and construction cost, so that energy-efficient building technologies can be affordable to society. Several studies have conducted to examine the design optimization of nZEH in sub-tropical climate to four season climate environment. Also, studies regarding the suitable design parameters for nZEH have been conducted in sub-tropical climate and the tropical climate [4] [5] [6] [7].

In order to have a comprehensive approach for optimizing the nZEH design parameters towards the lifecycle cost of a house, the implementation of the nzEH design parameters in a house should also be studied. This paper discusses the objective of the study, which is to understand how nZEH design parameter has been implemented and give impacts on architectural form of housing design, particularly in small housing development.

2. Literature review

Generally, the concept of Zero Energy Building (ZEB) is a building that is able to meet its energy needs from inexpensive, clean, easy to obtain, and renewable source. Other definition states a ZEB concept generates renewable energy on site as much as or even exceed its annual energy consumption. A zero energy building can be defined in several different ways, usually depending on the project objectives. Therefore, it explains more for the building owners to think about energy consumption. Torcellini, et al. [8] explained four different types of commonly used definitions, namely: net zero site energy, net zero source of energy, net zero energy cost, and net zero energy emissions. Furthermore, Haslam and Farrell [9] stated an energy-plus buildings or buildings that consume slightly more energy than they produce are called near-zero energy buildings (nZEB). This could also be applied to housing, where the term would change to near-zero energy house (nZEH).

2.1. Examples of nZEH Practices

There are several nZEH practices throughout the world. Pearl River Tower in China adopted nZEH concept which is equipped with wind turbines, radiant slabs, microturbines, ground heat pump, ventilated facades, BIPV, and natural lighting control system in its building concept. Based on the number of floors and space available, efforts were made in this buildig to bring the concept of Zero Energy Building (ZEB) by reducing the energy account for the use of building lighting and air conditioning system. In order to reduce the energy consumption, the building was designed by combining site orientation, shape of the building, natural lighting and automated building control systems. Also, one of the NZEH designs is Ecoterra nZEH in Canada which is equipped by Building-Integrated Photovoltaics (BIPV) [10]. Moreover, in Germany there are also Canadian Solar and Green City Energy that gather the installation of solar power plants with a capacity of 1 MWp on 16 rooftops, with a total area of 17,000 square meters in the city Haertensdorf, Germany. These renewable energy technologies (RET) support the implementation of ZEB concept [11].

In Indonesia, Listijono's designated house in Pluit Residential House was an ASEAN-winning Energy Efficiency and Conservation (EE & C) Best Practice Competition in Buildings in ASEAN Energy Awards - 2014. Pluit Residential House occupies a site area of about 250 square meters and was completed in March 2013 whereby equipped with three storeys and one flat roof for Solar cell, one cold water tank, one hot water tank, one Virtual Routing and Forwarding (VRF) Air Conditioning unit with hot water generation, three parabolas, one Pyramid meditation, two outlets voids, one sky lighting, Green roof and one Gazebo for leisure area. The total gross floor area of the building are 781 square meters designed by Aria Architect and John Budi H Listijono as its M&E Engineer [12].

2.2. nZEH Design Parameters

According to Boeck et al [13], several design parameters for nZEH used in previous studies are as follows: (1) Renewable Energy Technologies, namely the use of PV (a) the ratio of the panel area - roof area and (b) Directions azimuth (orientation) of PV panels; (2) The shape of the building (a) building orientation, (b) the slope of the roof; (3) building envelope that affect insulation; (4) fenestration (windows and doors) with the components of the window glazing type, window-to-wall ratio, type window sills, window direction, internal and external shading; (5) Heating-Ventilating-Air Conditioning (HVAC) which affected Natural Ventilation and mechanical Cooling / heating; (6) Lighting.

Boeck et al [13] also conducted a study to collect 65 scientific journals that discussed about energy efficiency in housing and apartment buildings. The results of this study were used as a reference in the preliminary variables comply to this research whereas resulted in 5 groups of variables other than the use of PV panels: building shape, building envelope, fenestration, HVAC, and lighting. Additional variables, i.e. PV panel, obtained from Bucking et al [10] who studied nZEH that uses PV panels.

To further validate the nZEH design parameters for tropical climate, some relevant studies were used as benchmarks. Development of the building design with Near Zero Energy House (nZEH) approach in tropical climate such as Indonesia is influenced by several variables. Based on the study, some variables which affect nZEH optimum design in Indonesia are building orientation direction, Panel Photovoltaic (PV), and fenestration (windows and doors) with window glazing type and window-to-wall ratio subvariables. Geographically, Indonesia lies in 6°North Latitude – 11°South Latitude and 95°Eastern Longitude – 141°Eastern Longitude, consequently this brought tropical climate to the nation. A prominent feature in the tropical climate area are high average daily temperatures and higher air humidity and relatively slow air flow for achieving thermal comfort. The use of energy in Indonesia focused on air cooling and dehumidification (reduction of air humidity) that are commonly pursued using passive design to obtain a suitable thermal confort inside the building [4] [7].

Furthermore, the case study of Pluit Residential House exposes the passive design concept. Passive design concept acknowledge climate in Jakarta Area and natural ventilation with description of building is facing south (facing the road) where all the doors and windows are located as the openings of natural ventilation and designed from the early stage of using sky lighting to create natural ventilation to induce air speed of >0.5m/s in each floor system and day lighting system for non-air and air conditioned areas. Other highlight to the design is the utilization of LED light that covers 98% of light. Using particular design make a big saving for having day lighting generated from the sky lighting, and two others void in this building whereby the light load in this building is only 3.4 Watt per square meter. As part of green technology, the solar PV panels contribute 25.7% of the total electric energy from the grid, therefore it is very important to monitor and maintain the performance of this solar PV Panel. Consequently, WIFI Plug installed in those 3 solar inverters to monitor and record the total electricity generated in these solar PV to keep the electric energy generated above 8 kWH/day with the average of 12 kWH/day [12].

Thus, from the experiments and case study, there are four design variables for nZEH that significantly suit the tropical climate and support energy efficiency, which are: building orientation (building azimuth), panel PV, fenestration dan passive design. These findings are also relevant with the studies of Wimmer [7] and Torcellini et al [8]. Firstly, to its tropical climate with high average temperatures caused by vertical positioning of the sun, Indonesia bound to have set up on building orientation toward the north 0° to get comfortable indoor temperatures and reduce expenditure due to electricity costs for air cooling system and lighting. Secondly, for tropical countries, such as Indonesia where the sun shines all year round, a renewable energy source derived from sunlight is seen as a potential energy. Solar radiation can be used to generate electricity (Photovoltaic) for building utilities, and heating the water (Photovoltaic Thermal). Thirdly, fenestration, materials and installation also contribute to the energy channeled through the windows, doors, or skylights, together with the airflow in the window components to the means of increased energy efficiency. Last but not least, passive design apt to be applied in tropical countries since the passive cooling system obtain a comfortable temperature (thermal comfort) in the building [7] [8].

3. Methodology

To further understand how nZEH design parameter has been implemented and give impacts on architectural form of housing design in tropical climate, this study used case study as the methodology. A cluster housing X in Tangerang, Banten is chosen as the case study. It was chosen based on the type of housing (small cluster housing) and located in the suburban area in Banten, as one of the growing population and economic areas in JABODETABEK, Indonesia. The research instruments were checklists, interview form and observation log book. Checklists were used for comparing the housing concept from the developer and the nZEH design parameters. Interview form and observation log book were used for guiding the surveyors during interview process with the developer respondents. There were three (3) respondents that provide the data for the case study. The respondents were the Project Manager, Design Staff and Marketing Manager of cluster housing X.

4. Result and discussion

There were three main aspects that were studied and observed in the case study, which are the building specification, construction specification, and the implementation of nZEH design parameters.

4.1. Building Specification

The surface area of one house in the cluster housing X was 72m^2, and its building area was 54m^2. The house consists of one living room/family room linked directly with dining room, two bedrooms, one bathroom, one kitchen, and one roof garden. The main living area of the house is on the first floor, whereas the second floor was utilised for roof garden area. The house can be occupied by 3 – 5 persons. When designing this house, the hourly average temperature and wind velocity were not specifically measured by the architect. For further illustration, Figure 1 shows the floor plan, views, and interior views of the house in cluster housing X.

IOP Conf. Series: Materials Science and Engineering **650** (2019) 012016 doi:10.1088/1757-899X/650/1/012016

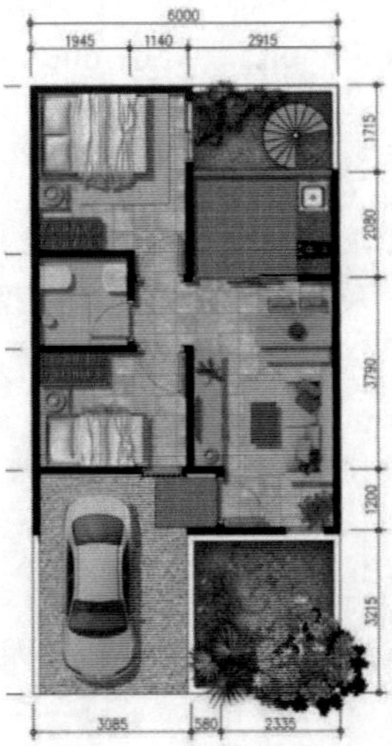

(a) Floor Plan

(b) Cluster Housing X-part of the area is still under construction

(c) Front view

(d) Sliding door that connects the family room and kitchen. This door uses Switchable Vision Panel technology, where the glass of the door can be switched into different level of shading and transparency, as needed by the occupants

(e) Skylight area viewed from the kitchen

(f) Roof garden with synthetic grass

Figure 1. Floor Plan and Views of Cluster Housing X (a-f) (on-site data, 2016)

4.2. Construction Specification

Regarding the materials, cluster housing X used light bricks for its external and internal walls due to the light weight, fast installation and heat absorbent. For the external and internal doors, aluminum were used for the frame and termite-protected wooden layers were used for the panel. As for the window, the frame was made of aluminum and the panel was made of glass. The floor tiles were made of ceramics, and the ceiling was constructed from concrete tiles. Due to confidentiality, the detailed construction cost of the house cannot be acquired. However, the respondents stated that the range for the construction cost of one house were approximately between Rp 500 million – Rp 750 million.

4.3. The Implementation of nZEH Design Parameters

As for the implementation of the nZEH design parameters, the aspects that were observed on the house were as follows:

- Renewable energy technology - Photovoltaic (PV) panel: The house did not utilise any PV panel. The reason is that PV panel is still expensive, and this house is targeted for the low – medium income consumers. Therefore, design parameters such as the panel/roof ratio and panel orientation were not considered and implemented.
- Building orientation (building azimuth): This aspect is used for the house design parameter, where the house was facing west or east for receiving the optimum sunlight.
- Passive design: Regarding its shape, the house was designed using the modern minimalist concept where it can be expanded in the future (for example: constructing additional room on the second floor). In terms of complying to the nZEH design parameters, this house did not fully implement them. The roof of the house is flat (no pitch / angle), and the building materials were mostly conventional. The natural HVAC could only be utilised through the sliding door which connects the family room and the kitchen, where the air circulation can be adjusted through that door. As described in Figure 1, the sliding door uses Switchable VisionPanel technology, where the glass of the door can be switched into different level of shading and transparency, as needed by the occupants. Moreover, for the lighting aspect, the house received natural lighting from the skylight area above the kitchen, and mostly used artificial lighting.
- Fenestration: For its windows and doors, there were no additional insulation, glazing, nor internal / external shading. The window-to-wall ratios were also not specifically designed to comply the nZEH requirement. The only component that adopted nZEH design parameter was the sliding door, which adopted the shading aspect of nZEH.
- Energy monitoring: There was no specific monitoring tool for the house's energyperformance / efficiency. The only metered tool available was the standard electricity monitoring tool from the Government National Electricity Company (PLN).
- Electricity consumption: From the respondents, the total electricity consumption per monthfor one house in this cluster was around Rp 300.000,- to Rp 500.000,-. This is still considered as normal for small houses. However, further approach to minimize the monthly electricity cost should be considered, especially for the low-medium income families.

From the results, it was found that most of the nZEH design parameters were not implemented yet in the house. The house only focused on the passive design (lighting and natural HVAC) and the building orientation for receiving optimum sunlight. Specific technology such as PV panel (RET), fenestration design, and daily energy monitoring tool – which mostly support energy efficiency for nZEH were not installed. Hence, the energy efficiency that can be derived from the house was only from the natural lighting (mostly during the day) and air circulation. Most of the house in the cluster housing X still uses air conditioning in their rooms.

Furthermore, it was also found that the developer of cluster housing X chose only minor parts of the nZEH design parameter in the house due to economical reasons. They opined that by having some parts of the Green Building aspects in the house, it could be a value added factor from the marketing point of view. Also, due to economical reason, the developer also viewed that utilizing RET,

IOP Conf. Series: Materials Science and Engineering **650** (2019) 012016 doi:10.1088/1757-899X/650/1/012016

fenestration design and energy monitoring tool are expensive which could increase the initial cost of the house. They believed that designing, constructing and selling a nZEH house would cost more than Rp 1 billion, and this is not quite suitable for small houses such as cluster housing X.

5. Conclusion

To conclude, this study showed that the cluster design housing X in Tangerang, Banten, poorly incorporate the nZEH design parameters. The lack of implementing the nZEH design parameters were due to economical reasons, which the developer viewed that most of the design parameters need higher material costs that lead to higher initial cost of the house. Although some studies found that nZEH can save more energy and has lower operational and maintenance costs, the developer still focused on having a reasonable initial cost, hence not prioritizing the full implementation of nZEH design parameters. The case study only focused on minor aspects of the nZEH design parameters, which are limited to passive design, fenestration, HVAC, lighting, and minor consideration to operation and maintenance.

The result of this study provide important feedback for further studies on nZEH, particularly in Indonesia. This case study is one finding of a small cluster housing in a suburban area in JABODETABEK. The nZEH design parameters that were implemented in this case study can further be used for optimization in order to increase the energy performance and achieve energy efficiency with optimum life cycle cost of the similar type of house. Moreover, similar approach can be conducted to other types of housing in Indonesia.

Acknowledgements

The authors would like to acknowledge Research Grant from Ministry of Research, Technology and Higher Education Republic of Indonesia (Kemenristek) for the funding support of the research project (Hibah Penelitian Unggulan Perguruan Tinggi 2016).

References

[1] Rattanongphisat W and Rordprapat W 2014 Strategy for Energy Efficient Buildings in Tropical Climate *Energy Procedia* Elsevier **52** pp 10 – 17

[2] Zhou A, Alleng G, and Byrne J 2001 Using Economic Incentives to Accelerate Development of Green Technologies *Green Building Workshop Sponsored by the Pacific Energy Center (China) and Environmental Market Solutions Inc.* ((U.S.) Shanghai China)

[3] Evola G, Margani G, and Marletta L 2014 Cost-effective Design Solutions for Low-rise Residential Net ZEBs in Mediterranean Climate *Energy and Buildings* **68** pp 7–18

[4] Koesalamwardi A B 2014 *Cost optimum design of near zero energy house using genetic algorithm* (Depok: Universitas Indonesia)

[5] Milajic A, Milosevic P, and Belkovic D 2013 Optimal Green Building Design Using Multi-Objective Genetic Algorithm *Reporting for Sustainability* pp 119-123

[6] Panão M J N, Rebelo M P, and Camelo S M L 2013 How low should be the energy required by a nearly zero-energy building? The load/generation energy balance of Mediterranean housing *Energy and Buildings* **61** pp 161–171

[7] Wimmer R 2013 Adapting Zero Carbon House for Tropical Climate – Passive Cooling Design in the Philipines *SB13 Sustainable Buildings, Infrastructure and Communities in Emerging Economies* (Manila) pp 1 – 8

[8] Torcellini P, Pless S, and Deru M 2006 Zero Energy Buildings: A Critical Look at The Definition *National Renewable Energy Laboratory* (Pacific Grove California)

[9] Haslam M P G and Farrell A 2014 Natural Ventilation Strategies in Nearzero-Energy Building: A design primer for students and professionals *Across:Architectural Research through to Practice: 48 Int. Conf. of the Architectural Science Assoc.* pp 619–630

[10] Bucking S, Athienitis A K, Zmeureanu R, O'Brien W J, and Doiron M 2013 Design Optimization Methodology for a Near Net Zero Energy Demonstration Home *Euro Sun*

2006 Int. Solar Energy Soc. Conf. Proc. Quebec
[11] Ibanez N 2010 Digital Construction [Online] http://www.constructiondigital.com/sectors/other-construction/german-solar-system- installed-ahead-schedule [Accessed: 20 November 2013].
[12] Listijono J B H 2014 Pluit Residential House *ASEAN Energy Efficiency and Conservation (EE&C) Best Practice Competition in Buildings* ASEAN Energy Awards Vientiane
[13] Boeck L D, Audenaert A, and Mesmaeker L D 2013 Improving the Energy Performance of Residential Buildings: A Literature Review *Hubrussel Research Paper*

IOP Conf. Series: Materials Science and Engineering **650** (2019) 012017 doi:10.1088/1757-899X/650/1/012017

Lean and sustainable construction: link between the sustainability report disclosure and the impact on profitable opportunities for investors

N C Kresnanto[1*] and W H Putri[2]

[1]Magister of Civil Engineering, Janabadra University, Jalan Tentara Rakyat Pelajar 57, Yogyakarta, Indonesia

[2]Accounting Department, Faculty of Economics, Janabadra University, Jalan Tentara Rakyat Pelajar 57, Yogyakarta, Indonesia

E-mail: *nindyo_ck@staff.janabadra.ac.id

Abstract. Lean management in the construction sector is a company-driven movement to gain various advantages in running the business process. By implementing lean management, the company is simultaneously able to achieve multiple goals, such as profit maximization, while being able to eliminate non-value added activities and improve its environmental and social performance. To communicate these activities, the company issues a sustainability report, whose quality of disclosure depends on what the company has done with its business processes. This study tries to examine whether the quality of disclosures in sustainability reports has been able to contribute to influencing the number of earnings per share, which in this case is essential information for investors to see opportunities for profitability. The findings of this study indicate that, overall, disclosures in sustainability reports have not been a critical reference for investors, especially in looking at their profitability opportunities, and they only position this sustainability report as an additional source of information to ensure their concerns about social and environmental responsibility. The implication of this research for the development of sustainable construction is that it functions as a note to increase stakeholder engagement, especially investors, so that they appreciate the efforts of management to achieve competitive advantages and not only profit oriented.

1. Introduction

Lean management strategies have been carried out by many companies since a while ago, but they are still initiating the improvement process towards sustainability as a form of accountability for their business processes. The triple bottom line concept has brought the company's paradigm to be more responsible and begin to pay attention to various aspects outside of finance [1]. As a form of continuous improvement, lean management is considered a concept that can solve multiple problems related to efficiency and productivity, reduce the number of non-value-added activities, and make the organization thinner and more flexible [2].

Moreover, in some companies that have paid more attention to improving the quality of their supply chain management into the sustainability paradigm, this lean management concept is becoming more appropriate because it is very easily associated with an eco-efficiency approach, which has two sides of influence, namely reducing costs and contributing to environmental performance.

Although in purpose, the concept of the value and the concept of waste management between lean and sustainability are different [3], both lead to an increase in the value of the company. Lean management prioritizes efficiency and productivity factors derived from business calculations while sustainability places more emphasis on how they carry out the process more responsibly for stakeholders. In the world of construction, the sustainability paradigm will be more visible when lean management designs are simultaneously chosen and harmonized with the sustainability concept, where investments that enter the company have taken into account the support and selection for efforts to improve environmental performance [4], such as reducing the use of hazardous materials, saving energy, and implementing an integrated waste management.

In addition to the consideration of improving environmental performance, construction companies are motivated to make sustainability efforts due to business value reasons [5]. If they can obtain a certain business value (for example, in environmentally friendly buildings, rent or unit-selling prices will be higher), they will take the opportunity and harmonize with the value chain they do. That is, economic motivation will tend to be a driver for sustainability compared to motivation to improve environmental performance. In this case, we can say that improving environmental performance is possible if it also simultaneously creates opportunities for strengthening economic/financial performance.

The progress of the construction sector in adapting the concept of sustainability has been demonstrated in several previous empirical studies. They realize that action is urgently needed to improve the quality of project management flows towards sustainability [6] and make it a benchmark for success. Efforts to develop sustainability evaluation indicators in the construction sector are made by basing on economic, social, environmental aspects [7], as well as in detail, entering internal and external efficiency indicators [8]. Even in its development, previous empirical studies have been able to compile indicators to evaluate Project Management Practices and Sustainable Infrastructure (SI). To satisfy the aim of our research, we found that one of the drivers of improving environmental performance was eco-efficiency [6]; therefore, the results of this study also confirmed the statements in the previous paragraph.

The common thread of the above explanation is that the presence of sustainability performance is an effort initiated by management [9][6], in the form of both lean and supply chain management, and is not a request from investors [8][10]. Besides, the disclosure of sustainable business processes in sustainability reports is more motivated by fulfilling the element of legitimacy and fulfilling community expectations compared to achieving the substance of sustainability [11]. Based on the results of previous studies, there are quite basic questions, i.e. how much the need or attention of investors to the presence of a sustainability report is and whether the existence of sustainability efforts initiated by management can influence the indicators that are of concern to investors. This study aims at presenting empirical findings of the quality of disclosure of sustainability reports, especially in the construction industry in several countries, and showing the effect of exposure of sustainability reports prepared using the Global Reporting Initiative/GRI G4 guidelines on earnings per share (EPS) variables.

2. Sustainable Construction and information disclosure in corporate sustainability reporting

The construction sector is an essential sector in the economy and takes a significant portion in providing employment as well as making various economic, environmental, and other social contributions. In response to sustainability issues, multiple initiatives to present the concept of sustainable construction have been applied in empirical practice, one of which is by developing the idea of green construction [12]. The objectives of the implementation of sustainable construction include maximizing wealth, maximizing utilities, minimizing resources, and minimizing impacts [13]. To make this happen, companies have a great interest in delivering information, among others, by improving the quality of disclosure in the sustainability report. Even though it is not regulated as mandatory, the quality of disclosure is an important matter to note to emphasize the commitment of sustainability and the company's efforts to fulfill its commitments. Many standards can be selected to guide the preparation of sustainability reports, such as the Global Reporting Initiative (GRI) and the International Integrated Reporting Committee (IIRC). This study uses the GRI G4 standard explicitly as a guideline for assessing the quality of disclosure of sustainability reports from construction companies selected as the research samples.

3. Information disclosure quality and Earning per Share

The existence of reporting that links sustainability implementation to financial parameters is currently becoming more prominent and transparent [14]. Besides, the information provided also has better quality, allowing various parties, including investors, to use sustainability reports as a profitability parameter for them. Investors are very interested in seeing how companies strive to be more profitable and more competitive whilst still considering stakeholders' interests more equitably. In addition, on the upstream side of production, the company is increasingly trying to be the best for its users while even taking into account various sustainability indicators.

The quality of disclosures in sustainability reports, especially economic indicators, is most often used to attract investors to the company. Investors easily find information related to financial performance and connect with expectations for profit, one of which is measured by the value of earnings per share / EPS. If the disclosure is following its expectations, investors will assume that the company is worthy of providing an opportunity for more profit. Thus, the first hypothesis (H1) is that the quality of disclosure of economic indicators affects the earnings per share / EPS. The better the disclosure of its financial indicators is, the better the internal business process is, making it able to promise higher profits.

On the other hand, the link between lean management and environmental performance is now increasingly connected because it is believed to be able to increase competitive advantage [15]. The slimmer a business process is, the more activities that are not value-added will be lost and replaced by increased productivity while accompanied by an increase in the prevention of environmental damage. Some of these advantages are very promising, particularly for investors, especially to create more opportunities to increase the value of their investments; hence, it can be assumed that more disclosures in the sustainability report about various production activities that are able to form environmental performance mean investors will be more interested because the company will be able to provide an ever increasing earnings value per share. Therefore, the second hypothesis (H2) of this study is that the quality of disclosure of environmental indicators influences earnings per share / EPS.

The popularity of socially responsible investment is also increasingly thought to be a motivator for companies to further improve their business processes according to the concept of sustainability. Although the results of previous studies related to the type of investors found that there were two motivations of investors when choosing the SRI instrument, namely finding opportunities to gain new financial benefits and avoiding environmental and social risks [16] [17], the tendency of investors to have other motivations cannot be ignored. Only does it appear in the type of investors who have more attention to the social implications that might arise from their investments in companies [18]. Based on the above phenomenon, the third hypothesis (H3) is that the quality of disclosure of social indicators influences earnings per share / EPS. The assumption built is that with the quality of disclosure of good social indicators in the accountability report, the company has made efforts to better mitigate social risk, allowing them to generate high profits, which impact on high EPS values.

4. Research Methods

The study we conducted aimed to observe the influence of the quality index of disclosure of sustainability reports in construction companies in several countries including Indonesia with the value of earnings per share (EPS) of the company's shares by compiling the index criteria as table 1.

In the first step, we conducted a content analysis on the quality of disclosures in the sustainability report and gave a score according to the specified index criteria. The population in this study was 28 construction companies in several countries, including Indonesia. Meanwhile, samples were taken through the purposive sampling method with construction companies that published their 2017 sustainability report as the sampling criteria.

We used a secondary data in the form of sustainability report during the period of 2017. The approach used was a direct approach, where the analysis starts from the theory and/or relevant research finding as a guide in converting content into code [19]. Data sources were obtained from the official website of each company and the National Center for Sustainability Reporting (NCSR) website at www.ncsr- id.org. The companies as the samples of this study consist of 26 Indonesian, Asia, Europe, and New Zealand company.

Table 1. Scoring index and criteria

Score Index	Details
1	Economic, environmental, and social indicators are fully disclosed according to the GRI G4 guidelines.
2	Economic, environmental, and social indicators are adequately disclosed according to the GRI G4 guidelines.
3	Economic, environmental, and social indicators are minimally disclosed according to the GRI G4 guidelines.
4	Economic, environmental, and social indicators are not disclosed.

These samples were then searched for earnings per share (EPS), and the effect on their EPS was observed on each of the disclosure indicators, namely economic, environmental, and social, using the multiple linear regressions. In short, the research model is as figure 1.

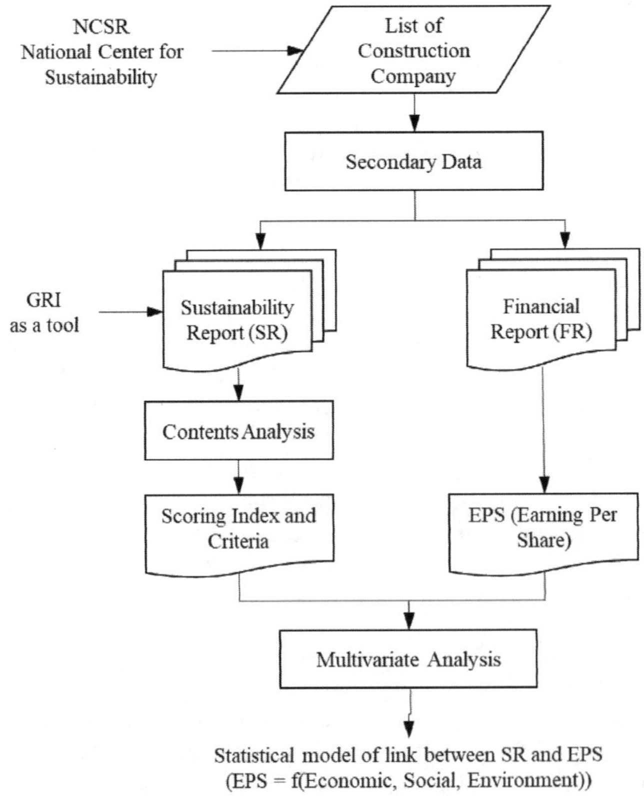

Figure 1. Research Model

5. Results and Discussion

The hypotheses were examined using a multivariate regression analysis in SPSS. Of 30 sample companies processed, four (4) data were outliers, resulting in 26 data to be processed with the following findings: Our statistical model is EPS = -263133.603 + 20597,297 Economic -105889.992 Environmental + 174047.592 Social. The results of hypothesis test indicate H1, disclosure of economic indicators does not affect EPS, H2, disclosure of environmental indicators does not affect EPS, and H3, disclosure of social indicators influences EPS (table 2).

Table 2. Hypothesis test result using multivariate regression analysis

Model	Unstandardized Coefficients		Standardized Coefficients	**t**	**Sig**
	B	**Std Error**	**Beta**		
(Constant)	-263133.603	160234.619		-1.642	0.115
Economic	20597.297	44954.909	0.108	0.458	0.651
Environmental	-105889.992	57928.392	-0.521	-1.828	0.081
Social	174047.592	74865.14	0.671	2.325	0.03

[a]Dependent Variable: EPS

From the three results of hypothesis testing, it appears that only the disclosure of social indicators affects EPS. This finding raises the suspicion that social disclosure is a concern of investors compared to the other two disclosures. In general, the response to growing public awareness of the importance of sustainable development may become the reason for this phenomenon [20]. However, this social disclosure is essential for investors to ensure that the company is free from potential disputes that arise due to social problems that may not be appropriately handled by the company.

Meanwhile, the findings on the disclosure of economic and environmental indicators in the sustainability report, which do not support the hypothesis, are thought to be due to the many factors that cause the opportunity for profit in the form of earnings per share / EPS to be unlikely seen from the disclosure in the sustainability report; instead, it can be obtained directly in the financial statements or annual reports. In this case, the quality of disclosure of economic and environmental indicators in the sustainability report is the least noticed information to determine the opportunities for the profitability of the company.

In addition, the results of simultaneous regression data indicate that the Sig value is at 0.107, which means that, simultaneously, the disclosure of the three indicators, economic, environmental, and social, does not affect EPS. This is also reinforced by the findings from the processed data that the Adjusted R Square value on the model is only at 0.134, indicating that the three disclosure elements only contribute 13.4% to the EPS variable whilst other factors outside the model have 86.6% influence. One of the possible allegations arising from this finding is that sustainability reports remain a complementary source of information because there is a tendency that an increase in reporting does not mean increasing transparency but merely releasing the company from its accountability obligations, instead; consequently, investors do not pay much attention to the quality of disclosure sustainability report [21].

These findings are in line with the previous research, in which sustainability reports are positioned as a tool to drive change and improve competitive advantage even though it remains in the conceptual stage. Specifically in the construction sector, the existence of sustainability reports still leaves challenges around data quality, stakeholder engagement, management reputation, and climate in the construction sector [14].

6. Conclusion and Limitation

Based on the data and hypothesis testing, the quality of disclosure of sustainability reports on sub-economic indicators does not support the proposed hypothesis. This finding is related to the fact that information about the potential amount of earnings per share / EPS cannot merely be accessed through sustainability reports, but instead, it can be seen in financial statements accessed through financial reports, prospectuses, and other information media. The same findings also exist in the environmental sub-indicators, which do not further support the research hypothesis. This is reasonably possible in many companies that still apply the command control approach that is as long as the company complies with the rules, this will not affect its potential profitability. However, on social sub-indicators, the results of this study indicate that the hypothesis is supported. In other words, the quality of social indicators in sustainability reports is positioned as a vital source of information to communicate the company's operational performance and can be a predictor of corporate profit opportunities. This research has specific practical implications for companies related to the need to increase stakeholders' involvement regarding improvements as valuable information for them. The limitation of this study is the non-robustness of the model, which restrains it from being generalized into a broader context.

References

[1] Putri W H and Sari N Y 2019 Eco-efficiency and eco-innovation: strategy to improve sustainable environmental performance *IOP Conference Series: Earth and Environmental Science* **vol 245** p 012049

[2] Bortolini M, Ferrari E, Galizia G, and Mora C 2016 A Rerence Framework Integrating Lean and Green Principles within Supply Chain Mnagement *Int. J. Soc. Behav. Educ. Bus. Ind. Eng.* **10** 876–81

[3] Johansson G and Sundin E 2014 Lean and green product development: Two sides of the same coin? *J. Clean. Prod.* **85** 104–21

[4] Hajmohammad S, Vachon S, Klassen R D, and Gavronski I 2013 Reprint of Lean management and supply management: Their role in green practices and performance *J. Clean. Prod.*

[5] Montabon F, Pagell M, and Wu Z 2016 Making Sustainability Sustainable *J. Supply Chain Manag.* **52** 11–27

[6] Xue B, Liu B, and Sun T 2018 What matters in achieving infrastructure sustainability through project management practices: A preliminary study of critical factors *Sustain.* **10**

[7] Shen L, Wu2 Y and Zhang X 2013 Key Assessment Indicators for the Sustainability of Infrastructure Projects *J. Constr. Eng. Manag.* **142** 1–12

[8] Meng J, Xue B, Liu B, and Fang N 2015 Relationships between top managers' leadership and infrastructure sustainability: A Chinese urbanization perspective *Eng. Constr. Archit. Manag.* **23** 237–60

[9] Martens M L and Carvalho M M 2017 Key factors of sustainability in project management context: A survey exploring the project managers' perspective *Int. J. Proj. Manag.* **35** 1084– 102

[10] Cormier D and Magnan M 2007 The revisited contribution of environmental reporting to investors' valuation of a firm's earnings: An international perspective *Ecol. Econ.* **62** 613–26

[11] Stubbs W, Higgins C, and Milne M 2013 Why do companies not produce sustainability reports? *Bus. Strateg. Environ.* **22** 456–70

[12] Presley A and Meade L 2010 Benchmarking for sustainability: An application to the sustainable construction industry *Benchmarking* **17** 435–51

[13] Ding G K C 2008 Sustainable construction--the role of environmental assessment tools. *J. Environ. Manage.* **86** 451–64

[14] Glass J 2012 The state of sustainability reporting in the construction sector *Smart Sustain. Built Environ.*

[15] Bai C, Satir A, and Sarkis J 2019 Investing in lean manufacturing practices: an environmental and operational perspective *Int. J. Prod. Res.* **57** 1037–51

[16] Renneboog L, Ter H J, and Zhang C 2008 The price of ethics and stakeholder governance: The performance of socially responsible mutual funds *J. Corp. Financ.*

[17] Chatzitheodorou K, Skouloudis A, Evangelinos K, and Nikolaou I 2019 Exploring socially responsible investment perspectives: A literature mapping and an investor classification *Sustain. Prod. Consum.* **19** 117–29

[18] Junkus J and Berry T D 2015 Socially responsible investing: a review of the critical issues *Manag. Financ.*

[19] Hsieh H F and Shannon S E 2005 Three approaches to qualitative content analysis. *Qual. Health Res.* **15** 1277–88

[20] Ehnert I, Parsa S, Roper I, Wagner M, and Muller-Camen M 2016 Reporting on sustainability and HRM: a comparative study of sustainability reporting practices by the world's largest companies *Int. J. Hum. Resour. Manag.* **27** 88–108

[21] Roberts J 2009 No one is perfect: The limits of transparency and an ethic for "intelligent" accountability *Accounting, Organ. Soc.* **34** 957–70

IOP Conf. Series: Materials Science and Engineering **650** (2019) 012018 doi:10.1088/1757-899X/650/1/012018

Factors Affecting Design and Management of Residential Community for Enhancing Well-being of Thai Early Stage Elderly

P Ruengtam[1]

[1]Interior Architecture Program, Faculty of Architecture, Urban Design, and Creative Arts, Mahasarakham University 44150 Thailand

*porntip.r@msu.ac.th

Abstract. This article presents factors related to environmental perception, residential community, and design of the healing environment, which affect Thai early stage elderly perceptions and requirements. Structural equation modeling was used to assess influences among the factors. Empirical data were collected from 419 samples of Thai early stage elderly. The research revealed that requirements of the residential community were classified into three groups including (1) general facilities and activities, (2) facilities related to health and security, and (3) facilities related to physical exercise. Results showed that factors of the elderly's requirements have direct effects on willingness to join in with project activities of the residential community. The result led to formulating policies in design and management of the residential community which enhanced the quality of life both physically and mentally for the Thai early stage elderly.

1. Introduction

The United Nations anticipates that by 2050, the number of elderly people, defined as older than 60 years, will reach 2 billion worldwide. This poses challenges for rich and poor countries alike. Thailand also faces this challenge as its population ages. As a result of modern medical technology and improved public health policy and planning the death rate in Thailand is declining, leading to an aging population. In its annual report for 2012, the Thailand National Committee on the Elderly [1] reported that the proportions of Thai elderly over 60 year in 2020, 2030, and 2040 will be 19.1%, 26.6% and 32.1% respectively. The key question is therefore how Thailand's government can address this scenario. In response, the National Committee on the Elderly under the Ministry of Social Development and Human Security Thailand [2] created a formal structure to define the condition of the elderly in Thailand: Firstly, those elderly people defined as having good living standards: are in good physical and mental health; have a happy family life; have access to social care friendly environment; have security and stability; have access to welfare and social services; lead valuable, independent lives with dignity; participate in the lives of family and community, and continue to follow news and information sources. Secondly, the family and the community are a key source of support for the elderly, acting as an efficient and enabling institution. Thirdly, the system of welfare and services guarantees a higher standard of living allows the elderly to participate fully in family and community life. Fourthly, all involved parties and sectors must work to ensure the accessibility of the welfare and services system so the elderly can use these services in a safe and supported manner with the protection of the system. Lastly, the correct approach shall be implemented in order to recognize the needs of all elderly people, including those facing difficulties, to ensure they are welcomed to participate fully in the community in all areas. This structure leads subsequently to the next issue which is that of where the elderly will live once they have retired and they are in the physical twilight of their lives. At this stage, they require help from their children, or other people, whenever they need to visit a hospital or seek medical assistance. The ideal solution may be the residential community to enhance elderly well-being. This researcher sought to examine a facility which would provide for all the needs of the elderly, which is well-managed, where the elderly receive medical care when they are sick, where they can enjoy a happy social life, where they can be a part of society, and where they can live conveniently and safely for the remainder of their lives. The aim of this study is to determine how such a place should operate. This will first require a study of the needs of the early stage elderly in Thailand in order to apply environmental design and management of residential community principles through the use of structural equation modeling (SEM). To achieve these goals, the study first of all determines the factors comprising the requirements of the elderly in Thailand along with a means of measuring these. Then an SEM is created to show the associations between these various factors along with the willingness of the elderly to join such a residential community. Recommendations can then be made to residential community design and management so that the well-being of the Thai elderly can be enhanced.

2. Research Background

2.1. Elderly

The National Committee on the Elderly under the Ministry of Social Development and Human Security Thailand [2] explained clearly that "the elderly are not a vulnerable nor social burden, but able to take part as the social development resources, so they shall be entitled to recognition and support by the family, community and the state to lead a valuable life with dignity and sustain their healthiness and living standards as long as possible". Meanwhile, the World Health Organization (WHO) [3] explained that throughout the world in general, it is commonly accepted that the elderly can be defined as those people aged 65 or older. However, this is a concept which may apply well in westernized nations but is less appropriate in Africa. Although the choice of number may seem arbitrary, it is also the age at which many countries begin to pay pensions. The UN offers no specific age definition, but has stated that 60 years is the agreed age at which the "older population" begins. The Act on the Elderly B.E. 2546 Thailand [4] sets out three different age groups as below:

- Early stage elderly (60-69 years): people are still able to look after themselves.
- Middle stage elderly (70-79 years): people become sick, weaken, and suffer from chronic disease or poor physical health.
- Late stage elderly (80 and older): people's health deteriorates further, including disability and degeneration.

This paper serves to build upon earlier studies [5], [6] which examined the factors associated with the environmental perceptions of such residential communities and the design of healing environments which are able to positively affect those Thai elderly residents in the early stage who are aged 60-69. This study focuses in particular on the early stage elderly.

2.2. Healing environment

The WHO [7] originally defined health in a manner which did not incorporate the notion of spirituality, but this was adjusted in 1998 to create a definition which covered the four dimensions of physical, mental, and social health, along with spiritual well-being. Meraviglia [8] defined spirituality as the expression and experience of the spirit in a manner which reflects faith in a supreme being or god, along with a feeling of self-connectedness integrating the minds, body and spirit. In elderly adults, this concept can be linked to an improved quality of life [9]. In Thai culture, this idea of spiritual well-being is connected to general respect for the elderly and will also involve Buddhism. It may involve an appreciation of religious practices such as pouring water upon the hands of respected older people, or requests for blessings and participation in other rituals or meditation [10].Barrera-Hernandez et al. [11] added that within the notion of the spiritual environment can be found the links between the physical and intangible elements which form a positive spiritual environment creating human well-being, environmental quality, and sustainable activities. The Cambridge Dictionary [12] gave the definition of a spiritual home as a place where an individual has a sense of belonging, despite not having been born in that place, probably as a result of the connections with the culture, people, and way of life. This definition fits the aims of this research to enhance elderly well-being in the spiritual and cultural environment of Thailand. Jonas and Chez [13] took the view that the future of medical management in the context of chronic diseases will lie in a focus on healing if a sustainable health care approach is to be employed. Healing can be defined as an ongoing process of repair and recovery, and forms the foundation of a new medical vision which brings together diverse techniques from all over the world to relieve suffering, treat chronic conditions, and improve well-being. Healing requires the creation of suitable attitudes and intentions on the part of both the recipient and the provider of the care, along with the use of self-care activities in the part of the elderly. The aim of to form healing relationships, making use of a deeper understanding of health promotion and maintenance, and suitable ways to effectively combine complementary and conventional medical techniques. Nelson et al. [14] explained that the "healing environment" can be considered similar to the therapeutic environment, which is typically designed to make use of the latest medical advances to provide high-quality patient care in a safe and secure setting, as well as inviting the family of the patient to collaborate in creating a psychosocially beneficial environment. From this description, it might be concluded that a spiritual environment enhancing the well-being of the Thai early stage elderly might incorporate many aspects of the therapeutic or healing environment. These concepts should be compared and further analyzed using the literature in the Thai context, and the findings can be incorporated in the design of the research questionnaire.

2.3. Environmental design
Plunz [15] described environmental design as a system of taking into consideration the surrounding environmental parameters as plans and policies are formulated to develop buildings, or products. Environmental design also plays a part in the arts and sciences when considering the human-designed environment. Such fields would include urban planning, interior design, and landscape architecture, as well as architecture and geography. Environmental design involves relating the physical surroundings to the needs of human activity, and therefore this context would include parks and buildings on a smaller scale, and whole community spaces on a larger scale. Environmental design can also be defined as the physical and constructed environment where people live their daily lives. It takes into account the experiences of the users of the designed space as well as their aesthetic perceptions of the environment created. In the context of this study, environmental design must be researched in order to meet the needs of the early stage elderly in Thailand.

2.4. Residential community
Paul et al. [16] explained that any community can be described as a social unit located in a particular area and sharing common values. It can be of any size, but will typically comprise a group of people who are connected for reasons other than family ties, and who value those connections in a social context. The WHO Regional Office for Europe [17] gave the definition of a community residential health facility as a non-hospital, community-based mental health facility which offers full-time accommodation and care for those personas who suffer mental health problems. Such facilities might include supervised housing in group homes without staff, group homes which have a number of residential staff or visiting staff, hotels which provide staff by day and night, homes or hostels which offer nursing staff full-time, or simply therapeutic communities or halfway houses. These facilities would include both public and private institutions, and some would seek profits while others would not. Perkins et al. [18] developed a textbook covering the main considerations in building design for elderly

people, and noted that there are a number of features which must be offered in skilled-nursing facilities for the elderly. These include a multipurpose room, library, coffee shop or snack bar, gift shop, outdoor seating areas, recreation facilities, art or activity rooms, clinics, and rehabilitation centers. The book added that where communities are designed for senior adults, it is important to create landscaped areas for walking, while facilities for leisure activities such as golf, lawn sports, fishing, gardening and so forth should be offered.

3. Research Method

3.1. Factor identification and questionnaire design

The related factors and their indicators were identified. A literature review was undertaken of related theories and concepts from many sources such as textbooks, research articles, on-line databases and annual reports. All factors were summarized and listed to compare similarities and differences in meanings of their components. The listed factors were saturated and categorized by the researcher, and constructed and prepared for analysis in the next section. A questionnaire survey was designed for data collection to confirm the factors or items of requirements in the residential community. The subjects comprised a group of Thai early stage elderly. The questionnaire was designed in three parts by referring to the listed factors. The first part requested general information of the respondents including gender, marital status, age, health condition, education, and economic status (6 questions). The second part concerned their expected requirement items in the residential community (30 questions), while the last part identified the elderly's willingness to participate in residential community activities (3 questions). The first part was measured by frequency (percentage) of respondents, while the second and third parts were assessed by a 5-level Likert scale from 'strongly disagree' to 'strongly agree'. The questionnaire items are listed in Table 1.

Table 1(a). Questionnaire items

Factor		Item
Elderly's requirements	Location	Q1: Calm and natural environment
		Q2: Near religious places
		Q3: Near hospitals
	Facility	Q4: Health food shop
		Q5: Convenience shop
		Q6: Beauty salon
		Q7: Laundry service
		Q8: Cleaning service
		Q9: Building maintenance service
		Q10: 24-hour security guards
		Q11: 24-hour medical center
		Q12: Care center for the elderly
		Q13: Sauna and spa
		Q14: Training center for improving quality of life
		Q15: Library
		Q16: Computer and internet room
		Q17: Karaoke lounge
		Q18: Swimming pool
		Q19: Fitness
		Q20: Outdoor stadium
		Q21: Garden and outdoor patio
		Q22: Indoor activities
		Q23: Religious place
		Q24: Sidewalks and bike lanes

IOP Conf. Series: Materials Science and Engineering **650** (2019) 012018 doi:10.1088/1757-899X/650/1/012018

Table 1(b). Questionnaire items

Factor		Item
		Q25: Fence and 24-hour gate guard
	Activity	Q26: Religious activity Q27: Recreational activities Q28: Important day activities Q29: Training activities Q30: Excursions Programs
Willingness		W1: Interesting in the community W2: Willingness to live in the community W3: Willingness to recommend the community to others

3.2. Validity and reliability

To ensure that the research items were appropriate for the elderly, interviews were held with five experts who had relevant experience in elderly behaviors and residential communities. The experts reviewed and suggested whether the items were accurate representations to measure the research model. They also suggested some items which were more appropriate in the context of the research. This exercise was useful to provide content validity and ensure that the items were neither ambiguous nor confusing. Cronbach's alpha was used to evaluate the reliability of the questionnaire. A pilot study was conducted on 30 target elderly to assess reliability. The coefficients for location, activity, and facility were 0.942, 0.953, and 0.850 respectively. The value for all items (Q1-Q30) was 0.964. All coefficients were above 0.7 and demonstrated that the questionnaire was reliable [19].

3.3. Data collection

Once the questionnaire was designed, a target group of Thai early stage elderly (age of 60 to 69 years) was selected using a convenient non-probability sampling technique. Almost all of the respondents lived in Bangkok while some resided in large cities surrounding the Bangkok area. The survey period was three months. Face-to-face interviews were conducted to explain the details of the questionnaire and ensure that the respondents understood the purpose of the survey. In total, 500 questionnaires were completed, with 81 rejected due to incomplete and bias responses. As such, 419 data sets were accepted as valid and used for analysis in the next section.

3.4. Exploratory factor analysis

Exploratory factor analysis (EFA) with varimax rotation was implemented to categorize the 30-items (Q1-Q30) and determine the underlying factor structure construct of the environmental design and management of the residential community. The Kaiser-Meyer-Olkin (KMO) measure of sampling adequacy was .97 (KMO>.7) [20], Bartlett's test of sphericity had a significant value of .000 (less than .05), approx. Chi-Square was 11805.305, and df was 435. Factor loading values of less than 0.5 were eliminated. The EFA output showed that three factors of the 30 items (Q1-Q30) were grouped as factor 1: Q2, Q4, Q6, Q7, Q8, Q9, Q13, Q14, Q15, Q16, Q17, Q23, Q26, Q28, Q29, and Q30; factor 2: Q1, Q3, Q5, Q10, Q11, Q12, and Q25; and factor 3: Q18, Q19, Q20, Q21, Q22, Q24 and Q27. Furthermore, the reliability of the questionnaire in terms of the three factors was assessed on the basis of Cronbach's alpha coefficient. According to Nunnally [19], Cronbach's alpha coefficients of 0.7 or higher are recognized as acceptable values. Thus, values of the coefficient were acceptable for all three factors, ranging 0.923, 0.939 and 0.962. The Cronbach's alpha value of all items (Q1-Q30) was 0.973. Outputs for all the questionnaire items are shown in Table 2.

Table 2. Factor loading of the EFA and Cronbach's alpha

Item	Factor loading 1	2	3	Cronbach's alpha	
Q28	.772	.213	.218	0.962 (activities and general facilities)	0.973 (Requirements)
Q29	.754	.157	.343		
Q14	.753	.156	.325		
Q15	.738	.193	.153		
Q2	.717	.426	.075		
Q30	.716	.120	.414		
Q13	.711	.207	.316		
Q26	.711	.346	.170		
Q23	.690	.401	.112		
Q4	.682	.308	.323		
Q16	.676	.119	.536		
Q9	.672	.326	.305		
Q17	.654	.079	.559		
Q7	.625	.319	.430		
Q8	.615	.324	.276		
Q6	.600	.209	.515		
Q11	.193	.845	.149	0.923 (health and security)	
Q10	.150	.831	.149		
Q3	.246	.735	.347		
Q12	.379	.721	.095		
Q25	.131	.719	.409		
Q1	.344	.673	.327		
Q5	.450	.553	.408		
Q19	.221	.287	.810	0.939 (exercises)	
Q18	.377	.204	.783		
Q20	.383	.368	.664		
Q22	.362	.483	.613		
Q27	.395	.445	.599		
Q21	.375	.496	.583		
Q24	.303	.531	.576		

In Table 2, EFA results indicated that requirements in the residential community of the Thai early stage elderly could be classified as three factors. These three factors were named by considering the majority of the included items as factor 1 'activities and general facilities', factor 2 'health and security', and factor 3 'exercises'.

3.5. Research hypotheses

To better understand the elderly's expected requirements in their residential community, research hypotheses were formulated. Once EFA had been performed, the three factors of the elderly's requirements in the residential community including 'activities and general facilities', 'exercises', and 'health and security' were treated as independent variables [21] which had direct effects on the dependent variable as the elderly's willingness to join in with the activities has of the residential community. This concept was presented as the conceptual research model (Figure 1). Three research hypotheses were developed as follows:

H1: The factor of activities and general facilities has an effect on the elderly's willingness to join in with the activities of the residential community.
H2: The factor of health and security has an effect on the elderly's willingness to join in with the activities of the residential community.
H3: The factor of exercises has an effect on the elderly's willingness to join in with the activities of the residential community.

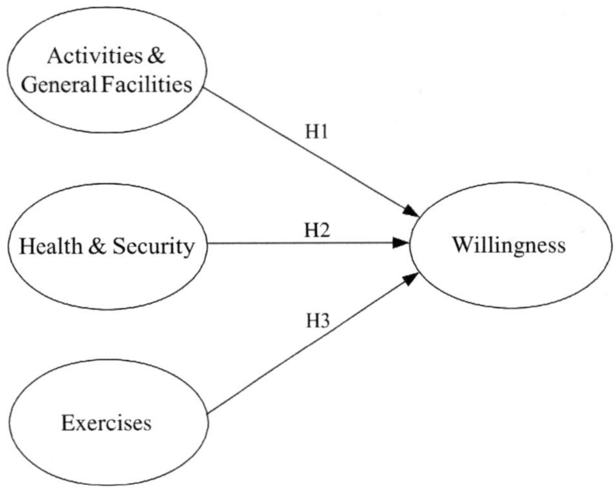

Figure 1. Conceptual Research Model

4. Analysis and results

4.1. Demographic information

Data profiles of the 419 Thai early stage elderly respondents were analyzed in terms of demographics as shown below in Table 3.

Table 3(a). Demographic information of the respondents

Description	Frequency	Percentage
Gender		
-Male	851	73.3
-Female	168	61.7
Marital status		
- Single	44	10.5
- Married	263	62.8
- Widowed	81	19.3
- Divorced	31	7.4
Age		
-60 -64 yrs.	265	67.1
-65 - 69 yrs.	154	76.1
Health condition		
-Strong, can take care of everything themselves	147	51.8
-Occasionally need some assistance	154	361.
-Mainly need assistance	81	4.7
-Need assistance all the time	4	8.8

Table 3(b). Demographic information of the respondents

Description	Frequency	Percentage
Education level		
-No education	86	7.1
-Primary school	75	1.4
-High school	112	78.1
-Bachelor degree	286	58.6
-Higher bachelor degree	17	5.5
Economic status (last income, before retirement)		
-Less than 10,000 Baht/month	56	87.4
-10,000 to 30,000 Baht/month	827	46.8
-30,001 to 50,000 Baht/month	35	17.9
-50,001 to 100,000 Baht/month	37	83.4
-More than 100,000 Baht/month	11	5.7

4.2. Structural equation modeling

Structural equation modeling (SEM) was used for data analysis. Byrne [22] stated that SEM can explain influences or effects between latent variables (factors) on other latent variables (factors) in the model. Here, once the EFA had been performed, the latent variable named 'willingness' with its three observed variables; W1 (interest in the community), W2 (willingness to live in the community) and W3 (willingness to recommend the community to others) was added to the model. The first output of the model analysis did not fit. The software suggested that some variables (items) should be deleted from the model and some covariance relations should be added. These suggestions were implemented. Then, the model was reanalyzed. The final output showed model fit by Chi-square = 57.627, df = 48, p = .161 (>.05), CMIN/DF = 1.201 (<3.5), GFI = .977 (>.9), RMSEA = .022 (<.08) [23] as shown in Figure 2.

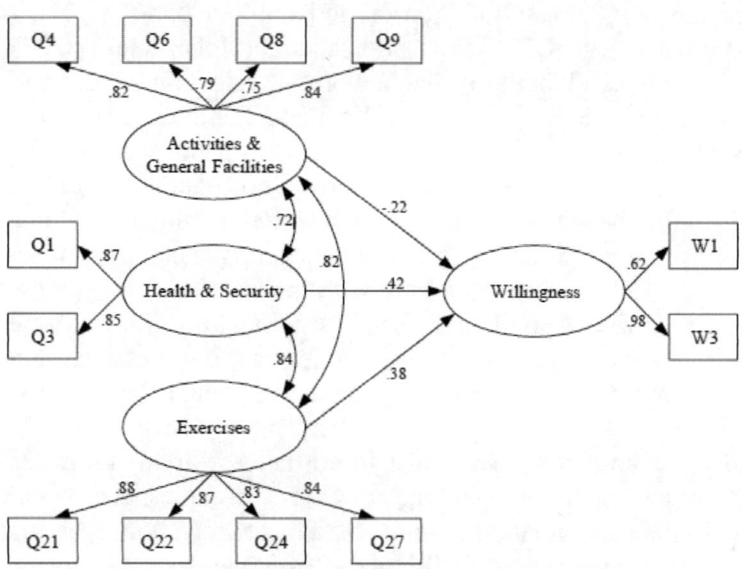

Chi-Square = 57.627, df = 48, p = .161 (>.05), CMIN/DF = 1.201 (<3.5),
GFI = .977 (>.90), RMSEA = .022 (<.08)

Figure 2. Structural equation model

The research hypotheses were tested using the outputs of the model as implemented above. Table 4 presents the test results of each factor and variables with significant effects (*p*-values < .05) between each other.

IOP Conf. Series: Materials Science and Engineering **650** (2019) 012018 doi:10.1088/1757-899X/650/1/012018

Table 4. Results of the tested hypotheses

Hypothesis	Relationship	Standardized Path Coefficient	Result	Significant (p)
H1	Activities and General facilities →Willingness	-.22	Supported	.019
H2	Health and Security →Willingness	.43	Supported	***
H3	Exercises →Willingness	.38	Supported	.005

Note: *** = $p < .001$

The SEM results in Figure 2 and Table 4 showed path coefficients in the standardized estimate of regression weight. All hypotheses (H1, H2, and H3) were significantly supported at the .05 level.

5. Conclusions

Factors related to environmental perception, residential community, and design of the healing environment which have effects on Thai early stage elderly requirements were studied. All of the elderly in this study resided in metropolitan areas. Profiles of the respondents are shown in Table 3. The requirements of the elderly were classified as three factors including ‘activities and general facilities’, ‘exercises’, and ‘health and security’. The factor of ‘activities and general facilities’ comprised four variables including the requirements of ‘building maintenance service’ (Q9), ‘health food shop’ (Q4), ‘beauty salon’ (Q6), and ‘cleaning service’ (Q8). The factor of ‘health and security’ comprised two variables as the requirements of ‘calm and natural environment’ (Q1), and ‘near hospitals’ (Q3). For the factor of ‘exercises’, four variables included the requirements of ‘garden and outdoor patio’ (Q21), ‘indoor activities’ (Q22), ‘recreational activities’ (Q27), and ‘sidewalks and bike lanes’ (Q24). In Figure 2, the factor of ‘activities and general facilities’ had a negative effect on “willingness”, thus, the Thai early stage elderly who required ‘activities and general facilities’ were not willing to join in with the activities of the residential community. The ‘health and security’ factor had a positive effect on ‘willingness’, and Thai early stage elderly who required ‘health and security’ were willing to join in with the activities of the residential community. Finally, the factor of ‘exercises’ had a positive effect on ‘willingness’, meaning that Thai early stage elderly who required ‘exercises’ were also willing to join in with the activities of the residential community. From the results, Thai early stage elderly with strong health do not require all facilities and activities in the residential community. The most important requirements in terms of ‘activities and general facilities’ were ‘building maintenance service’ and ‘health food shop’. The most important requirements in terms of the ‘health and security’ were ‘calm and natural environment’ and ‘near hospitals’. The most important requirements in terms of ‘exercises’ were ‘garden and outdoor patio’ and ‘indoor activities’. In the aspect of willingness to join in with the activities of the residential community, the factor which had the highest effect on the elderly was not ‘activities and general facilities’ or ‘exercises’ but ‘health and security’.

6. Acknowledgement

This research was supported funding by the research project grant provided by the National Research Council Thailand through the Mahasarakham University, Thailand (No. 5805042/2558).

References

[1] The Thailand Nation Elderly Committee 2012 *The Yearly Report of Thai Elderly Situation 2012* (Bangkok: The Ministry of Social Development and Human Security)

[2] The National Committee on the Elderly Thailand 2009 *The 2nd National Plan on the Elderly (2002-2021)* (Bangkok: The Ministry of Social Development and Human Security Thailand)

[3] World Health Organization 2019 *Definition of an older or elderly person* [cited 2019 Jan 15] Available from http://www.who.int/healthinfo/survey/ageingdefnolder/en/

[4] Department of Older Persons 2003 *The Act on the Elderly B E 2546 (2003 A D)* (Bangkok: Department of Older Persons, the Ministry of Social Development and Human Security Thailand)

[5] Ruengtam P 2016 Modelling Design and Management of Residential Community for Enhancing Elderly Well-Being in Thai Cultural and Spiritual Environments *Journal of Environmental Design and Planning Nakhara* Vol **12** pp 1-12

[6] Ruengtam P 2017 Factor Analysis of Built Environment Design and Management of Residential Communities for Enhancing Elderly Well-Being Elderly People *Journal of Procedia Engineering* 180 pp 966-974

[7] World Health Organization 1946 *Constitution of the World Health Organization* (Article80) (Geneva: WHO)

[8] Meraviglia M 1999 Critical Analysis of Spirituality and Its Empirical Indicators: Prayer and Meaning in Life *Journal of Holistic Nursing* Vol **17** No 1 March 1999 pp 18-33

[9] Gaskamp C, Sutter, R, and Mervigilia M 2006 Evidence-based guideline promoting spirituality in the older adult *Journal of Gerontological Nursing* **32** (11) pp 8-13

[10] Suchada T and Sawalai B 2010 *Value of Elderly from the Eyes of Thai Society* 1st ed (Nakhonpathom: Institute for Population and Social Research, Mahidol University)

[11] Barrera-Hernandez L F, Corral-Verdugo V, Tapia-Fonllem C O and Fraijo-Sing B S 2015 Psychological Meanings of "Positive Spiritual Environment" and "Place to Communicate with God" *European Scientific Journal* vol **11** No 26 pp 357-369

[12] Cambridge dictionary 2019 *Meaning of spiritual home* [cited 2019 Jan 15] Available from http://dictionary.cambridge.org/dictionary/english/spiritual-home

[13] Jonas W B and Chez R A 2004 Toward Optimal Healing Environments in Health Care *The Journal of Alternative and Complementary Medicine* Vol **10** Supplement 1 pp S-1–S-6

[14] Nelson C, West T and Goodman C 2005 *The hospital built environment: what role might funders of health services research play?* (Rockville MD: Agency for Healthcare Research and Quality 2005 Aug Contract No 290-04-0011 AHRQ Publication No 06-0106-EF)

[15] Plunz R 1982 *Design and the Public Good* (Cambridge: MIT Press)

[16] Paul J, Nadarajah, Yaso, Karen H and Victoria S 2012 *Sustainable Communities, Sustainable Development: Other Paths for Papua New Guinea* (Honolulu: University of Hawaii Press) p 14

[17] WHO Regional Office for Europe 2008 *Policies and Practices for Mental Health in Europe* (Copenhagen: WHO) p 71

[18] Perkins B, Hoglund J D, King D and Cohen E 2004 *Building Type Basics for Senior Living* (New Jersey: John Wiley & Sons, Inc.)

[19] Nunnally J C 1978 Psychometric Theory 2nd ed (New York: McGraw-Hill Inc.)

[20] Tabachnik B G and Fidel L S 2001 *Using multivariate statistics* 4th ed (Massachusetts: Allyn & Bacon)

[21] Xiong B, Skitmore M and Xia B 2015 A critical review of structural equation modeling applications in construction research *Automation in Construction* **49** pp 59-70

[22] Byrne B M 2010 *Structural Equation Modeling with AMOS, basic concepts, applications, and programming* (New York: Routledge Taylor & Francis Group)

[23] Arbuckle J L 2011 IBM SPSS Amos 20 User's (Crawfordville: Guide IBM Corporation)

IOP Conf. Series: Materials Science and Engineering **650** (2019) 012019 doi:10.1088/1757-899X/650/1/012019

The effect of heavy equipment management on the performance of the construction project and the construction company

A S Nugraha[1], L S Putranto[1]

[1]Tarumanagara University, Department of Civil Engineering Jakarta 11440, Indonesia

* anggriis@yahoo.com; lexy_putranto@yahoo.co.id

Abstract. The appropriate use of heavy equipment will affect the progress of a construction project. An inappropriate choice of heavy equipment will result in low productivity and at the end will incur additional costs. Previous studies have been observing separately the effect of heavy equipment management on the performance of the construction project and the effect of heavy equipment of the performance of the construction company. In this present paper, using structural equation modelling (SEM), both relationships will be studied simultaneously and additionally direct effect of performance of construction project on the performance of construction company will be observed as well. It was found that both relationships were significant and positive.

1. Introduction

According to Tatari and Skibniewski [1], construction equipment management is associated with procurement, retirement, replacement, operation, logistics, and maintenance of the equipment. The business objective of heavy equipment company is to minimize operation, maintenance and repair costs whilst keep maximizing the utilization of the equipment. These responsibilities can be classified into two groups, i.e. operational and strategic. These responsibilities are held by the project manager to determine the equipment requirement within a specific time period.

In Canada, according to Lucko [2], Fan et al [3], and Hendi [4] maintenance, operation, repairment, and replacements were factors affecting heavy equipment in construction. Gunawardena [5] conducted a study on 22 heavy equipment companies in Sri Lanka to find factors affecting heavy equipment management. The result was equipment choice, equipment finance, equipment maintenance equipment replacement and equipment record. In India Anbhule and Kumthekar [6] developed a design concept of equipment system for toll road construction project. On their system, appropriate design, equipment choice, equipment, installation, operation, and replacement became important factors in construction heavy equipment management.

Performance of construction project can be affected by the characteristics to the stakeholders such as owner, contractor and project managers. These characteristics usually consist of several factors, measurements, and indicators to describe whether construction project performance is going in the right direction. Besides costs, quality and time, according to Yeung et al. [7], Toor & Ogunlana [8], Cha & Kim [9] LaBarre & El-adawaty [10] health and safety of construction workers were also factors affecting the performance of the construction projects. Additionally, according to

Chan & Chan [11], Cheung et al. [12], Ugwu & Haupt [13] and Cha & Kim [9], environment and safety and finally according to Chan & Chan [11], Toor & Ogunlana [8] and Yeung [7] stakeholders satisfactions became important factors to measure the performance of construction projects.

Larsspm et a; [14] examined the importance of hard project management and team motivation in process performance in construction. Hard project management is based on rigorous planning and control. An analysis using structural equation model was conducted, with and without a mediating link between hard project management and process performance, based on data from a survey of 2,175 respondents, representing contractors and clients from 109 Swedish construction projects. The results confirm that hard project management is best executed through teams to improve process performance. "Path analysis," using the model with the mediating link, verifies that neglecting team motivation can undoubtedly weaken process performance.

This present paper was prepared in order to revealed whether in the Greater Jakarta there were relationships between:

1. The heavy equipment management and the performance of the construction project.
2. The performance of the construction project and the performance of the construction company.
3. The heavy equipment management and the performance of the construction company.
   Additionally, we also wanted to reveal whether the performance of the construction project can mediate between the heavy equipment management and the performance of the construction company.

2. Method

The current research was conducted in the Greater Jakarta. We interviewed 104 respondents, consist of project managers, general managers, site managers, equipment managers and competent engineers who were knowledgeable on heavy equipment management. The questionnaires to be used to assist the interview process were consist of 4 parts, the general data, the perception of respondents regarding the level of importance of heavy equipment management, the perception of respondents regarding the importance of factors affecting the performance construction project and finally, the perception of respondents regarding the importance of factors affecting the performance of the construction company. The general data consist of the name, the age, the position in the construction company, type of construction company (private, state-owned, foreign and others), employment length and education attainment of the respondents.

Questions regarding the level of importance of heavy equipment management consist of four sub- parts, i.e.:

X.1. The choice and ownership of heavy equipment:

1. Choice of the type of heavy equipment used based on the type of project.
2. The plan for heavy equipment choice.
3. The plan of the choice of usage duration of heavy equipment.
4. The plan of the choice of usage of heavy equipment type.
5. The plan of the choice of the number of heavy equipment used.
6. The choice of method of heavy equipment ownership (buy, rent, lease).
7. The project location (seaside, plateau, etc).
8. The field condition at the project.
9. The limitation of heavy equipment choice (traffic regulation, cost, and demolition).
10. The choice of heavy equipment vendor.
11. The final evaluation of heavy equipment choice.

X.2. The maintenance and repair of heavy equipment:

1. The pre-determined schedule of heavy equipment maintenance.
2. The preparation of all required resources to conduct efficient maintenance
3. The maintenance strategy (preventive steps at a heavy equipment maintenance plan).

4. The cost estimate of heavy equipment maintenance (additional costs, additional technicians, etc.).
5. The necessary repair during heavy equipment operation.
6. The experience of technician conducting repair and maintenance.
7. The communication between the technicians, the planners' group and the strategic managers.

X.3. The operation of heavy equipment:

1. The use of heavy equipment according to the project characteristic.
2. The availability of the heavy equipment for the operation
3. Determination of the number of hours of daily use of heavy equipment.
4. The working method of the heavy equipment used.
5. The skill of a heavy equipment operator.
6. The appropriate recruitment of heavy equipment operator.
7. The development of heavy equipment operator through training.
8. The communications between heavy equipment operators

X.4. The Replacement of heavy equipment:

1. The evaluation of overall owned heavy equipment.
2. The justification of the replacement of heavy equipment (inefficient heavy equipment, economical age of heavy equipment has been last.
3. The budget preparation for heavy equipment replacement.
4. The managerial planning for unused heavy equipment.
5. The replacement of obsolete heavy equipment types.

Questions regarding the level of importance of the performance of a construction project:

Y.1.1. Cost factors:

1. The cost control of every construction project which uses heavy equipment.
2. The manpower cost related to heavy equipment.
3. The operation cost related to the heavy equipment (fuel, lubricant, tires, etc.)
4. The cost control towards contingency cost (heavy equipment

Y.1.2. Quality factors:

1. The achievement of the working quality of the heavy equipment used.
2. The number of errors conducted during the heavy equipment operation.
3. The routine heavy equipment quality control.
4. The number of reworks.

Y.1.3. Time factors:

1. The schedule control for each construction work using heavy equipment.
2. The punctuality of the heavy equipment delivery time to the project.
3. The time lost due to a temporary break (heavy equipment repair, accident, etc.)
4. The bad plan of the working schedule.

Y.1.4. The health and safety factors:

1. The number of accidents happens.
2. The number of fatal accidents involving workers.
3. The number of light injury in the accident involving the workers.
4. The number of serious injury in the accident involving the workers.
5. The costs of the accident in the project.
6. The time lost due to the accident.

Y.1.5. The environment factors:

1. The project does not cause air pollution surrounding the project.
2. The project does not cause water pollution surrounding the project.
3. The project does not cause dangerous waste for the environment.
4. The sound level due to the heavy equipment.
5. The project does not cause damage due to the heavy equipment (access road, effects of heavy equipment vibration.

IOP Conf. Series: Materials Science and Engineering **650** (2019) 012019 doi:10.1088/1757-899X/650/1/012019

Y.1.6. Stakeholders satisfaction factors:
1. The owner satisfaction of the construction project.
2. The consultant satisfaction of the construction project.
3. The contractor satisfaction of the construction project.
4. The supplier satisfaction of the construction project.
5. The supervisors' satisfaction with the construction project.

Questions regarding the level of importance of the performance of a construction company:

Y.2.1. Client factors:
1. The number of clients using the service from the construction company.
2. The client satisfaction affecting the construction company performance.
3. The involvement of the clients to the construction company.

Y.2.2. Financial factors:
1. The income of the construction company.
2. The profit of the construction company.
3. The number of assets owned by the construction company.
4. The small and appropriate ratio of company total debt to the income.
5. The policy of the leader of the construction company in managing expenses.

Y.2.3. Manpower factors:
1. The knowledge development through the training to the company employees.
2. The development of technological knowledge to the company employees.
3. The competence of the company employees in understanding heavy equipment management.
4. The attitude of the company employees in servicing the company clients.

Y.2.4. The company internal condition factors:
1. The number of employees in the company.
2. The growth of the number of employees in the company.
3. The increase of the employee incomes in the company.
4. Employee satisfaction in the company.

Y.2.5. The company competitiveness factors:
1. The image and credibility of the company in the eyes of the clients.
2. The innovation or technology advancement used by the company.
3. The appropriate company marketing.
4. The appropriate social relation with the stakeholders.

The collected data were then analyzed using structural equation model.

3. Data Summary

Almost all of the mean of each observed indicators was above 3.00. The only exceptions were Y6.4 (2.94) and Y6.5 (2.88). Even these mean values were still higher than 2.5 (the departure from not important to important). Therefore, in general, in the perception of the respondents, all of the observed indicators were important. The followings were the data summary:

- Mean values of X1 was 3.51 and ranged between 3.03 and 3.89.
- Mean values of X2 was 3.32 and ranged between 3.04 and 3.53.
- Mean values of X3 was 3.34 and ranged between 3.14 and 3.71.
- Mean values of X4 was 3.13 and ranged between 3.03 and 3.28.
- Mean values of Y11 was 3.35 and ranged between 3.17 and 3.61.
- Mean values of Y12 was 3.18 and ranged between 3.06 and 3.33.
- Mean values of Y13 was 3.22 and ranged between 3.11 and 3.27.
- Mean values of Y14 was 3.16 and ranged between 3.09 and 3.34.
- Mean values of Y15 was 3.40 and ranged between 3.09 and 3.70.
- Mean values of Y16 was 3.16 and ranged between 2.88 and 3.51.
- Mean values of Y21 was 3.46 and ranged between 3.32 and 3.58.

IOP Conf. Series: Materials Science and Engineering **650** (2019) 012019 doi:10.1088/1757-899X/650/1/012019

- Mean values of Y22 was 3.43 and ranged between 3.12 and 3.79.
- Mean values of Y23 was 3.25 and ranged between 3.05 and 3.68.
- Mean values of Y24 was 3.30 and ranged between 3.01 and 3.62.
- Mean values of Y25 was 3.32 and ranged between 3.01 and 3.75.

4. Results and Discussions

The estimation result of the model is shown in Figure 1. It can be seen that at least there were 2 criteria for model fitness achieved. The χ^2=72.22 (more than 0.05) and root mean square error of Approximation (RMSEA)<0.001 (less than 0.05). This implies that the empirical data fit with the model. All standardized loading factors (SLF) were more than 0.7. This implies that each indicator could reflect the relevant constructs. All of the construct reliabilities (CR) were more than 0.7 reflecting good reliability. All of the average variance extracted (AVE) were more than 0.5, showing adequate convergence. The structural model consists of 2 equations:

$$Y1=0.7545\ X$$
$$Y2=0.5949\ X + 0.3618\ X$$

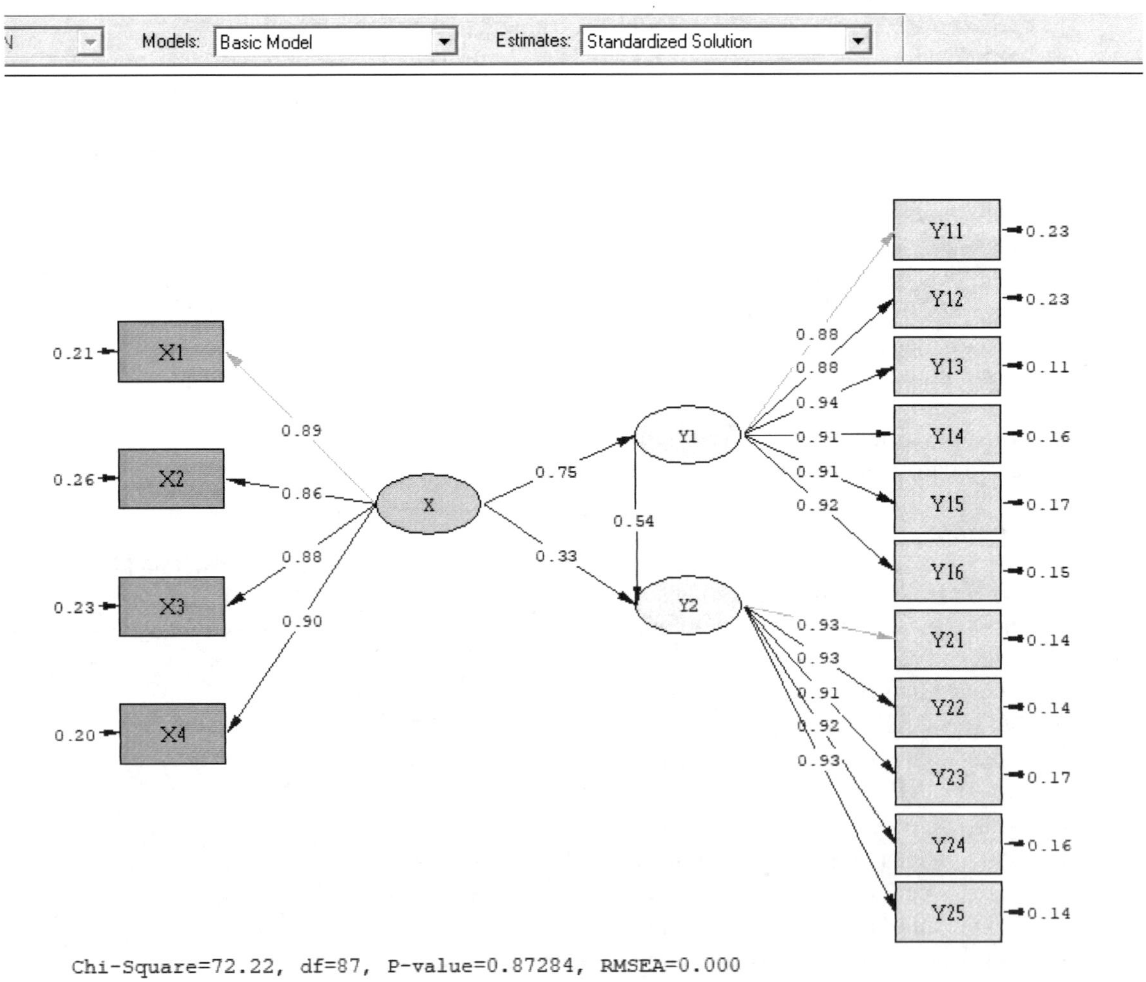

Figure 1. The structural equation model

IOP Conf. Series: Materials Science and Engineering **650** (2019) 012019 doi:10.1088/1757-899X/650/1/012019

5. Conclusions and Recommendations

Based on the analysis, it was found that:

- The heavy equipment management (X) was affecting significantly and positively to the performance of the construction project (Y1).
- The performance of the construction project (Y1) was affecting significantly and positively to the performance of the construction company (Y2).
- The heavy equipment management (X) was affecting significantly and positively to the performance of the construction company (Y2).
- The performance of the construction project (Y1) was mediating significantly and positively between the heavy equipment management (X) and the performance of the construction company.

Based on the results, we suggest that the construction company in the Greater Jakarta should pay sufficient attention to the heavy equipment management as it was proved that, it was a significantly and positively related with heavy equipment management, both directly and through the performance of the construction project.

References

[1] Tatari O & Skibniewski M 2006 Integrated agent-based construction equipment management: Conceptual design *Journal of Civil Engineering &Management* **12(3)** 231–6.

[2] Lucko G 2003 A statistical analysis and model of the residual value of different types of heavy construction Equipment. *Http://Scholar.Lib.Vt.Edu/Theses/Available/Etd-12032003-122642.*

[3] Fan H, AbouRizk S, and Kim H 2005 Building intelligent applications for construction equipment Management *Proceeding of the ASCE International Conference on Computing in Civil Engineering, Cancun, Mexico, 12-15 July 2005* 192-9.

[4] Hendi A 2007 Decision support system for equipment selection in construction projects *Master Thesis, University of Alberta, Edmontn, Alberta, Canada.*

[5] Gunawardena N D 1992 Construction equipment-management practices *Annual Transaction of IESL* 58-72.

[6] Anbhule Y R & Kumthekar M B 2013 3D equipment management system for highway construction projects: conceptual design IOSR Journal of Mechanical and Civil Engineering, **1(1)** 103.

[7] Yeung J Chan, A Chan D Chiang, Y & Yang H 2013 Developing a benchmarking model for construction projects in Hong Kong *Journal of Construction Engineering and Management* **139(6)** 705-16.

[8] Toor S & Ogunlana S O 2010. Beyond the 'iron triangle': Stakeholder perception of key performance indicators (KPIs) for large-scale public sector development projects. *International Journal of Project Management* **28(3)** 228–36.

[9] Cha H S & Kim C K 2011 Quantitative approach for project performance measurement on building construction in South Korea *KSCE Journal of Civil Engineering* **15** 1319–28.

[10] Labarre P S & El-adaway I H 2014 Project benchmarking : tool for mitigating conflicts, claims, and disputes through improved performance *Journal of Legal Affair & Dispute Resolution in Engineering Construction* **6(1)** 1–7.

[11] Chan A P C & Chan A P L 2004 Key performance indicators for measuring construction Success *Benchmarking* **11(2)** 203-21.

[12] Cheung S O Suen H C H & Cheung K K W 2004 PPMS: a web-based construction project performance monitoring system *Automation in Construction* **13** 361–76.

[13] Ugwu O O & Haupt T C 2007 Key performance indicators and assessment methods for infrastructure sustainability-a South African construction industry perspective *Building & Environment* **42** 665–80.

IOP Conf. Series: Materials Science and Engineering **650** (2019) 012019 doi:10.1088/1757-899X/650/1/012019

[14] Larsson J Eriksson P E & Pesämaa O 2018 The importance of hard project management and team motivation for construction project performance *International Journal of Managing Projects in Business* **11(2)** 275-88.

Hendi, A. (2007). *Decision support system for equipment selection in construction projects,*

IOP Conf. Series: Materials Science and Engineering **650** (2019) 012020 doi:10.1088/1757-899X/650/1/012020

Improving measureable external factors influence on construction project duration estimation

Anondho B[1], Latief Y[2] and Mochtar K[3]

[1]Lecturer at Civil Engineering Universitas Tarumanagara, Jakarta, Indonesia
[2]Profesor at Civil Engineering Department Universitas Indonesia. Depok, Indonesia
[3]Profesor at Civil Engineering Department Institut Teknologi Indonesia, Serpong, Indonesia

Abstract. The purpose of this study was to determine whether the use of building function variables as dummy variables affected the duration of construction projects, together with identified measureable external factors. Multivariate regression is a technique that estimates a single regression model with more than one outcome variable. When there is more than one predictor variable in a multivariate regression model, the model is a multivariate multiple regression. Goodness of Fit (R^2) of the equation regression analysist used as sensibility benchmarks. Variables are transformed to index number for unit synchronize. If the result is unsatisfied for predicting a project duration, an implementation of dummy variables could be tried if there any improvement of R^2. The trial on this research, 43 data show that in first stage, five influence factors: Education Index, Technology Absorption Index, Labor Experience Index, New Technology Availability Index, and Innovation Index generate regression equation without full fill the mathematical requirement. At the second stage, by adding dummy variables based on type of building, apartment, office, hotel and other then the three type together with the five external factors, afford a good R^2.

Keywords: measurable measureable external factors, duration of estimation, factor analysis, multivariate regression analysis, dummy variable.

INTRODUCTION

Some researchers have been tried to identified project successful factors up to now. Lists of variables have been identified shown in some literatures but no general agreement can be made (Chan et al, 2004). Experiences and previous research in the construction industry has shown that variation in production rate values for the same construction item is attributed to the effects of project conditions which are commonly called influence factors. Ng et al. (2001) found that no substantial changes were observed in Bromilow's observations in 1969, but there was a significant increase in the K coefficient value. The relationship is in different geographical regions such as in Australia (Ireland, 1985; Ng et al., 2001), Hong Kong (Chan et al., 1999); United Kingdom (Kaka et al., 1991), Malaysia (Chan, 2001), Nigeria (Ogunsemi et al., 2006), and the United States (Hoffman et al., 2007). Chan et al. (1999) also identified a significant relationship between the duration of construction projects and macro project factors such as floor area, and number of floors. In the study of Ng et al. (2001) conducted a categorization test on several qualitative attributes through variance analysis. The results show that the types of projects substantially have different influences, therefore, this study concludes that there are different groups of variants in the sample. In Australia, Love et al. (2005) The results of his research suggest construction duration determinants also include linear combinations of dirty floor area and number of floors. Since the success of a construction project could be seen based on the timeliness that has been scheduled, the accuracy of quality standards, and the accuracy of the budgeted costs, a framework concept was developed to identified influence variables on construction project success by Chan et al (2004). Then, Lee et al (2009) used some measureable external factors in his research to predict project duration in a relatively new environment. A similar previous studies showed that some measureable external factors influenced the duration of construction building project in Jakarta. Important factors in project success is measureable external factors, in addition of internal factors, where the measureable external factors are also affect the duration of the construction project and could cause project delays. Many researchers usually used factors analysis method to obtain dominant variables of sets of variables. It is about finding relationships between whole sets of variables, and finding the strength of those relationships. After this step, a dependent variable could be correlated with some independent variables to describe the relationship among them, known as influence independent variables to dependent variable. Regression analysis method is a statistical measurement used that attempts to determine the strength of the relationship between one dependent variable (usually denoted by Y) and a series of other changing variables (known as independent variables). Both methods refered a difference purpose between the method to identification of influencing factors and the method to calculated the magnitude of those influence factors. This caused an inconsistency results of factor analysis method, that the aim is to reduces the number of variables, and regression method which requires level of relationship strength. Since measureable external influence factors fluently composed of three main latent variables, which are: Economics, Resources and Technologies, it is difficult to established strong relationships between one of the three construction project success variables with those reduced measureable external factors (Sousa et al.2014). This paper describes a used indicator variables or dummy variables (Santoso: 2001) model to improved that relationships stated by increasing regression coefficient or goodness of fit (R^2)

METHODOLOGY

In explanatory variables, sometimes there are qualitative data that affect regression calculations (such as education, gender, etc.). Sousa et al. (2014) state that the duration of construction projects not only influences quantitative but also qualitative variables. Variable indicators, commonly referred to as dummy variables, are analyzes that can change qualitative variables into quantitative ones. This variable includes the binary variable. Qualitative factors can be included in regression analysis using variable indicators. The qualitative variables analyzed assume a value of 1 or 0 (Santoso: 2001). Using the binary coding pattern (1.0), this dummy variable is a dichotomous variable. The simple variable indicator regression formula with one independent quantitative variable and one dummy variable is (Kuncoro: 2001) n explanatory variables, sometimes there are qualitative data that affect regression calculations (such as education, gender, etc.). Sousa et al. (2014) state that the duration of construction projects not only influences quantitative but also qualitative variables. Variable indicators, commonly referred to as dummy variables, are analyzes that can change qualitative variables into quantitative ones. This variable includes the binary variable. Qualitative factors can be included in regression analysis using variable indicators. The qualitative variables analyzed assume a value of 1 or 0 (Santoso: 2001). Using the binary coding pattern (1.0), this dummy variable is a dichotomous variable. The simple variable indicator regression formula with one independent quantitative variable and one dummy variable is (Kuncoro: 2001)

$$y = \beta_0 + \delta_0 d + \beta_1 x + u$$

If dummy vaiable d = 1 or d = 0, then the formula becomes:

If d = 0, then y = $y = \beta_0 + \beta_1 x +$
If d = 1, then y = $y = (\beta_0 + \delta_0) + \beta_1 x + u$

IOP Conf. Series: Materials Science and Engineering **650** (2019) 012020 doi:10.1088/1757-899X/650/1/012020

Dummy variables are often used in regression and discriminant analysis. Where the coefficient has a direct interpretation. While the use of dummy variables in other multivariate techniques is more limited, especially in analyzes that rely on correlation patterns such as factor analysis because the correlation of binary variables is not well presented by the Pearson correlation coefficient (Hair et al.: 1998).

The steps of the study should be:

1. **Identification measureable external influence factors by factor analysis**
   A number of variables influencing the success of project implementation were identified following a thorough review from several journals (Chan et al 2004). This qualitative selection include measureable external environment to be selected. Analyze dominant measureable variables and collect data measure variables for further regression analysis.
2. **Determined the correlation equation of duration and those measureable external factors**
   Multivariate regression is a technique that estimates a single regression model with more than one outcome variable. When there is more than one predictor variable in a multivariate regression model, the model is a multivariate multiple regression.

3. **Calculate regression relationship strength**
   The strength of relationships stated by goodness of fit (R2) of the regression equation

$$y = \beta_0 + \beta_1 x + u$$

   R-squared is a goodness-of-fit measure for linear regression models. This statistic indicates the percentage of the variance in the dependent variable that the independent variables explain collectively. R-squared measures the strength of the relationship between indepemdent variables model and the dependent variable on a convenient 0 – 100% scale.

   $R^2 \leq$ dari 0,5 weak correlation
   $R^2 >$ dari 0,7 good correlation
4. **If $R^2 \leq 0{,}5$ choose dummy variables**
   If R2 $\leq$ 0.5, a regression analysis should be added a dummy variable/s, then the equation become:

$$y = \beta_0 + \delta_0 d + \beta_1 x + u$$

   A Dummy variable or Indicator Variable is an artificial variable created to represent an attribute with two or more distinct categories/levels. Dummy variables are created in this situation to 'trick' the regression algorithm into correctly analyzing attribute variables. It usually a latent variable/s

RESULT AND DISCUSSION

First Stage: Relationships without dummy

A study of the factors of measureable external influence on duration, identified several influence variables extracted from several journals. Kog and Loh (2012) identified Critical Success Factors that had an effect on each project actor, namely for architecture, civil work and the M & E division. Dai et al. (2009) was also conducted with a qualitative approach in his research. Identification of latent factors that influence labor productivity is carried out by a factor analysis approach. The Result of the previous study on influence factors on project duration conducted by qualitative method thorough questioner. Distribution of questionnaires and ongoing technical data collection projects carried out in the Jakarta and surrounding areas. Gathering 43 data showed that measureable external factors consist of: Inflation, Material Price Index, GDP, Exchange Rate, Interest Rate, Availability of Labor, Work Experience,

Education Level, Health Level, New Technology Availability, Technology Absorption Rate, Innovation, Inflation. After factor analysis, a set of 5 dominant factors was identified which are: education (X1), technology absorption index (X2), workforce experience (X3), availability of technology (X4), Innovation index (X5) as independent factors and Duration (Y) as dependent variable. The data comes from: BPS (work experience), WEF (education, technology absorption, availability of new technology), and the Global Innovation Index (innovation). Factor analysis and multiple linear regression analysis were conducted to test the relationship of measureable external factors to the duration of estimation of construction projects. In this study, the Goodness of Fit (R^2) of the equation regression analysist used as sensibility benchmarks. Variables are transformed to index number for unit synchronize.

The output of regression analysis in this first stage as seen in table 1

Table 1 *Model Summary*

Model	R	R *Square*	*Adjusted R Square*	*Std. Error of the Estimate*
1	,648[a]	,420	,341	88,70622462

a. *Predictors*: (Constant), X5, X2, X3, X4, X1

b. *Dependent Variable*: Y

The number of R^2 is 0.420. This means that only 42% of the duration can be explained by variables X1, X2, X3, X4, X5. While the remaining 58% is explained by other causes. This result is unsatisfied for predicting a project duration.

Second Stage: Analysis of Relationships with Dummy Variables

Analysis of relationships with dummy variables can be done if the indicator variable group has met the requirements of multivariate regression analysis. In this study, an analysis was performed using the dummy variable in the second stage. This dummy variables, 4 new independent variables are taken from the types of buildings, namely apartments (X6), offices (X7), hotels (X8), and other buildings (X9). Grouping for dummy variables can be seen in table 2. Because the number of dummy variables used for analysis is (k-1), then other building types are used as base groups (D1, D2, and D3 are set as zero) and are not included in multivariate regression analysis. The coding pattern that will be used in multivariate regression analysis can be seen in table 3.

Table 2 Variable Dummy Grouping Tables

Building Type	**Variabel *Dummy***
Apartment	D1 = 1, selain D1 = 0
Offices	D2 = 1, selain D2 = 0
Hotel	D3 =1, selain D3 = 0
Other Buildings	D4 =1, selain D4 = 0

Table 3 Variable Dummy coding patterns

Coefficient / Dummy Variable	**D1**	**D2**	**D3**
Apartment	1	0	0
offices	0	1	0
Hotel	0	0	1
Other Buildings	0	0	0

After the dummy variable was included in the regression analysis along with the Education Index, Technology Absorption Index, Labor Experience Index, New Technology Availability Index, and Innovation Index as independent variables. The results of multivariate regression analysis using dummy variables in the second stage will be described as follows.

Table 4 *Model Summary*

Model	R	*R Square*	*Adjusted R Square*	*Std. Error of the Estimate*
1	.886[a]	.785	.727	57.15459942

a. *Predictors*: (Constant). X9. X8. X7. X2. X6. X5. X3. X4. X1

b. *Dependent Variable*: Y

The ENTERED VARIABLES by tools shows that there are no variables removed, or in other words the nine independent variables are included in the regression calculation. The R square number is 0.785. this means 78.5% of the duration can be explained by the free variables entered. While the rest (100% - 78.5% = 21.5%) is explained by other reasons. Statistically a R^2 0.785 is a good relationship. The analysis is also shown that the standard error of Estimate is 57,145 (the unit used is the dependent variable, or in this case the duration). Note in the previous descriptive analysis, that the standard deviation of duration is 109,296, which is greater than the Standard Error of Estimate which is only 57,145. Because it is smaller than the standard deviation duration, the regression model is better at acting as a predictor of duration, rather than the average duration itself.

CONCLUSION

The analysis showed that by adding variable/s indicator or dummy variable improved the relationships equation determination by goodness of fit of its relationships. In this study the used of dummy variables increase the determination up to 86.9%. Although R^2 is used as benchmark, another regression model requirement should be noticed also to support the result. In this study, Education Index, Technology Absorption Index, Labor Experience Index, New Technology Availability Index, and Innovation Index together with building type, which are apartment, office, hotel and other then the three type, could be used as influence factors in predicting construction project duration.

However, an obstacles occur in implementation of dummy variables in setting type of the dummy variables which is usually a latent variables and should be so. Another work added in analyze with regression which are the standard mathematical requirement, but by using computer program such like SPSS® and so on.

REFERENCES

[1] Akanni, P O, A E Oke and O A Akpomiemie 2015 *Impact of Environmental Factors on Building Project Performances in Delta State, Nigeria* (HBRC Journal) **11**, 91-97

[2] Chan, Albert P C 1995 *Modelling Building Durations in Hong Kong* (Construction Management and Economics) **17**, 189-196
[3] Chan, Albert P C 2001 *Time-cost Relationship of Public Sector Projects in Malaysia* (International Jornal of Project Management) **19**, 223-229
[4] Chan, Albert P C, David Scott and Ada P L 2004 *Factors Affecting the Success of a Construction Project* (Journal of Construction Engineering and Management) **130**, 153-155
[5] Chan, D W M and M M Kumaraswamy 2002 *Compressing Construction Duration : Lesson Learned from Hong Kong Building Projects* (Journal of Construction Engineering and Management) **20**, 23-25
[6] Dai J, Goodrum, Paul M, Maloney, William F and Srinivisan C 2009 *Latent Structures of the factors affecting Construction Labor Productivity* (Journal of Construction Engineering and Management) **135**, 397-406
[7] Hair, Joseph F, et al 1998 *Multivariate Data Analysis* (New Jersey: Pretince-Hall inc)
[8] Kaka A and Price A D F 1991 *Relationship Between Value and Duration of Construction Projects* (Journal of Construction Management and Economics) **9**, 383-400
[9] Kog Y C and Loh P K 2012 *Critical Success Factors for Different Components of Construction Projects* (Journal of Construction Engineering and Management) **138**, 520-528
[10] Kuncoro M 2001 *Metode Penelitian Kuantitatif* (Yogyakarta: UPP AMP YKPM)
[11] Lee H, Shin J, Park M and Ryu H 2009 *Probabilistic Duration Estimation Model for High-rise Structural Work* (Journal of Construction Engineering and Management) **135**, 1289-1298
[12] Ng S T, Mak, Micahel M Y, Skitmore M, Lam K C and Varnam M 2001 *The Predictive Ability of Bromilow's Time–cost Model* (Journal of Construction Management and Economics) **19**, 165-173
[13] Santoso S 2001 *SPSS Versi 10: Mengolah Data Statistik secara Profesional* (Jakarta: PT Elex Media Komputindo)
[14] Sousa V, Nuno M A and Luis A D 2014 *Role of Statistics and Engineering Judgment in Developing Optimized Time-cost Relationship Models* (Journal of Construction Engineering and Management) **140**, 04014034

Progressive Infrastructure Projects in Indonesia: Views on Potential Growing Geoforensic and Dispute Cases

Chaidir Anwar Makarim[1]

[1] Professor in Geotechnical Engineering at Tarumanagara University, Indonesia. Specializing in geoforensic, construction disputes, arbitration, and value engineering.

E-mail: chaidir259@gmail.com

Abstract. Recently, progressive government infrastructure projects on: roads and bridges, harbour, oil and gas, water and sanitation, railroad, electricity, city transportation and information technology have reached big records of government projects since the independence of Indonesia – A country with 17.504 islands and projected population in 2030 of 350 million. Every construction project carries its years of serviceability usually between 50 – 100 years. During and before that time, cost of risk due to the unfulfillment of engineering standard of design, construction, management, usage and maintenance can cause problems. Cases of past project failures that bring consequences in terms of cost, disfunction/ failure and sometimes people's lives in some of the past projects are analyzed from the geoforensic point of view as cases to be avoided from the early beginning when the project was physically decided to be constructed in a certain geophysical area in specific parts of Indonesia.

Keywords: Infrastructure projects; Serviceability; Engineering standard of design, construction, management, usage, cost; Disfunction/ failure; Geoforensic point of view.

1. Introduction

Since the last 5 years, Indonesia government had adopted a progressive infrastructure development project in all part of Indonesia at the amount of more than 4.75 Trilliun Rupiah or $328,215,475.19 USD. In the 3rd quarter of year 2018, GDP development in general remained stable at 5.2% year-on-year. Government stimulation not only in Java island with its highly densed population, but spreaded out in Sumatra island to Papua/ West Irian as well with the intention to keep Indonesia people economically and politically stable and connected well. Construction sector is noted to be one of the business sectors that has scored positive performance amid slowing economic growth. The role of the construction sector on Gross Domestic Product (GDP) has also increased. Based on data from the Central Bureau of Statistics, annual economic growth in the 3rd quarter of year 2018 recorded a slight increase to 5.79% in 3rd quarter of year 2018 compared to 2nd quarter of year 2018 to 5.73% [9]. Those projects consisted of roads and bridges, harbor, oil and gas, water and sanitation, railroad, electricity, public transportation and information technology. According to London based publications recently – Pricewaterhouse Coopers, published its long-term global growth projections until 2050 within 32 largest economies in the world. This projects China to lead the four largest economies in the world, followed by India, the United States and Indonesia.

IOP Conf. Series: Materials Science and Engineering **650** (2019) 012021 doi:10.1088/1757-899X/650/1/012021

On April 20, 2019, at the infrastructure summit 2019, the Indonesia national arbitration center (BANI) was invited to give presentation on arbitration related to construction dispute cases where cost and quality control involved in. The event was conducted jointly by PT. Infrastructure Guarantee Fund (IIGF) and university network for Indonesia infrastructure development (UNIID), held at the university of Padjajaran in Bandung, West Java. On that occasion, as an arbiter with geoforensic experience, I have the honour to present – Based on my knowledge and experience in handling arbitration cases in construction disputes for the last seven years.

In the earlier Indonesia construction law No. 18 of 1999, arbitration and litigation were the alternative to solve dispute. Only recently, in the new Indonesia construction law No. 2 of 2017, dispute board (DB) is preffered before arbitration.

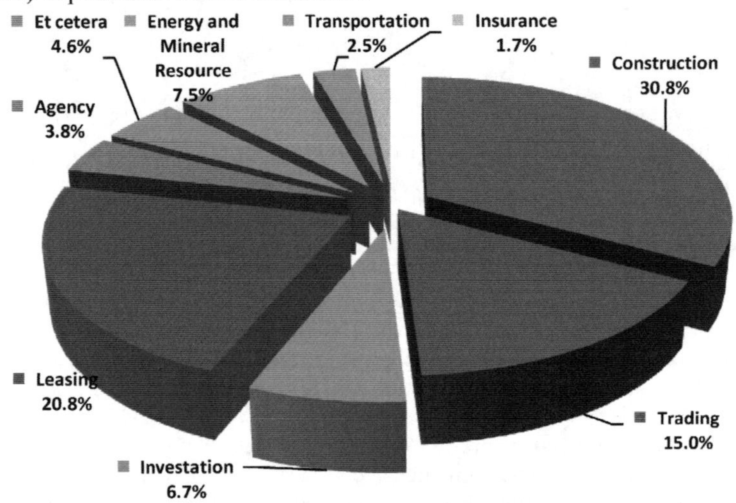

Figure 1. Registered Arbitration Case in BANI during a period of 2009-2013. [8]

2. The Role of Experts

Figure 2 shows the map of KPPIP (Komite Percepatan Penyediaan Infrastruktur Prioritas) – Accelerating Priority Infrastructure Provision Committee priority projects spreading out in Indonesia.

Figure 2. Ongoing infrastructure projects spread out in Indonesia. [5]

The big reason behind all of these is the huge commodities transportation and public transport costs between regions on national scale. For example, transportation of commodities from China is cheaper than from Java to Sumatra. If per kg from China is priced at Rp 10,000,-; from Java to Sumatra, it can cost more than Rp. 30,000,-. Because of these conditions, the government has made massive improvements and development of supporting infrastructure to all regions of Indonesia. This is done by the government as a form of investment to encourage economic growth in the future. The government hopes that the availability of adequate infrastructure will trigger economic growth in Indonesia, which impacts on the growth of the social life of the community, especially in areas close to the built infrastructure. Figure 3 below shows that infrastructure budget increased rapidly by the year 2016 to 2017.

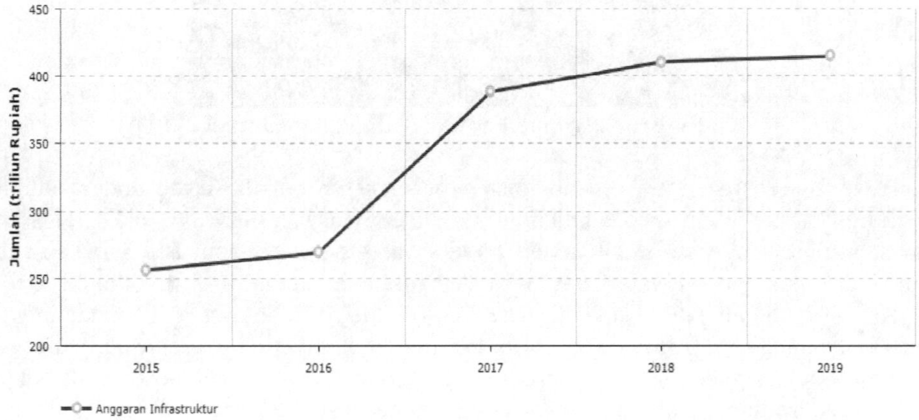

Figure 3. Infrastructure budget chart by the year 2015 to 2019. [5]

Infrastructure development is a continuous process that covers all aspects of people's lives - including social, economic, political and cultural aspects with the main goal of improving the welfare of the citizens as a whole. Therefore, the role of education in infrastructure development in Indonesia is very strategic. Infrastructure development should involve the active role of academics such as universities, experts in the fields of engineering and legal aspects, and research institutions in the process of design, implementation, development, and review. By involving the role of academics, it will create a domino effect for future development of infrastructure that will be carried forward by future generations. Research will develop which impact on the growth of quality human resources over time. In addition, assistance provided by certified experts will increase the reliability of infrastructure projects built in Indonesia.

2.1. Construction Problems: Failures or Disputes?

Infrastructure holds an important aspect for economic growth and for the welfare of society, however we often encounter various damage or failure within the projects. Is it “failures” or “dispute”?

2.1.1. Typical Causes of Failures

Unexpected factors or “Act of God” type failure as mentioned by the insurance policy in general are distinguished then failures that can be predicted (predictable) or human error type of failure. Geotechnically, most building and construction failures in Indonesia are caused by a lack of understanding of problematic land [6]. There are at least two buildings / construction collapsed in Jakarta due to the fault of designing very soft soil foundations [2]. Some land in Indonesia is very soft and expansive. An understanding of soil-water interactions when construction near the river or sea edge is also usually not seriously considered by most geotechnical engineers [8]. Uniquely, this is what often escapes design and impacts on problems later on.

Figure 4. Bridge Failure, Kalimantan/Borneo. [11]

In this case, the designers missed not only workmanship but the basic understanding of typical offshore design where hydro-geotechnical multi disciplinary comes into play. This incident may caused by scour that happened on top layer of soil around the pile. Scour is the loss of soil because of erosion in river bottoms or in waterfront areas. A hydraulic evaluation forms the basis for estimating the scour zone, which can extend several meters below the river bottom. All foundations must extend below any potential scour zone, so deep foundation (usually piles) are usually necessary. These foundation must be designed to safely support the design loads even if the upper soils are lost to scour.

Scour also reduce the lateral load capacity of piles because the upper soils are no longer present to provide lateral support. In other word, the distance from the structure to the ground surface becomes greater, so the moments in the pile increases. Therefore, it becomes necessary to use larger pile sections or battered piles to support the lateral loads.

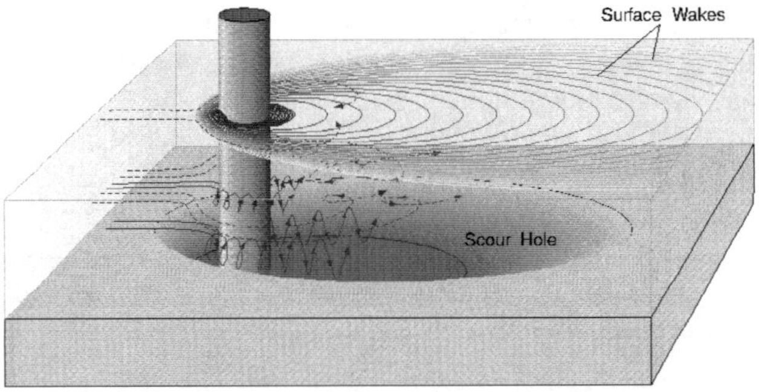

Figure 5. Illustration of scouring. [15]

Another case history failures involving that is diversion Tunnel is located below the spillway at an elevation of + 164.00 meter, construction type selected in this section is circular reinforced concrete lined, reinforced with rock bolts and steel ribs on the body of diversion tunnel. The Tunnel $\pm$ 550 meter length and 10 meter of diameter. Diversion tunnel is expected to drain the water discharge plan for 3200 m^3/second. Dam project is the result of cooperation of four governmental contactors in Indonesia.

This case position is related to a construction claim of flood incident and roof collapsed in a diversion tunnel that both of them took place in year 2010, at Dam Project-West Java. The successful Novum are found in both of cases. In the flood incident, the successful Novum was : the extreme climate anomaly of La Nina caused the increasing rainfall intensity, and eventually impact to water entered the diversion tunnel and causing extensive damage.

Figure 6 shows questions on its design, analysis and common practice in the world. Figure 7 shows blasting works by the spillway contractor that affecting works of contractor handling the diversion tunnel.

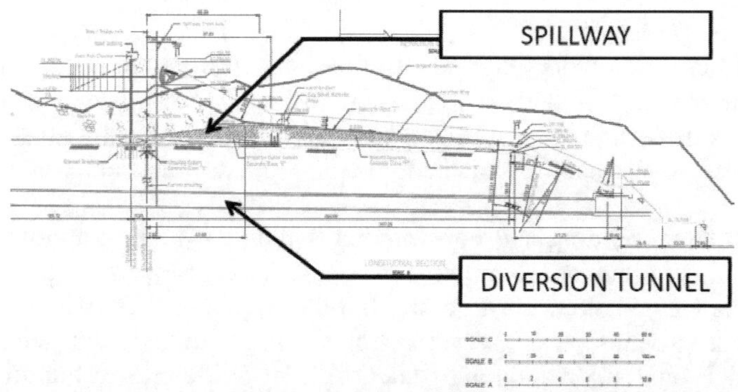

Figure 6. Drawing of spill Way Plan, Section and Detail. [11]

Figure 7. Simultaneous Blasting on Spill Way. [11]

According to Noon, R.K. (2001), Forensic Engineering is the implementation of engineering principles, knowledge, skills, and methodologies to answer questions of fact that have legal ramifications [10]. Initially, we only know the end results. This could be a building failure. Then the expert has to reconstruct it in order to know what causes it. Questions that should be answered is who, what, where, when, why, and how it happened.

Since last decade, Forensic Geotechnical Practice in Indonesia had gained National and International recognition through its involvement in solving construction disputes with: Contractors, International Insurances, Government, Developers/ Owners, and/or other Private Sectors. The growing number of construction cases solved by non-litigation institution in 2014 is estimated of more than 100 million USD which was believed still increases by 10-20% each year. Most building collapsed and construction failures in Indonesia did not happened by accident or Act of God. It started already when the designer composed the design or when the contractor interpreted the design to build the construction. Negligence and/or material defect could also contribute to the problem. Geotechnically, most of the building and construction failure in Indonesia were caused by inadequate understanding on Indonesia problematic soil.

Other building or construction failure in Indonesia were caused by less understanding on the nature and behavior of peat soil in Kalimantan and clay shale rapid degradation behavior in West Java. Understanding of soil-water interaction as the construction get closer to the riverside or ocean-side region were usually not considered seriously by most Geotechnical designer. Bridge collapsed, riverside retaining wall failures, or even road/infrastructure failures were mostly caused by lack of understanding on the soil behavior in those area.

2.1.2. Typical Causes of Disputes

The word "forensic" explains works that has to do with investigation. The result of the investigation should be bundled in such a way that could relate to law and courts. Based on Indonesia Construction Law No. 18/1999, construction failure is categorized as failure during construction, while building failures is failure after construction final handover for up to ten years where building insurance is usually still ongoing.

Indonesia has its own building and construction standard which is almost close the one adopted internationally such as the standard & practice in: USA, Japan, Malaysia, etc plus a few adjustment as dictated by Indonesia Geo-Ecological Structure. In other terms: if there is some criteria that is missing in the Indonesia code, a valid yet conservative related international code will be acceptable. This is agreed by ±27 professional association and stated in their guidelines of building failure and has been published as a guideline by the Indonesia Development Board of Construction, (LPJKN/Lembaga Pengembangan Jasa Konstruksi Nasional). LPJKN is stated in the law as the national authority for National Construction Regulation. It is also known that almost all professional association, just like HATTI (Indonesia Association of Geotechnical Engineer), were long time member of a worldwide international association namely The International Society for Soil Mechanics and Geotechnical Engineering (ISSMGE). HATTI professional member were mostly educated in Indonesia and/or abroad such as: USA, UK, France, Singapore, Japan, Malaysia, Australia, etc. which make them familiar with the international code of practice. It means that engineering practice in Indonesia is "friendly" and fully familiar with national or international construction standard, criteria, and specification.

2.1.3. Dispute Solution

Infrastructure construction failures that occur often cause disputes, both between contractors, insurance, owners, government, or the private sector. The law that regulates this is Indonesia Law No. 30 of 1999 ch. 11 on Alternative Dispute Resolution (ADR).

> "Dispute of differences of opinion that are not of a criminal native in nature may be resolved by the parties through Alternative Dispute Resolution (ADR) based on their good faith, by waiving such resolution by litigation in district court." (Umar, M.H., 2013) [7].

In that case, dispute in construction is usually solved through consultation, negotiation, mediation, conciliation or expert assessment, before it goes to arbitration. Litigation has long been the last resort due to its high cost and much longer time. Most of construction cases in Indonesia were not brought to litigation. There are 2 (two) solution that is usually chosen: nonbinding and binding ADR [1]:

(i) Non-Binding Dispute Resolution
- Conciliation: Where both parties appoint a conciliator who meets with parties separately in an effort to resolve differences. The main goals is to conciliate or seeking concessions.
- Mediation: Where both parties agreed to appoint a mediator (one or more) to solve differences in a private and confidential atmosphere in order to reach a win-win solution.
- Executive Tribunal: Where representatives of parties which previously has not been involved in the dispute, usually top management represent each party to negotiate for a settlement.

(ii) Binding Dispute Dispute Resolution
- Adjudication: This is a legal process where the arbiter or judge comes to a decision after reviewing evidence and argumentation including legal reasoning. Types of the disputes solved are: disputes between public officials or public bodies, between private parties and public officials or between private parties (individuals/ corporations).
- Negotiation. This is discussion for the purpose of exploring and arriving at an agreement in a legal way.

IOP Conf. Series: Materials Science and Engineering **650** (2019) 012021 doi:10.1088/1757-899X/650/1/012021

- Arbitration. It is a technique for the resolution of disputes outside the courts by a third party to review all evidence in the case and impose a decisions that is legally binding on both sides and enforceable in the courts.
- Expert Determination: This is when the parties agreed beforehand to be bound by decisions of an independent expert.
- Litigation. The process of solving dispute through the public court.

Finally, the arbitration award shall be final and binding upon both parties to the disputes [12].

2.1.4. Dispute Boards or Arbitration?

The first construction law was introduced on May 7th 1999, i.e.: Law No. 18 of 1999. At that time construction dispute will be solved through non-litigation such as mediation, conciliation, and arbitration, or through litigation (Criminal or Civil Law). It addresses potential construction failures due to four related parties i.e.: The Owner, The Designer, The Contractor, and the management construction with the consequences cases could be brought to Civil/Criminal Law. Parties could also get involved for other reason under the potential "state lost" (kerugian negara). In this case, criminal law/misdemeanor could not be avoided, where responsible party will be brought to justice through litigation.

In the new law, no government agent should be involved since dispute will be solved amicably through mediation, conciliation, Dispute Board or arbitration [13]. The new law was issued in 12 January 2017 i.e. Indonesia Construction Law No.2 of 2017 on Construction Services, addressed solving disputes through mediation, conciliation, arbitration, as well as through Dispute Adjudication Board. The latest is a completely new thing for Indonesia while actually the philosophy was deep rooted in Indonesia cultural heritage i.e. solving difference through "musyawarah" (peaceful understanding/ agreement).

When FIDIC published the "Red Book", it includes the traditional English system for the pre-arbitral resolution of disputes by the engineer. This practice is well adopted in common law countries – but unknown in civil law Countries.

Based on the old law, construction or building failures could possibly cause by faulty design, faulty function, faulty construction, or faulty management. Investigation related to construction failure (during construction process) or building failure (after construction completed) is the typical work done by government personnel including police and expert.

In government as well as in privately owned construction facilities, police has the right to investigate and bring the case to the court. Investigation is done by the help of certified forensic expert. Most of the time, fault was punished in front of court, for proven mismanagement, which could bring people to jail as one of the punishment. In other case, both parties agreed to bring the case to arbitration.

Recently, we often heard on the suggestion that the role of "Dispute Boards" be included in an Multi-hundred million dollar international projects contract. This suggestion should be made carefully towards a specific project as the experts said. Dispute Boards has already exists in Indonesia Construction Law No. 2 of 2017 for almost 2 years but not yet followed by government ordinance, ministerial decree, and so forth. No wonder the author found in some foreign financed project which in its contract include DAB (Dispute Adjudication Board) as a dispute resolution board (not yet formed) – both parties prefers to solve the case through arbitration.

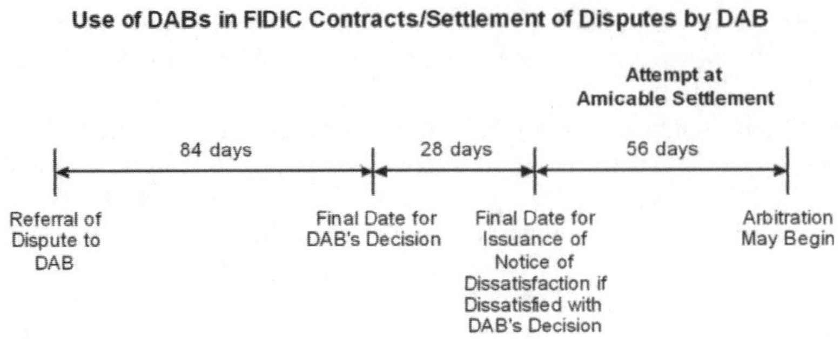

Figure 8. Timeline of FIDIC Contracts/Settlement of Disputes by DAB. [7]

As far as cost of DABs concern, the following is a quote from Christopher S, Sepällä presentation (2015) [14]:

- For multi-hundred million dollars project, a three-member DAB can cost much less than 1% of project cost, whereas international arbitration may cost 10 to 15% or more of project cost.
- Ad-hoc DABs are less costly than permanent DABs but, arguably, are not true DABs as they cannot perform a dispute prevention function.

However, since Dispute Board practice is something new for Indonesia, intensive training and procedures not to mention communication as well, should be introduced in detail by professional association – including practicing in a DAB case studies. With the existence of Dispute Boards, it is expected that the arising of Construction problems can be solved through mediation since Dispute Boards have been followed to guarding the cooperation project between the two parties from the beginning. To ensure the execution process of the Dispute Board decision, it is recommended that if the FIDIC Red Book is used, the parties must agree to a statement where if one of the parties fails to fulfill the DAB's decision, it is entitled to be able to go to arbitration (anyway), without referring to the previous dispute.

Finally, the author is in the opinion that Indonesia needs more time before applying Dispute Boards system as one of an alternative to resolve dispute in construction area. Intensive training/short courses and thorough understanding on the FIDIC Contract besides remuneration for the DAB Expert should be considered in a very detail manners.

3. Conclusion

Finally, this paper reach a conclusions as follows:

- Since the last 5 years, Indonesia government had adopted a progressive infrastructure development project of infrastructure in all part of Indonesia. It is hoped that the form of investement will trigger economic growth in Indonesia. However, every construction project carries its years of serviceability. Not so often, some cases happened – during that time, due to the unfulfillment of engineering standard of design, construction, management and usage. This brings consequence of cost, disfunction/ failure and sometimes lives. However, this problem can be avoided from the early beginning – before the project start, by applying geoforensic engineering practice to determine the project to be in a certain geophysical area in specific parts of Indonesia. The word "forensic" explains works that has to do with investigation which comes as a result of relation to law and courts.
- Failures on construction that occurred often causes disputes, both between contractors, insurance, owners, government, or the private sector. The law that regulates this is Indonesia Law No. 30 of 1999 ch. 11 on Alternative Dispute Resolution (ADR). Based on this law, disputes can be resolved through Alternative Dispute Resolution (ADR) based on their good faith, by waiving such resolution by litigation in district court. The Indonesian Arbitration Center (BANI) and ADR open its door to international as well as national arbitration. The

arbitration award shall be final and binding upon both parties to the disputes. The new law – Indonesia Construction Law No.2 of 2017 on Construction Services, addressed solving disputes through mediation, conciliation, arbitration, as well as through Dispute Adjudication Board. Dispute Boards has already exists in Indonesia Construction Law No. 2 of 2017 for almost 2 years but not yet followed by government ordinance, ministerial decree, and so forth. Dispute board practice is considered to be more efficient since it costs much lesser than international arbitration. However, since Dispute Board practice is something new for Indonesia, intensive training and procedures not to mention communication as well, should be introduced in detail by professional association – including practicing in a DAB case studies. Indonesia needs more time before applying Dispute Boards system as one of an alternative to resolve dispute in construction area.

References

[1] Charett, D. September 2012. *Dispute Boards in Australia – The Story So Far*, Build Law Issue No.15.

[2] Hulme, T. W., Law, Y. S. 1997. *Foundation Failure, Proceedings of the International Conference on Foundation Failures, 12-13 May 1997*. Singapore.

[3] Indonesia Investments. 2017. *Infrastruktur di Indonesia*. https://www.indonesia-investments.com/id/bisnis/risiko/infrastruktur/it-em381?.

[4] Kastratoke. 2014. *Tantangan Pembiayaan Pembangunan Infrastruktur Indonesia*. https://kajianstrategishms.wordpress.com/2014/09/25/tan-tangan-pembiayaan-pembangunan-infrastruktur-indonesia/, 2014.

[5] Komite Percepatan Penyediaan Infrastruktur Prioritas. 2016. *Priority Projects*. https://kppip.go.id/en/.

[6] Makarim, C. A and A. P, Paulus. 2008. *Peta Prakiraan Tanah Ekspansif di Pulau Jawa serta Dampak Perilakunya terhadap Konstruksi, Pusat Studi Geoteknik dan Lingkungan*. Jakarta: Universitas Tarumanagara.

[7] Makarim, Chaidir, A. 2013. *Role of Dispute Board in Indonesia – A Perspective Point of View*, BANI, 2013.

[8] Makarim, C. A. 2014. *Forensic Geotechnical Engineering in Indonesia: Cases and Trends of Future Development, Proceedings 18th Annual National Conference on Geotechnical Engineering, 11-12 November 2014*. HATTI. Jakarta.

[9] Maulana, R. 5 November 2018. *Sektor Konstruksi Makin Menggeliat*. https://ekonomi.bisnis.com/read/20181105/45/856681/sektor-konstruksi-makin-menggeliat.

[10] Noon, R. K. 2001. *Forensic Engineering Investigation*. CRC Press: London.

[11] Project Report 2012. *Laporan Investigasi Penyelidakan Kasus Kegagalan*.

[12] Republik Indonesia. 1999. *Undang-undang No. 30 Tahun 1999 tentang Arbitrase dan Alternatif Penyelesaian Sengketa*. Lembaran Negara RI Tahun 1999, No. 138. Sekretariat Negara. Jakarta.

[13] Republik Indonesia. 2017. *Undang-undang No. 2 Tahun 2017 tentang Jasa Konstruksi*. LL Sekretariat Negara RI Tahun 2017. Sekretariat Negara. Jakarta.

[14] Seppälä, Christopher R. 2015. *Fidic and Dispute Adjudication Boards (Dab(S)), International Federation of Consulting Engineers*.

[15] Wikipedia. Last edited on 4 March 2019. *Bridge scour*. https://en.wikipedia.org/wiki/Bridge_scour.

IOP Conf. Series: Materials Science and Engineering **650** (2019) 012022 doi:10.1088/1757-899X/650/1/012022

Correlation of geological conditions and levels of damage in Palu earthquake

R Hidayat*

Balai Sabo, Puslitbang Sumber Daya Air, Badan Litbang, Kementerian PU Sopalan, Maguwoharjo, Yogyakarta-55282

* rokhmathidayat33@yahoo.com

Abstract. The earthquake was occurred in Palu with magnitude 7.4 SR on September 28, 2018. The earthquake caused damage in infrastructure, settlements and casualties. The earthquake also caused liquefaction in Petobo and Balaroa. One of the factors that influence to the level of damage is geological conditions such as geological structures (faults) and rock types. The presence of faults determines the earthquake zone, so that it greatly affects the level of damage. The type of rock in Palu generally consists of allufial deposits and sedimentary rocks, make the level of damage that occurs has different levels. The results from survey showed that the level of damage to buildings was strongly related to the earthquake zone and rock conditions. Earthquake induce damage more intensively in locations with rocks in the form of soft allufial deposits and closed to the earthquake zone. This condition can be taken into consideration in earthquake disaster mitigation actions. The arrangement of the city of Palu must be carried out with careful planning because the Palu area is on a high tectonic route. To reduce the risk of disasters, structuring and development need to consider a map of earthquake-prone zones of damage.

1. Introduction

An earthquake with magnitude 7.4 SR had occurred on September 28 2018 at 18.01 WITA. The epicenter was 26 km north of Donggala and 80 km northwest of the city_of Palu with a depth of 10 km. Earthquake shocks were felt in Donggala Regency, Palu City, Parigi Moutong Regency, Sigi Regency, Poso Regency [1]. Disasters caused by earthquakes often result in many damage of property and casualties, so that mitigation efforts related to the disaster are needed. Mitigation is all actions in order to minimize disaster risk. The potential of geological resources as part of natural resources plays a role in mitigation efforts. Analysis of horizontal displacement and profile across the Palu fault shows the complexity of the fault zone, an average displacement of 3-5 meters, the total length of the fault zone is more than 140 km. The earthquake mechanism shows the direction of fault is north-south [2].

According to Pakpahan et al. 2015 [3], the Palu-Koro Fault is a very active fault. This fault has a location and source mechanism which is divided into several segments. In spatial planning, to reduce casualties due to earthquakes it is necessary to map vulnerable zones damaged by earthquakes. Geological and

geotechnical studies and analyzes need to be carried out that will direct suitability for settlement and land use [4]. The success of the damage reduction after the earthquake is proportional to the accuracy

IOP Conf. Series: Materials Science and Engineering **650** (2019) 012022 doi:10.1088/1757-899X/650/1/012022

of land use decisions. Geological data that allows reduction of earthquake damage and analyzed in each section separately. Geological data that needs to be examined include soil conditions, rocks and geological structures, especially rock layers and faults [5].

Sulawesi island is located at the colission of three Australian, Philipine and Sundanese plates (the Sunda plate is the eastern part of Eurasia). The Australian and Philippine plates move down the Sunda plate with speeds of 75 and 90 mm/year. In the south of Sulawesi, the fault zone extends from eastern Indonesia accommodates relative movements between the Australian and Pacific plates with transresive faults and block rotations [6]. The Australian plate subduction under the Sunda plate takes place in the Sunda-Banda Arc where it transitions to the Java Trench. Ada gerakan 42mm/tahun blok relatif pada zona patahan Palu-Koro [7].

Based on Minister of Public Works Regulation No. 21 of 2007 [8] concerning spatial planning guidelines for areas prone to volcanic and earthquake eruptions, stated that the level of vulnerability of an area to earthquakes is influenced by the type of rock, the slope, presence of fault structures, and earthquake strength. Various geological factors become land control that affect the level of damage, and can be developed into a micro zoning map of earthquake vulnerability. The micro zoning map is very important to support the improvement of the building code (building code) and land use management in earthquake- prone areas [9].

During 2000-2015, every year in the world there was an earthquake with magnitude 7.0 or more amounting to 11-28 times [10]. The area along Palu Bay is an area prone to earthquakes, caused by its geological condition which is in the front arc part of the subduction / collision zone, between the Continent plate in the north, and the Ocean plate in the south; (Hamilton, W., 1979) [11,12].

The strength of the Palu earthquake, which is above 7 in Richter's, has had a huge traumatic impact. The long time of the earthquake and the damage caused caused a great trauma. These natural disasters cause damage in all things. In fact the level of damage caused by the earthquake is not evenly distributed. Many buildings are had little cracks, but in other places there were buildings that were severely damaged and even collapsed. The earthquake also caused landslides, such as the 2006 Yogyakarta earthquake, which triggered landslides (Figure 1).

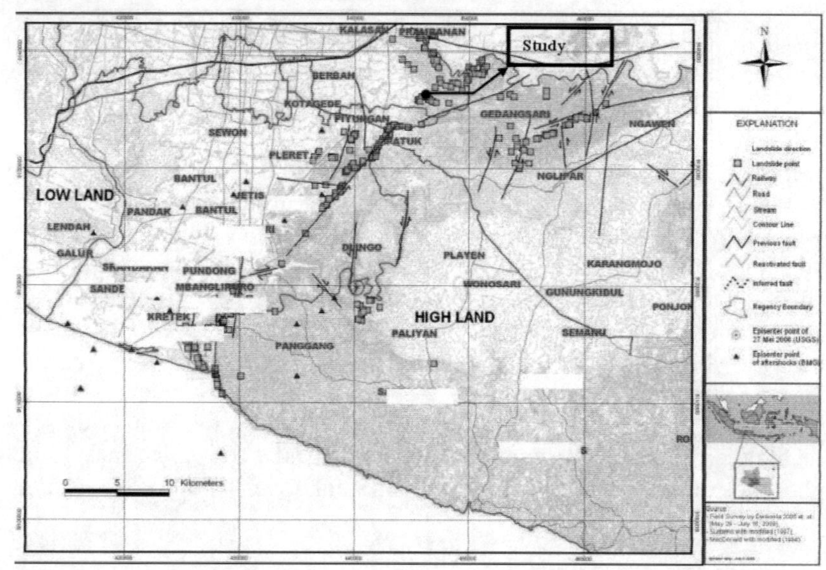

Figure 1. Point of Location of Mass Movement of Rocks due to Yogyakarta Earthquake 2006 [13]

2. Methods

This research was conducted through field surveys in several area in the city of Palu and its surroundings. Observe some field phenomena, namely the level of damage to buildings and soil cracks that occur,

IOP Conf. Series: Materials Science and Engineering **650** (2019) 012022 doi:10.1088/1757-899X/650/1/012022

then connected with local geological conditions. Observation and examination of field conditions is carried out with a descriptive approach, through measurement and recording of the level of damage in infrastructure.

To be able to properly identify and analyze the relationship of geological conditions and the level of damage at the research site, an investigation was conducted which included several activities in the form of interpretations and studies of regional geological data (previous studies). Based on this data it can be seen a description of the condition of the distribution of rocks and their geological structure. Based on field investigations, we can determine the level of damage of each survey location. The level of damage is grouped into three classes. That's mild, moderate and severe damage. The location of the survey must spread according to the geological conditions of the field.

Furthermore, geological analysis was carried out, correlation the level of damage caused by the earthquake and the geological conditions. So we got an overview of the geological factors dominant with the level of damage occurred. From the results of this investigation and analysis, expected to be able to recommend anticipatory steps in the development at the time of data collection, especially in regional spatial planning.

3. Results and discussion

Rock Condition

Based observations on the Geological Map Review of the Palu Sheet Scale 1: 250,000 (Rab Sukamto, et al., 1973) the study area generally consists of two rock formations, Aluvium-sediment beach and Molasa Celebes-Serasin.

a. Alluvium -sediment beach (Qap)

This unit consists of sedimentary material including gravel, sand, mud, and coral limestone. Formed in a river, delta and shallow sea environment, is the youngest sediment in this area. The deposits are of Holocene age. The rock conditions in this formation are generally material that has not undergone compacting (loose material). The rock structure generally consists of sand at the top, silt and clay at the bottom. Groundwater depth ranges from 0.5-16m below the surface of the ground.

b. Molasa Celebes and Sarasin (QTms)

This rock is spread in high altitudes, on the sides of mountains in the east and west of Palu City. This formation contains races derived from older rock formations, consisting of conglomerates, sandstones, mudstone, coral limestone, and marl, all of which are compacted. The rocks towards the sea turn to become finer grained clastic rocks.

Geological Structure

Based on the results of studies conducted by the Indonesian earthquake map revision team (Irsyam et al., 2010) active geological structures (fault) that pass through Palu City is Palu Koro Fault and Matano Fault, both of them are active faults, found in the valley of Palu. The main Palu-Koro fault is in the North-South direction while some of them (fractured section) are Southwest-Northeast-oriented. The active fault, which has a North-South direction, is an active fault due to the rejuvenation of an old reactivated structure, while the fracture trending Southwest-Northeast is a very active fault structure today. (Figure 2).

From the geological conditions, the physiography of the City of Palu is closely related to the condition of the fault structures that occur as well as the types of rocks that make up the City of Palu. The left and right sides of the city of Palu are the main fault zone (Palu-koro fault) and consist of rock that is harder than the constituent material.

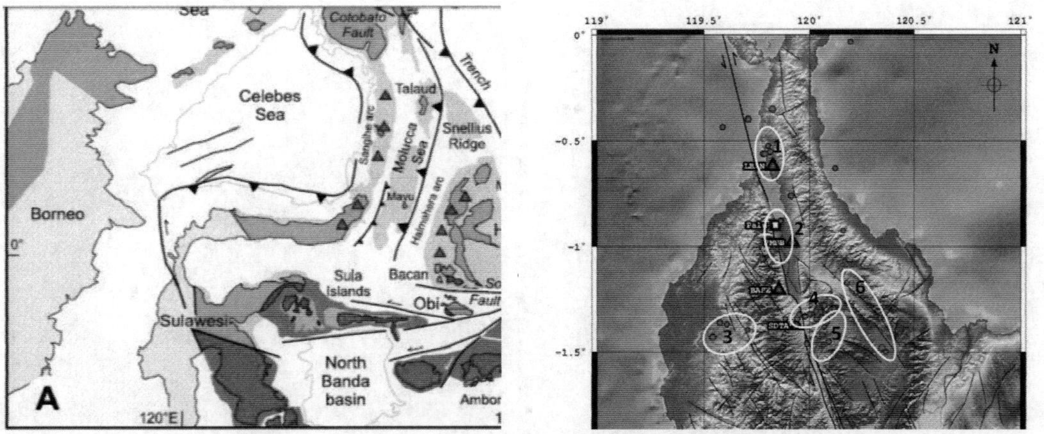

Figure 2. Regional tectonic system of Sulawesi Island and its surroundings [14] (left), Distribution of earthquake clusters around Palu Koro [3] (right).

The geological and topographic conditions of the Palu area really control the distribution of damage caused by earthquakes. In valleys area which is consisting of soft sediments, there is an increase in earthquake vibrations which results in increased levels of damage. Strengthening earthquakes or earthquake vibrations due to geological conditions have an important impact on the level of damage to buildings that occur. Field observations indicate that the level of damage to the building was caused not only by the level of earthquake strength, but also by the low community's understanding of the risk earthquake. The majority of buildings which is damaged do not follow regional regulations or standards for earthquake-prone areas. The majority of building structures are not designed to be earthquake resistant, so that was damaged.

In some region of Palu, there are large cracks and landslides. Landslides occur as a result of shear stresses that occur when vibrations from earthquakes exceed the maximum shear stress that can be received by the mass of rock / soil on the slope. while liquefaction is the process of changing the saturated granular soil from solid conditions to a liquid state due to the decrease in the effective stress of the soil. The decrease in effective stress causes the soil to lose its shear strength, so that the soil behaves like a liquid [15].

Figure 3 shows the surface displacement in the earthquake area of Palu. The main horizontal displacement process is north and south direcyion. West block moves to the south, while east block has a northward shift. Horizontal displacement is greater than 4 m along the fault zone. While the right picture shows the epicenter centers during earthquake peering occurred during the period 28 Sep - 12 October 2018.

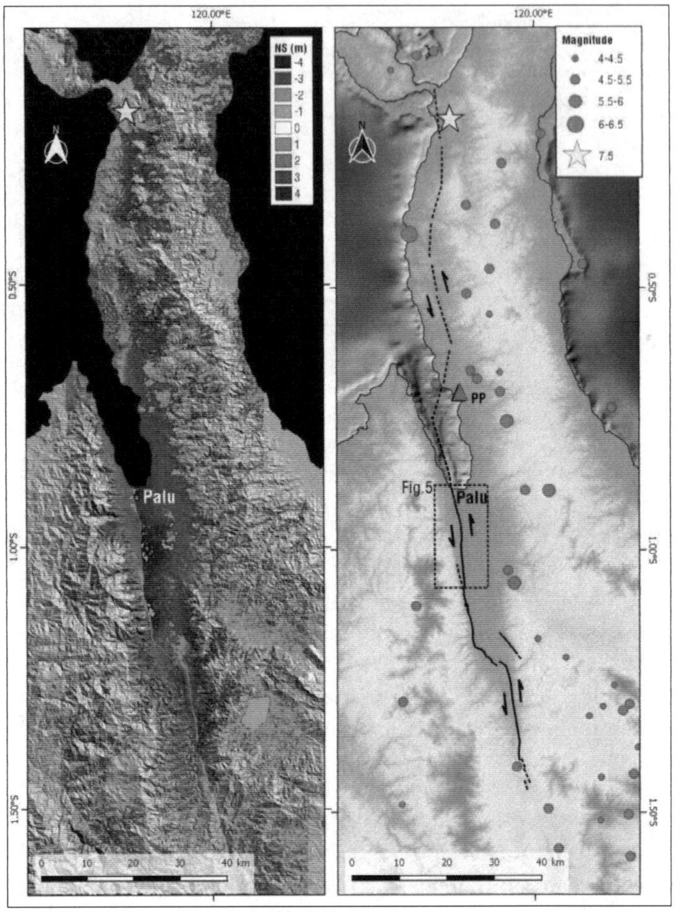

Figure 3. Horizontal Displacement in North and South Direction, The Red Area Are North Moves; The Blue Area Are South. Horizontal Displacement is Greater Than 4 M Along The Fault Zone. Picture Right, Describes The Fault Zone. The Red Circle Indicates USGS Epicentres (M> 4) For The Period Sep 28 - Oct 12 2018 (Valkaniotis Et Al, 2018).

Overall the damage caused by the Palu earthquake was quite severe. Field observations at several locations showed that many buildings were damaged, such as fractures and land subsidence. Many earthquakes cause mass movements. In some areas there are many damage in road bodies. In Petobo and Balaroa there is liquefaction or "dissolution" of rock / soil masses. The change in rock mass that is not compact (not consolidated), saturated with water into a melting state. Liquefaction occur mainly in alluvium regions. Earthquake waves cause ground shocks under certain conditions and one of them can cause liquefaction.

Liquefaction in Petobo and Balaroa

The liquefaction event is the behavior of the soil under a cyclic load that occurs due to a short cyclic load, the soil mass suddenly experiences a transition from a solid state to a liquid state or has a main consistency such as a liquid. Liquefaction in earthquake can be marked by the presence of ground movement in the horizontal direction, seepage of water out of the fracture of the ground, the movement of sloping or descending buildings, land subsidence, and the occurrence of landslides on dikes and slopes. Buildings that are in a simulated area can result in collapse, tilting or moving sideways

IOP Conf. Series: Materials Science and Engineering **650** (2019) 012022 doi:10.1088/1757-899X/650/1/012022

or even resulting in the collapse of the building. From a number of records of past events, it is known that liquefaction events have the potential to occur in the consistency of saturated granular soils that are detached from drainage properties in poor soil. If cyclic loads occur as during an earthquake, loose sand tends to decrease in volume, this results in an increase in pore water pressure and a decrease in the effective shear strength of the soil. The Palu earthquake caused liquefaction in Petobo and Balaroa (Figure 4).

The amplification factor in earthquakes is the ratio of maximum acceleration at ground level to bedrock. The frequency and amplitude that radiates from the bedrock to the surface of the earth will change as it passes through the soil sediment layer. This process causes severe damage to the structure of the building. Earthquake waves that spread on the ground will reflect and spread on the border of the soil layer which has different characteristics. Zones with thick soft sedimentary layers have vibration amplification / amplification 4-5 times [16]. The petobo area is quite far from the earthquake zone but experiences liquefaction due to the thick soft soil layer.

The potential of liquefaction in a soil deposit will be determined by a combination of several components, including the value of the soil properties index, such as dynamic modulus, humidity, fill weight, grain gradation, relative density and soil structure. Also environmental factors such as land formations, earthquake characteristics, geological conditions, groundwater position and effective soil stress [17]. Based field survey, the level of damage is grouped into three classes. That's mild, moderate and severe damage (table 1).

Figure 4. Likuefaction in Petobo (upper) and Balaroa (bottom)

Table 1. Damage Level Criteria.

No	Building Damage	Damage Level
1	Thin crack, still decent for residence	3
2	Large cracks (more than 2cm), not decent for residence	2
3	Severely damaged, not decent for residence	1

Figure 5. Plot the Location of Field Survey in Geological Map.

The level of damage caused by the Palu earthquake-based survey results, then plotted on a GIS-based geological map (Figure 5). Although there is no real correlation in the level of damage to buildings, most of the buildings that were severely damaged (damage classified as three levels) were concentrated in the central plains of Palu and around the fault line. The level of damage to buildings is greatly affected by the position of the building. In addition to being influenced by the distance to the earthquake zone, the level of damage also affected by soil / rock conditions. Lowland areas with geological conditions in the form of sediment deposits or loose land are severely damaged, while areas are rather high with geological conditions in the form of more compact sedimentary rocks experiencing minor damage. Balaroa and Petobo, although a little far from the earthquake zone but have loose soil conditions, experience the most severe levels of damage. While the Watusampu and West Ova regions with positions in hard rock zones, suffered minor damage.

Table 2. Damage level of each observation location

No	Location, Damage Level	Damage Class
1	Watusampu, the level of light damage, the building has a thin crack, it can still be inhabited	3
2	Balaroa, the level of damage is very heavy, experiencing liquefaction, destroyed buildings, some buried	1
3	Pengawu, the level of damage is moderate, the road has a crack of 5cm, the cracked building can still be inhabited	2
4	Baliasih, the level of damage is moderate, the ground has a crack of 10cm, light cracked buildings can still be inhabited	2
5	Gumbasa, the level of damage was very heavy, irrigation canals were cracked up to 25 cm, irrigation buildings were severely damaged	1
6	Petobo, the level of damage is very heavy, experiencing liquefaction, buildings destroyed, some buried	1
7	Horse racing, moderate damage, cracked soil, light cracked buildings can still be inhabited	3
8	Ova Barat, the level of damage is mild, light cracked buildings can still be inhabited	3

4. Conclusions

Based on the results of the study of the influence of geological conditions to the level of damage to buildings is very significant, especially fault zones and rock conditions. Development in Palu city of Palu must be done immediately so the communities activities can normal immediately as usual. Development in Palu city requires careful planning because the Palu area is on a high tectonic track. To reduce the risk of disasters due to population growth and increasing settlement needs, structuring and development need to consider a map of earthquake-prone zones of damage. The geological conditions, especially the distribution of rock types and fault zones, must be taken into consideration in the development of the development of Palu City in the future. Socialization is needed for the community to raise awareness that the Palu area and its surroundings are earthquake-prone areas. For this reason, efforts are needed to improve human quality / increase human resources, with knowledge about earth and earthquake. The level
of damage to buildings is closely related to the structure of the building, so it is necessary to socialize earthquake resistant buildings.

References

[1] Arnani M 2018 Gempa 7,7 Guncang Donggala Sulteng, Berpotensi Tsunami ed B Galih *https://nasional.kompas.com/read/2018/09/28/17173901/gempa-77-guncang-donggala-sulteng-berpotensi-tsunami*

[2] Valkaniotis S, Ganas A, Tsironi V and Barberopoulou A 2018 *A preliminary report on the M7.5 Palu earthquake co-seismic ruptures and landslides using image correlation techniques on optical satellite data, Report submitted to EMSC on 19 October 2018 12:00 UTC*, https://www.researchgate.net/publication/328414705

[3] Pakpahan S, Ngadmanto D and Masturyono 2015 Analisis Kegempaan di Zona Patahan Palu Koro,Sulawesi Tengah, Seismicity Analysis in Palu Koro Fault Zone, Central Sulawesi *Journal of Environment and Geological HazardsJurnal Lingkungan dan Bencana Geologi* **6** No.3 Desember 2015

[4] Westra P 2010 Kontribusi Geologi Dalam Pembangunan Kota Wilayah Bengkulu Paska Gempa Bumi *Jurnal Penelitian UNIB* **XVI** No.1 Januari 2010 pp 21-27 21

[5] Tudes S 2012 *Correlation Between Geology, Earthquake and Urban Planning* (https://www.intechopen.com/books/earthquake-research-and-analysis-statistical-studies- observations-and-planning/correlation-between-geology-earthquake-and-urban-planning)

[6] Rosa and J Taugaloidi 2004 GPS and seismological constraints on active tectonics and arc-continent collision in Papua New Guinea: Implications for mechanics of microplate rotations in a plate boundary zone *J. Geophys. Res.* **109** B05404, doi: 10.1029/2003JB002481

[7] Socquet A, W Simons, C Vigny, R McCaffrey, C Subarya, D Sarsito, B Ambrosius and W Spakman 2006 Microblock rotations and fault coupling in SE Asia triple junction (Sulawesi, Indonesia) from GPS and earthquake slip vector data *J. Geophys. Res.* **111** B08409 doi: 10.1029/2005JB003963

[8] Peraturan Menteri Nomor 21/PRT/M/2007 *Pedoman Penataan Ruang Kawasan Rawan Letusan Gunung Berapi Dan Kawasan Rawan Gempa Bumi* Kementerian PU

[9] Husein S, Karnawati D, Pramumijoyo S and Ratdomopurbo A 2007 Kontrol Geologi terhadap Respon Lahan dalam Gempabumi Yogyakarta 27 Mei 2006 upaya pembuatan peta zonasi mikro di daerah Bantul *Seminar Nasional 2007 Geotechnics for Earthquake Engineering Universitas Katolik Parahyangan* Bandung

[10] USGS (United States Geological Survey) 2017 *Tectonic Summaries of Magnitude 7 and Greater Earthquakes from 2000 to 2015*, Reston, Virginia

[11] Baeda and Yasir A 2011 Seismic and Tsunami Hazard Potential in Sulawesi Island, Indonesia *Journal of International Development and Coorperation* **17** No. 1 pp 17-30

[12] Daryono M, Natawidjaja D H and Sapiie B 2013 Paleoseismology of Palukoro Fault, Historical Records of 1907 & 1909, and recently 2012 Earthquake Events in Central Sulawesi, Indonesia. *Paparan Penelitian dalam 3rd International Symposium on Earthquakes and Disaster Mitigation* Yogyakarta

[13] Karnawati D 2006 The Mechanism Of Rock Mass Movements As The Impact Of Earthquake *Geology Engineering Review And Analysis dinamika TEKNIK SIPIL* **7** Nomor 2 Juli 2007 pp 179–190

[14] Hall R. 1999 *Neogene history of collision in the Halmahera region, Indonesia, Proceeding Indomesia Petroleum Assosiation 1999.*

[15] Yeats R S, Sieh.K and Allen C R 1997 The Geology of Earthquake **568** pp New York; Oxford University Press

[16] Setiani N, Zaenudin A and Hidayati S 2014 Amplifikasi Gempabumi Daerah Subang Jawa Barat, Jurusan Teknik Geofisika Universitas Lampung, https://media.neliti.com/media/publications/244755- amplifikasi-gempabumi-daerah-subang-jawa-afd1bf73.pdf

[17] Warman H and Jumas D Y 2013 Kajian Potensi Likuifaksi Pasca Gempa Dalam Rangka Mitigasi Bencana Di Padang *Jurnal Rekayasa Sipil*

[18] Badan Nasional Penanggulangan Bencana (BNPB) 2012 *Perka BNPB No.2 Tahun 2012 Tentang Pedoman Umum Pengkajian Bnecana, BNPB* Jakarta

[19] Kaharuddin M S, Hutagalung R and Nurhamdan 2011 Perkembangan Tektonik dan Implikasinya Terhadap Potensi Gempa dan Tsunami di Kawasan Pulau Sulawesi, Proceeding JCM Makassar 2011, 1-10, Makassar: *The 36th HAGI and 40th IAGI Annual Convention and Exhibition* pp 26-29 September 2011.

[20] Supartoyo, Sulaiman, C and Junaedi D 2014 Kelas Tektonik Patahan Palu Koro, Sulawesi Tengah, Tectonic class of Palu Koro Fault, Central Sulawesi *Journal of Environment and Geological Hazards, Jurnal Lingkungan dan Bencana Geologi* **4** No.2 pp 111-127

IOP Conf. Series: Materials Science and Engineering **650** (2019) 012023 doi:10.1088/1757-899X/650/1/012023

Evaluation of Ground Anchor Design on Man-made Cut Slope of Cemented Sand: A Case Study

E Bahsan[1], B S Sopandji[1] and F H Sagitaningrum[1,2*]

[1]Department of Civil Engineering, Faculty of Engineering, Universitas Indonesia, Depok 16424, Indonesia

[2]PhD Candidate, Razak Faculty of Technology and Informatics, Universiti Teknologi Malaysia, 54100 Kuala Lumpur, Malaysia

*sagitaningrum1994@graduate.utm.my

Abstract. A system of ground anchor and horizontal drain on a steep cut-slope in Bogor, West Java was designed to stabilize the slope below a bridge structure spanning across a river. Due to the difficulties in the construction of the ground anchor, the design of the ground anchor was evaluated and analyzed for its effect on the stability of the slope. The slope is assumed to have three layers of soil consist of sandy clay (medium stiff), sandy silt (hard), sandy silt (medium stiff), and cemented sand gravel. Four scenarios were evaluated to examine the slope stability: initial stage, the addition of horizontal drain, construction of bored piles with horizontal drain, and addition of ground anchor after the construction of bored piles and horizontal drain. Analysis were done with evaluating the Factor of Safety (FS) of the slope. It was found that the addition of horizontal drain played a big part in stabilizing the slope from initial stage (FS = 0.933 to FS = 1.546), whereas the addition of ground anchor and bored piles were not apparent after the addition of horizontal drain (FS = 1.545 and FS = 1.555), indicating that the main problem of the slope was the high groundwater level. Other than the constructed solutions, shortcrete should also be given at the critical slip surface at the top and surface of the slope in order to reduce erosion by run-off water.

1. Introduction

A prestressed grouted ground anchor is known as a stabilization method for various structures in geotechnical engineering. It is a structural element installed to transmit an applied tensile load into the ground of soil or rock. The ground anchor is installed in grout filled drill holes with basic components such as anchorage, free stressing length, and bond length. The combined system is capable of transmitting the prestressing force from the prestressing steel to the ground surface of the supporting structure [1].

One of the uses of a ground anchor is for a vertical cut-slope stabilization. In this case study, a system of ground anchor and horizontal drain was designed to stabilize the cut-slope below a bridge structure located in Bogor, West Java. However, when the construction had taken place, the designed specification was not fitted to the situation on the field. It was found that the diameter of the ground anchor specified was too large and not suitable for boring as the slope material was very hard to be penetrated with the available equipment. The aim of this report is to evaluate the slope condition and analyze the effect of the ground anchor to the stability of the slope.

2. Site Condition

The slope in this case study is located at one of the abutments of a tollway bridge construction which spans across a river in Bogor, West Java, Indonesia (the exact location will not be mentioned here due to confidentiality). Literature study and field survey was conducted in order to understand the site condition for this report.

A literature study was conducted by first identifying the slope's location to get its geological information. The geological information that was obtained from the Bogor Geological Map [2] inferred that the location is described as older sediments, lava, basalt and andesite with oligoclases-andesine, labradorite, olivine, pyroxene, and hornblende. From this information, it can be concluded that the location is dominated with vulcanic igneous rock due to its location near Gede-Pangrango mountain. It is also known that there wasn't any fault found at the slope location. At the surface of the slope location, it was found that the weathered rocks are classified as red latosol, andosol, and weathered sandstone [3]. The field survey was done in March 2018 to understand the geometry and soil material visually. At that time, the geometry of the slope was already modified by the cutting works. The abutment of the bridge is located at the top of the slope whereas the pier of the bridge is located at the toe. A very steep slope from the cutting works with an angle of $> 70^o$ and a height of 26 m was located near the pier of the bridge. The surface material of the cut section from the slope could be seen to be quite hard with several holes from the attempted ground anchor construction and water flowed out of the horizontal drain holes (Figure 1). From the other side of the slope, there was flowing water from the upper part of the slope, but it wasn't known whether it was from seepage of the slope or surface runoff (Figure 2).

Figure 1. The sight of cut slope material with holes from attempted construction for ground anchor and horizontal drain.

Figure 2. Water flowing at the other side of the slope from an unknown source.

2.1. Secondary Data Evaluation

There were no field and laboratory test done at this reviewed stage of construction due to the completion of the bridge structure at that time, and the situation in which the equipment was difficult to install. Analyses were made based on data from the previous tests done before the construction. The data included:

- Bridge Plan
- Longitudinal section of the bridge
- Boring log of ABT-1 (located at the abutment)
- Boring log of P-1 (located at the pier of the section)
- Boring log of NDB-17 (located near the river)

- Boring Record of Abutment-1 Bored Pile
- Boring log of Horizontal sub-drain
- Boring log of Ground Anchor

As mentioned before, the slope was located between the abutment and pier of the west side of the bridge. The boring logs of ABT-1, P-1, and NDB-17 were used to create the soil stratigraphy, whereas the longitudinal section was used to define the slope geometry. The initial section and the soil stratigraphy from the three boring logs can be seen in Figure 3. From this estimated section, a cut-slope construction was done between the P-1 boring point and P-1 location in order to construct the pier of the bridge at P-1 location. The initial section before the cutting works can be seen from the red full line of the geometry, whereas the dashed red line is the condition of the slope after the cutting works were done. The ground anchor and horizontal drain constructions at the point of the cut-slope near the P-1 location were known to be done after the bridge was completed. At the toe of Pier-1, a river flows through the bridge structure, which is around the NDB-17 point. The groundwater level was extracted from the boring log of P-1 and NDB-17 but was estimated at ABT-1 due to the unavailable data of the groundwater elevation from the boring log.

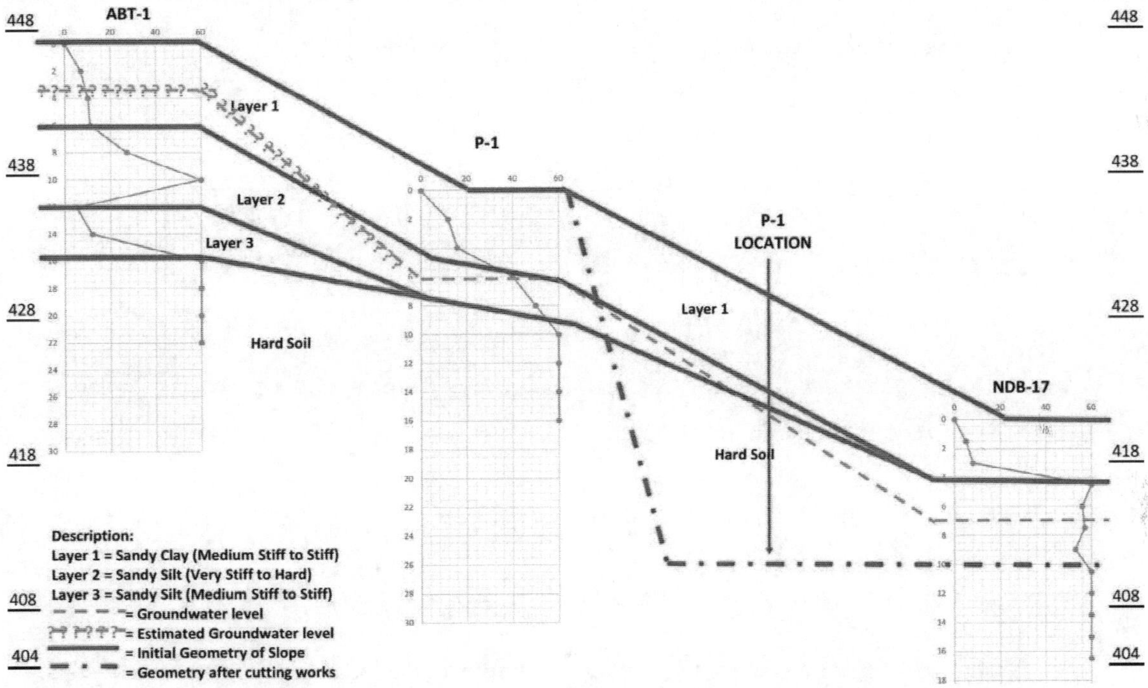

Figure 3. Estimated longitudinal section of the bridge consisting of the three deep boring logs and the location of the bridge's abutment (ABT-1) and pier-1 (P-1). The section shows the natural geometry before the construction of cutting and filling were executed.

In order to validate the interpretation for the stratigraphy in Figure 3, boring logs from the boring record of the bored pile in Abutment-1 was taken into account. It was known that the bored pile of 36.5 meters long was constructed from the elevation of +448.0 m from the top of the slope. At the boring log, elevation +448.0 m to +443.0 m was followed by red clay, which ended at boulders and hard soil at +443.0 m to +418.0 m, and boulders at elevation +418.0 m and 411.5 m. These data were then double-checked with the boring log for the horizontal drain construction at elevation +415.0 m at the face of the slope where breccia stone and andesite boulders with diameter > 35 cm and cemented clayey sand breccia was found. It can be inferred that the soil stratigraphy from the boring logs of ABT-1, P-1, and NDB-17 was correct according to the validation from the boring records.

2.2. Slope Stability Evaluation

Slope stability analysis was done by calculating the factor of safety (FS) of the slope which is basically the ratio of shear strength of the soil to the shear stress experienced by the soil (e.g. [4]). The shear strength is basically affected by cohesion (c) and internal friction angle (φ), whereas shear stress is affected by the weight of the soil itself or external loads at the surface. Some literature stated that a slope is safe when its FS is around 1.25 – 1.5 (e.g. [4] and [5]). Another source, which is the standard for geotechnical engineering construction in Indonesia (SNI 8460:2017) explained that the FS is based on the uncertainty of analysis, repairment cost if the structure failed, and the additional cost of designing the slope in a more conservative way [6]. From all the standards taken into account, this case study took the critical FS value of 1.5.

The slope analysis was done using the Finite Element Method (FEM) with Plaxis 8.2 2D software. The slope was assumed to be in plane strain condition (as seen in Figure 4) with the material model of Mohr-Coulomb.

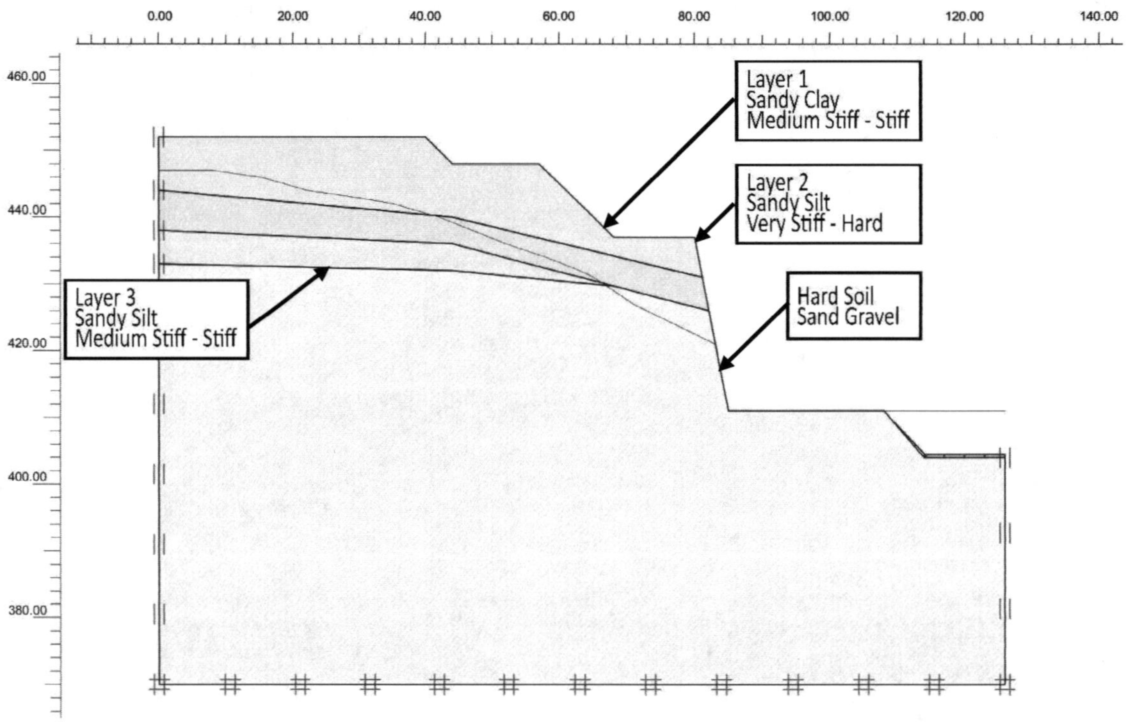

Figure 4. FEM model for the slope according to to the identified soil stratigraphy.

From the soil stratigraphy and N-SPT value from Figure 3, a correlation was done from the literature (e.g. [7]) in order to assume the needed parameters for the FEM model. The values used in the model can be seen in Table 1.

Table 1. Material parameters for each soil layer of the slope.

Soil Properties	Layers			
	Layer 1: Sandy Clay Medium Stiff - Stiff N-SPT = 9,5	Layer 2: Sandy Silt Very Stiff-Hard N-SPT = 33	Layer 4: Sandy Silt Medium Stiff-Stiff N-SPT = 8,5	Hard Soil (Sand Gravel) N-SPT = 60
Model Material	Mohr-Coulomb	Mohr-Coulomb	Mohr-Coulomb	Mohr-Coulomb
γ_{unsat} (kN/m^3)	17	13	17	18
γ_{sat} (kN/m^3)	19	17	19	21
E (kN/m^2)	20000	30000	30000	80000
υ (poisson ratio)	0.325	0.25	0.30	0.30
Cohesion (kN/m^2)	30	45	25	150
Friction Angle (°)	25	28	27	37

There are two stages of analysis in PLAXIS 8.2 for this case, the first is the initial stage of having the soil's own weight taken into account and the latter is the strength reduction (phi-c reduction) analysis in order to find the FS and the critical slip surface of the slope. Strength reduction itself is a method in which the cohesion and friction angle of the material was reduced with the estimated FS calculated through Finite Element Method until the value is non-convergent [8]. The FS value from the analysis will be the 'true' FS value for the situation and can be considered as the FS for the slope.

The previously described analysis was done to four stability scenarios according to the construction phases for the slope after the cutting works were finished. The first scenario represented the initial state after the cutting of the slope. The second scenario represented the condition after the horizontal drain was constructed. The third scenario represented the condition of the slope after the foundation for the pier and abutment were constructed, with the addition of horizontal drains. Lastly, the last scenario represented the condition the same as the third scenario with the addition of ground anchors.

2.3. Ground Anchor Modelling in PLAXIS 8.2

Ground Anchor in PLAXIS 8.2 was specifically modeled as a node-to-node element with the end of the structure as geogrid (tensile) elements. The node-to-node anchor modeled the free length of the ground anchors. As is implied in the name, the anchor was connected at both ends to nodes in the structural element, as if there was no contact of the anchor body to the surrounding soil. As for the geogrid (tensile) element, the model was designed to accommodate the bond length of ground anchors. In practice, the bond length was normally pressure grouted so that the soil can be completely in contact with the grouted body [9].

Although the precise stress state and interaction of the soil, in reality, is much more complex and cannot be modeled in 2D with precision, this estimation of stress distribution, deformations, and the stability of the structure on a global level is sufficient. It is also important to note that the grout body was assumed to not slip relative to the soil. Another consideration was that this type of model could not evaluate the pullout force of the ground anchor [10].

3. Results and Discussion

3.1. Finite Element Results

The results and analysis of the four scenarios are given below.

3.1.1. Initial Condition. The initial condition of the slope was reviewed with high Groundwater Elevation (GWL) starting from elevation +443.0 m at the top of the slope to around +420.0 m at the face of the slope, continued by the surface of the water from the river. These elevations were taken from the boring log of the three boreholes and can be seen in the software in Figure 5. The result for this condition was a factor of safety (FS) of 0.933 and potential of steep slight movement at the bottom of the slope as seen in Figure 6. This value indicated that at the intial condition, the slope was not safe as the calculated FS value was lower than 1.5.

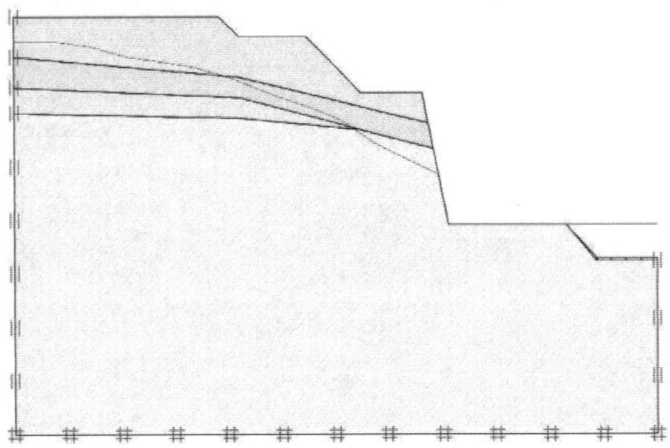

Figure 5. Finite Element representation of the initial groundwater level condition of the slope.

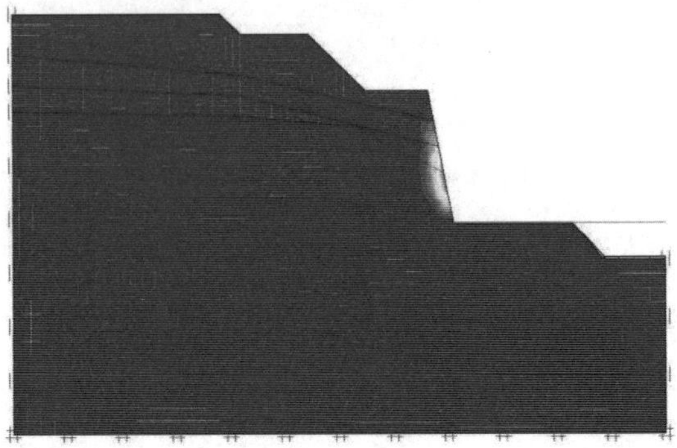

Figure 6. Simulation result for the initial groundwater condition (FS = 0.933).

3.1.2. Groundwater Condition Altered due to Horizontal Drain.

The second scenario was reviewed after the horizontal drain was constructed. The specification of the horizontal drain was 16 m long with the inclination degree of 5° towards the horizontal surface and was constructed at elevation +415.0 m as seen in Figure 7. This construction lowered the groundwater level at the surface slope which raised the FS value to 1.546. When compared to the initial condition, it can be seen that the slip surface of the slope shifted from the surface of the slope to the top of the slope which happened to be at the medium stiff to stiff sandy clay layer and very stiff to hard sandy silt layer. Since the value of FS for the slope at this condition is above 1.5, the slope is considered to be safe. The potential movement of the critical slip surface illustration of the slope can be seen in Figure 8.

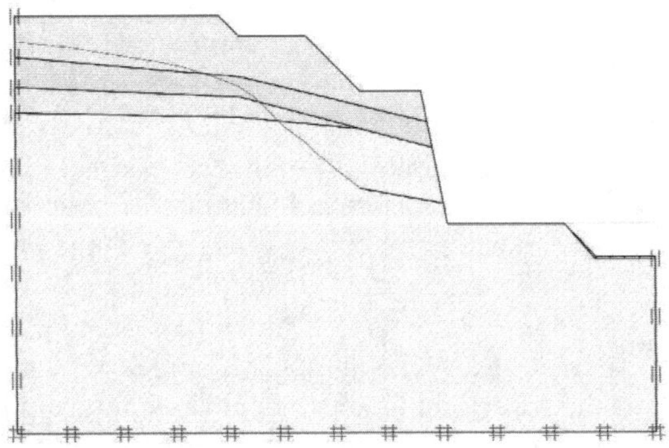

Figure 7. Finite Element representation of the altered groundwater condition due to horizontal drain construction.

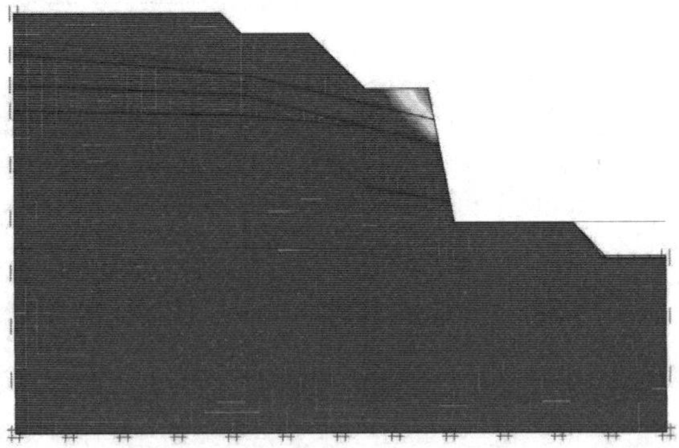

Figure 8. Simulation result for the altered groundwater condition due to horizontal drain construction (FS = 1.546).

3.1.3. Condition after bored pile construction.

In the construction process, the bored pile construction was done before the construction of horizontal sub-drain. However, this research wanted to emphasize the bored pile as a structure to stabilize the slope rather than only as a structure of the bridge itself. On the other hand, the horizontal sub-drain construction on this research was meant to emphasize the role of groundwater level for the stability of the slope.

IOP Conf. Series: Materials Science and Engineering **650** (2019) 012023 doi:10.1088/1757-899X/650/1/012023

The specifications for the bored pile were fc' = 29 MPa and Diameter = 1.2 m, whereas the length of the bored pile was 35 m for abutment-1 and 24 m for pier-1. The situation of the bored pile can be seen in Figure 9. The result of this condition can be seen in Figure 10. The FS value for this condition is 1.545. This value is not far off from the FS value in the horizontal drain condition which indicates that the bored pile construction didn't contribute much to the safety of the slope. Theoretically, the bored pile should contribute to the stability of slope when it is constructed at the failure area of the slope. From Figure 10, it can be seen that the critical slip surface of the slope was located at the tip of the steep cut slope, whereas the position of the bored piles was located at the top of the slope and at the toe of the steep cut slope. The critical slip surface location indicated that the bored pile's location was not in the critical slip surface of the slope, which is the reason why the FS value didn't increase from the bored pile constructions.

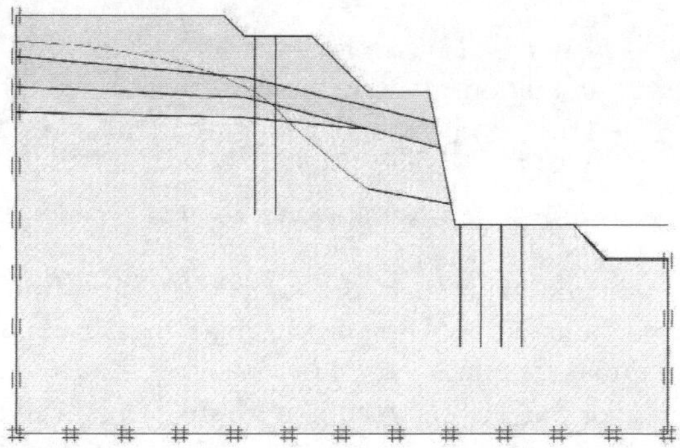

Figure 9. Finite Element representation of the slope condition after the construction of bored pile and horizontal drain.

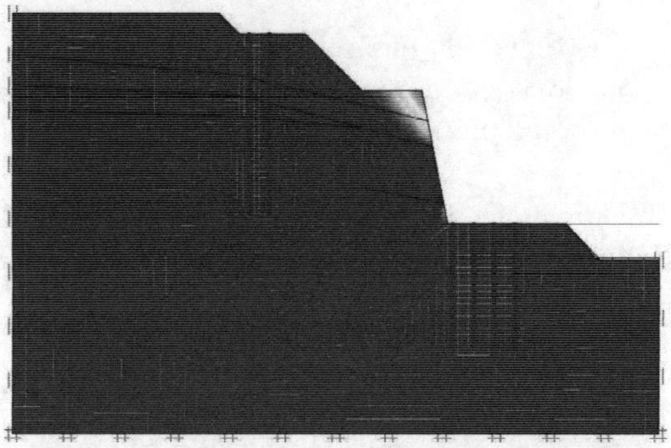

Figure 10. Simulation result for the slope condition after the construction of bored piles and horizontal drains (FS = 1.545).

3.1.4. Slope condition after ground anchor construction.

The construction of the ground anchor was done after the bored piles and horizontal drains were installed. The ground anchor specification in this research was the result of recalculating same bond strength of the ground anchor into a different diameter and spacing specification. The specification was altered due to the failure of penetrating the soil with even the highest grade of the drill bit. The aim of this analysis phase was to decide whether the new specification will still stabilize the slope in its constructed condition. The specification for the ground anchor in this condition was 25 m long with the diameter of 100 mm, the bond length of 6 m, and spacing of 2 m. The ground anchor was constructed at elevation +417.5 m and +421.5 m with 25° inclination from the horizontal axis (as seen in Figure 11). The pull out strength of the ground anchor was 600 kN. The simulation resulted with a critical slip surface around the top of the steep cut-slope surface, where the sandy clay and sand silt layers were located, with the FS value of 1.555 (as seen in Figure 12). This value of FS is not far off from the FS from the previous conditions. The FS value above 1.5indicated that the design was safe enough for stabilizing the slope.

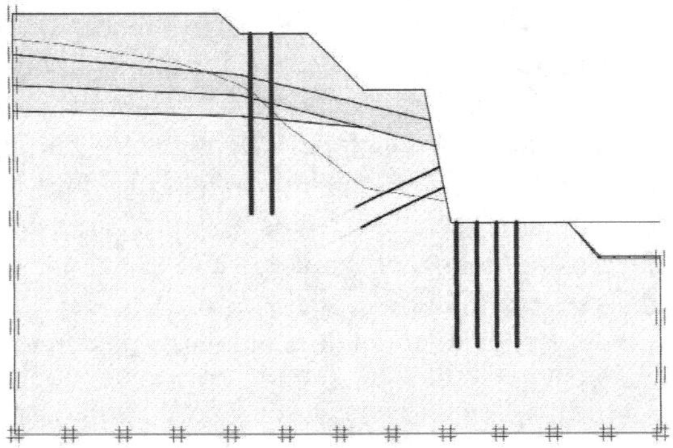

Figure 11. Finite Element representation of the slope condition after the construction of the bored pile, horizontal drain, and ground anchor

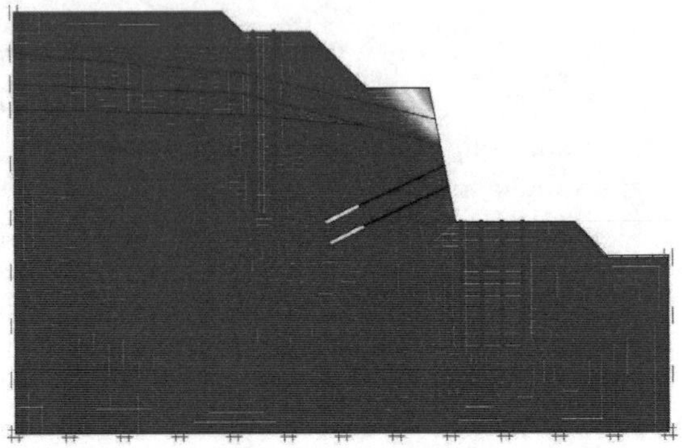

Figure 12. Simulation result for the slope condition after the construction of bored piles, horizontal drains, and ground anchors (FS = 1.555)

3.2. Discussion

According to the soil classification, it can be seen that the material at the steep cut slope itself was already quite hard. The data from the boring logs and field survey had shown that the material at the face of the slope was mostly cemented which indicated that the cohesion value of the soil is indeed quite high. But, this data should be reviewed with the Factor of Safety (FS) from the Finite Element Method where the summarized results can be seen in Table 2 below.

Table 2. Finite Element Method Results.

Conditions	Factor of Safety
Initial Condition	0.933
Addition of Horizontal Drain	1.546
Construction of Piles with the addition of Horizontal Drain	1.545
Construction of Ground Anchor with Piles and Horizontal Drain	1.555

From the Finite Element Method results, it can be seen that the natural condition's FS was under 1.5, which can be considered as unsafe. In most cases, cemented condition for the hard soil of a slope would be enough to maintain its stability without any stabilization method. Yet, the unsafe value of FS indicates that the high groundwater level of the slope needs a stabilizing measure in order to make the slope safe. From the horizontal drains construction condition, it can be seen that installing the horizontal drain has a big effect in which it lowered the groundwater level within the slope. The installation of horizontal drains successfully raised the FS value into a safer condition (FS > 1.5). In contrast, the construction of bored piles and ground anchors did not result in a significant increment of FS value, which validates that the main problem of the slope was not the shear strength of the material, but instead was the groundwater level of the slope.

It should be taken into account that near the constructed slope with the horizontal drain, there were running water on the surface of the slope which is shown in Figure 2. This condition indicates that surface water should also be taken into account in stabilizing the slope. Due to that reason, other than applying ground anchor and horizontal sub drains at the surface of the steep cut slope, shortcrete should also be added as a covering for the surface in order to prevent erosion of the slope from surface runoff. Not only that, since the critical slip surface was located at the top of the cut-slope, shortcrete should also be applied there in order to prevent erosion which could lead to landslide at the top of the slope.

IOP Conf. Series: Materials Science and Engineering **650** (2019) 012023 doi:10.1088/1757-899X/650/1/012023

4. Conclusion

In this study, the stability of a very steep cut-slope of cemented sand was evaluated. The slope is located at a tollway bridge spanning through a river in Bogor, West Java and had gone through cutting works. Several conditions for stabilization of the slope were evaluated using the Finite Element Method in order to understand the effects of the stabilization to the safety of the steep cut-slope.

It can be concluded that the initial state of the slope after the cutting works was not safe with the FS value of 0.933 and an additional possibility of a shallow failure on the slope surface from the displacement shading of the FEM model. After the installation of horizontal drains, the slope became quite stable due to the altered elevation of the groundwater level at the surface of the slope with the FS value of 1.546 and an additional slope failure area which moved to the top of the surface instead. Afterward, construction of ground anchor was evaluated in order to know the effect of the ground anchor to the stability of the slope. It was known that the ground anchor doesn't significantly add the stability of the slope with the FS value of 1.555 and the slope failure area still at the top of the slope. A similar condition of stability was seen when bored piles from the abutment and pier of the bridge were constructed, whereas the existence of the bored piles didn't affect the stability of the slope in general with the FS value of 1.545 and the slope failure surface was still at the top of the slope, which is the same condition with the addition of horizontal drain.

From the conditions, it can be inferred that the main problem of the stability of the slope was the high groundwater level at the slope due to the rise of FS value from 0.933 to 1.545. For further protection, shortcrete should be applied to the surface of the slope and at the critical slip surface, in addition, the existing construction to reduce the possibility of erosion of the surface of the slope.

References

[1] Sabatini P J, Pass D G and Bachus R C 1999 *Geotechnical Engineering Circular No. 4: Ground Anchors and Anchored Systems* (Washington: FHWA) p 281

[2] Effendi A C, Kusnama and Hermanto B 1998 *Peta Geologi Lembar Bogor, Jawa* vol 2 (Bandung: Geological Research and Development Centre)

[3] Wesley L D 2010 *Geotechnical Engineering in Residual Soils* (New York: John Wiley & Sons, Inc.)

[4] Duncan J M and Wright S G 2005 *Soil Strength and Slope Stability* (New York: John Wiley & Sons, Inc.)

[5] Abramson L W, Lee T S, Sharma S and Boyce G M 2001 *Slope Stability and Stabilization Methods* (New York: John Wiley & Sons, Inc.)

[6] SNI 8460:2017 2017 *Persyaratan perancangan geoteknik* (Jakarta: Badan Standardisasi Nasional)

[7] Budhu M 2010 *Soil Mechanics and Foundations* vol 3 (New York: John Wiley & Sons, Inc.)

[8] Griffiths D V and Lane P A 1999 Slope stability analysis by finite elements *Geotechnique* **49** No.3 pp 387-403

[9] Gouw Tjie-Liog 2014 Common Mistakes on the Application of Plaxis 2D in Analyzing Excavation Problems *Int. J. Appl. Eng. Res.* **9** No.21 pp 8291-8311

[10] Al-Khoury R, Bakker K J, Bonnier P G, Brand P J W, Broere W, Burd H J, Soltys G, Vermeer P A and Den Haag DOC 2002 *PLAXIS 2D – Version 8 Manual* (Netherlands: A. A. Balkema Publishers)

IOP Conf. Series: Materials Science and Engineering **650** (2019) 012024 doi:10.1088/1757-899X/650/1/012024

ANALYSIS OF THE STABILITY OF DAM BUILDING AFTER COUNTERWEIGHT ADDITIONS (A Case Study: Gonggang Dam, Magetan Regency)

P Kuntoro*, N Surjandari and R Hadiani

Department of Civil Engineering, Universitas Sebelas Maret, Surakarta, Indonesia

*pranatasetyo24@gmail.com

Abstract. The Gonggang Dam was first planned to be a rock filled dam, but due to the economic consideration, in the construction, the materials were changed from rock piles to random heaps with the fixed degree of slope. After being operated for a year with the observation on the possibility of seepage, no seepage was found. Therefore, to increase the stability of the dam body, a counterweight in the form of random pile completed with a filter and rip-rap was added to the lower part of the body of dam. The height of counterweight was determined based on the safety needs from the flow (surface) of the seepage line and the stability of the dam. The calculation results of stability analysis of the dam body with counterweights was considered effective as the downstream safety. Safety factors obtained from empty conditions, flood water levels and rapid drawdown without seismic load show the equation: security factor ≥1.2, which was in accordance to the permitted safety factor: sf ≥1.2. Meanwhile, the calculation result of the analysis of stability of the dam body with Operation Base Earthquake (OBE) and Maximum Design Earthquake (MDE) in empty condition and flood water level, the safety factors obtained were still above the permitted limit, ≥1.2. However, for rapid drawdown conditions with the effect of the MDE seismic load was not yet safe (collapse), so it was necessary to modify the design of the upstream leg section. The design modification was carried out by adding a random land fill structure that blends with the body of the main dam. The method was expected to be able to withstand the tension occurring upstream, thus increasing the stability of the dam body. This was proven to be effective based on the calculation results obtained by the safety factor of 1.115. The number was still included in the permitted Maximum Design Earthquake (MDE) safety number of load effect (sf ≥1).

1. Introduction

The Gonggang Dam that began operating in 2011 is located on the Gonggang River in Poncol Village, Poncol Sub-district, Magetan Regency, East Java. The Gonggang Dam is a central core embankment dam, with the height of 60.00 m, peak length of 226.00 m, peak width of 10.00 m, upstream slope of 1: 2.50 and downstream slope of 1: 2.00. The Gonggang Dam was at first planned to be a rock filled dam, but due to economic considerations, in the realization, the material was

IOP Conf. Series: Materials Science and Engineering **650** (2019) 012024 doi:10.1088/1757-899X/650/1/012024

changed from rock piles to random piles with a fixed slope. After being operated for a year with the observation on the possibility of seepage, no seepage was found. Therefore, to increase the stability of the dam body, a counterweight in the form of random pile completed with a filter and rip-rap was added to the lower part of the body of dam. The height of counterweight was determined based on the safety needs from the flow (surface) of the seepage line and the stability of the dam.

2. Safety Factor Requirements

The assessment of minimum safety factors for stability of embankment dam is based on *Pedoman Konstruksi dan Bangunan Sipil Analisis Dinamik Bendungan Urugan* (Construction and Civil Building Guidelines of Embankment Dam Dynamic Analysis) No 27/KPTS/D/2008 dated 31 January 2008, page 115. The following is the table of requirements of the safety factors.

Table 1. Requirements for minimum safety factors for the stability of the embankment dam (a)

No	Condition	Pore Water Pressure	SF Without earthquake	SF With earthquake
1.	Finished construction depends on: 1. Development schedule 2. Relationship between Pore pressure and time U / S and D / S slopes With an earthquake without damage used 50% coefficient Earthquake design	The increase in pore water pressure in piles and foundations is calculated using laboratory data and instrument monitoring	1.30	1.20
		Items just without instrument supervision	1.40	1.20
		Only on heaps without lab data. and with / without supervision of instruments	1.30	1.20
		Without instrument	1.30	1.20
2.	Lasting flow depends on: 1. Normal water level next to the bumpkin 2. U / S and D / S slopes. With an earthquake without damage used 100% coefficient. Earthquake design	From seepage analysis	1.50	1.20

IOP Conf. Series: Materials Science and Engineering **650** (2019) 012024 doi:10.1088/1757-899X/650/1/012024

Table 1. Requirements for minimum safety factors for the stability of the embankment dam (b)

No	Condition	Pore Water Pressure	SF Without earthquake	SF With earthquake
3	Operation of the reservoir depends on: 1. Maximum water level Upstream 2. Minimum water level upstream (dead storage). Slope U/S must be analyzed for fast receding conditions	Fast recede from normal water level to minimum water level. U / S and D / S slopes.	1.30	1.10
		Fast recede from maximum MA elevation to minimum MA elevation. The effect of the earthquake was taken 0% of the coefficient Earthquake design	1.30	-
4.	Emergency conditions depend: 1. Stacking the drainage system 2. Low tide because the use of water exceeds needs 3. Fast receding emergency needs	Fast recede from maximum water level to the lowest elevation of expenditure buildings. The effects of the earthquake are ignored	1.20	-

Note: check the standards about the method of dynamic gravity dam analysis
*) For OBE; while **) for MDE, FK $\geq$ 1

3. Imposition

According to Suyono Sosrodarsono, in the embankment dam, there are three loading conditions in calculation of the stability of the dam body, namely:

3.1. Initial condition of the dam

It is condition where tension and strain occur due to the load of the dam before being filled with water.

3.2. Dam condition after being filled with water

It is condition where tension and strain occur due to dam and water loads (maximum load of water at surface).

3.3. Rapid drawdown conditions

It is condition where tension and strain occur due to a decrease in reservoir water on the dam.

4. Method

The data collection is basic means to determine the resolution of a problem scientifically. The data collected comprise primary data and secondary data.

4.1. Primary data
Primary data refer to data obtained from direct observations of researchers at research locations such as:

- Survey on the location with the aim of observing the situation of the research location
- Taking photos of research locations for observation and analysis.

4.2. Secondary data
Secondary data refer to data obtained from laboratory at the time the research is carried out on the location. The secondary data cover information on soil, such as Soil Properties and Soil Engineering. Soil Properties include soil content weight (γ), water content (w), void ratio (e), porosity (n), specific gravity (Gs), Dry density, Saturade density, Poisson ratio, Young's modulus, Cohesion, Fristion angle, Dilatancy angle, Horizontal permeability, and Vertical permeability, while Soil Engineering comprise the results of Direct Shear Test, Triaxial Test, and Unconfined Test.

4.3. Image Data
Image data cover the image of floor plan, image of cross sections of the dam body, and image of elongated pieces of the dam body.

4.4. Technical Data
Technical data comprise data of main dam and inundation area. Data of main dam include type of dam, height of dam, length of peak, width of peak, elevation of peak of dam, elevation of river bed, slope of upstream, slope of downstream, and body volume of the dam. Data of inundation area consist of water catchment area, river length, maximum water level elevation, minimum water level elevation, dam reservoir volume, sediment reservoir volume, and inundation area.

Stage 1 – collecting data in the forms of:

- properties of soil obtained from the results of the Gonggang Dam Lab,
- map,
- floor plan,
- cross section, and
- elongated pieces.

Stage 2 – calculating the safety factor of the dam body without seismic load or with seismic load in a condition when:

- the dam is empty
- the dam is in flood water level, and
- the dam is in rapid drawdown condition.

Stage 3 – calculating stability, which is:

- calculating stability $SF \geq 1.2$

Stage 4 – with calculated stability before < after the implementation of Counterweight and Stability is safe, if it is not safe then recounting is done with design modifications.

Stage 5 – Determining the decision.

IOP Conf. Series: Materials Science and Engineering **650** (2019) 012024 doi:10.1088/1757-899X/650/1/012024

Below is the flow chart of the stages above.

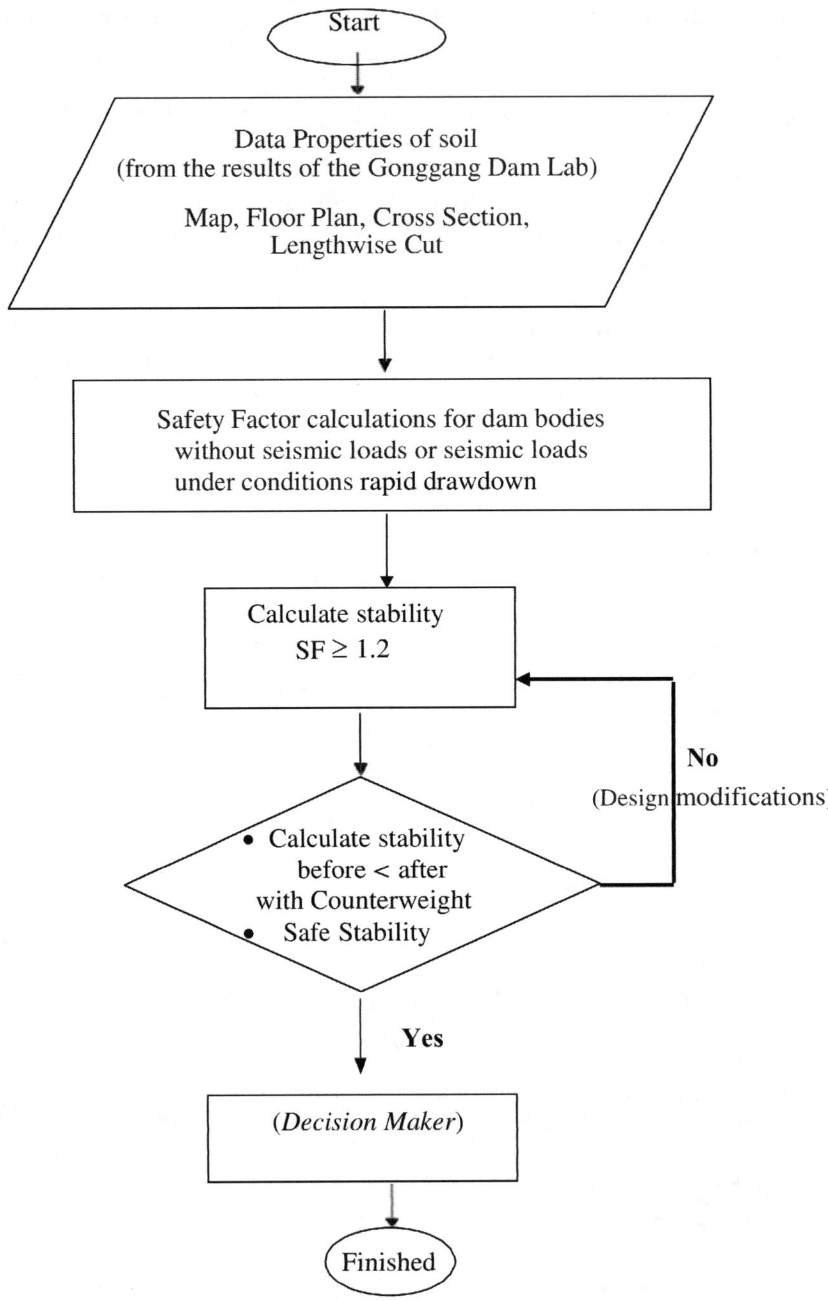

Figure 1. Chart of Research Flow

IOP Conf. Series: Materials Science and Engineering **650** (2019) 012024 doi:10.1088/1757-899X/650/1/012024

5. Result and Discussion

The following are the results and discussion of the calculations of the three loadings, with and without seismic loads.

Table 2. Calculation of safety factor in rapid drawdown condition

No	Condition	SF With earthquake		Total Displacement	
		Before	After	Before with counterweight	After with counterweight
1.	Rapid Draw down	1,184 (OBE)	1,109 (OBE)	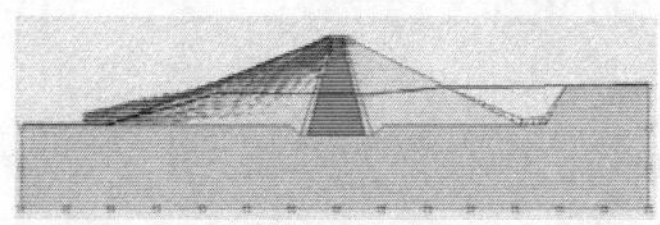 Picture 1	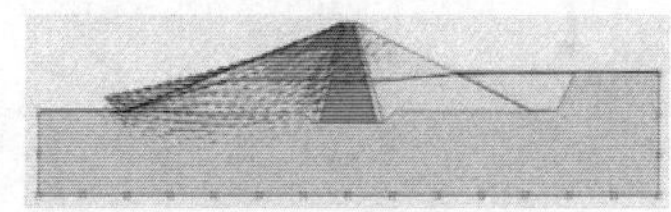 Picture 2
				Caption: • Conditions at the time of the dam are Rapid Drawdown • Filtration flow is in the center of the dam's body • The collapse mechanism is shown in red arrows, which are at the top of the dam's body towards the lower upstream • The tension does not reach the foundation	Caption: • Conditions at the time of the dam are Rapid Drawdown • Filtration flow is in the center of the dam's body • The collapse mechanism is shown in red arrows, which are at the top of the dam's body towards the lower upstream • Tension due to seismic loads, to foundation parts • With the presence of counterweight the greater the tension that occurs, especially on the foundation
		Collapse (MDE)	Collapse (MDE)	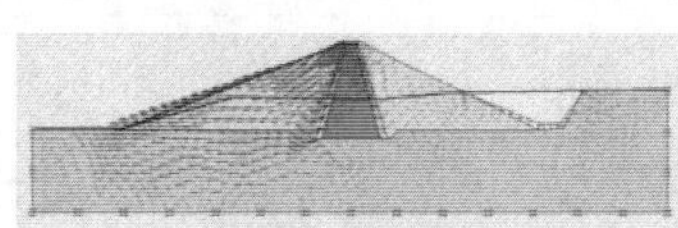 Picture 3	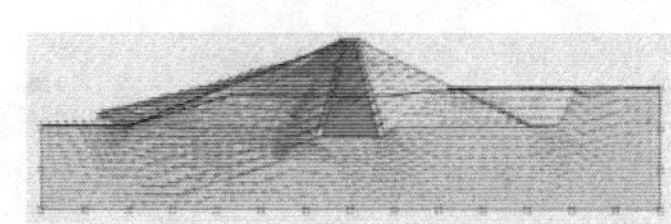 Picture 4
				Caption: ➢ **With the MDE earthquake the tension generated reaches the foundation**	Caption: ➢ **With the MDE earthquake the tension generated on the foundation is greater than without counterweight**

IOP Conf. Series: Materials Science and Engineering **650** (2019) 012024 doi:10.1088/1757-899X/650/1/012024

From the results of analysis of the Safety Factor that has not met the conditions in Rapid Drawdown with earthquakes Maximum Design Earthquake (MDE), a design modification is needed to increase the stability of the dam body. The design modification is focused mainly on the upstream part of the dam body, by adding a random embankment structure that blends with the existing heap.

The following is a modification of the dam body design calculation on the condition of Rapid Drawdown with seismic load Maximum Design Earthquake (MDE), to find the safety number required by using Plaxis Method.

Table 3. Calculation of safety factor in rapid modification condition

No.	Conditions of MDE Seismic Load		Captions
	Before with Counterweight	After with Counterweight	
1	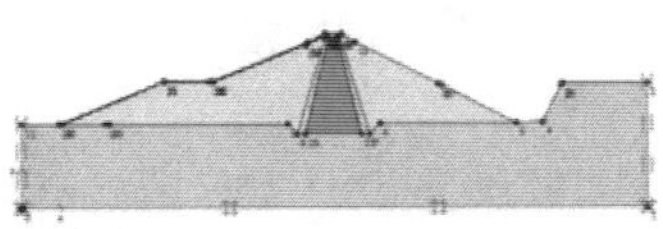		Depiction of design modification, by adding landfill upstream
	Figure 5. *Standart fixities*		
2	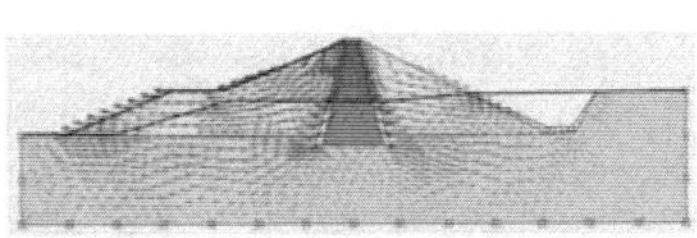	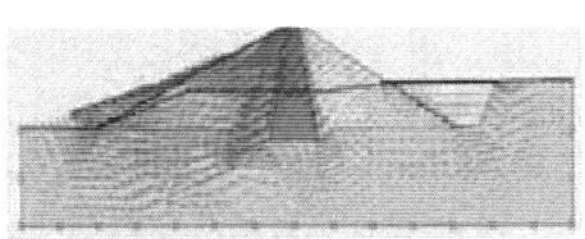	The calculation result of safety factor with the counterweight is higher than without counterweight, which is 1.115 > 1.00
	Figure 6. *Total deformation after design modification and without counterweight*	Figure 7. *Total deformation before design modification*	The tensions which occur without counterweight, with design modification is higher both upstream and downstream
		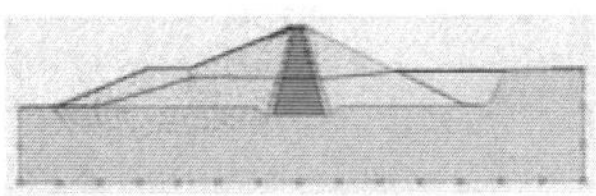 Figure 8. *Total deformation after design modification*	The tensions which occur after design modification is lower than before design modification.

6. Conclusion

Conclusion from the Safety Factor Calculation Table above can be elaborated as follows:

1. Counterweight is considered effective when there is no seismic load in all three conditions. It can be seen from the increasing safety factor.
2. Counterweight is effective when loading occurs downstream, in the presence of seismic loads.
3. Counterweight does not give any role at the time of loading which occurs upstream (rapid drawdown condition), especially in MDE earthquake conditions. Conversely, with the presence of counterweight, there is a large tension on the foundation.
4. In the operation of the reservoir, the speed of water level reduction is set to prevent rapid drawdown (the pattern of reservoir operations contains at least the procedure for removing water from the reservoir in accordance to the conditions of water reservoir volume and elevation and the water capacity and river capacity downstream of the dam)

7. Recommendation

To follow up this research, further research is needed regarding the development of the theme and methodology. The suggestions for further research include:

1. The necessity to do more detailed calculations of design modifications both in analysis and modeling in the laboratory, so that the best and most efficient alternatives can be determined to be applied in the field,
2. The necessity to study further about the technical implementation of design modifications.

References

[1] Astuti Y, Masrevaniah A, Marsudi and Suwanto 2012 analisa rembesan bendungan bajulmati terhadap bahaya piping untuk perencanaan perbaikan pondasi. *Journal of Teknik Pengairan Universitas Brawijaya Malang.* **3**(1) pp 51–60

[2] Beheshti A, Amirabbas, Kamanbedast, Akhbari and Hossein. 2013. seepage analysis of rock-fill dam subjected to water level fluctuation: a case study on gotvand-olya dam *Iranica Journal of Energy & Environment* **4**(2) pp 155-160

[3] Brinkgreve R B J 2007 Plaxis 2-D versi 8. *Delft University of Technology Civil Engineering Dimension* **14(**2) pp 100-109

[4] Departemen Pekerjaan Umum. Dirjen SDA. 2008. *Pedoman konstruksi dan bangunan sipil. analisis dinamik bendungan urugan.*

[5] Agustawijaya D S and Sukandi 2012 *The stability analysis of the Lusi Mud volcano embankment dams using FEM with a special reference to the Dam Point* (Bandung D.Dharma) P 10.

[6] Dharmayasa, Putu I G N, Redana I W, Putra and Tjok Gde Suwarsa 2014 analisis keamanan lereng bendungan utama pada bendungan benel di kabupaten jembrana. *Journal of Spektran*, **2**(2)

[7] Hasani H, Mamizadeh J and Karimi H 2013 stability of slope and seepage analysis in earth fills dams using numerical models (case study: ilam dam-iran) *World Applied Sciences Journal* **21**(9) pp 1398-1402

[8] Kamanbedast A & Delvari A 2012 Analysis of earth dam: seepage and stability using ansys and geo-studio software. *World Applied Sciences Journal* **17**(9) pp 1087-1094

[9] Loebis and Joesron 1987 *Banjir rencana untuk bangunan air.*

[10] Mori and Kiyotoka 1999 *Hidrologi untuk pengairan* (Jakarta: PT. Pradnya Paramita)

[11] Nanda, Ferry T N, Hamdhan and Noer I 2016. *Analisis rembesan dan stabilitas bendungan bajulmati dengan metode elemen hingga model 2D dan 3D.* Online Journal of Institut Teknologi Nasional **2** p 4

[12] Noori, Bahzad M A and Ismaeel K L 2011 *Evaluation of seepage and stability of Duhok Dam.* College of Engineering University of Duhok. Paramita. Jakarta. Pekerjaan Umum. Jakarta. Plaxis. Belanda.

[13] Sastraatmadja and Soedrajat 1982 *Anggaran biaya pelaksanaan* (Bandung:Nova)

[14] Soedibyo 1993 *Teknik bendungan.* (Jakarta: PT. Pradnya Paramitha)

[15] Soemarto C D 1999 *Hidrologi teknik.* 2nd edition. (Jakarta: Erlangga)

[16] Sosrodarsono S 1977 *Bendungan type urugan.* 4th edition (Jakarta: PT. Pradnya Paramitha)

Soft Soil Stabilization With Rice Husk Ash and Glass Powder Based on Physical Characteristics

D P Kusumastuti[1*], I Sepriyanna[1]

[1]Sekolah Tinggi Teknik PLN, Menara PLN Jalan Lingkar Luar Barat, Duri Kosambi, Cengkareng, Jakarta Barat,Indonesia

*dyah.pratiwi@sttpln.ac.id

Abstract. Soft soil has poor characteristics when used as part of construction. This is because soft soil generally has low bearing capacity, large changes due to settlement and compression and high water content. Efforts to improve soil, especially soft soil using chemical stabilization methods or by adding materials, are one of the best and most applied solutions. The selection of the good material will affect the results of stabilization so that the objectives to improve physical characteristics, especially water content and consistency limits on soft soil will be fulfilled. Rice husk ash and glass powder as stabilizers in soft soil can change physical characteristics, especially water content and consistency limits. Based on the results of the study, rice husk ash and glass powder were able to reduce water content, liquid limit and plastic limit and increase the soft soil shrinkage limit test results.

1. Introduction

Soil is useful as a construction material in various types of Civil Engineering work, besides that the soil also functions as a support for the foundation of the building [1]. Soil is a material that is very influential on the stability of building structures, such as building structures, highway pavements, tunnels, dams, etc [2]. However, not all soil can be directly used in Civil Engineering work, this is because some soil has a low bearing capacity, large deformations or changes and a large enough water content that affects the consistency of the soil.

Soft soil is one type of soil that has a low bearing capacity [3], if given a load on it will result in a very significant decrease in the occurrence of a non-uniform decline [4]. To overcome or reduce the problem of construction that is erected on soft soil is to make soil improvement. The method of soil improvement that is generally applied to soft soil is stabilization. Soil stabilization is an attempt to change or improve the technical characteristics of the soil to meet certain technical requirements [5]. In general there are two ways of soil stabilization, namely by mechanical stabilization and chemical stabilization [6].

Cheap and effective stabilization in soft soil is to add certain chemicals [7]. Chemicals commonly used are cement, lime, fly ash [5] and other materials characterized by pozzolan so that it can produce cementation processes [8]. In this study, chemical stabilization of soft soil was carried out using rice husk ash and glass powder. The selection of rice husk ash as stabilization material because rice husk ash can fill the cavities that exist between the droplets of soil [9] and contains a lot of silicate elements (SiO_2) and aluminate (Al_2O_3) so it is categorized as pozzolan [10]. While the glass powder used is waste from glass fragments which are then crushed into powder. Broken glass which becomes

IOP Conf. Series: Materials Science and Engineering **650** (2019) 012025 doi:10.1088/1757-899X/650/1/012025

powderhas silica content (SiO_2) [5][11], Na_2O and CaO which is quite large which is more than 70% [11]. Addition of husk ash and glass powder as a soft soil stabilator in this study will be reviewed changes in physical characteristics, especially water content and consistency limits, because generally soft soils have a high water content.

2. Research Methods

Research is experimental with a laboratory scale so that research is carried out at the Soil Mechanics Laboratory, Civil Department, Sekolah Tinggi Teknik PLN. The soil used is an example of soft soil disturbed. Soft soil samples were treated with the addition of rice husk ash with a percentage 0%, 2,5%, 5%, 7,5%, 10% and glass powder with a percentage 10%. The stages of research are generally divided into 3 stages, namely:

2.1. Preparation

The first thing to do at the preparation stage is to look for information through literature studies that fit the research topic with research needs, then prepare laboratory equipment used, namely measuring water content and consistency limits tests. Preparing the material used, namely soft soil as an example of disturbed soil, rice husk ash and glass powder.

Samples of disturbed soft soil used in this study were taken at a depth of 20 - 100 cm with a little waterlogged conditions in the rainy season. The sampling location for soft soil is in the PLN Substation, Duri Kosambi, Cengkareng, West Jakarta, Indonesia.

2.2. Implementation

The implementation stage consists of making and testing the test object. Making test specimens is carried out in the laboratory by preparing initial soft soil and soft soil stabilized with rice husk ash and glass powder. The specimens were made each variation of the addition of stabilator material on soft soil and soft initial soil as many as 3 pieces. Each specimen will be tested by the same method and treatment.

Testing is done to see the physical characteristics of soft soil, through testing moisture content and consistency limits. Testing the consistency limits of soft soil consists of testing the liquid limit, plastic limit and shrinkage limit..

2.3. Analysis

Stages of analysis are the final stages of the research conducted. Stages of analysis are carried out after a series of tests and data collection have been completed. Changes in physical characteristics, especially water content and the consistency limits of the initial soft soil and soft soil stabilized with rice husk ash and glass powder will be presented in the form of tables and graphs..

3. Result and Discussion

To facilitate the reading of the test results in the study, each soil sample is given a simple code that is sample 1 for original soil or initial soil, sample 2 for initial soil + 10% glass powder, sample 3 for initial soil + 10% glass powder + 2,5% rice husk ash, sample 4 for initial soil + 10% glass powder + 5% rice husk ash, sample 5 for initial soil + 10% glass powder + 7.5% rice husk ash and sample 6 for initial soil + 10% glass powder + 10% rice husk ash.

3.1. Water Content Test, w

Water content in the soil becomes a factor that affects the consistency and plasticity of the soil [12]. Water content is the ratio between the weight of the water and the weight of the soil. The weight of water used in the calculation of water content testing is the difference from the weight of the grain before and after drying in the oven for 24 hours. The results of water content testing are shown in Table 1.

Table 1. Water Content (*w*) Test Results

Code Sample	Water Content, *w* (%)
Sample 1	55.928
Sample 2	46.460
Sample 3	45.243
Sample 4	43.476
Sample 5	42.220
Sample 6	40.534

Based on Table 1, it can be seen that the addition of glass powder stabilizer and rice husk ash to soft soil affects changes in water content. The more stabilator ingredients are added, the lower the water content. The initial soil water content is reduced by 9,468% when 10% of rice husk ash is added, and is reduced by 15,394% when added by 10% rice husk ash and 10% glass powder.

3.2. Consistency Limits Tests

The consistency of soft soil, especially in clay soil, will change with changes in water content. Clay will become softer if the water content increases and conversely it will harden and crack if the water content decreases [13]. Testing the consistency limits is carried out through 3 stages, namely liquid limit testing, plastic limit testing and shrinkage limit testing. The results of testing the consistency limits are shown in Table 2.

Table 2. Consistency Limits Test Results

Code Sample	Liquid Limit (%)	Plastic Limit (%)	Shrinkage Limit (%)
Sample 1	72.878	61.544	18.54
Sample 2	67.314	47.198	22.95
Sample 3	67.08	44.537	24.74
Sample 4	66.743	44.506	25.14
Sample 5	65.172	44.237	33.11
Sample 6	62.413	40.652	35.15

Based on Table 2, there is a decrease in liquid limit and plastic limit along with the addition of stabilizers, namely rice husk ash and glass powder, where the lowest liquid limit value is 62,413% and the plastic limit is 40,652% in the original soil sample with the addition of 10% glass powder and 10% rice husk ash. In contrast to shrinkage limits, the more stabilizers (rice husk ash and glass powder) are added to the soft soil, the shrinkage limit value increases. The biggest shrinkage limit is 35.15% which is obtained by adding 10% stabilizer of glass powder and 10% rice husk ash or increasing by 16.61% compared to the initial soft soil.

The increase in shrinkage limit value is caused by the decrease in the content or moisture content in the soft soil due to the addition of stabilizers, namely glass powder and rice husk ash. Rice husk ash and glass powder absorb water which initially fills the pore space in soft soil, resulting in soft soil becoming more dense so that mercury is spilled during testing the shrinkage limit is more or more heavy.

IOP Conf. Series: Materials Science and Engineering **650** (2019) 012025 doi:10.1088/1757-899X/650/1/012025

3.3. Discussion

Based on the results of testing the physical characteristics of soft soil, it is seen that the addition of rice husk ash and glass powder greatly influences physical characteristics, especially water content (Figure1) and consistency limits (Figure 2).

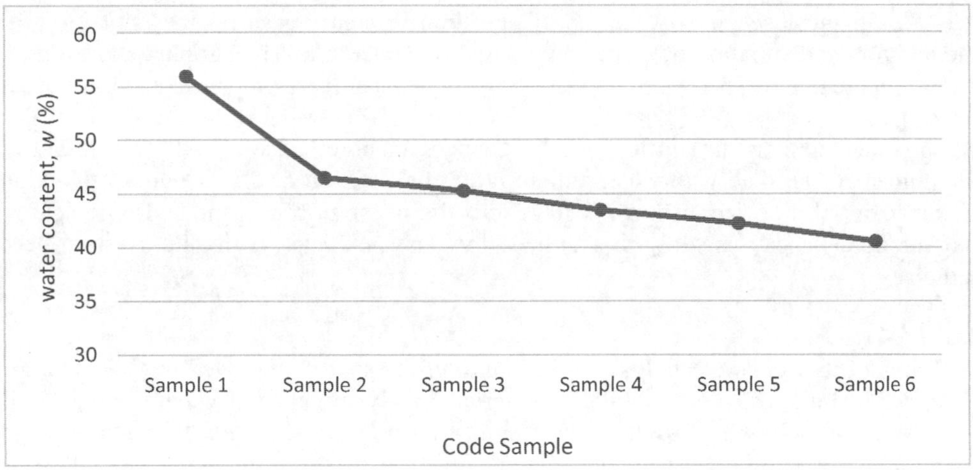

Figure 1. Water Content (w) Test Results

Based on Figure 1, that soft soil water content decreases with increasing and varying the stabilator material. The initial soft soil water content is 55,928% and the lowest water content is 40,534% which is obtained by adding 10% glass powder and 10% rice husk ash or decreasing by 27.52% compared to the initial soft soil moisture content. Glass powder and rice husk ash added to soft soil can function to absorb water so that soft soil characteristics become more rigid and dense.

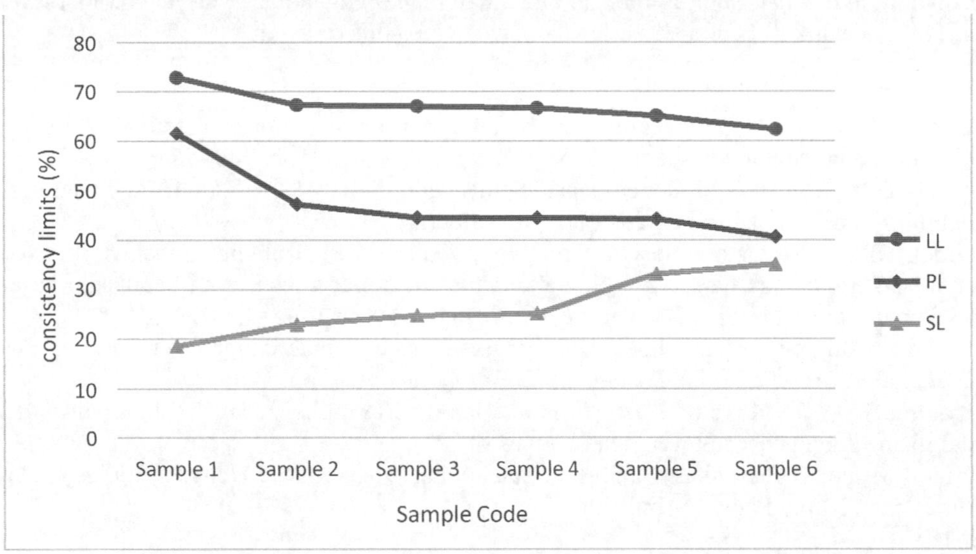

Figure 2. Consistency Limits Test Results of Stabilization Soft Soil with Rice Husk Ash and Glass Powder

Based on Figure 2, the addition of stabilizer materials, namely glass powder and rice husk ash greatly influences the soft soil consistency limits. Liquid limits and plastic limits decrease with increasing and varying stabilizer materials. This also strengthens changes in water content (Figure 1) which occurs in soft soil stabilized by glass powder and rice husk ash. In addition to reducing water content, glass powder and rice husk ash as stabilizers can also reduce liquid limits and plastic limit soft soil.

Conversely from shrinkage limit test result, the addition of rice husk ash and glass powder stabilizer material will increase the soft soil shrinkage limit. Increasing shrinkage limit in soft soil along with the addition and variation of glass powder and rice husk ash. The shrinkage limit in initial soft soil of 18.54% increased to 35.15% or increased by 89.59% compared to shrinkage limit of initial soft soil.

Increasing shrinkage limit is also influenced by changes in pore space in soft soil. Initially the initial soft soil pore space is filled with water and air, after the addition of the rice husk ash and glass powder stabilizer material, the rice husk ash will absorb the water stored in the soil pore space and urge the air to escape. This causes soft soil stabilized by glass powder and rice husk ash to become more rigid and dense.

4. Conclusion

Based on the results of the previous analysis and discussion, it can be concluded that the addition of glass powder and rice husk ash as a stabilator material greatly affects the water content and the consistency limits of soft soil. As a result of the stabilization of soft soil with the addition of glass powder and rice husk ash causes water content, liquid limit and plastic limit to decrease. Where the lowest result of water content, liquid limit and plastic limit on the stabilized soft soil was obtained at the addition of 10% glass powder and 10% rice husk ash. Conversely due to the stabilization of soft soil with glass powder and rice husk ash, the soft soil shrinkage limit increases. The biggest shrinkage limit was obtained by adding 10% glass powder and 10% rice husk ash.

Acknowlegments

Acknowledgments to Kemenristekdikti for assistance through the Penelitian Dosen Pemula 2019 Scheme, thanks to the Chair and Lembaga Penelitian dan Pengabdian Masyarakat (LPPM) of the Sekolah Tinggi Teknik PLN and all those who have helped this research.

References

[1] Karaseran A J *et al.* 2015 Pengaruh bahan campuran arang tempurung terhadap konsolidasi sekunder pada lempung ekspansif *J. Sipil Statik* vol. **3** no. 8 pp. 543–553

[2] Raswitaningrum T R J 2017 Penambahan Semen dan Abu Sekam Padi Untu Pengningkatan Stabilitas Tanah in *Semnastek 2017* no. November, pp. 1–12.

[3] Y. Firdaus, Reza Novianda; Suryo, Eko Andi; Zaika, "Pengaruh penambahan kadar kapur terhadap karakteristik dan daya dukung tanah lunak di kecamatan grati kabupaten pasuruan, jawa timur (," *J. Mhs. Jur. Tek. Sipil Univ. Brawijaya*, vol. 1, no. 3, 2018.

[4] Saleh A and Anggraini M 2019 Metoda Perbaikan Tanah Lunak dengan Penambahan Pasir in *Seminar Nasional Pakar ke -2 Buku 1: Sains dan Teknologi* pp. 1–4.

[5] Hutagalung S W R Kajian Kuat Tekan Bebas Stabilisasi Tanah Lempung dengan Stabilizing Agents Serbuk Kaca dan Semen.

[6] Sudarma I M Indeks Plastisitas Tanah Lempung Yang DIistabilisasi Dengan Varias Campuran Limestone , Kapur Padam, Abu Sekam dan Semen *Gradien* vol. **10** no. 1 pp. 96– 111, 2018.

[7] I. Yunus, Muhammad; Rauf, "Pengaruh Penambahan Kapur Terhadap Nilai Plastisitas Tanah Lempung Di Kabupaten Fakfak Provinsi Papua Barat," *J. Log.*, vol. 18, no. 1, pp. 26–31, 2018.

[8] A. Gunawan, Wiliam NIco; Manoppo, Fabian J; Sarajar, "Analisis Stabilitas Tanah Rawa Terhadap Embankment Jalan Tol Manado Bitung Dengan Menggunakan Semen Yang Dipadukan Dengan Abu Terbang (Fly Ash)," *J. Sipil Statik*, vol. 6, no. 3, pp. 189–198, 2018.

[9] Setiawan D A L S 2015 Studi Dan Analisa Campuran Tanah Lempung Dan Abu Sekam Padi Terhadap Nilai Permeabilitas Dengan Alat Falling Head *JRSDD* vol. **3** no. 3 pp. 493-506

[10] F. J. Pinasang, Denny Boy; O.B.A Sompie, "Analisis Campuran Kapur-Fly Ash dan Kapur-Abu Sekam Padi Terhadap Lempung Ekspansif," *J. Ilm. Media Engineeing*, vol. 6, no. 3, pp. 157–178, 2016.

[11] Karwur H Y, Tenda R, Wallah S E, and Windah R S J 2013 Kuat Tekan Beton Dengan Bahan Tambah Serbuk Kaca Sebagai Substitusi Parsial Semen *J. Sipil Statik* vol. **1** no. 4 pp. 276–281

[12] E. Yuda, Himmes Fitra; Zakaria, Zufialdi; Sukiyah, "Implikasi Kadar Air Tanah Terhadap Daya Dukung Tanah di Wilayah Gambir dan Sekitarnya," *Bull. Sci. Contrib.*, vol. 15, no. 1, pp. 95–100, 2017.

[13] Ardiyanti R, Faizien Z, and Sulistyorini D 2012 Stabilisasi Tanah Lempung Ekspansif Dengan Campuran Abu Sekam pp. 27–39

IOP Conf. Series: Materials Science and Engineering **650** (2019) 012026 doi:10.1088/1757-899X/650/1/012026

Effect of Compaction to Increase the Critical Height of a Slope without any Support

A.J Susilo[1], I Sumarli[1], G.S. Sentosa[1], A Prihatiningasih[1] and E Wongkar[1]

[1] Department of Civil Engineering, Tarumanagara University, Jakarta, Indonesia

[*] alfred@ft.untar.ac.id ;indas@ft.untar.ac.id ; gregoriuss@ft.untar.ac.id ; aniekp@ft.untar.ac.id ; edwinfw69@gmail.com

Abstract. Embankment is a common way to raise ground level of any construction. Raising the ground level will also help structure from being flooded, flood especially in major city such as in Capital City of Indonesia, Jakarta, is a yearly problem. On the other hand, embankment will also be a problem if not handle carefully. Clay and sand can be used as materials. Properties of material that should be investigate before the materials being used are cohesion and angle of friction. Beside selection of high quality materials, compaction can also help embankment from failure. When the materials get well-compacted the chance of sliding should be lower. This study analyse the correlation whether well-compacted material can have higher critical height and the result can be imply on the construction site as not every embankment need retaining structure to be stable but only by compaction the embankment can withstand by it self

1. Introduction

Flood is a problem that happened in many cities in Indonesia, especially Jakarta. Many new constructions raise their ground floor level to accommodate this situation. Elevating the base of construction means construct embankment using fill material. Fill material according to Federal Highway Administration (FHWA) usually consist of soil but may also include aggregate, rock, or crushed paving material; with coarser material placed at the bottom (sand and gravel) followed by more fined material (silt and clay) and get compacted to provide better support for any structure on top. Beside material, gradation also is very important in the construction of embankment in order to provide a firm foundation to withstand any loading condition without deflection or undesirable movement. The properties of soil such as unit weight, water content and specific gravity also influence the capability of embankment.

A certain amount of water is mixed during compaction proctor test defined by ASTM standard (ASTM D698 and ASTM D1557) in laboratory. Using proctor test, optimum water content and maximum dry density can be achieved.

Beside water content, according to FHWA [4], a reference of testing procedures stated by ASTM standard has been established considering the condition of soil material (Table 1).

IOP Conf. Series: Materials Science and Engineering **650** (2019) 012026 doi:10.1088/1757-899X/650/1/012026

Table 1. Embankment or fill material test procedures [4]

Property	Test Method	Reference
	Particle Size Analysis of Soils	ASTM D422
Gradation	Sieve Analysis of Fine and Coarse Aggregate	ASTM D135
	Unit Weight and Voids in Aggregate	ASTM D29
	Specific Gravity of Soils	ASTM D854
Unit Weight and	Relative Density of Cohesionless	ASTM D2049
Specific Gravity	Maximum Index Density of Soils Using a Vibratory Table	ASTM D4253
	Minimum Index Density of Soils and Calculation of Relative Density	ASTM D4254
Moisture of Density	Moisture-Density Relations of Soils and Soil-Aggregate Moistures Using 5.5 lb (2.49 kg)	ASTM D698
Characteristic	Rammer and 12 in (305 mm) Drop	(Standard)
	Moisture-Density Relations of Soils and Soil-Aggregate Moistures Using 10 lb (4.54 kg)	ASTM D1557
	Rammer and 18 in (457 mm) Drop	(Modified)
	Density of Soils in Place by the Sand Cone Method	ASTM D1558
Compacted Density	Density and Unit Weight of Soils in Place by the Rubber Baloon Method	ASTM D2167
(In-Place Density)	Density of Soils and Soils-Aggregate in Place by Nuclear Methods (Shallow-Depth)	ASTM D2922
	Density of Soils in Place by the Sleeve Method	ASTM D4564
	Unconsolidated Undrained Compressive Strength of Cohesive Soils in Triaxial Compression	ASTM D2950
	Direct Shear Toast of Soils Under Consolidated Drained Conditions	ASTM D3080
Shear Strength	Consolidated Undrained Triaxial Compression Test on Cohesive Soils	ASTM D4767
	One-Dimensional Consolidation Properties of Soils	ASTM D2435
Compressibility	One-Dimensional Consolidation Properties of Soils Using Controlled-Strain Loading	ASTM D4168
	One-Dimensional Swell on Settlement Potential of Cohesive Soils	ASTM D4546
	California Bearing Ratio (CBR) of Laboratory-Compacted Soils	ASTM D1883
Bearing Capacity >	Bearing Ratio of Soils in Place	ASTM D4429
Permeability	Permeability of Granular Soils by Constant Head	ASTM
	D2434 pH of Soils For Use in Corrosion Testing	ASTM G51
Corrosion	Field Measurement of Soils Resistivity by Using the Wenner Four-Electrode Method	ASTM G57
Resistance	Pore Water Extraction and Determination of the Soluble Salt Content of Soils by Refractometer	ASTM D4542

Many main contractors regularly use fill material of clay to raise the level of ground floor. As seen in Figure 1, contractor used soil to raise about 2.5 meter from existing soil layer.

Figure 1. Construction site with clay as embankment material to raise ground level

To prevent failure from sliding, fill material should be well-compacted (Figure 2). At construction site, compaction equipment such as sheepsfoot compactor, smooth drum, vibratory, rubber tire, jumping jacks, plates and trench compactors can be applied.

Figure 2. Soil compactor equipment to provide uniform compaction

IOP Conf. Series: Materials Science and Engineering **650** (2019) 012026 doi:10.1088/1757-899X/650/1/012026

Figure 3. Embankment with uniform compaction

As seen on Figure 3, compactor equipment is riding and milling on the top of fill material to compact the clay expecting a uniform density of the materials. In the pictures, it can be seen that ground floor will be risen as high as the fence or approximately 2.5 meters.

2. Slope Stability

Ground with uneven level is vulnerable to soil movement; the higher differences of level will risk the possibility of soil movement. The movement will be similar to water flows from higher level to lower elevation. According to Cruden and Varnes (1996) [3] there are five major categories of soil movement: Fall, Topple, Slide, Spread and Flow.

All those possibilities will not happen when the soil has high shear strength, compared to shear stress developed along the potential failure surface of the soil. Soil Shear strength is dependent to cohesion and internal friction angle. Higher shear strength can be achieved by:

1. Using high quality soil material
   - Clay: higher cohesion
   - Sand: higher angle of friction
2. Compaction will also increase the value of soil shear strength

Figure 4. Example of fill material failure

Figure 4 showing an example of slope failure. Embankment on this case using a mixture of clay and sand, the percentage of sand is dominant. The density is a major factor that can cause the failure of the embankment. In this case the embankment seems did not get enough compaction. Although a retaining wall was constructed, however, the shear stress was risen higher and the wall could not withstand the pressure. After an investigation, it was found that the factor of rain water run-off causing the weakening of the soil density, the drainage system was not designed properly resulting in higher lateral pressure to the retaining wall. In this paper, the rain water problem will not be discussed widely.

3. Common Solution for Slope Protection

To prevent embankment from failure, typically around the perimeter a retaining wall construction can be installed. Types of retaining wall used depend on how high lateral pressure generated behind the wall, equipment and material availability, and fill material quality.

Gravity wall, piling wall, cantilever wall and anchored wall are some types of retaining wall used by engineer to prevent soil from sliding. When designing retaining wall structure, engineer should calculate the safety factor against overturning, sliding, foundation pressure and the possibility failure of overall stability.

In this paper, a recommendation is proposed to reduce the construction cost by eliminating the cost of constructing retaining wall, replacing it by well-constructed embankment by compacting the soil layer by layer with tight control.

IOP Conf. Series: Materials Science and Engineering **650** (2019) 012026 doi:10.1088/1757-899X/650/1/012026

4. Critical Height Correlation to Compaction

Culmann's Analysis [3] (Figure 5) assumes the failure of a slope occurs along a plane when the average shearing stresses tending to cause the slip rather than the shear strength of the soil. Also, the most critical plane is the one that has a minimum ratio of the average shearing stress that tends to cause the failure of shear strength of the soil. A simple approach to estimate the critical height of a slope can be calculated by using the equation (1) [3]

$$H_{cr} = \frac{4c'}{\gamma}\left[\frac{\sin\beta\cos\phi'}{1-\cos(\beta-\Phi')}\right] \quad (1)$$

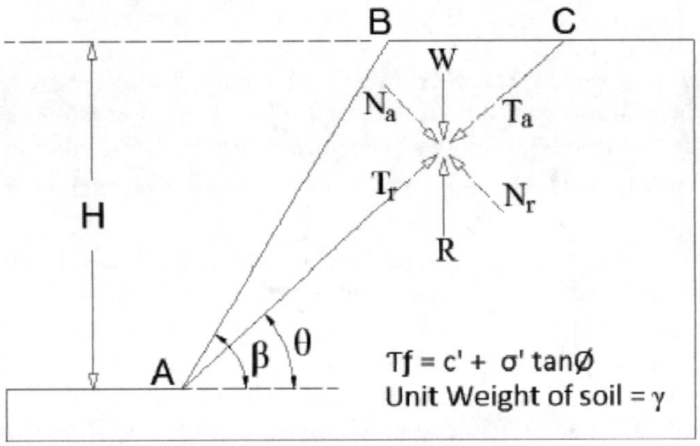

Figure 5. Finite slope

Where c' = Cohesion

Ø' = angle of internal friction

ɤ = Unit weight of soil

β = Slope angle with horizontal

H_{cr} = The maximum height of the slope for which equilibrium occurs

To use equation (1) properly, it is important to find parameters, unit weight, cohesion and angle of internal friction. Sometimes to find directly those parameters need costly effort. In this paper an alternative approach is proposed by using CBR (California Bearing Test) value, pocket vane shear value and empirical equation to estimate the approximately critical height.

The values of CBR and pocket vane shear can be found by testing soil samples in the laboratory. The ASTM standard (D1883) can be used to get CBR value. In order to get values of cohesion and angle of internal friction, the tested samples of CBR were cut in five layers and each layer was tested using pocket vane shear equipment. Average values from pocket vane shear test then can be defined.

This approach is taken from Wongkar (2018) study which correlated the value friction angle with CBR value. From pocket vane shear, value of soil strength (Su) can be found. Using equation 2 (Sowers, 1979) to get N-SPT value and using correlation of friction angle and N-SPT value from Meyerhoff (1956) as seen in table 2, friction angle can be predicted.

$$Su = (0.102 \text{ to } 0.179)\ N \quad (2)$$

IOP Conf. Series: Materials Science and Engineering **650** (2019) 012026 doi:10.1088/1757-899X/650/1/012026

Table 2. Correlation between N-SPT value with friction angle

SPT N (Blows/0,3m)	Soil Packing	Relative Density (%)	Friction Angle (°)
< 4	Very Loose	< 20	< 30
4 to 10	Loose	20 to 40	30 to 35
10 to 30	Compact	40 to 60	35 to 40
30 to 50	Dense	60 to 80	40 to 45
> 50	Very Dense	> 80	> 45

5. Case Study

One example to use this approach is taken from an embankment construction located at Tangkil, Citeureup, Bogor, West Java, Indonesia. An amount of soil sample for embankment construction was taken from the field and tested in the laboratory. The parameters result of laboratory testing was used for design embankment height by using equation (1). The basic important parameters are γ (unit weight) = 16 kN/m^3.

In the laboratory some soil samples were also tested to find CBR values with several defined values (CBR = 2.5, 5.0, 7.5, and 10). Table 3 shows the result of laboratory study to correlate the CBR values with angle of internal friction (Ø') values. The Ø' values are calculated using the empirical equation (2) and Table 2 after getting the values from pocket vane shear tested in the laboratory. Those values are shown in Table 3, column (4). The critical height, H_{cr}, shown in column (6) was calculated using equation (1). By using those soil parameters and assuming slope angle with horizontal, β = 90 degrees, then critical height can be determined and being compared.

To show how the equation works, below is sample calculation for CBR value =2.5:

Safety Factor = 3 (Assumption)

Assume β = 90°

c/SF = 30.64/3

= 10.212 kN/m^2

Ø/SF = 39.40°/3

= 13.13°

$$H_{cr} = \frac{4 \times 10.212}{16}\left[\frac{\sin(90°)\cos(13.13°)}{1-\cos(90°-13.13°)}\right]$$

= 3.22 m

Tabel 3. The Correlation of CBR values with Angle of Internal Friction (Wongkar (2018) and result of critical height for each CBR values

CBR value	Position of point samples tested using pocket vane shear		Cohesion C (kN/m^2)	Ø' average (degree)	γ (kN/m^3)	H_{cr} (m)
(1)	(2)		(3)	(4)	(5)	(6)
2.50	TOP		30.64	39.4	16.00	3.22
	SIDE	TOP				
		MIDDLE				
		BOTTOM				
	BOTTOM					
5.00	TOP		36.04	43.12	16.00	3.87
	SIDE	TOP				
		MIDDLE				
		BOTTOM				
	BOTTOM					
7.50	TOP		40.15	45.19	16.00	4.37
	SIDE	TOP				
		MIDDLE				
		BOTTOM				
	BOTTOM					
10.00	TOP		48.00	50.00	16.00	5.37
	SIDE	TOP				
		MIDDLE				
		BOTTOM				
	BOTTOM					

The graph correlation between CBR values and critical height is shown in Figure 6

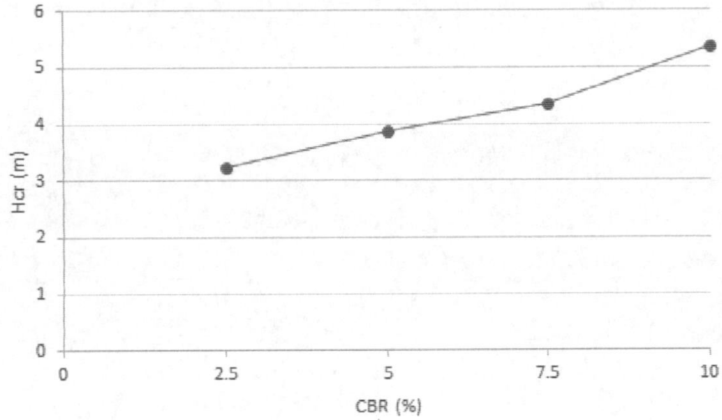

Figure 6. Correlation of CBR value versus Critical height

6. Conclusion

Based on calculation above shown that critical height increase corresponding with the increase of CBR value. The recommendation from this result is density of embankment play a significant role in slope stability failure. When soil material is not well compacted protection using retaining wall might be inevitable and vice versa.

References

[1] Nichols H L 1976 *Moving the Earth* (New York: McGraw-Hill Publishing Company)
[2] U.S. Bureau of Reclamation 1991 *Earth Manual* (Washington, DC)
[3] Das B M 2010 *Principles of Geotechnical Engineering*, 7th Edition
[4] Federal Highway Administration (FHWA) 2016 *User Guidelines for Waste and Byproduct Materials in Pavement Construction* Retrieved from https://www.fhwa.dot.gov/publications/research/infrastructure/structures/97148/app4.cfm
[5] Wongkar E 2018 *Korelasi nilai CBR terhadap tegangan vertical dan tegangan horizontal pada tanah lempung di daerah sentul*
[6] Sowers G F 1979 *Introductory Soil Mechanics and Foundations: Geotechnical Engineering*. Edition 4. (New York: Mac Millan)

IOP Conf. Series: Materials Science and Engineering **650** (2019) 012027 doi:10.1088/1757-899X/650/1/012027

The Permeability of Portland Cement-Stabilized Clay Shale

W Diana[1*], E Hartono[1] and A S Muntohar[1]

[1]Civil Engineering Departement, Universitas Muhammadiyah Yogyakarta, Jl Brawijaya Tamantirto Kasihan Bantul DI Yogyakarta

* willis.diana@umy.ac.id

Abstract. Research on the effect of Portland cement on the improvement of problematic soil has been extensively studied in many years and many places. Soil stabilization using cement has improved the strength and durability, enhanced the performance of the problematic soil such as clay shale. However, most studies in the field of Portland cement-stabilized soil have only focused on strength and durability improvement and the information on permeability characteristic has not adequately provided. This study aimed to investigate the effect of Portland cement stabilization on the consistency limits and permeability of the clay shale. The Portland cement content was varied from 2% to 10% by weight of dry soil. The stabilized clayshale specimens were cured for seven days before the testing. Constant head permeability method (method A) of the ASTM D5084 was used to determine the coefficient of permeability. The result shows that both the liquid limits and plastic limits decrease with increased Portland cement content, hence the plasticity index of the soil-PC mixing decrease. The permeability coefficient of the soil mixed with 2% of cement increased about 14 times greater than the untreated soil. Increase in cement content tends to decrease the coefficient of permeability.

1. Introduction

Shales are thinly laminated fine-grained pelite clastic rock composed predominantly of siliciclastic materials by the granulometric composition of mixtures of clays and particles size of powder or silt. The clay shale was originated from 50% to 70% of sedimentation rock on earth crust. Clayshale is formed majorly by clay mineral, claystone, soil, dust or rock with cementation, consolidation and sedimentation processes. This process results in a compact and overconsolidated structure. However, clayshale behaves brittle and low durability [1]. The clayshale is susceptible to weathering and rapidly degrade the geotechnical properties, strength, and durability [2]–[5]. Thus, the behavior generates problems such as unstable fill material, reduction in bearing capacity of shallow and deep foundation, slope stability, piping, and subsidence. The availability, mechanical, chemical properties, and the cost of the Portland Cement (PC) were the advantageous which favored and most used material among the chemical stabilizer. Many researchers have evaluated the properties of PC-soil mixing [6]–[11]. The hydrations reaction of the PC with soil water and pozzolanic reaction were the main factor in the improvement of PC-soil mixture [11]. Previous studies in PC-stabilized soil, generally, focused on strength and durability improvement, but lessen studies investigated the permeability characteristic. The permeability of a soil is its ability to allow water flow under a hydraulic head. To analysis subsurface flow, the permeability is a critical parameter [12].

Measuring and predicting soil permeability properties are still tricky. The porosity, the fabric, the density, and the composite of sediment are the influencing factors of the soil permeability [13].

IOP Conf. Series: Materials Science and Engineering **650** (2019) 012027 doi:10.1088/1757-899X/650/1/012027

Measuring the soil permeability was tough, time-consuming, and expensive [14]. There was lessen published data of measured permeability for mudrocks [13]. The study conducted by Neuzil [12] and Dewhust et al. [15] measured the permeability of the intact mudrocks samples. The study indicated a notable contribution to a better understanding of the permeability of the mudrock material. Casey et al. [13] tried to predict the permeability of the mudrocks based on its porosity and liquid limit. Reece et al. [16] quantified that small variation in fabrics and composition of the mudstones significantly affected the permeability, although the soil has similar grain-size distributions and porosity. Ewy [17] stated that testing protocols on shale were needed for mechanical testing, which should be designed around the permeability coefficient value of the shale.

Recently, literature has emerged a contradictory finding of the permeability on the stabilized-soil. A study on the permeability of a treated tropical residual soil using lime found that the permeability of the treated soil with 2% lime increased the coefficient of permeability of a brown saprolitic Brazilian soil, whereas contrary as for red lateritic Brazilian soil. Further addition of lime decreased the coefficient of permeability [18]. Alhassan [19] found that the permeability at specified lime content decreased to the lowest value at 6% RHA content, and then the coefficient of permeability slightly increased with further increase in RHA. Study on stabilized soil-cement found that the permeability of the soil decreased as cement content increases [6]. Mousavi and Wang [20] studied the effect of cement-peat ash-silica sand on the permeability of clay. The study found that at an identical void ratio, the coefficient of permeability decreases drastically by the addition of 18% cement. In contrast, Alhasan [19] and Elsharief [21] studied the stabilization of three tropical soil using lime and found the coefficient of permeability increased substantially at optimum lime content. The increased of the permeability properties of the stabilized soil with an increase in the percentage of MgO or lime [22].

It is now well established that the strength and the durability of the soil generally increased with the increasing percentage of the stabilized agent. However, the influence of stabilized agent such as PC on the permeability characteristic has remained discussable. The literature has emerged that offers contradictory findings of the influences of stabilized agent (lime, PC, RHA, etc.) content on the permeability characteristics. A large discrepancy among the research in the permeability of the stabilized soil was because of various soil and mineralogy, testing method, apparatus, and loading condition, etc. The results and recommendation in literature can be adapted as a general guide. The influence factors need to be verified for individual sites. Furthermore, the permeability of the stabilized soil should be carefully assessed in each case. Hence, an objective of this study was to investigate the effect of PC content on the plasticity and permeability characteristics of the clay shale.

2. Materials

The disturbed clay shale samples were collected from cutting slope of the Semarang-Bawen Toll Road at a depth of 1 m – 5 m from the surface. The ASTM D4318 [23] procedure was adopted to determine the Atterberg limits of the samples. The particle size analysis, include sieve analysis and hydrometer testing, was determined by the ASTM D422-63 [24]. The soil classification was according to the Unified Soil Classification System determined using the ASTM D2487 [25] procedure. The maximum dry density (MDD), and the optimum moisture content (OMC) of the untreated clay shale were determined using standard Proctor compaction test, the test based on the ASTM D698 [26]. The density and water content were determined for mixing the stabilized specimens. **Figure 1** shows the grain size distribution of the clay shale sample. The soils properties are presented in **Table 1**. The sample was classified as heavy clays (CH materials) according to the Unified Soil Classification System (USCS). The slake durability index (I_{d2}) was determined using ASTM D4644 [27] procedure. The clayshale has $I_{d(2)}$ = 1.47%, which was classified as very low durability [28]. The cement was Portland Composite Cement (PCC) which was produced by Holcim Cement Company.

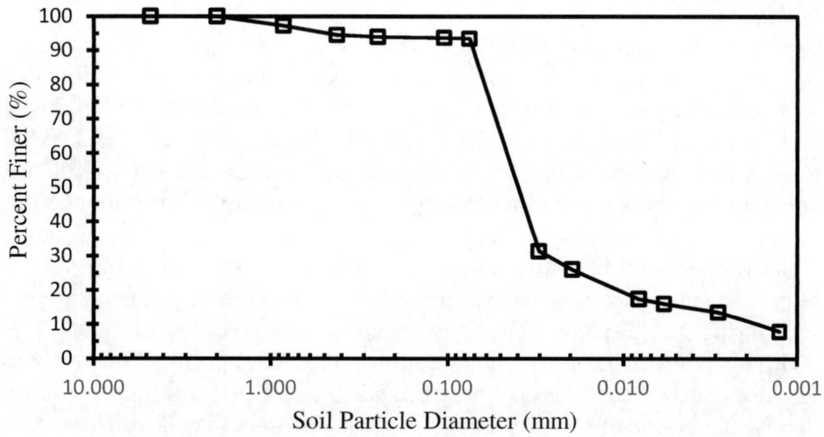

Figure 1. Grain size distribution.

Table 1. Engineering properties of soil sample

Properties	Value
Specific Gravity	2.65
Liquid Limit	51.04%
Plastic Limit	34.27%
Plasticity Index	16.77%
Percent passing more than no.200 sieve	93.46%
Optimum Moisture Content (OMC)	16.99%
Maximum Dry Density(MDD)	16.33 kN/m^3

3. Method

3.1. Sample Preparation

The clay shale sample was pulverized and sieved through no. 4 test sieve. The specimens were prepared in five different cement content (PC), which was 2%, 5%, 7%, and 10% by dry weight of the soil. The specimens were compacted statically in rigid steel mold with 35.2 mm in diameter and a height of 70 mm. All the specimens prepared at the MDD and OMC of the untreated soil. For the permeability test, after compacting each specimen was put in a plastic bag and cured for seven days in the room temperature about 28°C.

3.2. The Atterberg limits test

The soil samples were pulverized to pass through the sieve no. 40 (425μm). The testing procedure as outlined in ASTM D4318 [23] for determination of liquid limit (LL) and the plastic limit (PL). Before the test, the soil samples were oven-dried. The tests were carried out on untreated soil and the treated soils with 2%, 5%, 7%, and 10% of cement. The soil-PC mixing procedure for the liquid limit and plastic limit was prepared as recommended by Miura et al. [29]. The soil-PC was mixed thoroughly with water for 10 minutes. Considering the setting time of cement (±120 minutes) and ion exchange in the treated soil, then the soil-cement slurry was kept in a plastic bag for 60 minutes. Afterward, the liquid limit and plastic limit test was performed.

3.3. The permeability test

The coefficient of permeability (also referred to as hydraulic conductivity), *k* can be defined as the water discharge rate under conditions of laminar flow through a unit cross-sectional area of the soil at 20°C temperature condition and a unit hydraulic gradient [30]. Constant head permeability tests were carried out by a flexible wall permeameter apparatus which referred to ASTM D5084 method A. The apparatus provided the specimen with porous stone at end pieces, enclosed by a flexible membrane sealed at the cap and base and subjected to controlled cell pressures. Two drainage line on the top and two on the base were used to remove air from soil specimens and saturation of the hydraulic system. The main procedures of the permeability test were following (i) confining pressure was applied by a small cell pressure; (ii) saturation process was applied by back-pressure. The degree of saturation was evaluated by measuring the B-value. The specimen was adequately saturated if the B-value was greater than 0.95; (iii) consolidation; (iv) permeation phase. The permeation test was conduction by five trial of head loss. At this step, the backpressure was maintained throughout the permeation phase. The head loss across the permeameter was kept constant. The quantity of inflow, outflow, and changes in the height of specimens was monitored, measured, and recorded. The permeation test was continued until at least four value of permeability coefficient was obtained over an interval of time as follows: (a) the rate was between 0.75 and 1.25 of the ratio of outflow and inflow, and (b) the coefficient of permeability was steady. The steady condition was gained if four or more consecutive permeability coefficient was within the minimum value of 25% and no significant changes in the plot of the permeability coefficient versus time. The calculation of the permeability coefficient as follows;

$$k = \frac{\Delta QL}{Ah\Delta t} \tag{1}$$

where:

k = permeability coefficient (m/s).
ΔQ = quantity of flow for interval time Δt, the inflow and outflow average (m^3),
L = specimen's length (m)
Δt = time interval ($=t_1-t_2$) over which occurs of the flow ΔQ (s),
t_1 = time at the start of permeation trial (s),
t_2 = time at the end of permeation trial (s),
A = specimens cross sectional area (m^2),
h = head loss of water across the specimen (average value) ($=(h_1+h_2)/2$) (m),
h_1 = head loss at t_1 (m),
h_2 = head loss at t_2 (m).

4. Result and Discussion

4.1. The Atterberg Limits

The results obtained from the Atterberg limits test of untreated and treated soils with PC is presented in **Figure 2**. Both the liquid limit and plastic limit slightly decreases by the addition of PC. The plasticity index reduces 48% from 16.77% to 8.61% by the addition of 10% PC. The plasticity index reduction was due to the decrease in liquid limit and plastic limit. The decrease in plasticity index and the liquid limit is an indicator of improvement [31]. Many works correlated the plasticity index with the engineering properties of soil. A lower plasticity index will enhance the shear strength, compressibility, durability, and workability of the admixed soil [10], [32]. The soil mineralogy is the primary factor that influences the changes in the plasticity properties of the stabilized soil. Cation exchange in cement-soil chemical reaction leads to produce physical change and mineralogy of the stabilized clay [18].

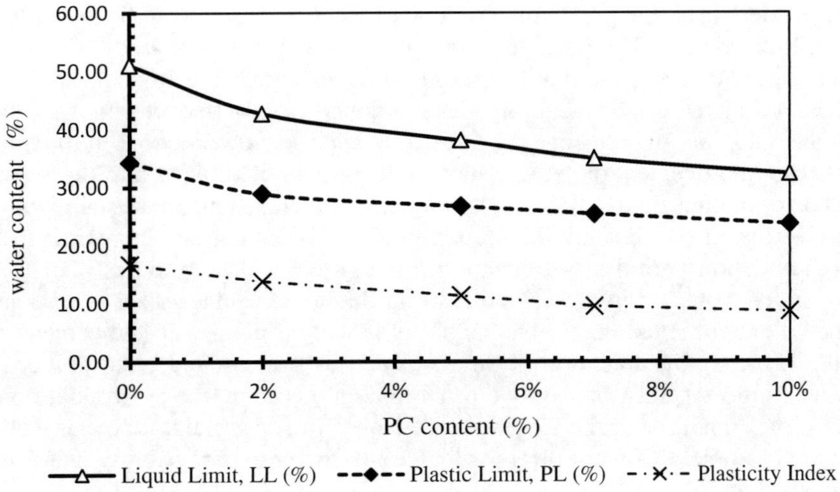

Figure 2. The Atterberg Limits of untreated and threated soils

4.2. Permeability of the Clayshale

Figure 3 shows the relationship of the permeability coefficient and PC content. The permeability coefficient was 9.65×10^{-5} m/s and 1.40×10^{-4} m/s respectively for untreated and 2% PC. The figure shows that the coefficient of permeability increased with the addition of 2-5% cement. The coefficient of permeability of the stabilized soil increases 12 to 14 times by mixing with 2% and 5% cement respectively. However, the addition of higher content of cement results in decreasing the coefficient of permeability. The permeability dropped to $2{,}9\times10^{-4}$ m/s by adding 7% of cement, and the value approximately closed to the untreated soil. This characteristic can be explained that the immediate chemical reaction was the soil agglomeration of the soil by addition of a low cement content e.g. 2-5%. The agglomeration causes a lager interparticle voids. As a consequence, the water rapidly flows in the void and results in substantial increases in the coefficient of permeability. Increase in cement content developed cementation product such as bonding gel which bound the soil particles and hinders the water flow in soil. Therefore, the coefficient of permeability decreases. The result is in agreement with the study conducted by Galvão et al. [18].

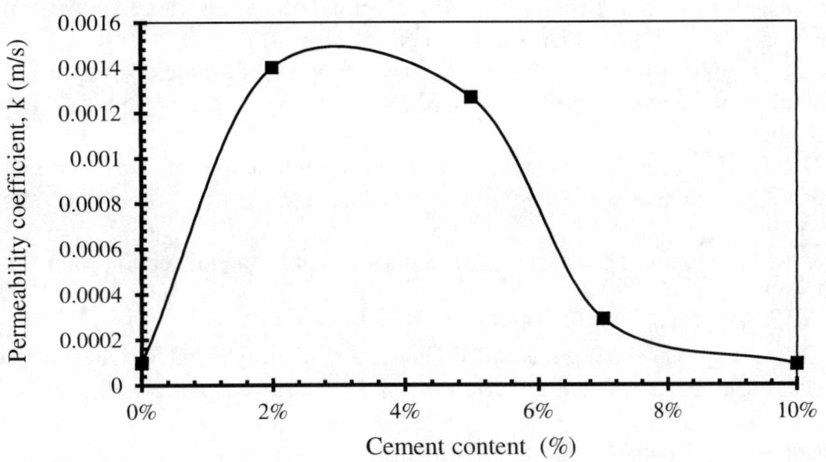

Figure 3. Variation of the permeability coefficient with the cement content.

Yi et al. [22] found that the hydration product of cement has a significant influence on the permeability of stabilized soil. The particles can fill the soil pores and lead to a decrease in the permeability. Besides, the cracks in stabilized soil were appeared due to the excessive hydration products and lead to increased permeability. The amount of macro-pores in the stabilized soil decreases with the addition of cement and causes a significant reduction in the coefficient of permeability [9]. The improvement of cement-stabilized soil can be divided into three zones which is active, inert, and deterioration zones [7]. In the active zone is the most effective stabilization, the pores are filled with the increased cementitious products. In this condition, the strength reaches the maximum value. At a short period of stabilization, the existence of large pores increases due to the presence of unhydrated cement particles. Solidification due to cementation gel is responsible for the decreasing of the volume of small pores. The large pores will be filled with the cementitious product and increase with the time. The condition of the pores causes strength and permeability development over time. However, unfortunately, the study did not take into account the determination of the micro and macropores of the samples and the mineralogy changes in soil-cement mixture. Notwithstanding the relatively limited sample, this work offers valuable insights into permeability characteristic of the clay shale-PC stabilization.

5. Conclusion

The consistency and the permeability characteristic of the clay shale samples mixed with cement has been successfully tested and discussed. The purpose of the current study was to assess the effect of cement on the plasticity properties and permeability of the treated and untreated soils. The following conclusions can be drawn :

1. The liquid limit, plastic limit, and plasticity index decreased with the cement content. The plasticity index reduces 48% from 16.77% to 8.61% by the addition of 10% PC
2. The coefficient of permeability increases about 12-14 times by the addition of 2%-4% cement. Further addition of cement, up to 10%, significantly decreases the coefficient of permeability of the stabilized soil.

References

[1] Haldar S K 2014 *Introduction to mineralogy and petrology* (Amsterdam: Elsevier)

[2] Agung P, Damianto B, Yuwono, & Isiatun. 2013 A Critical State Approach to Stability of Clay Shale for Design Structures of the Sentul Hill, West Java, Indonesia. *Geotechnical Engineering* 7–13 (National University of Singapore)

[3] Alatas I M, Kamaruddin S A, Nazir R, Irsyam M, & Himawan A 2015 Shear Strength Degradation of Semarang Bawen Clay Shale Due to Weathering Process. *Jurnal Teknologi,* **77(11)** 109–118.

[4] Widjaja B 2008 Engineering characteristics of Bukit Sentul clayshale based on laboratory and in situ tests *Geotechnical and Geophysical Site Characterization – Huang & Mayne (eds)* (pp. 1231–1237).

[5] Putera A M A, Pramusandi S, & Damianto B 2017 Identification and classification of clayshale characteristic and some considerations for slope stability *African Journal of Environmental Science and Technology* **11(4)** 163–197.

[6] Akiije I 2014 Strength and Permeability Characteristics of Selected Laterite Stabilized Using Powermax Cement *International Journal of Scientific & Engineering Research* **5(12)** 695–702.

[7] Horpibulsuk S, Rachan R, Chinkulkijniwat A, Raksachon Y, & Suddeepong A 2010 Analysis of strength development in cement-stabilized silty clay from microstructural considerations. *Construction and Building Materials* **24(10)** 2011–2021.

[8] Horpibulsuk S, Rachan R, & Suddeepong A 2011 Assessment of strength development in blended cement admixed Bangkok clay. *Construction and Building Materials* **25(4)** 1521–1531.
[9] Sasanian S, & Newson T A 2014 Basic parameters governing the behaviour of cement-treated clays. *Soils and Foundations* **54(2)** 209–224.
[10] Pakbaz M S, & Alipour R 2012 Influence of cement addition on the geotechnical properties of an Iranian clay *Applied Clay Science* **67–68** 1–4.
[11] Pakbaz M S, & Farzi M 2015 Comparison of the effect of mixing methods (dry vs. wet) on mechanical and hydraulic properties of treated soil with cement or lime *Applied Clay Science* **105–106** 156–169.
[12] Neuzil C E 1994 How permeable are clays and shales? *Water Resources Research* **30(2)** 145–150.
[13] Casey B, Germaine J T, Flemings P B, Reece J S, Gao B, & Betts W 2013 Liquid limit as a predictor of mudrock permeability. *Marine and Petroleum Geology* ***44*** 256–263.
[14] H. Ranaivomanana, A. Razakamanantsoa, and O. Amiri 2017 "Permeability Prediction of Soils Including Degree of Compaction and Microstructure," *Int. J. Geomech.* vol. 17 no. 4 p. 04016107
[15] Reece J S, Flemings P B, Dugan B, Long H, & Germaine J T 2012 Permeability- porosity relationships of shallow mudstones in the Ursa Basin, northern deepwater Gulf of Mexico: MUDSTONE PERMEABILITY-POROSITY BEHAVIOR *Journal of Geophysical Research: Solid Earth* **117(B12)**
[16] J. S. Reece, P. B. Flemings, B. Dugan, H. Long and J. T. Germaine 2012 "Permeability-porosity relationships of shallow mudstones in the Ursa Basin, northern deepwater Gulf of Mexico: mudstone permeability-porosity behavior " *J. Geophys. Res. Solid Earth* vol. 117 no. B12
[17] Ewy R T 2018 Practical approaches for addressing shale testing challenges associated with permeability, capillarity and brine interactions *Geomechanics for Energy and the Environment* **14** 3–15.
[18] de Brito Galvão T C, Elsharief A, & Simões G F 2004 Effects of Lime on Permeability and Compressibility of Two Tropical Residual Soils. *Journal of Environmental Engineering* **130(8)** 881–885.
[19] Alhassan M 2008 Permeability of Lateritic Soil Treated with Lime and Rice Husk Ash. *Technical Report* **6**
[20] Mousavi S E, & Wong L S 2016 Permeability Characteristics of Compacted and Stabilized Clay with Cement, Peat Ash and Silica Sand *Civil Engineering Infrastructures Journal* **49(1)** 149 – 164.
[21] Elsharief A M 2013 Lime Stabilization of Tropical Soils From Sudan For Road Construction. *International Journal of Geomate*.
[22] Yi Y, Liska M, & Al-Tabbaa A 2014 Properties of Two Model Soils Stabilized with Different Blends and Contents of GGBS, MgO, Lime, and PC *Journal of Materials in Civil Engineering* **26(2)** 267–274.
[23] ASTM D4318 2000 *Standard Test Methods for Liquid Limit, Plastic Limit, and Plasticity Index of Soils,*. (West Conshohocken, PA: ASTM International.)
[24] ASTM D422-63 2002 *Standard Test Method for Particle-Size Analysis of Soils.* (West Conshohocken, PA: ASTM International.)
[25] ASTM D2487 2000 *Standard Practice for Classification of Soils for Engineering Purposes (Unified Soil Classification System).* (West Conshohocken, PA: ASTM International.)
[26] ASTM D698 2000 *Standard Test Methods for Laboratory Compaction Characteristics of Soil Using Standard Effort (12 400 ft-lbf/ft3 (600 kN-m/m3)).* (West Conshohocken, PA: ASTM International)
[27] ASTM D4644 2004 *Standard Test Method for Slake Durability of Shales and Other Similar Weak Rocks.* (West Conshohocken, PA: ASTM International.)

[28] Erguler Z A, & Shakoor A 2009 Quantification of Fragment Size Distribution of Clay-Bearing Rocks after Slake Durability Testing *Environmental and Engineering Geoscience*, ***15*(2)** 81–89.
[29] Miura N, Horpibulsuk S, & Nagaraj T S 2001 Engineering Behavior of Cement Stabilized Clay at High Water Content. *Soils and Foundations* **41(5)** 33–45.
[30] ASTM D5084 2000 *Standard Test Methods for Measurement of Hydraulic Conductivity of Saturated Porous Materials Using a Flexible Wall Permeameter,*. (West Conshohocken, PA: ASTM International.)
[31] Basha E A, Hashim R, Mahmud H B, & Muntohar A S 2005 Stabilization of residual soil with rice husk ash and cement *Construction and Building Materials* **19(6)** 448–453.
[32] Victor R.,Too J T, Kaluli W 2014 Cement effects on the physical properties of expansive clay soil and the compressive strength of compressed interlocking clay blocks *European International Journal of Science and Technology* **3(8)** 74–82.

IOP Conf. Series: Materials Science and Engineering **650** (2019) 012028 doi:10.1088/1757-899X/650/1/012028

Durability of Reactive Powder Concrete

Widodo Kushartomo[1*] and Octavivia[1]

[1] Civil Engineering Undergraduate Study Program, Universitas Tarumanagara, Indonesia

*widodo@untar.ac.id

Abstract. Concrete is a generally used material in building construction. The construction will be built on the ground and could also be in places exposed to sea water. Sulfate contained in ground water and sea water could cause sulfate aggression to concrete itself. Therefore, the research used Reactive Powder Concrete which is one of the concretes that is currently popular and has a high durability. The research was done by doing a constant repetition of saturated sulfate submersion and specimen abstersion. The varieties of specimen used along with the research such as Type I, Type II, and PCC variations of cements, also varieties fas 0.2 and fas 0.25. Research shown that Reactive Powder Concrete which is contained type II cement has the best resistance because the compressive strength does not reduced to 30% and the reduced weight is less than 1% when exposed to high concentration of sulfate.

1. Introduction

Concrete is a material that's mostly used in buildings. Construction will be build in direct contact with soil that contain groundwater. The construction that's built close to the sea can led the construction be in contact with sea water. Ground water and sea water contained substances - chemicals that are reactive to the elements contained in concrete and can eventually lead to the disintegration of the concrete [1]. One of these chemicals is sulfate. Therefore, the concrete should have a high durability in order to minimized the sulfate attack.

One type of concrete that can meet the need for high durability is Reactive Powder Concrete (RPC). Reactive Powder Concrete is an ultra high-strength and high ductility by optimizing the aggregate with the micro size and proper gradation of all the particles contained in the concrete mix to produce the maximum density [2].

RPC is composed of very fine powders (cement, sand, quartz powder and silicafume), steel fiber (optional) and superplasticizer**.** The superplasticizer, used at its optimal dosage, decreases the water to cement ratio (w/c) while improving the workability of the concrete**.** . A very dense matrix is achieved by optimizing the granular packing of the dry fine powders. This compactness gives RPC ultra-high strength and durability. [3]

Microstructure enhancement of RPC is done by heat curing. Heat curing is performed by simply heating (normally at 90°C) the concrete at normal pressure after it has set properly.

IOP Conf. Series: Materials Science and Engineering **650** (2019) 012028 doi:10.1088/1757-899X/650/1/012028

This considerably accelerates the pozzolanic reaction, while modifying the microstructure of the hydrates that have formed [1].

The most common concrete attack is the sulfate attack on concrete. Murdock (1999) mentions the chemical sulfate attack on concrete will cause damage on concrete and can cause cracks on concrete itself [4]. According to Masruri (1993), sodium sulfate and magnesium sulfate have more detrimental effect than the loss caused by gypsum (calcium sulfate), because these salts are not only more soluble, but also resulted in a greater concentration of sulfate in groundwater and minerals react with cement, thus causing total damage to the cement paste [5].

The combination of water with cement will form a cement paste by the hydration process. The cement paste are combines aggregate, fill the empty space, and free flowing. Sulfate attack occurs by attacking calcium hydroxide which is formed from the cement hydration process. Cement hydration process can be seen as follows:

$$C_3S + H \rightarrow C\text{-}S\text{-}H + CH$$

Tricalcium silicate + Air → Calcium silicate hydrate + Calcium hydroxide

$$C_2S + H \rightarrow C\text{-}S\text{-}H + CH$$

Dicalcium silicate + Air → Calcium silicate hydrate + Calcium hydroxide

Mindess and Young (1981) describes the sulfate attack into 3 stages [6]:

1. Penetration of sulfate ion
   The first process is the diffusion sulfate ions into the pores of the concrete, whice is controlled by the permeability coefficient and the diffusion coefficient of the sulfate ions.

2. Gypsum corrosion
   In its initial stages, gypsum corrosion may actually be beneficial, since gypsum is more soluble than calcium hydroxide and the dissolution-crystallization reaction will allow gypsum to crystallize without expansion. Equation of gypsum corrosion is as follows:

$$CH+SO_4^{2-}+2H_2O \rightarrow C\bar{S}H_2+2OH^-$$

Calcium hydroxide + Sulfate + Water → Gypsum + Hydroxide

3. Sulfoaluminate corrosion
   As sulfoaluminate corrosion causes internal cracking, the diffusion coefficient of the concrete will be increased, thereby accelerating further sulfate attack. Ettringite in this case is formed from monosulfoaluminate can be seen below.

$$C_4A\bar{S}H_{12} + 2C\bar{S}H_2 + 16H \rightarrow C_6A\bar{S}_3H_{32}$$

Calcium Aluminate Hydrate + *Gypsum* + Water → Ettringite

C3A in the cement compounds also affect the sulfate reaction which affected by gypsum. The main initial reaction of C3A and gypsum reacts with water to form trisulfoaluminat / ettringite. If the sulfate is all consumed before C3A completely hydrate, then ettringite transform to another calcium sulfoaluminate hydrate containing less sulfate called monosulfoaluminate.

$$C_3A + 3C\overline{S}H_2 + 26H \rightarrow C_3A.3C\overline{S}_3.32H$$

$$C_3A + 3C\overline{S}H_2 + 26H \rightarrow C6A\overline{S}_3H_{32}$$

$$2C_3A + C_6A\overline{S}_3H_{32} + 4H \rightarrow 3C_4A\overline{S}H_{12}$$

At higher concentrations of sulfate, gypsum corrosion can contribute greater. Magnesium Sulfate can be even more aggressive because of the possibility of additional corrosive reaction due to the presence of magnesium ions, which can decompose both C-S-H and the calcium sulfoaluminates [6]:

$$C_3S_2H_3 + 3M\overline{S} \rightarrow 3C\overline{S}H_2 + 3MH + 2SH_x$$
$$C4A\overline{S}H_{12} + 3M\overline{S} \rightarrow 4C\overline{S}H_2 + 3MH + AH_3$$

This study aims to determine the force and weight of concrete due to sulfate attack in aggressive environments. The problems of this study was to determine the strength of concrete due to sulfate attack in aggressive environments and to determine the value of the weight of concrete due to sulfate attack in aggressive environments.

2. Research method

Materials that used in this research are 1. Water. 2. Sand with a maximum grain size of 0.6 mm. 3. There are 3 types of cement that being used: Type I cement, Type II cement, and PCC.
4. Silcafume with a grain size of 40 µm. 5. Quartz powder with 1200 mesh. 6. Superplasticizer used is Sika Viscocrete-10. 7. MgSO4 salt.

By using the material of No. 1-6, the mix design can be made. In this study, there are two types of mix design. The design are as follows:

- Water cement ratio (f_{as}) = 0,2 and 0,25
- Fine aggregate = 1,5 mass of cement
- *Silicafume* = 25% mass of cement
- *Super plasticizer* = 3% mass of cement
- *Quartz powder* = 25% mass of cement

The study begins with an examination of fine aggregate [7] based on standards that have been determined. After all the examination of fine aggregate finish, then make cylinders concrete specimen with a diameter of 75 mm and a height of 150 mm in accordance with the mix design that has been determined in advance. The slump test for this study is inverted slump test. After the concrete mixture is ready, then pour it into the mold. For the curing, its use steam curing method. For sulfate treatment, treatment used is based soundness test, ASTM C88 [8]. In this study sulfate used was a saturated solution of sulfate. MgSO4 salts dissolved in water until the salt can't dissolve anymore. Specimen will be tested when it's already 28 days.

IOP Conf. Series: Materials Science and Engineering **650** (2019) 012028 doi:10.1088/1757-899X/650/1/012028

3. Result and analysis

Based on the research, the compressive strength of concrete exposed to sulfate is lower than the concrete not exposed to sulfate. The results of compressive strength of concrete with and without immerse in sulfate can be seen in Table 1, Figure 2 and Figure 3.

Table 1. Concrete Test Result.

f_{as}	Cement Types	Compersive Strength			Mass
		Without Sulfate (MPa)	With Sulfate (MPa)	Difference (%)	Difference (%)
0,2	I	91,8100	59,4933	35,1995	1,2415
	II	85,9133	62,5400	27,2057	0,6234
	PCC	71,7367	47,6800	33,5347	0,8314
0,25	I	78,3700	37,2767	52,4350	1,4313
	II	76,1733	53,4367	29,8486	0,9732
	PCC	71,5300	42,9533	39,9506	1,0208

Concrete with immersion in sulfate mass reduction will occur as can be seen in Table 1. The mass reduction can also be seen in the specimen image. In Figure 1 it can be seen that there is still no holes are formed due to sulfate attack. Holes due to the formation of ettringite compounds that are expanding can be seen in Figure 1.

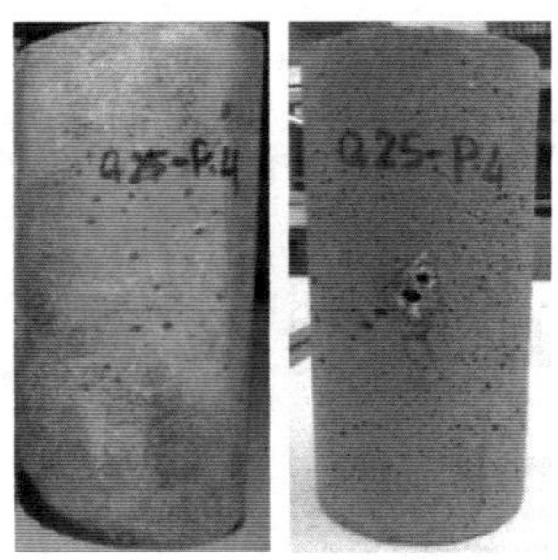

Figure 1. Specimen before (left) and after (right) immersion in sulfate.

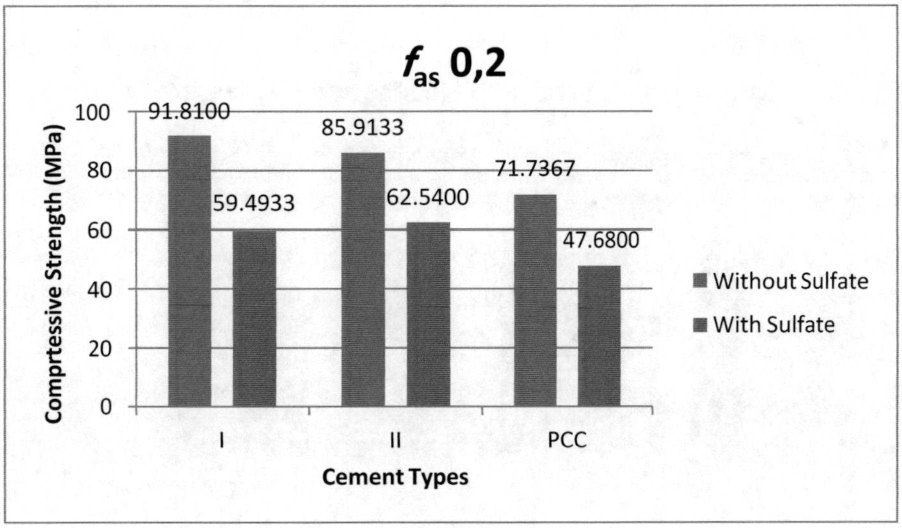

Figure 2. Compressive Strength with f_{as} 0,2 diagram.

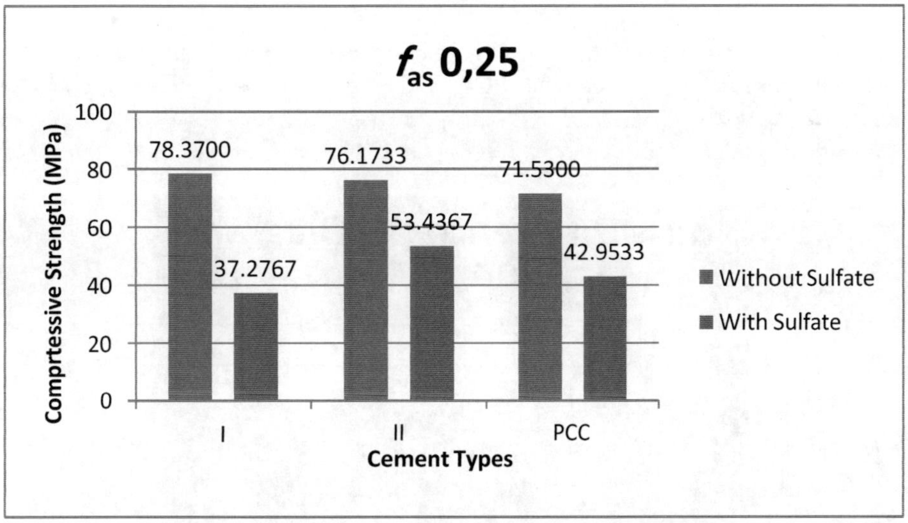

Figure 3. Compressive Strength with f_{as} 0,25 diagram.

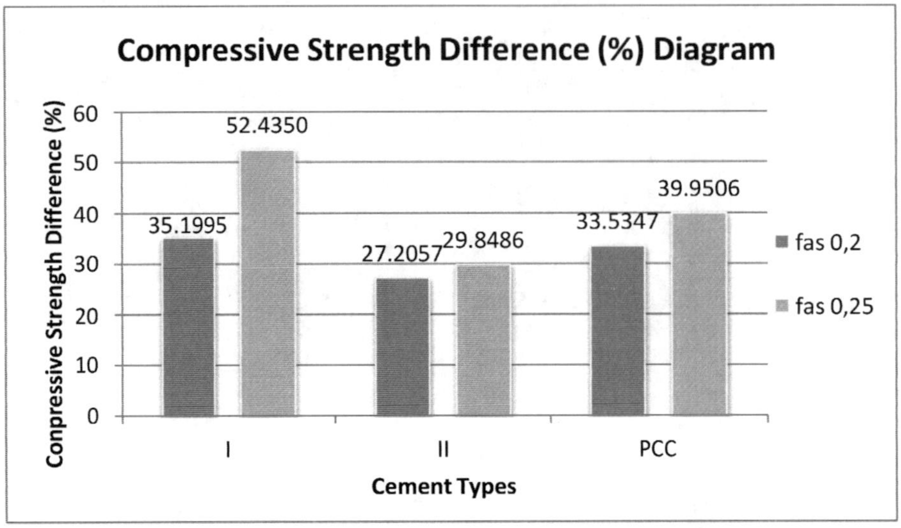

Figure 4. Compressive strength difference (%) diagram.

From Figure 4, if its sorted from the largest to the smallest compressive strength difference, that are specimen with cement type I, cement PCC and cement type II. This applies to specimen with f_{as} 0,2 as well as f_{as} 0,25.

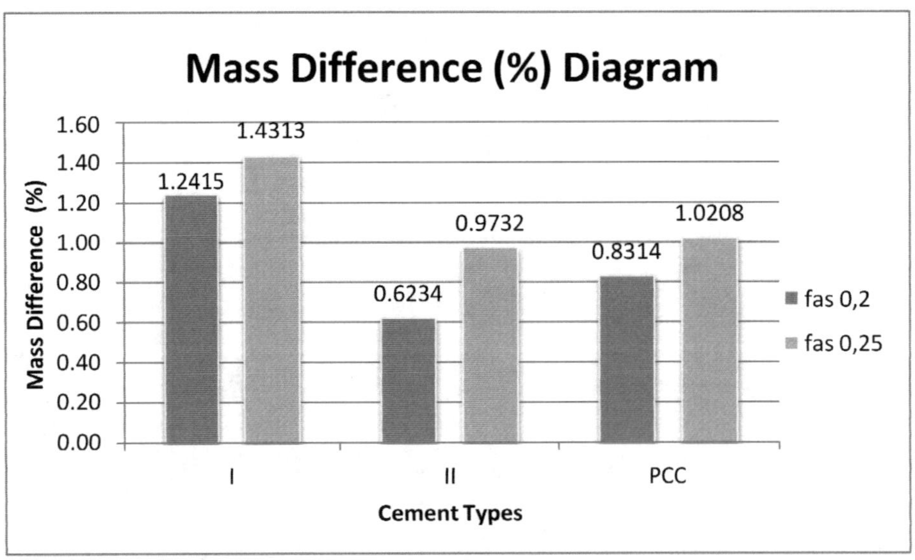

Figure 5. Mass Difference (%) diagram.

The less percentage difference in mass means the less mass from the specimen are missing. Based on Figure 5. the specimen with the least reduction in specimen mass is the specimen with Type II cement.

In the composition of the cement, C3A compound is a compound that was attacked by sulfate. Composition of C3A in type II cement is the least as much as 6%, so the cement Type II said to be good for sulfate attack. If C3A met with gypsum, it form ettringite that cause the crack expands and causes the compressive strength of concrete decreased.

IOP Conf. Series: Materials Science and Engineering **650** (2019) 012028 doi:10.1088/1757-899X/650/1/012028

It can be seen from the results of compressive strength test, specimens with Type II cement mixture produces % reduction in compressive strength at least compared to cement Type I and PCC. So it can be said that the Type II cement is the cement that has good resistance to sulfate compared with Type I cement and cement PCC.

4. Conclusion

Based on the research, we can conclude some of the following:

- Reactive Powder Concrete with a mixture of cement type II has a good resistance against sulfate, because the compressive strength only decreased by 27.2057% on the highly concentrated (saturated) sulfate attack.
- Reactive Powder Concrete is good for resistance to sulfate attack due to the weight loss of the test specimen is not up 1.5% to sulfate attack that has a high concentration.
- Based on this research, Type II cement is the cement that has good resistance to sulfate compared with Type I cement and cement PCC.
- Reactive Powder Concrete is best used on structures that are directly related to ground water and sea water, for examples foundation and harbor.
- In responding to sulfate attack, Reactive Powder Concrete with PCC cement more efficient than type II cement.

References

[1] Putra Dharma 2006 Penambahan Abu Sekam Pada Beton Dalam Mengantisipasi Kerusakan Akibat Magnesium Sulfat Pada Air Laut *Jurnal Ilmiah Teknik Sipil* **10**

[2] Richard P and Cheyrezy M 1995 Composition of Reactive Powder Concrete *Cement and Concrete Research* **25** p 1501

[3] Bonneau O, Vernet C, Moranville M and Aitcin P C 2000 Characterization of the granular packing and percolation threshold of reactive powder concrete *Cement and Concrete Research* **30** p 1861

[4] Murdock L J 1999 *Bahan dan Praktek Beton* (Ciracas: Erlangga)

[5] Masruri N 1993 Pengaruh Garam Sulfat terhadap Beton dan Cara Pencegahannya *Jurnal Pemukiman* **9** pp 11

[6] Mindess, Sidney and Young J F 1981 *Concrete* (America: Prentice-Hall)

[7] Laboratorium Konstruksi dan Teknologi Beton 1997 *Buku Petunjuk Praktikum Konstruksi dan Teknologi Beton* (Jakarta: Universitas Tarumanagara)

[8] American Society for Testing and Materials 1999 *Standard Test Method for Soundness of Aggregates by Use of Sodium Sulfate or Magnesium Sulfate* (Philadelphia: ASTM)

IOP Conf. Series: Materials Science and Engineering **650** (2019) 012029 doi:10.1088/1757-899X/650/1/012029

Numerical study of two low rise eccentrically braced steel frames under Indonesia seismic load using ELM and DAM

H Purnomo[1], M Orientilize[1]*, S A Rahim[1] and A Y Priyadi[1]

[1]Civil Engineering Department, Faculty of Engineering Universitas Indonesia

*mulia@eng.ui.ac.id

Abstract. Indonesia steel design code 2002 was revised in 2015. The latest version refers to AISC 360-10 which convert steel stability analysis to direct analysis method (DAM). DAM is an elastic analysis which includes second-order effect where P-delta and geometric imperfection are directly calculated during the analysis. Imperfection is modelled as notional load applied horizontally which is determined as 0.002 of gravity load. Numerical study using SAP software had been conducted to compare DAM and the previous method ELM on eccentrically braced frames (EBFs) under seismic load. It is known that EBF is one of earthquake resistant structural systems of steel structure. Numerical study was validated against experimental results performed by other researchers. Similar EBF frames used in the experiment were reanalyzed by ELM and DAM and coefficient of notional load is varied as 0.002 and 0.003. Seismic load applied in the analysis were varied based on three different seismic zones and soil conditions.

1. Introduction

Indonesia steel design code SNI 1729:2015 is a translation of AISC 360-2010 [1,2]. One of the significant revision of AISC 2010 is changing the stability analysis to Direct Analysis Method (DAM) substituting the preceding method Effective Length Method (ELM). The 2010 code uses DAM as its main analysis and ELM as an alternative. DAM is more accurate for stability analysis since second order effect is calculated explicitly in the analysis. Moreover, residual stress and out-of-plumbness effects may also be included in the analysis, therefore effective length factor can be omitted by taking effective length factor K as one.

As an alternative, ELM is still allowed to be used as explained on Appendix 7 of AISC 2010, if the ratio of maximum second-order drift to maximum first-order drift in all stories is equal to or less than 1.5 [1]. ELM includes the P-delta effect through amplification factor. To simplify the instability effect due to buckle, K factor is introduced as effective length where the value is determined according to relative rigidity between girder and column. Since the rapid development of ICT, P-delta effect and buckling can be directly calculated in the analysis. Hence, this method is named as direct analysis method, DAM.

Both ELM and DAM determine nominal capacity of individual column by using column interaction equations [3].

Initial imperfections, such as out of-plumbness and fabrication tolerances, result destabilizing effects. As explained in Section 2.2 AISC 2010, it can be included in the analysis through direct modeling or represented as notional loads. Notional loads are applied as lateral loads at each level as an addition to other lateral loads. The load value shall be 0.002 of gravity load of each level. According to clausal 2.b of Section C AISC 2010, the coefficient of 0.002 equals nominal initial story out-of-plumbness ratio as 1/500. However, the load coefficient can be adjusted proportionally, if necessary.

Several study have been done to compare DAM and ELM to structural steel design results [5- 7]. Analysis conducted on 1-storey with 1 bay structure result higher stress ratio when using ELM [3]. DAM has its advantages which are simpler to applied and more accurate for case which have significant second order effect. It is confirmed by study conducted on 1 story with 3 bays steel frame [5]. However, study which are compare DAM and ELM are rarely conducted on steel brace frame.

A numerical study to compare ELM and DAM on one and three-storey eccentric steel braced frame have been conducted. Preliminary research had been performed on one-story EBF, comparing flexural link and shear link [6]. The study was extended to three story EBF. Similar with former research, experimental study conducted by other researcher was reanalyzed using ELM and DAM. The frames were simulated against lateral loads represented seismic load from three seismic zones in Indonesia. Four stability methods were employed: ELM with first and second-order analysis, and DAM with notional load coefficients as 0.002 and 0.003.

2. Numerical Model of EBF and Validation

Numerical study was conducted using SAP2000 [8] software. Validation of numerical model was initially performed against two frames used in former experimental study conducted by other researcher, Shi [9] and Shujun [10] as shown in Figure 1 and 2. As described in AISC 341 [11] the two frames have the same link characteristics known as shear link.

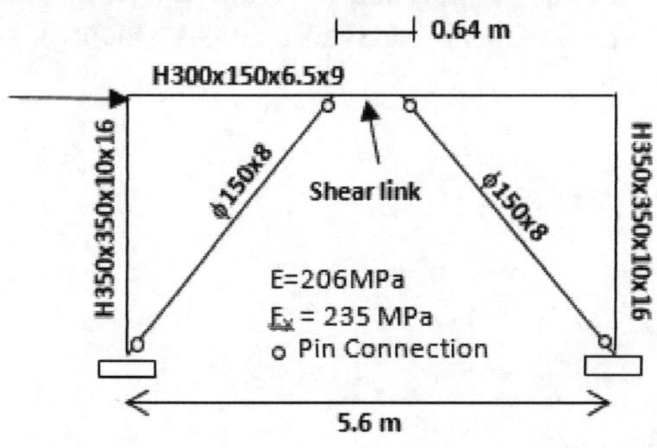

Figure 1. One-storey of EBF 1 [9].

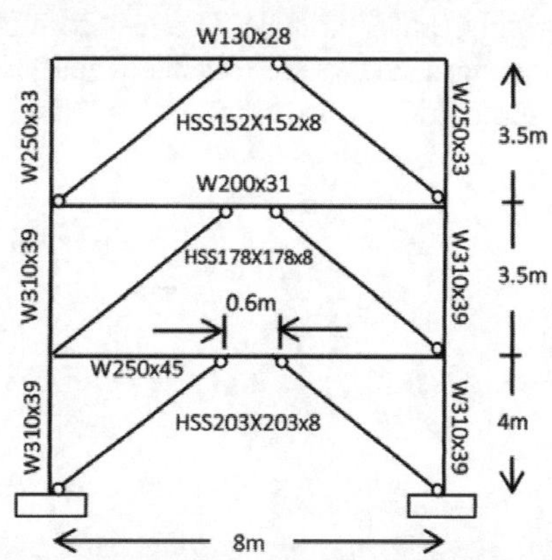

Figure 2. Three stories of EBF 3 [9].

Frame element are selected for beam, column and bracing. Following the experimental set-up, rigid and pin connections were assigned on connection of beam to column and brace connection, respectively. Moment curvature according FEMA 451 is assigned to link elements. [12]. To simulate the incremental loading, pushover analysis was carried out. Non linear of geometry and material were employed in the analysis. Plastic hinges were assigned on beam and column end, link and bracing.

The results are presented on load versus horizontal displacements. Both experiment and numerical results are plotted on the same graph as shown in Figure 3 and 4. As shown in Figure 3, experiment and numerical simulation of one storey frame is agree well. Furthermore, results from SAP analysis of three stories frame is not as good as one storey frame as presented in Figure 4. Initial stage of its load deflection curves agrees with experiment, where in the initial stage, the stiffness of SAP model is faintly higher compared to experimental result. After the plastic hinge was formed, the SAP result is above the test results. The maximum forces from SAP (365.85kN) is 10.9% higher than experiment (330kN). However, the first plastic hinge of SAP model occurred at 304.23kN which is late than experiment at 284.38kN.

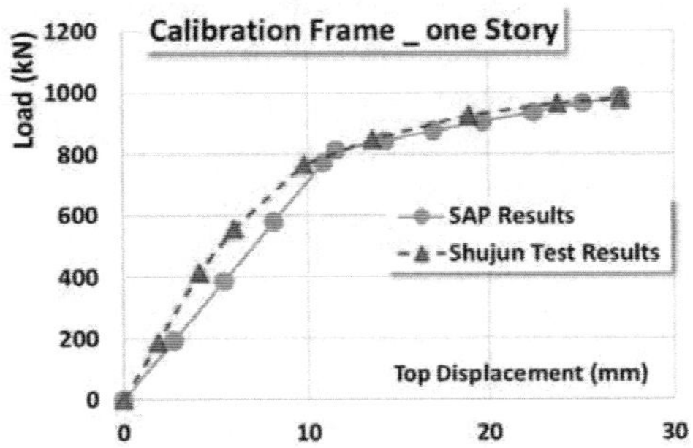

Figure 3. Validation of Numerical Model on one storey EBF.

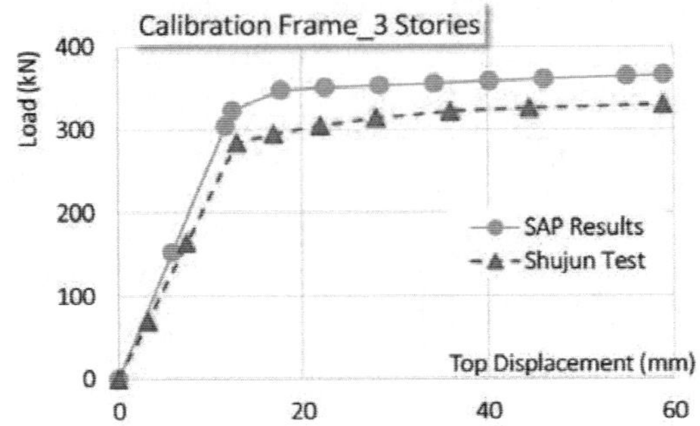

Figure 4.Validation of Numerical model on three stories EBF.

Figure 5 shows collapse mechanism of both frames. As shown, similar mechanism was detected as the plastic hinge form at shear link. The results confirm numerical results conducted by Shi et al. [9] where shear yielding occurs in mid-link. It can be concluded that numerical simulation using SAP software can simulate the experiment. Therefore, numerical models are considered valid and further study can be carried out using those models. The test results are used as reference in comparing ELM and DAM and named as advanced analysis. Advanced analysis is known as non-linear inelastic analysis where non-linearity in geometric and material are considered in the analysis.

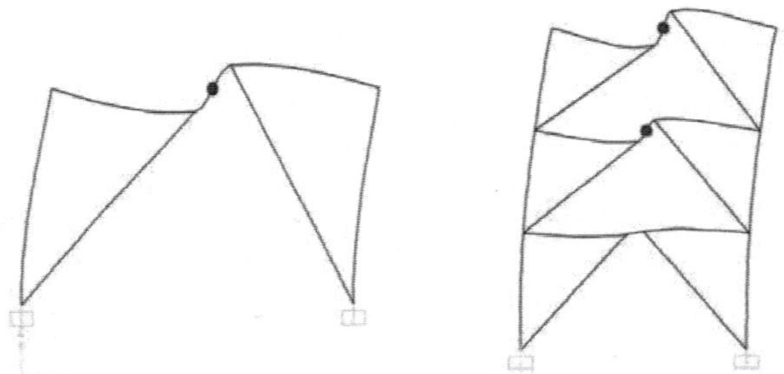

Figure 5. Plastic hinge on One-story and three stories EBF.

3. Results and Discussions

3.1. Comparison of DAM and ELM

Comparison study between DAM and ELM as steel stability analysis is conducted according to SNI 1729:2015 which accord to AISC 2010. ELM is the method adopted in the preceding design code, SNI

IOP Conf. Series: Materials Science and Engineering **650** (2019) 012029 doi:10.1088/1757-899X/650/1/012029

03-1729-2002. DAM was initially introduced as an alternative method besides ELM in AISC 2005. Refer to AISC 2010, DAM is the main method in SNI 2015 and ELM is one of alternative methods. Unlike SNI 2002 where ELM with first order elastic analysis is the only method used, SNI 2015 introduces second order elastic analysis of ELM. The difference between first and second order analysis is in determining the P-delta effect. As described on Appendix 8 AISC 2010, first order analysis calculates P-delta effect through amplification factor whereas second order analysis includes the effect in the analysis. Requirement details of DAM and ELM can be found on Chapter C and Appendix 8 of AISC 360-10, respectively. In addition, besides imperfection, DAM can also consider reduction of stiffness as 20% than initial condition, except for displacement calculation.

As explain in Section C2.2b, initial imperfections of column can be directly modeled in the analysis or represented as notional loads [1]. The loads are determined as follow:

$$Ni = 0.002\alpha Yi \quad (1)$$

where α =1.0 (LRFD); α = 1.6 (ASD)

Ni = notional load applied at level i, kips (N)

Yi = gravity load applied at level i from the LRFD/ASD load combination, kips (N)

The constant of 0.002 in equation (1) is determined based on amount of nominal initial story out-of-plumbness ratio of 1/500L and is allowed to adjust proportionally [1]. Two EBFs shown in Figure 1 and 2 were reanalyzed by two different approach of ELM, first order and second order approach. DAM with two notional load constant number of 0.002 and 0.003 were used. The loads were laterally applied according to equation 1. Hence, four stability methods named as ELM first order, ELM second order, DAM_0.002 and DAM_0.003 are compared.

The study aims to evaluate different notional load values which represent the column plumbness. The second objective is to see how effective the amplification moment factor can represent the P-delta effect on ELM analysis. Pushover analysis were carried where the maximum force is limited to ratio of second-order to first-order analysis as 1.5. Henceforth, one-story and three-stories EBF were loaded by 1940,09 kN and 1043,25 kN, respectively.

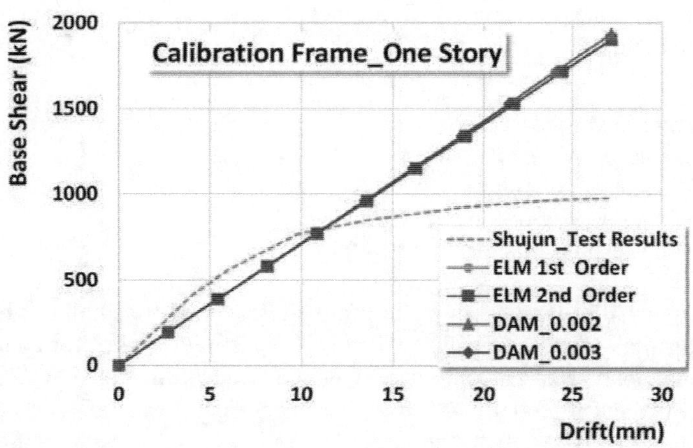

Figure 6. Comparison Results of one-storey EBF analyzed by ELM and DAM.

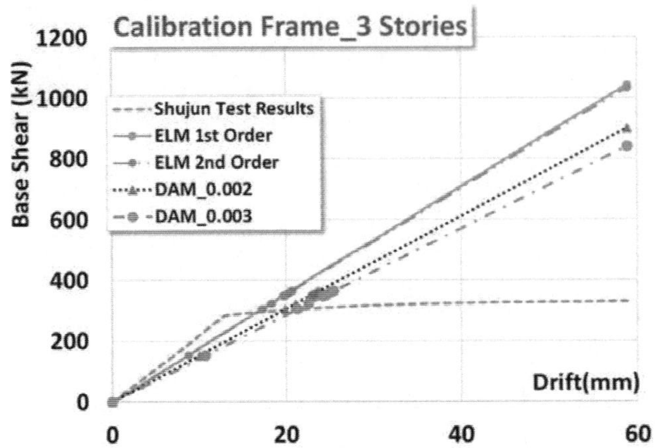

Figure 7. Comparison Results of Three-Storey EBF analyzed by ELM and DAM.

Figure 6 and 7 shows results of pushover analysis. As shown, stability analysis using ELM or DAM is not significant on one-storey EBF as can be seen from the four graphs that close each other. Slighly different are found on three-storey EBF as can be seen on Figure 7. At the same base shear, DAM_0.003 has the highest drift followed by DAM 0.002 and ELM. The results are due to lateral forces applied at DAM 0.003 is higher since it is added by notional loads. The difference is not significant since the loads is a function of gravity loads which come from EBF's selfweigth. Thought that the effect is not significant, ELM second order is less stiff than ELM first order. The result is predicted since ELM second order include the effect of P-delta in the analysis.

3.2. EBF subjected to Indonesian Seismic Zone

Study of ELM and DAM is proceeded by comparing PM ratio of EBFs as a results of seismic simulation. The EBFs were loaded by earthquake load according to Indonesia design code: SNI 1726-2012 [13]. The frames were simulated in 3 seismic zones, Samarinda, Jakarta and Padang which considered as light, medium and strong seismic zone, respectively. Soil types were varied as hard, medium and soft soil which were denoted as SC, SD and SE, respectively. Earthquake loads applied are determined as follows [14]:

$$F_x = C_{vx} V \tag{2}$$

where: F_x : Seismic load for x-storey
C_{vx} : Distribution factor for each floor
V : Base shear of structure, $V = C_s.\, W$
W : Effective Weight of the structure,
C_s : Seismic response factor

As discusses earlier, two EBFs used in this study are those tested by Shujun where the frames were loaded horizontally. Similar load case was applied in this study where horizontal load consisted of earthquake and notional load. Hence, combination of loads according to ASCE [15] are: (C1) 1.4D + 1.4ND; (C2) 1.2D + 1.0E and (C3) 0.9D + 1.0E, where D is dead load, ND is notional dead load, and E is earthquake. Combination C2 result the highest PM interaction for both EBFs, hereafter, only that combination result is discussed here. Structural responses are presented as PM interaction where it is determined based on ultimate load of axial and flexural compared to its each nominal capacity as follow:

IOP Conf. Series: Materials Science and Engineering **650** (2019) 012029 doi:10.1088/1757-899X/650/1/012029

For ratio of $P_u / \phi P_n \geq 0.2$ PM interaction : $P_u/\phi P_n + 8M_u/(9\phi P_n) \leq 1$ (3)

For ratio of $P_u / \phi P_n < 0.2$ PM interaction: $P_u/(2\phi P_n) + M_u / \phi P_n \leq 1$ (4)

Results of numerical simulation modeled based on experimental test is assumed as advanced analysis since the model can simulate the experimental test. Advanced analysis is a non-linear inelastic analysis considering non-linearity in geometric and material. Hence it can be used as reference to compare ELM and DAM.

Table 1 and 2 shows PM interaction results of EBFs. The value presented here were taken from column at the lowest floor. Similar with previous result, ELM and DAM have no different effect in term of PM ratio of one story EBF. As presented on Table 1, PM ratio of ELM and DAM are same. Minor effect of ELM and DAM is observed on PM ratio of three stories-EBF. The four methods, ELM first order, ELM second order, DAM_0.002 and DAM_0.003, have ratio below advanced analysis. PM-interactions obtained for all methods are below 0.5.

Table 1. PM Interaction of One-Storey EBF.

Seismic Zone	Soil Type	Earthquake Loads (kN)	PM Interaction of One-Storey EBF				
			Advanced Analysis	ELM 1st Order	ELM 2nd Order	DAM _0.002	DAM _0.003
Samarinda	SC	12.54	0.007	0.007	0.007	0.007	0.007
	SD	16.73	0.009	0.009	0.009	0.009	0.009
	SE	26.13	0.014	0.014	0.014	0.015	0.015
Jakarta	SC	72.74	0.04	0.040	0.04	0.04	0.040
	SD	82.77	0.046	0.046	0.046	0.046	0.046
	SE	94.06	0.052	0.052	0.052	0.052	0.052
Padang	SC	145.68	0.081	0.081	0.081	0.081	0.081
	SD	145.68	0.081	0.081	0.081	0.081	0.081
	SE	131.30	0.073	0.073	0.073	0.073	0.073

Table 2. PM Interaction of Three-Stories EBF.

Seismic Zone	Soil Type	Earthquake Loads (kN)	PM Interaction of Three-Storey EBF				
			Advanced Analysis	ELM 1st Order	ELM 2nd Order	DAM _0.002	DAM _0.003
Samarinda	SC	123,69	0,066	0,057	0,057	0,057	0,057
	SD	164,97	0,088	0,076	0,076	0,076	0,076
	SE	257,73	0,138	0,119	0,118	0,118	0,118
Jakarta	SC	717,38	0,384	0,330	0,329	0,329	0,328
	SD	816,33	0,437	0,375	0,374	0,374	0,374
	SE	927,65	0,497	0,427	0,425	0,425	0,425
Padang	SC	1436,77	0,769	0,661	0,658	0,659	0,658
	SD	1436,77	0,769	0,661	0,658	0,659	0,658
	SE	1295,00	0,693	0,595	0,593	0,594	0,593

Table 3. Drift of One-Storey EBF.

Seismic Zone	Soil Type	Earthquake Loads (kN)	Drift of One-Storey EBF				
			Advanced Analysis	ELM 1st Order	ELM 2nd Order	DAM _0.002	DAM _0.003
Samarinda	SC	12,54	0,18	0,18	0,18	0,18	0,18
	SD	16,73	0,23	0,24	0,24	0,24	0,24
	SE	26,13	0,37	0,37	0,37	0,37	0,37
Jakarta	SC	72,74	1,02	1,04	1,02	1,03	1,02
	SD	82,77	1,16	1,18	1,17	1,17	1,17
	SE	94,06	1,32	1,34	1,32	1,33	1,32
Padang	SC	145,68	2,04	2,07	2,05	2,05	2,05
	SD	145,68	2,04	2,07	2,05	2,05	2,05
	SE	131,30	1,84	1,87	1,85	1,85	1,85

allowable drift = 60mm

Table 4. Drift of Three-Stories EBF.

Seismic Zone	Soil Type	Earthquake Loads (kN)	Drift of Three-Storey EBF				
			Advanced Analysis	ELM 1st Order	ELM 2nd Order	DAM _0.002	DAM _0.003
Samarinda	SC	123,69	5,950	5,670	5,858	5,770	8,551
	SD	164,97	7,936	7,562	7,813	7,793	11,405
	SE	257,73	12,399	11,815	12,207	12,338	17,819
Jakarta	SC	717,38	-	32,886	33,977	34,857	49,598
	SD	816,33	-	37,422	38,663	39,705	56,439
	SE	927,65	-	42,525	43,935	45,159	64,135
Padang	SC	1436,77	-	65,865	68,048	70,102	99,334
	SD	1436,77	-	65,865	68,048	70,102	99,334
	SE	1295,00	-	59,365	61,333	63,156	89,532

allowable drift = 254mm

Comparison of horizontal displacement at top story called as drift are displayed on Table 3 and 4. The value of allowable drift are determined based on SNI 1729-2012 and ASCE 7-16, section 12.12. The EBFs are categorized as all other structures and classified in risk category II. Thus allowable drift is $0.02h_x$ where the limit for one storey is 60mm and 254mm for three stories. The EBFs drift are meet the requirement which is below the maximum value. Similar results are also found that ELM and DAM do not results significant effect to drift of both EBFs.

As presented on Table 4, DAM with notional load factor as 0.003 have the highest drift followed by DAM_0.002, ELM second order and ELM first order. As discussed earlier, this is due to addition of horizontal forces comes from notional loads and second order effect. EBFs with ELM analysis were only loaded by earthquake without notional load.

4. Conclusions

Comparing different methods of steel stability analysis, ELM and DAM, to two EBFs gives almost similar results on one storey frame. The difference of instability calculated by the two methods does not appear significantly. Hence, the buckling factor "K" between ELM and DAM seems similar. In term of PM ratio, three stories EBFs also have similar results. The effect of ELM and DAM appears evidently on three stories where DAM_0.003 has the highest drift. Initial imperfections of column as 0.002 (1/500L) and 0.003 that represented on notional loads slightly change the EBF responses. This is because of less dead load load worked on the structures. However, notional load is a function of gravity load. Both EBFs have enough strength against seismic loads in three seismic zones in Indonesia.

Acknowledgment
This works is supported by Hibah PITTA 2018 funded by DRPM Universitas Indonesia No. 2451/UN2.R3.1/ HKP.05.00/2018.

References

[1] AISC 2010 *Specification for Steel Structural Building*
[2] *Badan Standardisasi Nasional (BSN), SNI* 1729:2015, *Spesifikasi untuk Bangunan Gedung Baja Struktural,* (2015)
[3] Prajzner J 2006 *Evaluation of the Effective Length Method and the Direct Analysis Method for the Design of Steel Columns in Frame Structures* (University of Maryland)
[4] J Rozon and S K 2008 *Study of Global Behaviour of Eccentrically Braced Frames in Response to Seismic Loads* The 14th World Conference on Earthquake Engineering in Beijing, China.
[5] Surovek A E and Ziemian R D 2005 *The Direct Analysis Method: Bridging the Gap from Linear Elastic Analysis to Advanced Analysis in Steel Frame Design (ASCE library)*
[6] M Orientilize, H Purnomo, S A Rahim and A Y Priyadi 2019 *IOP Conference Series Materials Science and Engineering Series Materials Science and Engineering* **473**(1)
[7] Purnomo Heru, Orientilize Mulia, A Rahim Sjahril1 and Firdaus A Zaki 2019 MATEC Web Conf **276**
[8] SAP 2000 Adv. V.15, 2011, Analysis Reference Manual, Computers and Structures Inc. (Berkeley California)
[9] Shi Y J, Xiong J and Wang Y Q 2010 Experimental Studies on Seismic Performance of Multi-storey Steel Frame with Eccentric Brace *Journal of Building Structures* **31**(2) pp 29-34
[10] Shujun Hu, Z W 2015 Practical Advanced Analysis for Eccentrically Braced Frames. *Advanced Steel Construction* **11** No.1 pp 95-110
[11] AISC 341-16 2016 *Seismic Provisions for Structural Steel Buildings* (Chicago: American Institute of Steel Construction)
[12] FEMA 451 2006 *NEHRP Recommended Provisions: Design Examples* (California: Federal Emergency Management Agency)
[13] Badan Standardisasi Nasional (BSN) SNI 1726:2012 *Tata Cara Perencanaan Ketahanan Gempa untuk Struktur Bangunan Gedung dan Non Gedung*
[14] AISC and Structural Steel Educational Council 2006 *AISC Seismic Design Manual* (AISC inc.)
[15] ASCE 2017 *"Minimum Design Loads and Associated Criteria for Buildings and Other Structures."* Standards ASCE/SEI 7-16.

IOP Conf. Series: Materials Science and Engineering **650** (2019) 012030 doi:10.1088/1757-899X/650/1/012030

Infiltration of chloride ion in the self compacting concrete with calcium stearate

Agus Maryoto[1*], Buntara Sthenly Gan[2], Nor Intang Setyo Hermanto[1] and Rachmad Setijadi[1]

[1]Universitas Jenderal Soedirman, Purwokerto, Central Java, Indonesia
[2]Nihon University, Koriyama, Sendai, Japan

*agus_maryoto1971@yahoo.co.id

Abstract. Corrosion rates in reinforced concrete can be detected by the amount of chloride ions present in concrete. This study aims to determine the effect of calcium stearate use on self compacting concrete on the infiltration of chloride ions. The calcium stearate content used is 0, 1, 5 and 10 kg per 1 m^3 of self-compacting concrete (SCC). The concrete quality tested includes 20, 30 and 40 MPa. The parameters tested in this study were compressive strength and infiltration of chloride ions. Compressive strength specimen is in the form of cylinder 15 mm in diameter and 150 mm in height, while chloride ion infiltration specimen is in the form of cubes with sides of 150 mm. The results show that calcium stearate concrete with 10 kg / m3 of concrete when compared to normal concrete, the compressive strength decreases by -6.69%, 16.16% and 18.74% for concrete 20, 30 and 40 MPa respectively. Whereas the infiltration of chloride ions in concrete on 10 mm depth from the concrete surface was 0.153, 0.061 and 0.012% in the concrete of 20, 30 and 40 MPa. These infiltration of chloride ion in concrete 20, 30 and 40 MPa are lower around 287, 252, and 626% compared to concrete without calcium stearate.

1. Introduction

Self-compacting concrete (SCC) is concrete that is capable of compacting itself without the help of external compaction. This type of concrete can flow only by relying on its own weight [1]. SCC is able to fill even the narrowest space between even tight concrete reinforcements [2]. This capability is not possessed by conventional concrete, consequently SCC is generally denser and has a higher compressive strength [3,4]. The high compressive strength of SCC is also due to the use of a low water cement ratio of around 30%. Because the usage of a low water cement ratio, SCC has the small number of capillaries and concrete pores. These capillaries and pores are formed during a hydration reaction process where the water needed to react with cement is only about 25% of the weight of cement. This means that only 25% of the total weight of water used evaporates and leaves pores and capillaries. Another advantage of SCC concrete compared to conventional concrete is that it is more waterproof, lower shrinkage and creep, more resistant to corrosion attack [5], lower permeability, better flexural strength and durability.

Reinforced and pre-stressed concrete can be used for building structures located in free and un-free from corrosion attacks and corrosive locations [6,7]. The concrete structure located in the corrosive environment has a shorter service life. Bridge structures of reinforced or pre-stressed concrete located in coastal areas generally have a service life that is shorter than the design. This is caused by corrosion attacks on concrete reinforcement by chloride ions. It can be infiltrated into concrete due to pressure, diffusion and absorption. When corrosion starts in concrete reinforcement, the service load of reinforced concrete structures decreases by the time [8]. Although in general, the reduction in service capacity of reinforced concrete can be improved [9], the carrying out of repairs is very complicated and the costs are expensive.

Maryoto et al. [10] have investigated the use of calcium in conventional concrete. The results show that the use of calcium stearate in conventional concrete can reduce the infiltration of chloride ions. The effect of using calcium stearate is to make the concrete more hydrophobic. Finally, the corrosion attack that occurs in concrete contained calcium stearate becomes smaller. On the other hand, SCC is concrete made by adding superplasticizer as a significant additive to increase the workability of fresh concrete. Until now, there are no researchers to investigate whether the use of calcium stearate is safe or unsafe in SCC. Moreover, this study aims to determine the effect of calcium stearate on infiltration of chloride and compressive strength ions in SCC 20, 30 and 40 MPa.

2. Method

2.1. Materials and equipment

The materials used in this study include Portland Composite Cement (PCC), sand, broken stone, water, superplasticizer (Viscocrete 1003), coating material, NaCl and calcium stearate. Meanwhile the main equipment used includes concrete mixer, slump equipment, oven, balance, sieve shaker, molding cylinder diameter 150 mm and 300 mm height, cube molding 150x150x150 mm, compression machine / universal testing machine (UTM), X-ray fluorescence (XRF) and drilling machine.

Based on the physical properties of fine aggregates and coarse aggregates, the mix proportion of SCC 20, 30 and 40 MPa as shown in Table 1. The number of test specimens, type of test and calcium stearate content used in SCC are shown in Table 2.

Table 1. Mixed proportion of SCC.

Material	Quality of concrete (MPa)		
	20	30	40
Cement, PCC (kg)	350	425	545
Sand (kg)	910	815	740
Crushed Stone (kg)	910	920	886
Water (kg)	175	186	183
Superplasticizer (kg)	1.8	1.9	2.7

Table 2. Number of specimen.

Code of specimen	Compression test			Infiltration of chloride ion test		
	20	30	40	20	30	40
C0S0.5	3	3	3	3	3	3
C1S0.5	3	3	3	3	3	3
C5S0.5	3	3	3	3	3	3
C10S0.5	3	3	3	3	3	3

In the table: C is representing calcium stearate, one number behind C shows calcium stearate content in each m^3 of SCC (in kg), S is representing superplasticizer, and two numbers behind S show superplasticizer content in percentage by the cement weight.

2.2. Infiltration of chloride ion

The infiltration of chloride ions test was carried out to determine the amount of chloride ions that ingress into the SCC. The test object used is a cube measuring 150x150x150 cm. Specimens are removed from the water soaking tub after 28 days of water immersion. The test object is dried by placing it in an open space but not exposed to direct sunlight for 24 hours. Cover the 5 surfaces of the specimen with coating material so that the water cannot infiltrate into the concrete as shown in Figure 1. Soak the specimen cube with 3% NaCl solution for 3 months. The depth of 3% NaCl solution is 3 cm to the surface of the surface of the specimen as shown in Figure 2. The position of the uncoated surface is located in the above position. It means that the 3% NaCl solution can infiltrate into the SCC. The next step is to remove the specimen from the immersion of 3% NaCl solution after 3 months and dry it for 24 hours. Drilling specimens is carried out on uncoated surfaces such as Figure 3. The drilling powder is put in a glass baker as shown in Figure 4. A drilling scheme with a distance of 1, 2, 4, 6 and 8 cm is illustrated in Figure 5.

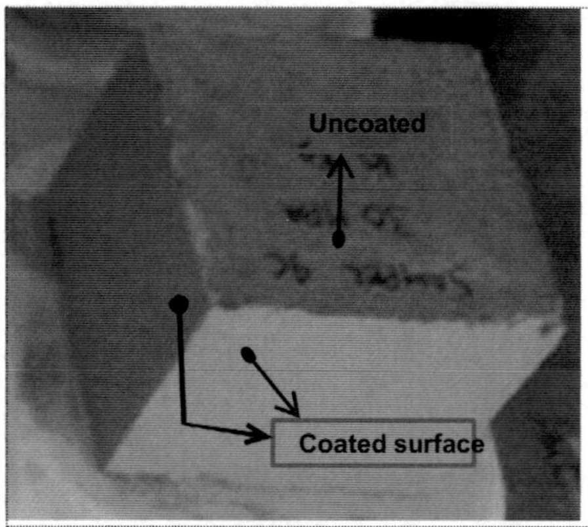

Figure 1. Coating protection.

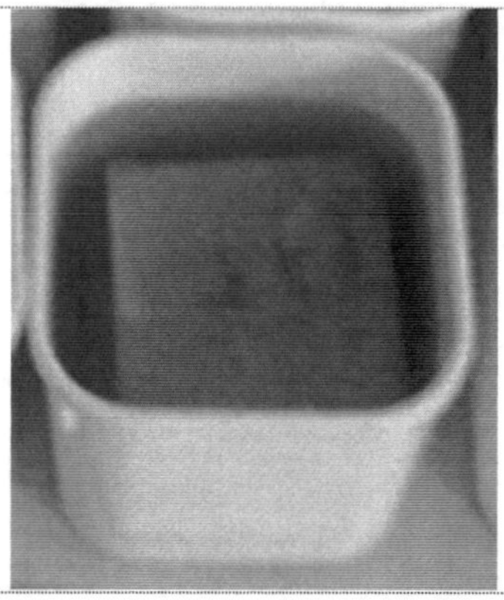

Figure 2. Soaking under NaCl solution.

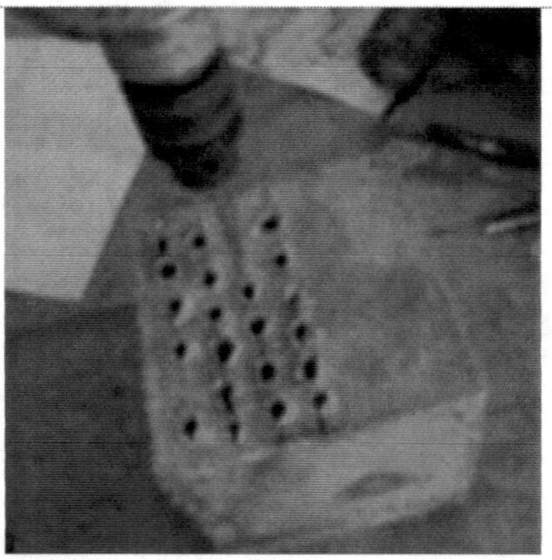

Figure 3. Taking powder of concrete.

Figure 4. Powder of concrete.

IOP Conf. Series: Materials Science and Engineering **650** (2019) 012030 doi:10.1088/1757-899X/650/1/012030

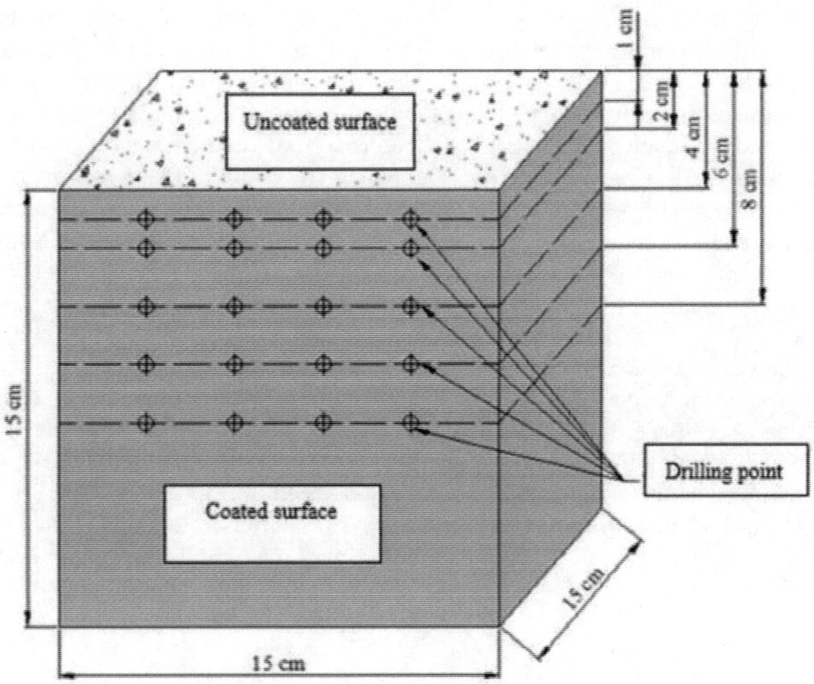

Figure 5. Schematic of drilling.

As for powder of concrete shown in the Fig. 4 is then analyzed to identify the chloride content using XRF. The working principle of chloride analysis using XRF is based on the collision of atoms on the surface of specimens by X-rays. Qualitative results are indicated by the peak of the spectrum representing the type of element in accordance with X-ray energy characteristics. Quantitative analysis is obtained by comparing the sample intensity with the standard. The steps for testing specimens with XRF are as follows Powder of concrete is finely ground by using a portable ball mill. Sieving powder of concrete so that its size is homogeneous at +/- 100 mesh. Enter the powder of concrete in the ring and weigh it to get the weight of the specimen. XRF and computer are turned on and wait for a while so the equipment is ready to be used to analyze the powder of concrete.

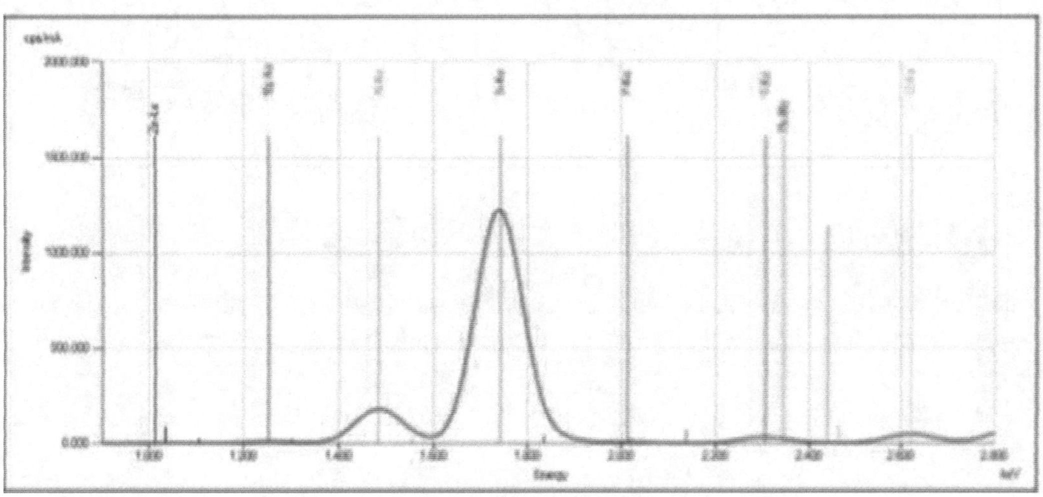

Figure 6. Spectogram

IOP Conf. Series: Materials Science and Engineering **650** (2019) 012030 doi:10.1088/1757-899X/650/1/012030

Specimens that are ready to be put in in XRF. Measurement variables are filled in accordance with Standard Operating Procedures of XRF. Energy to analyze is regulated. Filters are arranged according to the elements to be analyzed in this case is the element Chloride and Sodium. XRF is closed and runs an analysis program that is on the computer and is related to XRF. Testing of specimen is conducted and the result is a spectrogram. Percentage of each element contained in powder of SCC is obtained. Save the test results data on a computer. The result of elemental analysis is a spectrogram as in Figure 6. Base on the graph, it can be observed the intensity and energy reflected by Cl (chloride). The high intensity shows how much content in the test sample. High energy is a specific characteristic of certain elements. The output of the XRF instrument can be put in a table of mass content of each element in percentage.

2.3. Compressive strength

The concrete compressive strength test is carried out when the concrete is 28 days old and has been placed in an open space to dry the surface of the specimen. Capping concrete surfaces is conducted in order to obtain a good distribution working load on the concrete surface. Put the specimen in the testing position and turn on the compression machine at a speed of 0.19 - 0.39 MPa. The compressive strength of concrete is obtained when the maximum force that causes the specimen fall out. The amount of compressive strength is calculated by dividing the maximum force with the surface area of the cylindrical specimen.

3. Result and Discussions

3.1. Infiltration of chloride ions

Figures 7, 8 and 9 show the results of ion chloride infiltration tests on SCC 20, 30, and 40 MPa contained calcium stearate. Meanwhile, Figure 10 shows the relationship between infiltration of chloride ion in 6 cm depth and calcium stearate content in concrete.

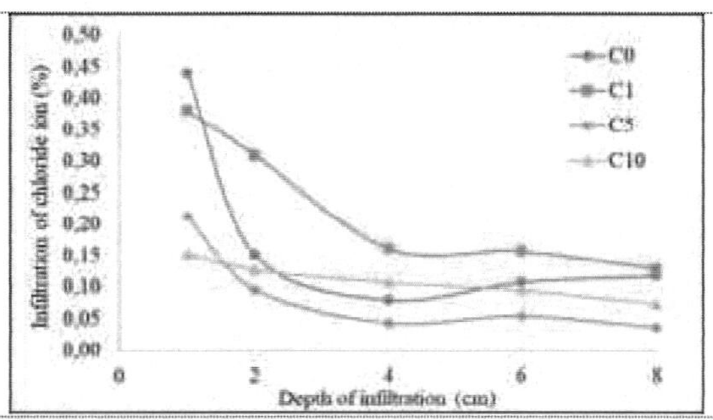

Figure 7. SCC 20 MPa.

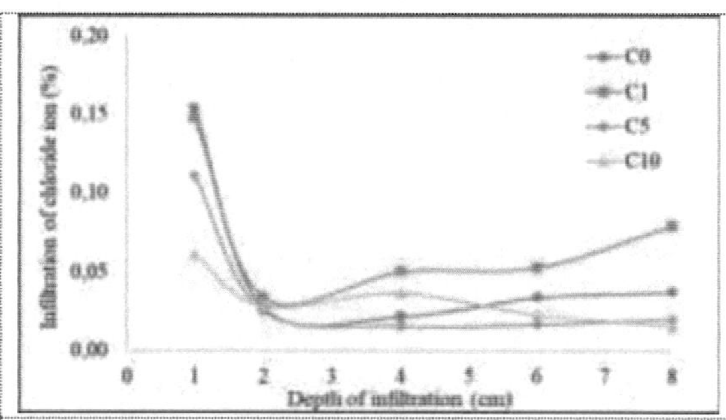

Figure 8. SCC 30 MPa.

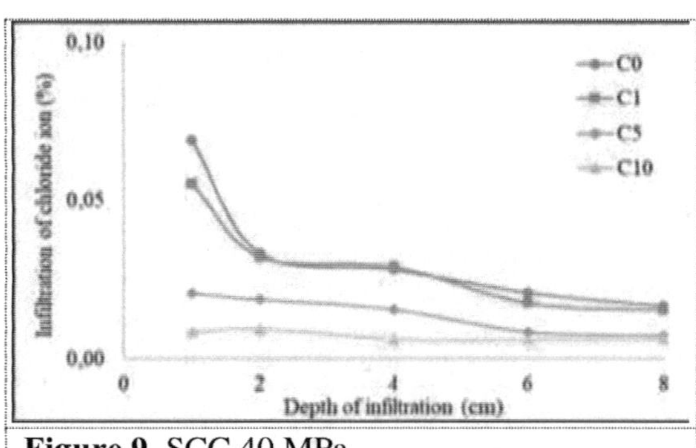

Figure 9. SCC 40 MPa.

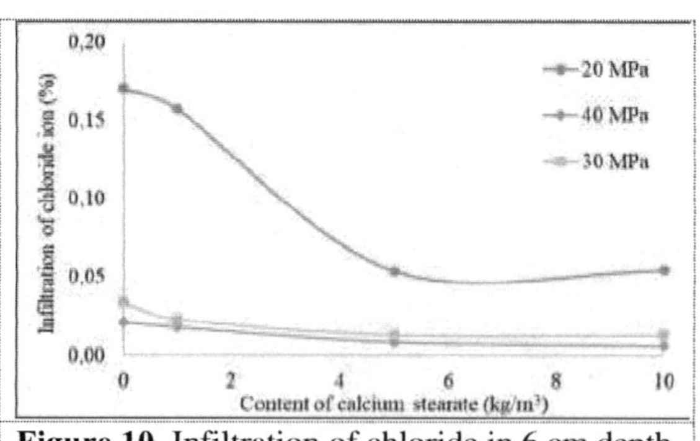

Figure 10. Infiltration of chloride in 6 cm depth.

IOP Conf. Series: Materials Science and Engineering **650** (2019) 012030 doi:10.1088/1757-899X/650/1/012030

In the figures 7, 8 and 9, C0, C1, C5 and C10 represent calcium stearate content in SCC. It is 0, 1, 5 and 10 kg per 1 m^3 of fresh SCC.

Figure 7 shows the relationship between the depth of the chloride ion infiltration and the percentage of chloride ion in the SCC 20 MPa. The deeper position of taking concrete powder from the surface, the infiltration of chloride ions decreases. The effect of this incident is that corrosion attacks on concrete reinforcement that placed deeper than the concrete surface also decreases. Furthermore, reinforced concrete located in a corrosive area, it would be safer if the size of the concrete cover is greater than 50 mm.

In addition, based on Figure 7, it can be found that SCC with the addition of calcium stearate, infiltration of chloride ion in the SCC is decreased. Higher the content of calcium stearate in SCC, the chloride ion is lower. The infiltration behavior of chloride ions in 30 and 40 MPa concrete was the same behaviour with SCC 20 MPa. Greater the addition of calcium stearate, the smaller the infiltration of chloride ions into concrete (Figures 8 and 9). This is understandable because calcium stearate in the concrete will react with cement to form the wax-like compound [10,11,12,13]. The wax-like have a hydrophobic property. This property cause the contact angle formed between the water and the concrete surface becomes large. Consequently, water will be difficult to ingress into the SCC.

Figure 10 shows the magnitude of chloride ion infiltration in SCC 20, 30 and 40 MPa at a depth of 6 cm from the surface of the SCC. Based on this figure, the interesting thing is that the higher the quality of the concrete used, the infiltration of chloride ions decreases. Infiltration of chloride ions at a depth of 6 cm in SCC 20, 30 and 40 MPa with an additional calcium stearate of 1 kg per 1 m^3 of concrete decreased by 8%, 32% and 15% compared to conventional concrete. This phenomenon occurs because the higher the quality of the concrete, the lower the water cement ratio used in the mix proportion of concrete. Based on Table 1, the water cement ratio used is 0.50, 0.44 and 0.34 for concrete grade 20, 30 and 40 MPa. Chemically, the water used by cement so that the hydration reaction can take place is 0.25 of the weight of cement. This means that there is still water left for 0.25, 0.19 and 0.14 of the weight of cement on concrete quality concrete 20, 30 and 40 MPa. The residual water evaporates and comes out of the concrete and the effect is leaving the pore and capillary. The bigger the water evaporates, the bigger the pore and capillaries are formed. This capillary network is used by chloride ions in and seep into the concrete.

3.2. Compressive strength

Compressive strength of SCC can be identified in the Figure 11. It can be observed that calcium stearate applied in SCC reduces the compressive strength. This decline of the compressive strength not only occurs in the SCC 20 MPa, but also in the SCC 30 and 40 MPa. This trend is matched to the usage of calcium stearate in the conventional concrete [10]. The effect of calcium stearate 1 kg per cubic meter in compressive strength of SCC is insignificant. Because of this, Calcium stearate is safe to be applied in the SCC. No negative effects in the SCC due to addition of calcium stearate. According to Maryoto, 2018 the reduction of compressive strength of SCC is caused by the bonding of wax-like constituent. Unlike the bonding of calcium silicate hydrate (C-S-H) in concrete, the bonding of wax-like constituent in SCC is very weak. As the influenced, the compressive strength of SCC containing calcium stearate is descend. The reduces of compressive strength is around -3,7%, 6% and 0.5% for SCC containing calcium stearate 1 kg per 1 m^3 of fresh concrete.

IOP Conf. Series: Materials Science and Engineering **650** (2019) 012030 doi:10.1088/1757-899X/650/1/012030

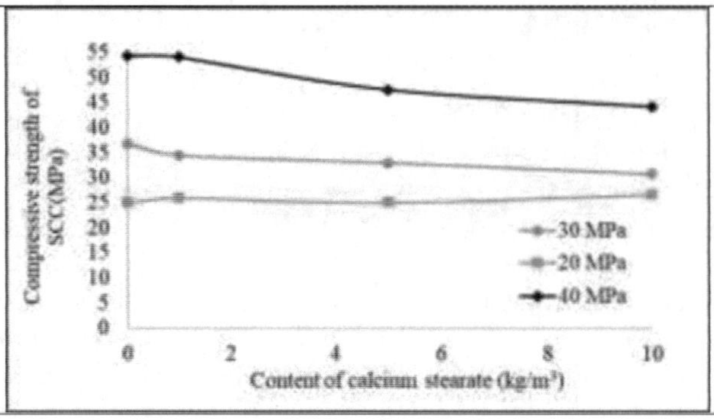

Figure 11. The average of compressive strength of SCC with calcium stearate

4. Conclusions

Infiltration of chloride ion in SCC is affected by containing calcium stearate. Additional of calcium stearate in SCC 20, 30 and 40 MPa reduces infiltration of chloride around 8%, 32% and 15% in 6 cm depth from surface of SCC. Lower infiltration of chloride ion, the resistance of SCC toward the corrosion attach increase. On the other hand, compressive strength of SCC 30 MPa with calcium stearate 1 kg per 1 m3 of SCC is reduced insignificantly. The decreasing of the compressive strength of concrete with calcium stearate is -3.7%, 6.1% and 0.525 for SCC 20, 30 and 40 MPa respectively. Even though the usage of calcium stearate on high content reduced the infiltration of chloride ion in the SCC significantly, it suggested that the safe usage of calcium stearate in SCC is 1 kg per 1 m^3 of fresh SCC.

Acknowledgement
Thank you to LPPM Universitas Jenderal Soedirman for funding this research through the 2019 IRC scheme so that this paper can be presented and published.

References

[1] Okamura H, Ozawa K and Ouchi M 2000 *Struc. Conc.* **1** pp 3-17
[2] Dines A, Harini S, Jasmine J P, Jincy J and Shagufta, J 2017 *Int. J. of Eng. Sci. & Res. Tech.* **6** pp 42-50
[3] Bradu A, Cazacu N, Florea N and Mihai P 2016 *Bul. Inst. Pol. Din Iasi* **62** pp 59-68
[4] Dhiyaneshwaran S, Ramanathan P, Baskar L and Venkatasubramani R 2013 *Jordan J. of Civ. Eng.* **7** pp 342-53
[5] Mittal M, Basu S and Sofi A 2013 *Int. J. of Civ.* Eng. **2** pp 61-6
[6] Kumar V, Singh R and Quraishi M A 2013 *J. of Mat. & Env. Sci.* **4** pp 726-31
[7] Loreto G, Benedetti M D, Iovino R, Nanni A and Gonzales-Nunez M A 2011 *Proc .of SPIE 7983* pp 1-11
[8] Neville A 1996 *Mat. & Struc.* **28** pp 63-70
[9] Haryanto Y, Gan B S and Maryoto A 2017 *Int. J. of Tech.* **8** pp 134-44
[10] Maryoto A, Gan B S, Hermanto N I S and Setijadi R 2018 *Int. J. of Corr.* pp 1-9
[11] Maryoto A, Gan B S, Hermanto N I S and Setijadi R 2018 *J. of Eng. Sci. & Tech.* **13** pp 3263-76
[12] Maryoto A, Gan B S, Hermanto N I S and Setijadi R 2017 *J. of Eng. & Appl. Sci.* **12** pp 7965-7
[13] Maryoto A, Gan B S, Hermanto N I S and Setijadi R 2017 *J. of Eng. & Appl. Sci.* **12** pp 5233-8

IOP Conf. Series: Materials Science and Engineering **650** (2019) 012031 doi:10.1088/1757-899X/650/1/012031

Road construction project waste material recommendations

Mega Waty [1]* and Hendrik Sulistio[1]

[1]Universitas Tarumanagara, Jl. Let. Jend. S. Parman No.10, Jakarta 11470, Indonesia

* mega@ft.untar.ac.id

Abstract. After risk identification and risk analysis done, so risk response is carried out. Risk response is handled by the service provider. Recommendation is the final result of risk management of road construction project waste material. The recommendation is the aggregate B and Ready Mix Concrete is necessary to pay attention to the waiting time and pay attention to the critical trajectory on material scheduling so that breakdowns and scheduling updates are made per week. Need to pay attention to rainfall and implementation methodology. Recommendations for lean concrete waste material are necessary to pay attention to the waiting time and pay attention to the critical trajectory on material scheduling so that breakdowns and scheduling updates are carried out per week. Need to pay attention to rainfall and implementation methodology. And the causes for minimizing waste material must be considered.

1. Introduction

Risk Analysis was carried out on 51 road construction projects which stated that Aggregate B, Concrete Ready Mix and Lean Concrete which produced a number of causes of waste material and most have a high risk rating is emphasis on time and cost emphasis (Waty and Sulistio, 2019)

And now will do recommendation for waste material on road construction project after risk response.

2. Respon risk

The Risk Response generates risks received by the contractor on the first worksheet and on the second worksheet results in transfers to the implementing contractor because the contractor as the service provider must accept the consequences if a problem occurs except natural disasters and the handling of the most important risks and who handles them. After the risk response is generated, recommendations are sought. Before recommendations are made, Focus Group Discussions are conducted.

3. Focus group discussion

Group Discussion is conducted to get the expected recommendations. There were seven FGD participants. Focus Group Discussion (FGD) is a way of several ways that are used to gather opinions or perceptions from experts in their fields, thus it can be used for input into a study. The discussion invites experts at a specific place and time. It starts with giving a presentation beforehand on the model that has been obtained then the analysis that produces the sources and causes of the waste material that has been produced, then a discussion is held to determine the right recommendations.

Expert Judgment
Expert Judgment are urgently needed at the meeting. There are about 35 questions that need to be made recommendations. For example, one of the questions is taken number one, which is information on road design drawings that are less complete.
Expert Judgment 1:
Recommendations for the causes of incomplete road design drawing information are necessary to verify before the auction is conducted.
Expert Judgment 2:
Recommendations are: It is necessary to do a typical picture with only long and cross section drawing. The recommendation is to disagree with the expert 1
Expert Judgment 3
Recommended factors for incomplete road design image information.
The recommendation is to agree with expert 1 but also agree with expert 2 because if expert 1 only discusses the auction and not too detailed. But agree with expert 2 because it is more detailed.
Expert Judgment 5
The recommendation is to agree with expert 1 and expert 2 by adding design details to avoid change orders.
Expert Judgment 6
Recommendation for the causes of incomplete road design image information is to agree with expert 5.
Expert Judgment 7
Recommendations The factors that cause image information to be incomplete are to agree with expert 1 and expert 2
Expert Judgment 4
Recommendations The factors that cause image information to be described as incomplete are those that do not agree with expert 5.

4. Result dan discussion

From the Focus Group Discussion produces recommendations, where there is one recommendation , namely Road Construction Project recommendations recommendations as listed in Table 1 below.

Recommendations for the Road Construction Project relating to material management which will be discussed below.

Table 1. Minimization of material waste for road construction project. (a)

Waste Material Aggregate B
Source: Design Causes: - Information on road design drawings that are incomplete Minimization recommendations: - Verification needs to be done before the auction is carried out, not only with typical long and cross section, but with detailed design so as to avoid change order.
Source: Material Procurement Order errors result in excess or lack of material for Minimization recommendations: - Orders must be made in accordance with auction requirements, needs and looking for responsible suppliers.
Source: Material Handling Causes: - Errors in the use of materials that need to be replaced, for example, do not comply with specified specifications such as abrasion and gradation and aggregate grain hardness - Material sent in less condition. - The amount of material needed is not clearly known because of imperfect planning.

Table 1. Minimization of material waste for road construction project. (b)

- Measurement of inaccurate field dimensions resulting in excess volume. Minimization recommendations: - Must be in accordance with the reference standards including SNI 03-4141-1996, SNI 1744: 2012 and SNI 7619: 2012 so that the aggregate used has a minimum immersion CBR value of at least 60% and a plasticity index of 0-10. - Find the appropriate borrow pit and avoid irresponsible suppliers. - Review design with re-measurement so that when shop 0%, a shop drawing can be made. Leveling of sub grade must be properly checked with Theodolite at each implementation and confirmed solid soil conditions.
Source: Behavior Control Factors that cause: - Material scheduling control deviations - Deviations in controlling material costs Minimization recommendations: - Need to pay attention to the waiting time and pay attention to the critical trajectory on material scheduling so that breakdowns and scheduling updates are carried out per week. Need to pay attention to rainfall and implementation methodology. It is necessary to pay attention to the Variant Schedule (SV) and Cost Variant (CV) because if SV and CV = negative then the work is finished late and costs higher than the budget. - Paying attention to the Work Breakdown Structure of each type of work so that if there is a delay in immediate handling, for example if there is a 5% delay, a meeting must be held and if the delay of 10% is necessary, the Show Cause Meeting should also take into account the weather and implementation methodology. It is necessary to pay attention to the Variant Schedule (SV) and Cost Variant (CV) because if SV and CV are negative so the work is finished late and costs higher than the budget.
Waste Material Lean Concrete
Source: Design Causes: - Order error due to choosing different product specifications Minimization Recommendation: - Must pay attention to specifications that are in accordance with the required terms of reference.
Source: Material Handling Causative factors - Material sent in less condition Minimization recommendations: Pay more attention to the sub contractor to get the right volume according to the specifications.
Source: Behavior Control Causes: - Material scheduling control deviations Minimization Recommendation: Need to pay attention to the waiting time and pay attention to the critical trajectory of project scheduling so that breakdowns and scheduling updates are made per week. Need to pay attention to rainfall and implementation methodology. It is necessary to pay attention to the Variant Schedule (SV) and Cost Variant (CV) because if SV and CV are negative so the work is finished late and costs higher than the budget.

Table 1. Minimization of material waste for road construction project. (c)

Waste Material Concrete Ready Mix
Source: Behavior Control Causes: - Deviations in controlling material costs
Minimization Recommendation: - Paying attention to the Work Breakdown Structure of each type of work so that if there is a delay in immediate handling, for example if there is a 5% delay, a meeting must be held and if the delay of 10% is necessary, the Show Cause Meeting should also take into account the weather and implementation methodology. It is necessary to pay attention to the Variant Schedule (SV) and Cost Variant (CV) because if SV and CV are negative so the work is finished late and costs higher than the budget.
Source: Material Handling Causes: - Damage to material in the place due to the slow cutting of concrete Minimization recommendations: - Immediately conduct a special team to cut before 8 hours. Source: Implementation Causes: - Work accidents in the field. - Unreliable equipment Minimization recommendations: - Applying the correct K3 in accordance with OHSAS or SMK3 with clear and precise PPE. - Repairing tools or using appropriate equipment in accordance with the requirements in the work reference frame of at least 90% of its novelty.

5. Recommendation

The recommendations for the causes of each of the most extreme materials are as shown below.for Road Construction Projects:

1. B Aggegate

Source: Behavior Control
Cause: Material scheduling control deviations
Recommendation: Need to pay attention to the waiting time and pay attention to the critical trajectory on material scheduling so that breakdowns and scheduling updates are carried out per week. Need to pay attention to rainfall and implementation methodology. It is necessary to pay attention to the Variant Schedule (SV) and Cost Variant (CV) because if SV and CV are negative so the work is finished late and costs higher than the budget.

2. Lean Concrete Source: Behavior Control

Cause: Material scheduling control deviations
Recommendation: Need to pay attention to the waiting time and pay attention to the critical trajectory on material scheduling so that breakdowns and scheduling updates are carried out per week. Need to pay attention to rainfall and implementation methodology. It is necessary to pay attention to the Variant Schedule (SV) and Cost Variant (CV) because if SV and CV are negative, so the work is finished late and costs higher than the budget.

3. Concrete Ready Mix Source: Behavior Control

Cause: Deviation in controlling material costs
Recommendation: Pay attention to the Work Breakdown Structure of each type of work so that if there

is a delay in immediate handling, for example, if a 5% delay occurs, a meeting must be held and if the delay of 10% is necessary, the Show Cause Meeting should also take into account the weather and implementation methodology. More concentration on the cost of implementation. It is necessary

to pay attention to the Variant Schedule (SV) and Cost Variant (CV) because if SV and CV are negative so the work is finished late and costs higher than the budget.

6. Conclusion

The causal factors that need to be considered to minimize waste material with risk management during the implementation phase of the road contruction project are as follows:

a. Waste Material B Aggregate Causes:

- Information on road design drawings that are incomplete.
- Order errors result in material overload or lack.
- Errors in the use of materials that need to be replaced, for example, do not conform to specifications specified such as abrasion and gradation and aggregate grain hardness.
- Material sent in less condition.
- The amount of material needed is not clearly known because of imperfect planning.
- Measurement of inaccurate field dimensions resulting in excess volume.
- Material scheduling control deviations.
- Deviations in controlling material costs.

b. Waste Material Lean Concrete Causes:

- Error ordering because choose a different specification.
- Material sent in less condition.
- Material scheduling control deviations.

c. Waste Material Concrete Ready Mix Causes:

- Deviations in controlling material costs.
- Damage to material in the place due to the delay in cutting concrete.
- Work accidents in the field.
- Unreliable equipment.

References

[1] Al-Bahar J and Crandall K 1990 Systematic risk Management approach for construction projects *Journal of Construction Engineering and Management* **116**(3) pp 533-546

[2] Bossink B A G and Browers H J H 1996 Construction waste: Quantification and source evaluation *Journal of Construction Engineering and Management* **122**(1) pp 55–60

[3] Choi H H and Mahadevan S 2008 Construction Project Risk Assesment Using Existing Data base and Project Specific Information *Journal Of Construction Engineering And Management* **134**(11) pp 894-903

[4] Ekanayake L and Ofori G 2000 Construction material waste source evaluation *Proc. Strategies for sustainable built environment* Pretoria pp 23-25

[5] Farmoso C T, Soibelman L M, Cesare C D and Isatto E L 2002 Material waste in building industry: Main causes and prevention *Journal of Construction Engineering and Management* **128** (4) pp 316–325

[6] Foley P, Howard G and Ian M 1984 *Future of the Sheffield Economy: an investigation using the Delphi Method* Departement of Town and Regional Planning, Faculty of Architectural Studies, University of Sheffield

[7] Gavilan R M and Bernold L E 1994 Source evaluation of solid waste in building construction *Journal of Construction Engineering and Management* **120**(3) pp 536 – 552

[8] Hanna A S, Thomas G and Swanson J R 2013 Constructions risk identification and allocation cooperative approach *Journal Of Construction Engineering And Management* **139**(9)

[9] Jaselskis E J and Russels J S 1992 Risk Analysis Approach To Selection of Contractor Evaluation Method *Journal Of Construction Engineering And Management* **118**(4) pp 814-821

[10] Loosemore M and Teo M 2001 A Theory of waste behaviour in the construction industry *Journal Construction management and economic* pp 741-751

[11] Polat G and Ballard G 2004 Waste in Turkish Construction – Need for lean construction techniques *Proceedings of the 12th Annual Conference of the International Group for Lean Construction IGLC-12*" August Denmark pp 488-501

[12] Waty M and Sulistio H 2019 Identifikasi Risiko Lanjutan Terhadap Sumber dan Penyebab Material Waste Proyek Konstruksi Jalan *Journal Media Komunikasi Teknik Sipil,* ID 21817 Semarang,Indonesia (Waiting Assigment)

[13] Waty M and Sulistio H Early Risk Identification On Source and Causes of Waste Road Project Construction Materials Tarumanagara International Conference On the Application of Technology and Engineering Jakarta Indonesia Nov 22-23 2018 *IOP Conf. Series: Materials Science and Engineering* **508** (2019) 012008, doi:10.1088/1757-899X/508/1/012008

[14] Waty M, Alisyahbana S, Gondokusumo O, Sulistio H, Hasyim C, Setiawan M I and Ahmar A S 2018 Modeling of Waste Material Cost on Road Construction Projects *International Journal Of Engineering & Technology* **7**(2) pp 474-477

[15] Waty M and Sulistio H 2019 Risk Analysis Waste Material on Road Construction Project " *International Review of Civil Engineering-Praise Worthy Prize*, ID 17207(In Review)

IOP Conf. Series: Materials Science and Engineering **650** (2019) 012032 doi:10.1088/1757-899X/650/1/012032

Comparison of the RSNI 1726:2018 and the SNI 1726:2012 design response spectra of 17 major cities in Indonesia

Suradjin Sutjipto and Indrawati Sumeru

Civil Engineering Department, Trisakti University, Jakarta, Indonesia

suradjin@trisakti.ac.id, indrawati@trisakti.ac.id

Abstract. Based on the 2017 Indonesian Seismic Map and the ASCE 7-16, the proposed Indonesian Seismic Design Code - RSNI 1726:2018, was developed to renew the current SNI 1726:2012 code. Taking a sampling of 17 major cities representing all regions in Indonesia, this paper presents the comparison of the RSNI 1726:2018 and the SNI 1726:2012 design response spectra. All cities located in high seismic areas have an anomaly characteristic, where the design response spectra of site class SE (soft soil) is less than that of site class SD (medium soil), or even less than that of site class SC (hard soil), contrary to what Structural and Geotechnical Engineers think so far, that the design response spectra of site class SE is always higher than that of site classes SD and SC.

1. Introduction

The construction industry always welcomes the revised Indonesian Seismic Design Code with great enthusiasm. They always want to know immediately, how much the load difference and the changes in the provisions relating to the design and construction. Commercial building developers may immediately take a decision to speed up the project design and construction process if it turns out that the seismic design load rises and/or the requirements are more stringent. On the contrary, they may postpone the project until the new code is implemented if the seismic design load decreases.

As a revision of the current Indonesian seismic design code SNI 1726:2012 [7], the proposed RSNI 1726:2018 [4] was developed based on the 2017 Indonesian Seismic Map [3] and the provisions of the ASCE 7-16 [1] which was adopted with several modifications adapting the conditions of Indonesia.

This paper is intended as a "preview" which shows the difference of the design response spectra according to the RSNI 1726:2018 and the SNI 1726:2012 of 17 major cities which are considered to represent the entire territory of Indonesia, based on verified data provided.

In addition, an observation on the possibility of an anomaly in which the condition of the design response spectra at short periods of the site class SC (hard soil), SD (medium soil) and SE (soft soil) do not increase sequentially, is also presented in this paper. This anomalous phenomenon occurs in earthquake-prone areas with relatively high spectral response acceleration at short periods (S_S) related to the application of the site coefficient F_a, which in fact decreases (de-amplifies).

2. Design response spectra

Figure 1 shows a schematic illustration of developing a design response spectrum according to the RSNI 1726:2018 or the ASCE 7-16. In principle, the spectral response acceleration values, S_S and S_1, are obtained from the MCE_R maps for 0.2 seconds periods (figure 2) and a 1-second period (figure 3), which are then multiplied by the seismic amplification factors F_a and F_v producing the spectral response acceleration parameters corresponding to the soil site class, namely S_{MS} and S_{M1}. The value of the design spectral response acceleration at short periods, S_{DS}, is 2/3 of the S_{MS} value, and the value of the design spectral response acceleration at a 1-second period, S_{D1}, is 2/3 of the S_{M1} value.

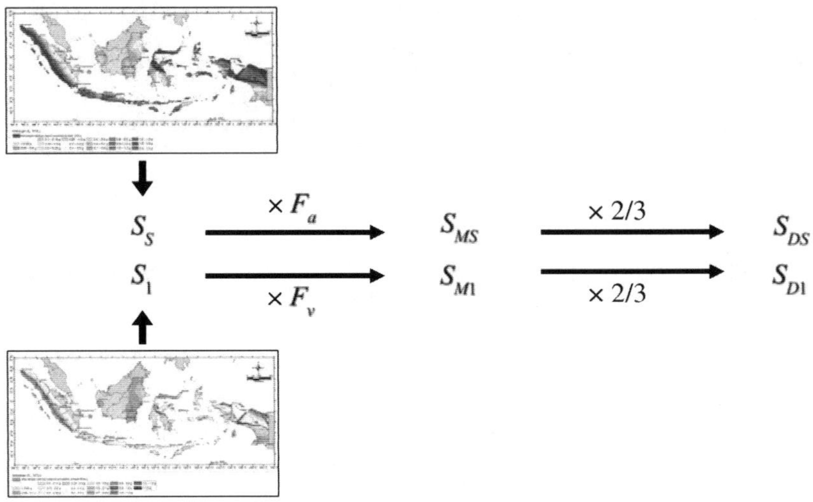

Figure 1. Developing design response spectra process.

The Risk-Targeted Maximum Considered Earthquake (MCE_R) spectral response acceleration maps with a 5% critical damping and a 1% probability of collapse within a 50 years, which were adjusted for the soil site class between SB and SC, for short periods (0.2 seconds) and a 1-second period according to the ASCE 7-16, as shown in figures 2 and 3, were developed based on hazard maps with a 2% probability of exceedance within 50 years (2,475-year earthquake return period) compiled by the National Earthquake Study Center Team in the 2017 Earthquake Source and Hazard Maps.

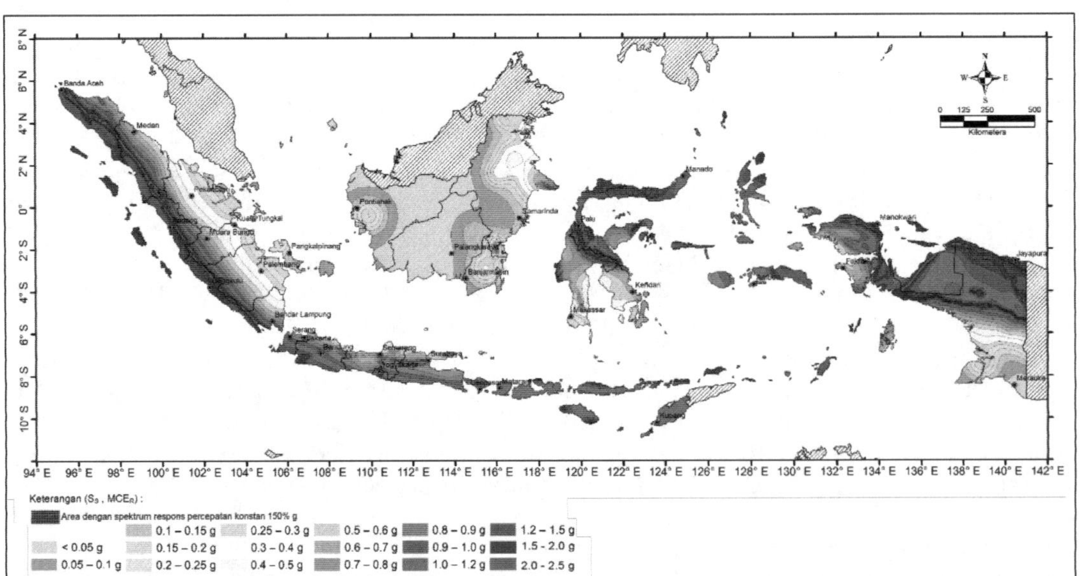

Figure 2. MCE_R map for 0.2-second spectral response acceleration (5% damping - site class SB/SC).

IOP Conf. Series: Materials Science and Engineering **650** (2019) 012032 doi:10.1088/1757-899X/650/1/012032

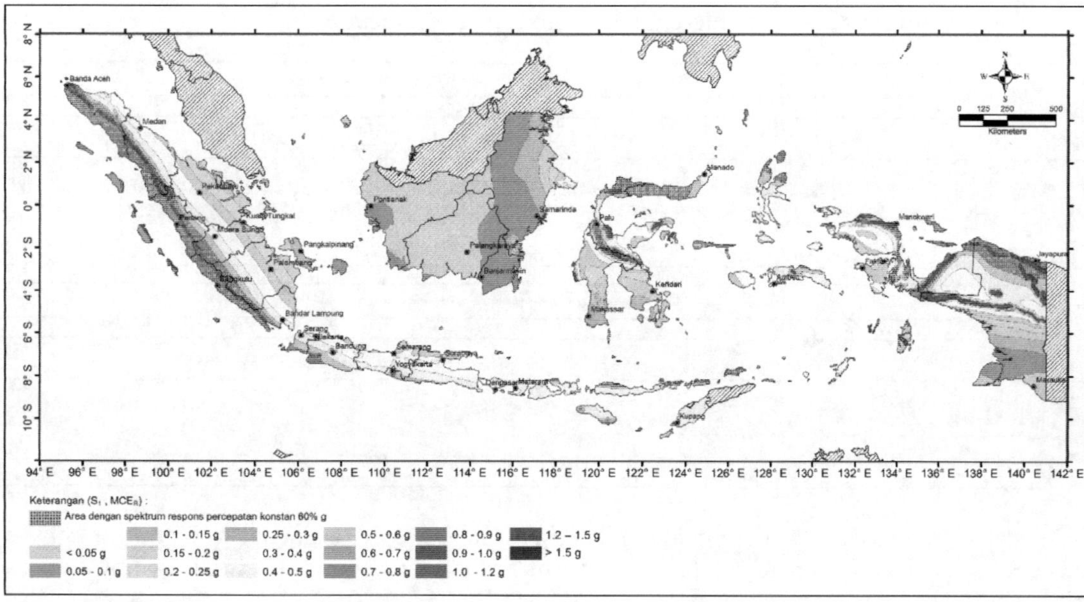

Figure 3. MCE_R map for 1-second spectral response acceleration (5% damping - site class SB/SC).

The MCE_R maps were developed with a grid interval of 0.1^0 and consist of approximately 96,000 data for each spectral response acceleration at short periods, S_S, and for spectral response acceleration at a 1-second period, S_1. In order to obtain S_S and S_1 values at given latitude and longitude coordinates, an interpolation calculation of 4 points on the closest grids is carried out. This MCE_R map will be refined and posted on the Puskim PU website to be freely used by public.

3. Site coefficient F_a and F_v

As shown in table 1 and table 2, the site coefficient values (which are also commonly known as the amplification factors) F_a and F_v were changed in RSNI 1726:2018 by adopting the recommendation from Pacific Earthquake Engineering Research (PEER) that are slightly different from the ones set on the ASCE 7-16. This is to avoid having to conduct site-specific response analysis for the soft soil (SE) site class when the values of spectral response acceleration $S_S \geq 1$ g, and $S_1 \geq 0.2$ g, as required in ASCE 7-16.

Table 1. Short-period site coefficient F_a.

Site Class	MCE_R Spectral Response Acceleration Parameter at Short Periods								
	$S_S \leq 0.25$			$S_S = 0.5$			$S_S = 0.75$		
	SNI 1726:2012	ASCE 7-16	PEER / RSNI 1726:2018	SNI 1726:2012	ASCE 7-16	PEER / RSNI 1726:2018	SNI 1726:2012	ASCE 7-16	PEER / RSNI 1726:2018
SA	0.8	0.8	**0.8**	0.8	0.8	**0.8**	0.8	0.8	**0.8**
SB	1.0	0.9	**0.9**	1.0	0.9	**0.9**	1.0	0.9	**0.9**
SC	1.2	1.3	**1.3**	1.2	1.3	**1.3**	1.1	1.2	**1.2**
SD	1.6	1.6	**1.6**	1.4	1.4	**1.4**	1.2	1.2	**1.2**
SE	2.5	2.4	**2.4**	1.7	1.7	**1.7**	1.2	1.3	**1.3**
SF	Site-Specific	Site-Specific	**Site-Specific**	Site-Specific	Site-Specific	**Site-Specific**	Site-Specific	Site-Specific	**Site-Specific**

Site Class	MCE_R Spectral Response Acceleration Parameter at Short Periods								
	$S_S = 1.0$			$S_S = 1.25$			$S_S \geq 1.5$		
	SNI 1726:2012	ASCE 7-16	PEER / RSNI 1726:2018	SNI 1726:2012	ASCE 7-16	PEER / RSNI 1726:2018	SNI 1726:2012	ASCE 7-16	PEER / RSNI 1726:2018
SA	0.8	0.8	**0.8**	0.8	0.8	**0.8**	0.8	0.8	**0.8**
SB	1.0	0.9	**0.9**	1.0	0.9	**0.9**	1.0	0.9	**0.9**
SC	1.0	1.2	**1.2**	1.0	1.2	**1.2**	1.0	1.2	**1.2**
SD	1.1	1.1	**1.1**	1.0	1.0	**1.0**	1.0	1.0	**1.0**
SE	0.9	Site-Specific	**1.1**	0.9	Site-Specific	**0.9**	0.9	Site-Specific	**0.8**
SF	Site-Specific	Site-Specific	**Site-Specific**	Site-Specific	Site-Specific	**Site-Specific**	Site-Specific	Site-Specific	**Site-Specific**

Table 2. Long-period site coefficient F_v.

Site Class	MCE$_R$ Spectral Response Acceleration Parameter at a 1-second Period								
	$S_1 \leq 0.1$			$S_1 = 0.2$			$S_1 = 0.3$		
	SNI 1726:2012	ASCE 7-16	PEER / RSNI 1726:2018	SNI 1726:2012	ASCE 7-16	PEER / RSNI 1726:2018	SNI 1726:2012	ASCE 7-16	PEER / RSNI 1726:2018
SA	0.8	0.8	**0.8**	0.8	0.8	**0.8**	0.8	0.8	**0.8**
SB	1.0	0.8	**0.8**	1.0	0.8	**0.8**	1.0	0.8	**0.8**
SC	1.7	1.5	**1.5**	1.6	1.5	**1.5**	1.5	1.5	**1.5**
SD	2.4	2.4	**2.4**	2.0	2.2	**2.2**	1.8	2.0	**2.0**
SE	3.5	4.2	**4.2**	3.2	Site-Specific	**3.3**	2.8	Site-Specific	**2.8**
SF	Site-Specific	Site-Specific	**Site-Specific**	Site-Specific	Site-Specific	**Site-Specific**	Site-Specific	Site-Specific	**Site-Specific**

Site Class	MCE$_R$ Spectral Response Acceleration Parameter at a 1-second Period								
	$S_1 = 0.4$			$S_1 = 0.5$			$S_1 \geq 0.6$		
	SNI 1726:2012	ASCE 7-16	PEER / RSNI 1726:2018	SNI 1726:2012	ASCE 7-16	PEER / RSNI 1726:2018	SNI 1726:2012	ASCE 7-16	PEER / RSNI 1726:2018
SA	0.8	0.8	**0.8**	0.8	0.8	**0.8**	0.8	0.8	**0.8**
SB	1.0	0.8	**0.8**	1.0	0.8	**0.8**	1.0	0.8	**0.8**
SC	1.4	1.5	**1.5**	1.3	1.5	**1.5**	1.3	1.4	**1.4**
SD	1.6	1.9	**1.9**	1.5	1.8	**1.8**	1.5	1.7	**1.7**
SE	2.4	Site-Specific	**2.4**	2.4	Site-Specific	**2.2**	2.4	Site-Specific	**2.0**
SF	Site-Specific	Site-Specific	**Site-Specific**	Site-Specific	Site-Specific	**Site-Specific**	Site-Specific	Site-Specific	**Site-Specific**

As stated by Crouse [2], the previous F_a and F_v values were considered to be too long have not been updated and are no longer compatible with the results of recent studies conducted by J Stewart et al [8]. This is the basis of the changes adopted by the RSNI 1726:2018.

According to the ASCE 7-16, the bed rock is no longer at the site class SB, but in between sites classes SB and SC. It should be noted that the MCE$_R$ maps presented in figures 2 and 3 have been adjusted to this criterion. Next, in contrast to the SNI 1726:2012, the S_S classification is added with "$\geq$ 1.5" and the S_1 classification is added with "$\geq$ 0.6 ", representing very high seismicity areas near the fault or earthquake source. It can be stated that F_a and F_v values variedly change, most of them rise, but some values remain the same as before, and some values go down, especially in the low seismicity areas. The occurrence of an anomaly in which the magnitude of the design response spectra for the hard soil (SC), medium soil (SD) and soft soil (SE) site classes did not increase sequentially as in the old standard due to fluctuations of F_a values and will be discussed further in the other section of this paper.

4. Design response spectra of 17 major cities in Indonesia

In order to see the design response spectra comparison of the RSNI 1726:2018 and the SNI 1726:2012, 17 major cities, which were considered to represent the entire territory of Indonesia from West to East, and from North to South, were chosen as samplings.

Comparison of the spectral response acceleration at the bedrock of the 17 major cities is shown in table 3. S_S and S_1 values derived from the MCE$_R$ map of the RSNI 1726:2018 had been verified by M Irsyam on May 23, 2018 Sub-Structures Technical Meeting in Jakarta, from the results of studies of W Sengara, Asrurifak and W Partono which were done separately.

From the observations of table 3, the values of spectral response acceleration at short periods (S_S) according to the RSNI 1726:2018 in ten cities, namely: Banda Aceh, Denpasar, Jakarta, Manado, Manokwari, Medan, Padang, Palembang, Pontianak and Surabaya, increase with ratios between 1.01 - 6.65 from the values of spectral response acceleration at short periods according to the SNI 1726:2012; one city, namely Jayapura, has no change; and six cities, namely Ambon, Balikpapan, Bandung, Kupang, Makassar and Semarang, in fact decrease with ratios ranging from 0.52 - 0.94.

The values of spectral response acceleration at a 1-second period according to the RSNI 1726:2018 in twelve cities increase with ratios between 1.01 - 2.14, namely: Balikpapan, Bandung, Denpasar, Jakarta, Kupang, Manado, Manokwari, Medan, Palembang, Pontianak, Semarang and Surabaya; two cities, Jayapura and Padang, have no change; and three cities, Ambon, Banda Aceh and Makassar, decrease with ratios of 0.79 - 0.93, compared to the values of spectral response acceleration at a 1-second period according to the SNI 1726:2012.

IOP Conf. Series: Materials Science and Engineering **650** (2019) 012032 doi:10.1088/1757-899X/650/1/012032

Table 3. Spectral response acceleration parameters at short periods and 1-second period based on MCE_R map of 17 major cities in Indonesia.

No.	City	Coordinate		S_S			S_1		
		Latitude	Longitude	SNI 1726:2012	RSNI 1726:2018	Ratio	SNI 1726:2012	RSNI 1726:2018	Ratio
1.	Ambon	-3.6554	128.1908	1.380	1.085	0.79	0.490	0.393	0.80
2.	Balikpapan	-1.2379	116.8529	0.235	0.123	0.52	0.082	0.083	1.01
3.	Banda Aceh	5.5483	95.3238	1.349	1.510	1.12	0.642	0.600	0.93
4.	Bandung	-6.9175	107.6191	1.450	1.176	0.81	0.486	0.510	1.05
5.	Denpasar	-8.6705	115.2126	0.977	0.984	1.01	0.360	0.397	1.10
6.	Jakarta	-6.1751	106.8650	0.664	0.779	1.17	0.293	0.379	1.29
7.	Jayapura	-2.5916	140.6690	1.500	1.500	1.00	0.600	0.600	1.00
8.	Kupang	-10.1772	123.6070	1.113	1.049	0.94	0.296	0.380	1.28
9.	Makassar	-5.1477	119.4327	0.317	0.222	0.70	0.142	0.112	0.79
10.	Manado	1.4748	124.8421	1.035	1.052	1.02	0.442	0.470	1.06
11.	Manokwari	-0.8615	134.0620	1.454	1.500	1.03	0.561	0.600	1.07
12.	Medan	3.5952	98.6722	0.526	0.652	1.24	0.332	0.360	1.08
13.	Padang	-0.9471	100.4172	1.398	1.481	1.06	0.600	0.600	1.00
14.	Palembang	-2.9761	104.7754	0.262	0.292	1.11	0.164	0.248	1.51
15.	Pontianak	-0.0263	109.3425	0.017	0.113	6.65	0.022	0.047	2.14
16.	Semarang	-7.0051	110.4381	1.098	0.880	0.80	0.364	0.379	1.04
17.	Surabaya	-7.2575	112.7521	0.663	0.710	1.07	0.247	0.315	1.28

Table 4 and table 5 present a comparison of the design response spectrum value for the site class categories of hard soil (SC), medium soil (SD) and soft soil (SE) of the 17 major cities in Indonesia at short periods (0.2 seconds), S_{DS}, and at a 1-second period, S_{D1}.

Table 4. Design response spectra at short periods (S_{DS}) of 17 major cities in Indonesia.

No.	City	S_{DS}									Anomaly		
		Hard Soil (SC)			Medium Soil (SD)			Soft Soil (SE)			1	2	3
		SNI 1726:2012	RSNI 1726:2018	Ratio	SNI 1726:2012	RSNI 1726:2018	Ratio	SNI 1726:2012	RSNI 1726:2018	Ratio	SD < SC < SE	SD < SE < SC	SE < SD < SC
1.	Ambon	0.920	0.868	0.94	0.920	0.771	0.84	0.828	0.746	0.90			•
2.	Balikpapan	0.188	0.107	0.57	0.251	0.131	0.52	0.392	0.197	0.50			
3.	Banda Aceh	0.899	1.208	1.34	0.899	1.007	1.12	0.809	0.805	0.99			•
4.	Bandung	0.967	0.941	0.97	0.967	0.807	0.84	0.870	0.752	0.86			•
5.	Denpasar	0.657	0.787	1.20	0.722	0.726	1.00	0.604	0.730	1.21		•	
6.	Jakarta	0.502	0.623	1.24	0.562	0.617	1.10	0.607	0.663	1.09	•		
7.	Jayapura	1.000	1.200	1.20	1.000	1.000	1.00	0.900	0.800	0.89			•
8.	Kupang	0.742	0.839	1.13	0.783	0.756	0.97	0.668	0.742	1.11			•
9.	Makassar	0.254	0.192	0.76	0.327	0.237	0.72	0.483	0.355	0.74			
10.	Manado	0.690	0.842	1.22	0.749	0.757	1.01	0.621	0.742	1.20			•
11.	Manokwari	0.969	1.200	1.24	0.969	1.000	1.03	0.872	0.800	0.92			•
12.	Medan	0.417	0.539	1.29	0.484	0.556	1.15	0.578	0.633	1.10			
13.	Padang	0.932	1.185	1.27	0.932	0.987	1.06	0.839	0.797	0.95			•
14.	Palembang	0.210	0.253	1.21	0.278	0.305	1.10	0.430	0.444	1.03			
15.	Pontianak	0.014	0.098	7.20	0.018	0.121	6.65	0.028	0.181	6.38			
16.	Semarang	0.732	0.704	0.96	0.777	0.673	0.87	0.659	0.702	1.07		•	
17.	Surabaya	0.502	0.576	1.15	0.561	0.583	1.04	0.607	0.646	1.06			

Table 5. Design response spectra at a 1-second period (S_{D1}) of 17 major cities in Indonesia.

No.	City	S_{D1}								
		Hard Soil (SC)			Medium Soil (SD)			Soft Soil (SE)		
		SNI 1726:2012	RSNI 1726:2018	Ratio	SNI 1726:2012	RSNI 1726:2018	Ratio	SNI 1726:2012	RSNI 1726:2018	Ratio
1.	Ambon	0.428	0.393	0.92	0.493	0.500	1.01	0.784	0.636	0.81
2.	Balikpapan	0.093	0.083	0.89	0.131	0.133	1.01	0.191	0.232	1.21
3.	Banda Aceh	0.556	0.560	1.01	0.642	0.680	1.06	1.027	0.800	0.78
4.	Bandung	0.426	0.507	1.19	0.491	0.609	1.24	0.778	0.741	0.95
5.	Denpasar	0.346	0.397	1.15	0.403	0.504	1.25	0.614	0.638	1.04
6.	Jakarta	0.294	0.379	1.29	0.354	0.485	1.37	0.552	0.628	1.14
7.	Jayapura	0.520	0.560	1.08	0.600	0.680	1.13	0.960	0.800	0.83
8.	Kupang	0.297	0.380	1.28	0.357	0.486	1.36	0.556	0.628	1.13
9.	Makassar	0.157	0.112	0.71	0.211	0.177	0.84	0.319	0.306	0.96
10.	Manado	0.400	0.470	1.17	0.459	0.573	1.25	0.707	0.708	1.00
11.	Manokwari	0.486	0.560	1.15	0.561	0.680	1.21	0.898	0.800	0.89
12.	Medan	0.325	0.360	1.11	0.384	0.466	1.21	0.591	0.614	1.04
13.	Padang	0.520	0.560	1.08	0.600	0.680	1.13	0.960	0.800	0.83
14.	Palembang	0.179	0.248	1.39	0.234	0.348	1.48	0.362	0.506	1.40
15.	Pontianak	0.025	0.047	1.89	0.035	0.075	2.14	0.051	0.132	2.56
16.	Semarang	0.348	0.379	1.09	0.406	0.485	1.20	0.617	0.628	1.02
17.	Surabaya	0.256	0.315	1.23	0.314	0.417	1.33	0.496	0.575	1.16

In accordance with the new site coefficient F_a, the S_{DS} values for the hard soil (SC) site class, 12 cities experience increases with ratios of 1.13 - 7.20, and 5 cities experience decreases with ratios ranging from 0.57 - 0.97. For the medium soil (SD) site class, 9 cities experience increases with ratios of 1.01 - 6.65, 2 cities have no change, and 6 cities experience declines with ratios of 0.52 - 0.97. For the soft soil (SE) site class, 9 cities experience increases with ratios of 1.03 - 6.38, and 8 cities experience decreases with ratios of 0.50 - 0.99.

With the application of the new site coefficient F_v, the S_{D1} values in the hard soil (SC) site class, 14 cities experience increases with ratios between 1.01 - 1.89, and 3 cities experience decreases with ratios ranging from 0.71 - 0.92. For the medium soil (SD) site class, 16 cities experience increases with ratios of 1.01 - 2.14, and 1 city decrease with a ratio of 0.84. For the soft soil (SE) site class, 9 cities experience increases with ratios between 1.02 - 2.56, 1 city have no change, and 7 cities experience decreases with ratios of 0.78 - 0.96.

5. Design response spectrum anomaly

So far, Structural and Geotechnical engineers have a mindset where the design response spectra of hard soil (SC), medium soil (SD) and soft soil (SE) site classes increases sequentially. In other words, the softer the soil, the greater the design response spectrum value. As shown in figure 4, the SNI 1726-1989 [5] and the SNI 1726-2002 [6] design response spectra also adhere to this rule. However, studies carried out by S Sutjipto [9, 10] indicate that the application of site coefficients F_a and F_v according to the SNI 1726:2012 changes these sequences so that it is then referred to as an anomaly. In fact, it also occurs in the RSNI 1726:2018.

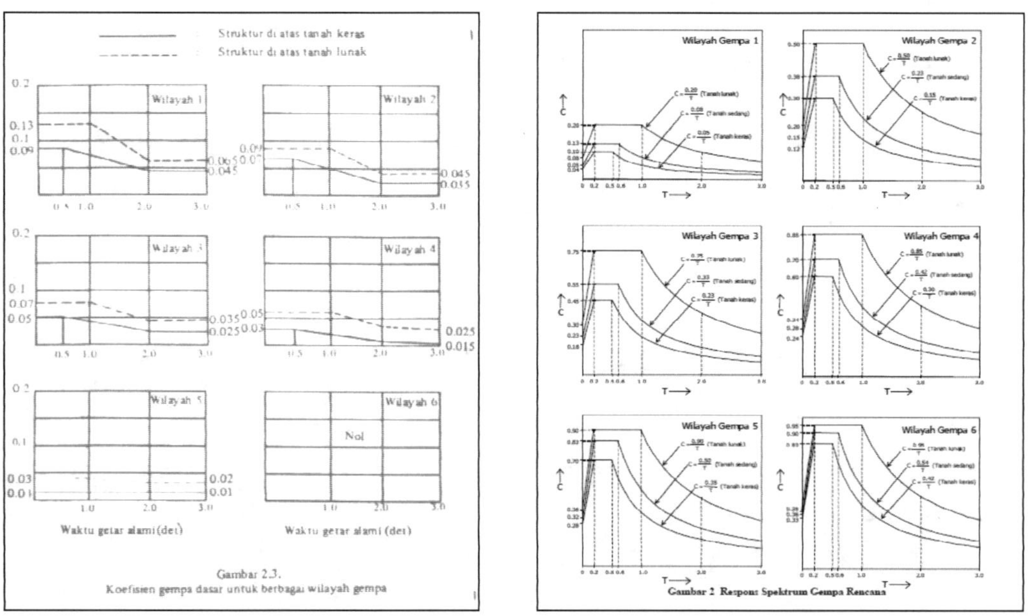

SNI 1726-1989 **SNI 1726-2002**

Figure 4. The SNI 1726-1989 and the SNI 1726-2002 design response spectra.

From table 4, it can be observed that the design response spectra at short periods (S_{DS}) of 17 selected cities experience 4 conditions and they can be grouped as follows:

- Normal : design response spectra SC < SD < SE
- Anomaly 1 : design response spectra SD < SC < SE
- Anomaly 2 : design response spectra SD < SE < SC
- Anomaly 3 : design response spectra SE < SD < SC

IOP Conf. Series: Materials Science and Engineering **650** (2019) 012032 doi:10.1088/1757-899X/650/1/012032

There are 6 cities with a normal condition, namely: Balikpapan, Makassar, Medan, Palembang, Pontianak and Surabaya. One city with Anomaly 1 condition is Jakarta. Two cities with Anomaly 2 condition are Denpasar and Semarang. And, 8 cities with Anomaly 3 condition are Ambon, Banda Aceh, Bandung, Jayapura, Kupang, Manado, Manokwari and Padang.

The observations in table 5 show the design response spectra at 1-second period (S_{D1}) of the 17 selected cities do not show anomalies, where the condition of the design response spectra SC <SD <SE. However, it should be noted that although the anomaly only occurs in short-periods response spectral, so it seems that it only affects the low-rise buildings, but in fact the impact can also affect the high-rise and medium-rise buildings due to the requirement of the minimum base shear $V = 0.044\ S_{DS}\ I_e\ W$ according to the RSNI 1726:2018.

6. Plot of 17 selected cities design response spectra

The design response spectrum plots based on the RSNI 1726:2018 and the SNI 1726:2012 for 17 selected cities are presented simultaneously in figure 5.

The design response spectra of the RSNI 1726:2018 are shown in solid curve lines, while the design response spectra of the SNI 1726:2012 are shown in intermittent curve lines. Green, yellow and red curves represent hard soil (SC), medium soil (SD) and soft soil (SE) site classes, respectively.

Plots without border show cities with Normal conditions (design response spectra SC < SD < SE). Plots with a single red line border indicate cities with Anomaly 1 condition (design response spectra SD < SC < SE), while plots with a double red line border indicate cities with Anomaly 2 condition (design response spectra SD < SE < SC), and plots with a triple red line border indicate cities with Anomaly 3 condition (design response spectra SE < SD < SC).

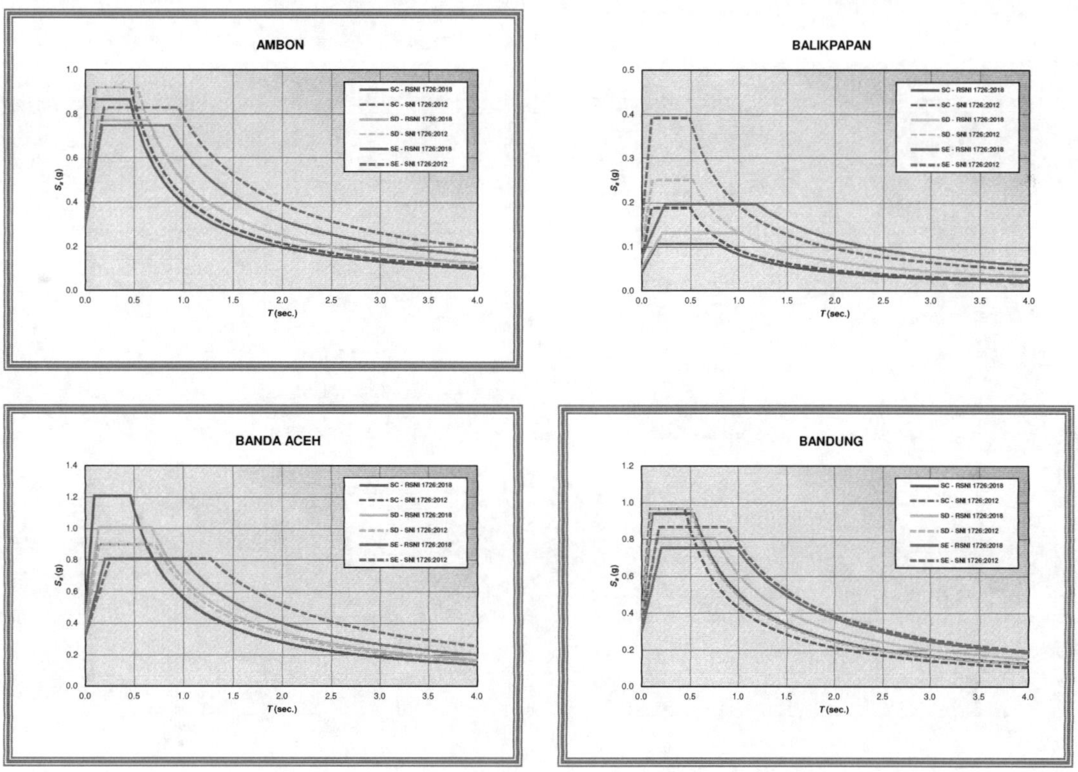

Figure 5. Design response spectra of 17 major cities in Indonesia.

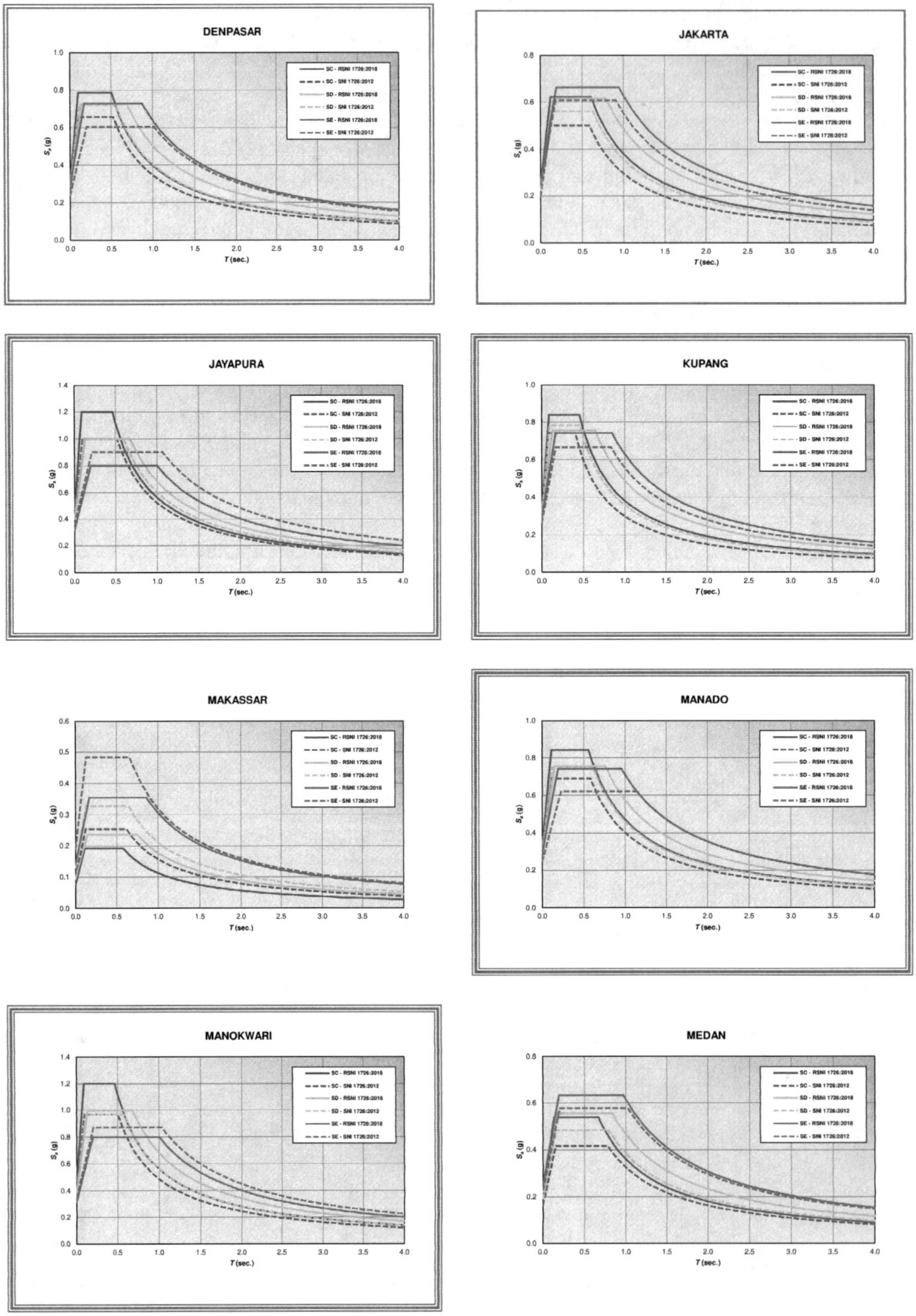

Figure 5. Design response spectra of 17 major cities in Indonesia (continued).

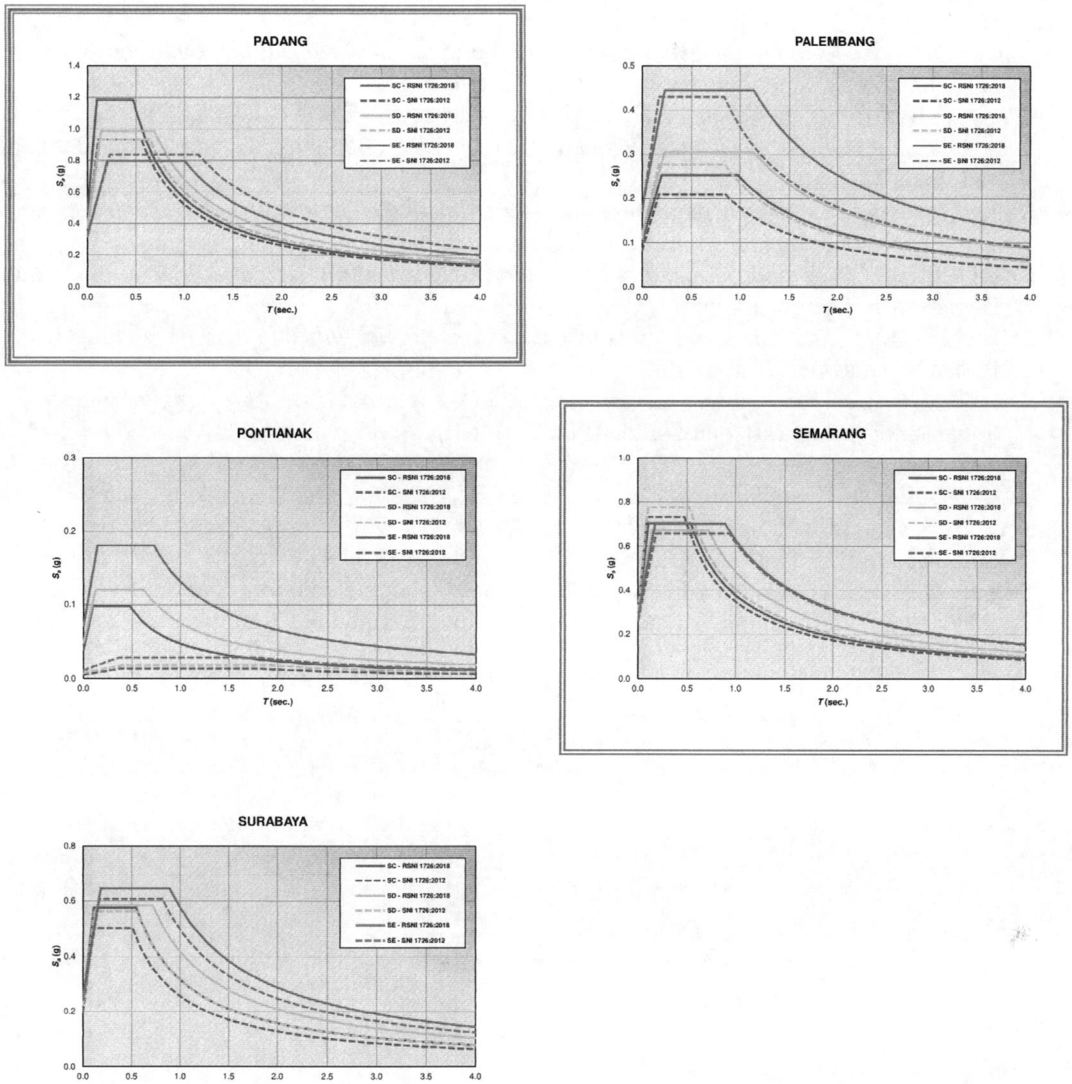

Figure 5. Design response spectra of 17 major cities in Indonesia (continued).

7. Conclusion

Based on the data presented in this paper, it can be concluded that:

- The design response spectra of the 17 selected cities based on the RSNI 1726:2018, compared to that of based on the SNI 1726:2012, at short-periods mostly increase 10% in average, with an exception of Pontianak which increase approximately 6.5 times, and at 1-second period mostly increase 17% in average, except Pontianak which increase approximately 2.3 times.
- Based on the RSNI 1726:2018 site coefficient F_a, three possibilities of anomaly conditions may occur in the cities located in earthquake-prone areas with a value of S_S above 0.75 g.

IOP Conf. Series: Materials Science and Engineering **650** (2019) 012032 doi:10.1088/1757-899X/650/1/012032

References

[1] ASCE/SEI 7-16 *Minimum Design Loads and Associated Criteria for Buildings and Other Structures* (VA: American Society of Civil Engineers)

[2] Crouse C B 2015 *Site Coefficients* F_a, F_v *and* F_{PGA} *Proposed for ASCE 7-16.* (CA: AECOM)

[3] *Peta Sumber dan Bahaya Gempa Indonesia Tahun 2017* (Bandung: Pusat Litbang Perumahan dan Pemukiman)

[4] RSNI 1726:2018 *Tata cara perencanaan ketahanan gempa untuk struktur bangunan gedung dan non gedung* (Jakarta: Badan Standarisasi Nasional)

[5] SNI 1726-1989 *Pedoman perencanaan ketahanan gempa untuk rumah dan gedung* (Bandung: Departemen Pekerjaan Umum)

[6] SNI 1726-2002 *Tata cara perencanaan ketahanan gempa untuk bangunan gedung* (Jakarta: Badan Standarisasi Nasional)

[7] SNI 1726:2012 *Tata cara perencanaan ketahanan gempa untuk struktur bangunan gedung dan non gedung* (Jakarta: Badan Standarisasi Nasional)

[8] Stewart J P and Seyhan E 2013 *Semi-Empirical Nonlinear Site Amplification and Its Application in NEHRP Site Factors* (CA: PEERC)

[9] Sutjipto S and Sumeru I 2015 *Anomali Spektrum Respons Desain SNI 1726:2012* (Jakarta: Seminar dan Pameran HAKI 2015)

[10] Sutjipto S 2016 *An Overview of Design Response Spectra in the Indonesian Seismic Code - SNI 1726:2012* (Bali: International Conference on Earthquake Engineering and Disaster Mitigation III)

IOP Conf. Series: Materials Science and Engineering **650** (2019) 012033 doi:10.1088/1757-899X/650/1/012033

Multi-storey Modular Cold-Formed Steel Building in Hong Kong: Challenges & Opportunities

Andy Prabowo Pho[1,2]
[1]PhD student, Department of Civil Engineering, The University of Hong Kong, Pokfulam Road, Hong Kong, China

*andyprb@connect.hku.hk

Abstract. Modular building construction has become widely popular across the continents, such as in America, Europe (United Kingdom), Asia (Singapore), and more recently in Australia. Promoting sustainability in construction, the application of modular construction has been successfully implemented in very broad range building functionalities. These include student residences, private housings, hotels, commercial buildings, military accommodation, health facilities, as well as other services units attached to the main buildings. Several low-rise buildings (6 to 10 stories) as well as high-rise buildings (20 to 30 stories) have used this type prefabricated off-site building system as an alternative construction method. Among those existing projects, modular steel building brings significant advantage on the manufacturing process and erection speed. These will be beneficial for Hong Kong which has very high pressure on housing demands. Due to its excellence on strength and durability, the modular integrated construction (MiC) in Hong Kong is looking to the direction of using cold-formed steel as structural component. Some challenges and opportunities are exist and these will be discussed in this paper.

1. Introduction

A modular building is constructed from stacks of building blocks. The blocks are fabricated in the workshop and they are delivered to the site as a 3D volumetric module [1]. Each module is assembled by prefabricated elements, such as floors, walls, roofs, mechanical services, and electrical fittings. Together with structural elements, one module may contain single unit or even more than one unit [2]. By obtaining 3D volumetric buildings, it will minimize the amount of on-site works such as large volume of concrete casting, steel erections, and major architectural and installation works [3]. Moreover, modular volumetric building is also suitable for projects which has limited construction space.

Among the existing construction material, steel structure gives relatively more benefit on this prefabricated off-site construction technique. The primary advantage of modular steel building is on the weight and flexibility which ease manufacturing process such that enhance the construction speed [4]. Technical issues and challenges together with proposed solutions along modular steel building fabricated by hot-rolled sections have been assessed in the technical note [5]. Nevertheless, there are some questions remaining on the durability (corrosive resistance and fire resistance) on this steel type.

[2]Presently on study leave at Department of Civil Engineering, Universitas Tarumanagara, Jakarta, Indonesia

Despite the positive future opportunities, much wider technical challenges on application of modular construction technology have been realized [2, 6]. There are still lack of confident on the structural behavior due to an inadequate design and construction guidelines as well as code requirements. Moreover, promoting of cold-formed steel as an alternative main structural element is remain scarce. Under the extreme loads, particularly very huge wind loads as in Hong Kong, the performance of modular building construction has been questioned. Nevertheless, there have been many studies on the performance and behavior of existing modular building structure. Those studies are summarized in the following section.

2. Literature Review on Modular Steel Building

The existing literatures provide examples on the practical structure design approach of the existing modular steel buildings. Due to lack of practical design codes on such type of structure, experimental and numerical tests have been conducted to assess the suitability and reliability of proposed system. Those existing studies confirm the promising performance of steel structure for modular integrated construction (MiC). Nonetheless, the studies have not been expanded to use cold-formed steel as primary structural elements.

Several tests of curtain wall structures from cold-formed steel acting as gravity load bearing element were conducted [7]. This was to investigate the possibility of secondary elements worked as primary structural element. It was shown from the tests if the composite action between wall board and its supporting structure would significantly increase the capacity of curtain wall structure. This technique was assessed for the application into high-rise building. Apart from that element test, the robustness of current practice for stability design method on steel structure from the BS 5950 was comprehensively observed. Several practical analytic approach for design recommendations to improve the design codes (BS 5950) were included. However, member imperfections that are associated with advanced analysis method have not been initiated.

Several case studies for success story of high-rise modular building construction projects in London were presented [3]. There were 3 case studies involved: 12-, 17-, and 25-story modular steel buildings. As the building heights were increased, concrete core wall was introduced as lateral force resisting system. This wall increased the overall structural stiffness such that building displacement due to wind loads can be controlled to meet a standard code [8]. Although the steel structure only carried the gravitation loadings, such reliable steel to concrete connection was capable to transfer the lateral loads to the concrete wall and to accommodate differential settlement that possibly arises [1]. Understanding on this mutual system, further study to dual cold-formed steel and concrete wall system is necessary.

Under an extreme loading, such as an earthquake, there have been some numerical studies performed [9, 10]. More recent study [11] focused on the behavior of 4-story building stabilized by concentric bracings. Detail regarding modelling approach of the bracing, vertical and horizontal connection between modules were evaluated. From the Incremental Dynamic Analysis (IDA) and Pushover Analysis, the simplification of such elements was verified and having consistent performance. Those studies have not seen a cold-formed steel as an alternative steel construction material.

As an additional alternative of structural system in a multi-story building, [12] propose an innovative flexible solution system using prefabricated modules. Rather than using dual system in [3], 100 mm concrete walls were spread out in both horizontal directions and along the building height provided as a stability element. This study proved that dual action between concrete walls attached to modular steel frames promoted flexible design of floor/room layout. However, the applicability for tall buildings is to be evaluated by further research. Another possible solution for modular high rise buildings is to use composite steel structure [4]. Using this composite action, it will also give benefit to the use of cold- formed steel. Nevertheless, this alternate composite system shall be investigated further.

Some extensive reviews of the past studies on the modular steel buildings have been provided

[13, 14]. Those cover the experimental investigations on component materials, connection systems, and other pertinent structural design aspects. From the last literatures, it is concluded that none of studies have involved cold-formed steel as main element to carry the structural imposed loads.

3. Alternative Modular Steel Building Systems

In general, there are 2 alternatives construction material conventionally used: concrete and steel. So far, modular steel building was formed by hot-rolled steel sections for main structural frame. Cold- formed steel sections dominate the non-structural element such as façade support, ceiling supports, and curtain walls. Using cold-formed steel as main structural element has not been explored. The advantages and disadvantages of using hot-rolled steel and cold-formed steel will be discussed in later chapter.

Based on [5], the types of steel modules can be classified into two categories which depends on how gravity loads will be transferred to the foundation:

- Continuously supported modules
- Corner supported modules

On those two types, only discussion about the vertical and lateral force resisting system will be presented. More details related with the manufacturing process, optional module sizes, accessories components, and construction strategies can be referred to [1, 5].

3.1. Continuously Supported Modules

This type is also known as 4-sided modules. The modules will be manufactured from a series of 2D-elements in form of panels. The fabrication process starts from floor cassette and then covers by curtain walls and ceiling panels. These curtain walls are designated to transfer the gravity loads in a module to the foundation. It is so-called load bearing wall system. An investigation of the capacity of vertical post of the wall elements due to stiffening effect of the enclosure board of the wall has been studied [7].

This module is also applicable for "partially open-sided" modules. Special attention should be addressed on the wall openings due to some windows and doors necessary. The openings will create disruption on the load transfer process. Lintel beams and columns around the opening shall be designed for structural elements. The structural elements of this 4-sided module are illustrated in **Figure 1**.

Figure 1. Typical structural view of 4-sided module (Image courtesy of Kingspan Off Site and Modular UK)

IOP Conf. Series: Materials Science and Engineering **650** (2019) 012033 doi:10.1088/1757-899X/650/1/012033

The stability of each module is provided by the vertical bracing along the wall and in-plane bracing under the floor cassette. The in-plane bracing acts as a floor diaphragm to transfer the lateral load in the building into vertical bracings. The module to module connection will assist the load transfer mechanism in addition to maintain structural integrity. Furthermore, the vertical bracing will capture the load and bring it to the foundation.

For low-rise buildings (6 to 10 stories height), the group of vertical bracings on several modules along the building perimeter can be relied as lateral force resisting element of the whole building. However, a reinforced concrete wall greatly improves the building stability in addition to vertical bracing in the modules of tall building [3]. In order to activate the concrete wall, the corner module inadjacent to the concrete wall shall be tied together. This connection must also consider a differential settlement between the core wall and steel modules.

3.2. Corner Supported Module

The corner supported module, which is also applicable for open-sided module, still relies on the hot-rolled steel frame. The frame works as primary vertical force resisting element and it provides stability of each module. Depending on the beam span, it is useful if an intermediate column is added to minimize the steel beam and corner column size. A picture of open-sided module is presented in the **Figure 2**.

Figure 2. Open-sided module (Image courtesy of Kingspan Off Site and Modular UK)

The infill wall is on separated plane from the main frame, but the wall is connected to the frame. Therefore, the stability of the module can be given through cross bracing in the wall plane. This module is often found on the buildings higher than 3 stories. The wall can be formed from lighter cold-formed steel sections whenever it does not contribute to the gravity load carrying system and stability system.

4. Applicability of Modular Building Construction in Hong Kong

According to [15], Hong Kong has been implementing prefabricated building technology since the middle of 90's. The project was Headquarter and Member Facilities of the Hong Kong Jockey Club. In the project, secondary beams and slab of 3-storey podium structure were constructed using precast element. The application of prefabricated elements in construction project in Hong Kong has been continuously extended until recent projects. Most of them use precast facades and prefab balconies to speed up the construction time. However, the application of modular volumetric

IOP Conf. Series: Materials Science and Engineering **650** (2019) 012033 doi:10.1088/1757-899X/650/1/012033

building technology has not been widely implemented. The modular building construction in Hong Kong is also called as Modular Integrated Construction (MiC) [8].

MiC is seen to be the future of construction industry in Hong Kong [16]. It is because Hong Kong is one of the most densely populated cities in the world. A housing shortage is one classical problem in this city. Therefore, Hong Kong authority for housing is pursuing prefabrication construction system to provide more public housings. It has been admitted if the prefabrication in construction is significantly increasing construction productivity.

There are some investments required to build the manufacturing facilities. Since the current modular building in Hong Kong were developed based on import product. Some other investments on the automation and robotic technology are paramount to support the prefabrication process. Nevertheless, the penetration of technology and innovation in Hong Kong construction industry will enhance the capability in providing modular building.

From the structural design view, a comprehensive design standard for this construction type must be developed. This is to protect the engineers in order to provide safe design. So far, the design practice still adopts the conventional design standard which stand independently. Stability systems and connection detailing are the key on the overall performance where these two have not been considered interactively. The advancement on construction materials are ready to support this construction to bring a more efficient and optimum structural design.

5. Benefits and Challenges Using Cold Formed

Although it uses as a secondary component, cold-formed steel member has been contributed to MiC projects. Four-sided module and continuously supported module use cold-formed steel for curtain wall element. It acts as primary structural component since the wall carry the gravity loads. In comparison, the open-sided module and corner supported module treat cold-formed steel members only for architectural supporting elements. The module relies on the hot-rolled steel profile at the corner positions or sometimes also at intermediate positions. The columns are predominantly from hollow steel section.

Table 1. Benefit and challenges of hot-rolled and cold-formed steel structure (a)

Comparison item	Hot-rolled steel	Cold-formed steel
Cross Section	- Local buckling is unlikely to happen	- Sections are relatively easy to experience local
Global Stability	- Due to relatively bulky section, the member imperfection will be small so that global stability can be maintained - Lateral torsional buckling tends to control the member design	- Plate imperfections will affect global stability - Members are likely to experience torsional or flexural-torsional instability
Residual stress	- Relatively lower	- It increases because of cold-forming process
Wind Load	- It will create relatively thick steel sections because the steel strength is relatively lower	- Since the yield strength is higher, it will create relatively slender structure and it will reduce the amount of steel used
Seismic Load	- It has good performance because of high ductility	- The performance is questionable because there is inadequate research in this field
Fire Engineering	- Strength degradation is gradually taken place on the temperature above 200oC [17, 18]	- The performance dips sharply when it exposes on temperature above 100oC [19]

Table 1. Benefit and challenges of hot-rolled and cold-formed steel structure (b)

Comparison item	Hot-rolled steel	Cold-formed steel
Durability	- Surface preparation and coating are required to protect from corrosion	- There is no requirement for coating because it can have corrosion resistance from galvanizing process
Fabrication Process	- Difficult to customize the cross sections. Market availability will govern the sections used	- Relatively easy to customize the section and it takes shorter time to fabricate the built-up sections
Steel Structure Weight	- Since the applicability of high strength steel is limited, the normal strength steel will give bigger	- Lighter structure weight is possibly gained because the sections will be smaller due to relatively high yield

Due to recent technology advancement, it is possible to form and roll thicker sections from cold- formed steel. It may be foreseen in the future the cold-formed steel is another option for main steel frame. Some benefits and challenges of selecting cold-formed steel in structural point of view have been summarized in **Table 1**.

[1] explains that the steel wall can be formed by C channels ranging from 70 to 100 mm deep. The channels are positioned in every 600 mm. The same sections are also used for ceilings and floors but with closer distance (within 400 mm). The vertical and horizontal bracing can use cold-formed or hot- rolled steel depending on the design. All those cold-formed steels are possibly made from light gauge steel. In the past, it was not easy to form thicker sections [20]. Therefore, cold-formed has not been exploited for the open frame structure.

As an example, it is presented the simple calculation of member size for corner column of 4-sided module. The column will be designed using hot-rolled steel and cold-formed steel to compare the results. It aims to use the optimum dimension. The building data consist of:

- Module size = 3.6 m x 7 m; Height = 3 m; Ultimate floor load 8 kN/m^2
- Number of stories = 10, so the total force on the column in the ground floor = 10 x 8 x 25.2/4
  = 504 kN (assume there are only 4 columns on the module)
- Hot-rolled steel specification: F_y = 460 MPa
- Cold-formed steel specification: Stainless Steel Lean Duplex (EN 1.4162) $F_{0.2}$ = 460 MPa
- All the sections are using Square Hollow Section (SHS)

5.1. Hot rolled-steel design

Trial section: SHS 100x100x10 (Effective Length, L_E = 3m)
According to the table B.2 [21]:

Bf = Bw = 100 mm; ri = 10 mm; t = 6.3 mm; A = 2320 mm^2; rx = ry = 38mm
I = 336*10^4 mm^4; S = 67.1*10^3 mm^3; Z = 80.9*10^3 mm^3
Since it uses S355:

$$\varepsilon = (\frac{275}{Fy})^{0.5} = (\frac{275}{460})^{0.5} = 0.77$$

Member Classification: $\frac{b}{tf} < 40\varepsilon \rightarrow 100 < \frac{40}{10} * 0.77 \rightarrow 10 \leq 30.93$ (semi-compact, class 3)

Slenderness ratio $\lambda = \frac{Le}{ry} = 78.95$ Assume K = 1
using and F_y then read table 24 in (BS5950, 2000) to get p_c
Compressive strength: p_c = 266.25 N/mm^2
Axial compressive capacity: $P_n = A*p_c$ = 617.7 kN → **unity** check= 504/617.7 = 0.82 < 1**(SAFE!)**

5.2. Cold-formed steel design

Trial section: SHS 100x100x6 (STALA, n.d.)

$B_f = B_w = 100$ mm; $r_i = 10$ mm; $t = 6$ mm; $t_i = 5.96$ mm (BMT)

Sectional properties of unreduced sections:

$b_w = B_w - 2(t + r_i)$	$b_w = 68$ mm
$b_{cw} = B_w - t$	$b_{cw} = 94$ mm
$b_f = B_f - 2(t + r_i)$	$b_f = 68$ mm
$b_{cf} = B_f - t$	$b_{cf} = 94$ mm
$r_c = r_i + 0.5t$	$r_c = 13$ mm
$u = 0.5\pi\, r_c$	$u = 20.42$ mm
$c_o = (2/\pi)\, r_c$	$c_o = 8.26$ mm
$A = t_i\,(2b_w + 2b_f + 4u)$	$A = 2107.93 mm^2$
$E = 205000$ MPa $\upsilon = 0.3$	

Local buckling:

Using effective widths concepts (BS5950, 1998), the properties of reduced section are as follow:

Slenderness limitation: $b_w / t_i = 11.41 < 500$

K factor: $h = b_{cf} / b_{cw} = 1 \rightarrow$ use curve 2: $K_1 = 7 - \frac{2h}{0.11+h} - 1.2h^3 = 3.998 \approx 4$

K_1

$p_{cr} = 0.904\, E\, K_1 (t_i / b_w)^2 = 5695$ MPa

$f_c / p_{cr} = 460 / 5695 = 0.08 < 0.123$

Effective width: $b_{eff} = b_w$ (fully effective)

Effective area: $A_{eff} = A_g = 2107.93$ mm^2 (Note: No unstiffened element on SHS section)

Axial capacity due to local buckling: $P_n = A_g * f_c = 969$ kN

Flexural buckling:

Since from local buckling check the full section is effective, then the other section properties are calculated:

$I_{corner} = 0.149\, r_c^3\, t_1 = 1951.02$ mm^4

$$I_x = I_y = 2 \quad \frac{b_f t_i^3}{12} + 2.\, b_f t_i [0.5 b_{cw}]^2 + 2 \quad \frac{t_i b_w{}^3}{12} + 4 I_{corner} + 4.\, \mu.\, t_i (0.5 b_w + C_o)^2$$

$I_x = I_y = 2399.37 + 1790527 + 312335.79 + 7804.1 + 869403 = 2.982\ 10^6$ mm^4

Mid-contour length: $p = 2(b_w + b_f) + 4u = 353.68$ mm

Enclosed Area: $A_p = b_{cw}\, b_{cf} - r_c^2 (4 - \pi) = 8690.03$ mm^2

Torsional Constant: $J = 4A_{p}{}_i^2 \frac{t}{p} = 5.09\ 10^6$ mm^4; warping constant = 0

$G = E / [2(1 + \upsilon)] = 7.885\ 10^4$ MPa

$r_x = \sqrt{(I_x/A)} = 37.6$ mm $\rightarrow r_y = r_x$ and x_o

$= 0$ $r_o = {}_x\sqrt{(r_y^2 + r_o^2 + x^2)} = 53.2$ mm

$\beta = 1 - (x_o/r_o)^2 = 1$

Check maximum allowable slenderness: $\frac{LE}{r_y} = \frac{3000}{37.6} = 79.79 < 180$ (Ok)

Elastic buckling stress under minor axis: $P_{EY} = \frac{\pi^2 E I_y}{L_E{}^2} = 670.38\ KN$

Perry coefficient: $\eta = 0.002 \left(\frac{L_e}{r_r} - 20\right) = 0.118$

Short strut capacity: $P_{cs} = A_{eff} F_{0.2} = 825KN \rightarrow \phi = ½ (P_{cs} + (1 + \eta)\, P_{Ey}) = 787.24$ kN

Flexural buckling resistance: $Pc = \frac{P_{Ey} P_{cs}}{\phi + \sqrt{\phi^2 - P_{Ey} P_{cs}}} = 529KN$

Torsional buckling:

Elastic flexural buckling resistance about major axis: $P_{EX} = \frac{\pi^2 EI_x}{L_E{}^2} = 670.38$ KN

Torsional buckling load : $P_T = \frac{1}{r_o{}^2}(G.J + 2\frac{\pi^2 \cdot E \cdot C_w}{L_E{}^2} = 141.81 \times 10^3$ KN

Torsional buckling load: $P_{TF} = \frac{1}{2\beta}\left[(P_{EX} + P_T) - \sqrt{(P_{EX} + P_T)^2 - 4 \cdot \beta \cdot P_{ExT} \cdot P}\right] = 670.38\ KN$

Since $P_{Ey} < P_T$, $\alpha = 1 \rightarrow \eta = 0.002\ (\alpha L_E/r_y - 20) = 0.118 \rightarrow \phi = 787.24$ kN

Torsional flexural buckling resistance: $P_c = \frac{P_{TF}P_{CS}}{\phi + \sqrt{\phi^2 + P_{TF}P_{cs}}} = 529KN$

The capacity of cold-formed steel column = min (local buckling; flexural buckling; torsional buckling)

$P_c = 529$ kN $\rightarrow$ unity check against $P_u = 504/529 = 0.95 < 1$ **SAFE!)**

Both design calculations give similar results. Using almost identical sections and same yield strength have given slightly different unity check value. Proposed section of SHS 100x100x6.3 (Hot-rolled) and 100x100x6 (cold-formed) can be used to carry the loads. However, the design process of cold-formed takes more effort. Finally, it can be concluded that the cold-formed section is applicable for main structural member of modular building.

6. Summary and Conclusion

The background theory including the development of current modular building structure have been presented. Due to technology advancement, MiC projects may promote manufacturing technique in construction. There are some benefits in term of time, cost, and sustainability can be gained with modular building.

In terms of structural design, further studies are necessary especially for the feasibility in high-rise buildings. The use of cold-formed steel for main structural member also remains valid for further investigations. Cold-formed steel has been used for wall-supporting structure but the application for open frame system remains scarce. However, the example calculation shows a greater possibility of using cold-formed steel for main column structure. The calculation is to provide quick check on the feasibility of using cold-formed steel structure.

References

[1] Lawson R M, Ogden R and Goodier C I 2014 *Design in modular construction*. CRC Press, Taylor & Francis

[2] Ramaji I M and Memari A M 2013 "Identification of Structural Issues in Design and Construction of Multi-Story Modular Buildings," presented at the 1st Residential Building Design & Construction Conference, Sands Casino Resort, (Bethlehem,PA)

[3] Lawson R M, Ogden R G and Bergin R 2012 Application of modular construction in high-rise buildings *Journal of Architectural Engineering,* **18**(2) pp 148-154

[4] Liew Z D J Y R and Chua Y S 2018 *Steel Concrete Composite Systems for Modular Construction of High-rise Buildings," presented at the 12th International Conference on Advances in Steel-Concrete Composite Structures (ASCCS 2018)* (València, Spain)

[5] Lawson R M *Building design using modules*. 2007 Ascot, (UK: The Steel Construction Institute)

[6] Jellen A C and Memari A M 2013 The State-of-the-Art Application of Modular Construction to Multi-Story Residential Buildings presented at the 1st Residential Building Design & Construction Conference, Sands Casino Resort, (Bethlehem, PA)

[7] Lawson R M and Richards J Modular 2010 design for high-rise buildings *Proceedings of the Institution of Civil Engineers* **163** pp 151-164

[8] Pan W, Su R, Cai Y and Young B 2018 Engineering modular systems for high-rise buildings: an update *Proceedings of the Institution of Civil Engineers - Civil*

Engineering **171**(4) pp 148-148

[9] Annan C D, Youssef M A and El Naggar M H 2009 Seismic Vulnerability Assessment of Modular Steel Buildings *Journal of Earthquake Engineering* **13(**8) pp 1065-1088

[10] Srisangeerthanan S, Hashemi M, Rajeev P, Gad E and Fernando S 2018 Numerical study on the effects of diaphragm stiffness and strength on the seismic response of multi-story modular buildings *Engineering Structures.* **163** p 25

[11] Fathieh A and Mercan O 2016 Seismic evaluation of modular steel buildings *Engineering Structures.* **122** pp 83-92

[12] Gunawardena T, Ngo T, Mendis P, and Alfano J 2016 Innovative Flexible Structural System Using Prefabricated Modules *Journal of Architectural Engineering* **22**(4)

[13] Lacey A W, Chen W, Hao H and Bi K 2018 Structural response of modular buildings - An overview *Journal of Building Engineering.* **16** pp 45-56

[14] Ferdous W, Bai Y, Ngo T D, Manalo A and Mendis P 2019 New advancements, challenges and opportunities of multi-storey modular buildings – A state-of-the-art review *Engineering Structures.* **183** pp 883-893

[15] Wong R W M, Hao J J L and Ho M F C 2018 *Prefabricated Building Construction Systems Adopted in Hong Kong*

[16] Wong J, Zhang J and Lee J A 2015 Vision of the Future Construction Industry of Hong Kong in *The 32nd International Symposium on Automation and Robotics in Construction.* (Qulu, Finland: International Asociation on Automation and Robotics in Construction)

[17] Outinen J, Kesti J and Mäkeläinen P 1997 Fire Design Model for Structural Steel S355 Based Upon Transient State Tensile Test Results *Journal Constructional Steel Research.* **42(**3) pp 161-169,

[18] Qiang X, Bijlaard F S K and Kolstein H 2012 Deterioration of mechanical properties of high strength structural steel S460N under transient state fire condition *Materials & Design.* **40** pp 521-527

[19] Chen J and Young B 2007 Experimental investigation of cold-formed steel material at elevated temperatures *Thin-Walled Structures* **45(**1) pp. 96-110

[20] Guo Y -J, Zhua A -Z, Pi Y -L, and Tin-Loi F 2007 Experimental study on compressive strengths of thick-walled cold-formed sections *Journal of Constructional Steel Research.* **63**(5) pp 718-723

[21] BS10210, *British Standard: Hot finished structural hollow sections of non-alloy and fine grain steels - Part 2: Tolerances, dimensions, and sectional properties.* London: British Standard Institute, 2006.

IOP Conf. Series: Materials Science and Engineering **650** (2019) 012034 doi:10.1088/1757-899X/650/1/012034

Development Precast and Prestressed Concrete Rigid Pavement System in Indonesia

Hari Nugraha Nurjaman[1], Lutfi Faizal[2], Nyoman Suaryana[3], Yudhi Dharmawan[4], Agus Wantoro[4], Purnomo[4], Suwito[5*]

[1]Indonesian Association of Precast and Prestressed Engineer, Jakarta, Indonesia
[2]Institute of Housing and Settlement, Ministry of Public Work and Housing, Bandung, Indonesia
[3]Directorate of Road Preservation, Ministry of Public Work and Housing, Jakarta, Indonesia
[4]Waskita Beton Precast Ltd, Jakarta, Indonesia
[5]Construction Engineering and Management Department, Faculty of Engineering, Podomoro University, Jakarta, Indonesia

*irsuwito@gmail.com

Abstract. Roadway pavements are important parts of transportation system and thus should be adequately strong and durable. Unfortunately, many pavements constructed in Indonesia fail to reach their design service life. Among major factors contributed to early damages to the pavements in Indonesia are unfulfilled specified pavement construction specifications, traffic overload and poor support condition. As an attempt to produce better pavement system and as part of an accelerated infrastructure program in Indonesia, new precast pavement system using unbonded posttensioned technology that has characteristic of self-centering has recently been developed. This self-centering concrete pavement is expected to reduce the impact of overload and be able to overcome imperfect subbase, subgrade and poor soil condition. The theoretical and experimental development programs of the new concrete rigid pavement system will be described in this paper. The developed pavement system is then implemented to see the effectiveness construction procedures and later on tested using Falling Weight Deflectometer (FWD) method to evaluate its structural performance.

1. Introduction

Since 2014, Indonesian government has conducted acceleration of infrastructure construction for all regions in Indonesia. Roadway or highway is essential part of infrastructure because it is responsible for transportation connectivity in supporting efficient economic activity. As such roadway pavements should be adequately strong and durable for their design service life. Early damages to pavements may cause traffic disruption and thus lead to negative user experiences and economic impact. Unfortunately, many pavements constructed in Indonesia fail to reach their design service life. Among major factors contributed to early damages to the pavements in Indonesia are unfulfilled specified pavement construction specifications dan traffic overload.

Traditional Highway Construction often faces difficulties such as supporting preparation work which is not optimum, shortage in asphalt delivery in flexible pavement case, highway obstructions

IOP Conf. Series: Materials Science and Engineering **650** (2019) 012034 doi:10.1088/1757-899X/650/1/012034

in case of conventional rigid pavement (jointed plain concrete pavement, JPCP, shown in figure 1), thus inability to meet quality specification in case of rigid pavement. In addition, the drainage system often time is addressed properly. This condition demands new pavement technology to be developed and implemented in pavement construction in Indonesia.

Figure 1. Conventional rigid pavement (JPCP)

A relatively recent innovative technology developed for highway pavement is precast concrete pavement. Since the precast concrete pavement system normally fabricated under controlled environment, it offers several potential advantages over traditional highway pavement system such as shorter construction time, higher quality (need less maintenance) and efficient use of materials. There have been several applications of precast concrete pavement system in Indonesia. It started with the use of Prestressed Precast Concrete Pavement (PCP) technology in a section of Cakung-Cilincing highway, Jakarta in 2007 (figure 2). It was then followed by the use of Precast Concrete Pavement Panel (PCPP) system in Suryacipta Industrial Estate (2013, figure 3) and Precast Prestressed Concrete Pavement (PPCP) in Kanci-Pejagan Toll Highway (2011, figure 4).

Figure 2. PCP in Cakung-Cilincing Highway 2007

Figure 3. PCPP in Suryacipta industrial estate, 2013

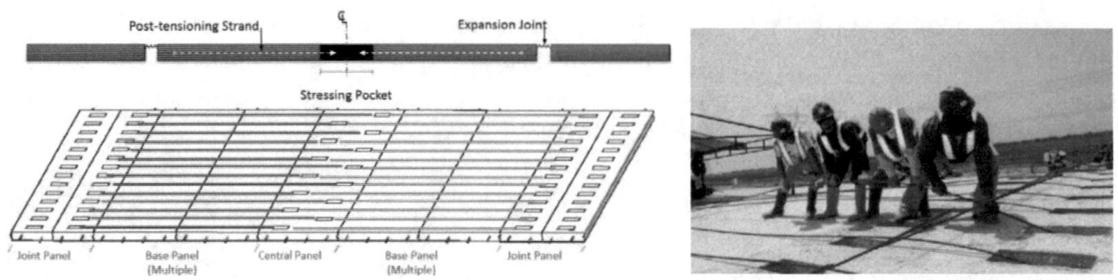

Figure 4. PPCP Kanci-Pejagan Tollway 2011

Study on various application of precast concrete pavement systems in construction condition Indonesia [8] reveals that Prestressed Precast Concrete Pavement technology demonstrates best performance. It is part of integrated development effort on highway rigid pavement technology based on precast prestressed concrete system initiated since 2015, by performing qualitative and quantitative research. Qualitative research was carried out on qualitative performance on the application of precast system in several highway constructions from 2010 to 2015, and in comparative study in United States in 2015. Quantitative research was carried out on performance of application of precast system in highway construction from 2010 to 2015, and in development experimental program from 2015 to 2017. This study will focus on the experimental/testing program developed to produce a precast pavement system that is durable and economics.

2. Program Development

2.1. System Development

The idea behind the system development of pavement is the re-centering capability of unbonded post- tensioned prestressing in resisting earthquake forces in buildings [7]. The use of re-centering capability in highway pavement would allow any imperfection in sub-base and sub-grade as well as overload condition [10]. This technology has been applied in several internal roads in precast plant (2015-2016) as seen in Figure 5. Design process of the system with finite element method which considered overload condition in standard subgrade specification and undersubgrade specification data can be seen on work conducted by Nurjaman et. al. [8].

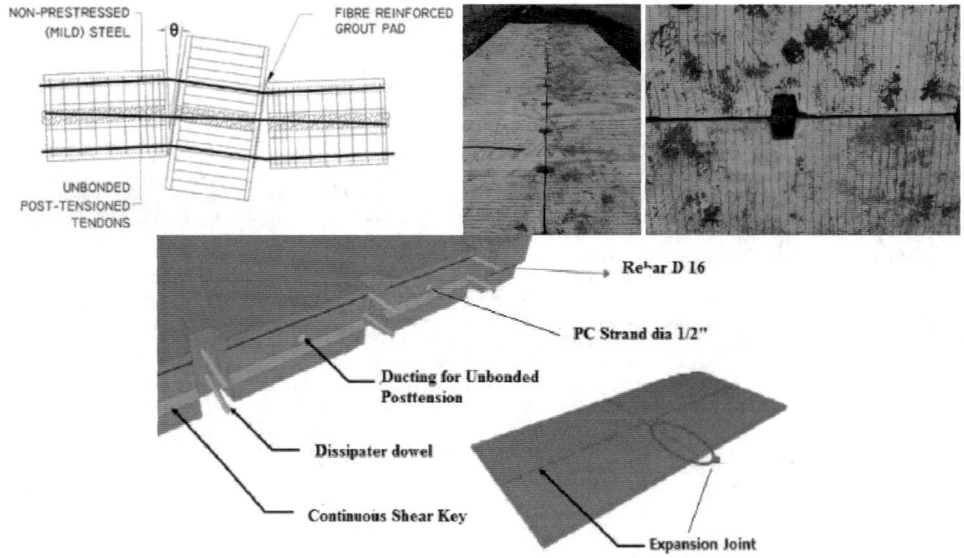

Figure 5. PCP unbonded post-tensioned technology with re-centering capability. Precast module is 12000x1800x180 and 12000x3600x180 with concrete quality fc' = 41.5 MPa

IOP Conf. Series: Materials Science and Engineering **650** (2019) 012034 doi:10.1088/1757-899X/650/1/012034

2.2. Experimental Program
Under the loading the critical part of a rigid concrete pavement system is at the connection system. The connection mainly undergoes internal flexural bending moment, shear forces as well as tensile force action. To develop a reliable precast and prestressed concrete system for rigid pavement, it is necessary to develop an experimental program that addresses those actions. Thus, in this study three (3) experimental tests had been programmed, i.e.: (1) Pull-Out Dowel Activator Test; (2) Flexural Test and
(3) Shear Test. Details of testing specimens and procedures will be given below.

2.2.1. Pull-Out Dowel Activator Test. The connection system between precast panels uses reinforcing bars called dowel activator. As the role of dowel activator is to connect between two precast pavement panel, it must be able to distribute forces due to traffic load. The ability of dowel to distribute forces is determined by conducting pull-out test. Specimens for dowel pull-out tests were prepared by embedding the dowel of diameter 16 mm into a concrete block to simulate the actual connection system. Two types of specimens were prepared, one for representing the precast panel thickness of 180 mm (SpRigWP_180) and the other for precast panel thickness of 200 mm (SpRigWP_200). The preparation and final form of specimens for pull-out dowel activator test is shown in figure 6.

Figure 6. Preparation and final form of specimens for pull-out test on dowel activator

The setup configuration for pull-out dowel activator test is shown in figure 7. Testing is carried out according to standard tensile test for reinforcing bar.

IOP Conf. Series: Materials Science and Engineering **650** (2019) 012034 doi:10.1088/1757-899X/650/1/012034

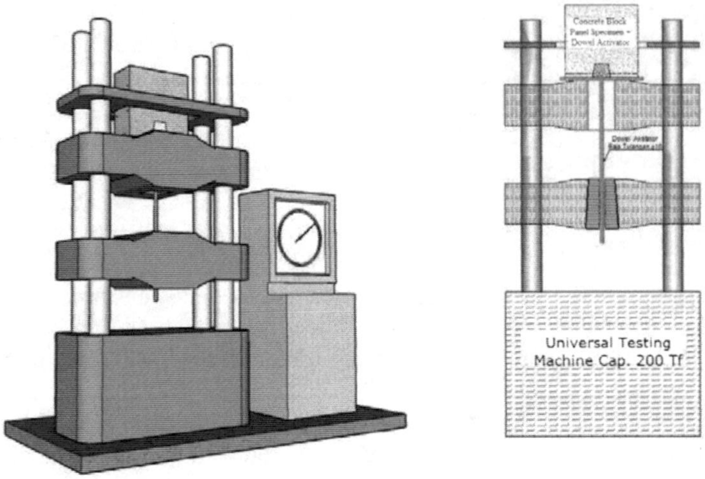

Figure 7. Setup configuration for pull-out dowel activator test

2.2.2. Flexural Test. The purpose of flexural testing on two precast concrete panel connected with dowel activator is find out the performance of connection system under flexural load. The setup configuration of flexural test is shown in Figure 8. The testing load will be applied and increased incrementally until the connection fails near the support where the maximum moment taken place. Two (2) specimens were prepared for the test, one with precast panel thickness of 180 mm (SpRigWP_180) and the other with thickness of 200 mm (SpRigWP_200).

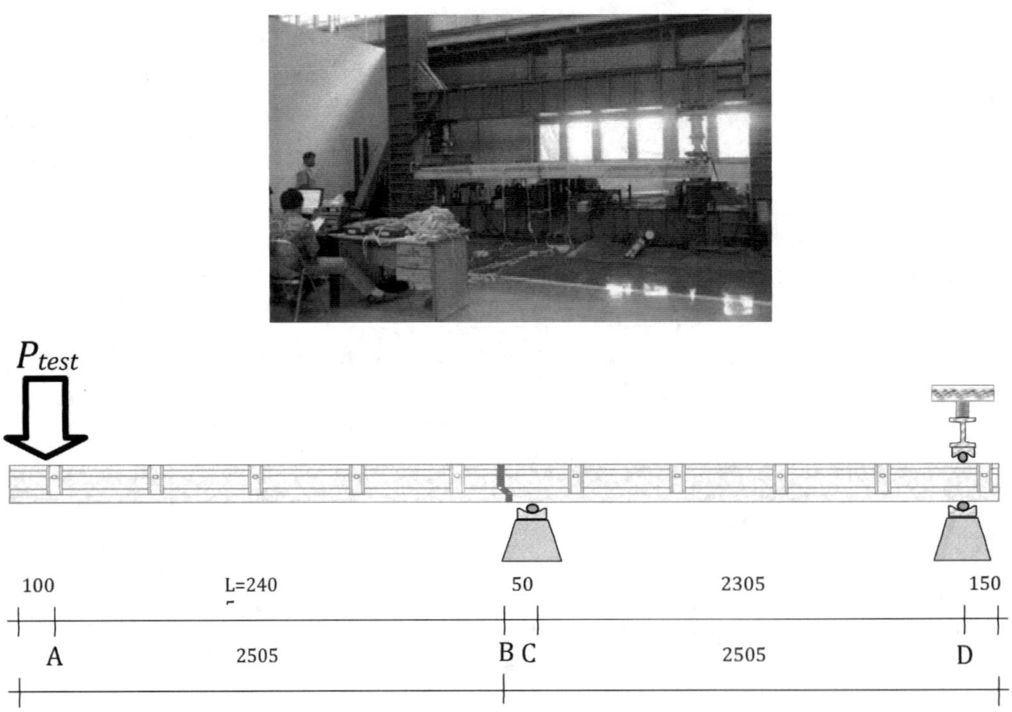

Figure 8. Picture and configuration for flexural test specimen

2.2.3. Shear Test. The Shear test was carried out on specimen that consists of two panel connected in the middle and supported by three (3) supports as shown in Figure 9 below. This test is conducted to determine the shear capacity of the connection by applying the vertical load on test specimen until failure load. Two types of specimens were prepared, one with the thickness of 180 mm (SpRigWP_180) and the other with thickness of 200 mm (SpRigWP_200). The dimension of specimen is of 5100 mm in length by 1800 mm in width.

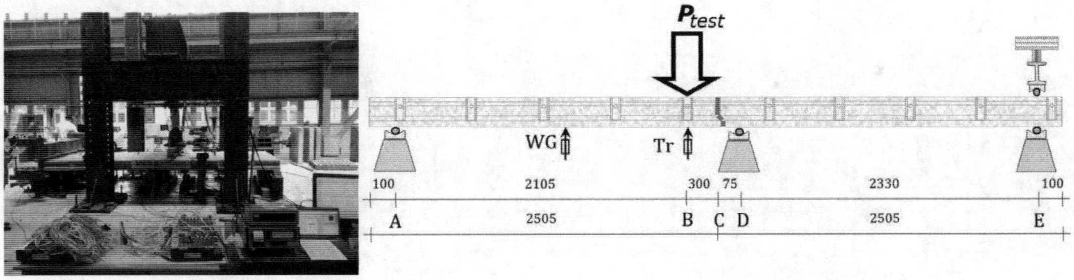

Figure 9. Picture and configuration for shear test specimen

2.3. Implementation and FWD Testing Program

As part of development program, the newly developed precast prestressed concrete pavement system was installed on segment of Margomulyo road (managed by Directorate of General Highway) in Gresik to evaluate execution aspect especially the effect of construction on traffic, and structural performance of the system in the actual condition. The road was selected because it is frequently suffered damage due to heavy traffic and heavy vehicles. The length of road segment is 48 m and the width is 23.4 m as shown in figure 10. The construction new pavement system was conducted in the month of December 2018. Structural testing was conducted using FWD method by Institute of Research and Development Road and Bridge, Ministry of Public Work and Housing, Bandung, Indonesia, 5 months after installation (May 2019). The locations of points in which FWD tests taken are shown in figure 10.

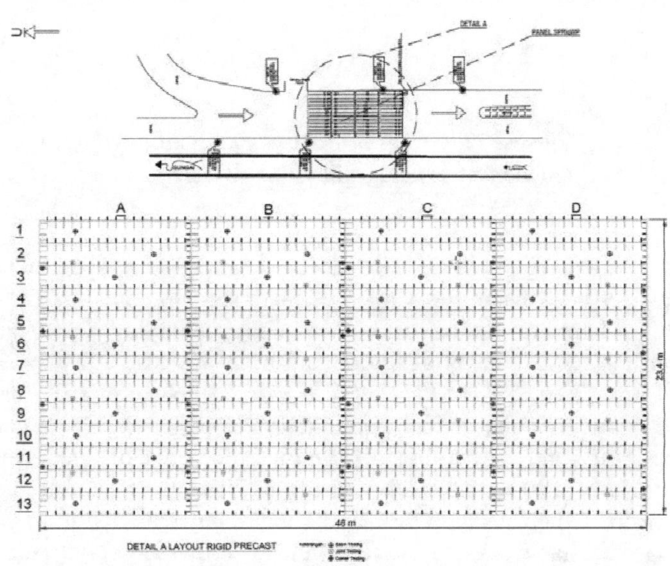

Figure 10. Road segment used for implementation of SpRigWP pavement and point locations for FWD test

3. Results and Discussions

3.1. Results of Pull-Out Dowel Activator Test

Figure 11 (a) and (b) show the relationship between load and elongation for specimens SpRigWP_180 and SpRigWP_200, respectively. The results show that the tensile behavior of dowel is similar to that of steel bar shown in figure 12. It shows that dowel anchorage system is able to mobilize the strength of dowel steel bar.

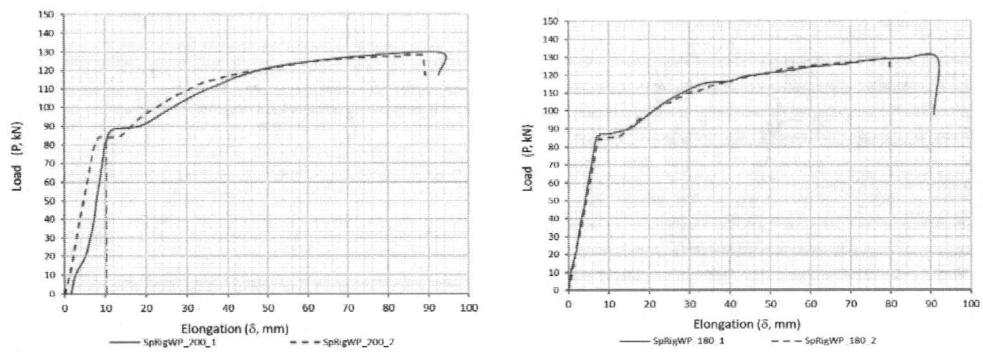

(a) SpRigWP_180 (b) SpRigWP_200

Figure 11. Graph of load vs elongation from pull-out dowel test

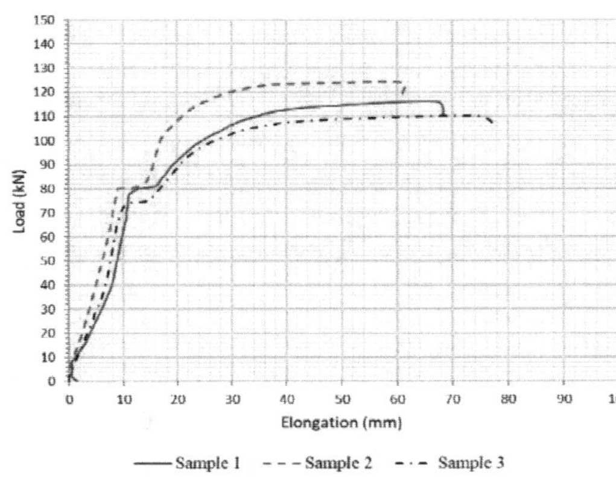

Figure 12. Graph of load vs elongation from tensile test steel bar diameter 16 mm

3.2. Results of Flexural Test

The flexural test results are presented in tables 1 and 2 for specimens SpRigWP_180 and SpRigWP_200, respectively. From the values of testing loads, one can derive the moment at the connection (Ptest·L, where Ptest is the testing load and L is the lever Arm shown in Figure 8). The moment capacity of tested specimens at yield condition are Mtest_180 = 18.624 kN·2.405 m = 44.79 kN·m and Mtest_200 =19.214 kN·2.405 m = 46.21 kN·m for specimens SpRigWP_180 and SpRigWP_200, respectively. Theoretical nominal moment for specimen can be determined from the section compatibility and the results of nominal moments of specimens SpRigWP_180 and SpRigWP_200 are of Mn_180 = 30.82 kN·m and Mn_200 = 37.18 kN·m, respectively. Comparing moment strength from experimental work and theoretical calculation shows that the values of moments from experimental work are higher than those of theoretical calculations (Mtest_180 = 44.79 kN·m > Mn_180 = 30.82 kN·m and Mtest_200 = 46.21 kN·m > Mn_200 = 37.18 kN·m). The results indicate that the specimens perform satisfactorily against flexural tests.

Design of reinforced concrete against bending moment requires that the moment capacity, ϕMn, is larger than or equal to ultimate moment, Mu (ϕfMn ≥ Mu), where ϕf is a strength-reduction factor, taken 0.9 for flexural moment, Mn is nominal moment strength. The moment capacity of specimens SpRigWP_180 and SpRigWP_200 are of 0.9·30.82 kN·m = 27.73 kN·m and 0.9·37.18 kN·m = 33.46 kN·m, respectively. The internal ultimate momen Mu resulting from the finite element analysis conducted on the specimens SpRigWP_180 and SpRigWP_200 under expected traffic load are of 5.76 kN·m for both of them. One can see that the specimens are safe against flexural moment (27.73 kN·m > 5.76 kN·m).

Table 1. Flexural Test Results of Specimen SpRigWP_180

No.	Specimen Condition	Load (P, kN)	Deflection (mm)	Strain (x 10^{-6})
1.	Self-weight	9.716	0.80	13
2.	Initial Crack	12.460	2.15	69
3.	Yield	18.624	15.87	1958
4.	Ultimate	23.475	227.56	13092
5.	Failure	19.428	320.95	No data

Table 2. Flexural Test Results of Specimen SpRigWP_200

No.	Specimen Condition	Load (P, kN)	Deflection (mm)	Strain (x 10^{-6})
1.	Selfweight	10.796	7.09	738
2.	Initial Crack	11.56	7.74	815
3.	Yield	19.214	17.47	1882
4.	Ultimate	27.681	156.17	21559
5.	Failure	21.752	396.85	No data

3.3. Results of Shear Test

Tables 3 and 4 show a summary of shear test results from specimens SpRigWP_180 and SpRigWP_200, respectively. The ultimate testing load, Ptest, for specimens SpRigWP_180 and SpRigWP_200 are of 382.96 kN and 415.52 kN, respectively. Using the approximation, nominal shear strength from test results, Vn_test, can be estimated using expression of Vn_test = 0.91Ptest. Thus, the nominal shear strength of specimens SpRigWP_180 and SpRigWP_200 are of 348.49 kN and 378.12 kN, respectively. Meanwhile, theoretical nominal shear force capacity of the concrete specimen can be determined using equation of $V_n = 0{,}17\lambda\sqrt{f_c'}\,b_w\,d$. The theoretical nominal shear strength of specimens SpRigWP_180 and SpRigWP_200 are then of 286.9 kN and 328.4 kN, respectively. It can be seen that for both specimens the nominal shear strength from test results are higher than those of from theoretical calculations. It can be concluded that the specimens perform satisfactorily against shear. Design of reinforced concrete against shear force requires that the shear capacity, ϕVn, is larger than or equal to ultimate shear force, Vu (ϕVn ≥ Vu), where ϕ is a strength-reduction factor, taken 0.75 for shear, Vn is nominal shear strength. The shear capacity of specimens SpRigWP_180 and SpRigWP_200 are of 0.75·286.9 kN = 215.2 kN and 0.75·328.4 kN = 246.3 kN, respectively. The internal moment shear force Vu resulting from the finite element analysis conducted on the specimens SpRigWP_180 and SpRigWP_200 under expected traffic load are of 80.54 kN for both of them. One can see that the specimens are safe against shear (215.2 kN > 144.97 kN).

Table 3. Shear Performance of Specimen SpRigWP_180

No.	Condition	Load (P, kN)	Deflection (mm)	Strain (10^{-6})
1.	Initial Crack	172,48	2,19	121
2.	Yield	309,63	4,41	1977
3.	Ultimate	382,20	9,29	2184
4.	Failure	301,84	20,29	14615
5.	Failure	157,78	33,29	No Data

Table 4. Shear Performance of Specimen SpRigWP_200

No.	Condition	Load (P, kN)	Deflection (mm)	Strain (10^{-6})
1.	Initial Crack	76,44	1,39	111
2.	Yield	398,96	7,89	1977
3.	Ultimate	415,52	8,59	3626
4.	Failure	370,44	14,49	19633
5.	Failure	129,36	35,59	No Data

3.4. Results of Implementation and FWD Testing
From impulses of FWD test conducted on the pavement are in the range of 50 – 150, which correspond to the deflection of 0,05 mm - 0,15 mm, respectively. According to circular number 15/SE/M/2015 about guide for implementation of precast prestressed concrete pavement, the allowable maximum deflection is 0.57 mm. This result shows that the developed precast prestressed concrete pavement performed extremely well.

4. Conclusions
With accelerated road infrastructure construction in Indonesia, new technology that is efficient in construction, materials and reliable structurally needs to be developed. Precast concrete rigid pavement offers rapid and efficient construction procedure, higher quality and better structural performance. New pavement system was developed based on precast and prestressed technologies that have re-centering capability.

Precast and/or prestressed components have been employed for rigid pavement in Indonesia since 2007 but the results are not always optimum since the design and construction were not integratedly carried out. As part integrated development effort on highway construction based on industrial manufacture that has been initiated since 2015, theoretical, experimental program that include dowel pull-out test, flexural test and shear test were conducted. To see the performance of newly developed pavement system in real world application, implementation was conducted and later on the constructe pavement was tested structurally using FWD method. Overall, the test results show satisfactory performance.

The developed pavement system offers an attractive alternative for the performance improvement of highway construction and maintenance in Indonesia. The improved performance will lead to better experiences for users. Reliable transportation system will enhance economic activities and contribute to sustainable development in which nowadays considered as a part of global movement to save the world from climate change.

Acknowledgments
This research is funded by Waskita Beton Precast Ltd. Also the writers extend their appreciation to the Indonesian Association of Precast and Prestressed Engineers and to the Research Institute of

Human Settlement, and Institute of Road Engineering Ministry of Public Works for providing testing facilities.

References

[1] AAHSTO 1993 *Guide for Design of Pavement Structures* (American : Association of State Highway and Transportation Office)
[2] ACI 325.12R-02 2002 *Guide for Design of Jointed Concrete Pavements for Streets and LocalRoad* (American Concrete Institute)
[3] Choubane B 2015 *Private communication*
[4] Faizal L 2016 *Applied Strategic Technology of Precast dan Prestress Concrete, in Continuous Program Development Conference Proceedings, ed Cakranegaraet al., Ministry of Public Work and Housing* (Jakarta: Indonesia)
[5] Gillen S L 2015 *Private communication*
[6] Nantung T E 2015 *Private communication*
[7] Nurjaman H N, Faizal L, Sidjabat H R, Hariandja B, Rivky R and Put Y 2014 Elsevier. *Procedia Eng. J.* **95** pp 82–86
[8] Nurjaman H N, Faizal L, Suaryana N, Hariandja B, Gambiro, Purnomo and Wicaksono S 2017 *Design, development, and application of precast and prestressed concrete system for rigid pavement in Indonesia. In: Proceedings of the 3rd International Conference on Construction and Building Engineering (*ICONBUILD)
[9] PCA 1995 *Thickness Design for Concrete Highway and Street Pavements* (Portland Cement Association)
[10] Suaryana N 2016 *Dimension and Load of Prestressed Rigid Pavement, Ministry of Public Work and Housing* (Bandung: Indonesia.)
[11] Tyson S S 2015 *Private communication*

IOP Conf. Series: Materials Science and Engineering **650** (2019) 012035 doi:10.1088/1757-899X/650/1/012035

Loading Test and Temperature Effect on Steel Arch Bridge

Sumargo[1] and Ahmad H S Harahap[2]

[1]Civil Engineering, Faculty of Civil Engineering, Universitas Jenderal Achmad Yani, Jalan Terusan Jendral Sudirman, Cimahi, Jawa Barat, Indonesia, 40525
[2]Civil Engineering, Faculty of Civil Engineering, Politeknik Negeri Bandung, Jalan Gegerkalong Hilir, Ciwaruga, Parongpong, Kabupaten Bandung Barat, Jawa Barat, Indonesia, 40559

smg.7ph1@gmail.com; ahmadharahap298@ymail.com

Abstract. Before the bridge is operated, it is necessary to do a test to find out whether the bridge has met the applicable requirements and regulations and works properly with the service load. Tests include static and dynamic loading of bridges on the Water Front City bridge, Bangkinang, Riau. Static load test in the form of measurement of deflection and strain of the bridge structure when carried out the load on the bridge in a controlled and measurable manner. The given load is at least equal to the bridge service load according to specifications and service defects that occur do not exceed the specified limits. While the dynamic load test is in the form of recording bridge behavior when receiving dynamic loads. Static loading is carried out by placing several truck vehicles with measured dimensions and loads on each wheel to determine the amount of deflection and stress at ¼ - ½ - ¾ spans. Static load test results on the structure of the bridge, the maximum strain that occurs in reinforcing steel and concrete in loading conditions with a load of 2 trucks (50 tons), 4 (100 tons), 6 (150 tons), 8 (200 tons), 10 (250 tons) and 12 trucks (300 tons) respectively 17.62 MPa, 52.82 MPa, 61.95 MPa, 51.08 MPa, 105.49 MPa and 128.48 MPa. In unloading conditions, the maximum strain is 51.31 MPa, 70.38 MPa, 52.47 MPa, 88.71 MPa,75.20 MPa and 13.21 MPa respectively. The temperature difference of 15°C in arch element gives camber in the middle of the span of 15 mm for the arc and 13 mm for the girder. The natural frequency of the WFC bridge is 1,127 Hz which occurs in the vertical direction at a span of 100 meters.

1. Introduction

Water Front City (WFC) bridge is located in Kampar District, Riau Province. The WFC Bridge located in Bangkinang City is a three-span 50 m - 100 m - 50 m arc bridge with an arc element and girder extending the box profile and all structural elements made of steel. The load from the longitudinal girder is transferred to the arc using a cylindrical vertical rod. Prior to the opening bridge to public, a vehicle load test is carried out to determine whether the bridge strength has been properly controlled and measured [1]. The given load is at least the same as the bridge service load in accordance with the design, meets the applicable requirements and codess and can function properly. Load testing on the bridge is one method of measuring the bridge structure response both through static and dynamic loading tests. Static load test in the form of measurement of deflection and strain of the bridge structure when the load is applied to the bridge gradually loading and unloading. While the dynamic load test is in the form of recording bridge behavior when receiving dynamic loads. Static load test will provide information in the form of distribution patterns of forces in the superstructure due to the load. Whereas in dynamic tests, it is useful to measure the natural frequency of bridges. If there is a shift in the natural frequency of the bridge, then it can be further calculated regarding the condition of the bridge.

The purpose of static and dynamic load testing will be used as an analysis of bridge capacity and database of initial bridge conditions that can be used as an evaluation of the condition of the bridge in the future.

2. Methods

The WFC Bridge with a total length of this arc bridge span is 200 meters can be seen in Figure 1.

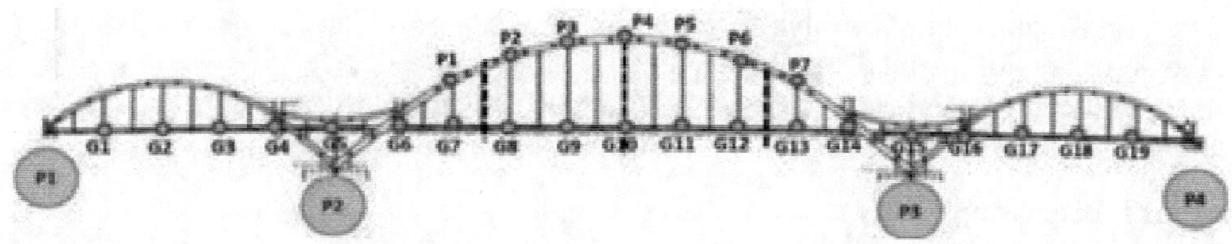

Figure 1. WFC Bridge and Sensors.

To find out the strain that occurs during loading tests [1-6], 10 strain gauges of TML type FLA-6-11 were installed in 10 different locations which can be seen in Figure 2. Only the 100 meter middle section bridge segment will be tested.

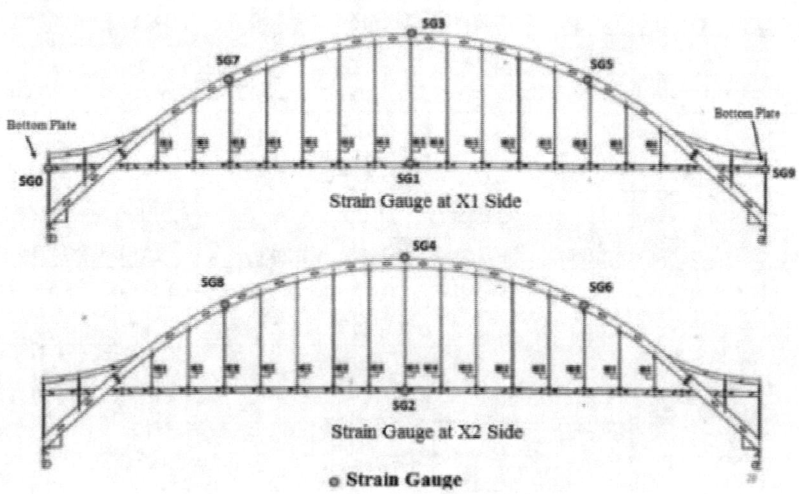

Figure 2. Location of Strain Gauges.

For deflection measurements, waterpass and total station were used instead of LVDTs [1-6] due to river conditions with deep and heavy currents do not possible. Waterpass (WP) and total station (TS) placement plan can be seen in Figure 3.

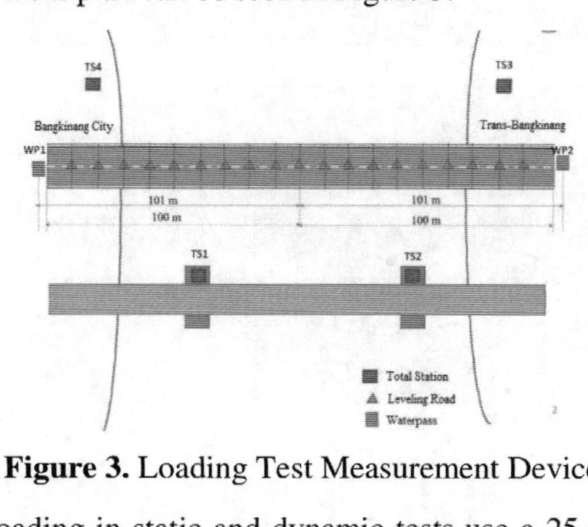

Figure 3. Loading Test Measurement Device

Figure 4. Truck Position on The Bridge.

Loading in static and dynamic tests use a 25 tonne load dump truck. In static testing, the total number of trucks used is 12 trucks, Figure 4. In static loading loads 0-50-100-150-200-250-300 tons and vice versa for unloading.

In dynamic tests, the truck is passed on a wooden track with a height of 200mm. Dynamic testing is carried out 5 times with 5 different excitation load locations, namely load excitation on the center line of the main span L/2 bridge, L/2 span A lane, L/4 bridge at center line, L/4 span A lane and on the 50 meter side span L/2 A lane.

The implementation of the bridge loading test carried out refers to the load test standard of the Association of American Society Highway Transport Organization (AASHTO).

3. Result and Discussion

Stages of static load test can be seen in Figure 5 and Table 1. The results of the measurement of bridge deflection for each resulting load configuration can be seen in Table 1. The longitudinal girder deflection curve and load- deflection graph in the girder can be seen in Figure 6.

Table 1. Girder Deflection in Long-Direction.

TABLE: Joint Displacements							
STA	U3 (0)	U3 (50)	U3 (100)	U3 (150)	U3 (200)	U3 (250)	U3 (300)
	mm	mm	mm	mm	mm	mm	mm
42,5	0,000	-0,277	-0,510	-0,430	-0,263	-0,198	-0,579
57,5	0,000	0,248	0,428	0,231	-0,110	-0,389	-0,281
75	0,000	5,679	6,025	-3,931	-6,680	-5,396	-5,004
100	0,000	-28,128	-34,750	-26,021	-19,731	-23,180	-20,614
125	0,000	5,200	4,764	-2,080	-4,098	-3,370	-4,944
142,5	0,000	0,183	0,300	0,193	-0,113	-0,329	-0,186
157,5	0,000	-0,210	-0,379	-0,377	-0,256	-0,177	-0,329

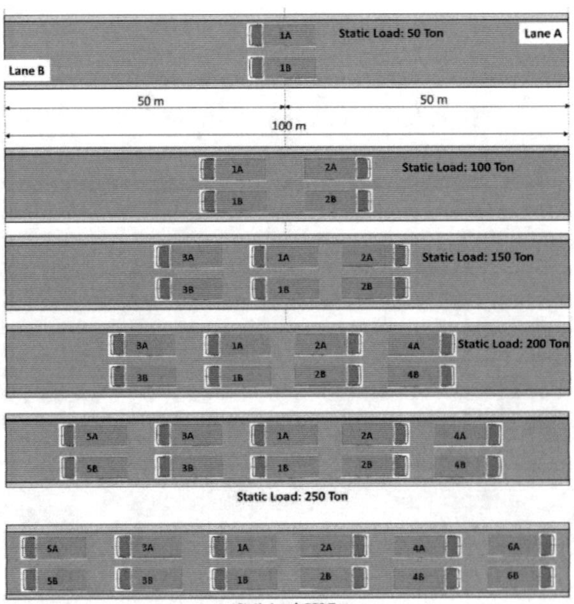

Figure 5. Loading Stages.

The results of measurement of arch deformation on the main span of the bridge can be seen in

Table 2 and Figure 7.

Table 2. Arch Deflection at the Main Span.

TABLE: Joint Displacements							
STA	U3 (0)	U3 (50)	U3 (100)	U3 (150)	U3 (200)	U3 (250)	U3 (300)
	mm	mm	mm	mm	mm	mm	mm
25	0,000	6,140	7,379	0,611	-2,242	-0,836	-0,662
50	0,000	-19,678	-25,499	-16,446	-10,947	-13,554	-11,847
75	0,000	5,380	5,753	2,092	0,010	0,824	-0,856

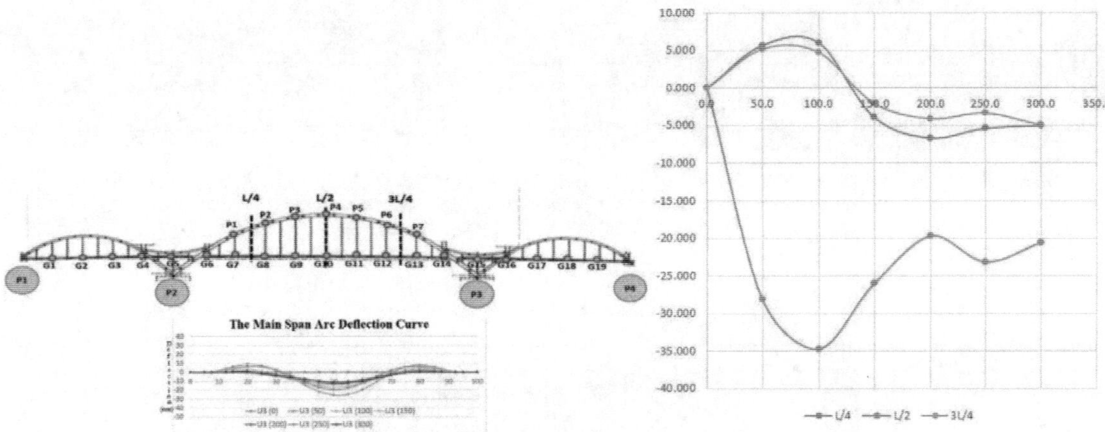

Figure 6. Load-Deflection Curve.

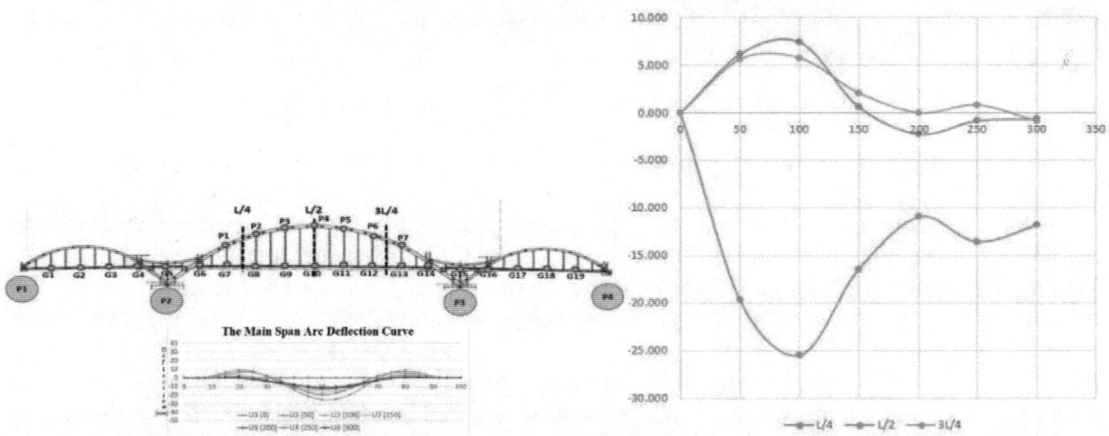

Figure 7. Deformation-Load Curve of Arch.

Designed deflection in girder and arches is 34,750mm and 25,499 mm, respectively, smaller than the allowable deflection L/800 = 125 mm. Deflection at L/2 shows a tendency that is opposite to deflection at L/4 and 3L/4 because the more the load is centered towards the L/2 the greater camber at L/4 and 3L/4. However, the more the load is evenly distributed towards the side, the deflection at L/2 also gets smaller, as well as deflection at L/4 and 3 /4.

Based on observations of all test equipment WP1, WP2, TS1, TS2, TS3,TS4 and comparing the suitability of the designed load-deflection curve with the actual test results, it was concluded that measurements based on TS2 and WP1 and WP2 were used for further analysis.

TS2 measured deflection at L/4, L/2 and 3L/4 sections of the main span bridge, and girder's actual and designed load-deflection curves are given, respectively in Figure 8, Figure 9, and Figure 10.

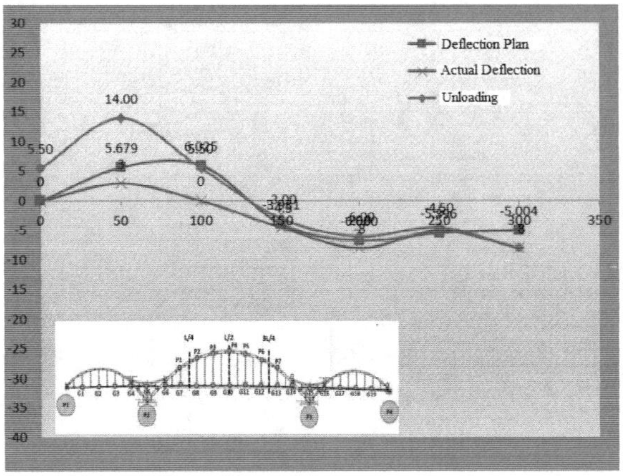

Figure 8. Load-Deflection Curve of Girder at L/4

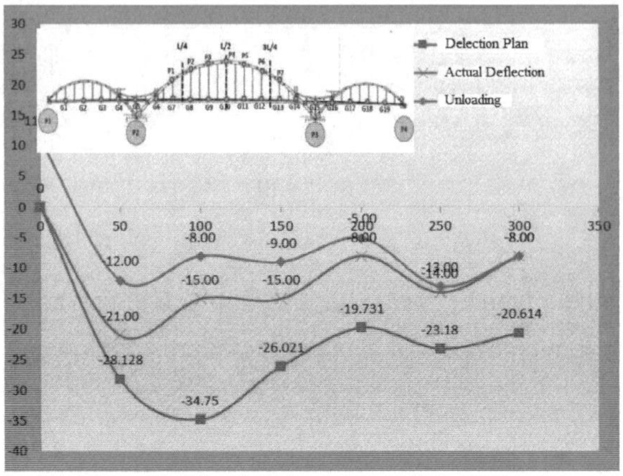

Figure 9. Load-Deflection Curve of Girder at L/2

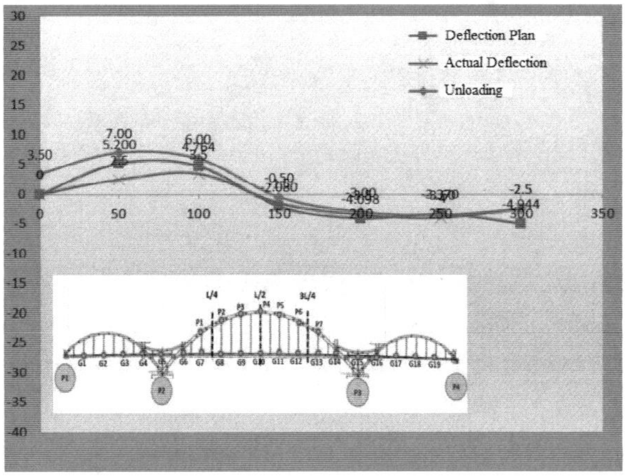

Figure 10. Load-Deflection Curve of Girder at 3L/4 - Actual and Design.

The curves conclude that the designed deflection at L/4, L/2 and 3L/4 gives a tendency that corresponds to the actual deflection and the actual deflection is smaller than the designed deflection.

The analysis shows that the temperature difference of 15˚C gives camber in the middle of the span of 15 mm for the arc and 13 mm for the girder or each temperature difference of 1˚C causes a camber of 1 mm. The difference in deflection in the arc and girder shows that due to the weight of the bridge itself, the hanging rod accepts the tensile force as designed. With the influence of temperature, the deflection that occurs during testing needs to be corrected by reducing the measured value during testing with a deflection opponent of 15 mm as shown in Figure 11.

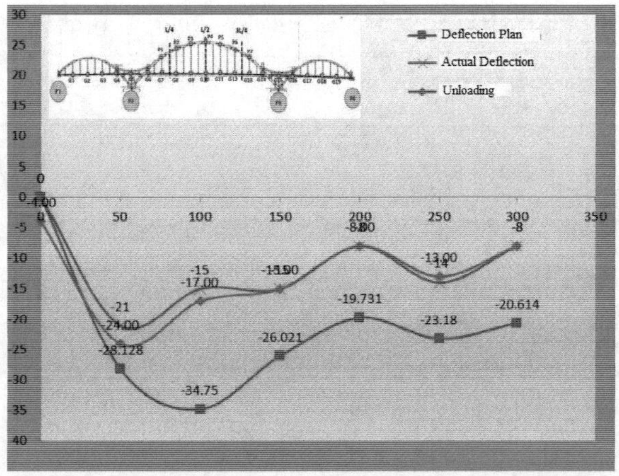

Figure 11. Corrected Actual and Design Load- Deflection Curve of Girder at L/2.

Maximum strain measurements on web of girder during loading stage can be seen in Table 3 and 4.

Table 3. Maximum Strain and Stress for each Static Test Loading Stage.

Loading Stage	Maximum Tension	Strain, με			
		SG 1	SG 3	SG 5	SG 7
50	17.62	-83.24	88.09	-24.57	-112.91
100	52.82	264.12	94.80	-80.83	-168.55
150	61.59	225.41	51.15	75.49	-143.55
200	51.08	180.77	-82.51	-84.85	-227.49
250	105.49	139.79	-58.47	7.82	-197.51
300	128.48	139.43	-41.13	-37.51	-5.83

The position of the strain gauge on the X1 and X2 side and the strain curve resulting from the static loading test can be seen in Figure 12 and 13, respectively.

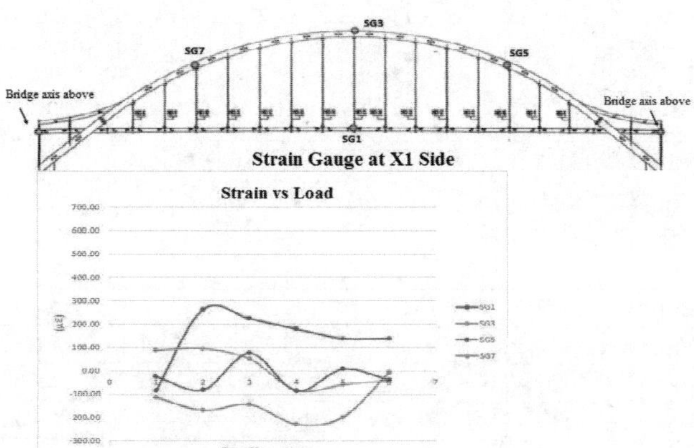

Figure 12. Position of Strain Gauge SG1, SG3, SG5, SG7 and Strain Curve of Static Test at Loading Stage.

IOP Conf. Series: Materials Science and Engineering **650** (2019) 012035 doi:10.1088/1757-899X/650/1/012035

Table 4. Maximum Stress and Strain Static Test at Loading Stage.

Loading Stage	Maximum Tension	Strain, με			
		SG 2	SG 4	SG 6	SG 8
50	17.62	-77.91	69.43	-28.91	-37.91
100	52.82	238.09	79.82	38.47	149.41
150	61.59	201.12	36.47	-12.21	309.76
200	51.08	183.83	-60.51	-50.83	255.38
250	105.49	58.18	-60.51	-38.83	527.47
300	128.48	173.86	6.50	24.84	642.41

Maximum strain measurements on web of girder during un-loading stage can be seen in Table 5 and 6.

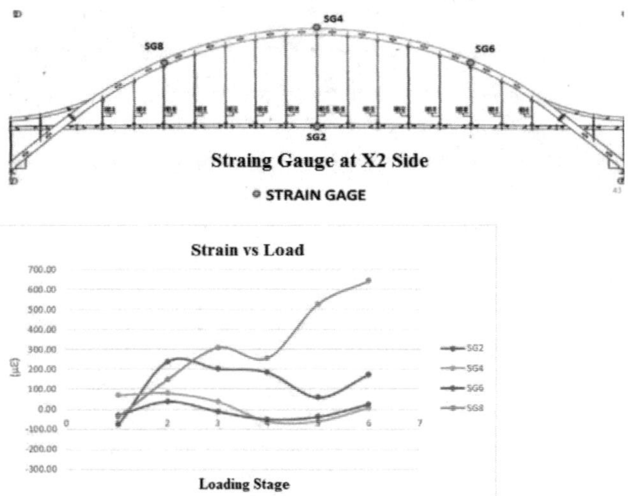

Figure 13. Position of Strain Gauge and Strain Curve Static Test at Loading Stage.

The position of the strain gauge on the X1 and X2 side and the strain curve of of the static test in the un-loading stage can be seen in Figure 14 and 15.

Table 5. Maximum Stress and Strain Static Test at Un-loading Stage.

Loading Stage	Maximum Tension	Strain, με			
		SG 1	SG 3	SG 5	SG 7
250	51.31	179.89	40.54	4.54	141.21
200	70.38	351.88	35.82	51.19	212.20
150	52.47	218.15	202.09	112.30	58.63
100	88.71	141.42	115.60	57.44	-110.70
50	75.20	118.21	187.76	102.97	130.14
0	13.21	-17.96	64.54	5.22	-67.69

IOP Conf. Series: Materials Science and Engineering **650** (2019) 012035 doi:10.1088/1757-899X/650/1/012035

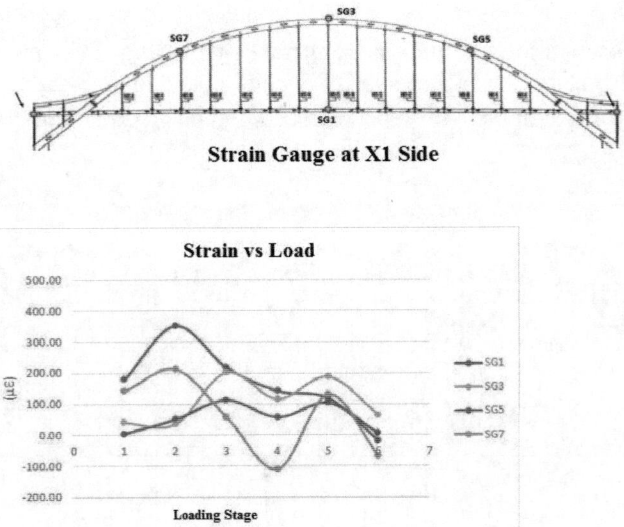

Figure 14. Position of Strain Gauge and Strain Curve Static Test Un-loading Stage

Table 6. Maximum Stress and Strain Static Test at Un-loading Stage.

Loading Stage	Maximum Tension	Strain, με			
		SG 2	SG 4	SG 6	SG 8
250	51.31	80.52	178.54	25.88	256.56
200	70.38	74.48	220.51	107.86	311.55
150	52.47	262.36	168.90	69.65	167.65
100	88.71	443.55	161.71	47.00	363.75
50	75.20	375.99	225.30	114.85	134.88
0	13.21	-32.07	-50.41	18.39	66.04

Five times dynamic testing were conducted with excitation load positioned at: (1) the center line of the main span at L/2, (2) the main span at L/2 bridge side, (3) center line of the main span at L/4, (4) the main span at L/4 bridge side, and (5) at midspan of the 50-meter span.

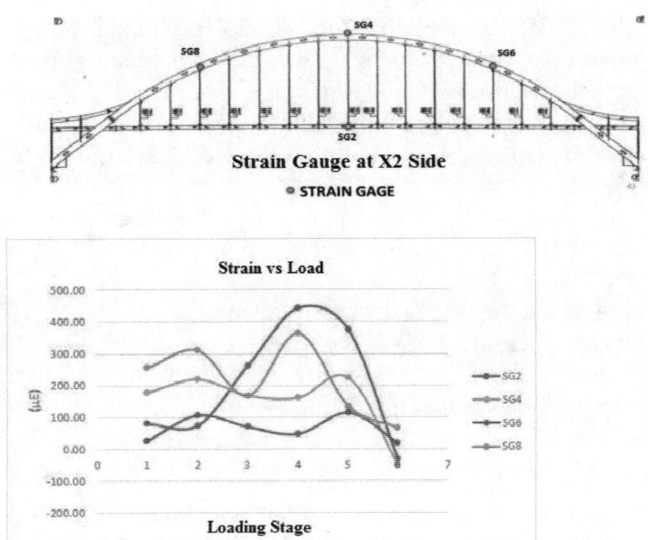

Figure 15. Position of Strain Gauge and Strain Curve Static Test at Un-loading Stage

The dynamic test uses impact excitation from trucks that weigh 25 tons, while the test devices consist of four accelerometer sensor units, with 3 units installed vertically, and one unit mounted laterally. Illustration of test and excitation parameters can be seen in Figure 16 and Figure 17.

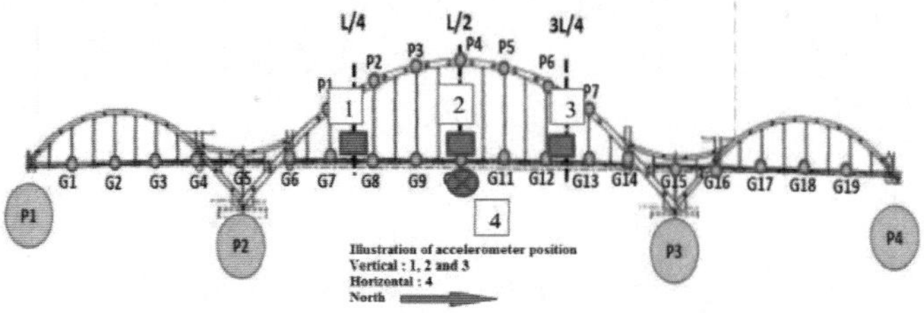

Figure 16. Position of Accelerometer.

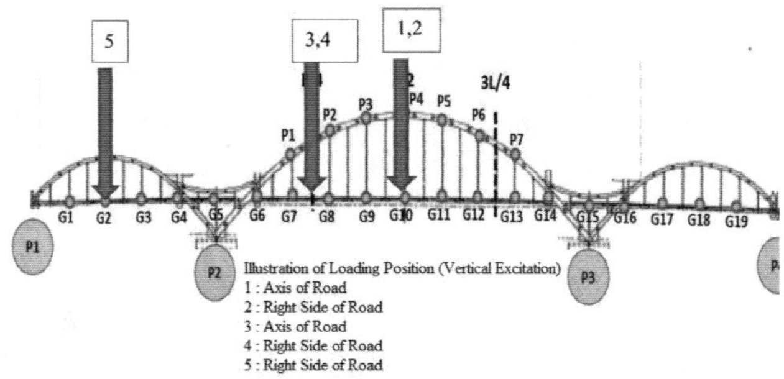

Figure 17. Position of Vertical Excitation Loading.

Based on the quality of the data, the test reference uses an accelerometer sensor in the position in the center of the vertical direction (accelerometer number 2).

In dynamic test 1, the reading frequency of the impact excitation can be seen in Figure 19. Graph of Frequency Domain can be seen in Figure 20. Frequency spectrum (Hz) for dynamic test 1 in the middle of the span and in the center of the bridge gives frequencies of 1.13485, 1.75385, 2.99187, 3.30137, 3.71404, 5.05522, 8.45976, 12.0706, 12.4833 and 13.2055.

At the second dynamic test, excitation impact was given at center line and on the right side of bridge. Frequency spectrums (Hz) for dynamic test 2 are 1.79214, 2.90451, 3.33709 and 10.0113.

In the third dynamic test, excitation impacts in a quarter and in the center of the bridge. Frequency spectrum (Hz) of dynamic test 3 shows values of 1.1327, 1.98222, 2.97333, 3.68126, 4.95555, 8.49522, 10.0527 and 12.4597.

In the fourth dynamic test, the excitation impact is on the quarter and on the right side of the bridge. Frequency spectrum (Hz) for dynamic test 4 shows values of 1.127, 1.97224, 2.95836 and 10.0021.

In the fifth dynamic test, excitation impacts at L/2 from the side arc (P1-P2) and on the right side of the bridge. The frequency spectrums (Hz) of the dynamic test 5 are 1.98273 and 2.79384.

The measurement results show that the natural frequency for a short span of 50 meters is 1.98273 Hz while for a span of 100 meters long is 1.127 Hz.

4. Conclusion

1. In testing with a symmetrical configuration loading, maximum stresses at steel web of girder for loading sequence of 50, 100, 150, 200, 250, and 300 tons are 17.62 MPa, 52.82 MPa, 61.59 MPa , 51.08 MPa, 105.49 MPa and
   128.48 Mpa, respectively. This value is still below the yield stress of the steel used which is 370

IOP Conf. Series: Materials Science and Engineering **650** (2019) 012035 doi:10.1088/1757-899X/650/1/012035

MPa.

2. The maximum stresses at un-loading stage of 250, 200, 150, 100, 50, and 0 tons are 51.31 MPa, 70.38 MPa, 52.47 MPa, 88.71 MPa,
75.20 MPa, 75.20 MPa and 13.21 Mpa, respectively.
3. Under unloading conditions the stresses do not return to their original value. For example, at a loading of 50 tons a stress of 17.62 MPa, while in an un-loading condition of 75.20 MPa or a stress difference of 57.58 MPa.
4. Testing under 70% of the actual load gives deflection smaller than the designed deflection. It means that the bridge is more rigid than it was designed.
5. The 15mm camber that occurs on the bridge girder due to its own weight without live load is caused by a temperature difference of 15°C on the lower and upper surfaces of the arc element. Because the four pedestals are in the form of pin connection, the bridge can only expand upwards and cause the girder to rise during the day and back down at night.
6. Strain and deflection values due to truck load test are smaller than required in AASHTO so that the bridge structure still behaves elasticly where the structure conditions return to normal (no residual structure response occurs) after un-loading.
7. The natural frequency of the WFC bridge is 1,127 Hz which occurs in the vertical direction at a span of 100 meters. However, with a middle span arch that reaches a height of 15 meters, the natural frequency is most likely to occur in this curve in the lateral direction which cannot be observed during testing.
8. The results of load test concludes that the Water Front City bridge is still elastic for 70% of designed load and safe to operate.

References

[1] Brownjohn J M, Moyo P, Omenzetter P and Lu Y 2003 Assessment of Highway Bridge Upgrading by Dynamic Testing and Finite-Element Model Updating *Journal of bridge engineering* pp 162-172

[2] Caglayan B O, Ozkagul K and Tezer O 2012 Assessment of a concrete arch bridge using static and dynamic load tests. *Structural Engineering and Mechanics*, **41** pp 83-93

[3] Nguyen V H, Schommer S, Maas S and Zurbes A 2015 *Static load testing with temperature compensation for structural health monitoring of bridges*. pp 1-30

[4] Riyono W A and Sukmara G 2014 Evaluasi Kapasitas Jembatan Sementara Tipe Modular Baja Dengan Metode Uji Pembebanan *Jurnal Jalan- Jembatan.* **31**(2) pp 111-123

[5] Setiati N R and Surviyanto A 2013 Analisis Uji Beban KendaraanTerhadap Jembatan Integral Penuh (Loading Test Analysis Of Full Integral Bridge) *Jurnal Teknik Sipil* pp 190-204.

[6] Wijayanto A, Nasution A and Zarkasi I 2017 Evaluasi Integritas Sistem Struktur Jembatan Dr. Ir. Soekarno. *Jurnal Teoretis dan Terapan Bidang Rekayasa Sipil*, pp 125 - 137

IOP Conf. Series: Materials Science and Engineering **650** (2019) 012036 doi:10.1088/1757-899X/650/1/012036

Shear capacity of fiber-reinforced concrete beams without transverse reinforcement

Wong Widjaja[1*], Daniel Christianto[2], Dennis Kurniadi[2], Lucyana Monica[2], Yuliana[2], Michelle[2]

[1]WME (South East Asia) Pte. Ltd
[2]Department of Civil Engineering, Tarumanagara University, Jakarta, Indonesia

*w.wong@wmeglobal.com

Abstract. This study studied the effect of using fiber on the shear strength of concrete beams without transverse reinforcement. The concrete mixture does not use coarse aggregate and uses stainless steel fiber with a diameter of 0.2 mm and a length of 36 mm. This study explains the effect of concrete quality on shear strength that can be carried by concrete beams with a ratio of 45 / 10.5, 25 / 10.5 and 20 / 10.5. The making of test specimens in this study was a beam of varying lengths of 110 cm, 70 cm and 60 cm while a width of 7 cm, and a height of 12.5 cm. From the test results, the shear stress value is calculated using the formula v / (b x d x √ (f'c)) > 2 in international units, in which the shear capacity value is bigger than the code states. Diagonal tension failure occurs on 110cm beam and shear tension failure on 70 cm and 60 cm beams.

1. Introduction

Today, many people research to produce concrete mixtures that have high strength. But with the development of the era, the development of concrete technology is also developing so that it has begun to enter the ultra-high performance-concrete generation. Reactive Powder Concrete (also known as RPC) is a concrete that has very high strength and is composed of cement, silica fume, and quartz powder [1].

One way to improve the quality of concrete as a construction material is to add fiber. The purpose of adding fiber is to increase the tensile strength of concrete. Fiber can also improve the concrete shear strength and can be used in planning in the design of reinforced concrete structures [2]. Fiber concrete is a composite material consisting of plain concrete and other materials in the form of fiber. Based on the material, fiber can be divided into four groups, namely steel fibers, glass fibers, artificial fibers, and natural fibers, the amount of fiber added to the concrete mix depending on the type of fiber used and the results to be achieved [3].

Adding fiber into the mixture will reduce workability quickly in line with the increase in fiber concentration and fiber aspect ratio [4]. So to get optimal results there are two things that must be considered carefully. The first thing is fiber aspect ratio, which is the ratio between fiber length (l) and fiber diameter (d). The second thing is fiber volume fraction (Vf), which is the percentage of fiber volume added to each unit of concrete volume [5].

In the ASCE Committee 426 report, the shear strength of concrete with or without reinforcement is the same, which is the value of the shear force that causes tilting cracks. In this case the shear reinforcement is considered to only withstand the excess shear force that can be held by the concrete without reinforcement. Crack in beams as shown in Figure 1 and cross section of a beam shown in Figure 2.

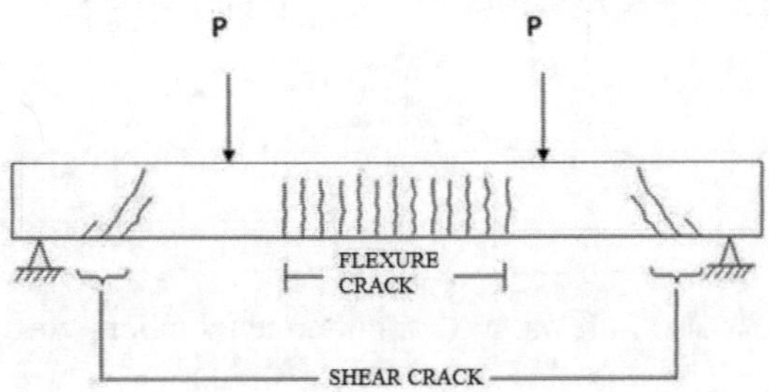

Figure 1. Cracks in beams

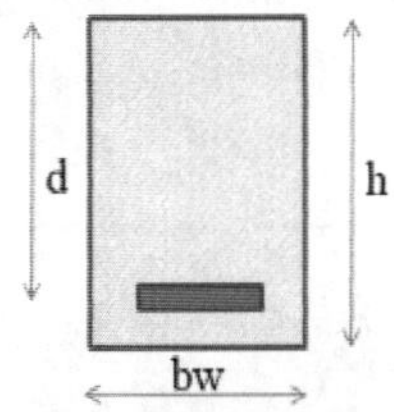

Figure 2. Cross section of a beam

We find that when comparing the relative strength of each beam (ru = Mu / Mfl), the transition point is a / d = 2.5 - 3 where the beam is critical to shear (the lowest bending moment in failure). Mu are the ultimate moment due to the shear force and Mfl is the ultimate bending moment. In this study examined the shear strength of concrete beams with ratios of a / d = 4.2857, 2.3809, and 1.9047 without transverse reinforcement where a is a shear span whereas d is the distance of the fiber pressed into the center of the reinforcing steel or the effective height of the beam with shape square cross section which can be seen in Figure 3. For a / d ratio < 2.5 it is known that it will be more resistant to shear forces that occur due to curved action (arch action). However, if the a / d ratio > 2.5 - 3 then the failure occurs simultaneously with the shear crack that appears [6].

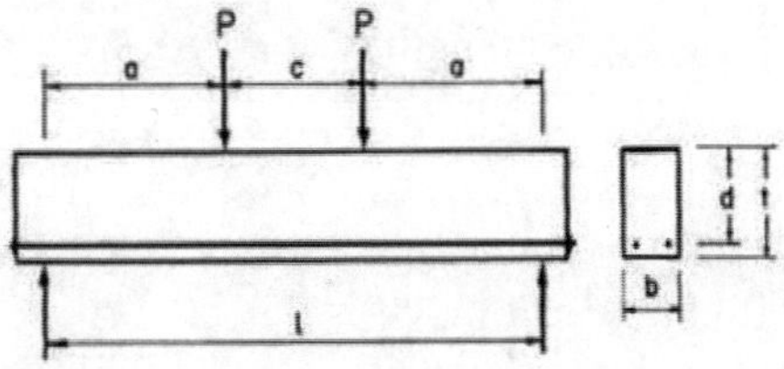

Figure 3. a/d ratio model (Source: Kani, 1967 [6])

Kani found out at the same time that the beam width did not have an important effect on shear strength [6]. He also concluded that shear stress decreases with increasing beam depth as seen in Figure 4.

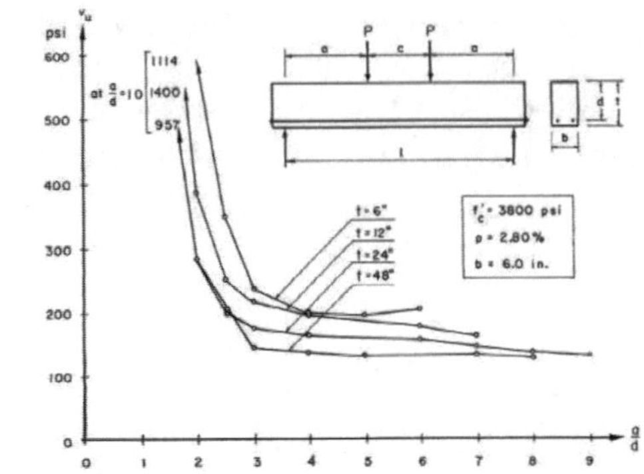

Figure 4. Shear stress at failure vs. a/d for concrete members with variable depth (Source: Kani, 1967 [6])

This study will compare the value of the shear force of the concrete beam with the ratio of a / d is different. This is done to evaluate the pattern of changes in shear force that occur in concrete and incorporate the results of the research obtained into the Derivation of Design Equation graph as shown in Figure 5.

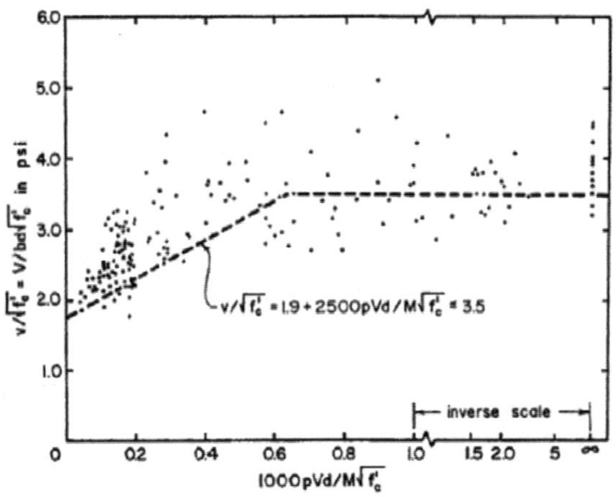

Figure 5. Derivation of design equation

The equation used to calculate the shear strength of concrete is:

$$Vc = \frac{1}{6}\sqrt{f'c}\, bw\, d \tag{1}$$

Where V_c is the shear capacity of normal concrete, then f'c is the compression stress of the concrete, b_w and d are stated previously in Figure 2.

2. Basic Theory

2.1. Concrete

Concrete is a mixture of portland cement, fine aggregate, coarse aggregate and water, with or without additional ingredients that form a solid period (SNI 03-2847-2002, Article 3.12). The main property of concrete is that it has a very high compressive load capacity, but not strong enough to withstand tensile stress. Because concrete is a composite, the quality of concrete is very dependent on the quality of each forming material.

IOP Conf. Series: Materials Science and Engineering **650** (2019) 012036 doi:10.1088/1757-899X/650/1/012036

2.2. Fiber-Reinforced Concrete

Fiber-reinforced concrete is a mixture of concrete added fiber into it evenly, both in the form of long fibers and short fibers. The basic idea of adding fiber is to reinforce concrete with steel fibers that are spread evenly into a concrete mixture with a random orientation. So that concrete does not experience cracks that are too early due to loading or heat hydration [7]. The main purpose of adding fiber to the concrete is to increase the tensile strength of the concrete. It should be noted that giving fiber does not add much to the compressive strength of the concrete, but only adds to the ductility of the concrete.

Adding fiber to the concrete mix can provide an advantage in improving some concrete properties [8], ie:

- Ductility (ductility), which is related to the ability of a material to absorb energy (energy absorption),
- Resistance to shock loads (impact resistance),
- The ability to withstand tensile and bending moments,
- Resilience to meltdown (fatique life),
- Resistance to the effect of shrinkage,
- Resistance to abrasion, fragmentation and spalling.

2.3. Reactive Powder Concrete

Reactive Powder Concrete (RPC) is a concrete that has very high strength and is composed of cement, silica fume, and quartz power [1]. The main principle in developing RPC according to Richard P. & Cheyrezy M., 1995 [9], is:

- Use of pozzolanic material in the form of silica fume.
- Increasing ductility by adding steel fibers to the concrete mix.
- Giving pressure to increase stir density.
- Perform treatment by heating method (steam curing) to achieve high strength.
- Use of superplasticizer from a type of polymer to improve performance.
- Increase homogeneity by removing coarse aggregates.
- Increase the density of the matrix by optimizing the aggregate grain size.

2.4. Shear Strength of Concrete

Beams are structural elements that support the external load and the weight itself, which results in a bending moment (M) and a shear force (V), as shown in Figure 6.

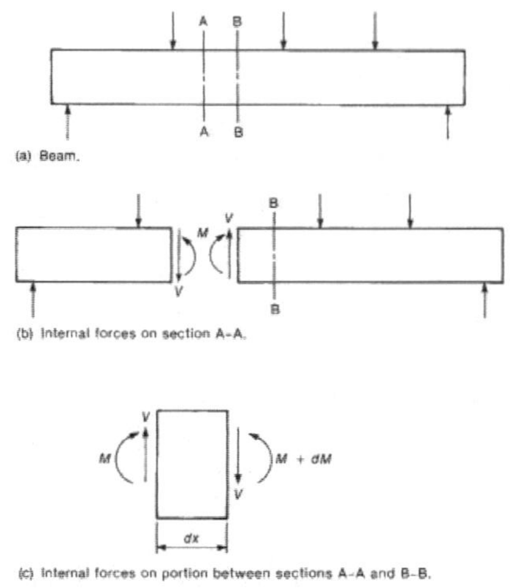

Figure 6. Internal forces on beam (Source: MacGregor and Wight, 2016 [3])

In the design of reinforced concrete structures, bending is always the first consideration to determine the moment of conditional resistance that can be used to ensure that a collapse can give a warning to the occupants. At the same time, the beam also withstands shear forces due to bending. The characteristic of a failure due to shear force on a reinforced concrete structural element is brittle, not ductile, and its collapse occurs suddenly without warning. This is because the strength of the reinforced concrete structure shear is mainly dependent on tensile strength and tension strength of concrete. Because the tensile strength of concrete is smaller than its compressive strength, the design of shear is very important in concrete structures. For reinforced concrete structural components, if the shear force that works is so large that it is beyond the ability of the concrete to hold it, additional reinforcement is needed to hold the shear.

In reinforced concrete beams stretching the elongated direction, steel reinforcements completely withstand the tensile force arising from bending. Meanwhile, if the working load continues to increase, tensile and shear stresses will also increase along with the load. While the steel reinforcement that is intended to withstand the pull in the beam is not located at the place where the diagonal tensile stress occurs. For this reason, stirrup reinforcement is needed to hold the diagonal tension in places needed.

Thus shear reinforcement has four functions [10] as follows:

- Bear a part of the planned external force (Vu),
- Prevents diagonal crack propagation so that it is not continuous to the tension section of concrete,
- Holding and tying the longitudinal reinforcement to its position so that the longitudinal reinforcement has a good capacity to carry flexure,
- Provide a kind of bond to the concrete area that is depressed if the stirrup is a closed stirrup.

The shear strength of the beam is influenced by the effective height factor (size effect). One of the researchers who conducted a study of this size effect was Kani. Kani conducted research at the University of Toronto, comparing four blocks without shear reinforcement with heights of 6, 12, 24, and 48 inch (153, 305, 610, and 1220 mm). The beam has concrete quality (f'c) = 3800 (26.2 MPa), flexural reinforcement ratio $\rho = 2.8\%$, and the ratio of a / d beams vary from 1 to 10.

We find that when comparing the relative strength of each beam, ru = Mu / Mfl (where Mu is the moment that causes the shear force and Mfl is the bending moment), that the transition point is a / d = 2.5 - 3 where the beam is critical to shear (bending moment lowest when a failure occurs. For the ratio below this limit, the beam will be able to withstand greater shear force due to arch action. However, if the a / d ratio is above the critical point, then concrete failure occurs along with the sliding crack that occurs.

Mu is the ultimate moment due to the shear force and Mfl is the ultimate bending moment. Then Kani made a graph based on the results of his research and called "valley of diagonal failure". The relationship between ru (relative strength) and the a / d ratio can be seen in Figure 7. Changes in the relative strength of the beam are clearly visible. Along with the effective increase in beam height, its strength will decrease.

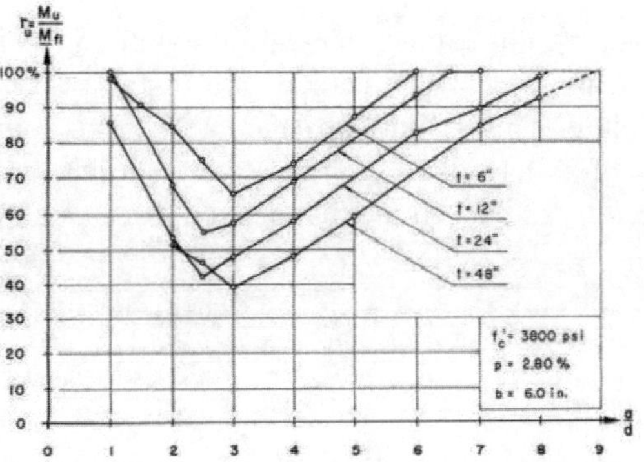

Figure 7. Relationship between the relative strength of beams with a/d ratio (Source: Kani, 1967 [6])

2.5. Types of Failure

There are several types of failures that can occur on a beam without stirrup reinforcement, namely:

1) Diagonal Failure: With increasing shear force the crack will widen and propagate until it penetrates the upper side of the beam until the collapse occurs. The combination of bending and shear forces is a basic cause of diagonal failure [11]. There are several types of diagonal failure that are often encountered, namely diagonal tension failure, shear tension failure, and shear compression failure:
    - Diagonal tension failure (Figure 8) occurs in beams with a / d ratios between 2.5 - 6. The crack that occurs spreads through the beam until it reaches the pressure area. When the beam experiences a critical point, the beam will fail due to the failure of the depressed concrete. This crack often occurs without warning and suddenly [11].

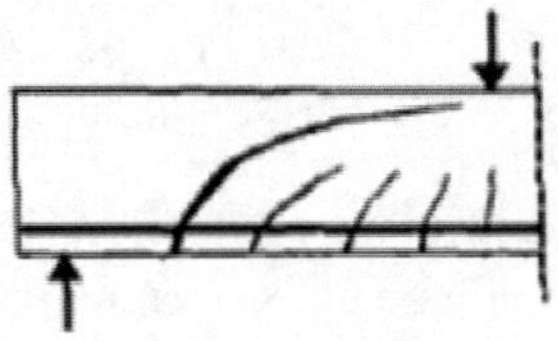

Figure 8. Diagonal tension failure (Source: Ziarra, 1993 [11]).

    - Shear tension failure is almost similar to the diagonal tension failure, but occurs generally in beams that are relatively short. Sliding cracks appear to spread on the beam but have not caused collapse on the beam. Further cracks occur around the longitudinal reinforcement of the last flexural crack and can cause loss of bond between longitudinal and concrete reinforcement (Figure 9). When the beam reaches a critical point, the beam will collapse splitting the concrete [11].

IOP Conf. Series: Materials Science and Engineering **650** (2019) 012036 doi:10.1088/1757-899X/650/1/012036

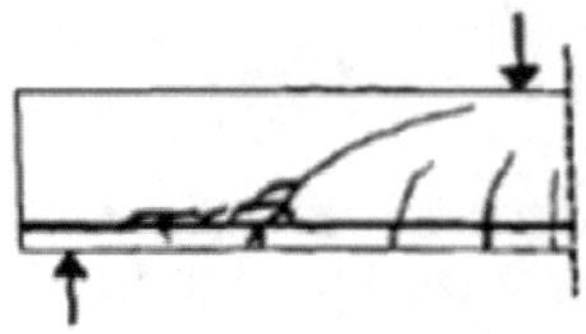

Figure 9. Shear tension failure (Source: Ziarra, 1993 [11])

- Shear Compression Failure spreads through the beam causing collapse when it reaches the pressure zone without any signs of a second crack as in the shear tension failure. This failure is called the shear compression failure (Figure 10). This type of failure occurs on relatively short beams. The ultimate load when a failure occurs can exceed the diagonal crack caused by arch action.

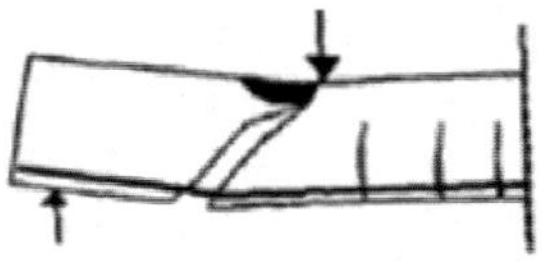

Figure 10. Shear compression failure
(Source: Ziarra, 1993 [11])

2) Flexural Failure: Flexural cracks occur due to moments and on long beams. Because of this, cracks get bigger at the point where the maximum moment value. Cracks are increasingly spread when the shear force on the concrete reaches the maximum tensile stress. Cracks that occur almost perpendicular and cause failure of the beam can be seen in Fig. 11 due to the two cases below [11]:
   - Under-reinforced beams: occur when flexural reinforcement has melted causing failure of the concrete press area.
   - Over-reinforced beams: occur when the concrete compressive area fails first before the flexural reinforcement melts and results in the concrete failure itself.

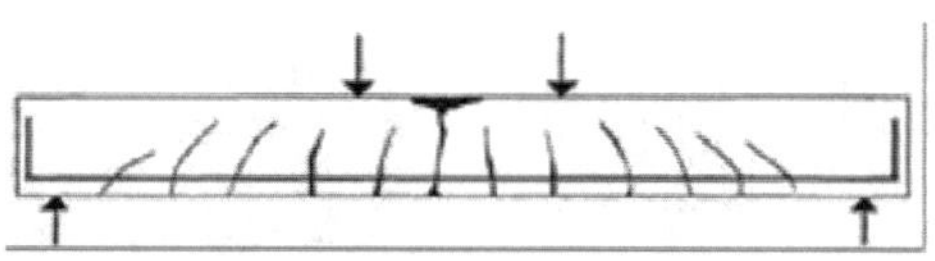

Figure 11. Flexural failure (Source: Ziarra, 1993 [11])

2.6. Derivation of Design Equation

In ACI-ASCE 326 a comparison graph has been made between with the parameter "ρVd / M" √ ("f'c") composed of 194 blocks of sample. In the graph there is also a line with the equation.

$$\frac{V}{bd\sqrt{f'c}} = 1.9 + (2500)\frac{\rho V d}{M\sqrt{f'c}} \qquad (2)$$

Formula (2) calculates the compressive stress in psi. The graph can be seen in Figure 12.

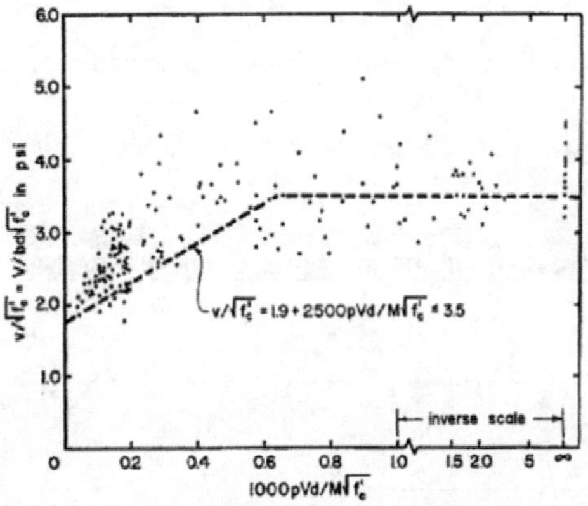

Figure 12. Derivation of design equation

3. Testing Model

Analysis of flexural reinforcement is calculated by planned f'c 90 MPa, the beam section used can be seen in Figure 13.

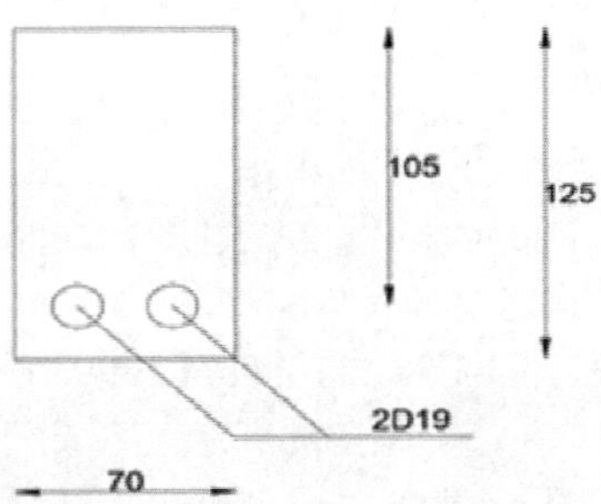

Figure 13. Cross section of beam

For modeling beam specimens, it can be seen in Figure 14 below:

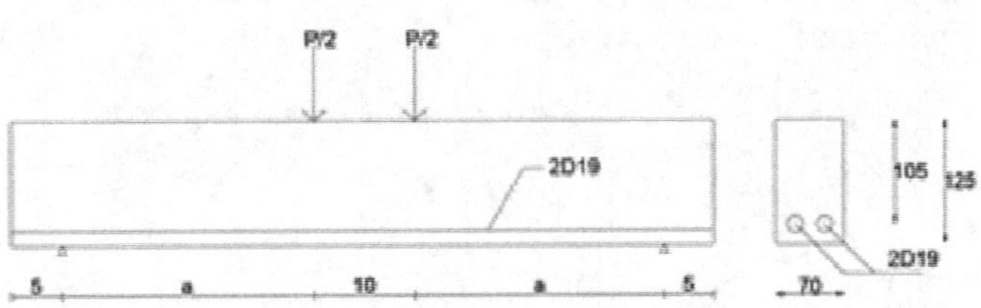

Figure 14. Modeling beam test objects

Test object size:

- Cylinder : Ø 10 cm x 20 cm (for strength test only)
- Beams : with a ratio 45/10,5, 25/10,5 and 20/10,5 Number of specimens: 2 cylinders 2 beams

The cement water factor used was 0.18 / 0.2 / 0.22 times the cement period

4. Results and Study

Based on the results of the research and calculations performed, it can be made a recapitulation of the analysis of the shear strength of the beam and how the effect of the increase in the quality of concrete on the shear strength of the beam as seen in Table 1 for fiberless reinforced concrete and Table 2 for fiber-reinforced concrete

Table 1. Recapitulation of shear strength analysis on non-fiber beam test objects

No.	Beam Length (m)	f'c (N/mm^2)	P (kN)	V_c (kN)	V_u (kN)	M_n (kNm)	M_u (kNm)	Failure
1	1,1	70,4	38,575	102,783	193,925	176,752	87,054	Shear
2	0,7	71,08	59,405	103,278	297,655	177,339	74,348	Shear
3	0,6	71,855	76,66	10,384	383,825	177,996	76,723	Shear

Table 2. Strong analyze recapitulation slide on fiber-reinforced beam test objects

No.	Beam Length (m)	f'c (N/mm^2)	P (kN)	V_c (kN)	V_u (kN)	M_n (kNm)	M_u (kNm)	Failure
1	1,1	92,485	48,77	117,807	24,49	19,417	109,992	Shear
2	1,1	69,1	39,79	10,183	20	175,597	89,787	Shear
3	1,1	67,945	39,245	109,752	197,275	174,533	88,561	Shear
4	0,7	90,865	69,71	116,771	34,918	190,583	87,229	Shear
5	0,7	85,72	59,815	113,417	299,705	187,728	74,861	Shear
6	0,7	76,98	46,58	107,479	23,353	182,001	58,317	Shear
7	0,6	83,605	93,645	112,009	46,875	186,542	93,708	Shear
8	0,6	79,64	68,305	109,321	34,205	183,877	68,368	Shear
9	0,6	63,535	43,88	97,643	219,925	170,116	43,943	Shear

Based on the data and test results obtained, calculate the values and then plotted into the derivation of design equation graph which can be seen in Figure 15 below.

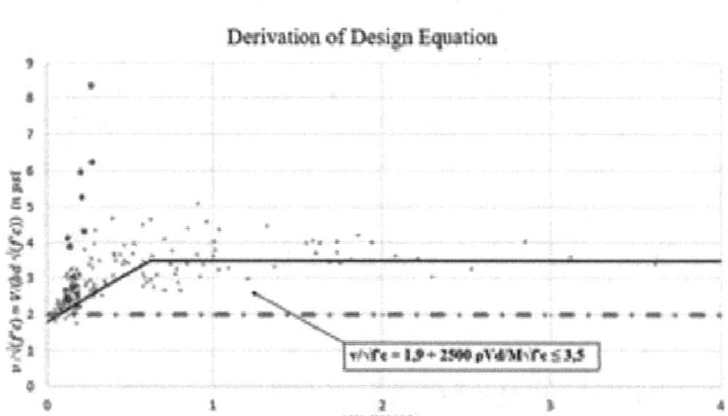

Figure 15. Derivation of design equation graph

IOP Conf. Series: Materials Science and Engineering **650** (2019) 012036 doi:10.1088/1757-899X/650/1/012036

From the results of the calculation, the shear strength graph against the a / d ratio based on changes in concrete quality can be seen in Figure 16.

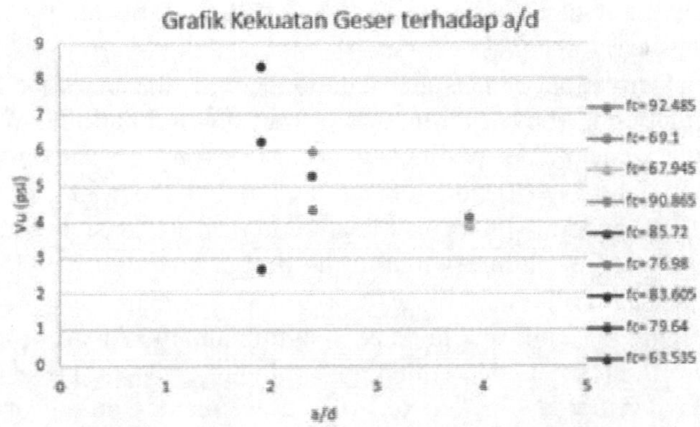

Figure 16. Graph of shear strength against a/d ratio

From Figure 16, we take the data which compression stress is at maximum from each type of the beams, then with comparing the formula (1) and (2) that has been calculated and plotted the values given in Figure 17.

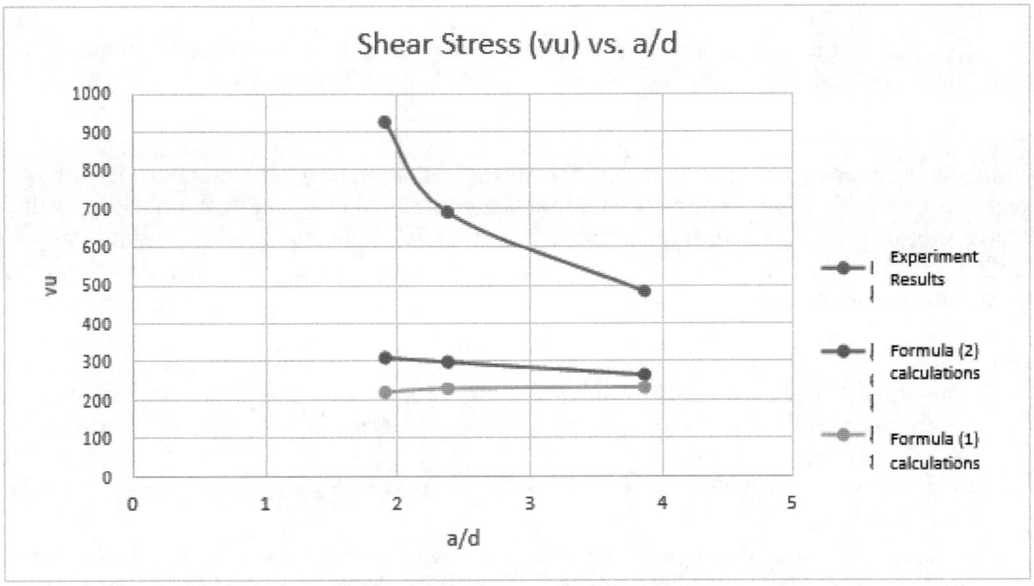

Figure 17. Graph of shear stress (vu) against a/d ratio

From the collected data, we can analyze these results as follows:

- Based on the graph of the shear strength of the a / d ratio, it can be seen that the pattern of changes in vu value due to changes in d resembles the results of research conducted by Kani. In addition, it can also be concluded that the smaller the a / d ratio will be more resistant to the shear force that occurs.
- Based on the Derivation of Design Equation graph, it can be seen that the point obtained from the calculation is above the minimum limit of > 2 and the greater the value of a / d the quality improvement does not have a large effect on the shear stress value.
- Concrete does not have flexible cracks and only shear cracks, can be seen from the calculation results where Mn <Mu. Shear failure in the design beam a / d = 45 / 10.5 is a type of diagonal tension failure while in the design beam a / d = 25 / 10.5 and 20 / 10.5 is a type of shear tension failure.
- Comparing Figure 15 and 16, there is a significant increase in shear capacity of concrete. This is because the formula that is being used in the code does resemble a regression from the data that they gathered. Based on the experiment, the shear capacity does increase significantly at one point of the shear stress beam.

5. Conclusion

Based on the results of laboratory testing and data analysis, conclusions are obtained, among others:

- Smaller the a/d ratio will be more resistant to the shear force that occurs.
- The greater the value of a/d, the quality improvement does not have a large effect on the shear stress value.
- Comparing the experiment and from the code, there is a significant increase in shear capacity of concrete.

Acknowledgements

I would like to express my deep gratitude for all the experts that have mentored me in building this project from the very beginning for the thorough guiding, patience, and support. Special thanks are also given to all parties who have contributed their time and energy in perfecting this experimental research work. I would also like to extend my thanks to the lab technicians who helped a lot so that the experiment can run smoothly.

References

[1] Kushartomo Widodo et.al 2013 *Pengaruh Penambahan Quartz Powder pada Reactive Powder Concrete terhadap Terbentuknya Kalsium - Silikat – Hidrat* (Jakarta:Universitas Tarumanagara)

[2] Christianto Daniel et.al 2019 *Effect of steel fiber on the shear strength of reactive powder concrete* TICATE Proceedings

[3] MacGregor J G and J K Wight 2016 *Reinforced Concrete Mechanics and Design* Seventh Edition.

[4] Christianto Daniel et.al 2014 *Pengaruh Volume Serat Lokal Terhadap Kuat Geser Beton Mortar* (Jakarta: Universitas Tarumanagara)

[5] Suhendro 1990 *Beton Fiber Lokal Konsep, Aplikasi, dan Permasalahannya* (Yogyakarta:PAU Ilmu Teknik, Universitas Gajah Mada)

[6] Kani G 1967 How Safe are Our Large Reinforced Concrete Beams *ACI Journal* **64** (3) 128- 141.

[7] Soroushian P and Bayasi Z 1987 Concept of Fiber Reinforced Concrete *Proceeding of The International Seminar on Fiber Reinforced Concrete* (Michigan: Michigan State University)

[8] Suhendro 1990 *Beton Fiber Lokal Konsep, Aplikasi, dan Permasalahannya* (Yogyakarta:PAU Ilmu Teknik, Universitas Gajah Mada)

[9] Richard P and Cheyrezy 1995 *Reactive Powder Conretes with High Ductility and 200-800 MPa Compressive Strength* (San Fransisco: ACI Spring Convention, April)

[10] Nawy E 1990 *Beton Bertulang Suatu Pendekatan Dasar* (Bandung: PT. Refika Aditama)
[11] Ziara M 1993 *The influence of confining the compression zone in the design of structural concrete beams* (Heriot-Watt University)

Nomenclatures

b,d	dimensions of beam	mm
bw	effective width of beam section	mm
d	effective height of beam section compression strength of concrete	N/mm^2
M	ultimate bending force	N/mm
Vc	shear strength concrete	N
Vu	ultimate shear force	N
ρ	ratio of reinforcement	

IOP Conf. Series: Materials Science and Engineering **650** (2019) 012037 doi:10.1088/1757-899X/650/1/012037

Experimental Study of Mortar Compressive Strength with Anadara Granosa Powder as a Substitute for Partial Use of Cement

A Sandjaya[1], Tavio[2], dan D Christianto[1]

[1] Bachelor of Civil Engineering Study Program, Universitas Tarumanagara, Tomang, Grogol Petamburan, West Jakarta, Indonesia
[2] Bachelor of Civil Engineering Study Program, Institut Teknologi Sepuluh Nopember, Keputih, Sukolilo, Surabaya, Indonesia

* arifs@ft.untar.ac.id

Abstract. Making cement produces carbon dioxide (CO_2) gas which results in the greenhouse effect. To reduce the use of cement, alternative materials are examined. One of the raw materials for making cement is limestone. Limestone is a sedimentary rock mainly composed of calcium carbonate. Blood clams (Anadara Granosa) are tried to replace cement because they have a high calcium carbonate mineral composition. Blood clams are crushed into powder and filtered using No.325 filter. Cement is replaced by weight by 5%, 10%, 15%, and 20%. Mortar specimens in the form of cubes with a size of 5 cm. Mortar compressive strength test was carried out at 7 and 28 days. Anadara Granosa powder reacts very weakly with water and increases the initial compressive strength of the mortar. The results of Anadara Granosa mortar compressive strength test at 7 days were higher than normal mortar. The most efficient composition is indicated by cement replacement of 20%.

1. Introduction

Cement is an industrial product with limestone as the main ingredient and clay or other substitute materials. The raw material is burned until it melts to form a clinker, which is then crushed and added to a certain amount of gypsum. 7% of the production of carbon dioxide (CO_2) emissions in nature comes from cement production and every 1 ton reduction in cement production results in a reduction of 1 ton of CO_2 gas emissions [1]. Because of this, research was carried out to reduce the use of cement so that cement production needs were reduced.

Egg shells contain high calcium which can be used to replace cement. Eggshell powder is filtered using a 90 μm filter. Eggshell powder can replace cement up to 10% in the concrete mixture, more than that the compressive strength and tensile will decrease. Another additive that can increase the concrete mixture with eggshell is fly ash [2].

Limestone is a sedimentary rock mainly composed of calcium carbonate in the form of calcite minerals. The main source of calcite is marine organisms. The mineral composition of blood clams (Anadara Granosa) of West Coast of Peninsular Malaysia is calcium carbonate (CaC) more than

IOP Conf. Series: Materials Science and Engineering **650** (2019) 012037 doi:10.1088/1757-899X/650/1/012037

98.7% and about 1.3% of composition are comprise of Magnesium (Mg), Sodium (Na), Phosphorus (P), Iron (Fe), Copper (Cu), Nickel (Ni), Beron (B), Zinc (Zn), and Silcon (Si) [3].

Research using Anadara Granosa ash as a substitute for Ordinary Portland Cement (OPC) of 4% by weight in concrete mixture showed that the resulting compressive strength was lower than normal concrete at 28 days, with curing ordinary water or brine for 7, 28, and 91 day [4].

Research using sea shells as a substitute for fine aggregates shows higher results of compressive, tensile and bending tests. The composition of the sea shell mixture is used as a substitute for fine aggregates of 20%, 40%, 60%, and 80%. From the results of the experiments, the optimum size of the fine aggregate substitute is 80%. When fine aggregates are replaced 100%, there is a decrease again [5].

Limestone should not be considered as a totally unreacted material when placed in a high pH environment with portland cement [6].

2. Research methods

Mixed plan based on SNI 03-6825-2002 Testing Methods of Compressive Strength of Portland Cement Mortar for Civil Works [7]. By using the weight of cement as a reference, the ratio of water to cement (W/C) is 0.48 and the ratio of sand to cement is 2.75. Cement used is Ordinary Portland Cement (OPC) type 1 and silica sand.

Two types of mixture are made, first is the weight of water maintained even though the weight of the cement decreases when replaced with Anadara Granosa powder and second is the weight of water adjusted to (W/C) with reduced weight of cement used when replaced with Anadara Granosa powder.

Anadara Granosa is obtained from seafood waste. Anadara Granosa is cleaned, dried (Figure 1), and crushed into powder. Then the powder is filtered using filter No. 325. The powder that passed the No. filter. 325 (<43 μm) was used instead of cement (Figure 2). The weight of cement replaced with Anadara Granosa powder is 5%, 10%, 15%, and 20% which is adjusted to the limits in EN-197-1 CEM II / A-M [8] where the composition of inorganic ingredients is 6-20%. Anadara Granosa does not go through the combustion process so it does not produce CO_2 emissions.

Figure 1. Anadara Granosa which has been cleaned and dried.

The mortar mixture is made into 5 cm cubes. Normal specimens (0%) and cement reduction test specimens as much as 20% each made 6 cubes were used as a comparison. The mixture of the first and second specimens for each composition was 6 cubes. The total specimens were 60 cubes. Test specimens were cured for 7 and 28 days.

Compressive strength test was carried out using ADR 3000 ELE machine with a maximum compressive capacity of 3000 KN made in England.

Figure 2. Anadara Granosa powder that passes the No. 325 filter.

3. Exprimental results

The results of compressive strength of normal test specimens are shown in Table 1. The average compressive strength for 0% at 7 days is 71.6 kN or 28.64 MPa and at 28 days is 108.7 kN or 43.48 MPa. The average compressive strength for (-20%) at 7 days is 58.1 kN or 23.24 MPa and at 28 days is 80.5 kN or 32.2 MPa.

Table 1. Compressive Strength Test Results for Test Specimens Normal.

0%				-20%			
7 days		28 days		7 days		28 days	
Weight (gram)	**Force (kN)**	Weight (gram)	**Force (kN)**	Weight (gram)	**Force (kN)**	Weight (gram)	**Force (kN)**
263,3	69,8	265,4	108,2	243,6	60,2	263,3	80,7
260,6	68,3	265,5	112,3	242,8	55,8	266,5	80,4
270,1	76,8	257,4	105,7	245,0	58,4	268,3	80,5

The compressive strength of the test specimen for mixture 1 is shown in Tables 2 and Table 3. The average compressive strength at 7 days is 47.4 kN or 18.96 MPa for 5%, 42.3 kN or 16.92 MPa for 10 %, 38.4 kN or 15.36 MPa for 15%, and 33.8 kN or 13.52 MPa for 20%. The average compressive strength at 28 days is 83.8 kN or 33.52 MPa for 5%, 76.9 kN or 30.76 MPa for 10%, 73.2 kN or 29.28 MPa for 15%, and 70.2 kN or 28.08 MPa for 20%.

Table 2. Test Results for Compressive Strength Test for Mixture 1 at 7 Days.

5%		10%		15%		20%	
Weight (gram)	**Force (kN)**	Weight (gram)	**Force (kN)**	Weight (gram)	**Force (kN)**	Weight (gram)	**Force (kN)**
252,5	47,6	257,3	40,3	252,5	37,8	256,3	35,2
251,4	47,8	253,2	41,5	254,1	38,8	252,8	30,8
252,4	46,7	255,1	45,2	252,0	38,5	250,5	35,4

Table 3. Test Results for Compressive Strength Test for Mixture 1 at 28 Days.

5%		10%		15%		20%	
Weight (gram)	**Force (kN)**	Weight (gram)	**Force (kN)**	Weight (gram)	**Force (kN)**	Weight (gram)	**Force (kN)**
254,9	84,4	255,6	74,7	259,4	70,6	253,6	69,7
256,6	85,3	251,7	77,3	251,4	74,5	256,5	70,4
255,6	81,8	250,1	78,8	254,4	74,4	253,8	70,5

The compressive strength of the test specimens for mixture 2 is shown in Tables 4 and Table 5. The average compressive strength at 7 days is 69.1 kN or 27.64 MPa for 5%, 67.8 kN or 27.12 MPa for 10 %, 66.9 kN or 26.76 MPa for 15%, and 64.0 kN or 25.6 MPa for 20%. The average compressive strength at 28 days is 93.9 kN or 37.56 MPa for 5%, 88.7 kN or 35.48 MPa for 10%, 85.5 kN or 34.2 MPa for 15%, and 82.6 kN or 33.04 MPa for 20%.

Table 4. Test Results for Compressive Strength Test for Mixture 2 at 7 Days.

5%		10%		15%		20%	
Weight (gram)	**Force (kN)**	Weight (gram)	**Force (kN)**	Weight (gram)	**Force (kN)**	Weight (gram)	**Force (kN)**
245,2	70,8	271,5	68,9	248,3	65,6	244,5	64,2
244,1	68,7	270,5	66,8	244,9	67,9	247,1	65,5
240,2	67,8	272,0	67,7	242,9	66,9	242,0	62,4

Table 5. Test Results for Compressive Strength Test for Mixture 2 at 28 Days.

5%		10%		15%		20%	
Weight (gram)	**Force (kN)**	Weight (gram)	**Force (kN)**	Weight (gram)	**Force (kN)**	Weight (gram)	**Force (kN)**
274,9	94,4	263,0	89,1	265,9	86,0	264,7	81,8
266,6	95,3	265,0	90,0	271,4	85,5	264,2	84,5
266,5	91,8	256,7	87,0	267,4	84,9	259,7	81,5

4. Discussion

The results of the compressive strength of normal specimens at 7 days were higher than those of mixture 1, but lower than the mixture 2. While the results of compressive strength at 28 days, only mixture 2 with a composition of 20% higher. The compressive strength test results from the test specimens for mixture 1 were lower than the mixture 2. Comparison can be seen in Figure 3 for the test at 7 days and Figure 4 for the test at 28 days.

Anadara Granosa powder will change color to gray if soaked in water for a long time.

IOP Conf. Series: Materials Science and Engineering **650** (2019) 012037 doi:10.1088/1757-899X/650/1/012037

5. Conclusion

In this study, the results of the compressive strength test of mixtures 1 and 2 showed that Anadara Granosa powder without burning did not react well with water. The results of the compressive strength tests of normal and mixed 2, Anadara Granosa powder were able to replace cement and be efficient at a composition of 20%. Anadara Granosa powder also increases the initial compressive strength of the mortar.

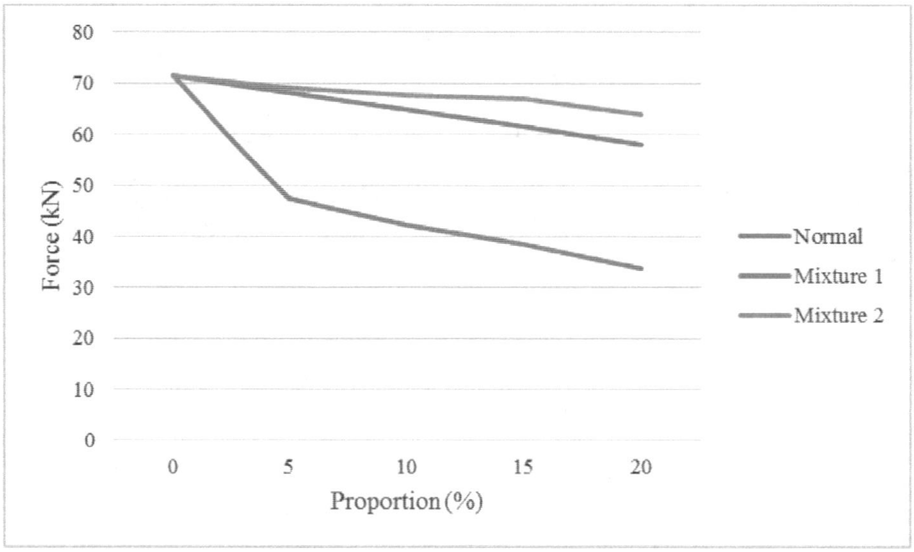

Figure 3. Graph Test Results Compressive Strength at 7 Days.

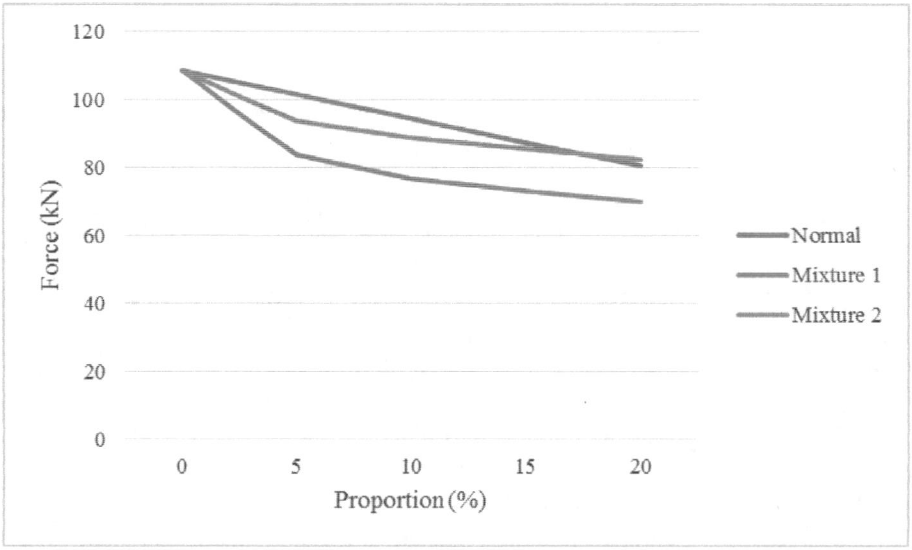

Figure 4. Graph Test Results Compressive Strength at 28 Days.

IOP Conf. Series: Materials Science and Engineering **650** (2019) 012037 doi:10.1088/1757-899X/650/1/012037

Reference

[1] Malhotra V M 1999 Making Concrete Greener with Fly Ash *Concrete International* **21 (5)** pp 61-66

[2] Yerramala A 2014 Properties of Concrete with Eggshell Powder as Cement Replacement *The Indian Concrete Journa*l October pp 94-102

[3] Awang-Hazmi A J, Zuki A B Z, Noordin M M, Jalila A, dan Norimah Y 2007 Mineral Composition of the Cockle (Anadara Granosa) Shells of West Coast of Peninsular Malaysia and It's Potensial as Biomaterial for Use in Bone Repair *Medwell Journals Journal of Animal and Veterinary Advances* **6 (5)** pp 591-594

[4] Tarisa E, Olivia M, dan Kamaldi A 2016 Durabilitas Beton Bubuk Kulit Kerang Di Lingkungan Air Laut Jom *FTEKNIK* **3 (2)** pp 1-6

[5] Mohanalakshmi M, Indhu S, Hema P, dan Prabba V C 2017 Developing Concrete Using Sea Shell as a Fine Aggregate *International Journal for Innovative Research in Science and Technology* **3** Issue 10 pp 282-286

[6] Paris J M dan Ferraro C C 2018 Evaluation of Particle Effects in Portland Cement Systems *Journal of Materials in Civil Engineering* **30** Issue 9 September

[7] Badan Standarisasi Nasional 2002 *Metode Pengujian Kekuatan Tekan Mortar Semen Portland untuk Pekerjaan Sipil SNI 03-6825-2002* (Indonesia)

[8] European Standard 2000 *Cement – Part 1: Composition, specification and conformity criteria for common cements EN 197-1* (Europe)

Rapid Visual Screening (RVS) for School Buildings after Earthquake in Lombok, West Nusa Tenggara, Indonesia

A Pujianto[1], H Prayuda[1], S A P Rosyidi[1], F Monika[1] and R Faizah[1]

[1]Department of Civil Engineering, Faculty of Engineering, Universitas Muhammadiyah Yogyakarta, Bantul, Daerah Istimewa Yogyakarta, Indonesia

* pujiantoasat@umy.ac.id ; hakasprayuda@umy.ac.id ; atmaja_sri@umy.ac.id ; fanny.monika.2007@ft.umy.ac.id ; restu.faizah@umy.ac.id

Abstract. This study discusses the results of the rapid assessment of school buildings in Lombok after an earthquake occurs. The earthquake in July 2018 with a high enough intensity caused a lot of damage to the building. The earthquake including the school building affected the entire area of Lombok Island. This study assessed 15 school buildings on Lombok Island. School samples were taken randomly from several districts that were rocked by the earthquake. The results showed that school buildings in all locations were part of a Non-Engineered Building where there were no results of analysis of earthquake resistant buildings. Overall, the condition of the building is categorized as being moderately damaged with non-structural components such as walls not having good connections with columns and beams. In addition, the structure and attributes of the roof are also damaged by a high percentage. In some buildings, it still does not follow good structural rules so that it is very dangerous for building users in the event of a large earthquake. Improvements using the retrofitting method are one of the most recommended ways to improve the performance of buildings that have mild and moderate damage.

Keywords—School Assessment, Earthquake, Rapid Visual.

1. Introduction

Geographically, Indonesia is in a very unstable location, namely in the ring of fire, a zone surrounded by very active plates. Indonesia is surrounded by the tectonic plates of Eurasia, Australia, and the Pacific, which are possible for earthquakes with higher intensity [1]. Earthquakes are natural phenomena that often occur in the world, which can cause a lot of loss of life [2].

After the earthquake, it is necessary to check the existing buildings to ensure the condition of the building is still safe or not to be used. The evaluation carried out depends on the method used so that the evaluation results will be concluded whether a building can be used, retrofitting must be done to increase the capacity to modify in order to reduce the danger for the future. In addition, an examination of the possibility of liquefaction is also carried out [3-5].

On 29 July 2018, there were reported earthquakes in the Lombok island region of Indonesia. The earthquake is reported to have a magnitude of 6.4 with vibrations felt up to several other regions in Indonesia. Many of the impacts resulting from the earthquake. It was reported that a lot of damage to homes, schools, health units, and public infrastructure occurred, so rapid evaluation was needed to be able to restore economic activity and conditions in the location immediately.

In this study, a visual assessment of school buildings in the area of Lombok Island will be conducted to check the structure and non-structural conditions. The structure must be safe from various types of disasters, so it does not affect operations and does not cause death to the users of the building [6]. Many cases reported damage to school buildings caused fatalities in various parts of the world due to natural disasters, some examples such as Pakistan [7], Philippines [8, 9], and Venezuela [10].

Many methods can be used to conduct rapid visual screening after an earthquake occurs. The most commonly used standards are FEMA 154 and FEMA 310 [11, 12] where FEMA 310 is more suitable for developing countries [13]. In this study, we will use a standard commonly used in Indonesia [14] given that the geographical conditions and conditions of most school buildings in Indonesia are part of the Non-Engineered Building. In this method, assessing the three main components of the building are vertical components, horizontal components, and foundation components.

In this study, an evaluation of 15 school buildings in Lombok Island, all schools evaluated were Muhammadiyah schools, which were suspected of being damaged by the earthquake. The school location shown in this study is a location that has been selected from several schools that have been surveyed as a whole. Through this research, it is expected to determine the condition of the building. In this case, the damage to the building is categorized into four, which are not damaged, lightly damaged, moderately damaged and severely damaged. Expected that through this research, it can be the basis of the method of proper repairs to buildings that have been damaged and become a consideration if the building will be used continuously after assessment.

2. Rapid Visual Screening

Rapid visual screening is one method for Evaluating Buildings Visually by Filling In The Forms That Have Been Made Before. The Questionnaire Contains Primary Data From The Buildings Reviewed. Vulnerability Assessment Of Buildings Using RVS Sheets, In General, Has Been Set In The Standard [11]. In General, A Number Of Things That Need To Be Considered According To FEMA 154 In Analyzing Building Vulnerabilities Include Location Seismicity, Where Spectral Acceleration Values Are Needed. In Addition, Data Is Also Needed On The Use Of The Building And Its User Population. The Things That Must Be Assessed Are Soil Types, Dangerous Non-Structural Elements That Collapse, Type Or Type Of Building, Number Of Floors, Vertical Irregularities, And Plan Irregularities And The Year The Building Was Built.

In This Study, Evaluations Were Carried Out Specifically For One Or Two-Story Simple Brick Buildings Due To The Earthquake Referring To The Standards As Outlined By The World Seismic Safety Initiative [14]. While The Level Of Damage To Each Component Consists Of:

a. There Is No Significant Damage.
b. Lightly Damaged, Damage To Structural Components Only Requires Repair Of The Cover And Non- Structural Components Need To Be Repaired.
c. Moderately Damaged, Where Structural Damage can still be repaired.
d. Severely damaged, the damage is too extensive if repaired will have a lot to be replaced or need to be rebuilt.

The evaluation carried out on the building includes vertical components, namely columns and walls, horizontal components namely beams and plates and also foundations. The level of damage to each component starts from 0 to 100 where the greater the value obtained shows the building will be more secure. Some things that must be considered when evaluating the structure and non-structural components include [11]:

a. Hair cracks on concrete that are less than 0.2 mm or invisible cracks indicate no significant damage.
b. Cracks on concrete components up to 2 mm wide are not considered dangerous and can be categorized as minor damage.
c. Cracks on structural components up to 5 mm wide indicate moderate damage.
d. Cracks on concrete components with a width greater than 5 mm indicate heavy damage and experience a significant reduction in strength.

e. Cracks or collapsing partitions do not indicate damage to the structure which is quite heavy.

3. Research Methodology

In examining the condition of the building, it uses the direct field survey method and records important data regarding the condition of the building being inspected. Before the inspection is carried out, it must be ensured that the building is safe for inspection and the observer team understands the evacuation route that must be followed if aftershocks occur. Some of the steps that must be followed in examining post-earthquake buildings are the following.

a. Prepare all equipment such as cameras, fill forms and GPS for the data collection process and determine the location of the building to be checked.
b. Make a location plan and floor plan for each floor. In the process of filling out this form, it is done for each floor, which then results are averaged each time.
c. Checked components consist of three parts, namely the foundation components, vertical components, and horizontal components. The foundation component consists of river stone foundation, beam foundation, and concrete tread foundation. While the vertical component consists of columns, column beam joints, diagonal cracked walls, cracked walls in beam or column joints, collapsed walls and partition walls. Whereas in horizontal components, checks are carried out only on beams, plates and roof truss.
d. In filling out the inspection form, the level of damage each is separated into 4 parts, which are not damaged, lightly damaged, moderately damaged and severely damaged. How to determine the level of damage has been determined in the standard that is used according to the width of the crack formed after the earthquake.
e. After completion of the inspection, the building is given a sign of whether it is suitable for use or must be repaired first or must be replaced with components that are not feasible with the new component.

4. Result And Discussion

The earthquake in Lombok began on July 29, 2018, and other major earthquakes still felt until August 19, 2018. Figure 1 shows the number and location of the earthquake occurring from July 29 to August 19. The big earthquake until 6.9 SR was at least felt to Bali Island and some areas in East Java. Because of this earthquake, there was a lot of damage to infrastructure and residential buildings to paralyze the economy of certain locations.

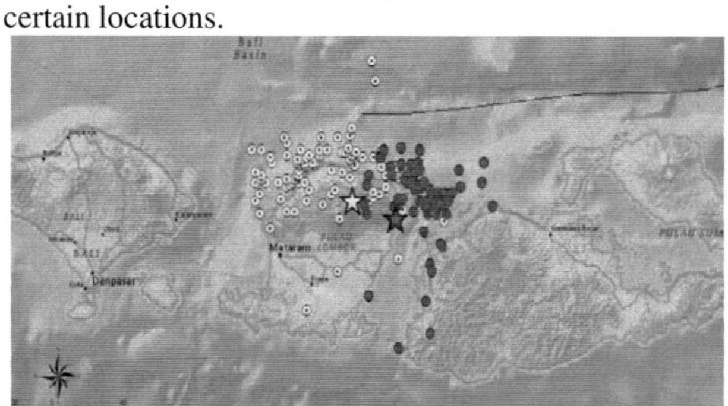

Figure 1. Number of points of the earthquake occurring from July 29 to August 19, 2018, on Lombok Island [15]

In Figure 2 it is Shake-map from the earthquake on 29 July 2018. Based on the shake-map it states that the areas of North Lombok and East Lombok feel shocks of SIG III-IV (VI-VII MMI). While the area of West Lombok, Mataram, Central Lombok, West Sumbawa, and Sumbawa Besar feel the shock with the intensity scale of the SIG-II (IV MMI). At the locations of Denpasar, Kuta, Nusa Dua,

Karangasem, Singaraja, and Gianyar, they felt a shock with the intensity scale of the SIG-II (MMI III-IV).

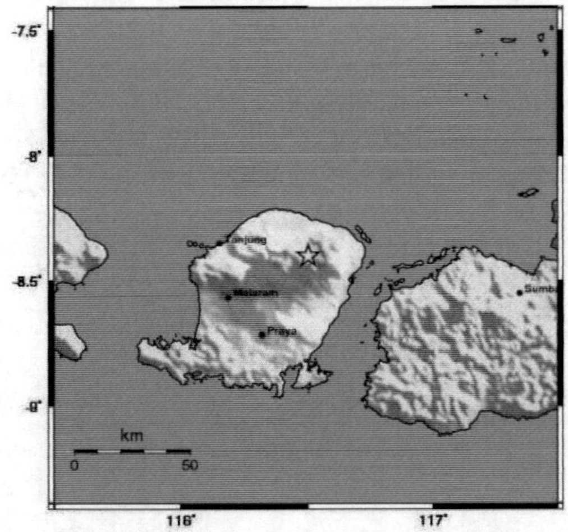

Figure 2. Shake map Lombok Earthquake on July 29, 2018 [15]

In this research, the three main components are divided into vertical components, horizontal components, and foundation components. In the vertical component, the analysis in the form of column conditions, walls, column and wall connections and beam, and column connections. Through the results of this analysis, it was found that only one school that obtained results did not experience significant damage or damage. While 14 other buildings suffered minor damage to moderate damage. The vertical component damage occurs on its special walls on column joints with walls, and there is no connection or reinforcement when there are doors or windows on the wall.

Figure 3 shows that 5 school buildings were moderately damaged, while 9 buildings had moderate damage to vertical components. The damage that occurs is not too damaging to the performance of structural components so that only non-structural components need to be improved. Improvements can use the retrofitting method.

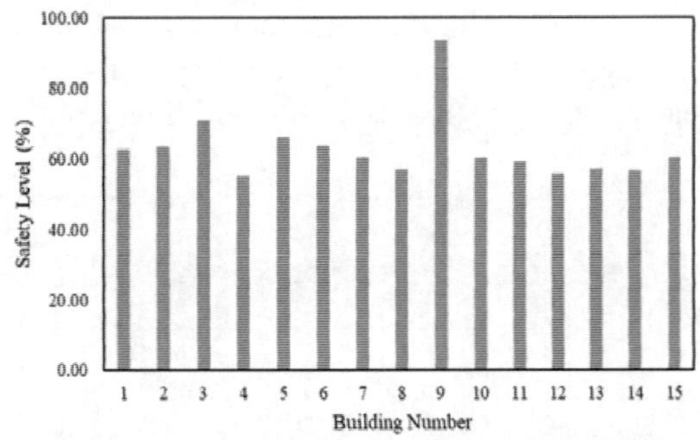

Figure 3. Building safety level in vertical components

As the material for future suggestions, it should be noted how good methods are in the process of building wall components. The damage caused by this earthquake is caused by a poor connection, so it is necessary to think about how good the connection between each building component.

Figure 4 shows the results of the building security level from the side of the horizontal component. This component assesses the feasibility of the beam, plate and roof structure. The results obtained showed that as many as two buildings did not experience significant damage, while 7 buildings suffered minor damage to the plates and roof. A total of 6 buildings were moderately damaged. Most damage is found on the roof components. Based on the results of direct observation in the field, the roof structure in general still uses wood construction that is quite old.

Apart from age, this damage is also affected by improper construction methods for roof structure work; it can be seen that there is no good connection between tile and easel made of steel structure so that in the event of a shock, tile is very easy to lose and fall. General damage that occurs in the horizontal component in the form of a ceiling drop due to not having a strong bond. While structural conditions only experience mild to moderate cracks, which are not expected to disturb structural conditions if quickly repaired.

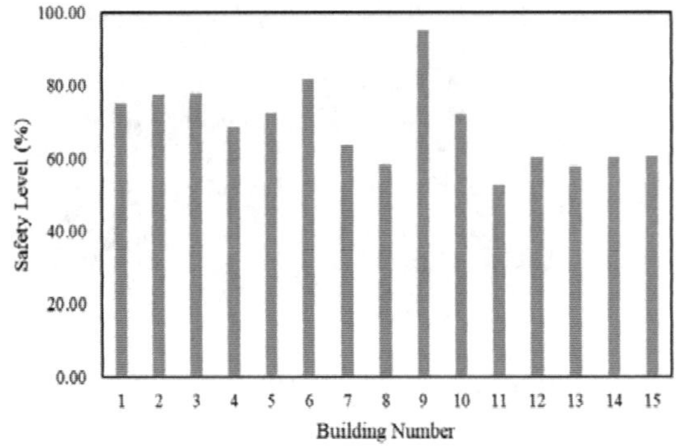

Figure 4. Building safety level in horizontal components

The foundation component shown in Figure 5 shows that most of the buildings did not experience significant damage. 11 buildings were not damaged on the foundation, and 4 buildings were damaged on the foundation. This damage occurs on the sloop, which appears to have a small crack. In some buildings that are quite old, the lack of maintenance on the foundation components causes some components to be long enough damage.

In examining these foundation components, there are quite a number of obstacles. Direct inspection methods that rely on visual vision are certainly not 100% can conclude the foundation is in good condition or not good. It is necessary to examine the foundation with better technology so that it can be known with certainty the foundation condition whether it is still functioning properly or not. Visual inspection can only detect damage if in the building area, especially in it there is a decrease in soil or sloop conditions that are clearly visible from outside the building.

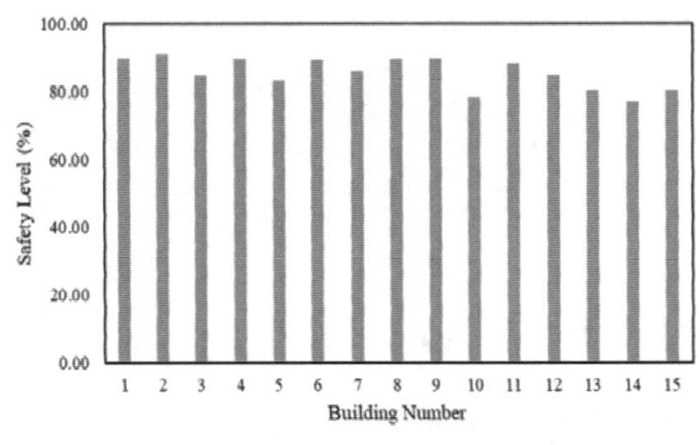

Figure 5. Level of building security on foundation components

Figure 6 describes the level of security of the building as a whole. Based on the results of the analysis, it was found that only 1 building was not damaged, while 6 school buildings suffered minor damage and 8 school buildings suffered moderate damage. There were no buildings that were severely

IOP Conf. Series: Materials Science and Engineering **650** (2019) 012038 doi:10.1088/1757-899X/650/1/012038

damaged. In Figure 6, it explains the percentage of damage to buildings, and this shows that 53% of school buildings were moderately damaged and 40% of buildings suffered minor damage.

The thing that really needs to be noticed in the field inspection is the level of damage to each component whether or not it can endanger the user. For example, some buildings do not experience significant damage to the foundation components but are experiencing moderate damage to vertical or horizontal components. Mathematically, it is possible that the building will be in a slightly damaged condition because it has a good level of security on the components of the foundation. Therefore, it needs to be taken into account if there is one component that is categorized as heavily damaged or damaged while eating the building has the potential to endanger its users.

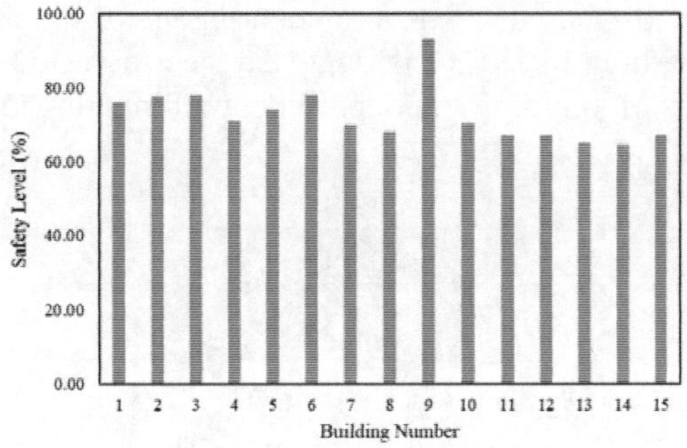

Figure 6. The overall building safety level

The level of damage obtained is one of the considerations of whether the building is suitable for use in the near future or not. Due to frequent aftershocks that make the building user feel traumatized. After obtaining the results, it needs to be explained to building users about a good method for maintenance and repairs that can be done, so that with these improvements, the building becomes safer and the durability of the building does not drop rapidly in a short time.

Figure 8 shows, the results of the percentage of damage that occurs in each component examined. 45% of the damage that occurs in all buildings comes from vertical components, while from horizontal components is 38% and from the foundation, components are 17%. This shows that vertical and horizontal components are very vulnerable to damage when an earthquake occurs.

The foundation component is the smallest component, which contributes damage to buildings due to using the rapid visual method. The level of difficulty obtained using this method is greater in assessing damage to the foundation. While the other components tend to be easier to assess because these components can be monitored properly.

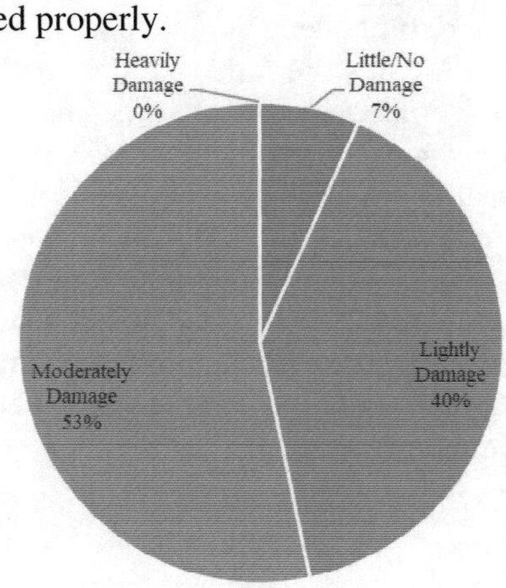

Figure 7. Percentage of damage to school buildings

IOP Conf. Series: Materials Science and Engineering **650** (2019) 012038 doi:10.1088/1757-899X/650/1/012038

As the biggest contributor to damage, it should be a concern during the construction process or maintenance of vertical components, besides being very vulnerable to damage, this component has the potential to damage and hoard humans in the event of a collapse. Even though a wall is one component of a building that is not considered to be a load restraint, if it does not have sufficient strength, the construction of a building can be categorized as not good.

In general, in addition to the building that has been assessed as old, it needs to be a concern also the method of development carried out. Many are found in buildings that do not use a connection system that can accommodate loads when an earthquake occurs. The knowledge that has not developed in several regions regarding good methods of development is one of the main obstacles that occur in the field. Thus, the government should strictly conduct searches to various earthquake-prone areas to be able to check the condition of buildings and provide training on how to repair and maintain the durability of these buildings.

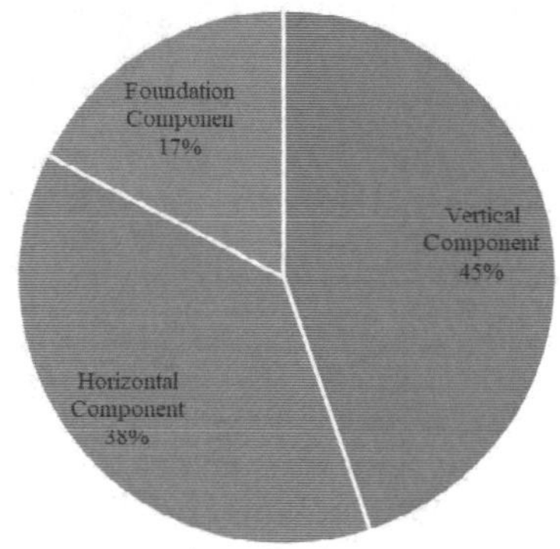

Figure 8. Percentage of causes of damage to buildings after the earthquake

Figure 9 is the percentage level of damage that occurs in each component that is carried out the assessment. Overall on the three main components, divided into 10 components. The results of the analysis show that the wall components that have diagonal cracks, wall connections with beams or columns and beam-column joints are the biggest part, which is damaged at 14% and 15% followed by damage to the roof structure with a damage rate of 14%.

It can be concluded that these four components are the components that are most prone to damage and can endanger the building user. Meanwhile, other components also contribute to damage in the event of an earthquake, structural components such as beams, plates, and columns contribute to damage of 9% to 10% and Sloop accounts for damage of 8%. This component of the structure also needs to be taken care of in the construction process.

The slab component can only be detected if the building is more than 1 floor, while on the ground floor of the building only damage occurs when the ceramic used is broken, and there are a few small cracks on the plate. Overall, hollow beam component cracks occur in the Drag section of the beam, and it can be concluded that shear and flexural damage often occurs together this is evident from the crack position that always occurs in the sliding location and bending location. Meanwhile, some column components experience sliding damage.

IOP Conf. Series: Materials Science and Engineering **650** (2019) 012038 doi:10.1088/1757-899X/650/1/012038

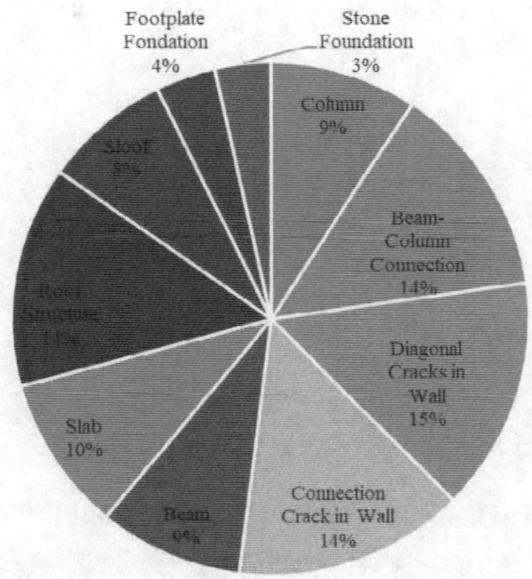

Figure 9. Percentage of damage to each component of a building

In Figure 10 through Figure 14 are some common examples of damage caused by earthquakes. In general, the damage that occurs in these locations.

Figure 10. Damage to wall components that do not have good connections with doors and windows

Figure 11. Damage to beam joints and columns

Figure 12. Damage to the roof frame structure

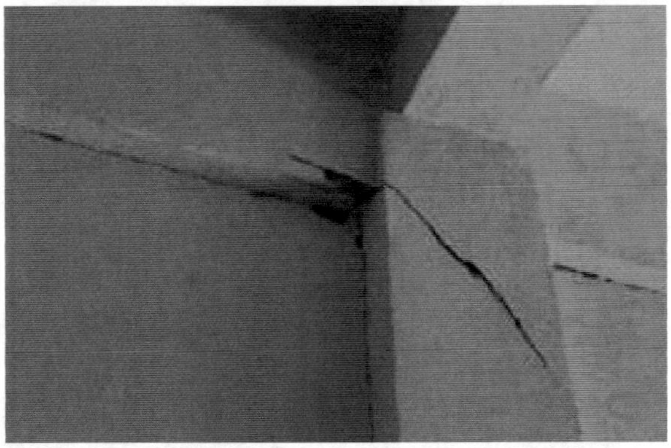

Figure 13. Damage to column components

Figure 14. Horizontal cracks occur on the wall

5. Conclusion

Based on the results of the analysis above, conclusions can be taken as follows:

1. Damage to 15 buildings consists of 53% moderately damaged, 40% lightly damaged and 7% not damaged.
2. The most damaged components are vertical components of 45%, horizontal components of 38% and foundation components as fast as 17%.
3. The most damage is on the wall, the wall connection with the beam or column, the connection on

the beam and column and on the roof structure.

Acknowledgements
The author would like to thanks the University of Muhammadiyah Yogyakarta especially Environment and Disaster Study Center that have supported this research. For the Muhammadiyah Disaster Management Center (MDMC) and thanks to the University of Muhammadiyah Mataram which facilitated the process of conducting the research.

References

[1] Prayuda, H., Zega, B. C., and Priyosulistyo, H. 2017 Prediction of Allowable Lateral Ground Acceleration (In-Plane Direction) of Confined Masonry Walls using Ambient Vibration (Microtremor) Analysis *Procedia Engineering: Sustainable Civil Engineering Structures and Construction Materials* (Indonesia) **171** pp. 1194-1203

[2] Allali, S. A., Abed, M., and Mebarki. 2018 Post-Earthquake Assessment of Building Damage Using Fuzzy Logic *Engineering Structures* **166** pp. 117-127

[3] ASCE/SEI 41-13 2014 Seismic Evaluation and Retrofit of Existing Buildings *American Society of Civil Engineers* (United States)

[4] Chakrabarti, A. 2007 *Handbook on Seismic Retrofit of Buildings* (India: Central Public Works Department and Indian Building Congress)

[5] El-Betar, S. A. 2018 Seismic Vulnerability Evaluation of Existing R. C. Buildings, Housing, and Building *National Research Center Journal* **14** pp. 189-197

[6] Ilumin, R. C., and Oreta, A. W. C. 2018 A Post Disaster Functional Asset Value Index for School Buildings *Procedia Engineering: 7th International Conference on Building Resilience; Using Scientific Knowledge to Inform Policy and Practice in Disaster Risk Reduction* (Thailand) **212** pp. 230-237

[7] Petal, M. 2008 Disaster Prevention for Schools Guidance for Education Sector Decision-Makers *International Strategy for Disaster Reduction Thematic Platform for Knowledge and Education* (Geneva)

[8] Shape, R L 1994 Luzon (Philippines) *Earthquake Tenth World Conference* (Rotterdam)

[9] United Nations International Children's Emergency Fund (UNICEF) 2014 *Philippines: Humanitarian Situation Report* (Philippines)

[10] Lopez, O. A., Hernandez, J. J., Re, G. D., and Puig, J. 2004 *Seismic Risk in Schools: The Venezuelan Project in Keeping Schools Safe in Earthquake* (Venezuela: Organisation for Economic Cooperations and Development)

[11] FEMA 2002 *FEMA 154: Rapid Visual Screening of Buildings for Potential Seismic Hazards – A Handbook* (United States :Federal Emergency Management Agency)

[12] FEMA 1998 *Handbook for the Seismic Evaluation of Buildings A Prestandard* (United States : Federal Emergency Management Agency)

[13] Rai, D. C 2005 *Review of Documents on Seismic Evaluation of Existing Buildings* (India:IITK-GSDMA-EQ03-V1.0)

[14] Boen, B 2017 Kajian Cara Cepat Keamanan Bangunan Tembokan Sederhana Satu atau Dua Lantai yang Rusak akibat Gempa dan Kajian Risiko Komponen Non-Struktural *World Seismic Safety Initiative* (Indonesia)

[15] Hidayati, N., Trisnawati, Sative, O., Wallansha, R., Sakti, A. P., Pramono, S., and Permana, D. *2018 Ulasan Guncangan Tanah Akibat Gempa Bumi Lombok Timur* Badan Meteorologi Klimatologi dan Geofisika, Indonesia.and Geoecology, Saint Petersburg, Rusia, 17-19 Mei 2017, 207-214.

IOP Conf. Series: Materials Science and Engineering **650** (2019) 012039 doi:10.1088/1757-899X/650/1/012039

Characteristics Comparison on Mechanical Properties of Mortars using Agriculture Waste as a Cement Replacement Materials

F Saleh[1], H Prayuda[1], F Monika[1] and M M A Pratama[2]

[1]Department of Civil Engineering, Faculty of Engineering, Universitas Muhammadiyah Yogyakarta, Bantul, Daerah Istimewa Yogyakarta, Indonesia
[2]Department of Civil Engineering, Faculty of Engineering, Universitas Negeri Malang, Malang, Jawa Timur, Indonesia

* dilla_vu@yahoo.com ; hakasprayuda@umy.ac.id ; fanny.monika.2007@ft.umy.ac.id ; mirza.abdillah.ft@um.ac.id

Abstract. Excessive use of cement makes the price of construction projects expensive. Mortar making using high amounts of cement also has the potential to increase the price of the construction. This study discusses the use of agricultural waste as a substitute for cement in mortar production. The waste used in this study was bagasse ash, fuel ash palm oil, and rice husk ash. Each waste is added as much as 25%, 50% and 75% which are then tested for mechanical properties such as water content, unit weight, absorption, and IRS. Compressive strength was tested when the mortar was 28 days old with a cube-shaped specimen with a size of 50 mm x 50 mm x 50 mm. From the test results, it was found that all of this waste can be used as a substitute for cement. To produce a compressive strength of 100 kg/cm2 can add each waste with a percentage range of 10-12% of the weight of cement.

Keywords: Mortars, Bagasse Ash, Fuel Ash, Rice Husk Ash

1. Introduction

Ordinary Portland Cement (OPC) is the basic cement which is currently the most produced in the world [1]. Higher demands for infrastructure and housing contribute greatly to the production of cement which is one of the ingredients of concrete manufacture [2]. As an effect, the release of CO2 into the environment free of cement production continues to increase causing the level of global warming to be increasingly difficult to control [2, 3]. It has been reported that around 4.1 billion tons of cement are produced every day in 2015, with each 1 ton of cement producing 0.7-1.1 CO2 emission [4, 5]. By continuing to increase gas production, this is certainly a special concern to maintain environmental conditions for the better and reduce the level of global warming.

In some tropical countries, agriculture is one of the reliable sectors. In Indonesia as a tropical country, it is very much developing the agricultural and plantation sectors. The plantation sector that develops is oil palm plantations, sugar cane plantations and rice farming which produce rice as one of the basic ingredients of food for the majority of the Indonesian people. In processing, the results of these plantations leave behind waste that has not been utilized so far. In general, plantation and agricultural waste in Indonesia has not been managed properly so that this waste will have a bad effect in the long term.

Mortar is a construction material component that is usually used not for structural components. The main material for making mortar consists of Portland cement, water and fine aggregate. In general, people in Indonesia use mortar as an adhesive for masonry on walls, whose function is not a structural component. Considering its function as an adhesive, the use of cement on mortar should be reduced or replaced with other materials that can reduce waste released into the wild so that it also has the potential to reduce the issue of global warming.

In the past few years, the use of pozzolanic materials in concrete or mortar constituents has become very essential and has experienced a significant increase [6]. Utilization of waste from agriculture as a mortar maker will certainly have advantages. Besides being able to reduce the composition of agricultural waste, the use of waste can also reduce the use of cement for material components that do not require high levels of strength. By utilizing these wastes it will produce new materials in the field of construction that are more environmentally friendly.

The use of bagasse ash as an added ingredient in the manufacture of concrete has been done before. Some researchers use this agricultural waste both on mortars and on concrete. Some of the studies that have been carried out include mechanical and durability properties of bagasse ash mortar [1, 7, 8], admixture in self-compacting concrete [9], high strength concrete with bagasse ash [6, 10]. In addition, the characteristics and microstructures of bagasse ash have also been carried out as a base for replacing Portland cement [11]. Based on several studies it can be concluded that this material has the potential to be used as a base for making mortar and concrete. However, with the condition of the origin of plants that are different locations with previous researchers, of course, they must be further investigated regarding capability of this material as an environmentally friendly mortar constituent.

In addition to bagasse ash, this study will compare the characteristics of other ingredients, namely palm oil ash and rice husk ash. In palm oil ash research has been carried out on its characteristics in making mortar [12-14], the use of palm oil ash as material for making concrete [15-17], and also making mortar with geopolymer systems [18, 19]. Through some of the results of the research, it is indicated that the material of palm oil ash has the potential to be used as one of the mineral substitute materials for cement. In addition to bagasse ash and palm oil ash, many wastes produced as a result of farming with rice husk ask are also not utilized optimally. Several studies on rice husk ash have also been carried out, including durability checks and microstructure rice husk ash mortar [20], physical, mechanical and chemical properties of rice husk ash mortar [21-24].

In this study will compare normal mortar without using added agriculture waste material against mortar by using each ingredient added by agriculture waste. Through this research, it is expected to produce optimum levels suitable for the utilization of waste agriculture as a mortar constituent. In addition, a fresh properties check will also be carried out and several mechanical properties checks such as water content, unit weight, and absorption value.

2. Materials and Test Method

The main materials used in this study are bagasse ash (BA), palm oil ash (POA), rice husk ash (RHA) and ordinary Portland cement (OPC). In addition, using water and fine aggregate with the same composition for all variations of the test object. Table 1 shows the results of examination of chemical properties for each of the main ingredients from previous studies. The ash baggage, palm oil ash and rice husk ash have a high $SiO2$ value that exceeds 50%, whereas in Portland cement has a higher CaO content of 50.76% followed by $SiO2$.

In this study using 12 variations where each of the added ingredients consisted of 4 variations. Variation used uses the percentage of volume, adding 25%, 50%, and 75% additional ingredients instead of cement. While 1 variation is used as a control using 100% Portland cement and without using added ingredients. In Table 2 shows the number of numbers for each of the 10 test items. In this study, testing of fresh properties and mechanical properties will be carried out on each variation of the test object used.

IOP Conf. Series: Materials Science and Engineering **650** (2019) 012039 doi:10.1088/1757-899X/650/1/012039

Fresh properties testing uses a flow table while mechanical properties are carried out by making specimens measuring 50 x 50 x 50 mm. Mechanical tests carried out in the form of examination of unit weight, water content, absorption, density and initial rate of suction (IRS). And compressive strength tested at the age of 268 days. Each test was carried out on 10 specimens which were then averaged to obtain results with standard deviations that met the specifications.

Table 1. Chemical properties of BA, POA, RHA, and OPC [2, 24, 25]

Materials	SiO_2	K_2O	Fe_2O_3	CaO	P_2O_5	MgO	Al_2O_3	TiO_2	Na_2O	MnO
BA	56.37	0.92	5.04	2.36	0.12	1.43	14.61	0.25	1.57	0.06
POA	64.16	8.25	6.33	5.81	5.18	4.87	3.73	0.19	0.18	0.18
RHA	93.8	0.12	0.3	0.89		0.32	0.74		0.28	
PC	23.86	0.92	2.19	50.76	0.12	1.36	5.77	0.25	0.91	0.06

Table 2. Mix proportion for each variation

Variation	Materials for 10 specimens (ml)					
	Cement	Water	Sand	BA	POA	RHA
100PC 0BA	600	300	1800	0	-	-
75PC 25BA	450	300	1800	150	-	-
50PC 50BA	300	300	1800	300	-	-
25PC 75BA	150	300	1800	450	-	-
100PC 0POA	600	300	1800	-	0	-
75PC 25POA	450	300	1800	-	150	-
50PC 50POA	300	300	1800	-	300	-
25PC 75POA	150	300	1800	-	450	-
100PC 0RHA	600	300	1800	-	-	0
75PC 25RHA	450	300	1800	-	-	150
50PC 50RHA	300	300	1800	-	-	300
25PC 75RHA	150	300	1800	-	-	450

3. Result and Discussion

The slump test is done before the concrete is put into the mold. In Figure 1 is the slump flow test results. The results show that the added value of the added material indicates that the spread value in the specimen is getting higher. In a variation of 75% added ingredients, rice husk ash has the smallest slump value, followed by palm oil ash and bagasse ash. However, all specimens meet the minimum specifications required for mortar manufacturing.

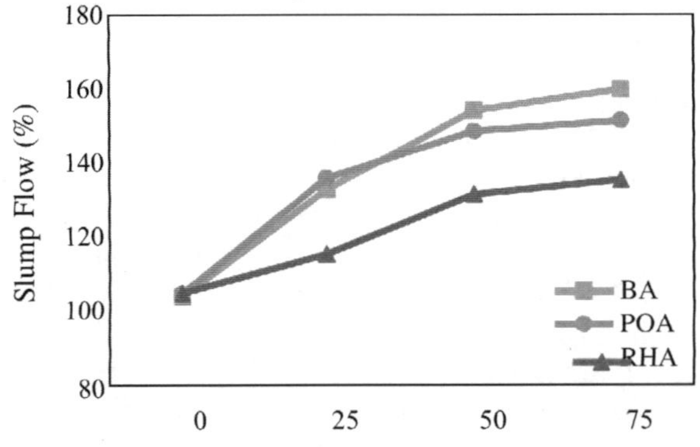

Figure 1. Flow test for each specimen

In Figure 2 shows the results of examination of fresh concrete, from these results it can be explained that in making mortar it is necessary to pay attention to the amount of water used. In this study using the same amount of water for each specimen. However, the more ingredients added, the more water will be needed. In Figure 2 it can be seen that the mortar is not plastic and tends not to fuse between parts. In addition, because of the higher pozzolanic properties, the mortar will react longer. During the stirring process, the water content in the ingredients is very small so that during the stirring process, this added material absorbs more water and has an effect on the value of the workability. It is recommended for the next study, by using different variations of the mixture, the amount of water used must also vary according to the type of material and the value of the expected workability.

Figure 2. Flow test specimen (a) 100% Portland cement; (b) 25% palm oil ash; (c) 50% palm oil ash; (d) 75% palm oil ash

The next test is the mechanical properties of mortar using ingredients added to agriculture waste as a substitute for cement. Examinations carried out in the form of unit weight, water content, absorption, mass density and IRS.

The results of unit weight testing as shown in Figure 3 (a) indicate that the more increasing the level of added material used will result in smaller unit weight or the mortar will be lighter. Mortars using this added material cannot be concluded with added ingredients such as what will produce lighter mortars because the results obtained tend to have a fluctuating pattern but with the longer trend will decrease. Mortar using bagasse ash has a tendency to decrease smaller than mortar using other added ingredients. Water content testing is also carried out with the results seen in Figure 3 (b). The results show that the higher the content of the added ingredients the higher the water content obtained in the mortar. This is because the more pores produced by the mortar because in this study using waste materials with a grain size not smaller than Portland cement. So that for further research it is necessary to pay attention to the grain size of the admixture material used. Increased water content by mortar from bagasse ash and rice husk ash tended to experience a similar increase, but the increase in water content in oil ash pump mortars tended to be smaller than the two previous materials.

The mass density test results show that for bagasse ash and rice husk ash tend to have a fluctuating pattern but will continue to decline the more added ingredients are used, while the mortar using palm

IOP Conf. Series: Materials Science and Engineering **650** (2019) 012039 doi:10.1088/1757-899X/650/1/012039

oil ash will decrease if more ingredients are added. Palm oil ash mortar has the smallest type of rice compared to mortar using other added ingredients. The results of the examination can be seen in Figure 3 (c). Basically, the lighter this material will be the better for construction material, but it needs to pay attention to its strength.

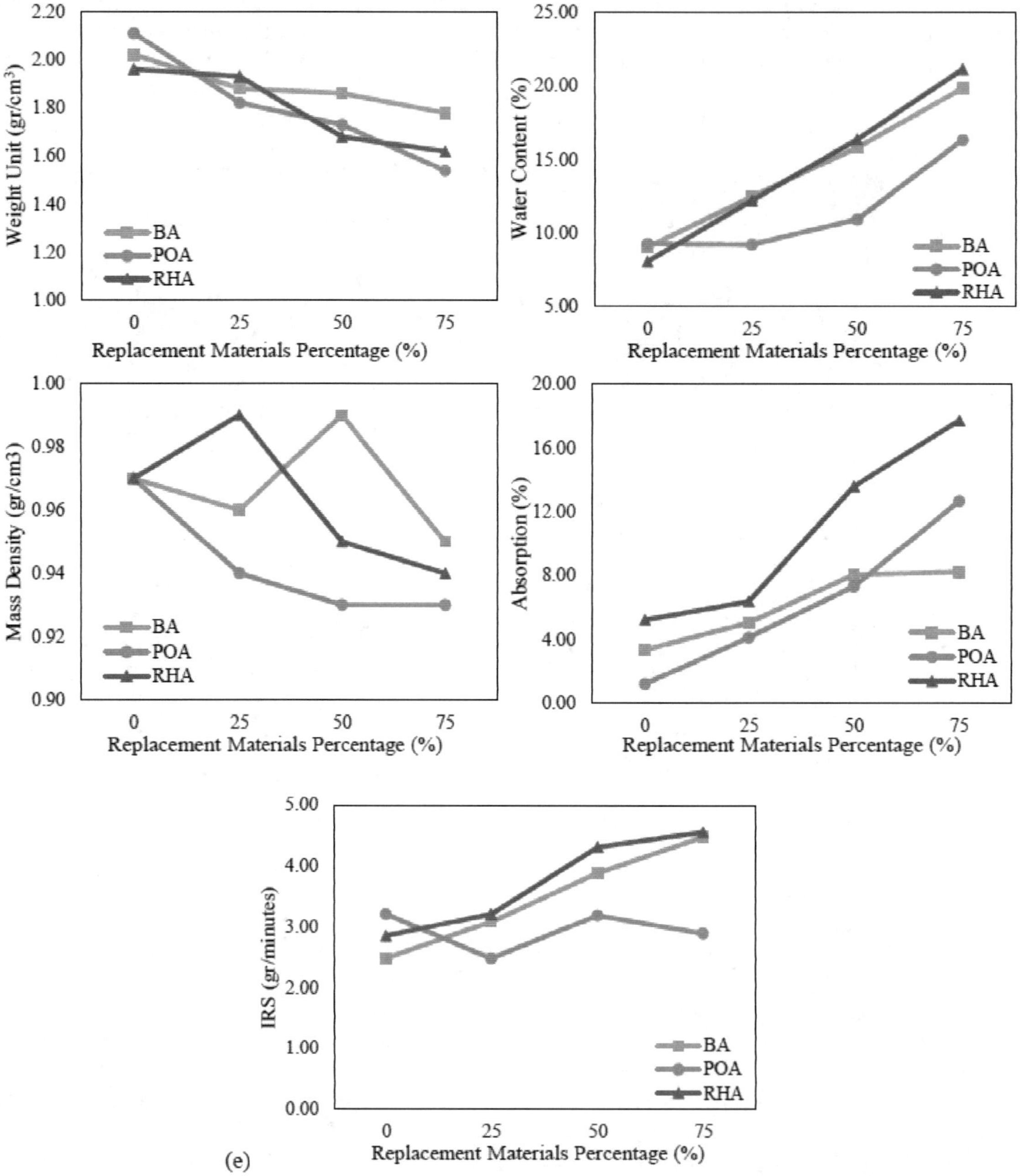

Figure 3. Mechanical properties of mortar (a) weight unit; (b) water content; (c) mass density; (d) absorption; (e) initial rate of suction

In Figure 3 (d) shows the results of absorption that occurs in mortar. The results show that the more added ingredients used will increase the absorption value, even if seen from the condition of the water content that is contained in the mortar also increasing. Of course the higher the absorption value indicates that the mortar will more easily absorb water. Mortar using rice husk ash admixture has the highest absorption value compared to other materials. The pore produced by this material is much higher than other materials, besides, the size of the grain also affects the absorption value produced. The more pore and the ease of material absorbing water will cause the mortar to absorb water more easily.

In Figure 3 (e) shows the IRS results where the more added ingredients are used, the higher the absorption rate (IRS) will also be. However, palm oil ash has a smaller IRS value compared to other higher materials even though the values are not much different. In addition, ingredients added to palm oil ash have a smaller increase in value compared to using other added ingredients which tend to increase faster.

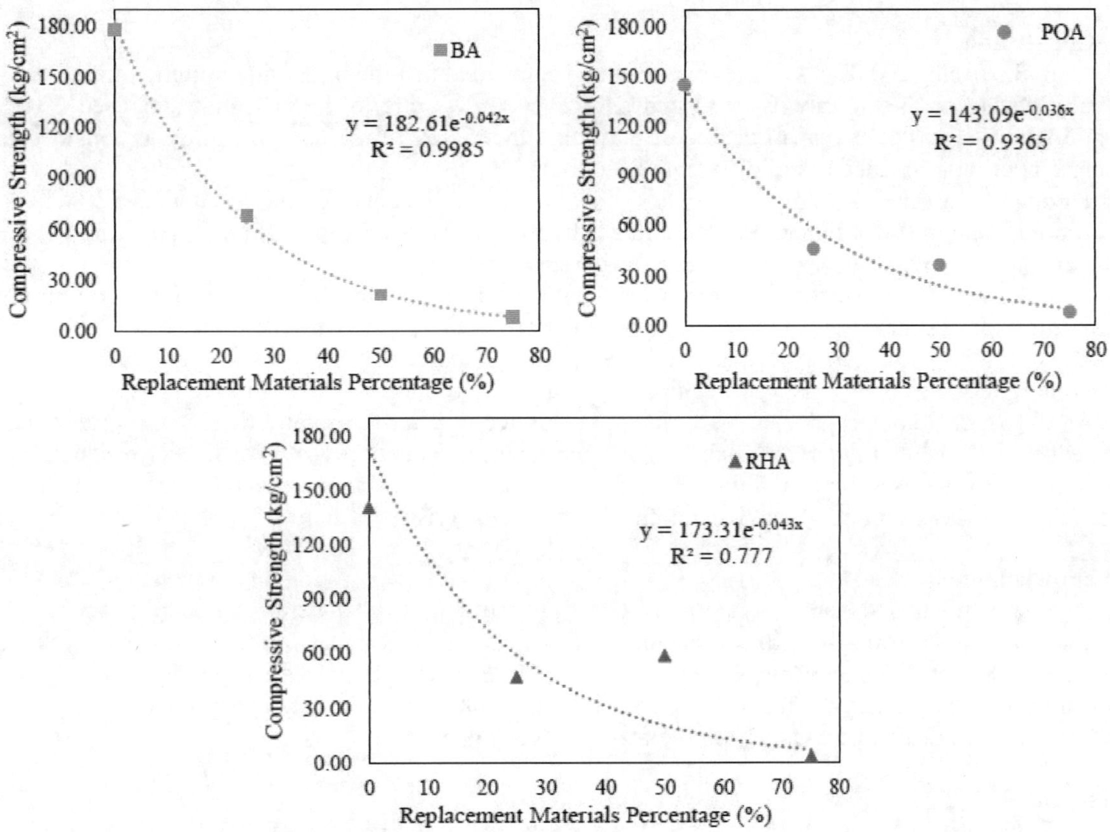

Figure 4. Compressive Strength (a) bagasse ash; (b) palm oil ash; (c) Rice Husk ash

In Figure 4 is the relationship between compressive strength at the age of 28 days with the percentage of added ingredients used. The results show that the results obtained are that the more added ingredients used, the less compressive strength produced. The highest compressive strength is produced by normal mortar without using ingredients added to agriculture waste. Table 3 is a standard mortar for buildings in Indonesia. Through this standard shows that to produce mortar does not have to use cement with a high compressive strength. So that it can be replaced with other pozzolan ingredients.

Through Figure 4, it can be calculated how many percent of the added material that can be used in accordance with the standard data in Table 3. Strengths that want to be produced adjust to the needs of the construction site. For example, if you want to produce mortar with B100 quality, which means compressive strength of 100 kg/cm2. So from Figure 4, using bagasse ash can replace cement by 12%, as well as using rice husk ash. While using palm oil ash will only be able to add about 10% of the total percentage of cement used. This will greatly help reduce the use of cement and can help reduce the cost of producing mortar in construction which is commonly used in some nonstructural components specifically in infrastructure.

Table 3. Mortar standard in Indonesia [26]

Class	Minimum Compressive Strength (kg/cm^2)		Maximum Absorption (%)
	Average for 5 sample	Each Sample	
B25	25	21	-
B40	40	35	-
B70	70	65	35
B100	100	90	25

4. Conclusion

Through the results and discussion above, it can be concluded that the increasing volume of material is added, the higher the flowability produced. However, this is due to the lack of water used, so the resulting mortar has a very small degree of plasticity. In the next study, it is necessary to consider the water binder ratio for each use of different added ingredients.

Through testing the mechanical properties show that the results obtained using materials are no better than mortars using 100% cement Portland, this is because mortars tend to lack water, resulting in large pores and no control on the size of the added material.

Through the results of the compressive strength it can be concluded that the more volume the admixture material increases, the smaller the compressive strength it will produce, but for some mixtures it still meets the specifications required by the applicable standards in Indonesia. In each ingredient added has the same characteristics of compressive strength produced at the age of 28 days.

At 100 kg cm2 compressive strength, it can be applied to bagasse ash and rice husk ash to replace cement by 12% while mortar with palm oil ash can replace cement by 10%. For further research, this percentage can be used by examining the effect of cement water factor and grain size of admixture material to further examine so that it can improve the characteristics of mortar

Acknowledgment

The author would like to thank the laboratory staff of the structural and material engineering laboratories for the Civil Engineering Department, Faculty of Engineering, Universitas Muhammadiyah Ygyakarta. And Lembaga Penelitian dan Pengabdian Masyarakat (LPPM) Universitas Muhammadiyah Yogyakarta as a funder in this research. The author would like to thank you to Department of Civil Engineering, Faculty of Engineering, Universitas Muhammadiyah Yogyakarta.

References

[1] Joshaghani. A., and Moeini, M .A. 2017 Evaluating the Effects of Sugar Case Baggase Ash (SCBA) and Nanosilica on the Mechanical and Durability Properties of Mortar. *Construction and Building Materials* **152** Pp. 818-831

[2] Noorvand. H., Ali. A. A. A., Demirboga. R., Noorvand. H., and Farzadnia, M. 2013 Physcal dan Chemical Characteristic of Unground Palm Oil Fuel Ash Cement Mortars with Nanosilica. *Construction and Building Materials* **48** Pp. 1104-1113

[3] Ranjbar. N., Mehrali. M., Behnia. A., Alengaram. U. J., and Jumaat, M. Z., 2014 Compressive Strength and Microstructural Analysis of Fly Ash/Palm Oil Fuel Ash Based Geopolimer Mortar *Materials and Design* **59** Pp. 532-539

[4] Part. W. K., Ramli. M., and Cheah, C. B. 2015 An Overview on the Influence of Various Factors on the Properties of Geopolymer Concrete Derived from Industrial by Products. *Construction and Building Materials* **55** Pp. 370-395

[5] Bosoaga. A., Masek. O., and Oakey, J. E. 2009 CO2 Capture Technologies for Cement Industry *Energy Procedia* **1** Pp. 133-140

[6] Rajasekar. A., Arunachalam. K., Kottaisamy. M., and Saraswathy, V. 2018 Durability Characteristics of Ultra High Strength Concrete with Treated Sugarcane Bagasse Ash.

Construction and Building Materials **171**. Pp. 350-356

[7] Arenas-Piedrahita. J. C., Montes-Garcia. P., Mendoza-Rangel. J M., Calvo. H. Z. L., Valdez-Tamez. P. I., and Martinez-Reyes, J. 2016 Mechanical and Durability of Mortars Prepared with Untreated Sugarcane Bagasse Ash and Untreated Fly Ash *Construction and Building Materials* **105** Pp. 69-81

[8] Praveenkumar. S., Shanmugasundaram. J., and Samynathan, B. 2017 Effect of Bagasse Ash in Properties of Cement Paste and Mortar. *International Journal of ChemTech Research*. **10(8)**. Pp. 219-225

[9] Moretti, J. P., Nunes. S., and Sales, A. 2018 Self-Compacting Concrete Indorporating Sugarcane Bagasse Ash *Construction and Building Materials* **172**. Pp. 635-649

[10] Jagadesh. P., Ramachandramurthy. A., and Murugesan, R. 2018 Evaluation of Mechanical Properties of Sugar Cane Bagasse Ash Concrete *Construction and Building Materials* **176** Pp. 608-617

[11] Mohamed. Y. A. A., Khalifa. S. A., Hegazi. S. E. F., and Elwahab, H. A. 2017 Effect of Sugarcane's Bagasse Ash Additive on Portland Cement Properties *International Journal of Sustainable Development Research* **3(6)** Pp. 85-89

[12] Yusuf. M. O., Johari. M. A. M., Ahmad. Z. A., and Maslehuddin, M. 2014 Shrinkage and Strength of Alkaline Activated Ground Steel Slag/Ultrafine Palm Oil Fuel Ash Pastes and Mortars *Materials and Design* **63** Pp. 710-718

[13] Lim. N. H. A. S., Ismail. M. A., Lee. H. S., Hussin. M. W., Sam. A. R. M., and Samadi, S. 2015 The Effect of High Volume Nano Palm Oil Fuel Ash on Microstructure Properties and Hydration Temperature of Mortar *Construction and Building Materials* **93** Pp. 29-34.

[14] Huseien. G. F., Ismail. M., Tahir. M. M., Mirza. J., Khalid. N H. A., Asaad. M. A., Husein. A. A., and Sarbini, N. N., 2018 Synergism between Palm Oil Fuel Ash and Slag: Production of Environmental Friendly Alkali Activated Mortars with Enhanced Properties *Construction and Building Materials* **170** Pp. 235-244

[15] Tunduba. Y. W., and Mirza, J. 2017 Mortars and Concrete Indorporating Palm Fuel Ash and Fly Ash *Regional Conference in Civil Engineering* (Indonesia)

[16] Prayuda. H., Saleh. F., Maulana. T. I., and Monika, F. 2017 Fresh and Mechanical Properties of Self- Compacting Concrete with Coarse Aggregate Replacement using Waste of Oil Palm Shell. *The 7th AIC-ICMR on Science and Engineering* (Indonesia)

[17] Hassan. I. O., Ismail. M., Forouzani. P., Majid. Z. A., and Mirza, J. 2014 Flow Characteristics of Ternary Blended Self-Consolidating Cement Mortars Indorporating Palm Oil Fuel Ash and Pulverised Burnt Clay *Construction and Building Materials* **64** Pp. 253-260

[18] Islam. A., Alengaram. U. J., Jumaat. M. Z., and Bashar, I. I. 2014 The Development of Compressive Strength of Ground Granulated Blast Furnace Slag-Palm Oil Fuel Ash-Fly ash Based Geopolymer Mortar *Materials and Design* **56**. Pp. 833-841

[19] Ranjbar. N., Mehrali. M., Alengaram. U. J., Metselaar. H. S.C., and Jumaat, M. Z. 2014 Compresisve Strength and Microstructural Analysis of Fly Ash/Palm Oil Fuel Ash Based Geopolymer Mortar under Elevated Temperatures *Construction and Building Materials* **65** Pp. 114-121

[20] Mohseni. E., Naseri. F., Amjadi. R., Khotbehsara. M. M., and Ranjbar, M. M. 2016 Microstructure and Durability Properties of Cement Mortars Containing Nano-TiO2 and Rice Husk Ash. *Construction and Building Materials* **114** Pp. 656-664

[21] Miyandehi. B. M., Feizbakhsh. A., Yazdi. M. A., Liu. Q., Yang. J., and Alipour, P. 2016 Performance and Properties of Mortar Mixed with Nano-CuO and Rice Husk Ash. *Cement and Concrete Composites* **74** Pp. 225-235

[22] Jamil. M., Khan. M N. N., Karim. M R., Kaish. A. B. M. A., and Zain, M. F. M. 2016 Physical and Chemical Contributions of Rice Husk Ash on the Properties of Mortar *Construction and Building Materials* **128**. Pp. 185-198

[23] Younes. M. M., Abdel-Rahman. H A., and Khattab, M M. 2018 Utilization of Rice Husk Ash and

Waste Glass in the Production of Ternary Blended Cement Mortar Composites *Journal of Building Engineering* **20** Pp. 42-50

[24] Adesina. P. A., and Olutoge, F. A. 2019 Structural Properties of Sustainable Concrete Developed using Rice Husk Ash and Hydrated Lime *Journal of Building Engineering* **25** Pp. 1-11.

[25] Jimenez-Quero. V. G., Leon-Martinez. F. M., Montes-Garcia. P., Gaona-Tiburcio. C., and Chacon-Nava, J. G. 2013 Influence of Sugar-cane Bagasse Ash and Fly Ash on the Rheological Behavior of Cement Pastes and Mortars *Construction and Building Materials* **40** Pp. 691-701

[26] Badan Standarisasi Nasional 1989 *SNI 03-0349-1989: Bata Beton untuk Pasangan Dindin* (Indonesia :Departemen Pekerjaan Umum)

IOP Conf. Series: Materials Science and Engineering **650** (2019) 012040 doi:10.1088/1757-899X/650/1/012040

Vulnerability Assessment on Non-Engineered Building in Earthquake Prone Area. Case Study: Klaten District, Central Java, Indonesia

M H Zufiar[1], H Prayuda[1] and F Monika[1]

[1]Department of Civil Engineering, Faculty of Engineering, Universitas Muhammadiyah Yogyakarta, Bantul, Daerah Istimewa Yogyakarta, Indonesia

* zulfiarheri@umy.ac.id ; hakasprayuda@umy.ac.id ; fanny.monika.2007@ft.umy.ac.id

Abstract. Most of the houses in Indonesia are still categorized as non-engineered buildings which do not use good structural analysis in the manufacturing process. The level of education and knowledge about the construction of simple houses is certainly the main cause. The geographical conditions of Indonesia which are quite prone to disasters such as earthquakes, eruptions and cause this condition is one of the real threats to the population of Indonesia. This study discusses the level of vulnerability of buildings in the Klaten area to earthquake hazards. This area is one of the areas affected by the earthquake in 2006, and the community living was rebuilt with funds from the Indonesian government. The results showed that at the time of the initial construction 95% of this building had been calculated to withstand earthquake loads, this was seen from the main building which had a fairly strong structural component and was equipped with a fairly good structural drawing. Along with the growth of the population, most of the buildings have undergone a change of shape, many people have added their own buildings without considering the strength of the construction to withstand earthquake loads. So that buildings that were initially categorized as resistant to earthquake forces are very dangerous if an earthquake occurs.

Keywords: Non-Engineered, Building Assessment, Earthquake, Klaten District.

1. Introduction

The earthquake occurred on 26 May 2006 with a strength of 5.9 SR and a depth of 11.3 km, the earthquake occurred in 5 seconds. As a result of the earthquake there was damage to the infrastructure and houses of the community, especially in the Bantul, Klaten and Yogyakarta areas. Based on the results of the investigation that non-engineered building is potentially the most severe damage occurs when an earthquake occurs because development is carried out without design analysis and supervision that meets the qualifications [1]. After the earthquake, the government of the Republic of Indonesia and various organizations provided assistance to reconstruct buildings that had collapsed due to the earthquake. During the process of building the house, the building was designed for earthquake resistant building houses with supervision provided by the government.

After more than 10 years of the earthquake, of course development continues to increase with an increase in population. In this study, we will assess buildings in an area affected by an earthquake in 2006, of course, the area has been educated on how to build constructions that can withstand earthquake forces. The purpose of this study was to assess the building of community houses, specifically houses that had been built after the earthquake from 2006 to 2007. The inspection was conducted to find out whether the building still met the earthquake resistant building rules and what changes had been made.

Methods for evaluating buildings vary greatly, several studies have been carried out using standard released by FEMA [2, 3], using statistical analysis methods [4], to use approach analysis using fuzzy logic programming languages [5-7]. This research will assess 100 houses that also combine several methods that apply in Indonesia and FEMA.

2. Research Method

This study uses an assessment with meted Rapid Visual Screening to identify buildings based on potential hazard levels observed. This research was conducted in Mlese Village, Klaten Regency, Central Java. The distribution sample can be seen in Figure 2 by taking 100 locations of houses that were built in 2006 and 2007 after the earthquake occurred

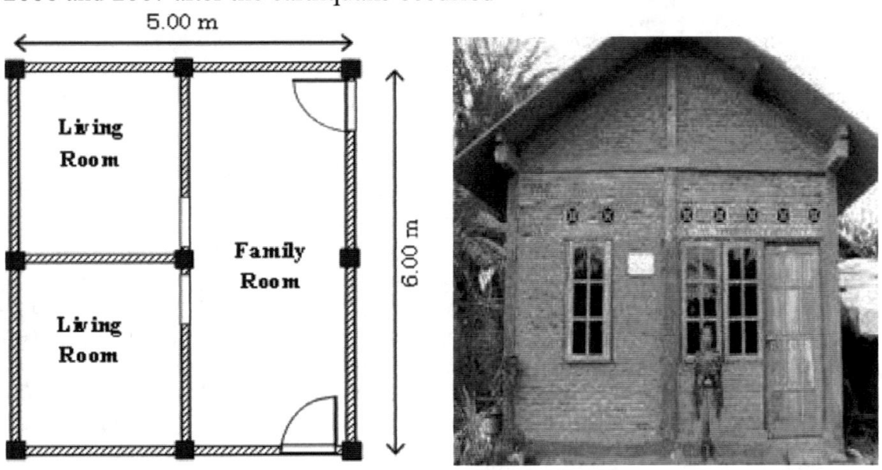

Figure 1. The initial design and shape of the building's initial construction in 2006

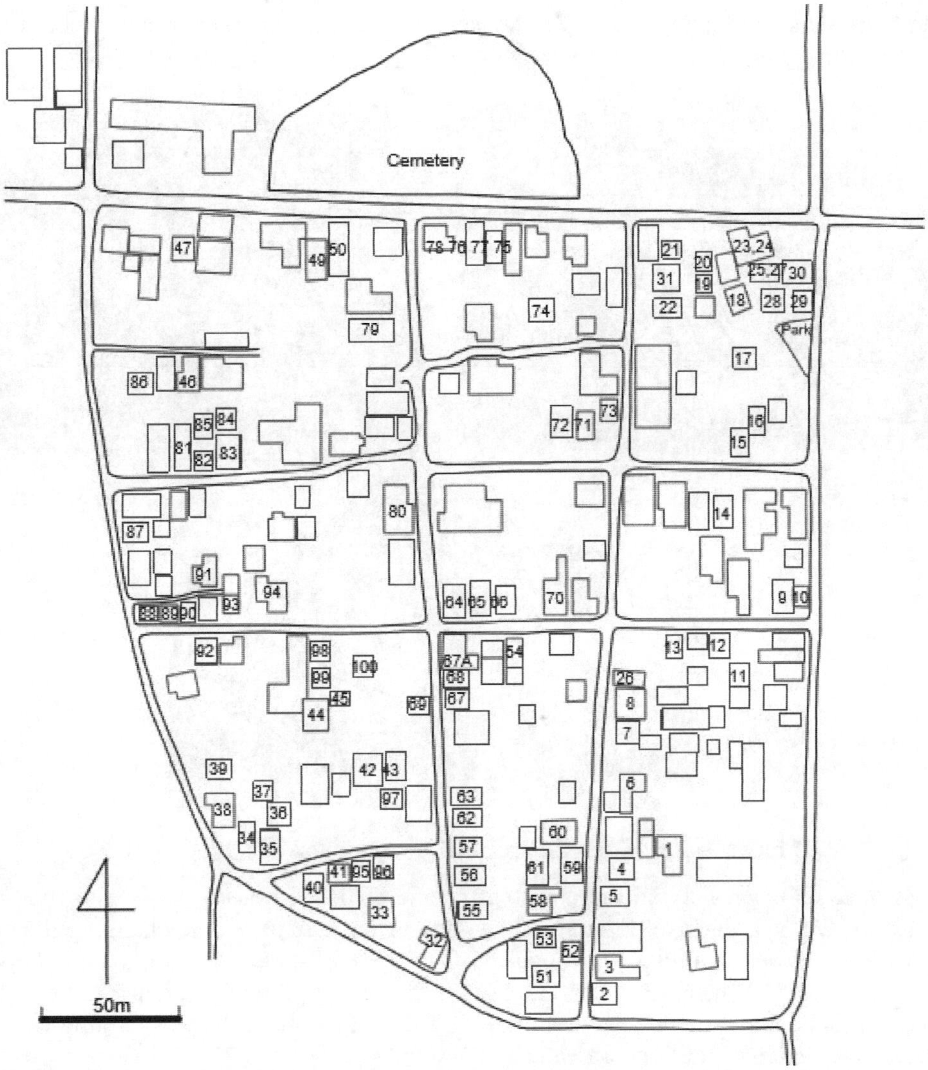

Figure 2. Plan location inspection and the number of positions of the building under study

In this study, we will examine houses built after earthquakes. At the beginning of its construction, the house was built on the condition of earthquake resistant buildings as shown in Figure 1. As time went on, the level of change that occurred in the 100 buildings will be checked. Next will be an examination of the level of building vulnerability both in terms of structural and non-structural components. The components examined were structural components, wall conditions, roof conditions, horizontal and vertical vulnerabilities.

3. Result and Discussion

3.1. Building Size

In 2007 the Mlese Urban Village carried out the reconstruction of 100 houses damaged after the earthquake. The construction of the house uses aid funds from the government in collaboration with several world organizations. Basically this development was designed following the rules of earthquake resistant houses made by the Ministry of Public Works of the Republic of Indonesia. The house was built with an area of 36 m2 complete with beam and column structures that have been designed to withstand earthquake loads. The house sample according to the rules can be seen in Figure 3. The size of the building used in 2006 was 5 x 6 m.

(a) (b) (c)

Figure 3. Reconstruction result after earthquake in 2006 until 2007

In 2017 an analysis of changes in the area that occurred in 100 houses was given funding to build houses after the 2006 earthquake. Figure 4 shows the percentage of people changing the condition of the building after 12 years of wider earthquakes. Only 20% of houses do not change the structure of the building so that it remains in accordance with the predetermined design. Whereas 19% of the houses have been added to the area of less than 50 m2, the addition of areas ranging from 50 to 100 m2 is around 48% and the addition of area above 100 m2 is 13%.

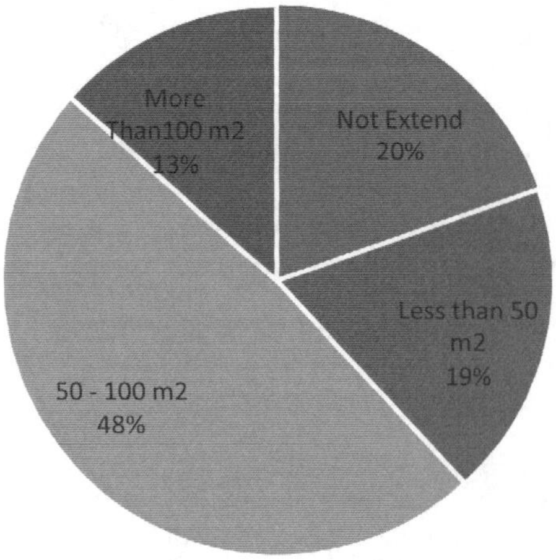

Figure 4. Buiding Size Information

In Figure 5 is a plan of one of the buildings that has increased the area. The building was built without using earthquake resistant building rules as regulated by the local government. So that it is necessary to do further examination regarding the level of vulnerability of buildings to earthquake forces that can occur without predictions beforehand. In addition, the addition of non-permanent extension material is also one of the most widely used components.

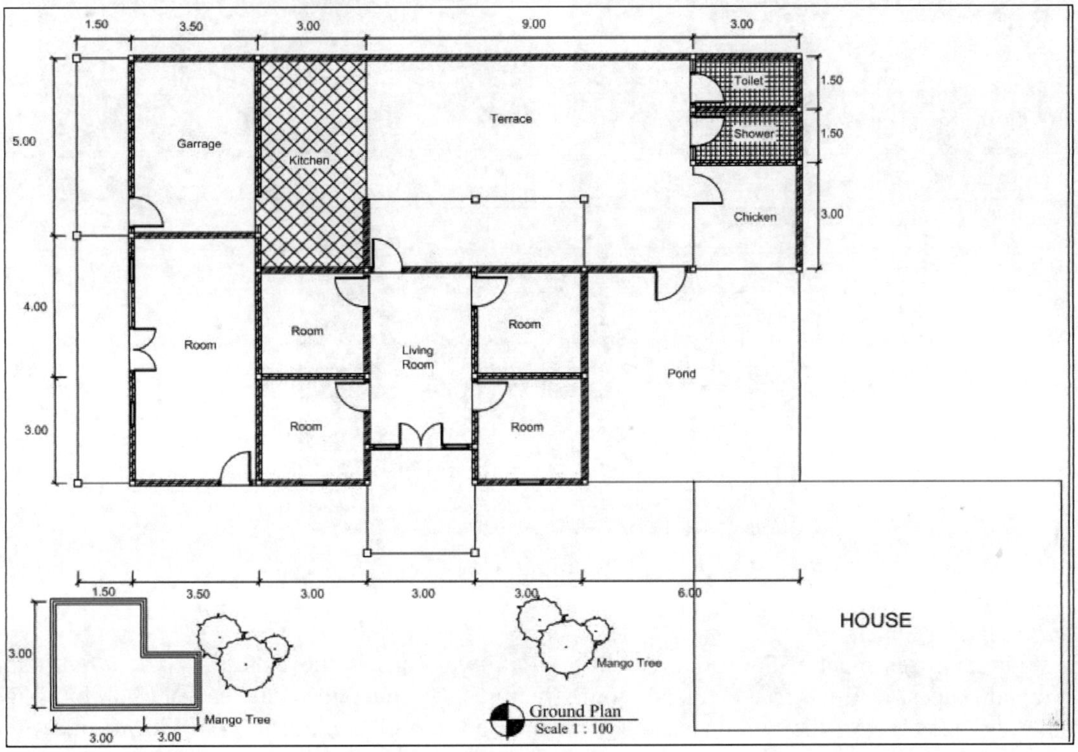

Figure 5. Example building with increase the size

(a) (b) (c)

Figure 6. Example building with increase the size

3.2. Building Assessment

In this study, an assessment of the conditions in the field was carried out regarding the configuration of horizontal shapes, wall material elements, roof structures to connections between old buildings and new buildings. 25.80% of the buildings built with the addition of area up to 2017 do not meet the horizontal configuration standard, the horizontal form of the building still does not take into account the applicable rules. This level of vulnerability needs to be taken into account because it will be predicted to continue to increase every year because growth will continue.

The addition of buildings is of course related to the use of wall elements, the survey results state that the owners of Mrese Village use red bricks without ties of 94.70%, of course this is not very good, although this component is a non-structural component. While the 5.30% addition of walls still uses wood or bamboo. Monitoring and training must be carried out in the development process, because these conditions are very important. The results of the post-earthquake investigation show that the high level of collapse in the wall components without bonding to the structural components will collapse due to an earthquake.

Whereas the extension structure system used 92.1% using practical columns and beams, 2.30% using wood reinforcement while using concrete frame structures only 5.70% of the 100 houses were inspected. Using practical columns and beams this is good enough and uses a fairly good shear and bending system. The overall roof truss structure is in the form of wood or bamboo construction, with 95.50% using tile roof coverings.

In addition, based on the results of the analysis, 95% of houses carried out by extension do not take into account the connections of old and new buildings, so the strength of the extension building needs to be analyzed in detail. Connections between old buildings and new buildings will be very dangerous in the event of an earthquake.

The analysis was also carried out using FEMA 154, based on the results of the analysis that had been carried out, the buildings were mostly categorized as C3 type (Concrete Frame with Unreinforced

IOP Conf. Series: Materials Science and Engineering **650** (2019) 012040 doi:10.1088/1757-899X/650/1/012040

Manson infill), with very varied Plan Irregularity levels and Type E Soil (1-3 stories). The results of the analysis show that 58.76% obtained a value of RVS of 0.3 and 41.28% obtaining a value of 0.8. Based on FEMA 154 all of these buildings do not include buildings that are safe from earthquake forces.

4. Conclusion

Based on the results of the examination and discussion above, it can be concluded that the level of vulnerability is increasing if the addition of building area continues without regard to the level of vulnerability to earthquake forces. In addition, the level of vulnerability of other buildings will continue to increase so that regular supervision needs to be carried out so that it will be able to control the development of the area.

Acknowledgements

Thank you for being conveyed to the Department of Civil Engineering, Faculty of Engineering, Universitas Muhammadiyah Yogyakarta. This research was conducted on the basis of collaboration with the Bandung Institute of Technology and Kyoto University. Thank you also to the local government and the entire research team.

References

[1] Pribadi. K. S., Zulfiar. M. H., and Foralisa, M. 2012 *Study of Construction Industry Contribution in Achieving Safer Building on Disaster. Report of Joint Research ITB and BP Construction* (Jakarta)

[2] E-Batar, S. A. 2018 Seismic Vulnerability Evaluation of Existing R.C. Buildings. *Housing and Building National Research Center Journal* **14** Pp. 189-197

[3] Perrone. D., Aiello. M. A., Pecce. M., and Rossi, F. 2015 Rapid Visual Screening for Seismic Evaluation of RC Hospital Buildings *Structures* **3**. Pp. 57-70

[4] Ningthoujam. M. C., and Nanda, R. P. 2018 Rapid Visual Screening Procedure of Existing Building Based on Statistical Analysis *International Journal of Disaster Risk Reduction* **28** Pp. 720- 730

[5] Allali. S. A., Abed. M., and Mebarki, A. 2018 Post-Earthquake Assessment of Buildings Damage Using Fuzzy Logic *Engineering Structures* **166** Pp. 117-127

[6] Sen, Z. 2010 Rapid Visual Earthquake Hazard Evaluation ef Existing Buildings by Fuzzy Logic Modelling *Expert Systems with Applications* **37** Pp. 5653-5660

[7] Demartinos. K., and Dritsos, S. 2006 First Level Pre-Earthquake Assessment of Buildings Using Fuzzy Logic *Earthquake Spectra* **22** Pp. 865-885

IOP Conf. Series: Materials Science and Engineering **650** (2019) 012041 doi:10.1088/1757-899X/650/1/012041

The Properties of Waste Rubber Tires in Increasing the Damping of Masonry Wall Structure

R Faizah[1*], H Priyosulistyo[2] and A Aminullah[3]

[1]Ph.D. Student of the Department of Civil and Environmental Engineering, Universitas Gadjah Mada, Yogyakarta, Indonesia.
[2]Professor of the Department of Civil and Environmental Engineering, Universitas Gadjah Mada, Yogyakarta, Indonesia.
[3]Senior lecturer of the Department of Civil and Environmental Engineering, Universitas Gadjah Mada, Yogyakarta, Indonesia.

* restu.faizah@umy.ac.id

Abstract. The accumulation of waste rubber tires causes environmental problems due to most of them cannot be recycled into new tires. Recently, this waste is gradually used as a material replacement in civil engineering such as in increasing damping properties. This study investigates the physical and mechanical properties of waste rubber tires including density (ρ), ultimate tensile strength (σ), elongation at break, hardness (Shore A), modulus of elasticity (E), and shear modulus (G). The specimens used were coded as A, B, C, and D to represent specified brand name. The testing method referred to ISO standards and was carried out in the laboratory of Center for Leather, Rubber, and Plastics (CLRP), and the structural laboratory of Department of Civil Engineering and Environmental, Gadjah Mada University, Yogyakarta, Indonesia. The result shows that the density of all brands is nearly the same which is around 1.1 gr/cm^3. The A rubber tire is indicated as the best damping properties since it has the lowest value on tensile strength, hardness, modulus of elasticity and shear modulus. However, the elongation at break is the highest value, compared to the other specimens. B rubber tire shows hard rubber, while C and D are high strength rubber. Therefore, B, C, and D rubber tires are appropriate to be used as barrier supports which must be able to withstand large forces, while the damping is not a priority.

1. Introduction

Economic improvement of the Indonesian community generates an increase in the number of car ownership. This reality also improves the number of waste tire feared to be an environmental problem. A car tire consists of rubbers, fillers, reinforcing materials, plasticizers, and chemicals. The amount of rubber is around 40% of all components, so that the rubber material is the highest volume in the waste tires [1]. Therefore, expanding the utilization of rubber components from tires towards higher benefits is a good solution for environmental problems.

Currently, the utilization of waste rubber tires in Indonesia is very limited, including rubber formed into fenders, ropes, sandals, trash containers, and table-chairs sets. In Civil Engineering fields, waste rubber tires are also applied as a material mixture of soil [2], concrete [3], mortar and asphalt [4], in the form of rubber tire crumbs. The previous research shows the great benefit of the waste rubber tires as the replacement materials in civil engineering. Various studies indicate that the additions of waste rubber tire can reduce unit-weight and increase damping in concrete/mortar. The additions of rubber tire crumbs in the mortar mixture can increase the damping ratio and decrease the unit-weight of mortar [5]. However, in concrete, it can increase the damping ratio [6], [7] and sound insulation, but decrease thermal conductivity [8]. Faizah, Priyosulistyo and Aminullah [5] also recommended the addition of rubber tire crumbs on mortar mixture to improve damping capacity of masonry walls.

Several studies [5] – [7] found that the addition of waste rubber tire in mortar/concrete mixture can increase the damping capacity of concrete/mortar. Meanwhile, mortar is very popular in Indonesian which is used in the masonry wall structures as a bed and head joints. Boen [9] also stated that the masonry wall infill is most widely used in the Indonesian house construction. Masonry wall damage under an earthquake often results in fatalities, so it is important to find a technique to increase the masonry wall resistance under earthquake. Using the high damping mortar as a bed joint or head joint on the masonry wall structure is a good idea to do.

The previous research took waste rubber tire crumbs from the waste collectors of mix tire brands. Each tire brand has specific properties that can indicate its damping properties. Therefore, properties investigation of each tire brand is critical to get the proper utilization. Using certain brands of rubber tire that have advantages in increasing damping, of course, will result in a more significant damping increase comparing by using mixed rubber tire brands.

There are many tire brands found in Indonesia, i.e., Bridgestone, Dunlop, Gajah Tunggal, Swallow, MRF, Goodyear, Achilles, Pirelli, Yokohama, Michelin, and others. Each of them has specific properties according to their service description and vehicle's requirements explored in term of load index (LI) and speed index (SI). The maximum load capacity of a passenger car tire is indicated by its Load Index (LI), while the maximum permissible speed (at full load) of the tire is indicated by its Speed Index (SI) [1]. In this study, 4 tire brands among those mentioned above are examined, and is coded by A, B, C, and D.

The properties tested in this study are density (ρ), ultimate tensile strength (σ), elongation at break, hardness (Shore A), modulus of elasticity (E), and shear modulus (G). These properties can be used to indicate the characteristics of certain brands of rubber tire related to the consideration in selecting them for specific uses. Edeskar [2] reported that the density of rubber tires was slightly higher than the density of water so that it will sink in the water. Besides, the compact density of rubber tires was higher than the bulk density, ranges from 1.08 – 1.27 gr/cm^3 for its compact and 0.45 – 0.99 gr/cm^3 for bulk. This result is in accordance with Bijarimi's result which has a rubber compounds density value of 1.14 – 1.15 gr/cm^3 [9].

Previous research also shows the results of rubber compound materials testing including tensile strength, elongation at break, and hardness. The rubber compound is a base material of rubber tire blended from natural rubber and chemical additive. Hendrawan and Purboputro [10] reported that the tensile strength of rubber compound was around 16.5 - 21.2 MPa, and the elongation at break was 289 - 576%. Besides, the hardness (Shore A) was around 61 – 71 [11] or 71 – 77 [10]. The hardness and elongation value indicates the elasticity of a material. The lower hardness or the higher elongation means the more elastic material. The elastic material has a good ability to return its shape to the original shape after loading. The rubber having high hardness is more resistant to abrasion and extrusion, while softer rubber will be easier to stretch and close on rough surfaces. The ratio of stress to strain in the elastic area represents a modulus of elasticity of the material. The material having a higher value of modulus of elasticity indicates a more rigid material [12].

2. Experimental Method

The experiment was carried out in the testing laboratory of Center for Leather, Rubber, and Plastics (CLRP), and the structural laboratory of Department of Civil Engineering and Environmental, Gadjah Mada University, Yogyakarta, Indonesia. The testing materials are waste rubber tires consisting of 4 brands coded by A, B, C, and D. All of them were from the car tires that is shown in Figure1. The physical and mechanical properties were given some tests including the density test (ρ), ultimate tensile strength test (σ), axial elongation test, hardness test (Shore A), modulus of elasticity test (E), and shear modulus test (G).

IOP Conf. Series: Materials Science and Engineering **650** (2019) 012041 doi:10.1088/1757-899X/650/1/012041

Figure 1. The waste rubber tires from the car tires

2.1. *Density Test*
The density test refers to ISO 2781: 2008 (Rubber, Vulcanized or Thermoplastic - Determination of Density) [13], using electronic densimeter having a precision of 0.01 presented in Figure 2. The beaker in the densimeter has a capacity of 250 cm^3. The specimen is a piece of the rubber tire having a mass of at least 2.5 gr. There are two methods for obtaining specimen densities, namely automatic and manual.

The specimen was placed on the lip of the beaker, and the digital monitor will shows the mass in the air (m_1). Furthermore, this specimen was inserted in a beaker containing water to obtain the mass in the water (m_2). Density (ρ) of the specimens was known automatically from the digital monitor, but it can be calculated manually by Equation 1.

$$\rho = \rho_w \frac{m_1}{m_1 - m_2} \quad (1)$$

where:

ρ = the density of rubber (gr/cm^3)
ρw = the density of water in the site (gr/cm^3)
m1 = the mass of rubber in the air (gr)
m2 = the mass of rubber in the water (gr)

Figure 2. The electric densimeter

2.2. *Hardness Test*
Hardness is a measure of the resistance to a reversible deformation of the rubber and shows the elasticity of a material. The lower hardness indicates the more elastic material. In this experiment, the hardness testing used the durometer (Shore A) method according to ISO 7619-1: 2010 (Rubber, vulcanized or thermoplastic Standard - determination of identification hardness - durometer method/Shore Hardness) [14]. The automatic durometer hardness (Shore A) shown in Figure 3 has a solid spherical indenter by the dimensions is presented in Figure 4.

The pressure load of durometer (Shore A) is 1 ± 0.1 kg or 12.5 ± 0.5 N by a scale of penetration depth on 0 to 100 during 3 - 15 seconds. The specimens have a thickness of approximately 6 mm, and a surface area to get at least 3 test points. The distance between points must be more than 5 mm, and the distance from the edge is more than 13 mm. The hardness value was obtained automatically from the digital monitor of the machine [12].

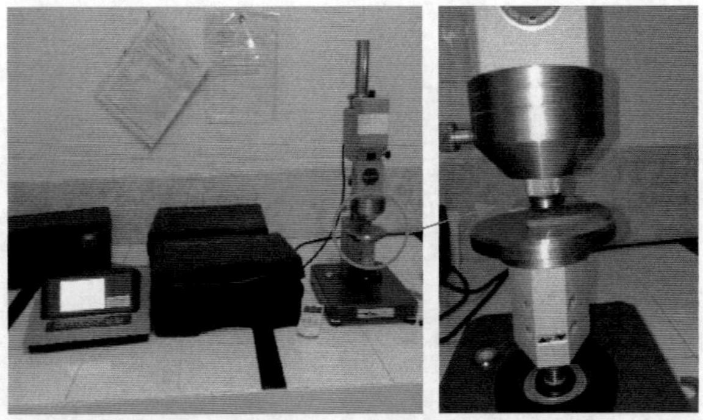

Figure 3. The automatic durometer hardness (Shore A)

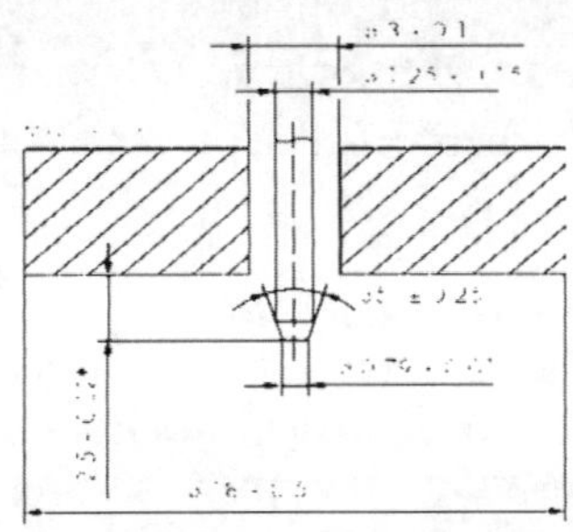

Figure 4. The dimension of indenter and presser foot durometer (Shore A)

2.3. Tensile Strength Test

Tensile strength is the maximum tensile stress that can be achieved in stretching pieces of specimens expressed by force per unit area of the initial cross-section of original specimens. The experimental procedure according to ISO 37: 2011 (Rubber, vulcanized or thermoplastic - determination of tensile stress-strain properties) [15]. The specimens are waste rubber tires in the form of dumb-bells as shown in Figure 5. It has 20 ± 0.5 mm of measurement length, and a thickness in a narrow section is 2 ± 0.2 mm. The experimental set-up is presented in Figure 6. The testing machine runs on 500 mm/minute grip speed until the specimen breaks.

Tensile strength (σ) is the maximum stress when the specimen broken expressed by force per unit area of the original cross section as mentioned in Equation 2.

$$\sigma = \frac{F}{A_0} \quad (2)$$

F is the nominal load applied to the specimen perpendicularly in units of Newton (N), and A_0 is a unit area of the original cross section before loading (m^2). The unit of tensile strength is MPa (SI) (where 1 MPa = 106 N/m^2).

In tensile strength test, there are three main parameters measured, namely tensile strength, elongation at break and modulus of elasticity of specimen. Elongation at break (%) is the maximum extension expressed as a percentage of extension at the break with original length before the sample is broken. Percent elongation at break is defined as Equation 3.

$$Percent\ elongation = 100\ \frac{L_B - L_0}{L_0} \quad (3)$$

L_0 and L_B denotation are the initial length of the tensile test specimen and its final length at rupture respectively [16]. Furthermore, the modulus of elasticity of specimen can be obtained by analyzing the strength-strain relationship curve resulted from the machine output automatically.

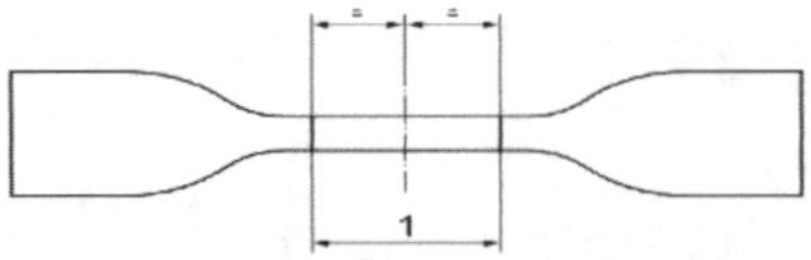

Figure 5. The dumb-bell specimens

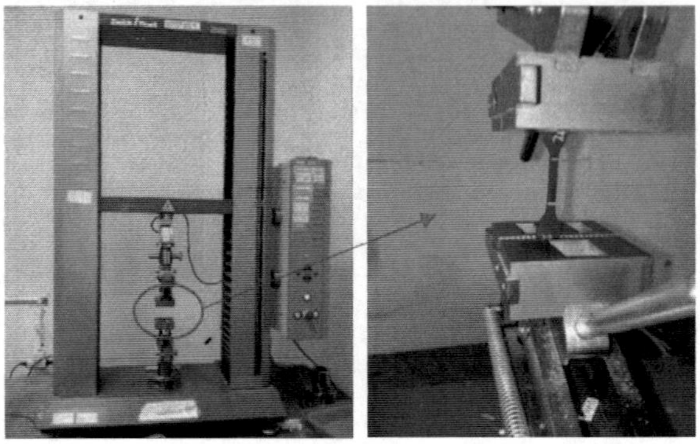

Figure 6. The experimental set-up of tensile strength test

2.4. Shear Modulus Test

Shear modulus or modulus of rigidity (G) which relates the components of the shearing stress and shearing strain is the coefficient of elasticity for a shearing force expressed by the ratio of shear stress to the displacement per unit sample length (shear strain). Shear modulus can be experimentally determined from the slope of a stress-strain curve created during tensile tests conducted on a sample of the material. The constant of proportionality relating shear stress and shear strain is the shear modulus. It is represented by G, in a unit of Newton per square meter (N/m^2) = Pascal (Pa) or pounds per square inch (psi) [16].

The specimen of the experiment is a waste rubber tire block with the length of a direction of withdrawal must be more than twice of the perpendicular direction. The set-up of the experiment is shown in Figure 7. The deformation of a rubber block if the forces (F) acting on the block is shown on the illustration in Figure 8, where F is the shearing load acting on the body, L is a height of the block, A is the area under shear and Δx is the deformation of the block. Furthermore, the shear stress induced in the body (τ), the shear strain (γ), and shear modulus (G) can be obtained using Equation 4.

$$G = \frac{\tau}{\gamma} = \frac{F/A}{\Delta x/L} = \frac{FL}{A\Delta x} \tag{4}$$

With G is a shear modulus, τ is shear stress, γ is a shear strain, *F* is acting forces, A is an area under shear, *L* is a height of the block, and Δx is the deformation of the block [16].

Figure 7. The experimental set-up of a shear modulus test

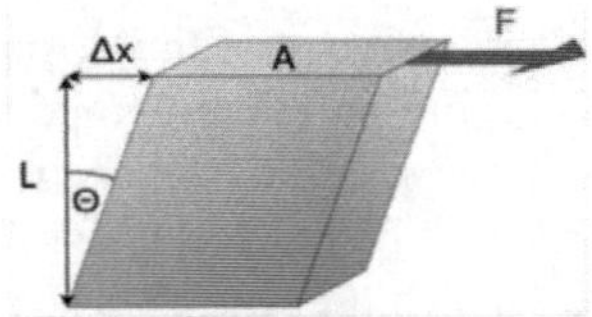

Figure 8. The illustration of deformation of rubber block due to shear force F

3. Result and Discussion

The physical and mechanical properties of waste rubber tires with a code of A, B, C, and D were investigated. The density test result shows that all specimens have a nearly same level of density, which is around 1.1 gr/cm^3. This value is in accordance with the results of Edeskar's research [2]. The density of waste rubber tire is lesser than the density of some materials such as sand, cement, lime, and asphalt. Therefore, the waste rubber tire might be the waste as a replacement material of sands in the concrete mixture to get the lighter concretes. Batayneh, Marie and Asi [17] confirmed that content increasing of the rubber tire crumbs in the concrete mix could reduce the slump and the unit-weight of the mixes. Faizah, Priyosulistyo, and Aminullah [5] also reported that the addition of rubber tire crumbs in the mortar mixture could reduce the unit-weight of the hardened mortar. Unit-weight of mortar with 40% and 60% rubber tire crumbs contents decrease up to 22% and 32% respectively. Besides that, the unit-weights of the mortar cubes containing 40% and 60% rubber tire crumbs can classify as a lightweight mortar because the unit weight is less than 1850 kg/m^3.

The specimen used in all of the experiment is prepared from a waste rubber car tire. Figure 9 shows a specimen of tensile strength test shaping a dumb-bells. The length of measurement is ± 20 mm, and a thickness in a narrow section is ± 2.0 mm. Some specimens break up in the middle, and some are slight to the edge under the tensile strength test, as shown in Figure 10.

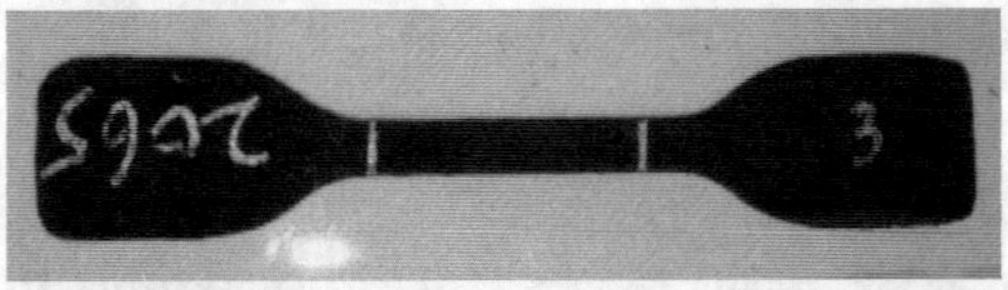

Figure 9. A specimen of tensile strength test (dumb-bells)

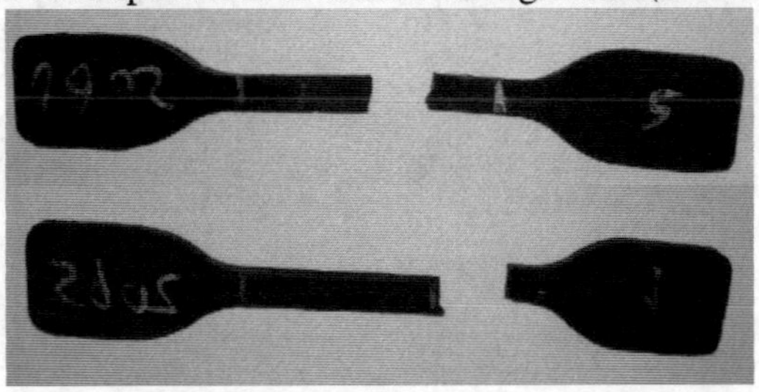

Figure 10. Specimens after a tensile strength test

The result of the tensile strength test given in Figure 11, shows that the value around of 10.64 – 14.358 MPa. B specimen has the highest value, but A specimen has the lowest value. These results exhibit that B specimen has the highest ability to withstand stress-loads which usually indicate a hard material, but A specimen is an opposite. The tensile strength result of this study is lesser than Bijarimis's result that was found around 16.5 - 21.2 MPa. Bijarimi investigated the tensile strength of the rubber tire compound [11]. The other report shows that the tensile strength of rubber compounds is around of 13.46 – 16.69 MPa [10]. The variation on this tensile strength value might be caused by the difference of rubber quality, chemical additive contents, and age of tires.

Elongation at break and modulus of elasticity were also obtained in conjunction with the tensile strength test given in Figure 12 and 13 respectively. The specimen can stretch up to 238 - 296% of its original length when forced stress. The A specimen has the highest elongation at break but the lowest on a modulus of elasticity. This result indicates that the A specimen is the softest and the most elastic one compared to other specimens. This finding is confirmed by the hardness of the test result given in Figure 14. The A specimen has the lowest value of hardness, while the B specimen has a highest, opposite to the tensile strength test result. The hardness test result is around 57.9 to 68.74, agrees with Sidabutar's report showed in Figure 15, which describes various elastomer hardness. The rubber has 20 – 90 hardness value and around of 60 for rubber tire cars. The compounds having higher durometer values indicate harder compounds [12].

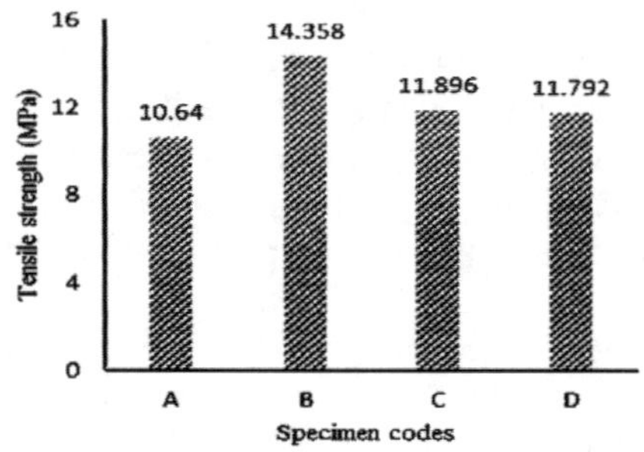

Figure 11. The tensile strength test

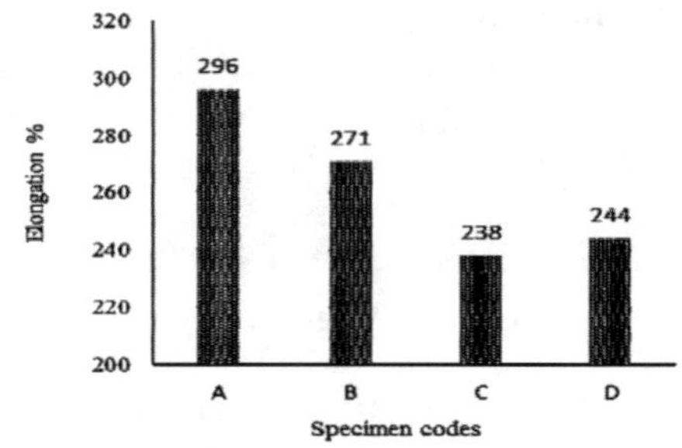

Figure 12. The elongation of the specimens

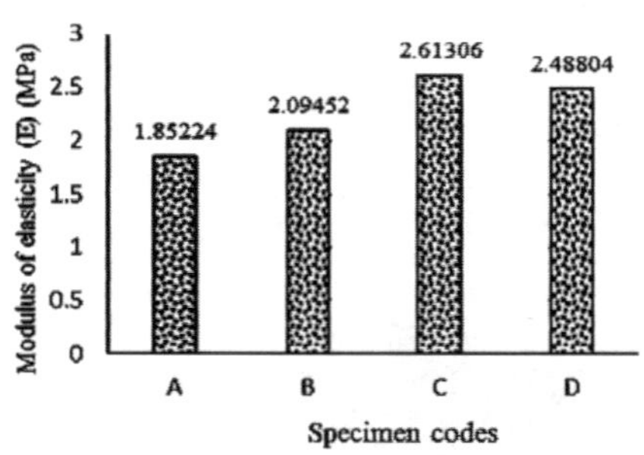

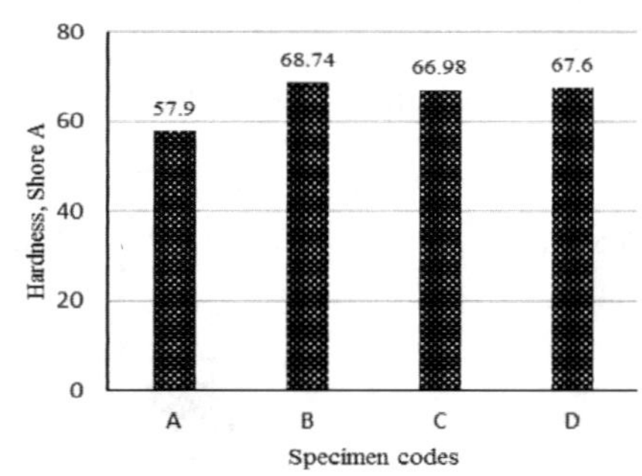

Figure 13. Modulus of Elasticity (E) **Figure 14.** Hardness (Shore A) of the specimens

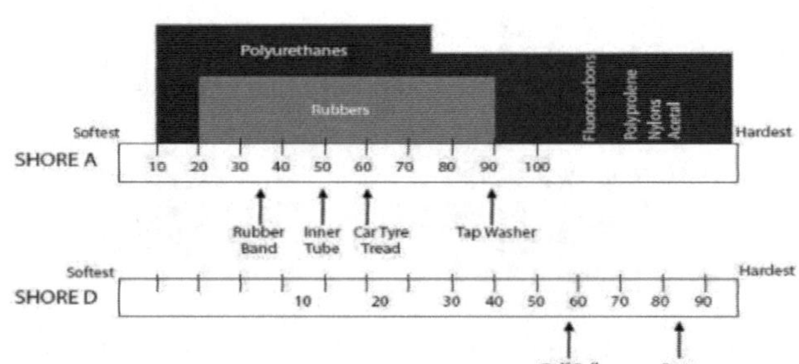

Figure 15. Various of elastomer hardness

The specimens were also examined by a shear modulus test, with an experimental set-up as represented in Figure 16. The specimen was attracted up to failure. The force was detected using a load-cell having a capacity of 1000 kg while a deformation was measured by a linear variable differential transformer (LVDT). After loading, the relationship curve between shear stress and shear strength can be described in Figure 17. From this figure, it can be obtained the shear modulus of all the specimens from the linear slope. The shear modulus of specimen is presented in Figure 18. The modulus of elasticity and shear modulus shows in-line results, where the highest to lowest values are owned by the C, D, B, A specimens respectively. The result of this study shows a good correlation between tensile strength, elongation at break, modulus of elasticity, hardness, and shear modulus of

IOP Conf. Series: Materials Science and Engineering **650** (2019) 012041 doi:10.1088/1757-899X/650/1/012041

specimens. All of the results are summarized in Table 1. The bold fonts show the minimum value, but the italic-underscore fonts show the maximum value.

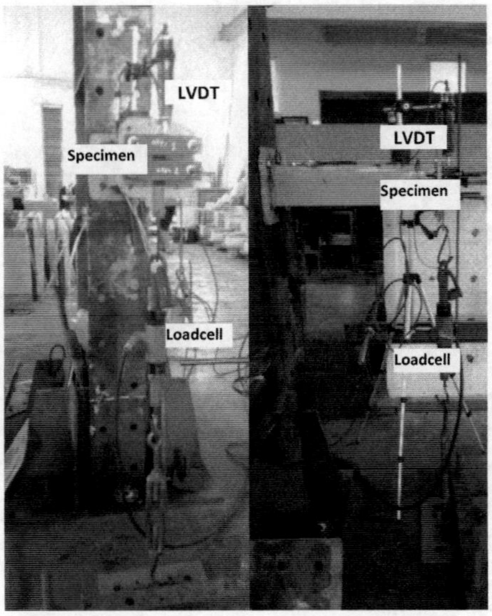

Figure 16. Experimental set-up of the shear modulus test

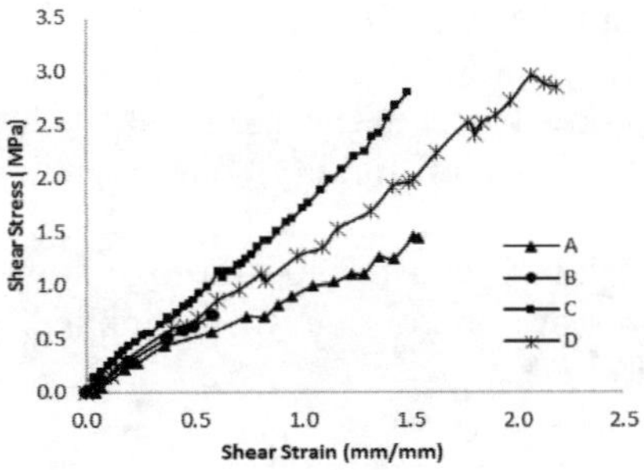

Figure 17. The relationships curve between shear strain and shear stress

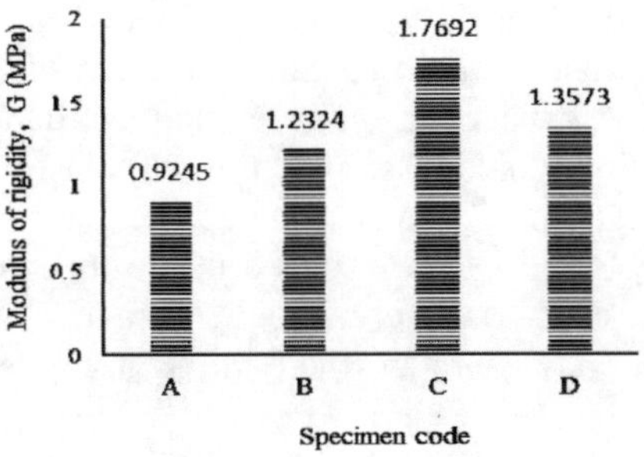

Figure 18. The shear modulus (G)

Table 1. The summary of the experimental result

No.	**The experiment**	**Specimen codes**			
		A	**B**	**C**	**D**
1	Density (gr/m^3)	1.112	1.090	1.120	1.136
2	Tensile Strength (MPa)	**10.640**	*14.358*	11.896	11.792
3	Hardness (Shore A)	**57.90**	*68.74*	66.98	67.60
4	Elongation at break (%)	*296*	271	**238**	244
5	Modulus of Elasticity, E (MPa)	**1.8522**	2.0945	*2.6131*	2.4880
6	Modulus of Rigidity, G (MPa)	**0.925**	1.232	*1.769*	1.357

Table 1 shows that the A rubber tire has the lowest value on tensile strength, hardness, modulus of elasticity, and shear modulus compared to the other specimens. However, the elongation at break was found to be the highest one. It can be analysed that the A rubber tire has soft and elastic properties since its tensile strength is the lowest one. This statement was confirmed by its elongation having the highest value. The high elongation at break indicates that the rubber tire is easy to stretch under stress-force. This characteristic is in accordance with the low value on a modulus of elasticity and shear modulus, representing the less stiff and less brittle materials. As a result, it has high stretch-ability under the loading. The materials that have a low value on a modulus of elasticity and shear modulus seem to indicate a high damping material [18]. This observation offers the finding that A rubber seems appropriate to be chosen as an additive or replacement material, which can increase the damping ability of the material.

The B rubber tire has the highest value on tensile strength and hardness indicating that B rubber does not easily change in shape. This rubber tire can withstand great force due to its hard property. Therefore, B rubber is appropriate to be used as a material that must resist a high impact as resisting axial load and abrasion coating materials.

The C rubber tire has the lowest elongation at break but is highest in modulus of elasticity and shear modulus. This property shows the good ability to keep the original shape when loaded by strength or shear forces. The C rubber tire might be appropriate to be used as an additive that prioritizes stiffness.

As shown in Table 1, the D rubber has an intermediate value on all properties compared to the other specimens. This fact indicates that the D rubber tire is an ordinary rubber having a general property. It does not have any specific property compared to the other rubber tires. Nevertheless, the D rubber tire has a secondary property after the C rubber tire as an additive that prioritizes stiffness. It is initiated by its value on elongation at break, modulus of elasticity and shear modulus which are almost the same with the C rubber tire.

4. Conclusion

Various tests have been carried out in this study with the specimens from waste rubber tire having A, B, C and D brands code. The findings of this study can be summarized as follows.

1. All waste rubber tires have a density of around 1.1 gr/cm^3.
2. The A rubber tire indicates the best damping properties since it has the lowest value on tensile strength, hardness, modulus of elasticity and modulus of rigidity, but having the highest elongation at break compared to the other specimens.
3. The B rubber tire shows a hard rubber material, while C and D shows a high strength rubbers materials. Therefore, the B, C, and D rubber tires are appropriate to be applied as a barrier supports materials, where the damping is not a priority.

Acknowledgment
The authors gratefully acknowledge the Research Directory of Gadjah Mada University which has given the Rekognisi Tugas Akhir (RTA) fund. The author also gave the highest appreciation to Universitas Muhammadiyah Yogyakarta as the main institution of the first author.

References

[1] Continental, Tyre Basics - Passenger Car Tyres. Continental Reifen Deutschland GmbH., 2013.
[2] Edeskär T 2006 *Use of Tyre Shreds in Civil Engineering Applications Technical and Environmental Properties* (Lulea University of Technology)
[3] Bisht K,and Ramana P V 2017 Evaluation of mechanical and durability properties of crumb rubber concrete *Construction and Building Materials* vol. **155**, pp. 811–817,
[4] FHWA-HIF 2014 *The Use of Recycled Tire Rubber to Modify Asphalt Binder and Mixtures* (US Department of Transportation Federal Highway Administration)
[5] Faizah R, Priyosulistyo H, and Aminullah A 2019 An Investigation on Mechanical Properties and Damping Behaviour of Hardened Mortar with Rubber Tire Crumbs (RTC) *MATEC Web of Conferences* vol. **258**, p. 05002
[6] A. Moustafa and M. ElGawady, "Damping Properties of High Strength Concrete with Scrap Tire Rubber," in Researchgate,net, 2015, p. 10.
[7] Youssf O, ElGawady M A, and Mills J E 2015 Experimental Investigation of Crumb Rubber Concrete Columns under Seismic Loading *Structures* vol. **3**, pp. 13–27
[8] Aliabdo A A, Abd Elmoaty A E M, and AbdElbaset M M 2015 Utilization of waste rubber in non-structural applications *Construction and Building Materials* vol. **91**, pp. 195–207,

[9] Boen T *Yogya Earthquake 27 May 2006, Structural Damage Report* p. 22.

[10] Hendrawan M A, and Purboputro P I 2015 Studi Karakteristik Sifat Mekanik Kompon Karet dengan Variasi Komposisi Sulfur dan Carbon Black sebagai Bahan Dasar Ban Luar, in *Simposium Nasional Teknologi Terapan (SNTT)* **3** p. 8.

[11] Bijarimi M, Zulkafli H, and Beg M D H, 2013 Mechanical Properties of Industrial Tyre Rubber Compounds *Journal of Applied Scince,*. [Online]. Available: https://scialert.net/fulltext/?doi=jas.2010.1345.1348. [Accessed: 23-Feb-2019].

[12] Sidabutar V 2013*Metoda Pengujian Sifat Fisik Barang Jadi Karet (Buku Ajar)* (Kementrian Perdagangan RI)

[13] ISO, ISO 2781:2008E (Rubber, Vulcanized or Thermoplastic - Determination of Density). 2008, p. 22.

[14] ISO 2010 *ISO 7619-1: 2010 (Rubber, vulcanized or thermoplastic Standard - determination of identification hardness - durometer method/Shore Hardness)*

[15] ISO 2011 *ISO 37:2011 (Rubber, vulcanized or thermoplastic-determination of Tensile Stress-Strain Properties).*

[16] F. P. Beer, 2007., Mechanics of materials, 6th ed. New York: McGraw-Hill, 2011.

[17] Batayneh M K, Marie I, and Asi I 2008 Promoting the use of crumb rubber concrete in developing countries *Waste Management,*vol. **28**, no. 11, pp. 2171–2176

[18] Siddique R, and Naik T R 2004 Properties of concrete containing scrap-tire rubber – an overview *Waste Management* vol. **24**, no. 6, pp. 563–569.

IOP Conf. Series: Materials Science and Engineering **650** (2019) 012042 doi:10.1088/1757-899X/650/1/012042

A joint choice decision model of intra-household interaction-based motorcycle mode and departure time in Yogyakarta, Indonesia

Muhammad Zudhy Irawan

Department of Civil and Environmental Engineering, Universitas Gadjah Mada, Yogyakarta, 55281, Indonesia

zudhyirawan@ugm.ac.id

Abstract. The motorcycle-based joint trip between family members in a household becomes a common phenomenon in responding to unreliable public transport. This study aims to model the joint choice decision between intra-household interaction-based joint trip using motorcycle and departure time. With the data collected from 416 workers using the motorcycles in Yogyakarta, Indonesia, a cross-nested logit model was developed to explore the individual and household characteristics, work-and travel-related characteristics influencing household-based joint trip and departure time. The model result reveals that motorcyclists are more likely to change their departure time rather than an intra-household-based joint trip. Motorcyclists departing on a peak hour have a lower elasticity due to the alteration of travel time and travel cost than motorcyclists leaving on an off-peak hour. In a case of the off-peak departure time, motorcyclists leaving home to work by driving alone are more elastic for a change of travel cost than motorcyclists departing to work with their family member.

1. Introduction

1.1. Background

Motorcycles become a primary mode of urban transportation in Southeast Asia, such as Indonesia, Thailand, Vietnam, and Malaysia. Mixed land use and narrow streets in Indonesia have caused motorcycles as a reliable travel mode to connect the point of origin to the point destination [1]. The number of motorcycles in Indonesia increases every year by reaching 105.15 million units in 2017, and it has significantly increased by 52.75% in the last five years. In 2016, motorcycles were accounted for 81.33% of all registered motorized vehicles [2]. Motorcycles are also indicated as one of the major contributors to a road accident, and it increases in accident fatality rates [3]. The Indonesian Ministry of Transportation recorded that 108,883 motorcycles were involved in traffic accidents in 2014 [4]. Meanwhile, forty-five percent of road accidents in Malaysia involved motorcycles, with 60% of them resulted in fatalities [5].

Understanding motorcycle ownership and usage in Asian countries have been carried out on many studies aiming to reduce motorcycle growth and to control its usage [6-9]. Several previous studies are also carried out concerning motorcycle fatalities [5,10,11]. However, there is a lack of studies on motorcyclists' behavior considering the departure time choice decision and the effect of intra-household interaction. It is crucial because intra-household interaction influences the mode choice

IOP Conf. Series: Materials Science and Engineering **650** (2019) 012042 doi:10.1088/1757-899X/650/1/012042

decision of family members [12]. In a case of the motorcycle-based joint tour between the family members in a household has become a way out due to the unreliable public transport service [13], analyzing the departure time choice and the intra-household interaction-based travel mode choice could be an approach to explore deeply the motorcyclists' behavior for a home-based work trip.

This study aims to model the joint choice of departure time and intra-household interaction-based ridesharing for the motorcyclist in Yogyakarta Urban Area and to understand the various factors influencing both of those choices. Some previous researches have shown that travelers choose travel mode and departure time simultaneously [14,15]. By understanding the interactions between the two dimensions, demand-side policies can be proposed aiming both to reduce the households' dependency on motorcycles and to increase the propensity of mode shifting from motorcycle to public transport. Understanding the departure time choice could also be a solution to decrease traffic congestion on peak hours or to spread traffic congestion outside peak hours, which remains one of the most common transport problems in urban areas. Based on random utility maximization theory, a cross-nested logit model within the theoretical framework of generalized extra value (GEV) class [16] was used to capture the relationship between alternatives for both dimensions of departure time and ridesharing mode.

The remainder of this paper is organized as follows. The next sub section gives a brief description of the study area. The second section reviews the main contributions to the choice of departure time and the effect of intra-household interaction on a joint trip. The third section describes the data collection. The fourth section defines a detailed formulation of the proposed models in this study. Furthermore, the fifth section presents detailed model estimation results and discussions. The last section provides the main conclusions and recommendations for further study.

1.2. Study Area: Yogyakarta Special Region

This section provides a brief description of the Yogyakarta Special Region (YSR) as an area of study. YSR is one of two special provinces in Indonesia with a population of 3.7 million in 2016 [17]. YSR consists of five regions: Yogyakarta as a capital city, Sleman, Bantul, Gunung Kidul, and Kulon Progo. As a capital city, a high number of motorized vehicles enter to Yogyakarta from its vicinity areas resulting in severe traffic congestion. The rise in the urban population linked to economic growth and motorization has led to an expansion of Yogyakarta city. Being adjacent to the city of Yogyakarta, Sleman and Bantul Regency have high urban growth attributing to residential projects. In 2016, YSR Agency for Land and Land Use introduced a Yogyakarta Urban Area (YUA) consisting the City of Yogyakarta, a part of Sleman and Bantul Regency [18]. YUA has an area of 19,651 hectares and includes 71 sub-districts in Yogyakarta, 15 sub-districts and 11 sub-districts in Sleman and Bantul, respectively.

Besides being a tourist destination, YSR is a city of education in Indonesia with more than 19 universities and colleges by the student population of about 351,293 in 2016. It nearly accounts for 9.5% of the YSR total population [17]. Almost all of the university students who come from the outside of YSR brought the motorcycle from their hometown, causing the number of the operated motorcycle is more than the registered motorcycle. There are 1,115,145 registered motorcycle in YSR in 2016 [19]. There is also a high number of elementary schools and high schools in YSR. Many students in elementary school and junior high school are escorted using motorcycle by their family members when they depart to/return from school. For senior high school students, most of them use a motorcycle because their age has met the minimum age threshold to have a driving license.

On the road, motorcycles are the dominant transport mode by around two-third of traffic. It also appears that more than one people occupy a motorcycle. Meanwhile, comparing the number of population and the number of registered motorcycles in YSR, the ratio value reaches 0.89 meaning that every people residing in Yogyakarta have a motorcycle mode. People living in YSR highly depend on a motorcycle in fulfilling their daily activities. They prefer to use a motorcycle compared to walking although on a trip less than 200 meters away. The high motorcycle dependence has also caused the disinterest on public transport use although the government has significantly improved the quality of public transport and the service coverage area. Survey result on public transport use (i.e., Trans-Jogja Bus) in 2017 reveals that the average load factor from 17 licensed bus routes reaches 17.35% only which consisted of 27.43% load factor on the 6 existing routes, and 11.84% load factor on the 11 new routes [20]. In YSR, purchasing a motorcycle is easy and cheap through a simple credit process. With a down payment of 500 thousand IDR (USD 34.45), people can own a new motorcycle with a price of around 20 million IDR (USD 1,378). In addition, since there is no policy of the motorized vehicle age limitation, the availability of the used motorcycles with a very affordable price makes it is much easier for people to own a motorcycle. Perhaps, when people decide to buy a used motorcycle by a credit system for the four-year duration, the monthly expenses to pay the motorcycle installment, tax, and operating costs, are insignificant different with the money that must be spent to buy a monthly ticket for using public transport.

2. Literature review

Studies on departure time have been extensively researched over the past decades in both dynamic user equilibrium analysis and econometric modeling. However, only a few studies incorporating the impact of intra-household interaction on the decision-making process of departure time. Yang *et al* [21] used the nested logit (NL) and cross-nested logit (CNL) model structures to estimate the joint choice of residential location, travel mode, and departure time for 8,900 households in Beijing. They pointed out that some factors such as house price, travel time, travel cost, working time flexibility, age, income, and car ownership was found as the influential variables on those three choice dimensions. They found that the departure time is the lowest substitutability and the residential location is the highest substitutability. It means that the travelers are more likely to change their departure time and travel mode rather than their residential location. In addition, travelers are willing to alter their departure time first rather than their travel mode if there is a change in travel time or other influential variables.

Ding and Mishra [15] applied the CNL model to the joint choice departure time and travel mode for urban commuting trips using a revealed preference data collected in the Maryland-Washington DC Region. Four travel modes: (1) drive alone, (2) transit, (3) shared ride, and (4) walk and bike, and two departures time: (1) peak and (2) off-peak were used in the choice decision model. The result found that males were more likely to drive alone and to use public transport, walk and bicycle relative to females, while young adults were less likely to drive alone. People who work in the CBD area tend to use public transport, walk and bicycle rather than drive alone. Also, individuals with flexible working time are more likely to depart to work on the off-peak period by using public transport, walk and bicycle, and drive alone.

Reviewing previous studies of motorcycle usage, it found that individual features such as socio-economic and demographics have a strong relationship with the motorcyclists' commuting behavior. By involving 503 students in six universities in Danang, Vietnam and applying a conditional logit regression model, Nguyen-Phouc *et al* [22] found that factor of travel time, travel distance, age, gender, income, bicycle and motorcycle ownership, driving license, location of school and living status have a strong correlation to mode choice decision. Surprisingly, travel cost does not influence mode choice. They also found that improving public transport service could potentially shift the motorcyclists to public transport users. Tran *et al* [23] revealed that household characteristics have a significant effect on the mode choice of motorcycle. They found that although 79% of 2605 respondents in Hanoi, Vietnam choose to walk from home to urban mass railway transit station for the trip distance less than 1 kilometer, 17% of them remain using motorcycles. Irawan *et al* [24] displayed that more than 70% of train users in Yogyakarta choose to use motorcycle from the train station to their destination place.

3. Data collection

Involving 416 respondents residing in YSR, an interview survey was conducted from March 6 to April 10, 2017. The selected respondent are people working in the City of Yogyakarta and riding a motorcycle during a trip to work. There are five parts in questionnaire form, as follows: (1) Individual characteristics: gender, age, and income, (2) Household characteristics: home location, number of motorized vehicle ownership, and number of family members, (3) travel-related features: travel time, travel cost, and number of time-constrained activities in a day, (4) work-related attribute: strict/flexible time to start working, and (5) trip chain.

Table 1 presents the descriptive statistics of the respondent. The sample consists of nearly 56% of men, and it is dominated by people aged 25-34 years old and people with income 1.5-3 million IDR by 38.46% and 42.55% respectively. The average number of family members in the household is 2.09, while the average number of motorized vehicle ownership is 2.29. Due to this, the ratio between the number of family members and the number of motorized vehicles in a household is 91.27% meaning that a family member in a household has one motorized vehicle. The sample also shows that only 18.03% respondents are living within the City of Yogyakarta (urban area), and 20.67% of them are residing on the outside of Yogyakarta Urban Area (rural area). The rest (61.30%) is residing in the suburban area (i.e., Yogyakarta Urban Area, but excluding the City of Yogyakarta). Taking into account the travel-and work-related characteristics, the average travel time from home to the workplace is 21 minutes, and the average trip cost is 5,348 IDR. Thirty-seven percent of respondents have two time-constrained activities, and 29% of them have more than or equal to three time-constrained activities in a day. Meanwhile, the proportion of respondents with work starting time flexibility is 27.64%.

Table 1. Descriptive statistics of respondents.

Variables		Mean	St. Dev.	N	%
Individual Characteristics					
Gender	Male	-	-	232	55.77
	Female	-	-	184	44.23
Age	20-24 years old	-	-	80	19.23
	25-34 years old	-	-	160	38.46
	35-44 years old	-	-	81	19.47
	45-54 years old	-	-	80	19.23
	55-64 years old	-	-	15	3.61
Income	< 1.5 million IDR	-	-	106	25.48
	1.5-3 million IDR	-	-	177	42.55
	3-6 million IDR	-	-	115	27.64
	> 6 million IDR	-	-	18	4.33
Household characteristics					
Number of a family member in a household		2.09	1.32	-	-
Number of motorized vehicle ownership		2.29	1.42	-	-
Home location	Urban	-	-	75	18.03
	Suburban	-	-	255	61.30
	Rural	-	-	86	20.67
Work-related characteristics					
Type of work starting time	Strict	-	-	301	72.36
	Flexible	-	-	115	27.64
Travel-related characteristics					
Travel cost (IDR per trip)		5,348	3,272	-	-
Travel time (minutes per trip)		20	11	-	-
Number of time-constrained activities in a day	1 trip	-	-	144	34.62
	2 trips	-	-	153	36.78
	≥ 3 trips	-	-	119	28.61

Considering the motorcyclists' trip chain, there are eight home-based trip chains to the workplace as shown in Figure 1 consisting of four single tours and four joint tours (ridesharing). The distribution of each home-based trip chain is displayed in Figure 2. It can be seen that 56% of motorcyclists' in

YSR (235 respondents) visit other places for escorting family members, eating, visiting friends, and shopping before departs to the workplace. From 56% of them, 179 respondents conduct a motorcycle-based ridesharing behavior. Concerning the trip chain of escorting family members first before leaving to the workplace, 64% of them (107 respondents) directly depart to the workplace after accompanying their family members, while 10% (18 respondents) and 26% (43 respondents) of them have to visit other places and going home, respectively, before departing to the workplace. Meanwhile, eleven respondents escort their family members after conducting other activities such as shopping, visiting friends, and eating. Referring to Figure 1 and Figure 2, there are 43% of motorcyclists (179 respondents) who jointly travel with their family member by using motorcycle consisting of 107 respondents with type C trip chain, 44 respondents with type E trip chain, 11 respondents with type G trip chain, and 17 respondents with type H trip chain.

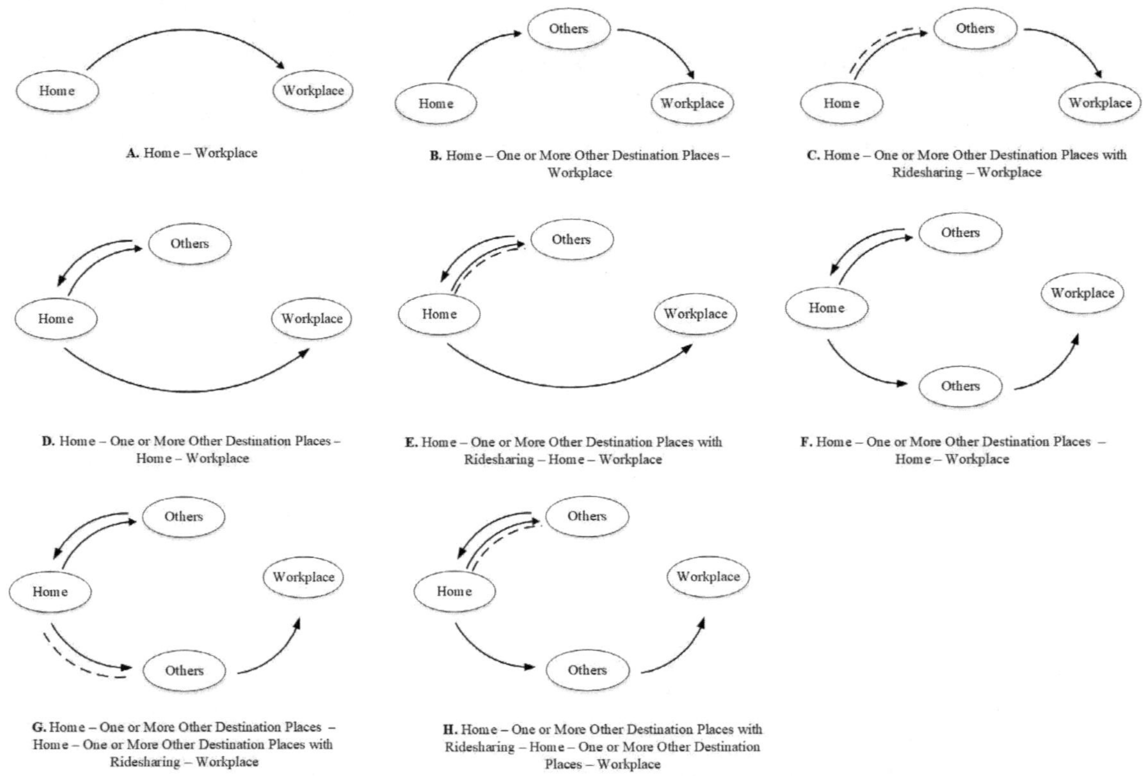

Figure 1. Types of motorcyclists' home-based tour in YSR.

Furthermore, Figure 3 shows the distribution of motorcyclists' departure time and total trip distance from home to the workplace. It reveals that the majority of motorcyclists depart from their home from 06.30 to 07.30 by 57.69%. Meanwhile, the percentage of motorcyclists' total trip distance from home to the workplace tends to be evenly distributed, with dominated by a trip distance ranging from three to four kilometers by 14%.

4. Proposed model

As previously explained, this study emphasizes the joint choice behavior of intra-household interaction-based joint trip using motorcycle mode and departure time for home-based work trips. There are two types of the joint trip using motorcycle mode choice: ridesharing and drive alone, while departure time choice subset consists of two alternatives: peak and off-peak. Due to this, the joint model of motorcycle-based joint trip and departure time creates four alternatives for each decision makers. It should be noted that in this study, peak hour is classified for motorcyclists departing from 06.30 to 7.30.

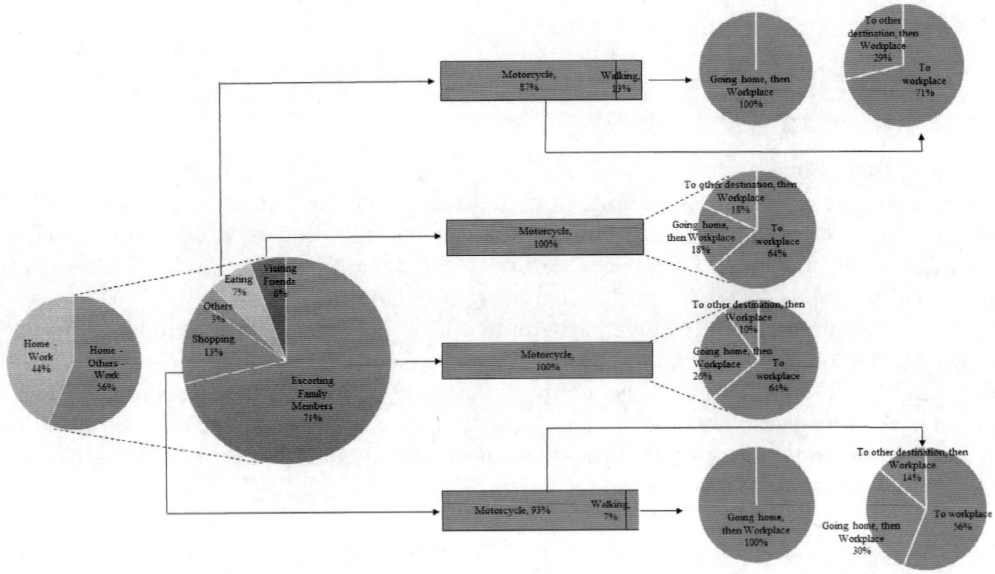

Figure 2. The distribution of motorcyclists' home-based tour in YSR.

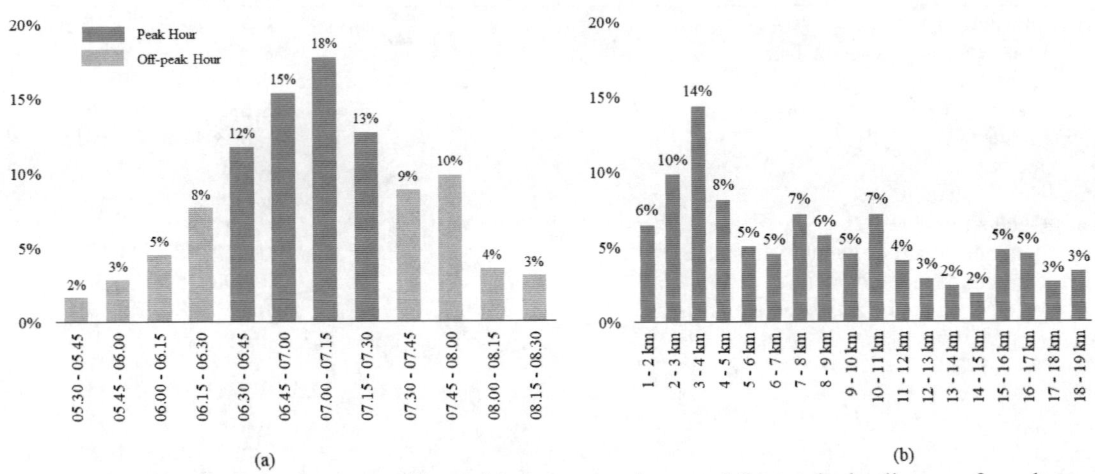

Figure 3. The distribution of motorcyclists' (a) departure time, and (b) total trip distance from home to the workplace.

Referring to the Generalized Extreme Value theorem [25-27], the choice probability of a cross-nested logit model alternative i (P_i) being chosen is defined in terms of conditional probabilities $(P_{i|m})$ and marginal probabilities (P_m) is shown as follows.

$$P_i = \sum_m P_{i|m} . P_m = \sum_m \left[\frac{\left(\tau_{im} . e^{V_i}\right)^{\frac{1}{\mu_m}}}{\sum_{j \in A_m} \left(\tau_{jm} . e^{Vj}\right)^{\frac{1}{\mu_m}}} . \frac{\left\{ \sum_{j \in A_m} \left(\tau_{jm} . e^{Vj}\right)^{\frac{1}{\mu_m}} \right\}^{\mu_m}}{\sum_l \left\{ \sum_{j \in A_l} \left(\tau_{jl} . e^{Vj}\right)^{\frac{1}{\mu_l}} \right\}^{\mu_l}} \right] \tag{1}$$

Where: $j \epsilon A_m$ is the set of all alternatives that belong to nest m, m and l are indices used to sum over all nests, τ_{im} are unknown parameters that characterize the portion of alternative i assigned to a nest in which τ_{im} are non-negative and must sum to one for every alternative. The parameters in Equation 1 are estimated based on the maximum likelihood method [28].

Furthermore, Wen and Koppelmen [26] formulated direct elasticities (*DE*) for cross-nested logit model is as follows.

$$DE = \left[(1-P_i) + \sum_m \left(\frac{1-\mu}{\mu} \right) \frac{P_{i|m}.P_m(1-P_{i|m})}{P_i} \right] \beta_k . X_{ik} \quad (2)$$

5. Result and discussion

The estimated parameters of cross-nested logit model are shown in Table 2, while the values of direct elasticity are displayed in Table 3. Direct elasticity is the variation in a decision maker's choice probability due to a change in one of the variables affecting that alternative. All the coefficients presented in Table 2 and Table 3 were estimated using Biogeme [29].

The model result shows that motorcyclists with flexible work starting time tend to depart during the off-peak period and to escort their family members before or on the way to the workplace. Taking into account the household characteristics, the number of a family member in a household shapes behavior of ridesharing. A more family member in a household increases the behavior of ridesharing by using motorcycles. Meanwhile, motorcyclists residing in the suburban and rural area are more likely to drive alone and to depart during the peak period.

Table 2. Estimation result of the cross-nested logit model.

Attributes/ data fit measures	Cross-nested logit	
	Parameter	t-statistic
Constant $_{\text{driving alone – peak period}}$	0.330	1.19
Constant $_{\text{ridesharing – off peak period}}$	0.198	1.84[c]
Constant $_{\text{ridesharing - peak period}}$	-0.069	-0.38
Age	0.632	4.75[a]
Gender	0.813	3.10[a]
Income	0.746	3.04[a]
Home location	0.460	2.43[b]
Occupation	0.765	3.04[a]
Number of family members in a household	0.521	7.19[a]
Number of time-constrained activities	-0.796	-3.56[a]
Number of motorized vehicle ownership	0.898	6.17[a]
Travel cost	-0.894	-7.47[a]
Travel time	-0.823	-2.30[b]
μ_{peak}	0.455	5.37[a]
$\mu_{\text{off peak}}$	0.636	2.44[a]
$\mu_{\text{driving alone}}$	0.359	10.87[a]
$\mu_{\text{ridesharing}}$	0.546	6.14[a]

Final LL = -567.334, Rho square = 0.315, AIC = 1176.667, BIC = 1261.311
Note: [a] Significance at the 1% level, [b] Significance at the 5% level, [c] Significance at the 10% level

Looking into the model fit in Table 2, the final log-likelihood value at the estimates as -567,334 with the rho square is 0.315. The Akaike and Bayesian information criterion (AIC and BIC) are 1176.667 and 1261.311 respectively. From Table 2, it also can be seen that, as expected, the coefficients for travel time (-0.823), travel cost (-0.894), and the number of the time-constrained activities (-0.796) have negative signs, and are statistically significant at levels of confidence well above the 95% limit. Flexible work starting time and living in an urban area contribute to the increase of utility function for workers using motorcycle mode by the coefficients of 0.765 and 0.460, respectively. Comparing among independent variables, variable of travel cost has a high coefficient value and the most significant variable influencing in both departure time and ridesharing behavior. Meanwhile, even though the variable of travel time also has a high coefficient value, its significant value at the confidence level does not reach above 99% limit. The value of constant is significant for driving alone only (0.198) that signify at 10% level, meaning that share driving by motorcycle and depart on the off-peak period produce the highest utility. However, it should be pointed out that the constant value for the behavior of driving alone on peak-period was fixed as one aiming to normalize the parameters.

From the value of dissimilarity parameters in the cross-nested logit, it shows that the dissimilarity parameter on ridesharing behavior (0.546) is smaller than departure time decision on the off-peak period (0.636), meaning the alternatives in the nest of ride-sharing behavior have high substitutability. If the utility variables alter, the motorcyclists will change their departure time first in a condition when they depart on the off-peak hour. Meanwhile, since the dissimilarity parameter on ridesharing behavior (0.546) is higher than departure time decision on peak hour (0.455), for motorcyclists departing on peak hour tend to change their joint tour into single tour first before changing their departure time. Furthermore, driving alone has the lowest dissimilarity parameter by 0.359 representing that when the utility variable change, they tend to shift their departure time in both peak and off-peak hour situation.

Table 3. Direct elasticities.

Departure Time	Driving alone		Ridesharing	
	Travel Cost	Travel Time	Travel Cost	Travel Time
Peak	-0.023	-0.008	-0.024	-0.008
Off peak	-0.041	-0.012	-0.027	-0.011

Concerning direct elasticities computing results as shown in Table 3, the average value of direct elasticities for travel time (0.010) are triple less than travel cost (0.029) meaning that motorcyclist has a higher sensitivity to travel cost than travel time. According to the results, a 1% increase in the travel cost for motorcyclists who are driving alone and departing on the off-peak hour to become available results in 0.041% decrease in the probability of driving alone and leaving home on the off-peak hour. Except on above condition, it can be seen that the values of elasticity in both travel time (0.008 and 0.011) and travel cost (0.024 and 0.027) for motorcycle-based ridesharing on peak period is slightly lower than driving alone (0.008 and 0.012 for travel time, and 0.023 and 0.041 for travel cost). Meanwhile, for the off-peak period, motorcyclists departing to work by escorting their family member are less elastic for a change of travel cost (0.027) than motorcyclists departing to work by driving alone (0.041). In addition, motorcyclists who depart on off-peak period are more sensitive to change in both travel time (0.012 and 0.011) and travel cost (0.041 and 0.027) than motorcyclists who depart on peak period (0.008 for travel time, and 0.023 and 0.024 for travel cost).

Taking into account the low elasticity values on both travel time and travel cost, it indicates that motorcyclists tend to be inelastic with respect to travel time and travel cost. Even though in this study does not consider the decision process of public transport choice, however, this situation is felt threatening the existing public transport (i.e. Trans-Jogja Bus) and the proposed rail-based public transport [30] in YSR. The transportation planners and policymakers must capture this situation to keep the existing demand and to increase the future demand for public transport in YSR.

6. Conclusion

Accommodating the effect of intra-household interaction on the decision-making process of departure time is a fundamental importance in understanding motorcyclists' travel behavior in YSR. By interviewing 416 workers as motorcyclists, this study has discussed the combined choice of intra-household interaction-based joint tour and departure time.

A cross-nested logit model was used to investigate the effect of individual characteristics (gender, income, age), household characteristics (number of family member, number of vehicle ownership, home location), work-related characteristic (flexible/strict work starting time), and travel-related characteristics (travel time, travel cost, number of time-constrained activities) on the decision process of motorcycle-based joint tour and departure time.

The result shows that a more family member in a household increases the tendency of a joint tour. Motorcyclists residing in an urban area with work starting time flexibility are more likely to depart on the off-peak hour by escorting their family member first. Comparing among alternative, a joint trip conducted on off-peak hour has the highest utility than other options. Travel cost becomes the most significant variable shaping the combined choice of intra-household interaction-based ridesharing and departure time. Motorcyclists also have a higher sensitivity to travel cost than travel time. Meanwhile, comparing between the choice of joint trip and departure time, if there is a change of influential variables such as travel time, travel cost, etc., individuals prefer to alter their departure time rather than a joint trip. In a condition when they have to depart to work on off-peak time duration, they are more sensitive regarding a change of travel time and travel cost than if they leave to work on peak time duration.

A future study could incorporate public transport mode into the existing joint model. Demand-side policies could be proposed aiming to influence the motorcyclists not only to change their departure time but also to shift them as public transport users. In addition, the emergence of motorcycle-based ride-sourcing must be anticipated by the policymakers whether it acts as a substitution, complement, or nothing for both motorcycle and public transport mode [31]. It would be also interesting to include electric motorcycle [1] in order to explore how the existing motorcyclist behaves and reacts due to the emergence of a motorcycle-based new alternative mode and further analyze the potential market share of this mode by reflecting on the failure of hybrid cars in replacing the gasoline-powered cars in Indonesia [32]. It would of interest to explore whether alternative travel mode provides additional insights regarding the decision process of departure time choice and intra-household interaction-based ridesharing behavior.

References

[1] Guerra E 2019 Electric vehicles, air pollution, and the motorcycle city: A stated preference survey of consumers' willingness to adopt electric motorcycle in Solo, Indonesia *Transportation Research Part D* **68** pp 52-64

[2] Central Bureau of Statistics 2017 Number of Registered Motorized Vehicle in Indonesia 1949-2016

[3] Ranney M L, Mello M J, Baird J B, Chai P R and Clark M A 2010 Correlates of motorcycle helmet use among recent graduates of a motorcycle training course. *Accident Analysis and Prevention* **42** pp 2057-2062.

[4] Indonesian Ministry of Transportation 2015 *Number of Traffic Accident in Indonesia* Jakarta

[5] Manan M M A and Várhelyi A 2012 Motorcycle fatalities in Malaysia. *IATSS Research* **36** pp 30-39

[6] Wen C H, Chiou Y C and Huang H L 2012 A dynamic analysis of motorcycle ownership and usage: A panel data modeling approach *Accident Analysis and Prevention* **49** pp 193-202

[7] Tuan V A and Shimizu T 2005 Modeling of household motorcycle ownership behaviour in Hanoi city *Journal of the Eastern Asia Society for Transportation Studies* **6** pp 1751-1765

[8] Wedagama D M P 2009 A multinomial logit model for estimating the influence of household characteristics on motorcycle ownership: A case study in Denpasar City, Bali *Journal of Civil Engineering* **29** pp 2-9

[9] Leong L V and Sadullah A F M 2007 A study on the motorcycle ownership: A case study in Penang State, Malaysia *Journal of the Eastern Asia Society for Transportation Studies* **7** pp 528-539

[10] Zhang J, Norton R, Tang K C, Lo S K, Jiatong Z and Wenkui G 2004 Motorcycle ownership and injury in China *International Journal of Injury Control and Safety Promotion* **11** pp 159-163

[11] Keng S H 2005 Helmet use and motorcycle fatalities in Taiwan *Accident Analysis & Prevention* **37** pp 349-355

[12] Irawan M Z and Sumi T 2011 Modelling departure time of student on public transport in Yogyakarta, Indonesia *Modern Applied Science* **5** pp 3-10

[13] Irawan M Z and Sumi T 2012 Motorcycle-based adolescents' travel behaviour during the school morning commuter and the effect of intra-household interaction on departure time and mode choice *Transportation Planning and Technology* **35** pp 263-279

[14] Bhat C R 1998 Accommodating flexible substitution patterns in multi-dimensional choice modeling: formulation and application to travel mode and departure time choice *Transportation Research Part B: Methodological* **32** 455-466

[15] Ding C and Mishra S 2015 Cross-nested joint model of travel mode and departure time choice for urban commuting trips: Case study in Maryland-Washington, DC Region *Journal of Urban Planning and Development* **141** pp 1-13

[16] Vovsha P 1997 The cross-nested logit model: Application to mode choice in the Tel-Aviv metropolitan area *Transportation Research Record* **1607** pp 6-15

[17] Central Bureau of Statistics 2017 Number of population in Yogyakarta Special Region

[18] Land and Land Use Agency 2016 *Delineation of Yogyakarta Urban Area* Yogyakarta Indonesia

[19] Financial and Assets Management Agency of the Yogyakarta Special Region 2017 Number of Registered Motorized Vehicle in Yogyakarta Special Province 2010-2016 Yogyakarta Indonesia

[20] Transportation Agency of Yogyakarta Special Region 2018 *Annual report the performance of TransJogja Bus in 2017* Yogyakarta Indonesia

[21] Yang L, Zheng G and Zhu X 2013 Cross-nested logit model for the joint choice of residential location, travel mode, and departure time *Habitat International* **28** pp 157-166

[22] Nguyen-Phuoc D Q, Gyimah R A, Tran A T P and Phan C T 2018 Mode choice among university to school in Danang, Vietnam *Travel Behavior and Society* **13** pp 1-10

[23] Tran M T, Zhang J and Fujiwara A 2014 Can we reduce the access by motorcycle to mass transit systems in future Hanoi? International Conference on Traffic and Transportation Studies (ICTTS) *Social and Behavioral Science* **138** pp 623-631

[24] Irawan M Z, Putri M K, Belgiawan P F and Dwitasari R 2017 Analyzing commuters' behavior on egress trip from railway stations in Yogyakarta, Indonesia *The Open Transportation Journal* **11** pp 53-66

[25] McFadden D 1987 *Modeling the choice of residential location, in Spatial Interaction Theory and Residential Location* edited by A. Kariqvist et al Amsterdam: North-Holland pp 75-96

[26] Wen C H and Koppelman F S 2011 The generalized nested logit model *Transportation Research Part B: Methodological* **35** pp 627-641

[27] Bekhor S and Prashker J N 2008 GEV-based destination choice models that account for unobserved similarities among alternatives *Transportation Research Part B* **42** pp 243-262

[28] Bierlaire M 2006 A theoretical analysis of the cross-nested logit model *Annals of Operations Research* **144** pp 287-300

[29] Bierlaire M 2016 *PythonBiogeme: a short introduction Report TRANSP-OR 160706* Transport and Mobility Laboratory, School of Architecture, Civil and Environmental Engineering, Ecole Polytecnique Fédérale de Lausanne

[30] Rezika W Y Priyanto S and Irawan M Z 2018 Estimation urban railway demand in Yogyakarta using Bivariate Ordered Probit Model *the 1st International Symposium on Transportation Studies for Developing Countries (ISTSDC 2017)* **181**

[31] Irawan M Z, Belgiawan P F, Tarigan A K M and Wijanarko F 2019 To compete or not compete: exploring the relationships between motorcycle-based ride-sourcing, motorcycle taxis, and public transport in the Jakarta metropolitan area *Transportation* Article in Press

[32] Irawan M Z, Belgiawan P F, Widyaparaga A, Deendarlianto, Budiman A, Muthohar I and Sopha B M 2018 A Market Share Analysis for Hybrid Cars in Indonesia *Case Studies on Transport Policy* **6** 336-341

IOP Conf. Series: Materials Science and Engineering **650** (2019) 012043 doi:10.1088/1757-899X/650/1/012043

A distribution strategy using a two-step optimization to maximize blood services considering stochastic travel times

Filscha Nurprihatin[1*], Refael Elnathan[1], Ruth Elisa Rumawan[1], Tasya Regina[1]

Department of Industrial Engineering, Faculty of Technology and Design, Universitas Bunda Mulia, Jakarta, Indonesia

*fnurprihatin@bundamulia.ac.id

Abstract. The blood service is a health service that utilizes human blood as basic material with humanitarian purposes, not for commercial one. Indonesian hospital ability in blood transfusions is generally still low, especially in terms of blood supply adequacy. In fact, there are still some provinces that experience excess blood supply while many other provinces experience a shortage of blood supply. The Blood Bank in Jakarta has the highest excess blood supply. Therefore, the blood can be transferred evenly from one province to another nearby province. The aim of this paper is to determine the allocation and the route of blood distribution to achieve the minimum travel times. Some variations in travel time are difficult to predict, so we take into account the stochastic properties of them. The effectiveness of blood distribution is very dependent on the accuracy of the target number of beneficiaries and the accuracy of the number of blood bags received in distribution activities. Meanwhile, the efficiency of blood bag distribution is measured by distribution routes that are directly related to transportation costs. This study uses a two-step optimization model to reach optimality. The first step is utilizing the transportation model to make sure the destination points are only the fastest to arrive. The second step is making use of the capacitated vehicle routing problem to ensure the routing is global optimal. This model successfully creates better blood demand fulfillment while minimizing transportation cost.

1. Introduction

Logistics management is part of supply chain management that plans, implements, and regulates the flow of products starting from the product is still in the form of raw materials to become finished products that are in the hands of consumers [1]. The focus of logistics activities is the efficiency of the distribution of goods from a supplier, through the manufacturing to be processed, to the point of consumption in a cost-effective manner, while still providing appropriate services for consumers [1]. In this context, efficient while effective supply chain planning and management of healthcare can result in human loss reduction [2], especially related to blood services.

Blood service is a health service that utilizes human blood as basic material with humanitarian purposes, not for commercial one [3]. Blood is a scarce, perishable and vital product, and no alternative has been found so far [2], [4]. Blood supply services in Indonesia are carried out by 421 Blood Banks throughout the country run by Indonesian Red Cross Society, and responsible to donor recruitment and collecting, securing, processing, storing, distributing and destroying blood

[2], [3]. The demand for blood, commonly called the Whole Blood (WB) and blood components are satisfied by the donation of eligible individuals [5]. The process of componentizing of WB yields three major components, such as Red Blood Cells (RBCs), Platelets and Plasma as shown in Figure 1 [6]. Every component has a different life span, making the problem more complicated. Blood supply management is different from a typical supply chain problem in three aspects [5]. First, the number of donations, and thus, the amount of blood supply, changes over time [5]. Second, the WB and blood products have short life spans, and there has to be a certain amount of time between two consecutive donations of a donor to replace the donated blood [5]. Third, in order to extract platelets, one of the most important blood products, donated WB has to be processed within 6 hours of donation [5].

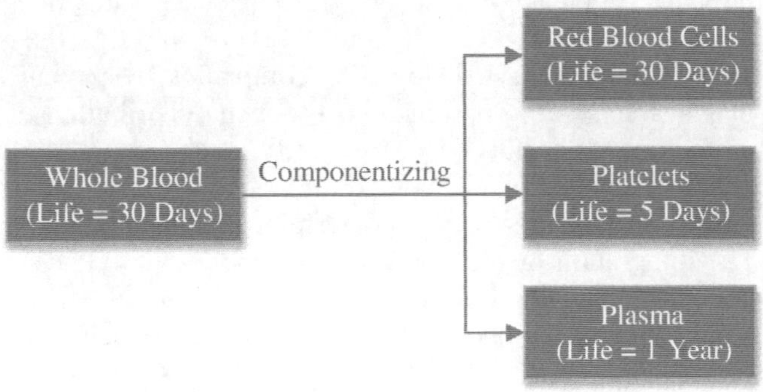

Figure 1. Componentizing of whole blood [6]

Indonesian hospital capability in blood transfusions is generally still low (average 55%), especially in terms of blood supply adequacy [7]. According to the World Health Organization (WHO) guidelines, the need for blood in an area is at least 2% of the population [3]. Unfortunately, Indonesia was only able to fulfill 81.2% of the blood demand in 2016 [3]. In other words, there is a shortage of 972,522 blood packages.

In fact, there are still some provinces that experience excess blood supply while many other provinces experience a shortage of blood supply. The wastage of blood products is a serious problem since a blood unit which is wasted at one particular blood bank may result in shortage at some other blood bank [6] Some blood banks do not perform blood collections by organizing blood donation camps but they order blood from a larger blood bank at regular intervals [6]. Table 1. shows that there are 5 out of 34 provinces fulfilled blood needs more than 100% while other 29 provinces did not. In other words, only 5 provinces meet the demand such as Jakarta (the highest: 302.7%), Yogyakarta, Bali, East Java and East Kalimantan. This table also shows the demand in Jakarta neighborhoods, such as West Java and Banten.

Table 1. Minimum Amount of Blood Needs and Blood Production in some Provinces in 2016 [3]

No	Province	Population	Minimal Blood Needs (Package)	Blood Production (Package)	Fulfillment (%)
1	Jakarta	10,277,628	205,553	622,136	302.7
2	Yogyakarta	3,720,912	74,418	113,390	152.4
3	Bali	4,200,069	84,001	104,362	124.2
4	East Java	39,075,152	781,503	901,658	115.4
5	East Kalimantan	3,501,232	70,025	70,152	100.2
6	West Java	47,379,389	947,588	589,999	62.3
7	Banten	12,203,148	244,063	143,522	58.8

The shortage and excess of blood supply in an area is a major problem. Inadequate number of

IOP Conf. Series: Materials Science and Engineering **650** (2019) 012043 doi:10.1088/1757-899X/650/1/012043

donations results in postponed surgeries, untreated patients and thus lost lives [2], [4], [5]. Having excess donations is not always better both because of donation/inventory related costs of outdated products and the ineligibility of the donor to donate for a certain amount of time [2], [5]. Excessive stock must be allocated appropriately to the needed regions to meet the blood demand. In addition, proper allocation and distribution can increase the effectiveness and efficiency of blood distribution [2]. Therefore, Blood Bank in Jakarta must be able to distribute blood effectively and efficiently so as to be able to help other areas that are lacking with low transportation costs.

Studies regarding the distribution effectiveness and efficiency have been carried out with various methods, including descriptive analytical methods, simulations, Genetic Algorithms (GA), Fuzzy Genetic Algorithms (FGA) and Simulated Annealing. The earlier study was conducted to measure the effectiveness and efficiency of distribution routes using analytical descriptive methods [8]. The simulation model was applied in previous work to improve the sustainability of the distribution network performance in British logistics companies by paying attention to exhaust emissions [9]. Another research tried to optimize costs for the distribution of fresh food products using GA, FGA, and Simulated Annealing [10]. Respectfully, the above studies did not solve the problem while ensuring the optimality as this paper did.

Previous research used the CVRP model to determine distribution routes while minimizing transportation costs. The integration of the transportation model and CVRP in this study aims to minimize the travel times of the blood distribution route from Jakarta which experiences an abundance of blood supply to the surrounding areas that are experiencing shortages. Travel times may be dynamic
[11] and vary exogenously due to traffic congestion, weather conditions, moving targets or mobile obstacles, affecting the efficiency and even feasibility of a delivery route [12], [13]. By minimizing the travel times of the distribution route, operational costs are assumed to be minimal as well in order to achieve distribution effectiveness and efficiency. In this study, there were 2 (two) types of locations considered, namely the source and the distribution points. The source point for distribution is defined as Blood Bank in Jakarta. Blood production distribution activities are carried out at Blood Bank in Jakarta that is overstocked with the closest Blood Banks in provinces that are experiencing shortages.

This study uses two-step optimization to reach out the distribution effectiveness and efficiency simultaneously. The first step is utilizing the transportation model to make sure the destination points are only the fastest to arrive. The second step is making use of the capacitated vehicle routing problem (CVRP) to ensure the routing is globally optimal.

2. Problem Formulation

The scope of this study is the blood distribution route in the Jakarta and its surrounding. Figure 2 shows one single source and 20 destination points, which is divided into two regions, West Java (shown in orange) and Banten (shown in blue). Some variations in travel time are difficult to predict, like traffic accidents, which cause congestion of which both the occurrence and delay are difficult to foresee [13]. Therefore, when it comes to the travel times we follow the assumption that they are continuous [12] and normally distributed, as demonstrated in the previous research in terms of speed of the vehicle [14]. We take the mean value as the input in the first and second step of the following optimization. The previous study takes into account the travel times from i to j and the other way around was equal [15], and successfully improved in this paper as in fact they are not.

Jakarta can distribute 416,583 blood packages while West Java and Banten need 357,589 and 100,541 blood packages respectively. The blood packages are assumed to be 450 ml each and are stored in a 35-liters cool box. In the distribution process, each cool box can store 77 blood packages and transports in a vehicle with capacity 22 cool box per trip. Therefore, each trip can transport 1694 blood packages and the trip is done once a week.

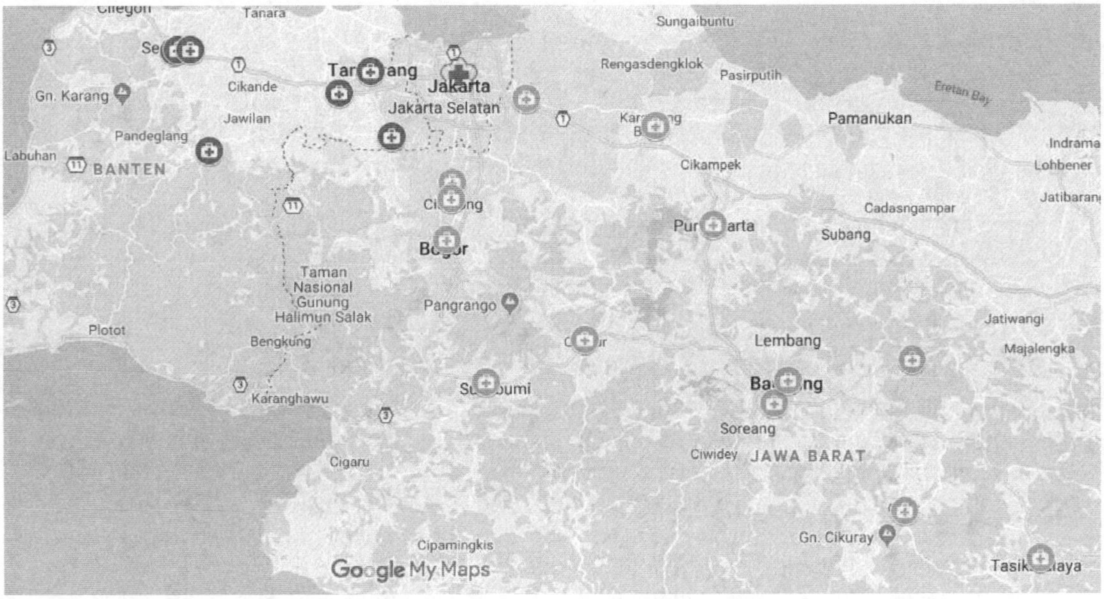

Figure 2. The Source and Destination Points

3. Model

3.1. Transportation Model

Transportation is the movement of products from one location to another, starting from the product still in the form of raw goods to finished goods distributed to customers in a supply chain system [16]. Transportation has the following characteristics, namely:

a. There are a number of specific sources and destinations. In this study, we use one single source and 20 destinations.
b. The number of goods to be distributed from each source and requested by each destination is varied. In this case, we consider the shortage of blood as the demand.
c. Goods are distributed from one source to a particular destination, considering the demand and capacity of the source.
d. The cost of shipping commodities from each source to each destination is different. This research contemplates the shipping cost as the travel times from the source to the destination.

The following shows the transportation model [16], [17].

Notation

n = number of source locations
m = number of demand points
c_{ij} = the cost of shipping one unit from i to j.
x_{ij} = quantity shipped from source i to destination j
a_i = the total amount to be shipped out of source i
b_j = the total amount to be shipped into destination j

Objective function

$$Minimize\ z = \sum_{i=1}^{m} \sum_{j=1}^{n} c_{ij} x_{ij} \quad (1)$$

IOP Conf. Series: Materials Science and Engineering **650** (2019) 012043 doi:10.1088/1757-899X/650/1/012043

Subject to

$$\sum_{j=1}^{n} x_{ij} \leq a_i,\ i = 1,2, \dots, m \tag{2}$$

$$\sum_{i=1}^{m} x_{ij} \geq b_j,\ j = 1,2, \dots, n \tag{3}$$

$$x_{ij} \geq 0, \forall i\ \forall j \tag{4}$$

Equation (1) minimizes the total cost (approximated by travel times) of shipping all units from *i* to *j*. Equation (2) ensures the quantity shipped cannot exceed the supply availability. Equation (3) makes sure the demand in destination *i* is met. Equation (4) ensures the non-negativity constraints.

3.2. Capacitated Vehicle Routing Problem (CVRP)[18]

Determination of distribution routes is the operational decision of each company with the ultimate goal is to minimize transportation costs which are approached by the length of the route [19]. CVRP has become the research material for combinatorial optimization. This research utilizing CVRP to minimize the total traveled times by the *m* vehicles, subject to [18]:

- Each route of a solution must to starts and ends at the depot, in this case depot means Blood Bank in Jakarta
- Each customer of a route is visited exactly once by exactly one vehicle.
- The total demand of each route does not exceed the capacity of vehicle.

Objective function

$$Minimize\ z = \sum_{i=0}^{m} \sum_{j=0}^{n} c_{ij} x_{ij} \tag{5}$$

Subject to

$$\sum_{j=0} x_{0j} = m \tag{6}$$

$$\sum_{j=0} x_{i0} = m \tag{7}$$

$$\sum_{j=1} x_{ij} = 1 \tag{8}$$

$$\sum_{i=1} x_{ij} = 1 \tag{9}$$

$$m \geq 1 \tag{10}$$

The objective function in equation (5) is to minimize total costs. The constraint (6) represents the number of modes of transportation used *m*. The constraint (7) guarantees that all modes of transportation departing from the distribution source return to the same place. Equations (8) and (9) guarantee that all distribution points will be visited.

4. Discussion

The blood is not distributed to all destinations, and only to 19 destination points instead, due to the demand is higher than the supply availability as shown in Figure 3. The destination point (node) 4

IOP Conf. Series: Materials Science and Engineering **650** (2019) 012043 doi:10.1088/1757-899X/650/1/012043

partially meets the demand since the capacity of supply while node number 6 is not supplied at all.

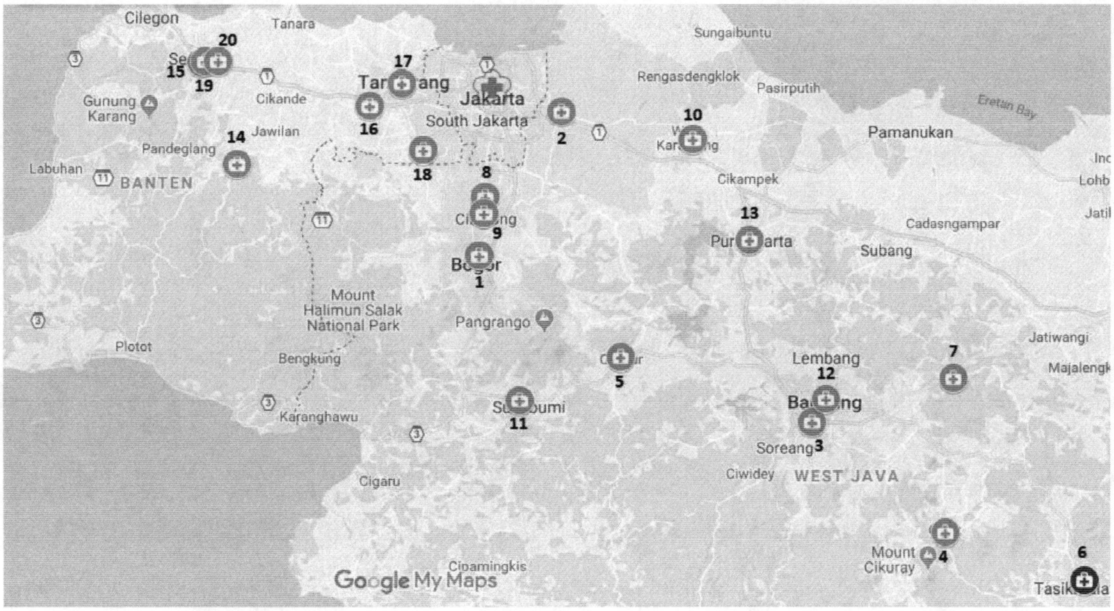

Figure 3. Visualization According to Transportation Model Results

The second model is CVRP which generates five trips from Jakarta to several distribution points every week, as shown in Table 2. With the determination of routes like this, it can be obtained the total travel times is 2,272.70 minutes.

Table 2. Distribution Route

Distribution Route	Times Traveled (min)	Blood Packages (per week)
Jakarta – 1 – 11 – 5 – Jakarta	434.90	1587
Jakarta – 2 – 8 – 9 – Jakarta	207.80	1587
Jakarta – 3 – 4 – 7 – 18 – Jakarta	810.96	1593
Jakarta – 13 – 12 – 10 – Jakarta	449.60	1587
Jakarta – 16 – 14 – 20 – 15 – 19 – 17 – Jakarta	369.44	1657
Total	2,272.70	8011

5. Conclusion

The results obtained from the two-step optimization with the transportation model and CVRP sequentially create better blood demand fulfillment. Blood Bank in Jakarta can distribute the blood to its surrounding, which is West Java and Banten, while minimizing the transportation cost. Figure 4 shows the better condition of blood demand fulfillment. This model escalates the demand fulfillment in Banten to 100%, while still maintaining the demand fulfillment in Jakarta also 100%. The performance in West Java is clearly satisfying enough since demand fulfillment is improved from 62.26% to 95.62%. Further research can use this model to determine the distribution routes of other regions that have excess blood supply to the surrounding area.

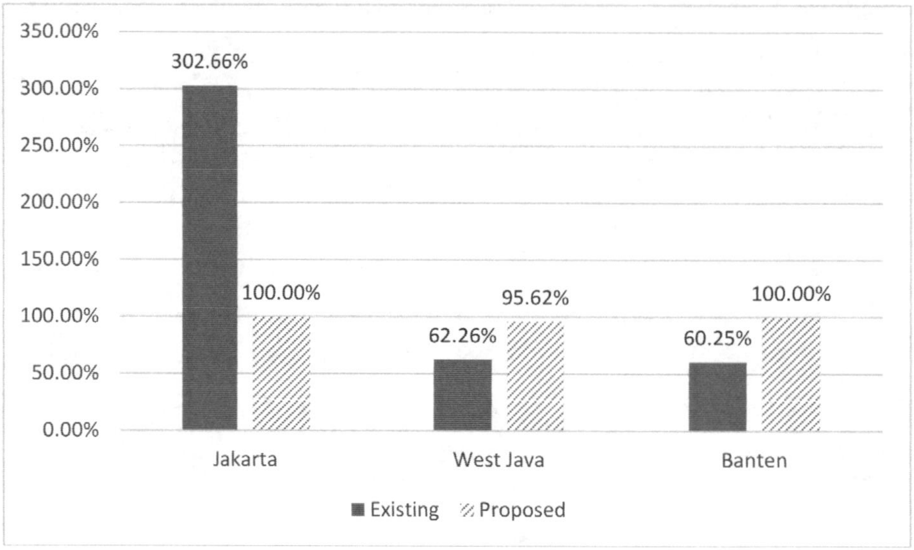

Figure 4. Percentage of Blood Demand Fulfillment

Further research can consider multi product perishable inventory decisions [6] as there is a relationship between storage duration and the length of stay in hospital and organ failure [20]. The next study may leads to minimize an objective function combining fuel consumption (or CO_2 emissions) and total route duration [12].

References

[1] Rushton A, Croucher P and Baker P 2014 *The handbook of logistics and distribution management: understanding the supply chain*, 5th ed. (New Delhi: Kogan Page Limited)
[2] Samani M R G, Torabi S A and Hosseini-Motlagh S M 2014 Integrated blood supply chain planning for disaster relief *Int. J. Disaster Risk Reduct* **27** pp 168–188
[3] Ministry of Health 2018 *Blood Service in Indonesia* (Jakarta)
[4] Zahraee S M, Rohani J M, Firouzi A and Shahpanah A 2015 Efficiency improvement of blood supply chain system using Taguchi method and dynamic simulation *Procedia Manuf.* **2** pp 1–5
[5] Ekici A, Özener O Ö and Çoban E 2018 Blood supply chain management and future research opportunities in *Operations Research Applications in Health Care Management* **262**
[6] Kahraman C and Topcu Y I 2018 *Eds.* Cham: Springer International Publishing AG, pp. 241–266.
[7] Lowalekar H and Ravichandran N 2013 Blood bank inventory management in India *Opsearch.* **51**(3) pp 376–399
[8] Ministry of Health 2015 *Strategic Plan of the Ministry of Health 2015-2019* (Jakarta)
[9] Karaxha H and Kristo I 2016 The logistics and management of distribution channels, the case of Kosovo *Ilir. Int. Rev* **6**(1) pp 37–48
[10] Danloup N, Mirzabeiki V, Allaoui H, Goncalves G, Julien D and Mena C 2015 Reducing transportation greenhouse gas emissions with collaborative distribution *Manag. Res. Rev* **38**(10) pp 1049–1067
[11] Nakandala D, Lau H and Zhang J 2016 Cost-optimization modelling for fresh food quality and transportation *Ind. Manag. Data Syst* **116** (3) pp 564–583
[12] Ritzinger U, Puchinger J and Hartl R F 2016 A survey on dynamic and stochastic vehicle routing problems *Int. J. Prod. Res* **54**(1) pp 215–231

IOP Conf. Series: Materials Science and Engineering **650** (2019) 012043 doi:10.1088/1757-899X/650/1/012043

[13] Gendreau M, Ghiani G and Guerriero E 2015 Time-dependent routing problems: A review *Comput. Oper. Res* **64** pp 189–197

[14] Spliet R, Dabia S, and Van Woensel T 2017 The time window assignment vehicle routing problem with time-dependent travel times *Transp. Sci* 52(2) pp 261–276

[15] Zeng W, Miwa T and Morikawa T 2016 Prediction of vehicle CO2 emission and its applicationto eco-routing navigation *Transp. Res. Part C* **68** pp 194–214

[16] Hiermann G, Puchinger J, Ropke S and Hartl R F 2016 The electric fleet size and mix vehicle routing problem with time windows and recharging stations *Eur. J. Oper. Res* 252(3) pp 995–1018

[17] Chopra S and Meindl P 2016 *Supply Chain Management*, 6th ed. (Harlow: Pearson Education Limited)

[18] Nahmias S. and Olsen T. L 2015 *Production and operations analysis*, 7th ed. (Long Grove: Waveland Press Inc)

[19] Liñán-García E, Villegas L C C, Dorantes P M and G M Méndez Metaheuristic hybridized applied to solve the capacity vehicle routing poblem in *Advances in Soft Computing*, O. Pichardo-Lagunas and S. Miranda-Jiménez, Eds. Cham: Springer International Publishing AG, 2017, pp. 377–387.

[20] Simchi-Levi D, Chen X and Bramel J 2014 *The logic of logistics: theory, algorithms, and applications for logistics management* vol 68 (New York: Springer Science+Business Media)

[21] Alshalani A, Howell A and Acker J P 2017 Impact of blood manufacturing and donor characteristics on membrane water permeability and in vitro quality parameters during hypothermic storage of red blood cells *Cryobiology* **80** pp 30–37

IOP Conf. Series: Materials Science and Engineering **650** (2019) 012044 doi:10.1088/1757-899X/650/1/012044

Comparative study of asphalt concrete characteristic between ideal gradation and laboratory tested gradation with crumb rubber

Asep Sundara[1*], Risna Rismiana Sari[2], Tanya Audia Balqis[3]

[1]Civil Engineering Department, Politeknik Negeri Bandung, Bandung, Indonesia
[2]Civil Engineering Department, Politeknik Negeri Bandung, Bandung, Indonesia
[3]Post-Graduate Student of Infrastructure Engineering Study Program, Applied Master Program, Politeknik Negeri Bandung, Bandung, Indonesia

*asepsundara24@yahoo.com

Abstract. Modifying asphalt concrete with using Crumb Rubber (CR) can provide environmental benefits. The method used in this study is an experimental laboratory with Mashall machine. The tests used two types of gradation, namely the gradations of Ideal and Laboratory Test. CR used in a mixture with a dry process method to replacement 0%, 5%, 15% and 20% of the weight aggregate at fraction No. 30 (0.6 mm). The results showed that the Ideal gradation mixture has a better characteristic than the mixture of laboratory tested gradations. Ideal gradations can increased stability into 10% from Laboratory Test gradations for mixture without CR. Addition of CR contect in AC mixture did not significantly effect for the characteristics. To ensure the effect of CR in mixture, further test must be taken.

1. Introduction

Asphalt concrete (AC) is a construction layer consisting of a mixture of asphalt and continuously graded aggregate, mixed, spread, and compacted at a specific temperature. Layers of asphalt concrete consist of mixture of three types namely Asphalt Concrete-Wearing Course (AC-WC), Asphalt Concrete- Binder Course (AC-BC), and Asphalt Concrete Base (AC-Base) [7].

Modified asphalt concrete wearing course (AC-WC) used for increases the strength of pavement structure with adding some additive such as chemicals, natural ingredients, and residual waste [4]. In many countries, crumb rubber (CR) are used as an addictive for flexible pavement cush as in US Department of Transportation Federal Highway Administration in America since 1986. The used of CR could reduce damage to flexible pavement due to weather and traffic load factors (AASHTO, 1982) [3]. Whereas in Indonesia the use of CR is quite high and contributes cush as large waste due to not utilized properly. According to the Indonesian Rubber Council at 2012, tire rubber production in Indonesia reached 15.4 million in 2011. Meanwhile, according to the Association of Tire Companies (2016) the use of tires reached 45 million tires/year.

By adding 50% CR on fraction No. 50 in AC mixtures can improve the resistance to permanent deformation due to ruts and provide better resistance to high temperatures and loads [9]. Further, AC mixture with adding 2% CR on mesh #80 resulted higher stability than mixture normal [3]. Moreover, there are still many of size and mix variations used CR to be researched in AC mixture. Therefore, this research mainly focused on CR content from 5.0%, 10%, 15%, and 20% by replacement weight of aggregate at fraction No. 30 (0.6 mm) on ideal and Laboratory tested gradations in AC mixture.

2. Literature Review

Asphalt mixture is asphalt made by mixing hard asphalt with an added ingredient. The strength of asphalt concrete is influenced by the properties of the mixed material. Modified polymer asphalt used in paved mixtures can improve better resistance to deformation, overcome cracks and increase resistance to aging damage. [4)].

Modified asphalt was also developed with the addition of used residual tire rubber. Tire rubber can be used as pavement material with two methods, dry process and wet process method. The first is called dry process, namely the method of adding used tire powder as a substitute for part of the aggregate in concrete asphalt mixture and not part of asphalt binder. Whereas for the wet process is the method of adding used tire powder to the asphalt as a binder where before asphalt is added to the asphalt concrete mixture [6].

The percentage of used CR added in dry processes varies shows that 3% -5% of used tire powder with aggregate weight is commonly used [1]. Research on the effect of mixture that used CR on AC mixture was carried out by Laos. The variation is 1%, 2% and 3% of CR with a mesh size of 40. But it produces uneven linear stability and flow between the addition of CR to the stability value of the mixture [5]. Then it was resumed by Gosali, also on AC mixture with 80 mesh size. The results showed that the addition of 1-2% CR can produce a better value of stability than the standard mixture [3].

3. Methodology

In general, this study consists of several preparations and tests, as follows:

3.1 Material

Materials cush as fine aggregates, coarse aggregates, stone ash (filler) used in this study came from Cagak, Subang, West Java. The bitumen from Pertamina asphalt with 60/70 penetration. Then the wasted crumb rubber were ordered from PT Pura Agung, Mojokerto, East Java. Figure 1. shows the materials.

(a) Aggregate (b) Crumb Rubber

Figure 1. Materials

IOP Conf. Series: Materials Science and Engineering **650** (2019) 012044 doi:10.1088/1757-899X/650/1/012044

3.2 Specimens

The specimens used in the study are two types of gradation, which are ideal and laboratory tested gradations. Asphalt content used from 5.0 % until 7.0 %, while the CR content from 5.0%, 10%, 15%, and 20% by replacement weight of aggregate at fraction No. 30 (0.6 mm). It means that the mixture is not modified asphalt because it's still used standard bitumen with 60/70 penetration.

3.3 Testing Methods

Marshall test was carried out to obtain mixture characteristic, such as Voids Mix Aggregate (VFB), Voids Fillied Bitumen (VFB) and Voids in Mixture (VIM), Stability, and Flow by using the machine as shown in Figure 2. The testion is carried out according to RSNI M-01-2003. The results of this test are compared with the specifications of Spesifikasi Umum Bina Marga, 2010 [2][8].

Figure 2. Mashall test

4. Result and Discussion

4.1 Aggregate Testing Result

The aggregate characteristics which used in Asphlat Concrete (AC) mixture are shown in **Table 1.** It can be seen that the aggregate used in this study has fulfilled the Specifications [2]

Table 1. Aggregate Characteristic

No	Testing	Method	Specification	Result
A	**Coarse Aggregate**			
1	Satured Surface dry Specific Gravity (SSD)	SNI 1969:2008	-	2.65
2	Bulk Specific Gravity	SNI 1969:2008	Min. 2,5	2.58
3	Apparent Specific Gravity	SNI 1969:2008	-	2.78
4	Absorption	SNI 1969:2008	Max 3%	2.77%
5	Los Angeles Abrasion (500 spin)	SNI 2417:2008	Max 40%	14,19%
6	Materials Finer than 75-µm (No. 200) Sieve	SNI 03-4142-1996	Max 2%	0.46%
7	Coating of Bitumen	SNI 2439:2011	Min. 95%	100%
B	**Fine Aggregate**			
1	Satured Surface dry Specific Gravity (SSD)	SNI 1969:2008	-	2.64
2	Bulk Specific Gravity	SNI 1969:2008	Min. 2,5	2.55
3	Apparent Specific Gravity	SNI 1969:2008	-	2.79
4	Absorption	SNI 1969:2008	Max 3%	3.32%
5	Materials Finer than 75-µm (No. 200) Sieve	SNI ASTMC117:2012	Max 10%	10.95%
6	Sand Equvalent	SNI 03-4428-1997	Min. 60%	79.10%

4.2 Asphalt Test Result
Asphalt test was conducted to determine the characteristics of asphalt in AC mixture. Based on Table 2. it can be seen that the characteristics of the 60/70 penetration asphalt used in this study have fulfilled the Specifications [2]

Table 2. Asphalt Pen 60/70 Characteristic

No	Test	Method	Specification	Result
1	Penetration	SNI 2456:2011	60-70	63
2	Kinematic Viscosity 135°C	SNI 7729:2011	≥ 300°C	500°C
3	Softening point	SNI 2434:2011	≥ 48°C	52.1°C
4	Dactility	SNI 2432:2011	≥ 100 cm	> 140 cm
5	Density	SNI 2441:2011	≥ 1	1.04

4.3 Gradations
The gradations used in this study are 2 types, Ideal gradation and Laboratory Test gradation. The difference between the two will be seen in the cumulative percentage of passing size for each filter size shown in **Table. 3** of the following:

IOP Conf. Series: Materials Science and Engineering **650** (2019) 012044 doi:10.1088/1757-899X/650/1/012044

Table 3. Aggregate gradation used

Sieve		Percentage passing of the size (%)			
(mm)	(Inch/No)	Lower limit	Upper limit	Ideal gradation	Laboratory tested Gradation
25	1	100	100	100	100
19	¾	100	100	100	100
12.5	½	90	100	95	100
9.5	3/8	77	90	81	87.4
4.75	No. 4	53	69	53	59.1
2.36	No.8	33	53	33.5	52.2
1.18	No.16	21	40	22	30.9
0.6	No.30	14	30	16	17.3
0.3	No.50	9	22	12	10
0.15	No.100	6	15	9.5	7
0.075	No.200	4	9	7	4

Based on Table. 3 show that laboratory tested gradations has percentage cumulative which tends to approach the low limit value. Laboratory tested gradations has more fine aggregate and less coarce aggregate than ideal gradations. Later in this paper, ideal gradations will be stated as A1 meanwhile laboratory tested gradations is A2.

4.4 Comparasion of Marshall Test Result

The test results obtained the compasion of Asphlat Concrate (AC) characteristic between two types of gradations can be seen in **Figure 3.** untill **Figure 7.**

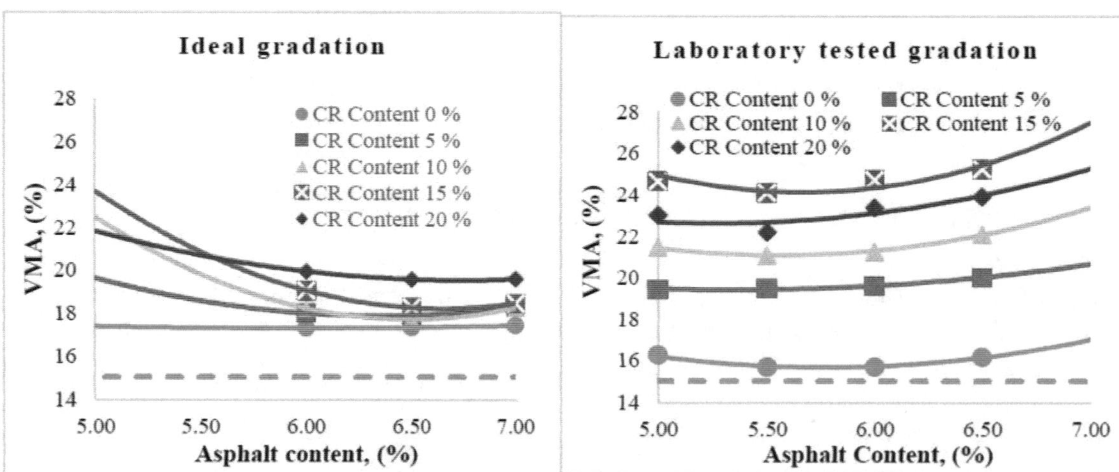

Figure 3. Comparison of VMA

Figure 3. Shows that the value of Voids Mix Aggregate (VMA). This figure shows the relationship between voinds on aggregate for 2 types of gradations. The result shows that the A2 has a quite a lot of voids gap in each addition of CR compared to the normal mixture. While A1 has fewer voids gap.

VMA values are higher in each contect of CR in both gradations but different things happen in 15% CR for ideal gradation. This mixture actually has a lower voids than 20% CR.

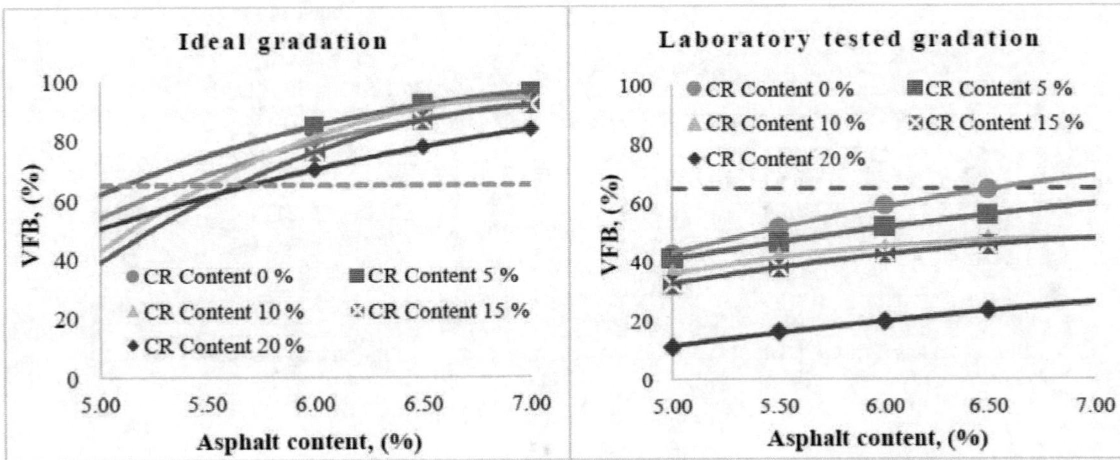

Figure 4. Comparison of VFB

Figure 4. Shows that the value of VFB, it shows that A1 is better than A2. It is laso seen that VFB of A1 value meets spesifications which is a niminum of 65%. While A2 for all mixures that did not meet the specifications. It happened because the voids of A2 are not fully filled with asphalt. A2 has smoother gradation, so that it has more voids. Therefore, asphalt liquid couldn't fill the gap of void easily.

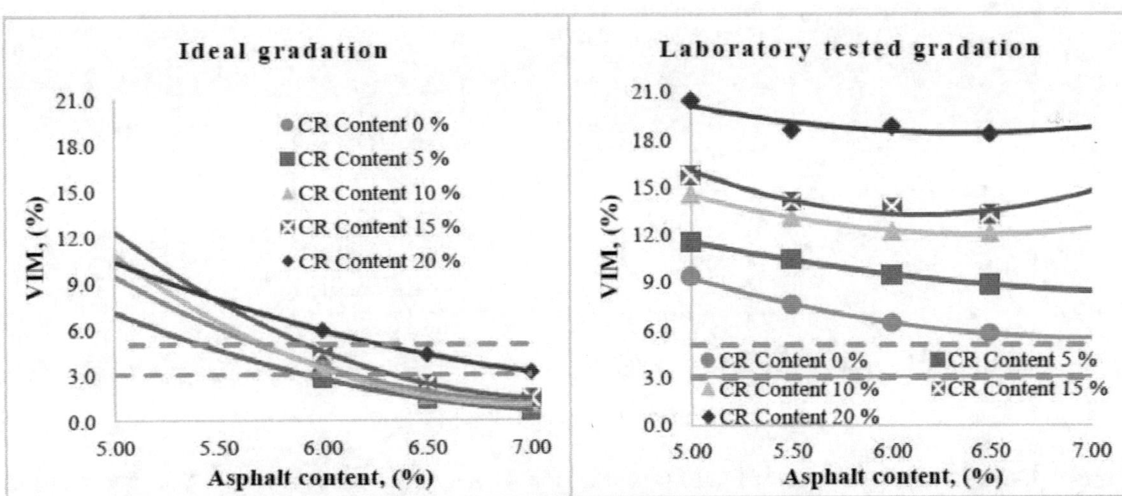

Figure 5. Comparison of Voids In Mix

Figure 5. shows the value of VIM between A1 and A2. Based on this figure, A1 is better than A2 because it can be seen from the VIM graph in A1 has meet specifications. Meanwhile, none of A2 entered the minimum VIM specifications, even the VIM in A2 reached a double value (21% than 12%). It happened, A2 has more voids due to the effect of more fine aggregate. For the more, the voids can't be filled with CR and increase the percentage of voids in mixture.

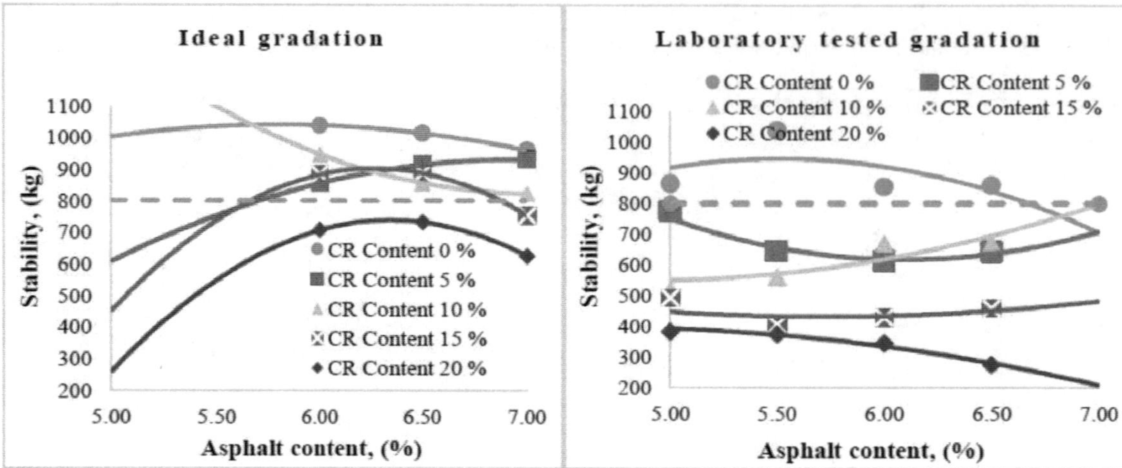

Figure 6. Comparison of Stability

Figure 6. Shows that the value of Stability between A1 and A2. It shows that the A1 is better than A2, excpet for CR 20% was 700 kg, below minimum specification. Meanwhile for A2, all mixtures contanting with CR did not meet any minimum specification. In A2 mixture with CR contenct, voids ware in huge amount so it results lower stability compared to mixture without CR [9].

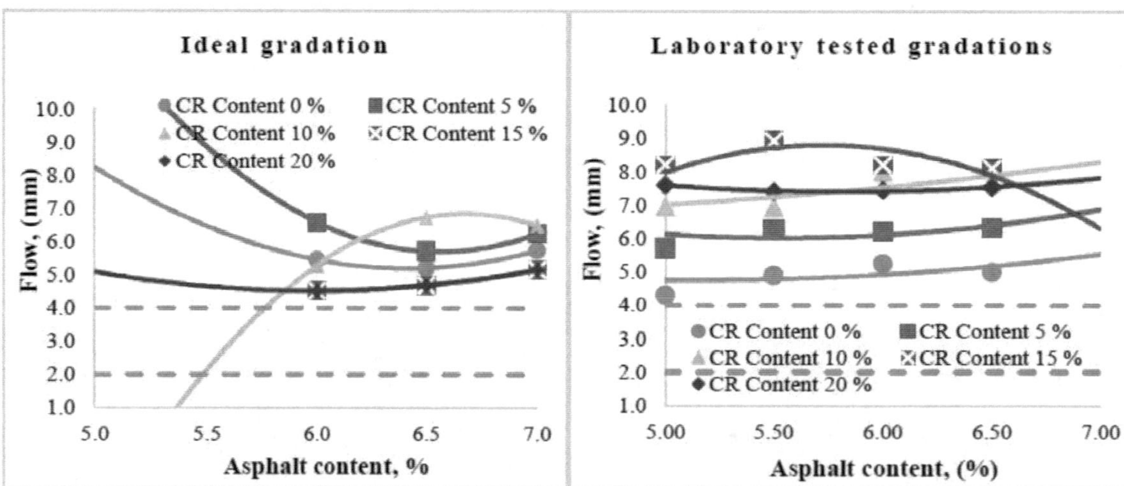

Figure 7. Comparison of Flow

Figure 7. Shows that the value of Flow. Based on the figure, flow values from both of gradations did not meet specifications. So that the best gradation cannot be taken between the two. It happens because the CR couldn't be able to maximally filled the voids, resulted in large amount of flow.

IOP Conf. Series: Materials Science and Engineering **650** (2019) 012044 doi:10.1088/1757-899X/650/1/012044

5. Conclusion and Recommendation

Based on the results of the study it was found that:

- Ideal gradations is better than Laboratory Test gradations based on characteristic of Marshall and is recommended for mixture with CR.
- Ideal gradations can increased stability into 10% from Laboratory Test gradations for mixture without CR.
- Addition of CR contect in AC mixture did not significantly effect for the characteristics. This could happen due to the size of CR could be bigger than it should be.
- To ensure the effect of CR in mixture, further test must be taken cush as test with different CR variations and sizes.

Acknowledgment

The author acknowledges to UPPM POLBAN that already support the funding for commencement the research and Bandung Research Center of Roads and Bridges that support the laboratory for the research.

References

[1] Ahmad Omar Asad 2015 *Using Crumb Rubber (CR) of Scrap Tire in Hot Mix Asphalt Design* (Jordan: Department of Civil Engineering Facullty of Engineering Technology Zarqa University)

[2] Direktorat Jenderal Bina Marga 2010 *Spesifikasi Umum Divisi 6* (Jakarta: BSN)

[3] Gosali Gavin dkk 2016 *Pengaruh Penambahan Serbuk Ban Karet Mesh #80 pada Campuran Laston untuk Perkerasan Jalan Raya* (Surabaya: Jurusan Teknik Sipil Universitas Kristen Petra)

[4] Isnanda Sofyan M Saleh and Muhammad Isya 2018 Pengaruh Subtitusi Polystyren (PS) dan Abu Arang Tempurung Kelapa Sebagai Filler Terhadap Karakteristik Campuran AC- WC *Jurnal Teknik Sipil Universitas Syiah Kuala, ISSN 2088-9321, E-ISSN 2502-5295*

[5] Laos Charly *2016 Pengaruh Penambahan Serbuk Ban Karet pada Campuran Laston untuk Perkerasan Jalan Raya* (Surabaya: Jurusan Teknik Sipil Universitas Kristen Petra)

[6] Mulyani Sri dan Hamdani Dani 2017 *Teknik Pencampuran yang Optimal antara Crumb Rubber dan Aspal Pen 60/70* (Bandung: Pusat Penelitian Jalan dan Jembatan)

[7] Spesifikasi Umum Jalan dan Jembatan 2006 *Campuran Bersapal Panas Buku V* (Jakarta: Kementerian Pekerjaan Umum)

[8] Standar Nasioal Indonesia No. M-01-2003 2003 *Metode Pengujian Campuran Beraspal Panas dengan Alat Marshall* (Jakarta: Departemen Pekerjaan Umum)

[9] Sugiyanto Gito 2008 Kajian Karakteristik Campuran Hot Rolled Asphalt Akibat Tambahan Limbah Serbuk Ban Bekas *Jurnal Teknik Fakultas Sains dan Teknik Universitas Jenderal Soedirman Purwokerto*

IOP Conf. Series: Materials Science and Engineering **650** (2019) 012045 doi:10.1088/1757-899X/650/1/012045

Temperature sensitivity effect in asphalt modification with natural rubber sir20

R Utami* and Suherman

Civil Engineering Department, Politeknik Negeri Bandung, Jl. Gegerkalong Hilir, Ds. Ciwaruga Kotak Pos 1234 Bandung 40012, Indonesia.

E-mail: *retnoutami@polban.ac.id

Abstract. The use of asphalt material is needed so that surface layer could be water-resistant. Asphalt material also improves tensile stress which means enhancing support capacity for traffic load. Asphalt modification can be done by adding certain materials, such as natural rubber, to improve asphalt performance. Experimental methods were taken in this study. Asphalt tests which carried out are specific gravity, softening point and penetration with natural rubber variation 2%, 4%, 6%, 7%, 8%, 9%, 11% and 13%. Results were shown that the higher addition of natural rubber SIR20, asphalt more hardened and more insensitive to changes in temperature. This hardness characteristic of asphalt could be the initial parameter for making asphalt mixtures that more resistant to rutting and permanent deformation.

1. Introduction

According to SNI 6749:2008, natural rubber asphalt is hard asphalt modified with elastomeric type of polymers. In Indonesia, the use of synthetic polymers to improve the quality of asphalt has been carried out, but the synthetic polymer is still imported so that the use of natural rubber as asphalt added material is expected to enhance the use of local product. SIR20 is an acronym for Standard Indonesian Rubber which originated from latex coagulant and one of Indonesia's export commodities.

The use of oil based asphalt in Indonesia is widely spread in across the nation. Most common mixture in Indonesia is Asphalt Concrete Wearing Course (AC-WC). There are many types of AC-WC modification to improve its performance. Modification of AC-WC with rubber material addition is a system to improve AC-WC performance in reducing deformation, increasing crack resistance and increasing asphalt attachment to aggregates [1]. However, the use of raw rubber in solid form as asphalt additive will require longer mixing time and higher temperature than without rubber, because the rubber still has long molecular chain so it is less desirable [2].

The problem of using solid natural rubber as asphalt additives can be handled by carrying out depolymerisation techniques on natural rubber [3]. Depolymerisation is the process of terminating a long chain of polymer molecules into shorter molecular chain, known as mastication stage or grinding process of natural rubber in an Open Mill machine.

In this study, the selection of SIR 20 natural rubber content variations was 2%, 4%, 6%, 7%, 8%, 9%, 11% and 13% determined based on variations that had not been done by previous researchers. With the addition of SIR 20 it is expected to obtain asphalt mixture with higher hardness characteristic. Therefore, this study will be carried out experimentally by doing tests on penetration, softening point and specific gravity from modified bitumen. This study aims to learn how much the

IOP Conf. Series: Materials Science and Engineering **650** (2019) 012045 doi:10.1088/1757-899X/650/1/012045

effect of SIR20 solid natural rubber on temperature sensitivity in asphalt 60/70 from the Penetration Index value.

2. Literature Review

Natural rubber as elastomer properties, acted as polymer modification which can change the performance of current conventional asphalt paving roads against temperature changes and traffic loading. Numerous studies have been carried out on natural rubber based on different aspects of compatibility and interactions between the polymers and the asphalt aiming to improve the performances of the modified materials. From the review, the natural rubber used in a form of latex concentrated, Ribbed Smoked Sheet (RSS) and Liquid Natural Rubber (LNR) show promise in improving the asphalt mixes. Therefore, further study is needed to make them applicable as rubberized asphalt binder, thus provide better performance and services on the asphalt pavement [4].

Research on mechanical depolymerization of natural rubber for asphalt mixture additives has been carried out by doing tests as follows [5]:

- Rubber asphalt quality test: penetration, softening point, flash point, viscosity and ductility.
- Aggregate and filler quality tests: gradation, specific gravity, abrasion, sand equivalent and stickiness.
- Marshall Test: Marshall stability, flow, and Marshall Quotien.

Another study on Mechanized Depolymerization of Natural Rubber for Asphalt Additives. This study wanted to know the effect of SIR 20 natural rubber depolymerisation on the mixing process, softening point and asphalt penetration. Besides that, the study also wanted to know the chemical properties of rubber after it was depolymerized and the quality of the mixture of rubber asphalt. The method used in this study is SIR 20 natural rubber mechanically depolymerized using open mill and chemical tools with the addition of 2% peptizer and 1% HNS. Rubber depolymerization results are stored for a certain period of time and then mixed with bitumen and measured the chemical properties of the rubber, such as Mooney viscosity, initial plasticity (Po) and plasticity. The results obtained are mechanical depolymerization of natural rubber can accelerate rubber solubility to bitumen, and accelerate the mixing process from 660 minutes to only 50 minutes and the concentration of rubber to bitumen which meets the requirements of softening point and penetration of polymer asphalt is 5% and 7%. However, to be widely applied, further research is needed on the testing of rubber asphalt as a whole as well as the calculation of its economic technology [3].

Another study with natural rubber mix with bitumen was conducted in Malaysia. It used Natural Rubber Latex on asphalt pen 60/70. It resulted that percentages of 7% Natural Rubber Latex or higher are considered for the optimum content as the softening point at 7, 9 and 12% are all higher than unmodified bitumen [6].

Adding natural rubber into bitumen was also conducted in India. The binders used in the study were PB 80/100 and Natural Rubber Modified Bitumen (NRMB) 60/70. Natural Rubber Modified Bitumen was mostly in situ mixing of natural rubber latex with bitumen. Mixing process of bitumen and natural rubber latex was using wet process in which natural rubber latex was added into hot boiling bitumen in 140°C with kerosene as viscosity modifier. Natural rubber latex was added at about 2% by weight of bitumen. From the study it was known that by adding natural rubber latex, penetration value is decreasing and softening point is increasing [7].

IOP Conf. Series: Materials Science and Engineering **650** (2019) 012045 doi:10.1088/1757-899X/650/1/012045

3. Test Methodology

3.1. Materials
Materials for this study were as follows:
- Natural rubber SIR20 with catalyzer Peptizerdi-(o-bezamidephenyl) disulfide and Hidroxilamin neutral sulphate.
- Asphalt Pen 60/70 from Pertamina.

3.2. Equipment
There were two main equipment involved in this study, to depolymerize natural rubber SIR20 and to test asphalt modification characteristics. An open mill and mixer were used for the former process, as shown in Figure 1 and 2. Meanwhile, a softening point and penetration test equipment was used for testing asphalt modification due to its sensitivity on temperature, as shown in Figure 3 and 4.

Figure 1. Open Mill Tool

Figure 2. Mixing Equipment

Figure 3. Softening Point Test

Figure 4. Penetration Test

3.3. Methodology
First stage was depolymerisation process of natural rubber SIR20 with its catalyser. As natural rubber depolymerized, mixing process were taken. Asphalt Pen 60/70 were heated at certain temperature and liquid natural rubber were added accordingly based on planned variations, which were 2%, 4%, 6%, 7%, 8%, 9%, 11% and 13%. Once asphalt and natural rubber SIR20 became one liquid solution, first stage was completed.

Second stage was testing asphalt modification characteristic. Due to its purpose to ensure asphalt modification's temperature sensitivity, two main tests were taken, penetration test (ASTM D5-97) [8] and softening point test (ASTM D36 – 95) [9]. Meanwhile, specific gravity test were taken to define asphalt modification purity against organic content. Once those tests were done, there were left analysing stage to calculate asphalt modification Penetration Index.

IOP Conf. Series: Materials Science and Engineering **650** (2019) 012045 doi:10.1088/1757-899X/650/1/012045

4. Results and Discussion

In addition to the 60/70 pen asphalt material, modified asphalt with elastomeric natural rubber was used as a mixture. The characteristics of mixing pen 60/70 asphalt and natural rubber are presented in Table 1.

Table 1. Asphalt Modification Characteristic

Characteristic	SIR20 Content								
	0%	2%	4%	6%	7%	8%	9%	11%	13%
Penetration	65	64	62	60	57	58	55	52	48
Softening Point	50.9	51	53	54.3	54.2	54.9	55.4	56.6	57.7
Spec. Gravity	1.037	1.033	1.034	1.035	1.0356	1.036	1.036	1.0363	1.0367

Based on the results as seen on Table 1, from the penetration value it can be concluded that the higher the level of rubber on the asphalt, the smaller the penetration value of the asphalt. The small penetration value indicates that the asphalt properties are getting harder. This shows that asphalt with a greater percentage of rubber will show a smaller penetration value, which means that the bitumen becomes harder.

After testing the softening point and obtaining the results, it can be seen that the greater the rubber content on the asphalt, the softening point that affects the asphalt hardness will be higher. In this study, 60/70 pen bitumen has a softening point temperature of 50.9°C. This temperature meets the hard asphalt point temperature specifications of ≥ 48 °C, based on the General Specifications of Bina Marga 2010 revision 3 [10].

Modified asphalt with natural rubber has a minimum softening point temperature requirement of 54°C. Thus, asphalt with a rubber mixture of 2% and 4% does not meet specifications with a softening point of 51°C and 53°C. However, on asphalt with a minimum rubber mixture of 6% - 13% the softening point temperature has met the specifications of 54.3 - 57.7 °C.

Specific gravity of asphalt as well as on asphalt modification was supposedly minimum 1.00. On table 1, it can be concluded that all asphalt modification is pure against organic content. Furthermore, the more natural rubber content on asphalt modification, specific gravity value gets closer value to original asphalt specific gravity (0% natural rubber content). It can be inferred that the more natural rubber addition could have more similar behaviour with natural asphalt.

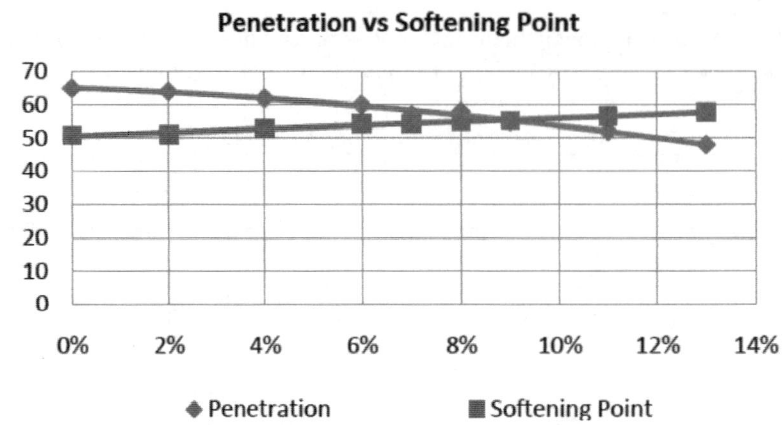

Figure 5. Penetration vs Softening Point

Figure 5 combined two main characteristic in defining temperature sensitivity, which are penetration and softening point. As we can see, penetration and softening point has different trend of curve, penetration value tend to get lower as adding the natural rubber content. Meanwhile, softening value tend to get higher as adding the natural rubber content. But, at 9% of natural rubber content, softening point was on higher value than penetration. This showed that at 9% asphalt modification condition reached its optimum hardness.

A study states that the addition of natural rubber into the asphalt can change the nature of the asphalt mixture especially in penetration and softening points so that it can reduce the sensitivity of bitumen to temperature [11]. To determine the sensitivity of asphalt to temperature, a certain parameter is needed, namely by calculating the penetration index. The penetration index (Penetration Index, PI) can be obtained by the following equation:

$$PI = \frac{(20-500\,A)}{(1+50A)} \tag{1}$$

$$A = \frac{(log\,800-log\,pen)}{(TL-25)} \tag{2}$$

Whereas,
TL = Softening point value
pen = penetration value
Log pen = logarithmic value of penetration

Table 2. Penetration Index in Asphalt Modification

Characteristic	SIR20 Content								
	0%	2%	4%	6%	7%	8%	9%	11%	13%
PI	-0.337	-0.352	0.056	0.275	0.12	0.324	0.301	0.423	0.459

Based on Table 2, it can be seen that the more rubber content in the asphalt, the penetration index obtained is closer to +1, which is 0.459. The penetration index that is getting closer to +1 indicates that the asphalt properties are getting harder and more insensitive to temperature. The nature of this asphalt is the desired properties of the asphalt mixture with rubber, because with decreasing sensitivity to temperature, the asphalt mixture will further increase the resistance of the asphalt mixture to permanent deformation due to the addition of the load.

IOP Conf. Series: Materials Science and Engineering **650** (2019) 012045 doi:10.1088/1757-899X/650/1/012045

5. Conclusion

Addition of rubber to solid SIR 20 on asphalt pen 60/70 adds hardness to the asphalt. This added value of hardness can be seen in the decrease in the penetration value to 48, the increase in the softening point value to 57.7°C and the Penetration Index value to 0.459. The nature of this asphalt is the desired nature of the asphalt mixture with rubber, because with decreasing sensitivity to temperature, the asphalt mixture will further increase the resistance of the asphalt mixture to permanent deformation due to the addition of load.

This increase in asphalt hardness can be used as an initial parameter to make asphalt mixtures that are more resistant to rutting and permanent deformation. However, further testing of asphalt mixes is needed to prove such resilience such as testing asphalt mixes with the refusal density method and Wheel Tracking Machine.

References

[1] Suroso T W 2007 Peningkatan Kinerja Campuran Beraspal dengan Karet Alam dan Karet Sintetis *Jurnal Jalan-Jembatan* **24** (1) 14 - 25

[2] Vichitcholchai N, Panmai J, dan Na-Ranong N 2012 Modification of Asphalt Cement by Natural Rubber for Pavement *Construction Rubber Thai Journal* 1 32-39

[3] Prastanto H 2014 Depolimerisasi Karet Alam Secara Mekanis untuk Bahan Aditif Aspal *Jurnal Penelitian Karet* **32(1)** 81 - 87

[4] NorFazira, NorHidayah, Jaya R P 2016 An Overview On Natural Rubber Application For Asphalt *Modification International Journal of Agriculture, Forestry and Plantation* **2** 212-218

[5] Prastanto H, Cifriadi A, and Ramadhan A 2015 Karakteristik dan Hasil Uji Marshall Aspal Termodifikasi dengan Karet Alam Terdepolimerisasi sebagai Aditif *Jurnal Penelitian Karet* **33 (1)** 75 - 82

[6] Shafii M A and Veng C L Y 2017 Effect of Blending Temperature and Blending Time on Physical Properties of NRL-Modified Bitumen *International Journal of Applied Engineering Research* **12 (13)** 3844-3849

[7] Krishnapriya M G 2015 Performance Evaluation Of Natural Rubber Modified Bituminous Mixes *Journal of Civil, Structural. Environmental, Water Resources and Infrastructure Engineering* **5 (1)** 1-12

[8] ASTM D5-97 *Standard Test Method for Penetration of Bituminous Materials* (West Conshohocken, PA: ASTM International)

[9] ASTM D36-95 *Standard Test Method for Softening Point of Bitumen* (Ring-and-Ball Apparatus) (West Conshohocken, PA: ASTM International)

[10] Kementerian Pekerjaan Umum 2010 Section 6.3 *Spesifikasi Campuran Beraspal Panas*

[11] Darunifah and Nurkhayati 2007 *Pengaruh Bahan Tambahan Karet Padat Terhadap Karakteristik Campuran Hot Rolled Sheet Wearing Course (HRS-WC)* (Semarang: Program Pasca Sarjana Universitas Diponogero Semarang)

IOP Conf. Series: Materials Science and Engineering **650** (2019) 012046 doi:10.1088/1757-899X/650/1/012046

Review study of irregularity track monitoring based on inertial measure unit and power spectral density analysis

Istiar and H Widyastuti

Civil Engineering Department, Institut Teknologi Sepuluh Nopember, Jl. Arif Rahman Hakim Surabaya 60111

E-mail: istiar@yahoo.com; h.w.dyas@gmail.com

Abstract. The maintenance of railway in Indonesia is determined by the track quality index (TQI). Analysed of TQI is based on track geometric inspection data. Currently, track geometric inspection uses a Track Geometric Vehicle (TGV) or Track Recorded Vehicle (TRV). The TRV's provide accurate geometry measurements. But, it was expensive to run, may disrupt regular train services during their operation, need for occupation of track limit the number of measurement operations undertaken annually and optical sensors are sensitive to the dirt, found in the railway environment. TRV's data was analysed by statistical function of the standard deviation to determine of TQI. TQI has some disadvantages, for example, it cannot represented the wavelength contents of the geometry defect, which is related to the particular issue in train-track interaction. Therefore, it is necessary to propose effective, efficient and reliable track quality assessment method. This paper will review some irregularity track monitoring methods that were probably used to track monitoring in Indonesia.

1. Introduction

The existence of railroad infrastructure must be able to guarantee the safety of train service. In 2018, most of train accident in Indonesia was derailment. Derailment was caused by irregularity track geometry [1]. Train accident would affect in disruption of train service. In order to keep the trains running without disruptions, an efficient maintenance policy based on risk assessment of the different components of the infrastructure is essential to anticipate problems before they occur [2].

The maintenance of railway in Indonesia is determined by the track quality index (TQI). Analysed of TQI is based on track geometric inspection data. Currently, track geometric inspection uses a Track Geometric Vehicle (TGV) or Track Recorded Vehicle (TRV). Track geometry was accurately measurements by TRV but are expensive to do, both in terms of operators side and the measurement system still worked corectly [3]. However, these vehicles may disrupt regular train services during their operation (O'Brien et al., 2016), high running costs and the need for occupation of track limit the number of measurement operations undertaken annually [4], and optical sensors are sensitive to the dirt, found in the railway environment [3]. Therefore, it is necessary to propose effective, efficient and reliable track quality assessment method.

IOP Conf. Series: Materials Science and Engineering **650** (2019) 012046 doi:10.1088/1757-899X/650/1/012046

Real-time automated monitoring systems are beneficial in terms of the effectiveness of condition monitoring in identifying maintenance issues in any structural health monitoring system (Malekjafarian et al. 2015 in [6]). Over last few decades railway vehicle dynamics has changed from being fundamental mechanical engineering discipline to one that utilizes sensors, electronics devices and computer processing [7]. The sensor reviewed on this paper is inertial measurement unit (IMU). Inertial measurement unit (IMU) usually used for track geometry monitoring system wasaccelerometer and gyro-meter. Accelerometer recorded moving acceleration an object in three axis. While gyro-meter measured orientation and angular velocity an object in three axis.

All recorded data by TRV was used for track quality index (TQI) analysing in Indonesia, track maintenance system commonly use the Track Quality Index (TQI) method, which is a statistical analysis of the standard deviation of each irregularity track. TQI has some disadvantages, for example, it cannot represented the wavelength contents of the geometry defect, which is related to the particular issue in train-track interaction [8]. So, PSD can be consider as a method to analyse track quality index in Indonesia. The application of Power Spectral Density in railway inspection is relatively new [8]. Various countries such as USA, China, France and Germany have modelled their own spectra of railway track irregularity. Various studies the use of IMU in-service railway vehicles and the use of PSD in the analysis of track irregularity have been proposed. Those studies are described in the next sections.

2. Monitoring track irregularity from in-service railway vehicles

An inertial measurement unit (IMU) is an electronic device that measures and reports a body's specific force, angular rate, and sometimes the magnetic field surroundings the body, using a combination of accelerometers and gyroscopes, sometimes also magnetometers.

Some researchers have been propose IMU as sensor to monitor track irregularity. [3] used accelerometer and gyro-meter to record dynamic response of vehicle. Two vehicles were measured the dynamic response, Tyne and Wear Metro and Class 175 in UK. Inertial measurement unit (IMU) were installed on axle-box and boogie. The detail location of IMU installation were showed on Table 1 and Table 2.

Table 1 Installed sensors equipment on Tyne and Wear trials [3]

Designation	Description	Range
$\ddot{z}_{L,\ axlebox}$	Vertically sensing accelerometer on rear-left axle-box	± 100 g
$\ddot{z}_{R,\ axlebox}$	Vertically sensing accelerometer on front-right axle-box	± 100 g
$\ddot{z}_{L,\ bogie}$	Vertically sensing accelerometer above rear-left axle-box	± 10 g
$\ddot{z}_{R,\ bogie}$	Vertically sensing accelerometer above front-right axle-box	± 10 g
Δz_L	Displacement (6a) to 8	± 38 mm
Δz_R	Displacement (7a) to 8	± 38 mm
$\emptyset_{bogie}$	Pitch-rate gyro installed on bogie	± 50° s-1
Θ_{bogie}	Roll rate gyro installed on bogie	± 50° s-1
Ψ_{bogie}	Yaw rate gyro installed on bogie	± 50° s-1
$\ddot{y}_{bogie}$	Laterally sensing accelerometer installed on bogie	± 10 g
$\ddot{z}_{body}$	Vertically sensing accelerometer over instrumented bogie installed on body	± 4 g 0 – 10 V
v	Voltage proportional to vehicle speed	0 – 23 m/s

IOP Conf. Series: Materials Science and Engineering **650** (2019) 012046 doi:10.1088/1757-899X/650/1/012046

Table 2. Installed sensors equipment on Class 175 [3]

Designation	Description	Range
$\ddot{z}$ bogie	Vertically sensing accelerometer (somewhat offset laterally) installed on bogie	± 10 g
$\ddot{y}$ bogie	Laterally sensing accelerometer installed on bogie	± 10 g
Øbogie	Pitch-rate gyro installed on bogie	± 50° s-1
Θbogie	Roll rate gyro installed on bogie	± 50° s-1
Ψ bogie	Yaw rate gyro installed on bogie	± 50° s-1
$\ddot{z}$ body	Vertically sensing accelerometer over instrumented bogie installed on body	± 4 g
v	Voltage proportional to vehicle speed	0 – 10 V 0 – 45 m/s

An accelerometer installed on a bogie or axle-box senses acceleration roughly perpendicular to the trajectory of the sensor through space, see Fig 1.

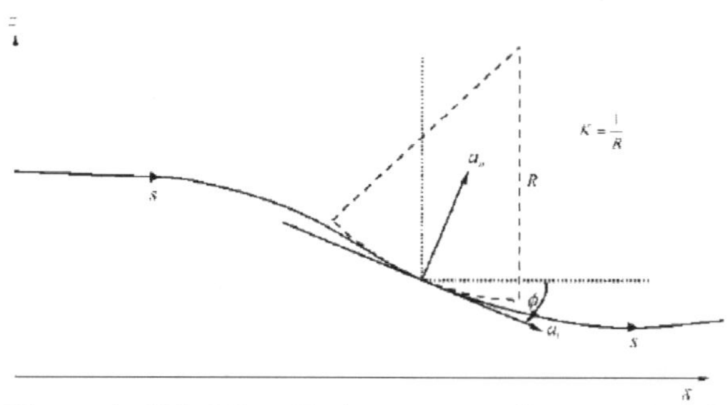

Figure 1. Side View Trajectory coordinate system [3]

The sensed acceleration arising from following the instantaneous curvature of the trajectory is showed by:

$$\alpha = \frac{v_x^2}{R} = v_x^2 K,\ K = \frac{a}{v_x^2} \tag{1}$$

where a is the (instantaneous) acceleration, R the radius of curvature, and K the curvature. In a similar way, assuming that the pitch of the bogie is basically the same as the angle from horizontal of the track under the middle of the bogie, it can obtain an approximate relationship between trajectory curvature and pitch-rate.

$$Ø = \frac{v_x}{R} = v_x K,\ K = \frac{Ø}{v_x} \tag{2}$$

where $\dot{\phi}$ is pitch-rate. The prediction of vertical displacement related with the curvature, is obtained by double integrating the curvature with respect to displacement along the track:

$$\hat{z} = \iint K\, ds\, ds \tag{3}$$

A plan-view trajectory can be described by the instantaneous (horizontal) curvature K (s) as a function of displacement s along the trajectory, as illustrated in Fig. 2. Lateral accelerometer installed on a bogie, inspects curvature through the centripetal force experienced by the accelerometer. Yaw rate gyro installed on a bogie inspects curvature from the turning rate of the bogie yaw. An accelerometer sensing acceleration perpendicular to a trajectory, inspects the instantaneous curvature according to:

IOP Conf. Series: Materials Science and Engineering **650** (2019) 012046 doi:10.1088/1757-899X/650/1/012046

$$\alpha = \frac{v_x{}^2}{R} = K^{\alpha} v_x{}^2, K^{\alpha} = \frac{a}{v_x{}^2} \tag{4}$$

where a is the lateral acceleration, R the radius of curvature, K^a the accelerometer-derived curvature, and v_x the vehicle speed along the track. A A yaw rate gyro offend the curvature of the trajectory through the bogie, ignoring the yaw and lateral resonances, according to:

$$\Psi = \frac{v_x}{R} = K^{\Psi} v_x, K^{\Psi} = \frac{\Psi}{v_x} \tag{5}$$

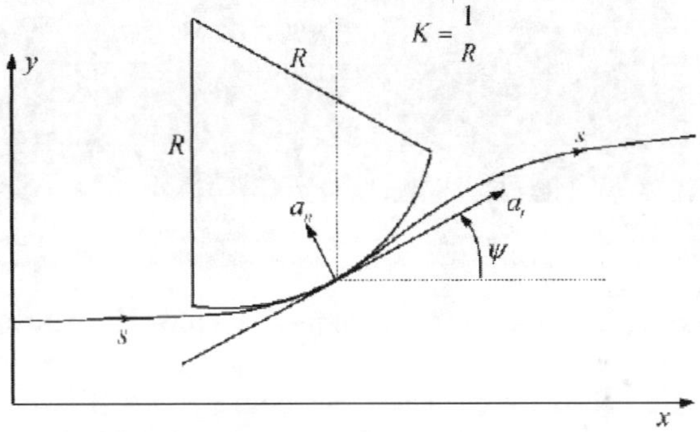

Figure 2. Horizontal coordinate system [9]

Results from [3] showed effective vertical irregularity monitoring, specially the ability to monitor vertical irregularity over a wide range of vehicle speeds down to about 1 m s-1, where vertically accelerometers combined with displacement transducers are unable to function correctly. In principle, either bogie lateral acceleration or yaw rate can be processed to give a prediction of mean lateral track irregularity, but a yaw rate gyro provides consistent prediction down to lower vehicle speeds than does an accelerometer and does not require compensation for the impact of bogie roll. An improved prediction can be obtained by inverting the dynamic relationship between the mean track alignment and bogie yaw movement.

Another track geometry parameter predicted by IMU data was cross-level. Cross-level is the height difference between the left and right rails on a track. In the research of [10], cross-level data was collected by a number of spring nest displacement (SND) sensors. The sensors equipment installed between the bogie and the axle box as shown in Figure 3 (b). The readings of these sensors were a combination of relative extension/ compression between the bogie and the wheel set, roll of the bogie and roll of the wheelset. This is illustrated in Figure 4.

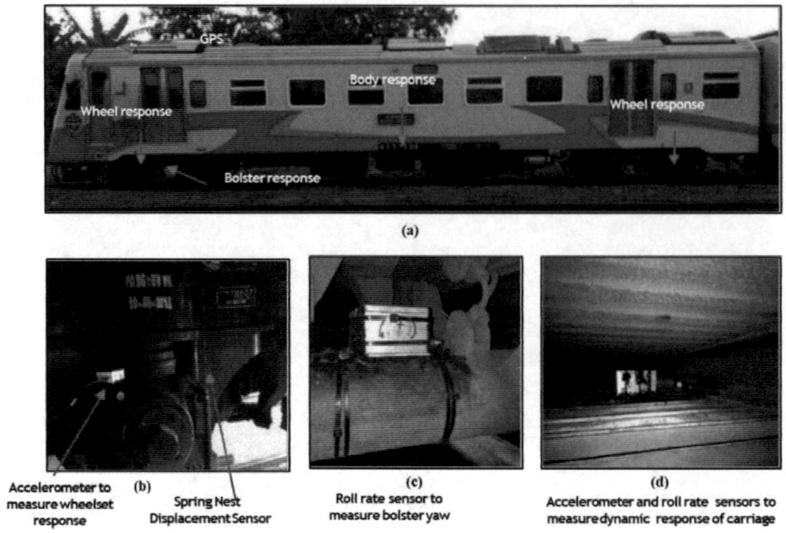

Figure 3. (a) Instrumentation of the passenger wagon to measure track condition and dynamic response. (b) Detailed view of the Spring Nest Displacement Sensor measuring primary suspension behaviour. (c) Roll rate sensor to measure the bogie roll. (d) Tri-axial accelerometers and roll rate sensors to measure the dynamic response of the wagon body [10].

The mean measurement on the left and right wheels of a wheelset of the SND sensor gives the measurement of the relative extension/compression between the bogie and the set (mean (SNDleft, SNDright)). The absolute roll (φ) of the bogie was the result of integration of the roll rate sensor measurement. The displacement contribution at SND from the bogie roll was determined by used the roll ($SNDBroll = w \times \sin(\emptyset)$). w is the distance from the centre of the bogie to the attachment point of the SND sensor. Knowing the roll of the bogie and the raw measurement of the SND sensor, the relative extension/compression between the bogie, the roll of the cross level between the left and right rail can be calculated as shown in Equation 6.

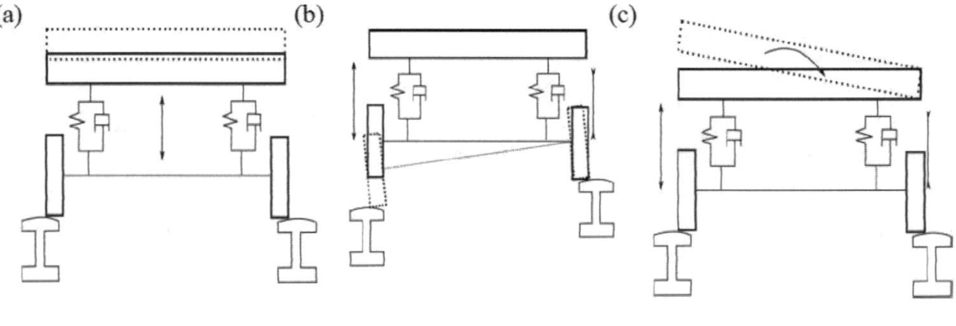

Figure 4. Primary suspension motion represented as a sum of (a) vertical compression/extension (b) Wheelset Roll (c) Bolster Roll [10]

$$CL = (SND\ Raw_{left} - mean(SND_{left}SND_{right}) - SND_{Broll}) + (SND\ Raw_{right} - mean(SND_{left}SND_{right}) + SND_{Broll}) \quad (6)$$

3. Power Spectral Density (PSD) Analysis

Several countries such as Britain, Germany, USA and China have made Comprehensive evaluations of the track irregularity spectrum (Zhipping and Shouhua, 2009 in [8]). Each country had specific characteristics of the track measured, which affect to analytical expressions of the PSD function. Those studies are described in the following sections.

3.1. The FRA PSD Standards (United States)

In USA, railway track was classified into 9 categories of track classes by the US Federal Railroad Administration (FRA). Classes 1 to 6 are designed for general tracks and Classes 7 to 9 are designed for high speed railways. One-sided power spectral density (PSD) function was used to described the random track irregularity for each track classes. The function is only applied to the wavelength range of 1.524 m to 304.8 m. Because the measurement equipment had limitation in the field measurement (Liu et al., 2011 in [8]). The empirical formula of PSD is as follows:
For vertical alignment:

$$S_{av}(\Omega) = \frac{k*Av*\Omega_c{}^2}{\Omega^2*(\Omega^2+\Omega_c{}^2)} \tag{7}$$

For lateral alignment:

$$S_{al}(\Omega) = \frac{k*Aa*\Omega_c{}^2}{\Omega^2*(\Omega^2+\Omega_c{}^2)} \tag{8}$$

For gauge and super elevation irregularity (cross level):

$$S_{gauge/cl}(\Omega) = \frac{4*k*Av*\Omega_c{}^2}{(\Omega^2+\Omega_c{}^2)*(\Omega^2+\Omega_c{}^2)} \tag{9}$$

where:

$S_{av}(\Omega)$	=	PSD of track vertical alignment irregularity [cm^2/(rad/m)]
$S_{al}(\Omega)$	=	PSD of track lateral alignment irregularity [cm^2/(rad/m)]
$S_{gauge/cl}(\Omega)$	=	PSD of track gauge or super elevation irregularity (cross level) [cm^2/(rad/m)]
Ω	=	spatial wave number (rad/m)
Ω_C, Ω_S	=	critical wavenumber (rad/m)
A_v, A_a	=	roughness coefficient related to the line grade (cm^2*rad/m)
k	=	a determined variable (≈ 0.25)

The spatial wavenumber (Ω) is related to the frequency per time unit f_h (Hertz) by the following relation $\Omega = 2 * \pi * f_h/v$. The parameters used in Equations (7) to (9) was presented in Table 3.

Table 3. The Coefficients of Power Spectral Density (PSD) function (Xia, 2002 in [8])

Line Class	Max Line Speed		Av (cm^2.rad/m)	Aa (cm^2.rad/m)	ΩC (rad/m)	ΩS (rad/m)
	Freight (km/h)	Passenger (km/h)				
1	16	24	1.2107	3.3634	0.8245	0.6046
2	40	48	1.0181	1.2107	0.8245	0.9308
3	64	97	0.6816	0.4128	0.8245	0.852
4	97	129	0.5376	0.3027	0.8245	1.1312
5	129	145	0.2095	0.0762	0.8245	0.8209
6	177	177	0.0339	0.0339	0.8245	0.438

Note: the coefficients of track classes 7 to 9 are not defined yet by the FR

IOP Conf. Series: Materials Science and Engineering **650** (2019) 012046 doi:10.1088/1757-899X/650/1/012046

3.2. German PSD Standard

Most of the European countries was used the German track PSD spectrum for dynamic simulations of railway vehicles (Zhiqiang et al., 2009 in [8]). The German track PSD spectrum is characterized by a single-sided spectrum. Irregularity track in the range of 0,01*2π to 0,4*2π (rad/m) was best representation of the German track PSD spectrum model (Zhang et al., 2010 in [8]). The PSD function is expressed by:

For longitudinal profile:

$$S_v(\Omega) = \frac{Ap*\Omega_c^{\,2}}{(\Omega^2+\Omega_\gamma^{\,2})*(\Omega^2+\Omega_c^{\,2})} \tag{10}$$

For lateral alignment:

$$S_{al}(\Omega) = \frac{Aa*\Omega_c^{\,2}}{(\Omega^2+\Omega_\gamma^{\,2})*(\Omega^2+\Omega_c^{\,2})} \tag{11}$$

For cross level or super elevation irregularity:

$$S_{cl}(\Omega) = \frac{\left(Ap*\frac{\Omega_c^{\,2}}{a^2}\right)*\Omega^2}{(\Omega^2+\Omega_\gamma^{\,2})*(\Omega^2+\Omega_c^{\,2})*(\Omega^2+\Omega_s^{\,2})} \tag{12}$$

S_v (Ω) = PSD of track longitudinal profile irregularity [m^2/(rad/m)]
S_{al} (Ω) = PSD of track lateral alignment irregularity [m^2/(rad/m)]
$S_{gauge/cl}$ (Ω) = PSD of cross level or super elevation irregularity [cm^2/(rad/m)]
Ω = $2\pi/\lambda$ denotes spatial wavenumber (rad/m)
Ω_C, Ω_γ, Ω_S = critical wavenumber (rad/m)
A = 0,75 m (one half of track gauge)
A_p = scale factor for longitudinal profile (m^2·rad/m)
A_a = scale factor for alignment (m^2·rad/)

In Table 4 was showed the parameters for the above equations, which represent the track irregularity with low and high levels of disturbation.

Table 4. Parameters of Track PSD (Lin *et al.*, 2004 in [8])

Parameters	Aa ($10^{-7}m^2$rad/m]	Ap ($10^{-7}m^2$rad/m]	Ω_C (rad/m)	Ω_γ (rad/m)	Ω_S (rad/m)
Low Disturbance	2.119	4.032	0.82	0.0206	0.438
High Disturbance	6.125	10.8	0.82	0.0206	0.438

3.3. The Standard of Chinese PSD

The Chinese Academy of Railway Science (CARS) had published various spectra of track irregularities (also referred as PSD Standards). The standards of evaluation and diagnosis of track quality was suitable for three different operational speed classes: 200 km/h, 160 km/h, and 120 km/h (Xianmai et al., 2008 in [8]). The spectrum range of different operational speed classes is given to accommodate the disparity of spectral amplitude that may vary from one track section to another. The track condition with a certain spectrum is usually fitted with one of these ranges. The higher the track quality, the closer the spectral curve of the track section to the lower limit of the standard value. Conversely, the lower the track quality, the closer the spectral curve to the upper limit of the standard value.

The Chinese PSD function is is described using single-sided spectrum, which depend on 6 coefficients as shown below:

$$S(f) = \frac{af^2+b}{cf^2+df^2+ef^2+k} \quad (13)$$

where S(f) denotes track irregularity PSD in unit [mm2/(1/m)] and f is wavenumber, often called spatial frequency of a wave, measured in 1/m.

The coefficients of spectral for Equation (13) are given in Table 5 to 7.

Table 5. The Parameters of Spectral for line speed design of 200 km/h (Xianmai *et al.*, 2008 in [8])

Track Irregularity		a	b	c	d	e	k
Gauge	Upper	362.2681	0.2393	15370.86	681.2174	10.267	-0.0007
	General	54.0439	0.0357	8254.682	365.8602	5.5139	-0.0004
	Lower	119.2536	0.0783	36295.99	1619.269	24.2936	-0.0018
Cross level (Super elevation deflect)	Upper	951.449	2.1747	47442.79	2121.78	25.473	0.0112
	General	35.4842	0.0811	6369.446	284.8838	3.4199	0.0015
	Lower	238.6205	0.5418	85347.97	3842.441	45.8306	0.02
Alignment	Upper	0.0	0.00699	0.0	1.0	0,01893	0.00003
	General	0.0	0.00194	0.0	1.0	0,01894	0.00003
	Lower	0.0	0.00097	0.0	1.0	0,01893	0.00003
Longitudinal Profile	Upper	0.0	0.00353	0.0	1.0	0.00752	0.0
	General	0.0	0.00098	0.0	1.0	0.00788	0.0
	Lower	0.0	0.00047	0.0	1.0	0.00783	0.0

Table 6. The Parameters of Spectral for line speed design of 160 km/h (Xianmai et al., 2008 in [8])

Track Irregularity		a	b	c	d	e	k
Gauge	Upper	612.3768	0.4046	8660.944	383.8376	5.7852	-0.0004
	General	213.1331	0.1408	10851.22	480.9504	7.2484	-0.0005
	Lower	187.2267	0.1238	31792.31	1407.659	21.23	-0.0015
Cross level (Super elevation deflect)	Upper	1890.022	4.2158	19981.09	984.2226	18.5928	0.0011
	General	94.9519	0.2118	3613.811	178.0026	3.3627	0.0002
	Lower	511.6737	1.1433	65036.82	3191.977	60.4676	0.0036
Alignment	Upper	0.0	0.01751	0.0	1.0	0.01893	0.00003
	General	0.0	0.00486	0.0	1.0	0.01893	0.00003
	Lower	0.0	0.00146	0.0	1.0	0.01893	0.00003
Longitudinal Profile	Upper	0.0	0.01016	0.0	1.0	0.00704	0.0
	General	0.0	0.0029	0.0	1.0	0.00758	0.0
	Lower	0.0	0.00084	0.0	1.0	0.0075	0.0

IOP Conf. Series: Materials Science and Engineering **650** (2019) 012046 doi:10.1088/1757-899X/650/1/012046

Table 7. The Parameters of Spectral for line speed design of 120 km/h (Xianmai et al.. 2008 in [8])

Track Irregularity		a	b	c	d	e	k
Gauge	Upper	640.74	0.4233	6524.507	289.1726	4.3582	-0.0003
	General	255.976	0.1691	7819.645	346.5745	5.2233	-0.0004
	Lower	325.929	0.2151	28425.14	1261.602	18.9956	-0.0015
Cross level (Super elevation deflect)	Upper	1830.68	7.3882	20908.35	1028.226	30.9382	0.008
	General	110.624	0.44649	2527.1	124.2566	3.73927	0.00097
	Lower	1077.07	4.3434	70234.04	3460.22	103.9562	0.027
Alignment	Upper	0.0	0.02622	0.0	1.0	0.01893	0.00003
	General	0.0	0.00874	0.0	1.0	0.01893	0.00003
	Lower	0.0	0.00306	0.0	1.0	0.01893	0.00003
Longitudinal Profile	Upper	0.0	0.01351	0.0	1.0	0.00687	0.0
	General	0.0	0.00478	0.0	1.0	0.00739	0.0
	Lower	0.0	0.00166	0.0	1.0	0.00721	0.0

Note: Although PSD provides a limit range of the spectral amplitude, this does not mean that the track spectrum cannot be lower or higher than the threshold limit value. This range is proposed based on the expected amplitude span of the chinese track irregularity spectra

3.4. SNCF PSD Standards (France)

SNCF proposed a PSD based on single-sided spectrum through an investigation on the railway track in France, which is suitable for vertical alignment. The equation used a function of cyclic wavenumber [cycles/m], described the track irregularities within the range of 2 m ≤ L ≤ 40 m [11)]. The SNCF model is as follows:
For vertical irregularity:

$$G_{\Gamma\Gamma}(n) = \frac{A}{(1+\frac{n}{n0})^3} \quad (14)$$

where:

A = Indicaation of the state of rail surface [m3] or [m2/(cycle/m)]

$$\begin{cases} 308\ X\ 0.509.10^{-6} \text{ for good state} \\ 308x1.790.10^{-6} \text{ for good state} \end{cases}$$

n_0 = Coefficient, equal to 0.0489 (cycle/m)

n = Cyclic wavenumber (cycle/m)

3.5. The Standard of Braun PSD

The International Organization for Standardization (ISO) formulate a uniform method to measure vertical surface of roads, highways, and off-road terrain. Braun, as cited by [11], has then adapted the road condition model from ISO to the context of the railway. The Braun PSD model has two value limits; upper and lower limit values, and it is based on single-sided spectra. The Braun model is described by the following equation:
For vertical irregularity:

$$G_{TT}(n) = G_{TT}(n_0) * (\frac{n}{n_0})^{-w} \quad (15)$$

where:

$G_{rr}(n_0)$ = scale factor of rail roughness [m³] or[m²/(cycle/m)]

$\begin{cases} 5\,.\,10^{-7} \text{ for the upper limit} \\ 1{,}0\,.\,10^{-7} \text{ for the lower limit} \end{cases}$

n_0 = 1/(2π) (cycle/m)

w = waviness, with values usually ranging from 1,5 to 3,5

n = cyclic wavenumber (cycle/m)

4. Conclusion

Nowdays, irregularity track is determined by statistical analysed, based on track geometriy vehicle measurement result. Statistical analysis of track geometriy vehicle data cannot reflect the wavelength contents of the geometry defect, which is inherently related to the particular issue in train-track interaction. Irregularity track could be determined by inertial measurement unit (IMU), installed at in-service vehicle. Inertial measurement unit (IMU) usually used for sensor dynamic response, were accelerometer, gyro-meter and spring nest displacement sensor. The collected data of IMU was analysed by Power Spectral Density (PSD). So, the probability of research topic that could be proposed, based on review study was analyze track geometric index based on inertial measurement unit by power spectral density.

5. References

[1] Liu X, Saat R, and Barkan C P R 2012 Analysis of Causes of Major Train Derailment and Their Effect on Accident Rates TRB: *J. Trans. Resc. Board* **2289** 154–163

[2] Jamshidi A, Roohi S F, Núñez A, Babuska R, De Schutter B, Dollevoet R, and Li Z 2016 Probabilistic defect-based risk assessment approach for rail failures in railway infrastructure *IFAC-PapersOnLine* **49(3)** 73-77

[3] Westeon P F, Ling C S, Roberts C, Goodman C J, Li P, and Goodall R M 2007 Monitoring Vertical Track Irregularity from In-service Railway Vehicles *Proc. IMechE* **221** 75-88

[4] OBrien E J, Bowe C, Quirke P, and Cantero D 2016 Determination of longitudinal profile of railway track using vehicle-based inertial readings *Proc. Inst. of Mech. Eng.* **231/5** 518-534

[5] OBrien E J, Quirke P, Bowe C, and Cantero D 2017 Determination of Railway Track Longitudinal Profile Using Measured Inertial Response of An In-service Railway Vehicle Struct. *Health Mon.* **17(6)** 1425-1440

[6] Quirke P, OBrien E J, Bowe C, Malekjafarian A, and Cantero D 2018 Estimation of Railway Track Longitudinal Profile Using Vehicle-Based Inertial Measurements *Int. Cong. And Exh. "Sust.Civ. Infr.: Innov. Infra. Geo*. 145-148

[7] Chudzikiewicz A, Bogacz R, Kostrzewski M, and Konowrocki R 2018 Condition Monitoring of Railway Track Systems by Using Acceleration Signals on Wheelset Axle- Boxes *J. Tran.* **33(2)** 555–566

[8] Berawi 2013 Improving Railway Track Maintenance Using Power Spectral Density (PSD) *PhD Dissertation* (Porto: Universidade Do Porto)

[9] Weston P F, Ling C S, Goodman C J, Roberts C, Li P, and Goodall R M 2007b Monitoring Lateral Track Irregularity from In-service Railway Vehicles *Proc. IMechE* **221** 89-100.

[10] Nadarajah N, Shamdani A, Hardie G, Chiu W K, and Widyastuti H 2018 Prediction of Railway Vehicles' Dynamic Behaviour with Machine Learning Algorithms Elect. *J. of Struct. Eng.* **18(1)** 38 – 46

[11] Broeck 2001 A prediction model for ground-borne vibrations due to railway traffic *PhD Thesis* (Leuven: Katholike Universiteit Leuven)

IOP Conf. Series: Materials Science and Engineering **650** (2019) 012047 doi:10.1088/1757-899X/650/1/012047

Model of relationship between car ownership growth and economic growth in Java

N C Kresnanto

Magister of Civil Engineering, Janabadra University, Jalan Tentara Rakyat Mataram 57, Yogyakarta, Indonesia

E-mail: nindyo_ck@staff.janabadra.ac.id

Abstract. Vehicle growth is a problem that is inseparable from sustainable transportation. In 2014, the number of private vehicle ownership in Indonesia reached more than 448 vehicles per 1,000 people. Specifically, for cars it reaches 324 vehicles per 1000 people. The growth of private vehicle ownership is strongly influenced by economic growth (one measure of the economic growth of the economy is Gross Domestic Product - GDP). It is necessary to look at the mathematical relationship between trends in economic growth represented by GDP and the growth of private vehicle ownership, with a model. This research is located on Java Island using provincial administrative based data (5 provinces namely Banten, Capital Region of Jakarta, West Java, Central Java, and East Java). The growth model of vehicle ownership on Java, in general, can be approached with the Gompertz Function. The Gompertz function is a type of mathematical model for a time series and is named after Benjamin Gompertz (1779-1865). It is a sigmoid function which describes growth as being slowed at the start and end of a given time period.

1. Introduction

Many developing countries have experienced high growth in motorized vehicles at the end of this decade, along with the development of their economic growth [1]. The results of the study on the sustainable mobility project (World Business Council for Sustainable Development, 2004) projected a five-fold increase in the number of vehicles in China and Latin America in 2050. Meanwhile, in Indonesia, the 2014 Central Bureau of Statistics noted that the number of motor vehicles had reached 114,209,266 units consisting of 12,599,138 passenger cars, 2,398,846 buses, 6,235,136 freight cars, and 92,976,240 motorbikes. The increase in the number of motorized vehicles per year in Indonesia is 14.55%, this increase is a very high increase compared to the increase in the average population in the range of 2010-2016 which is equal to 1.36%. If the population of Indonesia in 2014 was 254.9 million (Susenas 2014), the choice of motorized vehicles in Indonesia reached more than 448 vehicles per 1.000 people. The number and growth of these vehicles is a separate problem related to sustainable transportation.

Increased ownership of motorized vehicles will certainly also affect economic growth. Economic growth in Indonesia, between the period 2000-2004, economic recovery occurred with an average growth of Gross Domestic Product (GDP) at 4.6% per year [2]. After that, GDP growth accelerated (with the exception of 2009 time, due to global financial shocks and uncertainty, Indonesia's GDP growth fell to 4.6%, a figure that is still amazing) and peaked at 6.5% in 2011. Nevertheless, after 2011 the expansion of the Indonesian economy began to slow down greatly. Between 2011 and 2015 Indonesia's economic growth slowed sharply (which is explained in more detail below). When compared with the growth of motorized vehicles, it can be concluded while the growth rate of high motor vehicles at the end of this decade actually has an impact on declining GDP.

By taking into account the background of the very rapid development of motorized vehicles and the possible impacts that arise as a result, it is necessary to examine the extent of the influence of ownership of motorized vehicles (especially cars) on the economic growth of a region or vice versa. Several previous studies have tried to build a model to predict the growth of motorized vehicles, which of course the results can be used as a reference for planning. The growth of motorized vehicles from various countries, especially developing countries, was concluded following the S-Curve pattern [3]. A study of several growth models of motorized vehicles related to socio-economic independent variables and other variables such as vehicle type, demand-supply, and others has also been done [4]. In Columbia, a model has been created to predict the increase in ownership of motorized vehicles (motorbikes and cars), and has tested the sensitivity of income levels to the ownership of these vehicles [5]. Based on the per capita GDP of the population, the growth of vehicles in China follows the S-Curve pattern [6]. From a number of studies submitted it can be said that there is a significant relationship between the growth of vehicles with socio-economic variables, especially GDP (income).

Given the importance of predicting the growth of motorized vehicles, this paper will discuss the relationship between the growth of motor vehicles and economic progress with the main indicator of GDP per capita. In the discussion, this paper takes a case study on Java and the type of vehicle to be studied is a car because this type of vehicle has a fairly high growth rate.

2. Methodology

2.1. Gompertz Model for Vehicle Ownership Predictions

The speed of expansion of vehicle ownership in developing countries is mainly driven by market demand, this means that an increase in vehicle ownership is closely related to the level of consumer wealth and intention to buy [6]. Income variables usually better reflect aggregate demand. Therefore, researchers can use per capita income as an economic factor in Gompertz's functions in developing countries. The Gompertz function is widely applied to build relationships between motor vehicle ownership and economic factors. The relationship of increasing motor vehicle ownership per 1,000 people was approached by the Gompertz Model as in equation 1.

$$V_t^* = \gamma \,.\, e^{\alpha . e^{\beta . GDP_t}} \tag{1}$$

where

- Vt : vehicle ownership per 1000 people in year t model
- GDP_t : Per capita GDP in year t
- γ : coefficient which describes the saturation condition of vehicle ownership per 1,000 residents (this figure is taken as the maximum value of vehicle ownership data per 1,000 residents, whichis 1,299 vehicles / 1,000 people)
- α : negative coefficient which describes the initial condition of the model
- β : negative coefficient which determines the curvature of the Gompertz curve

In general, the Gompertz model will form an S-Curve pattern, as in Figure 1a.

IOP Conf. Series: Materials Science and Engineering **650** (2019) 012047 doi:10.1088/1757-899X/650/1/012047

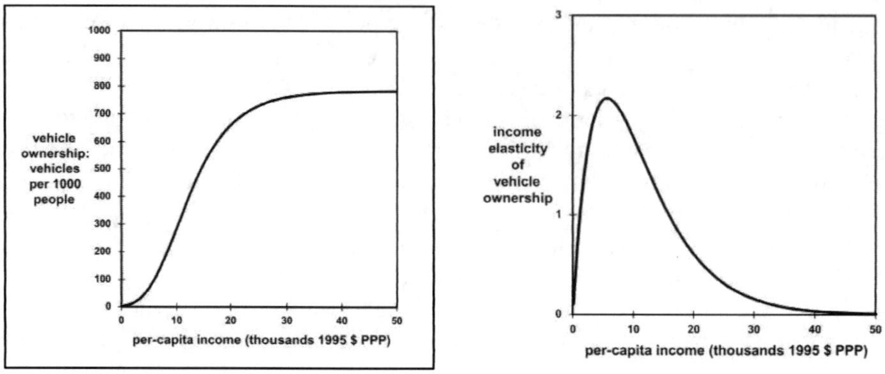

(a) Gompertz Function *(b) Gompertz Function Elasticity*

Figure 1. Illustration of Gompertz Function and Income Elasticity

2.2. Gompertz Model Sensitivity

The sensitivity of the model uses internal equations [3], as follow:

$$\mu_t{}^{LR} = \alpha.\gamma.GDP_{t}.e^{\beta.GDP_t} \tag{2}$$

This elasticity is all positive at all GDP levels because $\boldsymbol{\alpha}$ and $\boldsymbol{\beta}$ are negative. The elasticity value increases from GDP = 0 to maximum at GDP = -1 / $\boldsymbol{\beta}$ then decreases to zero asymptote line (as in Figure 5b).

3. Discussion

3.1. Growth of Motor Vehicles, Population and Road Network in Indonesia

The number of motorized vehicles in Indonesia in 2015 reached 121.39 million units [7]. Data from the Central Statistics Agency (BPS) shows that of these figures the most were motorbikes with 98.88 million units (81.5 percent). Passenger cars followed with a total of 13.48 million units (11.11%), then freight cars 6.6 million units (5.45%), and bus cars with 2.4 million units (1.99%) of the total vehicle. While this data was collected from BPS Indonesia, the growth of motorized vehicles (passenger cars and motorbikes) from 1949-2015 can be seen in Figure 2.

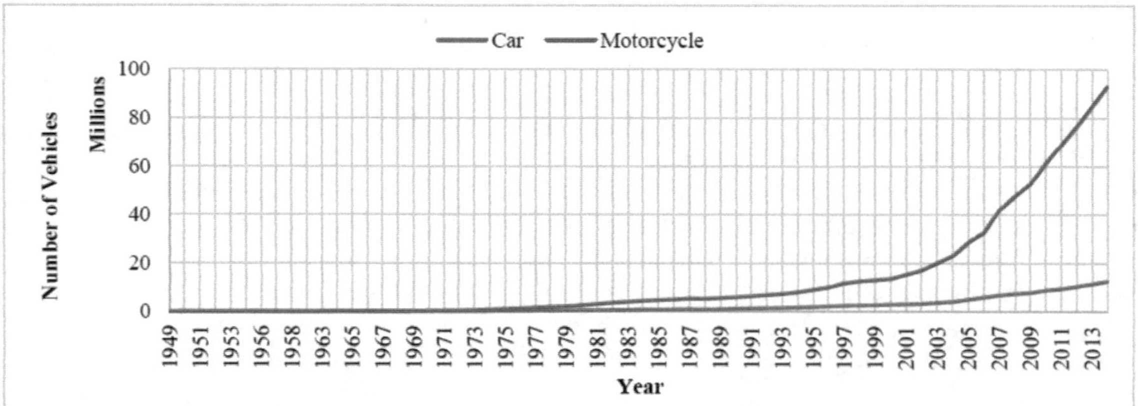

Figure 2. Growth in the number of motorized vehicles (passenger cars and motorbikes) in Indonesia

IOP Conf. Series: Materials Science and Engineering **650** (2019) 012047 doi:10.1088/1757-899X/650/1/012047

The annual growth rate of motor vehicles from 1949-2014 was 11.96% / year for passenger car vehicles and 19.59% / year for motorcycle vehicles. The growth of these vehicles also far exceeds the growth of Indonesia's own population which in the range of 2010-2016 was 1.36%.

Furthermore, what needs to be considered related to the growth of motorized vehicles is supporting infrastructure for motorized vehicles, namely in the form of road networks. Data on paved road networks in Indonesia from 1957-2016 can be seen in Figure 3. The average development per year is 6%. This will be a big problem if the rate of increase is far below the growth of motor vehicles. Impacts that occur such as congestion, pollution, etc. will be more quickly achieved. Another derivative impact is that economic conditions will get worse because, for example, the cost of travel is higher due to congestion, health costs due to pollution, stress costs due to road conditions, and other costs.

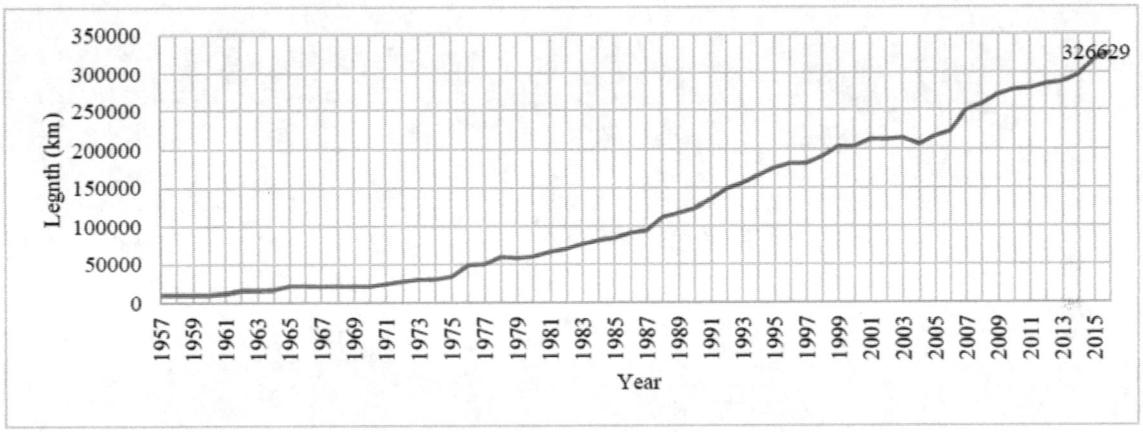

Figure 3. Development of the length of paved roads in Indonesia

3.2. Indonesia's Gross Domestic Product (GDP) as an Indicator of Economic Progress

In the economy of a country there is an indicator that is used to assess whether the economy is going well or badly. Indicators in assessing the economy must be used to find out the total income earned by everyone in the economy. The right and appropriate indicator in making these measurements is the Gross Domestic Product (GDP) / Gross Domestic Product (GDP). And GDP per capita which is the amount of GDP when compared to the population in a country is a better tool that can tell us what happens to the average population, the standard of living of its citizens.

3.3. Relationship Between Vehicle Growth and Economy

Referring to several previous studies, that there is a very strong relationship between economic growth (with GDP indicators) and vehicle ownership per 1000 population [6]. In the Java research location Banten Province, DKI Jakarta Province, West Java Province, Central Java Province and East Java Province, the relationship between vehicle ownership and Per capita GDP is shown in Figure 4. There is a very significant relationship between the two variables.

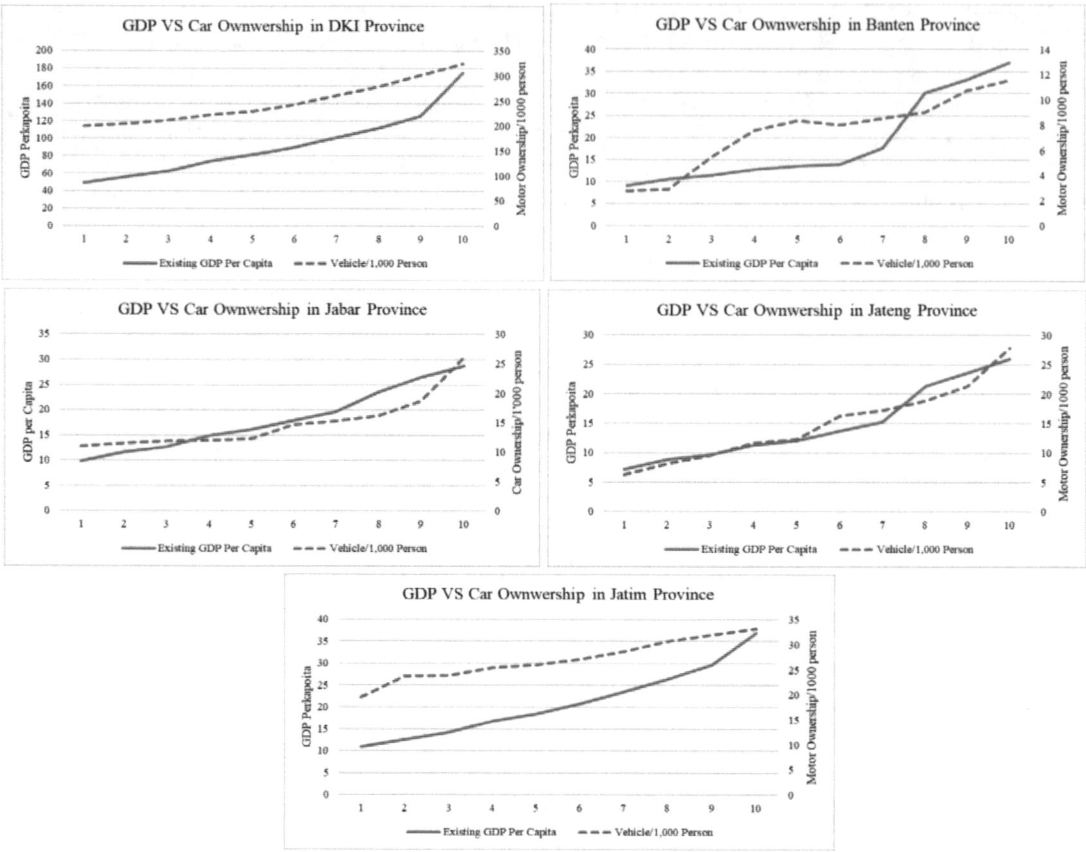

Figure 4. Relationship between motor vehicle ownership and annual GDP in several provinces in Java

3.4. Calibration Results of the Relationship Model of Car Ownership with GDP of Per capita in Java Island

Furthermore, the Gompertz Model was applied in the Car Growth Model for case studies in Java using data from 5 provinces, namely the Special Capital Region (DKI) of Jakarta, Banten Province, West Java Province (West Java), Central Java Province, and Java Province. East (East Java). The data used is socio-economic data in the form of GDP per capita data and vehicle ownership data from 2005 to 2014.

Based on the Gompertz model, the initial value of the α coefficient describes the initial condition calculated using the formulation approach used by [9]. If in year t = 1, the GDP value is 0 and $V_t^*= 1$ then equation (1) will b= 1

$$V^*_1 = \gamma . e^{\alpha . e^{\beta . o}} \quad (3)$$

$$V^*_1 = \gamma . e^{\alpha} \quad (4)$$

So,

$$\alpha = \log \frac{\gamma}{V^*_1} \rightarrow \alpha = \log \frac{324}{1} = 5{,}021 \quad (5)$$

IOP Conf. Series: Materials Science and Engineering **650** (2019) 012047 doi:10.1088/1757-899X/650/1/012047

γ, coefficient which describes the saturation condition of vehicle ownership per 1000 residents (this figure is taken as the maximum value of vehicle ownership data per 1000 inhabitants, namely 324 vehicles / 1000 people). So that there is only a coefficient perlu that needs to be calibrated. With a simple calibration model (R2), with various applications of the value of β, the best value of β is 0.0325 with an R2 value of 0.9456. The result of the calibration coefficient iteration β can be seen in Figure 5.

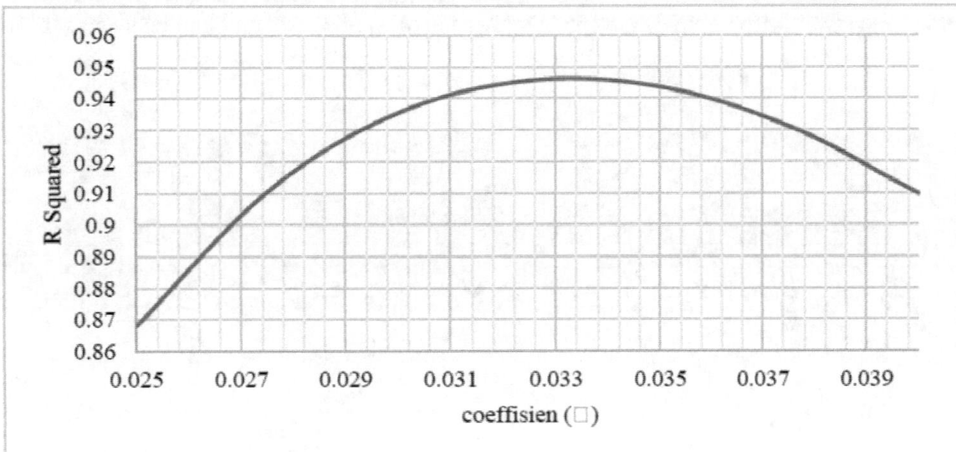

Figure 5. β coefficient calibration results

Plotting existing data compared to the model can be seen in Figure 6. The growth model of cars based on GDP obtained is:

$$V^*_1 = 1299.\,e^{3{,}11361.e^{0{,}023.GDP_t}} \quad (6)$$

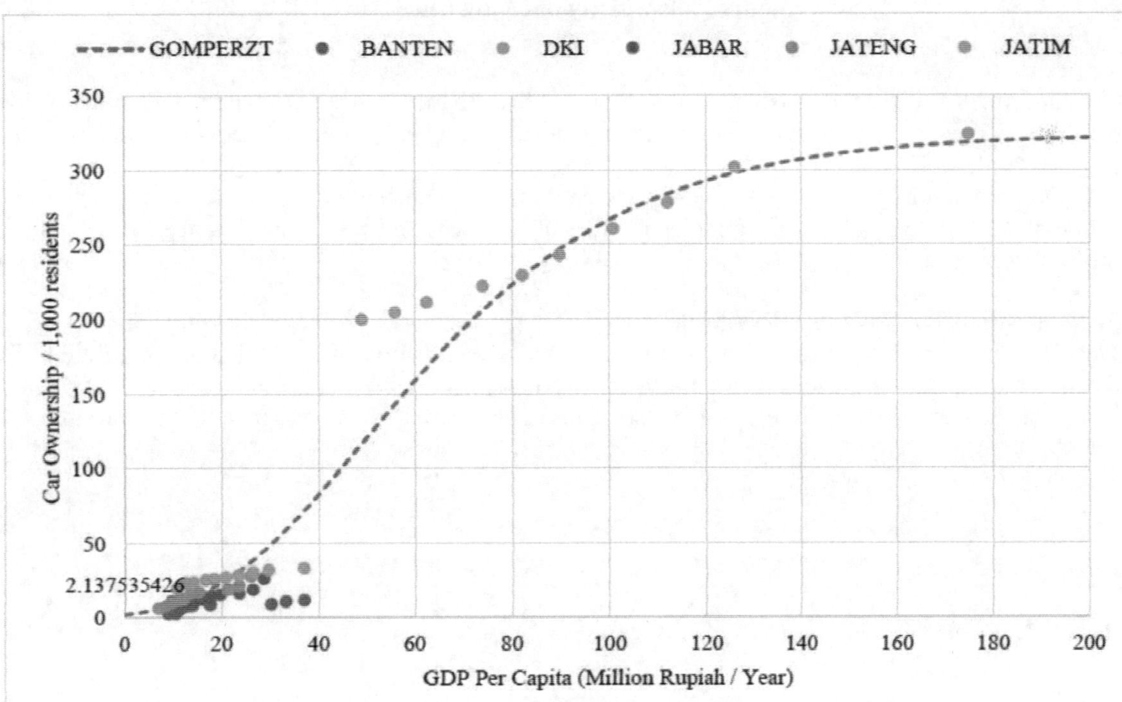

Figure 6. Comparison of model results with existing data

IOP Conf. Series: Materials Science and Engineering **650** (2019) 012047 doi:10.1088/1757-899X/650/1/012047

It is seen that the growth pattern of the car follows the pattern of the Gompertz model with asymptotes in 324 cars per 1,000 inhabitants. The province with the highest level of ownership is DKI Jakarta Province, while the other provinces are at the beginning of the increase in function. The provinces of Central Java and East Java have growth rates above the S-Curve Gompertz while West Java and Banten are at the bottom of the curve.

The results of the model sensitivity were obtained that the highest peak sensitivity was at 30 million GDP / year with a sensitivity value of 1.8465. The results of total sensitivity can be seen in Figure 7.

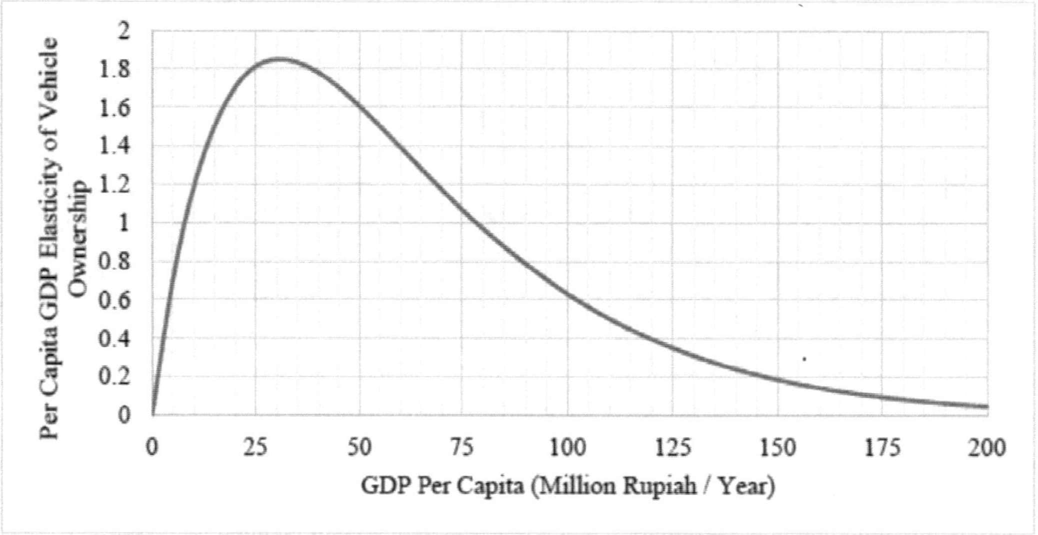

Figure 7. Sensitivity of car ownership to GDP

4. Result

The car growth model on Java Island can be approached with the Gompertz function. This function is a negative exponential function with the asymptote used is the highest value of ownership of four-wheeled vehicles in the DKI Jakarta Province, namely 324 four-wheeled vehicles per 1,000 residents. If all provinces are assumed to follow the Gompertz pattern, it can be estimated that at one time per 1,000 people there will be 324 cars, meaning the number of four-wheeled vehicles will range from 32.4% to the total population.

5. Suggestions for Advanced Research

The Gompertz function for predicting motorized vehicle ownership can be applied to other regions by recalibrating its forming coefficients. Further research is also needed regarding the value of vehicle ownership asymptotes in each region and also needs to be developed with other economic variables.

IOP Conf. Series: Materials Science and Engineering **650** (2019) 012047 doi:10.1088/1757-899X/650/1/012047

References

[1] Acevedo J 2013 *Studying Car and Motorcycle Ownership Levels in Developing Countries Using Individual Income* 1–18
[2] Indonesia Investments 2017 *Produk Domestik Bruto Indonesia*
[3] Dargay J, Gately D, and Sommer M 2007 Vehicle Ownership and Income Growth, World wide: 1960--2030 *Energy J.* **28** 1–32
[4] Gerard de Jong, James Fox A D 2009 *A comparison of car ownership models Gerard Bilingualism* **110** 115–22
[5] Gómez-Gélvez J A and Obando C 2013 Modeling Car Ownership in Urban Areas of Developing Countries: Case Study of Bogotá, Colombia *J. Transp. Res. Board* **2394** 111–8
[6] Wu T, Zhao H, and Ou X 2014 Vehicle ownership analysis based on GDP per capita in China: 1963-2050 *Sustain.* **6** 4877–99
[7] KATADATA 2016 *Berapa Jumlah Kendaraan Bermotor di Indonesia?*
[8] Badan Pusat Statistik 2018 *Jumlah Kendaraan Bermotor (Unit)*
[9] Rządkowski G, Głażewska I, and Sawińska K 2015 The Gompertz Function and Its Applications In Management *Found. Manag.* **7** 185–90

IOP Conf. Series: Materials Science and Engineering **650** (2019) 012048 doi:10.1088/1757-899X/650/1/012048

Study of willingness to pay the Jakarta-Bandung highspeed train: a case study of Argo Parahyangan train passangers

A L Putri and H Widyastuti*

Department of Civil Engineering, Institut Teknologi Sepuluh Nopember Surabaya.

E-mail: *hera.widyastuti@yahoo.co.uk

Abstract. Currently the increase in vehicle volume in the cities of Bandung and Jakarta has grown very rapidly. This is accompanied by an increase in the volume of vehicles traveling from Jakarta to Bandung and vice versa. High Speed Train (HST) becoming the government's choice as a form of mass transportation modernization in Indonesia in building inter-city connectivity and regional development. Study or research on willingness to pay (WTP) is needed to find out the prices that are desired by prospective users. The method used in this study are distribution of questionnaires and interviews with stated preference techniques. The result that the highest probability value is in scenario 1 frequency with a probability value of 96.00%. Or it can be interpreted that as many as 96.00% of respondents of Argo Parahyangan train users will use the HST if the facilities offered are the frequency of train departures once an hour. The biggest WTP is in the scenario conditions offered, namely intermodal accessibility connected with the LRT (Light Rail Transit) mode, and also once a hour frequency scenario where the WTP value reaches.

1. Introduction

Currently the increase in vehicle volume in the two cities has grown very rapidly. This is accompanied by an increase in the volume of vehicles traveling from Jakarta to Bandung and vice versa. Construction of the Cipularang toll road (Cikampek-Purwakarta-Padalarang) has caused the road to become more burdened and has a negative impact without increasing the performance of other modes of service. Even the Cipularang toll road is currently unable to accommodate the large volume of vehicles passing through the toll road. Based on data from the Badan Pengatur Jalan Tol (BPJT), the Jakarta-Cikampek Toll Road is the highest volume toll road in Indonesia, with the highest volume reaching 205,111,304 in 2017. While in normal conditions the traffic is 140 thousand vehicles per day. In addition, the Jakarta - Bandung toll road segment is currently being accelerated with the construction of the Jakarta - Cikampek elevated project. This condition makes the distance of Jakarta-Bandung along 142.3 Km which should be reached in 2-2.5 hours to 3-4 hours in normal conditions, in conditions of weekends.

In normal conditions, Bus and travel are very widely used by users of public transportation from Bandung to Jakarta or vice versa. However, at present, many bus users and travel users switch to using trains because the transportation modes are scheduled and free of traffic jams. Based on PT. KAI's Annual Report in 2017 the Jakarta-Bandung corridor railroad namely the Argo Parahyangan Train can only serve 2,000-2,500 passengers per day. Currently the occupancy of users of the Argo Parahyangan

IOP Conf. Series: Materials Science and Engineering **650** (2019) 012048 doi:10.1088/1757-899X/650/1/012048

Jakarta - Bandung train tends to be high, even at certain times (peak season), the PT KAI provides additional (facultative) trains to accommodate the large number of occupant passengers.

Therefore the government need to harmonize the infrastructure development of the Jakarta-Bandung corridor with the movement of economic development that occurs around Jakarta and Bandung. Under these conditions, the HSR became the government's choice as a form of mass transportation modernization in Indonesia in building inter-city connectivity and regional development. PT Kereta Cepat Indonesia–China (KCIC) is a company that operates the Indonesian fast train network which is planned to be built with the Jakarta-Bandung route with a distance of 142.3 km. This company is a joint project with Pilar Sinergi BUMN Indonesia (a consortium of four Indonesian state-owned enterprises: Kereta Api Indonesia (KAI), Wijaya Karya (Wika), PTPN VIII, and Jasa Marga) with China Railways.

The Jakarta-Bandung Highspeed Train is a new alternative mode that will later compete with existing modes of transportation. With the new transportation, the thing that must be considered is the purchasing power and WTP the tariff of the service user. Where this ability can be a willingness based on user perceptions, namely WTP which can be defined as the WTP service users for compensation for an item or service enjoyed [1]. The WTP concept is useful in visualizing the perspective of system users [2] in [3].

2. Literature review

Research studies aim to complement the basic theories taken from literature studies, so that they can contribute to further research. Previous research studies that were related were in scope (High Speed Rail), probability of using new mode, WTP, interview/survey interview, stated preference techniques and regression analysis. The summary of previous research is as follows:

HSR (High Speed Rail) is an excellent alternative mode because of its ability to capture a relatively large market share, especially for medium distances [4]. In Japan (as Pioneer HSR since 1964) there was a rapid decline in air transportation recorded after the extension of the HSR network. According to Taniguchi (1992), HST is more competitive with a distance of less than 438 miles due to higher frequency services, cheaper rates, proximity to the city center and service and security reliability of [5].Recently, in 2009, the Spanish AVE (HSR Spain) enjoyed 85% market share in Madrid - the Seville Route, more than 70% on the Madrid - Malaga route, and about 50% in the Madrid - Barcelona corridor in 2009. But this advantage is reduced for the Barcelona-Seville route, considering the distance is more than 400 miles. Thus, AVE only enjoys a 30% market share of seats on this Barcelona-Seville route.

The Seoul-Jeju route analyzed as an empirical case study, in which the construction of an undersea tunnel to connect Seoul and Jeju city by HSR has been considered. This study analyzes passengers' choice of the mode of transportation when air transportation is in-competition with HST. The study also included two new variables, ‘safety of transportation’ and ‘availability of duty free shopping’ in addition to traditional transport choice variables such as travel time, travel costs and frequency of service to reflect special characteristics of the market [6].

Using Stated-Preference, this study discusses changes in air transportation demand with the entry of 'LCC' into domestic services, and with the beginning of the operation of the Linear Chuo Shinkansen [7]. To create a choice model, it is estimated to use the Nested-Logit model parameters by two types of travel objectives to consider differences in price sensitivity between business and non-business. Travel attributes used include:

- Total Travel Costs (TotalFares)
- Total Travel Time (Total Travel Time)
- Frequency and others (Frequency and other)

Analyzed competition from (HST) with several alternative modes in the Madrid- Zaragoza-Barcelona corridor. This analysis is based on the estimation of the disaggregate demand model using Revealed Preference (RP) and a mixture of database Revealed Preferences (RP) and Stated Preferences (SP). Model specifications aim to explain changes in demand for HST as a result of changes in travel time, travel costs, outbound access, reliability and convenience in all modes competing in this corridor [8].

IOP Conf. Series: Materials Science and Engineering **650** (2019) 012048 doi:10.1088/1757-899X/650/1/012048

3. Method

The data needed is in the study is primary data. Primary data, is data obtained directly in the field by distributing questionnaires. The questionnaire contains questions related to the characteristics of modal users, and the characteristics of travel. This questionnaire was designed to find out the characteristics of the respondents who traveled in the Jakarta-Bandung corridor by asking about age, gender, occupation, travel frequency, the most frequently used transportation modes, travel time, purpose of the trip, and the cost of one trip.

3.1. Designing the Stated Preference Questionnaire

In designing the questionnaire that will be used is a stated preference technique. Stated preference techniques are approaches to respondents to find out their responses to different situations. Each individual is asked for his response if they are faced with the situation given in the actual situation (how they prefer the choice offered). In the literature study that has been done, [6], [7], [8] and [11] show that the variables or attributes that most influence the choice of modes are rates, travel time, frequency and accessibility. When building a new route, applying new variables that reflect actual market conditions will help understand the passenger mode of behavior choices in a more realistic way [5]. The new variable here is the frequency of train departure or headway and intermodal accessibility because this variable will characterize the advantages of the HST compared to other modes.

Respondents will be faced with several scenario choices, which they will choose whether they are willing to use the HST with a number of conditions offered. The tariff scenario offered for HST uses assumptions based on the comparison of Argo Parahyangan Priority train rates, which have relatively the same facilities. Argo Parahyangan Priority train tariff is Rp. 240,000 so that in this study a price range starting from Rp. 240,000 - Rp. 300,000.

Of the four variables used (time, cost, frequency and accessibility) the scenario formed:

- Scenario 1 (Rp. 300k, 30 min)
  Scenario 1 offers a tariff of Rp. 300.000 (flat tariff) with a travel time of 30 minutes
- Scenario 2 (Rp. 280k, 40 min)
  Scenario 2 offers a tariff of Rp. 280.000 (flat tariff) with a travel time of 40 minutes
- Scenario 3 (Rp. 260k, 50 min)
  Scenario 3 offers a tariff of Rp. 260.000 (flat tariff) with a travel time 50 minutes
- Scenario 4 (Rp. 240k, 60 min)
  Scenario 4 offered a tariff of Rp. 240.000 (flat tariff) with a travel time 60 minutes
- Scenario 5 (frequency once an hour)
  Scenario offered is the frequency of train departures once an hour
- Scenario 6 (frequency every 2 hours)
  Scenario offered is the frequency of train departures every 2 hours
- Scenario 7 (frequency every 3 hours)
  Scenario offered is the frequency of train departures every 3 hours
- Scenario 8 (Connect with LRT)
  Scenario offered are intermodal accessibility connected with LRT mode (Light Rail Transit)
- Scenario 9 (Connect with Bus)
  Scenario offered are intermodal accessibility connected with Bus mode

IOP Conf. Series: Materials Science and Engineering **650** (2019) 012048 doi:10.1088/1757-899X/650/1/012048

3.2. Number of Samples

Determination of the number of samples in this study is using Slovin formula 1.

$$n = \frac{N}{1 + N_e^{\ 2}} \quad (1)$$

where
n : Sampel size
N : Population size
e : Percent of allowance for inaccuracy due to sampling errors. Assuming that the population is normally distributed.

The population size is obtained from the number of Argo Parahyangan Railway passengers based on data from the PT. KAI Annual Report, which is 2,500 passengers per day, so the calculation of sample size is as follows

$$n_{KA} = \frac{2500}{1+2500\,(0.1)^2} = 99.960015 \sim 100 \text{ Respondents} \quad (2)$$

So the minimum number of samples needed in this study is 100 respondents.

4. Data analysis

The analysis used to determine the passenger characteristics of the Jakarta-Bandung corridor used descriptive statistical methods. By using the statistical method, it can be seen the description of the collected data in the form of the highest percentage percentage of each passenger characteristic of the Jakarta-Bandung corridor. The next analysis is binary logistic regression analysis. Binary logit regression analysis (binary logistic regression) is used to determine the probability percentage of respondents who choose to be willing to use the (HST). In addition, binary logistic regression analysis can also be used to determine the dominant variables / attributes affecting respondents to choose the mode [9]. To determine the approximate value of probability mode selection can be calculated using the following formula

$$\text{In}\frac{p}{1-p} = \beta_0 + \beta_0 + \beta_1 X_1 + \beta_2 X_2 + \cdots \beta_K X_K \quad (3)$$

$\beta_0 + \beta_1 + \beta_2 + \beta_K$: Binary data with two responses, for example not willing to use the HST (0), are willing to use the HST (1). Binary data is a dependent variable
(X): Is an independent variable, which has an influence factor in the choice of mode.

After this calculation is complete, the exponential (exp) value is obtained which will be used to calculate the estimated probability value using the following formula:

$$P_{(HST)} = \frac{exp^{logit(p)}}{1+exp^{logit(p)}} \quad (4)$$

$$P_{(others)} = \frac{1}{1+exp^{logit(p)}} \quad (5)$$

To determine the factors that influence the independent variable with the dependent variable, it can be determined by comparing the significant value, i.e. if the value is sig. < $\boldsymbol{\alpha}$, with a tolerance value of $\boldsymbol{\alpha}$=0.05%, then the variable has an effect, but if the value is sig. > $\boldsymbol{\alpha}$ then the variable has no effect.

5. Result of analysis

The data collection process is done by distributing questionnaires at Bandung Station and Gambir Station. The survey was conducted at that location because it was to directly interview passengers who actually traveled with that mode. Based on the calculation of the number of samples, the minimum sample needed for this study was 100 respondents. The results of the survey in the field obtained samples from argo parahyangan train passengers as much as 177. This because for the analysis needed more

IOP Conf. Series: Materials Science and Engineering **650** (2019) 012048 doi:10.1088/1757-899X/650/1/012048

samples to minimize respondents' answers that are incomplete / valid and to get a more valid model in the analysis. From the primary data obtained from the survey in the field it is known the characteristics of passenger trains. As well as to find out how much the probability of train passengers who will move using HST are then analyzed using stated preference techniques. To analyze the data using the SPSS program (Statistical Package for the Social). The characteristics of train users and the results of logistic regression testing for each variable can be seen in Table 1

Table 1. Logistic regression test results for each variable (Scenario 2)

Characteristics	Willingness to Move	
	Yes	No
Male	20	63
Female	35	59
Sig. value 0.61 (> value α 0.1) then the variable not significant		
17 to 30 years	44	87
31 to 45 years	11	32
Sig. value 0.621 (> value α 0.1) then the variable not significant		
Government Employees	19	46
Private Employees	10	50
Entrepreneur	10	9
Sig. value 0.962 (> value α 0.1) then the variable not significant		
< Rp 3 to 5 mio	36	59
Rp 5 to 7 mio	14	43
> Rp 7 mio	5	20
Sig. 0.104 (> value α 0.1) then the variable not significant		
Consideration of ease of mode and mobility	7	7
Comfort and safety	18	41
Consideration of time and speed	28	68
Sig. 0.470 (> value α 0.1) then the variable not significant		
Work	20	43
Family Visit	11	42
Tourism	16	27
Others	5	6
Sig. 0.310 (> value α 0.1) then the variable not significant		
Once a month	19	34
2 times a month	6	30
3 times a month	10	6
Others	20	52
Sig. 0.014 (> value α 0.1) then the variable not significant		

5.1. Logistic regression test for scenario 2 (Rp.280k, 40 min)
The dependent variable in this section is the willingness to move using the HST with the scenario offered, which is 40 minutes travel time with a ticket price of Rp.280,000. The test results of each independent variable with binary logistic regression can be seen in Table 2.

Table 2. Logistic regression test results for frequency variable

		B	S.E	Wald	df	Sig.	Exp(B)
Step 1*	Frequency			10.6	3	0.01	
	Frequency (1)	-1.02	0.531	3.743	1	0.05	0.05
	Frequency (2)	1.09	0.591	3.424	1	0.06	2.982
	Constrant	-0.582	0.286	4.127	1	0.042	0.559

- Logit (P) for the frequency of travel 2 times a month

$$\text{Logit (P)} = \ln\frac{p}{1-p} = \beta_0 + \beta_1$$

-0,582 - 1,028 (2 times a month)
= -1,61

- Logit (P) for the frequency of travel 3 times a month

$$\text{Logit (P)} = \ln\frac{p}{1-p}\ \beta_0 + \beta_1$$

-0,582 + 1,093 (3 times a month)
= 0,511

5.1.1. Determining the probability value
The equation of modeling the test results of each variable free frequency of travel 1 month 3 times

$$\text{Logit (P)} = \ln\frac{p}{1-p}$$

= -0,582 + 1,093 (3 times a month)
= 0,511

Then the probability of respondents with a frequency of 3 times a month to choose to be willing to use the HST with scenario 2 (Rp.280k, 40 min)

$$P\text{ (3 times a month)} = \frac{exp^{logit(p)}}{1+ exp^{logit(p)}}$$

$$P\text{ (3 times a month)} = \frac{exp^{0,511}}{1+ exp^{0,511}}$$

= 0,625041
= 62,50%

5.1.2. Testing the logistic regression model
Before the model is declared feasible, the model must be tested statistically. Testing the binary logistic regression model uses the Hosmer and Lemeshow Test with the assumption:

- Ho: The model has been quite able to explain the data (*Goodness of Fit*)
- H1: The model does not adequately explain the data

Criteria:

- Acceptence Ho if *p-value* sig. > 0,1
- Reject Ho if *p-value* sig. < 0,1

IOP Conf. Series: Materials Science and Engineering **650** (2019) 012048 doi:10.1088/1757-899X/650/1/012048

Table 3. Hosmer and Lemeshow Test based on frequency

Step	Chi-square	df	Sig.
1	0.0000	1	1.0000

It can be seen from Table 3 of the Hosmer and Lemeshow Test for the regression results of frequency variables that the chi square value of 0.00 <2.71 chi square table value (df = 1 with significance 0.1) is 0,000 <or sig value 1.00> 0, 10 so that the decision is Ho accepted, with a confidence level of 90% can believe that the logistic model used is sufficient to explain the data / according to testing. This proves that the logistic regression model is feasible to interpretation.

Table 4. Classification plot for frequency variable

			Predicted		
			Scenario 2 (Rp. 280k,40 min)		Percentage Correct
	Observed		No	Yes	
Step 1	Scenario 2	No	116	6	95.1
		Yes	45	10	18.2
	Overall percentage				71.2

Table 4 Classification plots for frequency variables show that the logistic regression used is quite good because it is able to guess correctly 71.2% of the conditions that occur.

5.1.3. Interpretation of the Binary Logistic Regression Model

The probability of Argo Parahyangan train passengers with a frequency of 3 times a month choosing to use the high speed is 62.50%.

The binary logit regression results or the probability of respondents using the HST for each scenario can be seen in Table 5.

Table 5. Summary of results of binary logit regression analysis

Variable	Category	Scenario	Probability
Trip Purpose	Tourism	1 (Rp. 300k, 30 min)	16,27%
	Family visit	1 (Rp. 300k, 30 min)	13,20%
	Others	1 (Rp. 300k, 30 min)	72,71%
Travel Frequency	2 times a month	2 (Rp. 280k, 40 min)	16,66%
	3 times a month	2 (Rp. 280k, 40 min)	62,50%
Income	Rp 5 to 7 mio	2 (Rp. 280k, 40 min)	28,07%
Income	> Rp 7 mio	2 (Rp. 280k, 40 min)	12,00%
	Rp 5 to 7 mio	5 (once an hours)	42,12%
Income			
	> Rp 7 mio	5 (once an hour)	96,00%
Income	>Rp 7 mio	6 (every 2 hours)	48,00%
Income	>Rp. 7 mio	7 (every 3 hours)	20,00%
Travel Frequency	2 times a month	9 (connect bus)	83,34%
	3 times a month	9(connect bus)	63,76%

IOP Conf. Series: Materials Science and Engineering **650** (2019) 012048 doi:10.1088/1757-899X/650/1/012048

Based on Table 5, the results show that the highest probability value is in scenario 5 (frequency once an hour) with a probability value of 96.00%. Or it can be interpreted that as many as 96.00% of respondents of Argo Parahyangan train users are willing to use HST if the facilities offered are the frequency of train departures once an hour. So it can be concluded that Argo Parahyangan train passengers are more likely to like the frequency departure facility as a consideration to switch to using the HST

5.2. Willingness to Pay (WTP)

WTP is the willingness of the user to issue a reward for the services they obtain. WTP analysis is the average expected rate, expected service priority and WTP more for service improvement. The approach used in WTP analysis is based on the user's perception of the tariffs of the public transportation [10].

Several travel choice scenarios are given to respondents, this is where stated preference methods start running. This is because the fast train has not operated at the time the respondent was given questions about rates, frequency travel time and accessibility. For the question of the scenario with the choice of frequency and accessibility, the respondent is given the opportunity to choose the rate that the respondent wants.

The WTP value obtained from each respondent, in the form of the maximum rupiah value that the respondent is willing to pay for the tariff for railroad services, is processed to obtain the mean value of the WTP value, using the formula:

$$MWTP = \frac{1}{n} \sum_{i=1}^{n} WTPi \tag{6}$$

where:
n : sample size
WTP_i : The maximum WTP value of respondents to i
MWTP : Average WTP
A summary of the WTP value can be seen in Table 6 and the graph in Figure 1.

Table 6. Summary of value of *Willingness to Pay*

Scenario	WTP Value	Cumulative % of respondents will use the HST
5 (once an hour)	Rp. 240.000	52,94%
6 (every 2 hours)	Rp. 210.000	45.83%
7 (every 3 hours)	Rp. 200.000	50,82%
8 (Connect LRT)	Rp. 240.000	55,29%
9 (Connect Bus)	Rp. 215.000	42,95%

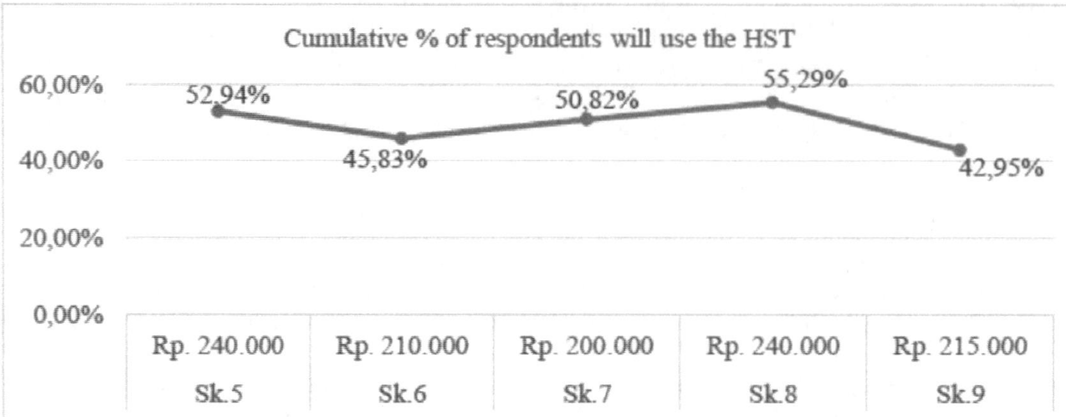

Figure 1. Summary graph of *Willingness to pay*

6. Conclusion

From the results of data analysis and discussion in this study, it can be concluded that:

- The analysis results obtained by the probability model of logistic regression between the Argo Parahyangan train and the HST, obtained the result that the highest probability value is in scenario 1 frequency with a probability value of 96.00%. Or it can be interpreted that as many as 96.00% of respondents of Argo Parahyangan train users will use the HST if the facilities offered are the frequency of train departures once an hour.
- The biggest WTP value is in the scenario conditions offered, namely intermodal accessibility connected with the LRT (Light Rail Transit) mode, and also once a hour frequency scenario where the WTP value reaches.

References

[1] Mazzulla E *2008 Willingness-to-pay of public transport users for improvement in service quality European Transport \ Trasporti Europei* n. 38 (2008): 107-118 Another reference

[2] Khisty C J and Lall B K 2003 *Transportation engineering, an introduction Third Edition* (New Jersey: Pearson Education, Inc.)

[3] Basuki J T 2009 Exploring the Willingness and Ability to Pay for Paratransit in Bandung, Indonesia *Journal of Public Transportation* **Vol. 12** No.2

[4] Román C, Espino R, and Martín J C 2007 Competition of high-speed train with air transport:the case of Madrid–Barcelona *J. Air Transp. Manage* **13** 277–284.

[5] Albalate D and Bel G 2012 *The Economics and Politics of High-speed Rail. Lessons from Experiences Abroad* (Lanham: Rowman and Littlefield Publishers (Lexington Books))

[6] Lee J-K 2016 A Study on travelers transport mode choice behavior using the mixed logit model: A case study of The Seoul-Jeju route *Jurnal Air Transport Management*

[7] Ono I 2015 Stated preference analysis to estimate the domestic transport demand following the future entry of LCCs and the inauguration of the Linear Chou Shinkansen in Japan *Journal of Air Transport Management*

[8] Román C, Espino R, and Martín J C 2010 Analyzing competition between the high speed train and alternative modes.The case of the Madrid–Zaragoza–Barcelona corridor *J. Choice Model* **3 (1)** 84–108.

[9] R C Jou 2011 *Airport ground access mode choice behavior after the introduction of a new mode: A case study of Taoyuan International Airport in Taiwan Transportation Research Part E* **47** 371–381

[10] P Chiambaretto 2013 Measuring the willingness-to-pay of air-rail intermodal passengers *Journal of Air Transport Management* **26** 50-54
[11] Widyastuti H and Willy K 2015 *Analisis Willingness To Pay Menggunakan Binary Choice Model (Studi Kasus : Rencana Re- Aktivasi Rute Kereta Api Jember-Panarukan)* p. 10

Estimation of Vehicle Stops Based on Modified Canadian Capacity Guide Formula Under Non-Lane Based Road Traffic Condition

A A Farabi[1]

[1]Lecturer, Department of Civil Engineering, University of Asia Pacific (UAP), Dhaka-1205, Dhaka, Bangladesh

* a.al.farabi.buet@gmail.com

Abstract: Vehicle stops estimation is one of the important parameters to assess the performance of a signalized road intersection. Canadian Capacity Guide provides a formula that can estimate number of vehicles that stops at least once due to the traffic signal. This study reviews the applicability of this formula for non-lane based traffic. The formula is segmented into two periods to check the estimation of stops during red period and green period. It is found that the formula underestimates the number of stops during red period. Also, the formula estimated number of stops is significantly higher during the green period. As traffic operation and vehicle maneuver of non-lane based traffic are much different from lane based disciplined traffic, the formula cannot predict vehicle stops accurately. Therefore, the estimation of vehicle stops by Canadian Capacity Guide formula is found to deviate considerably from field observed number of stops of vehicles. Thus, a modified regression formula is provided that can estimate the number of stops of vehicles for non-lane based traffic operation. The modified formula fits fairly good with the local traffic condition.

Keywords—Canadian Capacity Guide, Number of Stops, Non-Lane based traffic, Regression Analysis

1. Introduction

Various parameters exist to assess the performance of a signalized intersection. Delay, Queue length, vehicle stops are some much used parameters to assess this performance. Similar to other parameters mentioned, vehicle stops can be directly measured by performing field survey. Alternatively, analytical models are also found that can estimate number of stops of vehicles in an approach. Among various models, Webster [1] developed relationship between stop and delay based on uniform arrival of vehicles. Webster and Cobbe [2] developed a model based on random arrivals that can estimate the numbers of stops of vehicles. Cronje [3] explained the number of stops from queue diagram and also developed a model to estimate vehicle stops based on markov process. Rakha [4] reviewed the existing models of vehicle stops and proposed two approaches for estimating vehicle stops at under saturated and oversaturated condition. Canadian Capacity Guide [5] provides a model that is developed based on Webster and Cobbe [2] provided model. The Canadian Capacity Guide model estimates the number of vehicles that stops at least one time due to signal. This model is developed from uniform queue build up model and a progression adjustment factor (P.F.) is multiplied to get the estimate of number of stops. However, traffic situation of Dhaka city of Bangladesh is much different from western lane based traffic. Traffic does not follow any lane behaviour in Dhaka city. This is why vehicle queue formation is different from lane based traffic. Vehicles are found to form cluster near approach. Lateral gaps between vehicles are very less during queue building as gaps are filled up by small sized vehicles like motor bike. During discharge of vehicles, irregular pattern of movement is found. Vehicles even tend to change direction laterally to move faster. As a result, the model does not predict the number of stops properly in case of non-lane based traffic. To estimate number of stops properly, Hadiuzzaman et al. [6] proposed two approaches in case of non-lane based mixed traffic operation. One approach computes vehicle stops based on microsimulation technique and other one is based on analytical approach. Both the approaches to estimate vehicle stops are quite useful for non-lane based traffic. Specially, the analytical approach deals to compute stops for oversaturated condition.

Yet, a simpler approach to estimate number of stops for both under saturated and oversaturated condition was necessary. This fact led the study of analysing Canadian Capacity Guide provided formula for number of stops of vehicles. Applicability of this formula is checked with field traffic condition. Field traffic condition includes both under-saturated and over-saturated condition. After analysing the formula and comparing it with observed field values, a modification is provided to this simple formula. The modified formula can be used to estimate number of stops of vehicles in case of non-lane based traffic. However, non-lane based heterogeneous traffic (Presence of Non-motorized vehicle) is beyond the scope of this study.

2. Methodology for the study

The formula provided by Canadian Capacity Guide [5] to estimate number of stops of vehicles for at least once during signal operation is as follows:

$$N_s = \frac{P.F.*t_e*v*(C-g)}{60*C*(1-y)} \tag{1}$$

Where,

N_s = Number of vehicles that stops at least once during the evaluation time. The maximum value must be less than the following:

$$N_{s(max)} \leq \frac{v*t_e}{60}$$

P.F. = Progression Adjustment Factor

v = Arrival Flow Rate (vehicles/hour)

C = Cycle time (seconds)

g = Effective green time (seconds)

S = Saturation flow rate (vehicles/hour)

$y = \frac{v}{S}$ capped at $y \leq 0.99$

te = Evaluation time (minutes)

Now, formula (1) can be simplified more if the evaluation time is considered same as the cycle time. Therefore, simplification of formula (1) while considering consistency of units produce the following version:

$$N_s = \frac{P.F.*v*(C-g)}{(1-y)} \tag{2}$$

Where,

v = Arrival flow rate (vehicles/hour)

C-g = r = Red signal phase period (hour)

Cronje [3] explained number of stops from the queue diagram. If there is an initial queue at the beginning and no overflow at the end of cycle, the total number of stops per cycle is as follows:

$$N_s = Q_B + r*v + \frac{(Q_B + r*v)*v}{S-v} \tag{3}$$

IOP Conf. Series: Materials Science and Engineering **650** (2019) 012049 doi:10.1088/1757-899X/650/1/012049

Where, Q_B = Initial Queue (vehicles)
If there is no initial queue at the beginning and no overflow at the end of cycle, then total number of stops per cycle:

$$N_s = r * v + \frac{(r * v) * v}{S - v} \tag{4}$$

The first term of the equation (4) represents number of stops during red period while the later term represents number of stops during green period. By simple mathematical operation and replacing r by (C-g), it can be shown:

$$N_s = \frac{v * (C - g)}{(1 - y)} \tag{5}$$

So, the equation (5) is simply the same as equation (2) except the term progression adjustment factor which is absent in equation (5). Thus, the formula provided by Canadian Capacity Guide in equation (1) is just an improved version of equation (4).

Therefore, the study focuses on the equation (4) as it splits the number of stops in two distinct periods. This equation is used to determine number of stops in red period and green period consecutively by applying appropriate values of traffic parameters. Parameters are found from performing a traffic survey on selected road intersection. The following diagram of Figure 1 provides a clear understanding of procedure followed for the research study (analysis and model development parts are not shown):

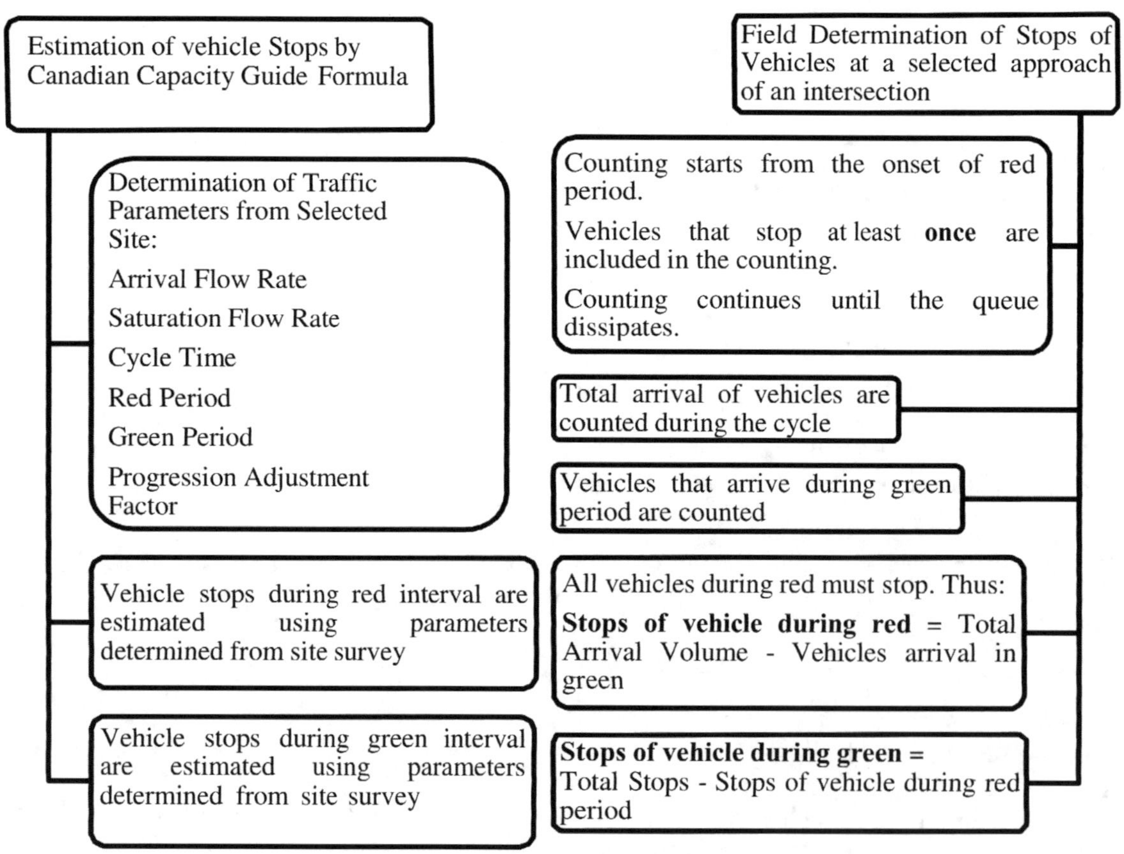

Figure 1. Diagrammatic representation of the methodology

The methodology explained above in Figure 1 is followed to conduct this study. At first, traffic parameters are determined form the site. Traffic parameters such as signal timings, arrival flow rate, saturation flow rate, Progression Adjustment Factor are measured as follows: Signal timings are measured directly from video recording. Average arrival flow rate is determined by Mcshane et al. [7] provided procedure. Vehicles that depart during the green period and form queue at the end of green period are the total arrival period during that cycle. This total arrival volume is divided by cycle time to get the average arrival flow rate (v) in vehicle per hour unit. Road Note 34 [8] is used to estimate saturation flow rate. Road Note 34 is applied to each cycle of traffic flow in terms of PCU per hour. PCE values required for estimation of saturation flows are taken from Hadiuzzaman et al. [9]. Saturation flow in PCU per hour is then converted into equivalent saturation flow in vehicle per hour unit using the procedure mentioned in Canadian Capacity Guide [5]. A total of thirty (30) cycles are used to estimate the average saturation flow (in vehicle per hour unit). Lost times are also determined by Road Note 34 [8]. Average of Initial lost times and final lost times from 30 cycles have been used. Average lost times are used to determine effective green time. Another important parameter is Progression Adjustment Factor (P.F.), it is determined by Strong et al. [10] provided formula. This formula is much better suited than using default values of Highway Capacity Manual 2000 [11]. After determining the above mentioned parameters, number of stops is estimated. Next, counting of number of stops is conducted that provide the actual number of stops of vehicles. After that, results of formula is compared with the observed results. Deviation of the formula from the observed value justifies the need of modification. Thus a modification is performed for the existing formula.

So, the rest of the paper explains data collection procedure followed by statistical analysis of vehicle stops. After that a linear regression model development is shown. The later portion describes the major findings and limitations of the study.

3. Data Collection

Data for this study is collected from Hotel Intercontinental intersection (Previously known as Hotel Sheraton intersection) of Dhaka City of Bangladesh. Few attributes of the intersection are shown in Table 1-

Table 1. Few features of the selected road intersection

Intersection Name	Approach Name	Road Width (meter)	Traffic Stream Type	Pedestrian behavior	Road Side friction
Hotel Intercontinental	East	12.21	Only motorized vehicle is present	Relatively disciplined	Low
	South	12.11	Only motorized vehicle is present	Relatively disciplined	Low
	North	10.23	Only motorized vehicle is present	Relatively disciplined	Low

Based on the information given on Table 1, it is understandable that this intersection has some advantages over other intersections of Dhaka city. Thus, data collection was performed on this selected site and it was done using video camera recording. Some techniques were needed to be followed such as: for long queue where a single camera cannot capture the entire queue, two video cameras were used simultaneously keeping a common overlapping portion. Also, some considerations were applied during video data transcription. As the East approach of the intersection has exclusive left turning facility, left turning vehicles were not supposed to stop due to signal. However, queued vehicles sometimes block the passage as they don't follow any lane behavior. These additional vehicle stops were not considered into the statistical comparison. This is due to the fact that the analysis is focused for queued vehicles only and the formula is not supposed to recognize the stops of vehicles of other lanes. This consideration is also applied for North approach and South approach of the intersection.

4. Comparative Analysis of Vehicle Stops

Data collected from site survey is used to compare between observed stops and formula stops. These data with corresponding traffic parameters are shown below in Table 2:

Table 2. Comparison between Formula estimated number of stops and Observed number of stops

Approach Name	Obs No.	Cycle Time, C (Seconds)	Effective green g (Seconds)	Arrival flow rate, *v* (veh/hour)	Saturation flow rate, S (veh/hour)	Degree of saturation, X	P.F.	Canadian Capacity Guide Formula: Number of Stops (vehicles)		Observed Stops in red (veh)	Observed Stops in green (veh)
								Stops in red (veh)	Stops in green (veh)		
(1)	(2)	(3)	(4)	(5)	(6)	(7)	(8)	(9)	(10)	(11)	(12)
East	1	306	130.57	1906	6062	0.74	0.99	92	42	92	36
	2	366	147.57	2056		0.84	1.18	147	75	145	40
	3	165	86.57	2444		0.77	1.44	76	51	73	7
	4	88	50.57	2169		0.62	1.3	29	16	28	7
	5	131	57.57	2226		0.84	1.38	63	37	61	2
	6	115	67.57	2693		0.76	1.33	47	38	45	3
	7	209	67.57	1809		0.92	1.12	79	34	79	0
	8	122	50.57	2597		1.03	1.11	57	43	58	3
North	1	140	51.23	900	3702	0.66	0.84	19	6	19	0
	2	229	94.23	2406		1.58	1	90	167	97	0
	3	146	12.23	642		2.07	0.95	23	5	23	0
	4	98	44.23	1690		1.01	0.93	24	20	24	0
	5	129	63.23	2289		1.26	1.22	51	83	56	0
	6	110	62.23	2455		1.17	1.07	35	69	36	0
	7	152	69.23	1493		0.89	0.49	17	11	18	0
	8	119	38.23	2269		1.91	0.98	50	79	57	0
	9	126	24.23	1486		2.09	1.04	44	30	54	0
South	1	207	41.38	661	2946	1.12	1.08	10	43	33	3
	2	65	12.38	554		0.99	0.86	2	9	7	0
	3	64	13.38	844		1.37	0.09	0	1	2	0
	4	36	9.38	600		0.78	0.43	1	3	2	0
	5	87	19.38	663		1.01	0.88	3	14	11	0
	6	78	11.38	554		1.29	1.16	3	15	12	0
	7	39	10.38	462		0.59	0.52	0	2	2	1
	8	73	15.38	592		0.95	1.27	3	15	12	0
	9	69	7.38	522		1.66	0.73	1	7	5	0
	10	55	29.38	982		0.62	0.52	2	6	4	3
	11	54	12.38	400		0.59	1.14	1	6	7	0

Approach Name	Obs No.	Cycle Time, C (Seconds)	Effective green g (Seconds)	Arrival flow rate, v (veh/hour)	Saturation flow rate, S (veh/hour)	Degree of saturation, X	P.F.	Canadian Capacity Guide Formula: Number of Stops (vehicles)		Observed Stops in red (veh)	Observed Stops in green (veh)
								Stops in red (veh)	Stops in green (veh)		
(1)	(2)	(3)	(4)	(5)	(6)	(7)	(8)	(9)	(10)	(11)	(12)
	12	92	35.38	587		0.52	0.51	1	6	5	0
	13	161	24.38	336		0.75	1.02	2	15	13	0
	14	239	31.38	377		0.97	1.11	4	28	24	1
	15	104	30.38	762		0.89	0.96	5	20	15	0
	16	208	63.38	606		0.68	0.63	4	19	16	1
South	17	79	16.38	730	2946	1.2	0.88	4	15	11	0
	18	136	18.38	371		0.93	1.07	2	15	13	0
	19	166	69.38	890		0.72	0.73	7	24	18	2
	20	59	14.38	733		1.02	1.21	4	15	11	0
	21	58	19.38	435		0.44	0.85	1	5	4	0
	22	148	24.38	633		1.3	1.1	7	31	24	2
	23	88	11.38	450		1.18	1.04	2	12	10	0
	24	172	20.38	419		1.2	0.86	2	17	15	1

Table 2 shows the summery of the collected data. Traffic parameters such as signal timings, arrival flow rate, saturation flow rate, Progression Adjustment Factor (P.F.) are measured by methods mentioned in the methodology. The first two columns describe the name of approach and observation periods. The column (3) provides cycle time which is determined from the survey. The effective green time (column 4) is determined by deducting lost times from Green + Amber time. Column (5) to column (8) show other traffic parameters necessary for number of stops calculation. Saturation flow listed in column (6) requires PCE factors. The following PCE values are used:

Table 3. Regressed PCE Values derived by Hadiuzzaman et al. [9]

Intersection Name	Approach Name	Pickup	Car	Bike	Auto-Rickshaw	Heavy vehicles
Sheraton	East	1.211	1.0	0.108	0.690	4.513
Bangla Motor	North	1.313	1.0	0.453	0.338	2.252

PCE values shown on Table 3 are determined by following synchronous regression method by Hadiuzzaman et al. [9]. In this study, East approach and South approach uses the same PCE values of East approach as both the approaches have right turning vehicles only and also both of them are car dominated as found during this study. Average percentage of vehicles for East approach is found to be: Car-44%, Auto-Rickshaw-37%, Motor Bike-13%, HV-1% and Other vehicles-5%. South approach of the same intersection have: Car-56%, Auto-Rickshaw-23%, Motor Bike-15%, HV-2% and Other vehicles-4%. Thus, PCE values of East approach can be considered approximately same for these two approaches. On the other hand, North approach of the selected intersection is just at the downstream of Bangla Motor intersection. These two approaches have approximately same road width. Approach width of Bangla Motor is 10.6 meter which is pretty close to North approach (Width is 10.23 meter) of

Hotel Intercontinental. Also, vehicle composition for both the approaches is close. Therefore, it can be considered a reasonable approximation for PCE values.

Using all the parameters, number of stops is determined which is shown in the following column (9) and column (10) of Table 2. Number of stops is split into red and green interval as mentioned previously. It helps to analyse and interpret the values more deeply. Column (11) and column (12) express the number of stops counted from video recordings directly.

5. Regression Model Development

Again, Table 2 shows that the formula estimated number of stops during red period is much close to the observed value in case of East approach and South approach. However, it underestimates much in case of South approach. Overall, the formula does not predict accurately in red period. By analysing data of Table 2, it can be found that the average value of underestimation during red period is 5.5 vehicles with a standard deviation of ±6 vehicles. On the other hand, vehicle stops in green period are over-estimated by the formula. Average value of over estimation during green period is 25 vehicles with a standard deviation of ±30 vehicles. When the segmented periods are analysed combinedly, it is found that the formula overestimates the number of stops in total. Average number of over-estimation by the Canadian Capacity Guide Formula is 20 vehicles with a standard deviation of ±30 vehicles. So, it is clear that the formula does not estimate the practical results properly.

As the estimations of the Canadian Capacity Guide formula do not exactly match with field data, it is necessary to ensure whether the variation is great enough to modify the existing formula. A statistical t-test is performed to ensure the modification justification. For the t-test, differences between formula estimated vehicle stops and field observed vehicle stops are determined at first. For the Canadian Capacity Guide Formula to be applicable for non-lane based traffic, the difference should be statistically zero. So, the mean of differences should be zero in order to accept the hypothesis that the formula is okay. Now, using the mean of differences, standard deviation and number of observations, a t-test is performed which is shown in Table 4:

Table 4. Summary of t-test

	Sample mean of differences	Sample standard deviation	Population mean of differences	Number of Observations	t-statistic	|Critical t| for 5% LOS	Comment
Stops in green period	-25.3	30.04	0	41	-5.39	-	-
Stops in red period	5.54	6.1	0	41	+5.81	-	-
Total number of stops	-19.76	30.27	0	41	-4.18	1.96	Hypothesis is rejected

Table 4 clearly explains that is the null hypothesis is rejected at 5% level of significance. So, the Canadian capacity guide formula does not predict accurately enough for non-lane based traffic. So, the formula needs to be modified to suit non-lane based mixed traffic operation. In this study, a linear regression model is provided that can be suitable to estimate the actual number of stops of vehicles. The formula model is as follows:

$$N_{S(Field)} = a_1 * N_{S(Formula)} + a_2 \tag{6}$$

Where,

a_1, a_2 = Regression Coefficients

$N_{S\ (Field)}$ = Field counted number of stops of vehicles

$N_{S\ (Formula)}$ = Formula predicted number of stops of vehicles

After performing this simple linear regression based on the data provided on Table 2, the following regression model is found which is shown in Table 5:

Table 5. Proposed Regression Model

Proposed Model				$N_{S(Field)} = 0.598 * N_{S(Formula)}$	
a_1	a_2	t_{a1}	t_{a2}	F_{value}	R^2
0.598	0	184	-	337.4	0.89

Table 5 provides a linear regression model that fits much well with the practical data. R^2 value obtained in this model is 0.89 that shows a good correlation with field data. Coefficient obtained from the study is 0.598. So, formula predicts 40% higher than the actual number of vehicle stops. Now, the calibrated regression formula is checked against another set of collected data from the same intersection. The relationship between modified formula predicted stops and observed field stops is shown in the following Figure 2.

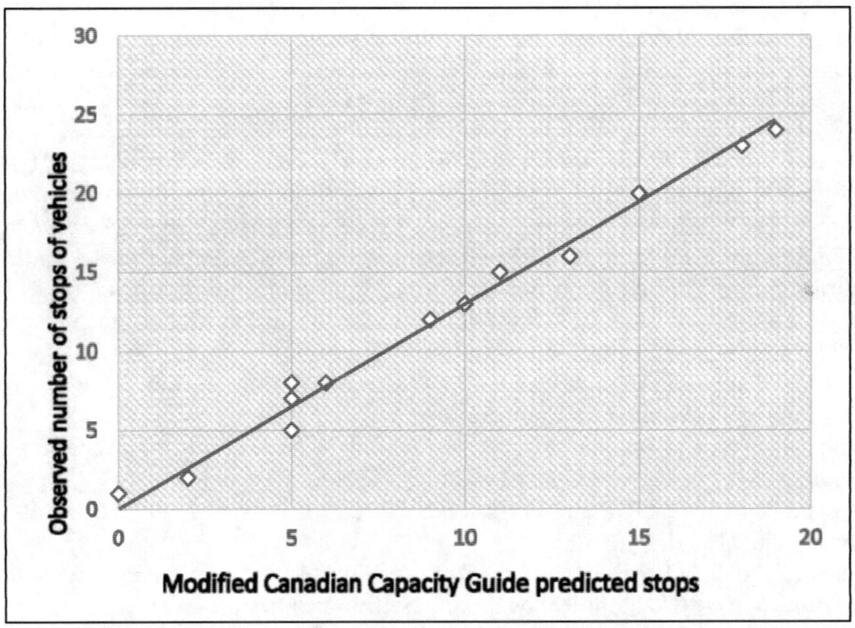

Figure 2. Correlation between Observed number of stops and Model predicted number of stops

Observed number of stops fits very well with the model predicted number of stops as shown in Figure 2. So, correlation between the observed stops and model predicted stops is much better. Therefore, the calibrated formula just provides a multiplication of an Adjustment Factor (A.F.) to estimate number of stops properly:

$$Ns = \frac{A.F. * P.F. * t_e * v * (C - g)}{60 * C * (1 - y)}$$

Where, A.F. = 0.598

This calibrated formula can be used to estimate number of stops of non-lane based traffic. However, the selected intersection is much well-disciplined as mentioned previously. Thus, it may not provide proper results for traffic with many external interruptions to movement.

6. Conclusion and Recommendation

The study provides some valuable findings in context of non-lane based traffic operation. It is found out that the number of stops of vehicles estimated by Canadian Capacity Guide formula does not fit with the field observed number of stops of vehicles. Moreover, the formula estimates lower number of stops during red period. On the other hand, it estimates much higher number of stops during green period. Overall, the formula cannot be used to estimate number of stops of non-lane based traffic. Thus, the existing formula provided by Canadian Capacity Guide is modified to estimate actual number of stops of vehicles. The modified regression formula fits well with field data. The modified formula just provides an adjustment term (A.F. = 0.598). Thus, it is much easier to predict the actual number of stops. Also, the correlation coefficient (R^2 value) is 0.89 which indicates a very good fit with field data. So, this modified formula can be used to estimate number of stops of vehicles for non-lane based traffic. However, the findings are based on three approaches of a single intersection. The intersection is relatively well disciplined and traffic movement incurs less friction from pedestrian and roadside activities. The intersection is free from such disturbances which are very common in other intersections of Dhaka city. So, more study is necessary to establish a general model that will provide accurate results for every intersection with non-lane based motorized traffic. Also, the study excludes non lane based heterogeneous mixed traffic which is predominant in the city. The modified formula may not be suitable at all in predicting stops under such mixed traffic condition. Nonetheless, the findings from this study can provide a valuable base for further research into this area.

References

[1] Webster F 1958 *Traffic Signal Settings* (London: Her Majesty's Stationary Office)

[2] Webster F V and Cobbe B M 1966 *Traffic Signals* (London: Her Majesty's Stationary Office)

[3] Cronje W B 1983 *Derivation of Equations for Queue Length, Stops, and Delay for Fixed-Time Traffic Signals* (Washington: Transportation Research Board) issue 905 pp 93–95

[4] Rakha H, Kang Y S and Dion F 2001 *Estimating Vehicle Stops at Under-Saturated and Over-Saturated Fixed-Time Signalized Intersections* Transportation Research Record **1776** 128-137

[5] ITE 2008 *Canadian Capacity Guide for Signalized Intersections* ed J W Gough (Canada: Institute of Transportation Engineers)

[6] Hadiuzzaman M and Rahman M 2009 *Estimating Vehicle Stops at an Oversaturated Signalized Intersection for Mixed-Traffic Conditions* ITE Journal 76-80

[7] McShane W R, Roess R P and Prassas E S 2004 *Traffic Engineering* (USA: Pearson Education International)

[8] Transport and Road Research Laboratory 1963 *A Method of Measuring Saturation Flow at Traffic Signals* (London: Her Majesty's Stationary Office)

[9] Hadiuzzaman M, Rahman M M and Karim M A 2008 *Saturation Flow Model at Signalized Intersection for Non-Lane Based Traffic* Canadian Journal of Transportation **2** 77-90

[10] Strong D W and Rouphail N M 2006 *Incorporating the Effects of Traffic Signal Progression into the Proposed Incremental Queue Accumulation (IQA) Method* (Washington D.C: Transportation Research Board)

[11] TRB 2000 *Highway Capacity Manual* (Washington DC: National Research Council)

IOP Conf. Series: Materials Science and Engineering **650** (2019) 012050 doi:10.1088/1757-899X/650/1/012050

Road Planning for Improvement of reinforced concrete at PT. Krakatau Bandar Samudra Cilegon-Banten Province

T Rosdiyani[1]*, G Noor[2] dan Dindin[3]

[1,2,3]Civil Engineering Department, Universitas Banten Jaya

* tellyrosdiyani004@gmail.com

Abstract. Road hardening is one of the elements of highway construction that is very important for smooth land transportation so as to provide comfort and safety for its users. The development of the economic and industrial sectors has increased, therefore the need for road transport facilities and infrastructure has increased and the development in PT. Krakatau Bandar Samudra, therefore the need for road repairs. Flexible hardening until rigid hardening, is expected to help improve surfing transportation services for the community and industry, so it needs to be well planned based on planning standards and applicable criteria in Indonesia. The method used in this planning uses Bina Marga T-14-2003, this plan includes to determine the thickness, dowel test diameter used and the dimensions of the drainage channel. In this planning requires primary and secondary data. Secondary data is obtained from the Public Works Agency in the form of location maps, previous technical specifications of data, and land data while the primary data is in the form of Average daily traffic data. The results of the analysis carried out obtained by the thickness of the concrete plate using Continuous Concrete Without Strengthening, the dowel test diameter and drainage channel dimensions that have fulfilled the requirements of safe numbers.

1. Introduction

Government Regulation No. 34 of 2006 and Law Number 38 of 2004 states "Roads are land transportation facilities that form a transportation network to connect the regions so that the wheels of the economy and development can rotate well". The increasing development of the economic and industrial sectors, Cilegon City has been given priority in road construction, especially in the Industrial Estate of PT. Krakatau Bandar Samudra is expected to help improve the economy with services, facilitate transportation for the community and industry.

PT. Krakatau Bandar Samudra which is located in industrial area, its heavy traffic which is economic support route for Java. Road of area at PT. Krakatau Bandar Samudra has a visually decreasing pavement structure with many deflections which can endanger safety and even hinder the rate of transportation. In conditions such as these rigid sidewalks make a choice in improving the road at that location. Rigid pavement planning is emphasized to determine the rigid pavement thickness so as being able to withstand the load of vehicles passing on the road, and to determine the diameter of the reinforcement and drainage dimensions required in this plan.

IOP Conf. Series: Materials Science and Engineering **650** (2019) 012050 doi:10.1088/1757-899X/650/1/012050

2. Research Methods

2.1 Data Collection Techniques

Planning for rigid hardening of the road at PT. Krakatau Bandar Samudra requires data to be processed as a reference to determine the hardening thickness, the diameter of the reinforced concrete plates in supporting the load of the vehicle that passes through these parts and the drainage dimension. Data collection techniques are carried out in the following ways.

2.1.1 Primary Data Conducting surveys directly in the Industrial Estate of PT. Survey of Krakatau Bandar Samudra vehicles to obtain data. Average daily traffic is conducted for two days representing workdays and holidays. The location of the study is shown in figure 1.

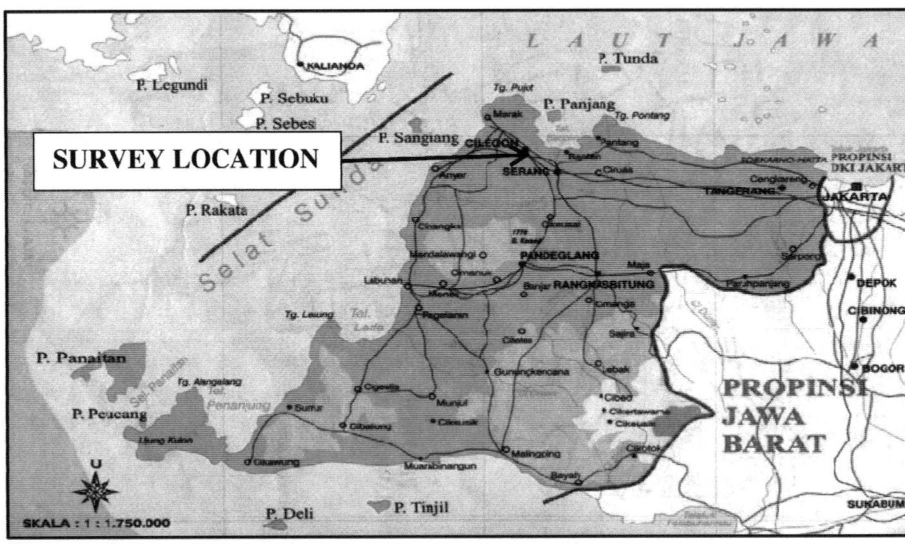

Source: Office of Highways

Figure 1. Survey Location

2.1.2 Secondary Data Secondary data is written data from objects in the form of maps, images, technical specifications, implementation methods, traffic data, land investigation data.

2.2Analysis methods

Planning road repairs in the PT Krakatau Bandar Samudra area using rigid hardening to determine the thickness and diameter of the reinforcement along with the dimensions of the drainage channel while the analysis step process is shown in Figure 1. As follows.

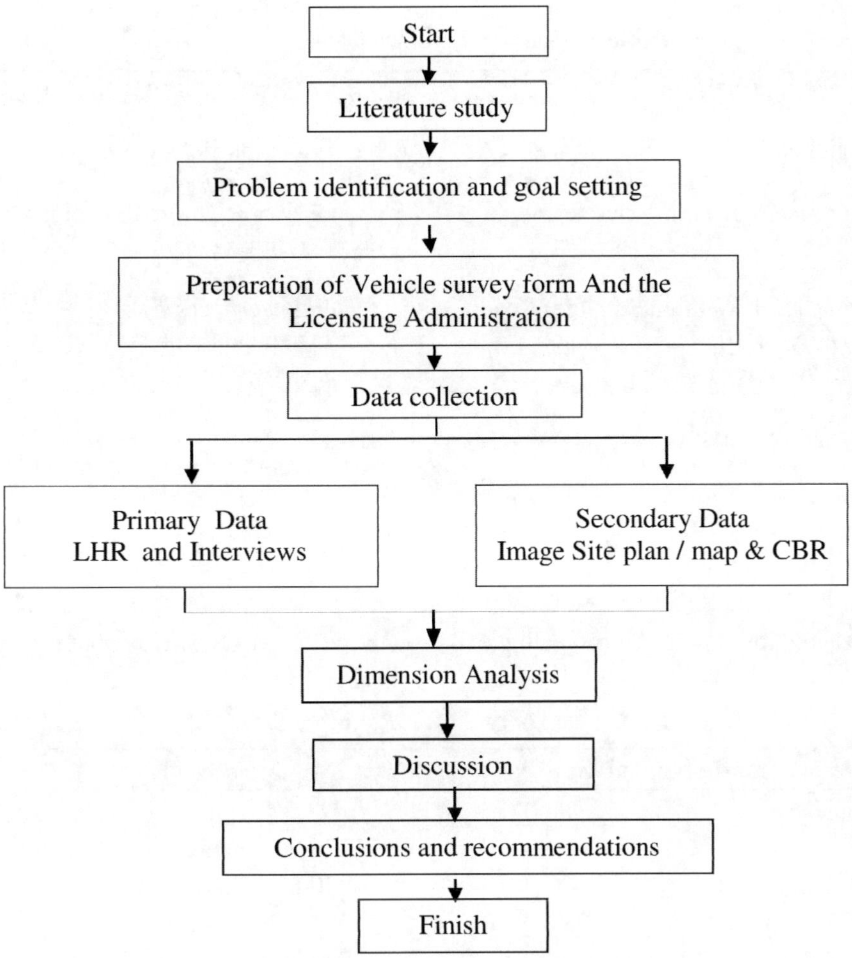

Figure 2. Planning Flow Chart

Based on Figure 2, it can be seen the stages in planning road improvement in the PT Krakatau Bandar Samudra area starting with the preliminary stage reviewing the literature review, the relevant literature with rigid pavement planning, then specifies the problem statement based on the background and identification of the problems that occur on the road segment so that it sets the objectives of the plan. Next set up formal and administrative licensing for the data collection phase. The data needed is in the form of primary data and secondary data. Primary data consists of vehicle average daily traffic data and data obtained by means of interviews, while secondary data obtained map / site plan data as well as land data in the form of CBR. After the data is collected then analysis of calculations to find concrete thickness, dowel test diameter and drainage channel dimensions and the next step to get the results can be concluded as well as recommendations.

IOP Conf. Series: Materials Science and Engineering **650** (2019) 012050 doi:10.1088/1757-899X/650/1/012050

3. Results And Discussion

3.1Data Planning Parameters

Table 1. Planning Parameter Data

a.	Role of Road: Special Road	b.	Road Type : 2 lane 2 direction, undivided
c.	The width of the road: 8, 00 meter	d.	Load security factor 1,1
e.	CBR subgrade: 5%	f.	Bending tensile strength (Fcf) : 4.0 MPa (F`c = 285 kg/cm²)
g.	Underfloor material: Thin concrete mixture	h.	Quality of Reinforcing Steel: Threaded steel reinforcement 24(fy: melting stress = 2400 kg/cm²
i.	Shoulder the road: Without Shoulder Road	j.	Ruji/dowel : Yes
k.	Average Daily Traffic Data: passenger car (2 ton)		260 vehicle / day
l.	Bus (8 ton)		170 vehicle / day
m.	Truck 2 small axle (6 ton)		200 vehicle / day
n.	Truck 2 big axle (13 ton)		170 vehicle / day
o.	Truck 3 big axle (20 ton)		150 vehicle / day
p.	Collaborative Truck (30 ton)		60 vehicle / day
q.	Traffic growth		5%
r.	Age of plan		10 year

Calculation of average daily traffic by projecting 5% growth and 10 years time is obtained: Average Daily Traffic = TOTAL VEHICLE X (1 + 1)n

Table 2. Average Daily Traffic Projected Growt 5%

Transportation type			Number of vehicles	
Pssenger car	2	Ton	423.54	vehicle
Bus	8	Ton	276.93	vehicle
Truck 2 small axle	6	Ton	530.6	vehicle
Truck 2 big axle	13	Ton	325.8	vehicle
Truck 3 big axle	20	Ton	244.35	vehicle
Collaborative Truck	30	Ton	97.74	vehicle

Source: Analysis results

3.2 Calculating the Amount of Commercial Vehicle Axes

Table 3(a). Vehicle Amount Recapitulation and Load Configuration

No	Transportation type	Axis Load Configuration				Number of Vehicles	Take the Axis per vehicle	Number of axes	Single Wheel Single Axis		Double Wheel Single Axis		Double Wheel Tandem Axis	
		Front wheel	Rear Wheel	Front Trailer Wheels	Rear Trailer Eheels				Axis Load (Ton)	Number of Axes (bh)	Axis Load (Ton)	Number of Axes (bh)	Axis Load (Ton)	Number of Axes (bh)

Table 3(b). Vehicle Amount Recapitulation and Load Configuration

1		2				3	4	5	6	7	8	9	10	11
1	Pssenger car	1	1	-	-	260	-	-	-	-	-	-	-	-
2	Bus	3	5	-	-	170	2	340	3	170	5	170	-	-
3	Truck 2 small axle	2	4	-	-	200	2	400	2	200	-	-	-	-
				-	-	-	-	-	4	200	-	-	-	-
4	Truck 2 big axle	5	8	-	-	170	2	340	5	170	8	170	-	-
5	Truck 3 big axle	6	14	-	-	150	2	300	6	150	-	-	14	150
							4	240	6	60	-	-	14	60
6	Collaborative truck	6	14	5	5	60	-	-	-	-	5	60	-	-
							-	-	-	-	5	60	-	-
Total								1620		950		460		210

Source : Analysis results

Number of commercial vehicle axis (JSKN) during the plan life:

JSKN = 365 x JSKNH x R (1)

= 365 x 1620 x 12.6

= 0.75×10^7

Where :

JSKN is the Number of Commercial Vehicle Axis

JSKNH is the Number of axes of the daily commercial vehicle when the road is opened

R is the growth rate. For 5% in the table on the guidelines Pd T-14 2003 was obtained 12.6 while the JSKN plan was obtained for:

JSKN plan = CxR xJSKNH x 365 (2)

= 3.7×10^7

Table 4(a). Axis Plan Repetition Calculation

Axis Type	**Axis Load (ton)**	**Amount of Axis**	**Proportion of Expenses**	**Poportion of Axes**	**Traffic Plans**	**Repetition that occurs**
(1)	**(2)**	**(3)**	**(4)**	**(5)**	**(6)**	**(7) = (4) x(5) x(6)**
Single Wheel Single Axis	6	210	0.22	0.59	3.7×10^7	0.48×10^6
	5	170	0.18	0.59	3.7×10^7	0.39×10^6
	4	200	0.21	0.59	3.7×10^7	0.45×10^6
	3	170	0.18	0.59	3.7×10^7	0.39×10^6
	2	200	0.21	0.59	3.7×10^7	0.45×10^6
Total Single Wheel Single Axis		**950**	**1.00**			
Double Wheel Single Axis	8	290	0.63	0.28	3.7×10^7	0.65×10^6
	5	170	0.36	0.28	3.7×10^7	0.37×10^6
Total Double		**460**	**1.00**			

Table 4(b). Axis Plan Repetition Calculation

Wheel Single Axis						
Double Wheel Tandem Axis	14	210	1.00	0.13	3.7 x 10^7	0.48 x 10^6
Total Double Wheel Tandem Axis		**210**	**1.00**			
		Kumulatif				3.7 x 106

Based on Table 4 above , it can be seen that the values the Single Wheel Single Axis, Double Wheel Single Axis or Double Wheel Tandem Axis in column 4 load are obtained by the following formula equation:

<u>Jumlah Sumbu</u> For example 210/950 = 0.22 as well as for Double Wheel Single Axis and Double Wheel Tandem Axis *Total STRT*

The proportion of axes in column 5 is obtained by the equation of division between the total number of axes and the number of axes based on the load with the following example:

$$\frac{Total\ STRT\ JS}{Total\ Jumlah\ Sumbu\ (JSKNH)} \quad (3)$$

As an example: 950/1620 = 0,59

3.3** Determining **the Bottom Base Layer

This value is obtained by aligning the graph on the Bina Marga 2002 guideline with the CBR (California Bearing Ratio) data of 5% and the number of axis repetitions from the analysis of 3.7 x 106, which is obtained

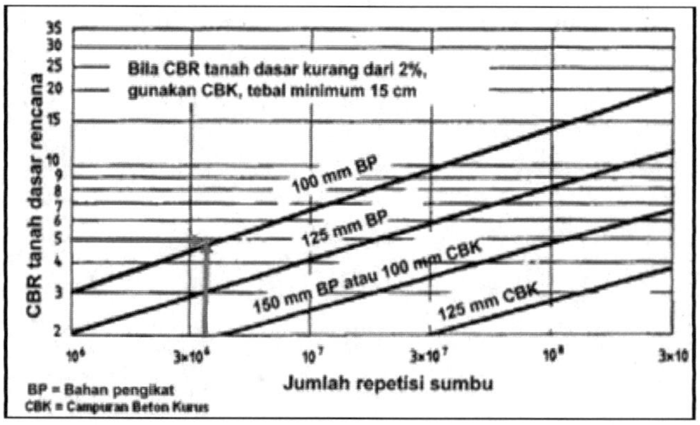

Figure 3. Minimum bottom layer thickness

Based on Figure 3, the bottom foundation layer thickness of 100 mm was taken and the effective CBR was 14%. JSK value is 0.75 x 107 with 4.0 Mpa concrete tensile strength, so that the estimated concrete thickness of 25 cm can be taken, the value shows that the total fatigue that occurs <100%, it can be concluded that the calculation is sufficient and the plate thickness is 25 cm can be used

3.4Determining Dowel / Ruji Diameter

After determining the estimated thickness of 250 mm, the test diameter can be determined based on the table in the Planning Guidelines for Cement Bina Marga Concrete Road Pd T-14 2003 as follows:

Table 5. Ruji's Diameter

No	Concrete Plate Thickness, h (mm)	Ruji's Diameter (mm)
1	$125 < h \leq 140$	20
2	$140 < h \leq 160$	24
3	$160 < h \leq 190$	28
4	$190 < h \leq 220$	33
5	**$220 < h \leq 250$**	**36**

Source : Bina Marga Pd-T-14-2003

Based on the table above the diameter of the test is 36 mm with a distance of 300 mm and a length of 450 mm

Coefficient of friction plate with foundation (μ) = 1,3 Distance Between Connections (L) = 4 m, transverse 5 m
Plate Thickness (h) = 200 mm
Steel Tensile Voltage (fs) = 4,0
MPa Concrete Quality (fc) = 350 kg/cm^2 Concrete Specific Gravity = 2400 kg/cm^2
Concrete Tensile Strength (f_{ct}) 0,4-0,5 = 40 kg/cm^2

MR Modulus Elastic Steel (Es) = 20000 kg/cm^2
Steel Melt Voltage (fy) = 3900 kg/cm^2
Modulus Elastic Beton (Ec) = 1400 $\sqrt{fc}$
Gravitasi (g) = 9,81 m/s^2

Transverse Reinforcement

$$As = \frac{\mu . L . M . g . h}{2fs} \quad (4)$$

Lengthening Reinforcement

$$P_s = \frac{1 \quad fc\,.(1{,}3-0{,}2.\mu)}{(fy-n.fc\,)} \quad (5)$$

here As is the area of reinforcement

Based on the data above transverse reinforcement obtained As = 48.46 mm2 with As min = 0.1% x 190 x 1000 = 282.76 mm2 then using D12 - 300 reinforcement, while longitudinal reinforcement can get As min = 0.1% x 200 x1000 = 200mm2 then use D12 - 400 mm.

Based on the crossing distance of 4.00 m and the distance of the longitudinal joint of 5.00 m in the form of a square, then for a single woven plain reinforcement taken an diameter of 8 mm with a distance of 200 mm which has a length of 251 mm2 and 251 mm2 transverse axles it is safe to be used as single woven plain reinforcement.

IOP Conf. Series: Materials Science and Engineering **650** (2019) 012050 doi:10.1088/1757-899X/650/1/012050

3.5 Planning dimensions of drainage channels

3.5.1 Calculation of Debit Flow Plans (Q)

Regional Data on Road Watering Area of PT Krakatau Bandar Samudra, Cilegon City Banten

Channel Length Plan = 300 m	Width of drainage area
I_1 Road Pavement = 4,0 m	I_2 roadside
I_3 Outside Road = 10 m	Area of Jetting
A_1 Road Pavement = 1.200 m^2	A_2 Road Shoulder Section
A_3 Outside Road = 3000 m^2	total area = 4.200 m^2 = 0,00420 km^2
Average Material Coefficient	0,70 - 0,95
Average Land Use Coefficie	0,60 – 0,90
Runoff Factor	1,2

$$C = \frac{C1.A1 + C2.A2 + C3.A3.FK}{A1 + A2 + A3} \quad (6)$$

Where :
C_1 = 0,70 Is the value of the flow coefficient for concrete roads
C_2 = 0 Is the value of the flow coefficient not to use the shoulder of the road
C_3 = 0,60 Is the value of the flow coefficient for the outer part of the regional road
Industry F_k = 1,2 for industrial runoff factors
A_1 = 1.200m^2 Extent of Pavement area.
A_2 = 0 m^2 The width of the road shoulder area.
A_3 = 3.000m^2 Extensive area of the outer part of the road.

Then Value C for Industrial Estate Road PT. Krakatau Bandar Samudra Cilegon Banten are: 0.71. with the concentration time (Tc) of this industrial area the road is 5,978 minutes.

3.5.2 Rainfall Data Analysis

From BMKG Data Class 1 Serang in 2017/2018 The average rainfall is 152.07 mm and the 10-year calculation return is 333 mm. Then the flow debit (Q) can be calculated by the equation:

$$Q = \frac{1}{3,6}\, C \times I \times A = 0,455 \text{ m}^3/\text{seconds} \quad (7)$$

Where :
C = average coefficient
I = Rain intensity (mm / hour)
A = Area of drainage area (m^2)

IOP Conf. Series: Materials Science and Engineering **650** (2019) 012050 doi:10.1088/1757-899X/650/1/012050

3.5.3 Drainage channel dimensions

Flow (Q) = 0,296 m^3 / seconds
Channel Basics (B) = 0,75 channel height
(H) Wide channel appearance Fs = B.H
= 0,75 H x H
=0,75 H^2

Around the Wet Channel Ps = B + 2 H
= 2,75 H
Radius hidrolik Rs = Fs / Ps
(0,75 H^2) / (2,75 H)
= 0,275 H
Formula Manning V= 1 / n $Rs^{2/3}$ $I^{1/2}$
= (1/0,014) $(0{,}273H)^{2/3}$ $0{,}02^{1/2}$
= 71,4 $0{,}273^{2/3}$ $0{,}15^{1/2}$ $H^{2/3}$
= 2,92 $H^{2/3}$
Debit Aliran Q = 0,455 m^3 / seconds Q
= Fs.V

From these calculations the dimensions of the channel dimensions are set to H = 0.99 m and B = 0.45 m.What was previously done was checking the channel dimensions that the Qdebit channel > 0.455 m3/ sec.

4. Conclusion

Increasing the road using rigid pavement on the roads in the PT Krakatau Bandar Samudra area can be concluded as follows:

a. The estimated thickness of the concrete plate is 250 mm or 25 cm
b. Dowel / Ruji diameter is 36 mm with a distance of 300 mm and a length of 450 mm while the longitudinal reinforcement joints use reinforcement rods 4D12 mm - 400 mm, transverse reinforcement per connection using reinforcement rods 4D12 mm - 300 mm, as well as for single woven plain reinforcement using D8 mm - 200 mm.
c. The dimensions of the drainage channel that can accommodate the discharge occur are B × H = 0.99 m × 0.45 m with 1SP: 4 Sandstone pair.

Bibliography

[1] Manu A I *Rigid Pavement* (Jakarta : 1995 Department of Public Works)
[2] Ministry of Public Works directorate general of clan community 2002 *Construction and Building Guidelines for Cement Concrete Road Construction* (Jakarta)
[3] Ministry of Public Works directorate general of clan community 2003 *Guidelines for Construction and Buildings for Cement Road Concrete Construction* (Jakarta)
[4] Directorate of Planning and Operation of Traffic in Urban Areas 1999 *Rekayasa Lalu Lintas* (Jakarta: Direktorat Jenderal Perhubungan Darat)
[5] Government regulations, Nomor 34 Tahun 2006 *about the Road* (Jakarta)
[6] Undang-undang Nomor 38 Tahun 2004 *about the Road* (Jakarta)

The Overloading Effect on the Design Life of Road and Thickness of Pavement Layer

A Rahmawati[1], E Adly[1], I Lutfiyanto[1] and A Syifa M[1]

[1]Department of Civil Engineering, Faculty of Engineering, Universitas Muhammadiyah Yogyakarta, Jl. Brawijaya, Geblagan, Tamantirto, Kec. Kasihan, Bantul, Daerah Istimewa Yogyakarta, 55183 Indonesia.

* anita.rahmawati@umy.ac.id ; emil@umy.ac.id

Abstract. Solo-Yogyakarta KM 9-15 Road is the national highway, which supports the traffic of goods and services, but many passing goods transport vehicles do not match the allowable load. As a consequently, the damage on the pavement occurred. Therefore, this study aims was to know the effect of overloading on design life and pavement thickness. This study used the method of Binamarga to analyze the effect of pavement thickness caused by overloading. The analysis was carried out by using the formula of design life decreasing with the value of Cumulative Equivalent Standard Axle (CESA) for overloading and standard load. Based on the 2013 Binamarga method, the design life is reduced by about eight years, from the previously determined is 20 years. Moreover, based on the analysis on pavement thickness of 20 years design life for a standard load, it showed that pavement thickness is needed for total surface layer is 100 mm consist of AC WC 40 mm and AC BC 60 mm. In addition, the base course layer is 145 mm (AC Base), and the subbase course layer is 300 mm (LPA Kelas A). At the same time, for overloading, the analysis result showed that it was required the pavement thickness for the surface layer is 100 mm, the base course layer is 180 mm, and the subbase course layer is 300 mm. These results indicate that overloading causes a reduction in the design life of the pavement and an increase in the thickness of the pavement.

Keywords: Bina Marga Method, Design Life, Overloading, Pavement Thickness

1. Introduction

The highway is one of transportation infrastructure that is intended for traffic and is very influential in the activities of daily life. Road plays an essential role in supporting services for various fields, one of which is the economy of a region. The increasing human needs in various aspects of life, it will impact on increasing traffic volume. This phenomenon will cause multiple problems on the road surface. The issues caused include damage to the road. Therefore, it will affect the level of comfort, safety, and smoothness in traffic.

Vehicles passing on roads, which bring load exceed the maximum specified load cause excessive loading on the pavement and affect the design life on the road [1]. Based on some previous studies, it can be concluded that the overloading can decrease the design life.

The study which using AASHTO 1993 and the calculation of cumulative ESAL (Equivalent Standard Axle Load) values, then the remaining pavement life was obtained. This section contains the

IOP Conf. Series: Materials Science and Engineering **650** (2019) 012051 doi:10.1088/1757-899X/650/1/012051

research background, purpose, contribution (the research benefit from theoretical and practical), research result and implication (Practical advice based on research result). The research result and implication in the introduction section are an only suggestion (not mandatory) [2].

Based on the design life of the 20-year plan, the cumulative ESAL value is 64,533,642 SAL, so there is a decrease in service life of eight years. The remaining life of the pavement plan is only 54.75%, and a pavement service life is reduced by 25.94%. Similar research on the effect of overloading were also not much different from previous research [3,4].

The research regarding the effect of overloading with slightly different methods showed that there was a reduction in service life from the design life and caused various road damage problems [5-14].

This study discussed the effect of overloading on design life and pavement thickness. Furthermore, this study used the method of Binamarga 2013 to analyze the decrease of service life and to analyze the effect of overloading on pavement thickness [15].

2. Literature Review

2.1. Axle Load and Overloading

The axle load of the vehicle is affected by a load of the vehicle and the axle configuration. Two similar vehicles will have different axle load because of the difference in load. Therefore, on road pavement planning, there needs to be a variation in axle loads so that that equivalent numbers can be determined for proper planning [16].

The amount of load received by the pavement depends on several things, i.e., the total weight of the vehicle, axle configuration, the contact area between the wheels and pavement and the speed of the vehicle. Overloading is a condition when the road receives a vehicle load that is not following the planned standard load. Researcher explained that the causes of road damage are overloading that repeatedly occurs, rainwater, air temperature, and poor quality pavement. The design of the road must be adequately planned and must also be adequately maintained so that it can serve the traffic load in accordance with the service life of the road [17].

2.2. Design Life

Based on the guidelines for flexible pavement of the Department Pekerjaan Umum (2013), the design life is the amount of time in the year calculated when the road is opened for vehicle traffic until heavy (structural) repairs need to be given, i.e., the new pavement layer (overlay). The design life used for the construction of new pavement is 20 years, and design life for road improvement is ten years [15]. The equation used to determine the Rest of Design Life (RDL) as follows:

$$\text{RDL} = \frac{\textit{Standard CESA Accumulation}}{\textit{Overloading CESA Accumulation}} \times DL \quad (1)$$

2.3. Serviceability

Serviceability can be calculated under some circumstances. These circumstances are the construction of the pavement has been completed, the traffic on the road has been achieved, and the traffic on the road has begun to be opened. Over time, the road serviceability will be reduced. The reduced serviceability depends on routine pavement maintenance [18].

2.4. Speed

Speed is a level of movement of vehicles or traffic expressed in kilometers per hour. The level of traffic density influences the speed and travel time of traveling from one place to the destination. By increasing the volume of traffic on the road, the level of comfort, safety, and smoothness when passing traffic will be affected. Therefore, an evaluation of the traffic condition needs to be carried out in order to maintain the level of smoothness of a road [6].

IOP Conf. Series: Materials Science and Engineering **650** (2019) 012051 doi:10.1088/1757-899X/650/1/012051

2.5. Equivalent Number of Vehicle Loads (E)

An equivalent number is a number that states the ratio of the level of damage to the pavement caused by a single axle vehicle load trajectory by a single standard load path of 8.16 tons, [15]. The equation used to determine the equivalent number of vehicles is as follows:

$$\text{ESA}_4 = \frac{L_{ij}}{SL} \quad (2)$$

$$\text{ESA}_4 = \left(\frac{\text{Single axle load(kg)}}{8160}\right)^4 \quad (3)$$

$$\text{ESA}_4 = 0.086 \left(\frac{\text{Double axle load(kg)}}{8160}\right)^4 \quad (4)$$

$$\text{ESA}_4 = 0.086 \left(\frac{\text{Triple axle load (kg)}}{8160}\right)^4 \quad (5)$$

2.6. Traffic Load

Traffic load can be calculated based on the equivalent number of the standard axle load. [15] The accumulation of traffic axle loads during the design life can be determined by the following CESA (Cumulative Equivalent Standard Axle) formula:

$$\text{ESA} = \sum_{Traktor-Trailer}^{MP} ADT\ x\ ESA_4 \times DL \quad (6)$$

$$\text{CESA}_4 = 365 \times ESA \times R \quad (7)$$

$$\text{CESA}_5 = CESA_4\ x\ TM \quad (8)$$

When :

CESA	= Cumulative Equivalent Standard Axle
ADT`	= Average Daily Traffic
DL	= Lane Distribution
365	= Number of days in a year
R	= Multiplier of traffic growth
TM	= Traffic Multiplier

2.7. Pavement Thickness

The thickness of each pavement material for each surface layer, the base course layer, and the subbase course layer can be determined by the following table [15].

Table 1. Selection Design Types of Pavement [15]

Pavement Structure	Design	ESA (million) in 20 years (rank 4 unless otherwise specified)
		>10 – 30
Rigid Pavement with heavy traffic (CBR ≥ 2,5%)	4	2
Rigid Pavement with low traffic	4A	-
AC WC modification or SMA modification with CTB (ESA rank 5)	3	2
AC with CTB (ESA rank 5)	3	2
AC thick ≥100 mm with granular fondation (ESA rank 5)	3A	2
AC or HRS thin with granular fondation	3	-
Burda or Burtu with LPA kelas A	5	-
Soil Cement	6	-
Unsurfaced	7	-

Table 2. Design Type 3A [15]

	FFF3	FFF4	FFF5	FFF6	FFF7	FFF8
Commulative Equivalent Single Axle for 20 years design (10^6 ESA5)	>4–7	>7–10	>10-20	>20–30	>30–50	>50–100
PAVEMENT LAYER THICKNESS (mm)						
AC WC	40	40	40	40	40	40
AC BC	60	60	60	60	60	60
AC *Base*	80	105	145	160	180	210
LPA Kelas A	300	300	300	300	300	300

3. Research Method

3.1. The Step of Research

In this step, the arrangement of planning and observing on the preliminary stage will be carried out in order to obtain the general description of the problems in the field. The preparation stage carried out in this study is shown in Figure 1.

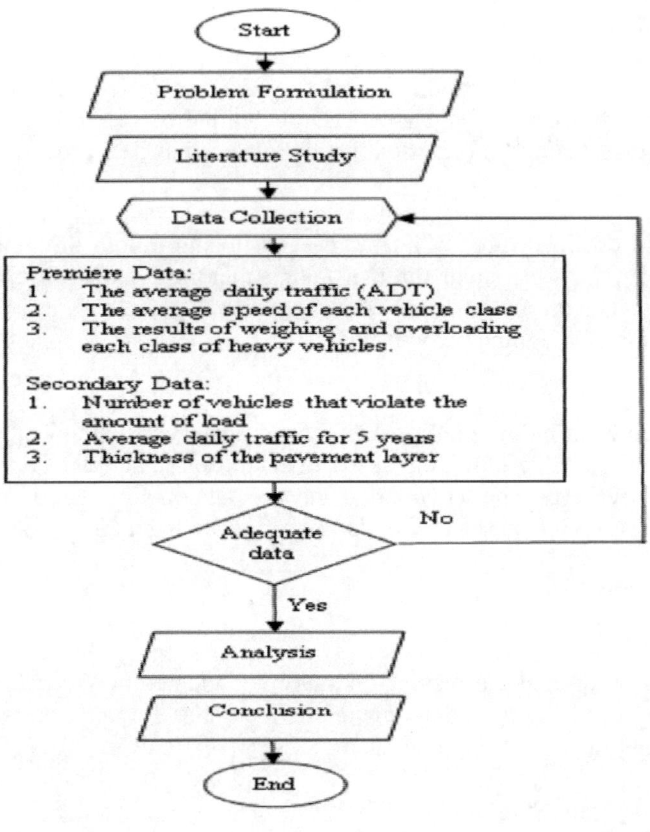

Figure 1. Chart of Research Flow

3.2. Research Location

The research location is on Solo-Yogyakarta KM 9-15 Road, Sleman. This road is a national road that connects two provinces, namely the provinces of Yogyakarta and Central Java.

Figure 2. Research Location

3.3. Data Collection Technique

In analyzing the effect of overloading on the design life of road in the Yogyakarta – Solo KM 9-15, primary and secondary data are required. The primary data came from the field observation and the secondary data obtained from related agencies. The technique of data collection in this study was by finding the primary and secondary information that will be used as data processing material.

3.4. Primary Data

The primary data were obtained directly in the research location both through observation or survey. The observation and survey were done through taking pictures of heavy vehicles, and conducting a survey on road surface conditions on Solo-Yogyakarta KM 9-15 Road, average daily traffic (ADT), vehicle speed, the results of weighing heavy vehicles.

3.5. Secondary Data

The data were obtained from indirect research at related agencies such as National Road Planning and Supervision Service of the Province of Yogyakarta Special Region and the Transportation Service of the Special Province of Yogyakarta. The secondary data were data on violation of overloading that occurred on Solo-Yogyakarta KM 9-15 Road. Average daily traffic data from 2014 to 2018.

4. Result and Discussion

4.1. Overloading Data

Based on the data collection and the result of a survey conducted at Weighbridge and Department of Transportation, it was obtained the data on violation number for an excessive load of the vehicle occurred at both weighbridge in Solo-Yogyakarta KM 9-15 Road in 2015. The number of violating vehicles and vehicles that do not violate every month and the total violations that occured within one year on the Kalitirto and Tamanmartani Weighbridges are shown in Table 3.

IOP Conf. Series: Materials Science and Engineering **650** (2019) 012051 doi:10.1088/1757-899X/650/1/012051

Figure 3. Overloading Heavy Vehicle that cross the Solo-Yogyakarta KM 9-15 Road

Table 3. The number of overloading violations on Kalitirto and Tamanmartani Weighbridge in 2015 [19]

Month	Kalitirto			Tamanmartani		
	The number of Vehicles	Violating Vehicle	Vehicles do not violate	The number of Vehicles	Violating Vehicle	Vehicles do not violate
January	18475	2979	15496	19221	3175	16046
February	16919	2789	14130	17905	2784	15121
March	18798	3095	15703	19788	3183	16605
April	19144	3123	16021	18499	3145	15354
May	20224	3233	16991	19867	3297	16570
June	19789	3306	16483	19614	3629	15985
July	10085	1757	8328	10132	1969	8163
August	19341	3121	16220	19335	3453	15882
September	18942	3232	15710	19868	3641	16227
October	20577	3512	17065	21165	3998	17167
November	19194	3340	15854	20301	3894	16407
December	17515	2857	14658	19001	3382	15619
Total	219003	36344	182659	224696	39550	185146

Source: DIY Transportation Agency for Land Transportation, 2015

Based on the recapitulation of overloading violations on Kalitirto and Tamanmartani Weighbridges, it can be seen that the number of overloading violations at Tamantani weighbridge is higher than at Kalitirto weighbridge.

The total number of vehicles that violated overloading that passed through the Kalitirto Weighbridge were 36,344 vehicles while those passing through the Tamanmartani Weighbridge were 39,550 vehicles. It means, the number of overloading vehicles that passed through Solo-Yogyakarta KM 9-15 Road which going to Solo were more significant than the vehicles went to Yogyakarta.

Table 4. Number of Violations for Each JBI Group on the Kalitirto and Tamanmartani Weighbridges [19]

No.	Group	Weighbridge	
		Kalitirto	Tamanmartani
1.	Group I	8650	11780
2.	Group II	13723	13894
3.	Group III	6992	7302
4.	Group IV	5564	5871

Source: Dishub DIY Bidang Angkutan Darat, 2015

From the data recapitulation of overloading violations occurred on Kalitirto and Tamanmartani Weighbridge, it can be seen the number of violations for each class of JBI (weight allowed). The total number of vehicles violating each JBI can be seen in Table 5. Whereas based on the results of the load survey that has been carried out in the Weighbridges, the average overloading is obtained in Table 5 as follows:

Table 5. Average Overloading of Each JBI Group

No.	JBI Vehicles	Overloading Average	Overloading Vehicles
1.	JBI < 8 ton	2220 kg	24 %
2.	8-ton ≤ JBI ≤ 14 ton	5079 kg	31 %
3.	14 ton < JBI ≤ 21 ton	14788 kg	35 %
4.	21 ton < JBI ≤ 28 ton	17118 kg	39 %

4.2. Traffic Growth (i)

The traffic growth estimation in Road internode Solo – Yogyakarta KM 9-15 was based on the total number of traffic volume which passing that highway every year starting from 2014 to 2018. These data were obtained from DIY National Road Planning and Supervision Office. The way to determine the traffic growth figures is done by using two methods, i.e., the Exponential Method and Linear Regression Method. Based on the result of the calculation of traffic growth figures by using the exponential method, it was obtained 8.26%. Meanwhile, based on the calculation by using the linear regression method, the traffic growth figure was 7.5%. From these two methods, the traffic growth figure that used was 7.5% from the linear regression method. The method of linear regression seems to be more accurate compared to the exponential method.

4.3. The Value of CESA of standard load and CESA of Overloading

The load caused by traffic can be calculated based on the equivalent method toward a load of standard axle and overloading. To determine the accumulation of traffic axle load during the design life, CESA4 (Cumulative Equivalent Standard Axle) formula can be used in equation 7. The result of the calculation of CESA4 for standard load and overloading can be seen in Table 6 and Table 7 as follows.

Table 6. The $CESA_4$ Value for Standard Load

JBI	Types of Vehicle	ADT 2018	ESA_4	ESA	$CESA_4$
I	Cars	33.977	0,00045	7,645	56.207,531
I	Non Bus	282	0,01762	2,484	18.262,852
I	Pick up, box cars	2.097	0,0069	7,235	53.193,131
I	Bus	378	0,05937	11,221	82.498,980
II	Large Bus	694	0,0439	15,233	111.995,986
III	Truck 2 Axle	1.702	0,7345	625,060	45.95.563,007
IV	Truck 3 Axle	809	0,6438	260,417	1.914.636,565
IV	Truck semi trailer	68	0,5465	18,581	136.611,135
IV	Trailer	79	1,7493	69,097	508.014,618
		Total			**7.476.983,805**

The TM value used to calculate $CESA_5$ is 2. The result of the calculation of CESA5 for standard load and overloading with the design life of 20 years can be seen in Table 8 as follows.

Table 7. The $CESA_4$ Value for Overloading

Gol. JBI	Types of Vehicle	ADT 2018	%	ADT*) 2018	ESA_4	ESA	$CESA_4$
I	Cars	33.977	100	33.977	0,00045	7,645	56.207,531
I	Non Bus	282	100	282	0,01762	2,484	18.262,852
I	Pick up, box cars	2.097	27,5	577	0,0490	14,129	103879,163
			72,5	1.520	0,0069	5,245	38.562,263
I	Bus	378	100	378	0,05937	11,221	82.498,980
II	Large Bus	694	100	694	0,04390	15,233	111.995,986
III	Truck 2 Axle	1.702	19,4	330	7,9278	1308,827	9.622.751,325
			80,6	1.372	0,7345	503,798	3.704.021,137
IV	Truck 3 Axle	809	15,5	125	5,1866	325,187	2.390.838,235
			84,5	684	0,6438	220,052	1.617.865,214
IV	Truck semi trailer	68	15,5	11	3,1151	16,417	120.700,985
			84,5	57	0,5465	15,701	115.436,814
IV	Trailer	79	100	79	1,7493	69,097	508.014,618
		Total					**18.491.035,103**

Table 8. the Value of CESA5 of Standard Load and CESA5 of Overloading

$CESA_5$ of standard load	$CESA_5$ of Overloading
($CESA_4$ x TM)	
$14{,}9 \times 10^6$.	$36{,}9 \times 10^6$.

4.4. The Decrease of Design Life
Based on the result of the calculation of standard load and overloading CESA above, it can be seen that the value of overloading CESA is more significant than the value of standard load CESA. This phenomenon that affects the service life on road pavement. To determine the rest of design life can be used the equation 1. From the calculation of the rest of design life for projection life of 20 years, it can be obtained the rest of the design life was eight years, and the decrease of design life was 12 years.

4.5. The Effect of Load on the Pavement Thickness Performance
Based on the result of the analysis on the decreasing of design life was caused by the overloading on Solo-Yogyakarta KM 9-15 Road, it was known that there was a decrease of design life from the design life that has been determined before. The analysis of the effect of the load against pavement thickness performance aims to know the pavement thickness performance that was caused by the average load and overloading. Table 1 is used to determine which number of pavement designs are appropriate. The type of pavement design is 3A. Pavement thickness is determined by plotting the CESA5 value in the Table 10 for standard load and overloading.

The analysis of pavement layer thickness due to the standard load and overloading by using the method of Binamarga 2013, it can be obtained the pavement layer thickness with a standard load for surface layer are 40 mm for ACWC and 60 mm for AC BC, the base course layer is 145 mm, and the subbase course layer is 300 mm. At the same time, for overloading, it can be obtained the surface layer are 40 mm for ACWC and 60 mm for AC BC, the base course layer is 180 mm, and the subbase course layer is 300 mm. The comparison can be observed in Figure 4.

Table 9 . Pavement Layer Thickness for Standard Load and Overloading[15]

	FFF3	FFF4	FFF5	FFF6	FFF7	FFF8
Commulative Equivalent Single Axle for 20 years design (10^6 ESA5)	>4–7	>7–10	>10-20	>20–30	>30–50	>50–100
	PAVEMENT LAYER THICKNESS (mm)					
AC WC	40	40	40	40	40	40
AC BC	60	60	60	60	60	60
AC *Base*	80	105	145	160	180	210
LPA Kelas A	300	300	300	300	300	300
			Standard load		**Overloading**	

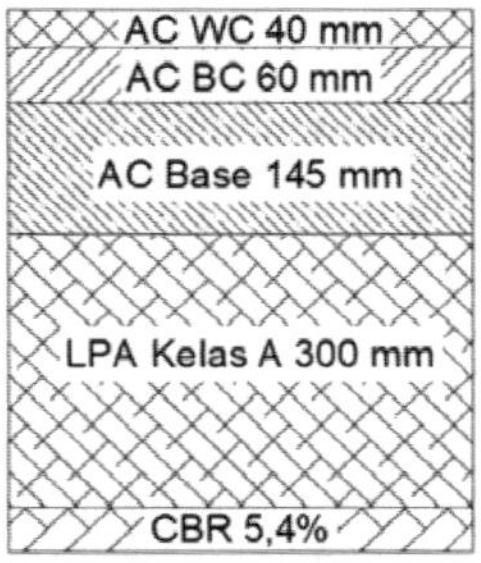

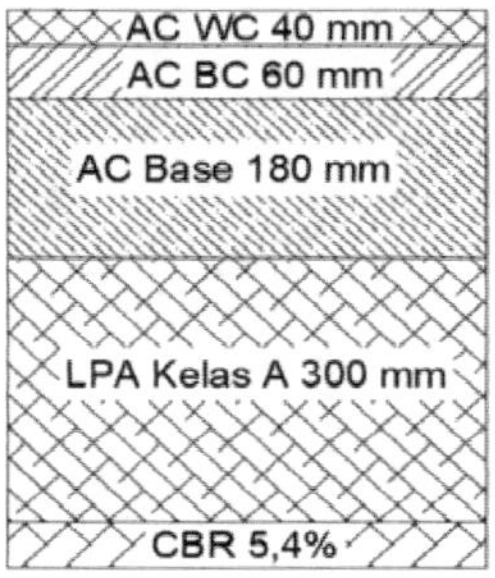

Figure 4. The Pavement Layer Arrangement due to the Impose of Standard Load and Overloading based on Binamarga Method.

5. Conclusion

Based on the result of the data analysis and the discussion on the effect of overloading and pavement thickness that have been carried out, then it can be concluded as follows:

a. From the calculation of $CESA_5$ value, it can be obtained the figure of $CESA_5$ Binamarga 2013 Method for a standard load of 14,9 × 10^6 ESAL and the figure of $CESA_5$ for overloading, i.e., 36,9 × 10^6 ESAL. From these calculations, there was a quite significant difference between the value of $CESA_5$ for the standard load and overloading. This phenomenon will decrease the design life of road pavement on Solo-Yogyakarta KM 9-15 Road.

b. Based on the analysis on the decrease of design life, it was obtained that the rest of design life for Binamarga 2002 Method was 8 years from the design life that had been determined before, i.e., 20 years.

c. From the result of the calculation of pavement thickness for standard load and overloading, it can be obtained that the pavement thickness of Solo-Yogyakarta KM 9-15 Road at the design life of 20 years as follows:
   - Due to the standard load, the thickness of the pavement for the standard load at the surface layer was 100 mm consist of AC WC 40mm and AC BC 60mm, base course layer was 145mm, and subbase course layer was 300 mm.
   - Due to the overloading, the thickness of the pavement layer with overloading found that the thickness of the pavement layer for the surface layer was 100 mm consist of AC WC 40mm and AC BC 60mm, base course layer was 180mm, and subbase course layer was 300 mm.

References

[1] Sari D N 2014 Analisa Beban Kendaraan Terhadap Deraja Kerusakan Jalan dan Umur Sisa *Jurnal Teknik Sipil dan Lingkungan* **2(4)** 615-620.

[2] Sentosa L, dan Roza A A 2012 Analisis Dampak Beban Overloading Kendaraan pada Struktur Rigid Pavemenet Terhadap Umur Rencana Perkerasan *Jurnal Teknik Sipil* **19(2)** 161-168.

[3] Morisca M 2014 Evaluasi Beban Kendaraan Terhadap Derajat Kerusakan dan Umur Jalan *Jurnal Teknik Sipil dan Lingkungan* **2(4)** 692-699.

[4] Suriyatno, Purnawan, dan Putri E E 2015 Analisis Tebal lapis Tambah dan Umur Sisa Perkerasan Akibat Beban Berlebih Kendaraan *Prosiding Annual Civil Engineering Seminar 2015, Pekanbaru, 4 November 2015,*164-176.

[5] Simanjuntak G I, Pramusetyo A, Riyanto B, dan Supriyanto 2014 Analisis Pengaruh Muatan Lebih (Overloading) Terhadap Kinerja Jalan dan Umur Rencana Perkerasan Lentur *Jurnal Karya Teknik Sipil* **3(3)** 539-551.

[6] Situmorang R A, Wartadinata P W, Setiadji B H, dan Supriyono 2013 Analisis Kinerja dan Perkerasan Lentur akibat Pengaruh Muatan Lebih (Overloading) *Jurnal Karya Teknik Sipil* **2(2)** 359-370.

[7] Atiya A F , Sari O D , Purwanto D, dan Setiadji B H 2014 Analisis Pengaruh Kinerja Jembatan Timbang Terhadap Kinerja Perkerasan dan Umur Rencana Jalan *Jurnal Karya Teknik Sipil* **3(3)** 662-673.

[8] Syafriana, Saleh S F, and Anggraini R 2015 Evaluasi Umur Layan Jalan dengan Memperhitungkan Beban Berlebih di Ruas Jalan Lintas Timur Provinsi Aceh *Jurnal Transportasi* **15(2)** 115-124.

[9] Pandey S V 2013 Kerusakan Jalan Daerah Akibat beban Overloading *Jurnal Tekno Sipil* **11(58)**

[10] Wandi A, Saleh S M, and Isya M 2016 Analisis Kerusakan Jalan Akibat Beban Berlebih *Jurnal Teknik Sipil* **5(3)**, 317-328.

[11] Saleh S M, Sjafruddin A, Tamin O Z, dan Frazila R B 2009 Pengaruh Muatan Truk Berlebih Terhadap Biaya pemeliharaan Jalan *Jurnal Transportasi* **9(1)** 79-89.

[12] Cahyono S D 2012 Pengaruh Beban Lalu Lintas Terhadap Kerusakan pada Jalan Raya Ngawi-Caruban *Jurnal Teknik Sipil* **13(1)** 66-73.

[13] Syaifullah, Wiguna I P A, dan Herijanto W 2013 Analisis Pengaruh Beban Berlebih Kendaraan Terhadap Pembebanan Biaya Pemeliharaan Jalan *Jurnal Teknik Sipil* **2** 1-8.

[14] Martina R, Saleh S M, dan Isya M 2018 Kajian Beban Aktual Kendaraan pada Konstruksi Jalan Menggunakan Weight In Motion (WIM) *Jurnal Teknik Sipil* **1(3)** 701-714.

[15] Departemen Pekerjaan Umum 2013 *Pedoman Perencanaan Tebal Perkerasan Lentur Direktorat Jendral Bina Marga* (Jakarta)

[16] Sukirman S 1999 *Perkerasan Lentur Jalan Raya* (Yogyakarta : Nova)

[17] Suwardo dan Sugiarto 2004 Tingkat Kerataan Jalan berdasarkan Alat Rolling Straight Edge Untuk Mengestimasi Pelayanan Jalan, *Prosiding Simposium VII FSTPT, Universitas Katolik Parahyangan, 11 September 2004,* 1-9.

[18] Hardiyatmo H C 2016 *Perancangan Perkerasan Jalan dan Penyelidikan Tanah* (Yogyakarta : Universitas Gajah Mada)

[19] Dishub DIY Bidang Angkutan Darat 2015 *Jumlah Pelanggaran Kelebihan Beban (Overloading) pada Jembatan Timbang Kalitirto & Tamanmartani Tahun 2015* (Yogyakarta)

The role of scrap rubber size against the characteristics of ballast layer

S A P Rosyidi[1], D M Setiawan[1], C Budiyantoro[2], and L N Bintari[1]

[1]Department of Civil Engineering, Faculty of Engineering, Universitas Muhammadiyah Yogyakarta, Bantul, Daerah Istimewa Yogyakarta, Indonesia
[2]Department of Mechanical Engineering, Faculty of Engineering, Universitas Muhammadiyah Yogyakarta, Bantul, Daerah Istimewa Yogyakarta, Indonesia

* atmaja_sri@umy.ac.id ; diansetiawanm@ft.umy.ac.id

Abstract. This research utilizes scrap rubber from motorcycle tires as a ballast layer mixture component. A compression test method was used with three test samples: ballast (sample I), ballast with 10% uniform graded scrap rubber (sample II) and ballast with 10% continuous graded scrap rubber (sample III). This research aims to analyze the elastic modulus, durability and vertical deformation of the ballast layer. It can be concluded that the elastic modulus of sample I, II, and III was 14.28 MPa, 8.53 MPa, and 3.10 MPa, respectively. Moreover, the abrasion of sample I, II, and III was 162.1 gr (3.24%), 54.2 gr (1.08%), and 50.1 gr (1%). Furthermore, at the same number of vertical deformation (5 mm), the sample I was able to receive the most considerable load that is 1980.9 Kg, followed by sample II by 1252.2 Kg and sample III by 443.7 Kg.

Keywords: ballast durability, elastic modulus, rail track, scrap rubber, vertical deformation.

1. Introduction

The ballast layer functions to hold the sleepers construction and distribute the load from the sleeper to the subgrade. In recent years, conventional track structure is the most widely used railroad type because of its low construction cost compared to the slab track. However, the performance of ballast layers needs to be improved to fulfill its function as a load distributor in the rail track structure [1]. The poor ballast material conditions can be a benchmark for the needs of maintenance work and the application of train speed restrictions [2, 3]. One effort to avoid the speed limitation is by strengthening the ballast structure. The use of scrap rubber as an added component in the ballast layer is a proposed method in this study to reduce the use of aggregate material. With the addition of rubber, it is also expected to be able to decrease the friction between aggregates and to minimize the ballast wear. The previous study has reported that the utilization of scrap rubber from waste motorcycle tires is not only the solution that can reduce the aggregate use, but also serves to increase the durability of the railroad structure [4].

Sol-Sanchez et al., (2014) have found that the optimum rubber content used as elastic material in the ballast layer was 10% [5]. Also, Hameed et al., (2016) have identified that the flexible mixture can also be used in sleepers to reduce cracks by 80-100% [6]. It is also known that the stiffness of the ballast layer is different for each layer. The greatest stiffness is in the bottom part of the ballast layer, and the flexible portion is found at the top of the ballast layer [7]. In another study, Navaratnarajah et al., (2017) have concluded that the use of rubber mixtures on ballast layers can reduce long-term degradation and deformation [8].

Most studies in rubber usage for rail track structure have only been carried out with crumb rubber and natural rubber. In this study, the author has used the scrap rubber from outer motorcycle tire with two different sizes as additional material in ballast layer. This study aimed to evaluate and validate the characteristics of modified ballast with scrap rubber mixtures through a compressive strength test by analyzing the elastic modulus, ballast material durability, and vertical deformation. Thus it can be seen the effect of adding scrap rubber from outer motorcycle tire on the ballast layer and the characteristics comparison between ballast with uniformly sized scrap rubber and ballast with graded-sized scrap rubber.

2. Research Method

2.1. Material Preparation and Testing

The ballast aggregates were obtained from Kulon Progo, while the scrap rubber was obtained from the outer tires of motor vehicles (Figure 1). At the stage of material testing, the aggregate physical properties were examined to identify the specifications of the material used. Regulations regarding testing aggregate specifications are based on the Indonesian National Standard [9, 10, 11, 12]. The physical tests performed on the ballast aggregate were the specific gravity and absorption tests, sieve analysis, mud content and abrasion with Los Angeles machines.

2.2. Mixture Planning and Sample Making

The specimens were made in the ballast box with a size of 40 cm x 30 cm x 20 cm. The test material prepared is a mixture of ballast and scrap rubber. The mixing process is accompanied by a manual compaction process with a pulverizer that has a load of 4.5 kg, a diameter of 6 cm and a falling height of 20 cm. Scrap rubber used in this study is divided into two (2) types. The first is scrap rubber with a uniform size of 3/8", while the second is scrap rubber with the graded size of 1", ½", ¾", and No.4, with the percentage of each type of scrap rubber is 10% from the total aggregate weight. The 10% of scrap rubber in the mixture is the optimum level of rubber utilization in this research based on the results of Sol-Sanchez et al. (2015) study [4]. After the aggregate material is known to meet the requirements, the specimens can be made according to the mix planned. There are three samples, which are sample I (ballast), sample II (ballast and 10% scrap rubber with a uniform size of 3/8"), and sample III (ballast with 10% scrap rubber with graded-size of 1", ½ ", ¾", and No.4) as shown in Table 1.

Furthermore, to identify the change in samples performance after the initial maximum loads, then each sample was tested twice, namely IA, IB, IIA, IIB, IIIA, and IIIB. Next, the mixture of ballast aggregate and scrap rubber is put into the ballast box per 1/3 the height of the mold, then the layer is compacted with a manual compactor 25 times for all sides, and the center of the layer with a falling height is approximately 20 cm. The same thing was done in the second and the third layers until the box is filled. In order to control the compaction, the compaction energy that produced by a manual compactor was transferred through a plate that covers the surface of the ballast layer.

Table 1. Specimens

No.	Configuration	Name
1.	Ballast	Sampel I (A and B)
2.	Ballast + 10% Scrap Rubber 3/8‖	Sampel II (A and B)
3.	Ballast + 10% Scrap Rubber 1‖, ½‖, ¾‖, dan No.4	Sampel III (A and B)

IOP Conf. Series: Materials Science and Engineering **650** (2019) 012052 doi:10.1088/1757-899X/650/1/012052

2.3. *Compressive Strength Testing*
At this stage, the sample was weighed then the compressive strength testing was conducted using a Micro-Computer Universal Testing Machine (UTM) tool with a load plate area measuring 30 cm x 15 cm (Figure 1). The results of this test are in the form of stress, strain, and elongation.

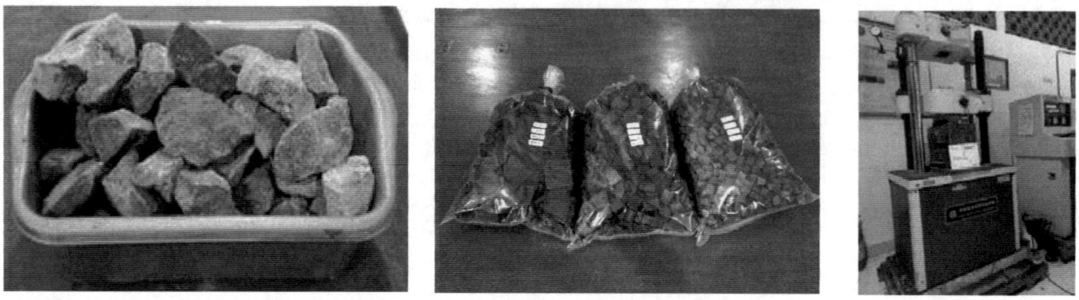

Figure 1. Ballast Aggregate, Scrap Rubber and Compressive Strength Testing Process with Universal Testing Machine

2.4. *Elastic Modulus Examination*
The elastic modulus (E) is obtained based on the compressive strength test data and is processed in the form of a stress-strain relationship curve with the trend-line approach.

2.5. *Vertical Deformation Examination*
Examination of vertical deformation is conducted based on the change of sample height that occurs due to the vertical loading process is given by the UTM Machine. The deformation value shows the level of stiffness and density of the ballast layer.

2.6. *Aggregate Abrasion Examination*
The method to find out the abrasion value of aggregate material in the mixture is based on the wear that occurs due to the compressive strength test process and by comparing the amount of deteriorating aggregate material after the compressive strength test process. The comparison is made by taking 5000 gr of ballast material from the total mixture weight before and after the compressive strength test of each mixture type. Furthermore, the ballast materials from each sample were compared using sieve analysis with the size of 1 inch to No.4. The value of material abrasion is obtained from the percentage of aggregate that has the grain size smaller than 25.4 mm. In another word, the material was able to pass the sieve with the size of a ¾ inch.

3. Result and Discussion

3.1. *Ballast Aggregate Checking Result*
In Table 2, the results of aggregate physical tests are shown. It can be concluded that the ballast met the specifications determined by the Indonesian National Standard (SNI).

Table 2. Aggregate Examination Results

No	Examination	Specification	Result	Unit
1.	Specific Gravity	≥ 2.6	2.63	-
	a. Bulk	≥ 2.6	2.66	-
	b. Dry	≥ 2.6	2.70	-
	c. Apparent	≤ 3.0	0.95	%
2.	Water Absorption	≤ 0.5	1.88	%
3.	Mud Content	≤ 25.0	17.66	%

IOP Conf. Series: Materials Science and Engineering **650** (2019) 012052 doi:10.1088/1757-899X/650/1/012052

3.2. Mixture Characteristics

Each sample has a different mixture characteristic that occurs because of the variety of constituent materials. Identifying the characteristics of the mixture was done to determine the volume of the material and the cavity that fills a ballast box — the more varied the size of scrap rubber mixed in the ballast modification, the smaller the volume of the cavity. This result is caused by the use of scrap rubber which serves to fill the cavities. The amount of cavities in each sample is presented in Table 3.

Table 3. Mixture Characteristics

Parameter	S.I	S.II	S.III
Volume Weight (gr/cm^3)	1.60	1.46	1.43
Scrap Rubber Volume Percentage	-	13.3%	14.5%
Ballast Volume Percentage	62.4%	49.5%	56.2%
Pore Volume Percentage	37.6%	37.2%	29.3%
Total	100%	100%	100%

3.3. Elastic Modulus

Based on the results of the analysis, it is known that the addition of scrap rubber to the ballast mixture can reduce the stiffness of the ballast layer. As presented in Table 4, the highest modulus of elasticity is shown by sample I (ballast without scrap rubber) in the first test (A). Whereas, the modulus of elasticity of ballast mixture with the addition of scrap rubber in samples II and III has a value of elastic modulus that is smaller than sample I. Samples II and III have more elastic properties due to the existence of scrap rubber that able to decrease the stiffness of the ballast layer, thereby reducing the modulus of elasticity. Furthermore, it can be concluded that sample III, ballast with graded-sized scrap rubber, has a lower modulus of elasticity compared to sample II, ballast with uniformly sized scrap rubber. This result is due to the addition of scrap tire rubber with various sizes that can fill the cavities in the ballast layer in sample III. This result is also in line with the results of Farhan et al. (2015), Setiawan and Rosyidi (2018), Signes et al. (2016), and Sol-Sanchez et al. (2014) which show that the addition of elastic material can increase deformation and reduce stiffness [13, 14, 15, 5].

Table 4. Elastic Modulus

No	Sample	Stress (kPa)	Strain (%)	E (MPa)	Changes in E (%)
1	Sampel IA	328.39	2.30	14.28	-10.08%
2	Sampel IB	337.64	2.63	12.84	
3	Sampel IIA	266.24	3.12	8.53	-9.26%
4	Sampel IIB	131.55	1.70	7.74	
5	Sampel IIIA	52.01	1.68	3.10	-37.42%
6	Sampel IIIB	39.25	2.02	1.94	

To find out the change in samples stiffness after given the first maximum loads, then each sample was tested twice, namely A and B. It can be concluded that in the second test (B), each sample experienced a decrease in the level of stiffness because it had received the previous maximum load. As shown in Figure 2 and 3, the reduction in elastic modulus in the second test due to the first maximum loads that occurred in sample I was 10.08% (from A = 14.28 MPa to B = 12.84 MPa), sample II 9.26% (from A = 8.53 MPa to B = 7.74 MPa) and sample III 37.42% (from A = 3.10 MPa to B = 1.94 MPa).

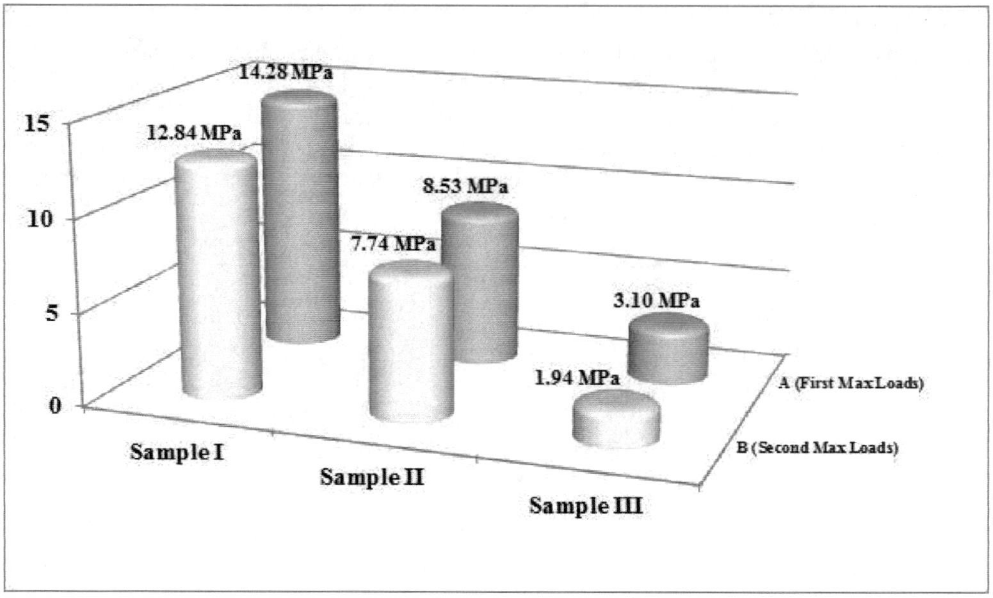

Figure 2. Elastic Modulus in the First and Second Max Loads

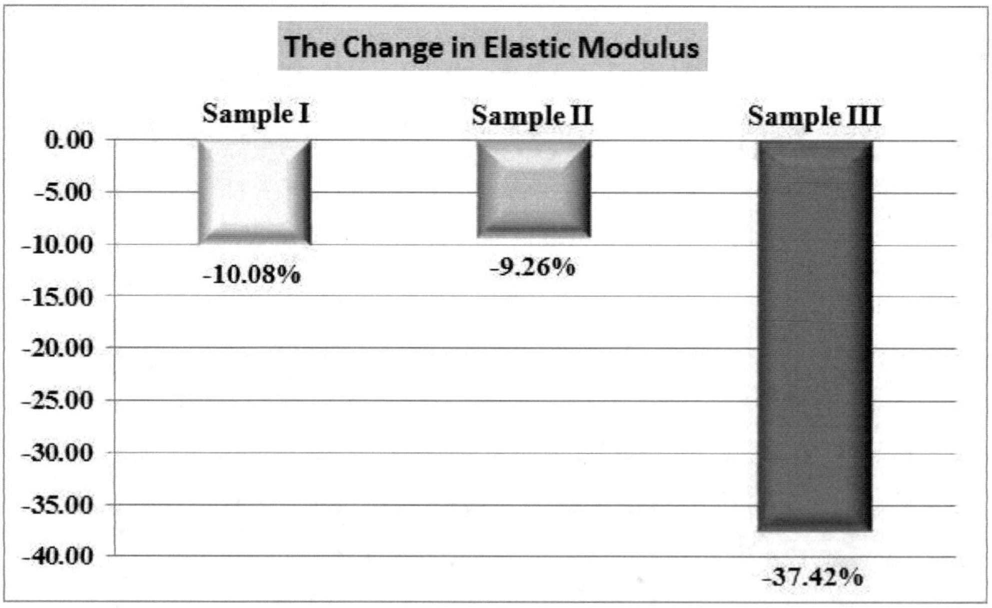

Figure 3. The Change in Elastic Modulus

3.4. Ballast Material Durability

Based on the following Figure 4 to 6, the results of sieve analysis tests in sample II were obtained after the testing which shows the changes in aggregate size distribution patterns. This result is due to the loads given during compaction using manual compactor and also the compressive testing process using UTM tools. The most significant changes in aggregate size distribution patterns occurred in sample 1.

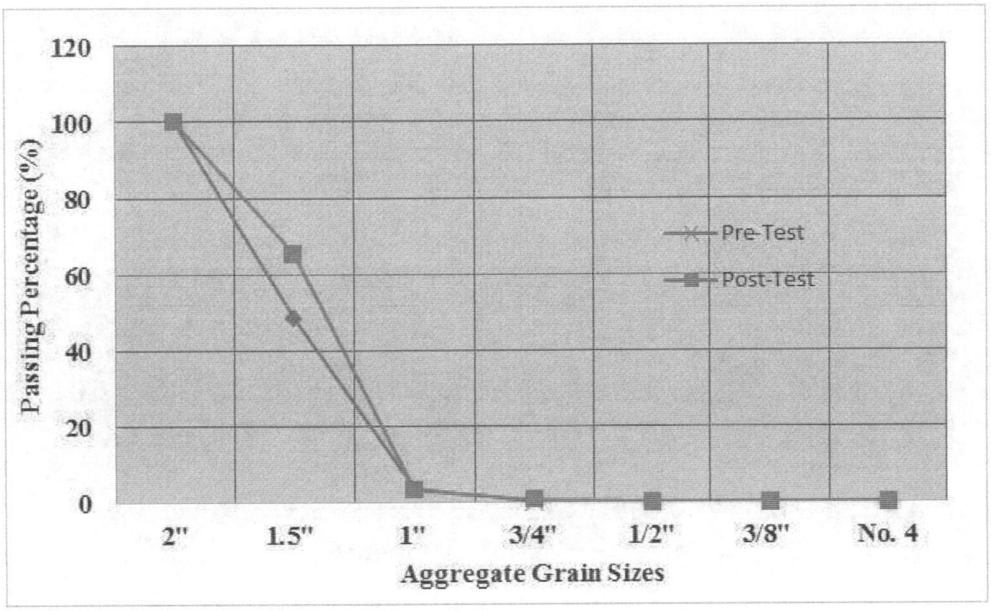

Figure 4. The Change in Aggregate Gradation on Sample I

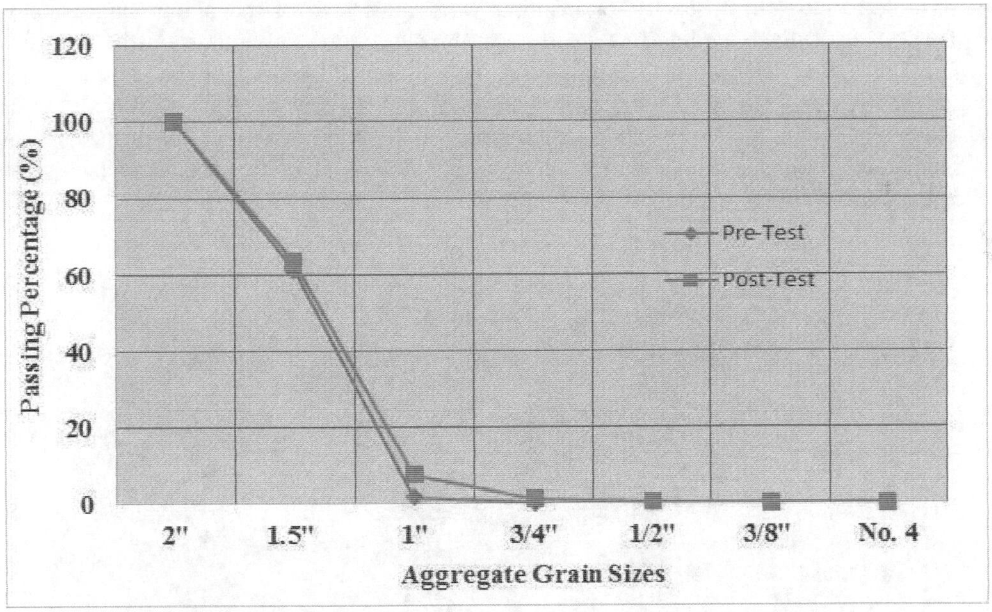

Figure 5. The Change in Aggregate Gradation on Sample II

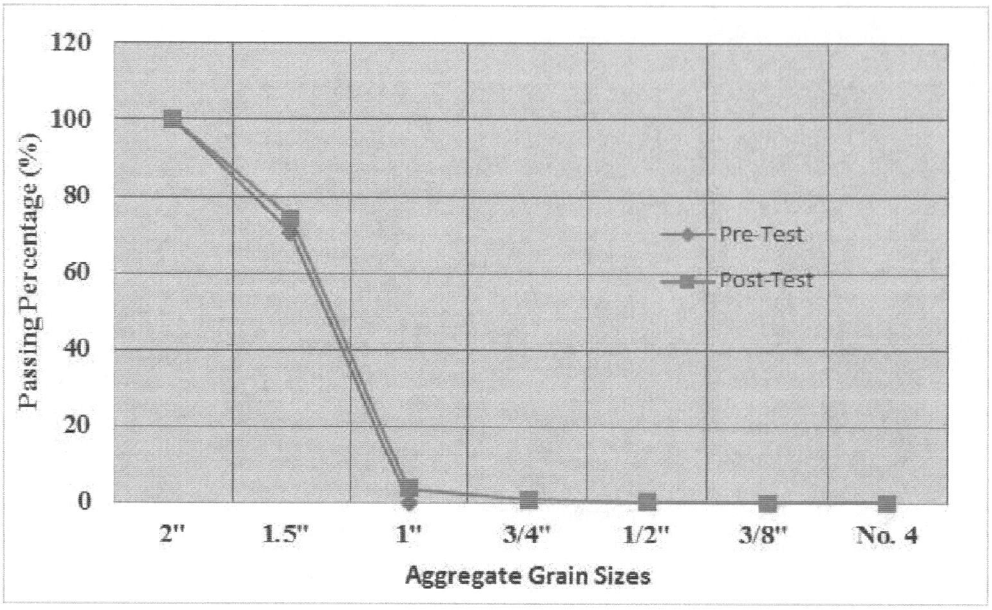

Figure 6. The Change in Aggregate Gradation on Sample III

The deterioration of ballast can be seen from the change in aggregate grain size that becomes smaller and able to pass the filter ¾", or in other words, the size of the ballast aggregate becomes smaller than 25.4 mm. Each sample has a different abrasion value depending on the modification of the mixture. The lower percentage of abrasion in the sample indicates that the sample has a higher degree of durability due to reduced friction or direct contact between the aggregates by the presence of the scrap rubber in the ballast layer. The scrap rubber can fill the cavity or pore in the mixture of the samples so that the modification of scrap rubber in the ballast layer can reduce the level of ballast material deterioration.

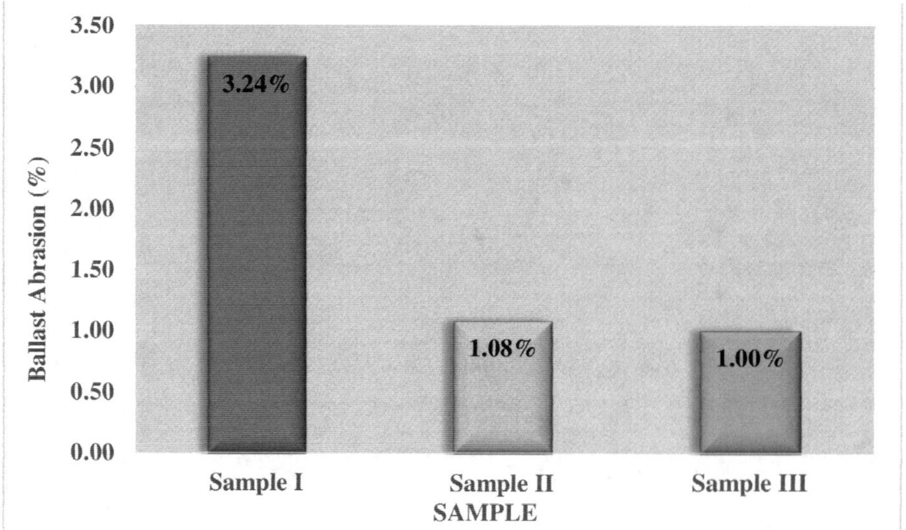

Figure 7. Weight and Percentage of Ballast Abrasion

Based on the information obtained in Figure 7, the highest aggregate abrasion percentage was found in sample I (ballast without scrap rubber) by 3.24% (162.1 grams) while the least was found in sample III (ballast with modified scrap rubber with graded-sized) by 1% (50.1 grams). Although sample II and sample III both used the scrap rubber, however, there were differences in results between these two samples. The abrasion value of sample III was lower than that of sample II due to the presence of graded-sized scrap rubber in the sample III which had a variety of rubber sizes so that it could fill the cavity in the ballast layer better and more evenly.

The addition of scrap rubber on the ballast layer can reduce the damage of ballast material. This is also in line with the research of D'Angelo et al. (2016), D'Angelo et al. (2017), Indraratna et al. (2014), Indraratna et al. (2017), Nimbalkar et al. (2012), and Setiawan and Rosyidi (2018) which state that the elastic materials in the ballast can minimize the movement of the ballast material and reduce the friction between aggregate so that material durability increases and material degradation decreases [16, 17, 18, 19, 20, 14].

3.5. Vertical Deformation

Vertical deformation values obtained based on the relationship between the loads and the change of samples height. Based on Figure 8 below, it can be seen that the height of a sample changes at a particular load. The test results show varying values. Ballast material without scrap rubber exhibits the highest stiffness. Based on the information obtained from Figure 8, at the same vertical deformation value of 5 mm, the sample I (ballast without scrap rubber) showed the highest level of stiffness because it was able to retain the loads with a value of 1980.9 Kg. Whereas sample III, ballast with additional graded-sized scrap rubber shows the lowest level of stiffness because it is only able to withstand the load of 443.7 Kg at 5 mm vertical deformation. Elastic material in the form of scrap rubber added to the ballast mixture causes an increase in the elastic characteristics of the mixture. Therefore, to achieve the same vertical deformation value, the load that can be retained by the ballast layer with scrap rubber mixture is smaller than the ballast without scrap rubber mixture. The elastic properties possessed by rubber cause a change in samples height when given the loads as in samples II and III. This statement is supported by the research of Setiawan and Rosyidi (2018) and Signes et al. (2016), where it is stated that the use of scrap rubber can increase permanent deformation in the ballast and sub-ballast layers [14, 15].

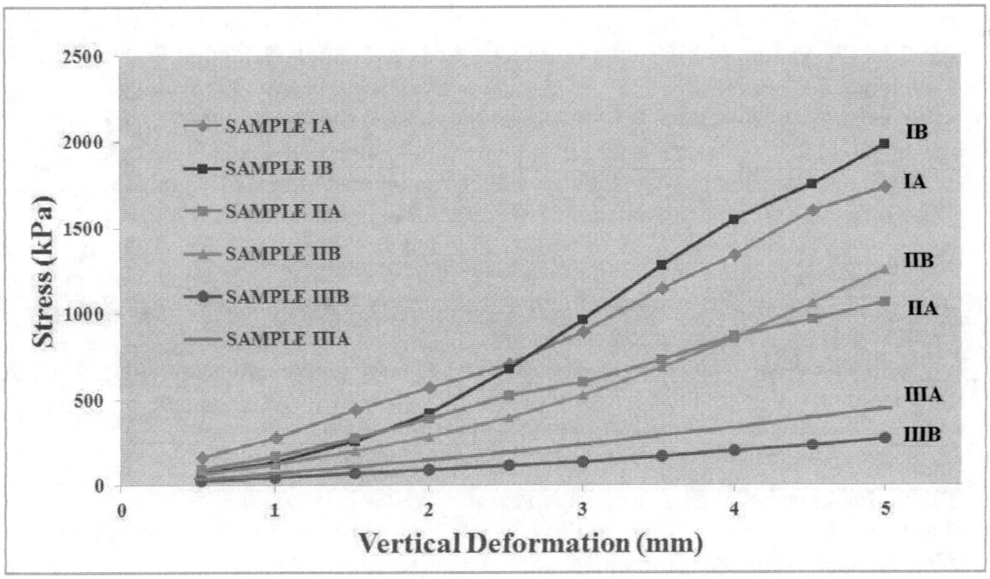

Figure 8. Vertical Deformation and Loads

4. Conclusions

From the results of the research conducted, the following results are summarized:

a. Ballast without scrap rubber mixture has a higher value of elastic modulus compared to ballast with scrap rubber mixture.
b. The addition of scrap rubber with uniformly sized on the ballast layer could produce greater elastic modulus compared to the addition of scrap rubber with graded-sized.
c. The decreasing on samples modulus of elasticity due to previous loading in ballast without scrap rubber (sample I) is 12.69%, ballast with 10% uniformly sized scrap rubber (sample II) is 6.23%, and ballast with 10% graded-sized scrap rubber (sample III) is 40.20%.
d. Ballast with scrap rubber mixture has a higher value of durability compared to ballast without scrap rubber mixture.
e. The addition of scrap rubber with graded-sized on ballast layer could produce higher ballast durability compared to the addition of scrap rubber with uniformly sized.
f. The ballast abrasion percentage in sample I is 3.24% (162.1 gram), sample II is 1.08% (54.2 grams), and sample III is 1% (50.1 grams).
g. At the same vertical deformation value of 5 mm, the sample I has the highest level of stiffness because it can withstand the loads up to a 1980.9 Kg. While sample III shows the lowest level of stiffness because it is only able to retain loads of 443.7 Kg.
h. Modification of the ballast layer by adding scrap rubber or elastic material can reduce the stiffness of the ballast layer which is characterized by a decrease in elastic modulus and an increase in the value of vertical deformation. But on the other hand, modification of ballast layers with scrap rubber can increase ballast durability because it can reduce the possibility of friction between aggregate particles so that it can reduce the ballast material damage.

Acknowledgements

The authors would like to thank Lembaga Penelitian, Publikasi & Pengabdian Masyarakat Universitas Muhammadiyah Yogyakarta (LP3M UMY) for the Research Funds in the scheme of Multi-Discipline 2018. The authors also would like to thank Robby Rahman and Nikmatusholihah for their contribution to this research.

References

[1] Setiawan D M, Muthohar I and Ghataora G 2013 Conventional and Unconventional Railway for Rail Ways on Soft Ground in Indonesia (Case Study: Rantau Prapat-Duri Railways Development) *Proceeding of The 16th FSTPT International Symposium, Universitas Muhammadiyah Surakarta, 1-3 November 2013* 610-620.

[2] Setiawan, D M 2016 Pembatasan Kecepatan Maksimum dan Kaitannya Terhadap Kapasitas Lintas Jalur Kereta Api Muara Enim–Lahat Sumatera Selatan *Prosiding Seminar Nasional Teknik Sipil ke-VI 2016. Surakarta, 25 Mei 2016* 36-46. ISSN: 2459-9727.

[3] Setiawan, D M, and Rosyidi, S A P ―Track Quality Index as Track Quality Assessment Indicator‖ *Proceeding of The 19th International Symposium of Indonesian Inter-University Transportation Studies Forum (FSTPT), Universitas Islam Indonesia, Indonesia, October 2016.* Ch. 2, pp. 197-207, ISBN: 979-95721-2-19.

[4] Sol-Sanchez M, Thom N H, Navaro F M, Gomez M C R, and Airey G D 2015 A Study into The Use of Crumb Rubber in Railway *Ballast Construction and Building Materials* **75** 19 - 24.

[5] Sol-Sanchez M, Navaro F M, and Gomez M C R 2014 The Use of Deconstructed Tires as Elastic Elements in Railway Tracks *Materials* **7** 5903-5919.

[6] Hameed A S, and Shashikala A P 2016 Suitanability of Rubber Concrete for Railway Sleepers *Perspectives in Science* **8** 32-35.

[7] Zakeri J A, and Mosayebi S A 2016 Study of Ballast Layer Stiffness in Railway Tracks *Gradevinar* **68(4)** 311-318.

[8] Navaratnarajah S K and Indraratna B 2017 Use of Rubber Mats to Improve The Deformation and Degradation Behavior of Rail Ballast Under Cyclic Loading *Journal of Geotechnical and Geoenvironmental Engineering* **143(6)** 1943-5606.

[9] BSN 1990 *SNI 03-1968-1990 Metode Pengujian Tentang Analisis Saringan Agregat Halus dan Kasar* (Jakarta: Badan Standardisasi Nasional)

[10] BSN 1996 *SNI 03-4141-1996 Metode Pengujian Gumpalan Lempung dan Butir-butir Mudah Pecah dalam Agregat* (Jakarta: Badan Standardisasi Nasional)

[11] BSN 2008a *SNI 1969-2008 Cara Uji Berat Jenis dan Penyerapan Agregat Kasar* (Jakarta: Badan Standardisasi Nasional)

[12] BSN 2008b *SNI 2417:2008 Cara Keausan Agregat dengan Mesin Abrasi Los Angeles* (Jakarta: Badan Standardisasi Nasional)

[13] Farhan A H, Dawson A R, Thom N H, Adam S, and Smith M J 2015 Flexural Characteristics of Rubberized Cement-Stabilized Crushed Aggregate for Pavement Structure *Materials and Design* **88** 897-905.

[14] Setiawan, D M, and Rosyidi S A P 2018 Vertical Deformation and Ballast Abrasion Characteristics of Asphalt-Scrap Rubber Track Bed *International Journal on Advanced Science, Engineering and Information Technology* **8(6)** pp. 2479-2484.

[15] Signes C H, Hernandez P M, Roca J G, de la Torre M E, and Franco R I 2016 An Evaluation of the Resilient Modulus and Permanent Deformation of Unbound Mixture of Granular Materials and Rubber Particles from Scrap Tyres to Be Used in Subbalast Layers *Transportation Research Procedia* **18** 384 – 391.

[16] D'Angelo G, Thom N, and Presti D L 2016 Bitumen Stabilized Ballast: A Potential Solution for Railway Track-Bed *Construction and Building Materials* **124** 118-126.

[17] D'Angelo G, Thom N, and Presti D L 2017 Optimisation of Bitumen Emulsion Properties for Ballast *Materiales De Construction* **67** 124-133.

[18] Indraratna B, Nimbalkar S, Navaratnarajah S K, Rujikiatkamjorn C, and Neville T 2014 Use of Shock Mats for Mitigating Degradation of Railroad Ballast *Sri Lankan Geotechnical Journal - Special Issue on Ground Improvement* **6(1)** 32-41.

[19] Indraratna B, Ngo N T, and Rujikiatkamjorn C 2017 Improved Performance of Ballasted Rail Tracks Using Plastics and Rubber Inclusions *Proceeding in Transportation Geotechnics and Geoecology, Saint Petersburg, Rusia, 17-19 Mei 2017* 207-214.

[20] Nimbalkar S, Indraratna B, Dash S K, and Christie D 2012 Improved Performance of Railway Ballast Under Impact Loads Using Shock Mats *Journal of Geotechnical and Geoenvironmental Engineering* **138(3)** 281-294.

IOP Conf. Series: Materials Science and Engineering **650** (2019) 012053 doi:10.1088/1757-899X/650/1/012053

Model of queuing in the railway level crossing (case study: Imam Bonjol railway level crossing in Blitar)

H Widyastuti[1*], A Utami[2], Z M Dzulfiqar[3]

[1]Department of Civil Engineering, Institut Teknologi Sepuluh Nopember, Surabaya 60111, East Java, Indonesia
[2]Department of Civil Engineering, Pertamina University, South Jakarta 12220, DKI Jakarta, Indonesia
[3]PT. Hutama Marga Waskita, East Jakarta 13840, Jakarta, Indonesia

E-mail: *hera.widyastuti@yahoo.co.uk

Abstract. A high number of accidents in railway level crossing causes loss of life and material. Furthermore, the losses are also experienced by users of traffic on the highway such as disturbance in the form of delay which can lead to increase the length of travel time where the vehicle will stop, and give impact to the emergency queue vehicle at railway level crossing. Considering those problems, there has been build two models which are: the first model is the queuing of traffic length and with considering the length and speed of the train, and the second model is the queuing of traffic with considering the gate closing time. The methodology applied was an observation on field and survey data and then calculation from Road Capacity Guidelines of Indonesia (PKJI 2014) and queue length calculation using the queuing analysis method. Data have been collected and analysed by making the model relationship using linear regression method. The result showed that there is a positive relationship between the train lengths with the duration of gate closing time. However, the train speed has a negative relationship. Moreover, the length of the train has a positive relationship with the length of the vehicle queue.

1. Introduction

Railway networks offer significant advantages for the transport of passengers and freight as they occupy 2-3 times less land per passenger than other modes of road transport [1]. In addition, the railway infrastructure can also be used to reduce traffic congestion so that society can move efficiently around the city and to connect the port area and the industry to the hinterland [2]. The increasing of vehicle volume and the rapid growth of population have affected the increasing of travel intensity resulting in an increased of direct-access to the city center [3]. On the other hand, the rapid growth of the population also raises the demand for access through large-capacity transportation modes through the railroad [4]. Those conditions give consequence to arise of new railway level crossings.

Railway level crossings create serious potential conflict points for collisions between road vehicles and trains. Safety at level crossings is a worldwide issue, which increasingly attracts the attention of relevant transport authorities, the rail industry and public [5]. Blitar is one of the cities in East Java where there are 18 railway level crossings both with and without barriers. Along with the

IOP Conf. Series: Materials Science and Engineering **650** (2019) 012053 doi:10.1088/1757-899X/650/1/012053

existence of railway level crossings in Blitar, the transportation office of Blitar plans to conduct an evaluation at the selected railway level crossings. Imam Bonjol railway level is selected to be experimental object due to high number of vehicles crossed this road. The high number of traffic is directly proportional to the risk of accidents at a railway level crossing. The closure of the railway level crossing on this road causes vehicle activities to be blocked and cause congestion. One of the congestion factors at the railway level crossings is the duration of closing time as the train passes through the crossing [6]. The uncertain duration of gate closing time cause road users to be undisciplined by break through the crossing gates while the train pass through the crossings [3]. Considering of those problem, it is necessary to make a model of relationship between closing of railway level crossing, length and speed of train and length of vehicle queue of in Imam Bonjol railway Level Crossings, Blitar.

2. Literature Review

2.1. Crossing/Intersection

Crossroads (intersection) are two or more sections of a road that meet each other, intersect or cross. The crossing between the road and the railroad is also called the intersection (crossing). The railway crossing is divided into two types. First type is Railway level crossing. According to the regulations of the director general of land transportation number: SK.770 / KA.401 / DRJD / 2005, Railway level crossing is the direct intersection between railway and road. The second type is Railway crossing, according to Road Traffic Planning Guidance with Railway Lane (No: 008 / PW / 2004), Railway crossing is crossing between railway and road in different level.

2.2. Non Signalized Intersections

Non Signalized intersections is an intersection without a Traffic Signal Equipment as a traffic controller. MKJI'97 called the intersection as non-signalized intersection. The non-signalized intersection is one of the most complex locations in a traffic system. Non-signalized intersection analysis must consider a wide variety of prevailing conditions [7].

2.3. Traffic Management

2.3.1. Road Capacity

The calculation of road capacity (C) is to determine the performance of a highway segment to accommodate the number of vehicles per hour (vehicle / hour) on the existing road [8]. Based on the Indonesian Road Capacity Guidelines, to calculate Road Capacity can use the formula [7]:

$$C = C0 \times FCLJ \times FCPA \times FCHS \times FCUK \tag{1}$$

where:

C : road capacity (vehicle / hour)
C0 : base capacity (vehicle / hour)
FCLJ : capacity factors related to lane width
FCPA : capacity factor related to separation of directions FCHS: related capacity factors Side barriers
FCUK : capacity factor related to City size

IOP Conf. Series: Materials Science and Engineering **650** (2019) 012053 doi:10.1088/1757-899X/650/1/012053

2.3.2. Degree of Saturation

Degree of Saturation is a main parameter used to determine the performance level of service in a road segment. Value of Degree of Saturation (DS) are vary between 0 to 1.According to the Indonesian Road Capacity Guidelines, the degree of saturation can be calculated using the formula [7] :

$$DS=Q/C \tag{2}$$

where:

DS : Value of degree of saturation
Q : Traffic volume (vehicle/ Hour)
C : Existing road capacity (vehicle/ Hour)

2.3.3. Analysis of Queue Length

At this stage the queue length will be calculated using Queuing analysis method, which is a derivative formulation of the shocked wave formula. In the calculation of the vehicle queue length some parameters will be obtained, those are: TQ (duration of the queue time), QM (vehicle queue), and queue length. The formula that will be used is [12]:

$$TQ = \frac{\mu.r}{\mu - \lambda} \tag{3}$$

$$QM = \frac{\lambda_r}{3600} \tag{4}$$

2.3.4. Regression

Regression analysis technique is a technique that can be used to generate relations in numerical form and to generate relations in numerical form and to determine how the variables are interrelated [11]. In a simple linear regression analysis the variables used are expressed in the general form:

$$Y = a+bx \tag{5}$$

3. Method

The data were collected using two approaches, those are: field video recordings at selected railway level crossings and direct traffic counting.

3.1. Experimental Objective

Railway Level Crossings safety is one of the most critical issues for railways. The uncertain duration of gate closing time cause road users to be undisciplined by break through the crossing gates while the train pass through the crossings. One of the congestion factors at the railway level crossings is the duration of closing time as the train passes through the crossing. For all these reasons, Imam Bonjol railway level crossing were selected for our field measurement campaign. For location detailed, can be shown in Figure 3.1.

IOP Conf. Series: Materials Science and Engineering **650** (2019) 012053 doi:10.1088/1757-899X/650/1/012053

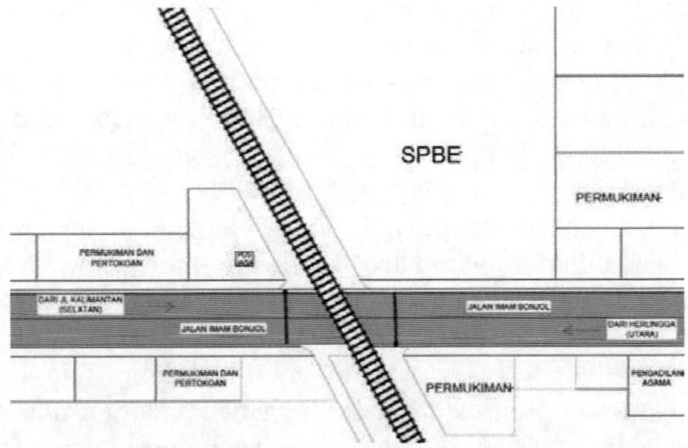

Figure 1. Location of Imam Bonjol Railway Level Crossing

3.2. Data Collecting

The data that will be used in this research is primary and secondary data. The secondary data to support the investigation come from dedicated database provided by PT. Kereta Api Indonesia, while primary data will be obtained by doing direct survey at Imam Bonjol railway level crossing. Table 1 shown types and function of data collected.

Table 1. Types and Function of Data Collected

Data Types	Function
Train Schedule (Secondary Data)	To determine the type of the train which pass through the railway level crossing that affected the closing of gate crossing
Geometric Data (Primary Data)	To determine the road performance in the railway level crossings
Length and Train Velocity (Primary Data)	To determine length and velocity of the train that affected the duration of gate of railway level crossing closing time
Closing time of Railway Level Crossing (Primary Data)	To determine the duration of gate closing as the train pass through the railway level crossing.
Traffic Data (Primary Data)	To determine the vehicle queue when the of gate of railway level crossing are closed

3.3. Data Analysis

In the analysis phase, an analysis of traffic data will be carried out before and after the train passes the railway level crossing. Then continued by analyse the capacity (C) and degree of saturation (DS) before and after the train passes. The length of gate closing time were recorded and continued by calculate the vehicle queue length by using shocked wave method [7]

When the railway level crossing closed, there will be vehicle delay which cause a queue [9]. So in the implementation of traffic analysis, the length and queue behavior are calculated when crossing begins to close until the crossing returns to open and the traffic flow returns to normal. The analysis used in queuing analysis is shocked wave method. In addition, the traffic analysis also calculated the road capacity by using a reference from the Indonesian Road Capacity Guide or PKJI 2014.

3.4. Data Analysis and Modelling

After the data are collected, data processing will be conducted and divided into two models. First model is relationship between the length of the train and the closing time of the railway level crossing, and the second model is relationship between the closing time of the railway level crossing and the length of the vehicle queue.

Multiple linear regression analysis is used to get the relationship of more than one variable while simple linear regression is used to get a mathematical relationship in the form of an equation between a single independent variable with a single independent variable [10]. Simple linear regression has only one variable associated with one non-independent variable, with the form $y = ax + b$ [11]. Formulation of model relationship between train length, train velocity and closing time of the railway level crossing are conducted by using Linear Regression with SPSS as a supporting software. Furthermore, to formulate the relationship model between the duration of closing Time and the Queue Lines using simple linear Regression.

3.5. Conclusion and Suggestion

Conclusion is based on the results obtained at the analysis stage. This research is expected to be a matter of consideration in designing early warning developments at railroad level crossings. Furthermore, this research is expected to be a reference material in conducting further studies and research related to early warning.

4. Analysis

4.1. Traffic Analysis

After amount of vehicle could be determined, calculation of light vehicle will be done, and so does the vehicle ratio, until the capacity of each direction and the saturation degree are found. From that analysis, the length of the vehicle queue also can be determined. The following Table 2 will show the capacity and DS at the Imam Bonjol railway level crossing.

Table 2. Capacity and Degree of Saturation

Data Types	Imam Bonjol Railway Level Crossing	
Time	06:15-07:15 (Weekday)	06:30-07:30 (Weekend)
Peak Hour (vehicle/hour)	1611	1576
Capacity (vehicle/hour)	2321	2321
DS	0.69	0.68

IOP Conf. Series: Materials Science and Engineering **650** (2019) 012053 doi:10.1088/1757-899X/650/1/012053

4.2. Traffic Analysis when Train Passing through Railway Level Crossing

Traffic analysis when the train passing through railway level crossing is about the same with the preceding traffic analysis, which is using PKJI 2014. But, this sub-chapter will show the result of data analysis calculation and the vehicle queue length when the series of train passing through the railway level crossing. SKR conversion on Imam Bonjol Railway Level Crossing are mentioned in Table 3.

Table 3. SKR Conversion Imam Bonjol

Road	Direction	7:27 Skr /c	cycle time	Skr/hr
	S	32	120	960
	U	37	120	1110
Imam Bonjol	Direction	10:22 Skr /c	cycle/time	Skr /hr
	S	21	81	933
	U	30	81	1333

4.3. Queuing Analysis

At this stage, it will calculate the length of the queue using the Queuing analysis method. In the calculation of the queue using queuing analysis will be obtained TQ (duration of the queue time), QM (vehicle queue), and queue length.

Example of Queue length calculation:

On Jalan Imam Bonjol Utara, March 14, 2018, at 7:27:

μ = 2321 vehicle/hour

r = 120 second

λ = 1110 vehicle/hour

So, the duration of the queue:

$$TQ = \frac{2321 \, x \, 120}{2321-1110} = 229.99 \text{ second}$$

The length of the vehicle queue in light vehicle units:

$$QM = \frac{\lambda r}{3600}$$

$$= \frac{1110 \, x \, 120}{3600} = 37 \text{ skr}$$

Where units of light vehicles are assumed to be private vehicles with a vehicle length of 3 meters so that the queue length =

QM = 37 x 3 = 111 meter

4.4. Comparative Analysis between Real Field Queue Length and Queuing Analysis Calculation

At this stage in measuring the queue length of the vehicle using 2 methods, namely manual or field survey using meter and recorded by surveyors and the second using queuing analysis. Then compared the results of the length of the field survey queue with the calculation of queuing analysis using the ANOVA method, namely:

Hypothesis test

Ho: The average result of the length of the field survey queue by calculating the formula is all the same.

H1: The average result of the length of the field survey queue with the calculation of the formula is not the same.

$HH_0 : AA_1 = AA_2$
$HH_1 : AA_1 \neq AA_2$
Comparing with the significant, alfa (α)= 0,05
if $F_{count} \geq F_{Table}$ } So, rejected HH_0, accepted H_1
P-value < α

From the results of the ANOVA above, the results of F count and also P-value are obtained. Both parameters will be used to determine which hypothesis will be accepted and will be rejected. F recapitulation and P value of all processes can be seen in Table 4.

Table 4. Recapitulation of F-count and P-value

Direction	Conclusion					Explanation
North	*Fcount*	0.184	<	F Table	4.15	Ho accepted, H1 rejected
	P-Value	0.671	>	*α*	0.05	
South	*Fcount*	0.148	<	F Table	4.15	Ho accepted, H1 rejected
	P-Value	0.703	>	*α*	0.05	

4.5. Model Equations

In this study a relationship model will be created using multiple linear regression methods. The relationship model that is made is between the closing time of the railway level crossing with the length of the train circuit and the speed of the train, as well as the model of the relationship between the length of the vehicle queue and the length of the queue formula and the length of the field survey queue. From the analysis results using the SPSS auxiliary program, the results in Table 5 are obtained:

Table 5. Model and significant equation output of Imam Bonjol

	Coefficients	Standard Error	t Stat	P-value
Intercept	-6.64911	16.12293	-0.4124	0.685879
Tram length	5.649409	1.333239	4.237357	0.000717
Velocity (Km/hr)	-0.867659	0.369582	2.347677	0.033027

From the results above, the model equation is obtained on the Imam Bonjol railway level crossing:

$Y = 5.649X_1 - 0.868 X_2 - 6.649$
Y = Length of railway level crossings gate closing
time X_1 = Train length
X_2 = Train speed

The meaning of the equation is that each increase in one train circuit increases the length of closing time by 5.649 seconds, and the meaning of 0.868 means that the increasing speed of 1 km/h will reducing 0.868 seconds the length of the railway level crossing gate closing time.

From the results of data analysis in Table 5 also obtained a significant test of data with the results of the P-value circuit length of 0.000717 smaller than alpha where the alpha used is 0.05 so it can be stated that the number of train circuits has a significant relationship to the length of railway level crossing gate closing time. While the train circuit speed with P-value 0.0133027 is smaller than alpha, it means that the train speed has a significant relationship to the length of railway level crossing gate closing time. The second model is the relationship model between the length of railway level crossing gate closing time and the length of the vehicle queue using field survey data between the length of the closing time and the length of the vehicle queue and also the calculation data formula queuing analysis queue length. In the equation used, x is the length of the closing time and y is the length of the vehicle queue. The linear regression results in the relationship between the length of the closure and the queue length of the formula for Imam Bonjol Railway Level Crossing can be seen in Figure 2.

The queue length on Imam Bonjol Railway Level Crossing needs to be reviewed because Imam Bonjol Street is a road that is quite heavy in traffic. With hourly traffic flow during peak hours it reaches 1611 vehicle / hour. So that when the train passes the crossing there will be a queue which results in a delay in the vehicle. The model of the vehicle queue length and railway level crossing gate closing time on weekday can be seen in Figure 2 and weekend condition can be seen in Figure 3.

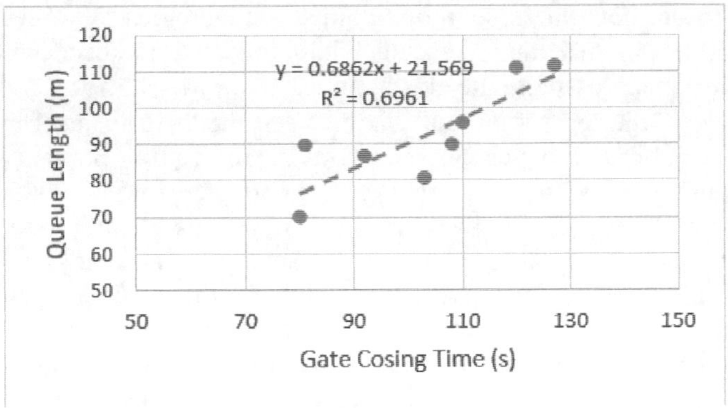

Figure 2. Regression chart Linear relationship Duration of gate closing time and the length of the queue (Weekday)

The model equation formed is y = 0.6862x + 21,569 with the value of R-Square is 0.6961. This equation implies the increasing of 1 second railway level crossing gate closing time, it will added 0.686 meters of vehicle queue length.

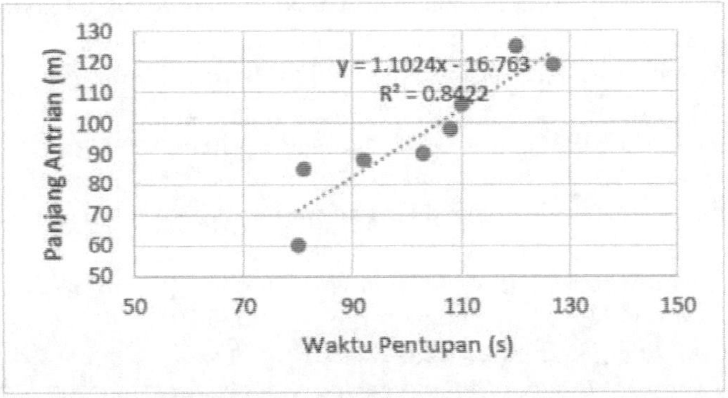

Figure 3. Regression chart Linear relationship Duration of gate closing time and the length of the queue (Weekend)

The model equation formed is y = 1.1024x - 16.76 with the R-Square value is 0.842. Y is vehicle queue length and x express the duration of railway level crossing gate closing time. This equation implies the increasing of 1 second railway level crossing gate closing time, it will added 0.686 meters of vehicle queue length.

5. Conclusion

This research examines that railway level crossings create serious potential conflict points for collisions between road vehicles and trains. The uncertain duration of gate closing time cause road users to be undisciplined by break through the crossing gates while the train pass through the crossings. Some conclusions obtained from some of the results made in this study are as follows.

- The results of the regression analysis of the relationship between the length of the train (number of trains / carriages) and the duration of gate closing time has a significant effect
- The results of the regression model of the relationship between the duration of gate closing time with considering the length and the speed of the train on Imam Bonjol Street have a positive influence on the relationship model that is on Imam Bonjol Street $y = 5.649X1 - 0.868\ X2 - 6.649$. With y is the duration of gate closing time, x1 is the train length and x2 is the train speed. Each increase in one train series increases 5.649 seconds of gate closing time, and -0.868 means that the increasing 1 km/hour of train speed will reduce 0.868 seconds of the railway level crossings gate closing time.
- The results of the regression analysis between the railway level crossings gate closing time with the length of the vehicle queue in Imam Bonjol Street (weekday) have a positive relationship. The equation model is $y = 0.6862x + 21{,}569$. This equation implies that the increasing of 1 second of railway level crossings gate closing time, it will add 0.6862 meters of vehicle queue length. The results of the regression analysis between the railway level crossings gate closing time with the length of the vehicle queue in Imam Bonjol Street (weekend) is $y = 1.1024x - 16.76$. This equation implies increasing 1 second of railway level crossings gate closing time, it will add 1.1024 meters of vehicle queue length.

References

[1] Lingamanaik S N, Thompson C, Nadarajah N, Ravitharan R, and Widyastuti H 2017 Using instrumented revenue vehicles to inspect track integrity and rolling stock performance in a passenger network during peak times *Procedia Eng* **188** pp 424–431

[2] Widyastuti H, Herijanto W, Kartika A A G, and Rahardjo B 2016 Railway infrastructure to support inter-modal transportation from port to hinterland (case Study-Manyar port) *Procedia - Soc. Behav. Sci.* **227** pp 181–185

[3] Utami A and Widyastuti H 2019 Model Panjang Antrian Kendaraan pada Perlintasan Sebidang Tanpa Palang Pintu (Studi Kasus: Perlintasan Sebidang Jl . Gayung Kebonsari Surabaya) *Jurnal Aplikasi Teknik Sipil* **17** no 23 pp 27–34

[4] Palar B, Timboeleng J A, and Rompis S Y R 2016 *Analisa Gelombang Kejut pada Persimpangan* **4** no. 9 pp 559–566

[5] Tey L, Ferreira L, and Wallace A 2011 Measuring driver responses at railway level crossings *Accid. Anal. Prev.* **43** no. 6 pp 2134–2141

[6] Gruyter G and Currie 2016 Rail Road Crossing Impact : An International Synthesis *Transp. Rev*. **36** pp 793–815

[7] Directorate General Bina Marga *1997 Highway Capacity Manual Project (Hcm)* **1** no. 1 pp 564

[8] Koloway B S 2009 Kinerja Ruas Jalan Perkotaan Jalan Prof Dr Satrio, DKI Jakarta *J. Perenc. Wil. dan Kota* **20** no 3 p hlm 215-230

[9] Irawati 2001 *Pengaruh Derajat Kejenuhan Jalan Terhadap Tundaan Dan Panjang Antrian Pada Pintu Lintasan Kereta Api Dengan Variasi Lama Penutupan Jalan Studi Kasus Pintu Lintasan 1, Jalan Pecindilan Surabaya* (Surabaya: Institut Teknologi Sepuluh Nopember)

[10] Supranto J 2000 *Statistik teori dan aplikasi jilid 1*

[11] Tamin O Z 2003 *Perencanaan dan Pemodelan Transportasi Edisi Kedua*

[12] Adolf D M 1990 *Traffic Flow Fundamentals* (Amerika: Prentice-Hall.Inc)

Evaluation of Technical Specifications of Gradation Course Base Subdrain Layer Structure on Road

E O Hutasoit[1], C A Prastyanto [1], U Lasminto [1]

[1]Department of Civil Engineering, Institut Teknologi Sepuluh Nopember Surabaya 60111 Indonesia

* caturarifp@gmail.com ; umboro.lasminto@gmail.com

Abstract. Subdrain layer is a type of pavement that can reduce excess water from rainwater so that it can seep into the ground. The purpose of this research is to find the relationship between the optimal water velocity to pass through the base layer and maximum carrying capacity. Sieve analysis was carried out to obtain aggregate gradation variations by modifying the Bina Marga Standards 2010. There are 5 different types of gradations used based on the reduction in the percentage of filler specifications of 8%, 5%, 4%, 3%, and 0%. Calculation of fillers is used to find out the weight of aggregate each sieve size. In addition, the compaction test was carried out on ASTM D Modified 1883 - 87 to obtain optimum water content by using from the maximum dry density curve. The results of compaction the maximum dry density varies with the range 1,911-1,981 gr/cm3. The results of the maximum dry density show that the optimum water content varies between 9-10%. Next, the test sample is reused for CBR testing which aims to determine the carrying capacity of the test sample. Results tend to decrease when the filler is reduced. Obtained CBR test results is in the range of 62-74%. The sample is reused for the infiltration test to find the speed of the water capable of passing the test sample. Based on the results of the infiltration test, the constant infiltration rate varies from 3.143 to 6.750 mm/minute. The constant speed tends to increase when the percentage of filler decreases.

Keywords: Gradation, CBR, Filler, Subdrain Layer, Compaction, Infiltration Rate

1. Introduction

Road damage is a common problem that often occurs in the rainy season. Indonesia has high rainfall so rainwater cannot flow properly because it is not supported by a good pavement structure. This causes puddles that cannot be avoided so that over time the road becomes damaged. Puddles that are above the pavement layer provide an opportunity for water to enter into the pavement layer. Water is known to be one of the causes of damage where water easily dissolves and break down materials between layers of material. Aggregate bond becomes loose due to water damage and cracks arise when the vehicle passes repeatedly [1].

Additionally, the road experiences further a decrease in structural function with increasing age and the number of heavy tonnage vehicles that cross the road [2]. Roads are damaged in a relatively short period of time from either newly built or repaired. Often road damage in Indonesia is unable to reach the age of the plan, although various efforts have been made (design, planning and maintenance) to overcome the damage to the road.[3] Based on this review, the effect of water on pavement damage that often occurs every year, encourages various stakeholders to carry out studies related to the characteristics and movement of water on the pavement structure [4].

Puddles that occur on the road are the main cause of damage to roads [5]. This puddle is created from rainwater that cannot seep into the pavement structure. Due to the fact that it is still not optimal to overcome the puddle. One of the most effective ways is through a translucent pavement or can be of overcoming called a subdrain layer [6]. The subdrain layer is a type of pavement that allows water to pass through the pavement layer allowing the water to seep directly into the soil [7]. The subdrainage layer systems becomes a good solution for water damage road. T pavement use drain moisture materials.

The purpose of this study is to find the most optimal gradation to be used as a foundation layer using Class A Aggregates. The gradation uses was the 2010 General Specifications gradation. The results shows that 4% filler gradation was the most optimum gradation to be used as subdrain layer with CBR value 67 % and has a constant infiltration rate of 3,667 mm/ minutes. The material used is material derived from the quarry of Watukosek Village, Ngoro District, East Java which a class A aggregate materials. Based on the results of this study, the most optimal gradation to be used as a subdrain layer is a 4% filler variety with a CBR value of 67%. The chosen filler variation is a gradation that can produce a high carrying capacity and infiltration rate allowing rainwater to penetrate into the ground or not be waterproof. In this research, sieving analysis will be carried out to obtain variations of grade aggregates according to the General Specifications of div. 5 Grained Bina Marga Class A. Base material must fulfil specification requirements or perform better than other construction materials. Specification of Roads published by Bina Marga 2010 is widely used for construction purposes in Indonesia.

The design will be used to determine the time and speed of water that can flow with new gradations on the foundation layer. It is then adapted to the bearing capacity of Bina Marga's road specification to achieve effective and durable pavement system. It is important to create material layers that allow for water to quickly and efficiently flow without moving particle materials in order to prevent premature pavement damage caused by puddle water.

2. Literature Review

Previous studies relating to scope drainage layer, subsurface drainage layer, infiltration rate, and performance of drainage layer include: thickness, type materials, aggregate gradation, and pavement camber. These studies focuses on design considerations of the base as a drainage layer which results in high strength and easily penetrable pavement structure.

Cardegen published a design guide for subdrainage pavements that has been widely used by engineers as the foundation for permeable pavement design since 1972. Cardegen's design considers the need for a pavement subsurface drainage system. Another research by Raymond and Mirian (1996) shows that subdrainage layer can be used as pervious pavement and geotextile can replace the subdrainage layer [8]. A subdrainage layer typically contains more gravel and presents a thicker layer. It has a strong infiltration capacity.

The subdrainage layer produce a positive pavement performance, especially in its capability as drainage pavement. Several guidelines use the Cedergen model as a requirement for subdrainage layer. FHWA has given a limit and time duration for puddles enter to the subsoil. According to the Federal Highway Administration (FHWA) model, the maximum time to drain is 300 m/day and is effected by the thickness of permeable pavement. The FHWA assumes approaches for the design of subdrainage layer to remove water without allowing the system to become fully saturated.

Based on FHWA, the infiltration water into pavement is most complex. Some water assumed by Darcy's law uses a velocity formula such as:

$$v = k.i$$

where: v is velocity, k is constant of permeability and I is the hydraulic gradient [7]

Based on Zaika and Djakfar, the model for permeable pavement is influenced by gradation.

IOP Conf. Series: Materials Science and Engineering **650** (2019) 012054 doi:10.1088/1757-899X/650/1/012054

The gradation samples must satisfy criterias such as:

1. Base materials must hava a water storage is used for entering to subsoil.
2. Base materials must have a high strength to supported load traffic.
3. Base materials must be able passed by water.
4. Materials must be able to strain so that subgrade, subbase and surface can be intact [9]

According to Integrated Corridor Management (ICM), the base course pavement is influenced by temperature of moisture content. The temperature element is one of the analysis for the design of drainage layer. Additionally, drainage layer also depends on the duration of rainfall to distribute the water into the pavement's structure. Materials with different combination such as fewer fines and more uniform aggregate and different size aggregates have shown to have higher infiltration rate.

3. Method

In this study using several stages of research. The research stage is used as the basic concept of development in facilitating the research process until to data analysis.

First, base material was prepared by sieving analysis. Sieve analysis, CBR, MDD, OMC and Infiltration rate values were obtained for this base material. A stock of filler particles was prepated by wieving a soil through No. 200 sieve. These sieved filler were incrementally added to the prepared base material to make 5 different variation soil samples. All tests relevant to base materials were preformed to each of these prepared samples.

3.1 Materials

Crushed stone aggregate taken from Watukosek Village, Mojokerto City were used in this study as aggregate Claas A. Standard ASTM C 136 was used to make grading of materials proposed for use as aggregate. The method was used to determine the grading of materials for use as subdrainage layer. The aggregate were dried at temperature of sunlight. When using sieve analysis, aggregate class A were retained for percentage passing and percentage retained were calculated. Spesification of Bina Marga 2010 have limitations on the maximum and minimum recommended particle size exist.

Aggregate Class A is typically produced by natural process and used as a pavement because of that materials have high permeability.

3.2 Sieve Analysis

The sieve used is size number 1.5;1; 3/8;4; 10;40;200. The purpose of sieve analysis is to obtain gradations of the test sample to be used. Sieved test materials using mechanical vibration machines where the sieve holes are smaller in sequence. In order for the sample to be representative, approximately 6 kg of each materials sample was taken for sieve analysis.

3.3 Compaction Test of Sample

Compaction of the base course is one of the important proses for construction of any types of pavement. At the beginning of the compaction process, the dry volume weight increases with increasing water content [10]. When the water content is gradually added with the same compaction effort, the weight of the soil grains also increases. At water levels greater than certain water levels, the water content becomes optimum and the increase in water actually reduces the weight of the dry volume. The relationship between the weight of the dry volume and the weight of the soil and water content depends on the effort given by the pounder. This research uses a Modified Proctor to determine the characteristics of its material density.

Compaction test is provide the determining water content which is selected in five layers into mold of given dimensions. Materials compacted by 56 blows with the hammer. Water content is to determine the relationship between dry unit weight and water content for the materials.the result must be plotted, represents Optimum Water content and Maximum Dry Denstiy. Figure 1 shows the the OMC change according to MDD.

3.4 California Bearing Ratio

Various material preparations and test equipment refer to ASTM D 1883 specifications. This method is used to determine the potensial strength of base course material for use in road pavement. In this research CBR testing was soaked. CBR in soaked condition is needed to know the performance of pavement the saturated materials condition. CBR values are values that state the quality of a material at the same penetration. Preparation and CBR test equipment refers to the ASTM D1883 standard. In this study using CBR soaked for 4x24 hours. Reading the proving ring for 10 minutes. The result of these tests are shown in Table 4.3. The quality and consistency are needed to determination the properties of base materials.

California Bearig Ratio is equation to compare stress of materials and depth penetration. Value of CBR shows strength of soils and CBR standard. Based on specification of Bina Marga 2010, the standart of values CBR is 90%. The formula to determine the value of can be shown as Eq. (1)

$$CBR = \frac{P}{Ps} \quad (1)$$

Where;
P ; stress of materials
Ps : California stress standard [11].

3.5 Infiltration Test

The infiltration test is used to determine the speed of water that can penetrate the test object. The test object used is a test object that has been compacted with the same optimum water content during the CBR test. Modified CBR has a pore on the bottom cover that functions as a water way out of the test object. Water reduction every 60 seconds is calculated and recorded. The falling water flow starts to be calculated using a stopwatch. The treatment of infiltration testing on other gradations remains the same.

3.6 Sample Test

The material sample tested in this research is a base course A material that comes from the quarry of Watukosek Village, Ngoro District, East Java.

3.7 Material Mixing Method

This material uses a method of reducing filler material in the DGH 2010 gradation. Filler reduction is based on the upper specifications and middle specifications, namely 8%, 5%, 4%, 3% and 0%. Based on table 4.1, the percentage of the sieve sizes shown after the filler reduction is shown. The purpose of this modification is to obtain a variable CBR value and infiltration rate so that the most optimal gradation is obtained in the subdrain layer.Finding approved base materials with varied percentages of filler is difficult. Therefore, it was decided to prepare soil samples, with satisfy all the Bina Marga Spesification Base material requirement.

IOP Conf. Series: Materials Science and Engineering **650** (2019) 012054 doi:10.1088/1757-899X/650/1/012054

Table 1 Variations in Subdrain Layer Research.

Sieve Number	Sieve Size (mm)	Bottom Spesification	Top Spesification	Sample 1 % Passing	Sample 2 % Passing	Sample 3 % Passing	Sample 4 % Passing	Sample 5 % Passing
2"	50	100	100	100.00	100.00	100.00	100.00	100.00
1 1/2"	37.5	100	100	100.00	100.00	100.00	100.00	100.00
1"	25.0	79	85	85.00	82.00	84.35	81.62	81.05
3/8"	9.50	44	58	58.00	51.00	56.17	49.97	48.42
No. 4	4.75	29	44	44.00	36.50	41.57	35.16	33.16
No.10	2.0	17	30	30.00	23.50	26.96	21.89	19.47
No.40	0.425	7	18	17.00	12.00	13.39	10.15	7.37
No.200	0.075	2	8	8.00	5.00	4.00	3.00	0.00
Filler	0	0	0	0.00	0.00	0.00	0.00	0.00

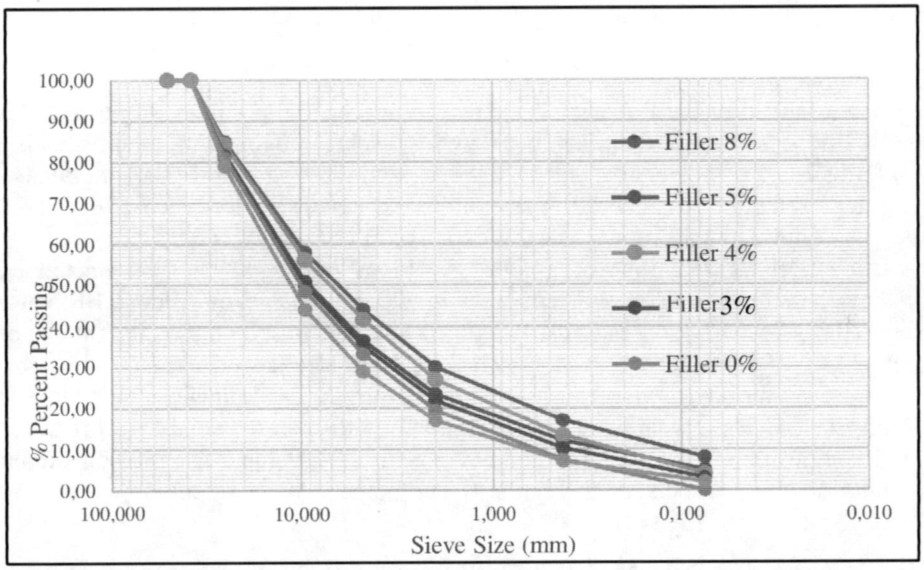

Figure 1.Particle size distribution curves of considered gradations from Table 1

4. Result and Discussion

4.1 Maximum Dry density (MDD)

Figure 2 shows the highest maximum dry weight of 8% filler gradation. The maximum dry density and the optimum water content of the base materials sampels were measured and shown with the compaction curve. The results of density testing are shown in figure 2 shows the maximum dry weight and optimum water content. From the figure is found that with a decrease filler in compaction the values of MDD decrease.

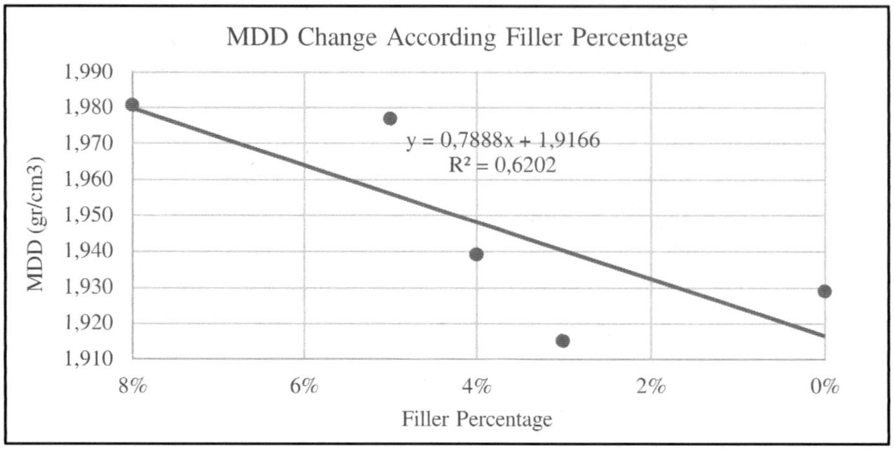

Figure 2. Relationship between MDD and Filler Percentage

4.2 *California Bearing Ratio*

In this paper we looked at the effect of filler percentage on CBR values due to a reduction in the percentage of filler into the subdrain layer. The CBR uses a proving ring reading with a maximum load of 7.492 lbs for 1 dvs or equivalent to 3.358 kg and the volume of CBR molds used is 2140.9 cm3. Tables 2 presents the result of CBR test in soaked condition. For each effect of filler samples were tested and performed load versus penetration curve have been calculated.

Figure 3 shows the CBR graph of all the gradations that have been subtracted based on the gradation characteristics based on the reduction in the percentage of filler. CBR values for each sample vary. This is because filler as a filler is getting smaller, Based on the graph it can be concluded that the percentage of filler decreases, the CBR value will decrease. Based on the test results, it is known that the maximum CBR that can be obtained from the percentage of filler is 74,654% for 4 days CBR immersion. The gradation is the strongest to withstand the road load for the subrain layer foundation. Relationship of Gradation of Aggregates with CBR values. The relationship between filler reduction and gradation can be seen in Figure 4. The percentage of fillers is arranged by the largest percentage of 8%, 5%, 4%, 3% and 0%

Table 2. The results of CBR values are arranged by Gradation characteristics.

Filler Percentage	Optimum Water Content (%)	Maximum Dry Density (maks. (gr/cm3)	CBR Value (%)
8%	9,00	1,981	74,654
5%	9,00	1,977	65,598
4%	10,00	1,939	67,345
3%	11,00	1,915	63,763
0%	9,00	1,929	62,400

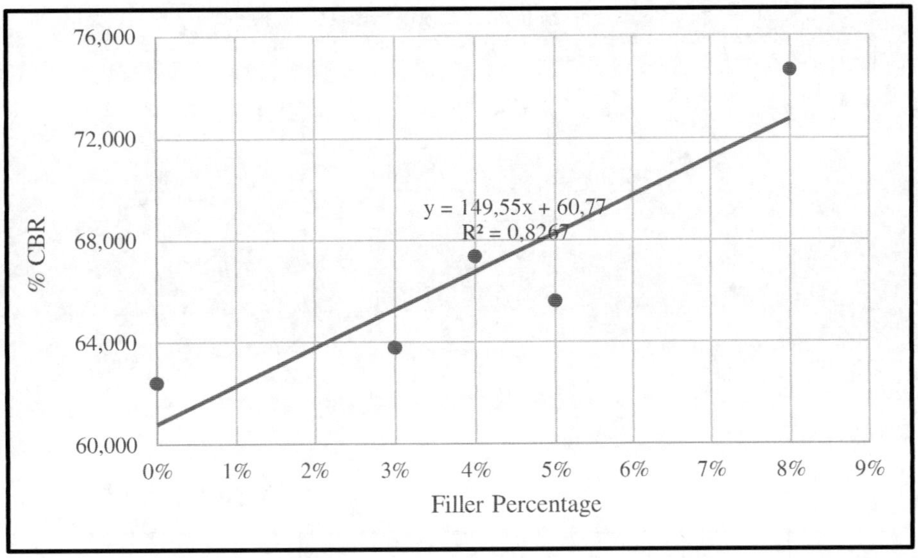

Figure 3. Graph of Filler Relationships with CBR

Based on Figure 2, it can be seen that the highest CBR value is in the 8% filler, which is 74.54%. This is because the percentage of fillers possessed by these gradations is the largest and has the best density, because the fillers fill empty cavities between coarse and fine aggregates. However, this gradation is the most watertight gradation. Analyzing these results clearly reveals that filler percentage has a close relationship to the bearing strength of the material. 4% filler gradation is the most optimum gradation because it has the optimum CBR because it has the most varied coarse and fine aggregate percentage.

4.3 Correlation of Aggregate Gradation to Infiltration Rate

The infiltration rate aims to determine the speed of water that can pass through a sample. In this research the test sample used was granular material, so the infiltration method was used. The infiltration rate test results are presented in Table 3.

Table 3. Infiltration Test Results

Filler Percentage	Steady Infiltration Rate mm/menit
8%	3,200
5%	3,667
4%	4,000
3%	4,667
0%	6,750

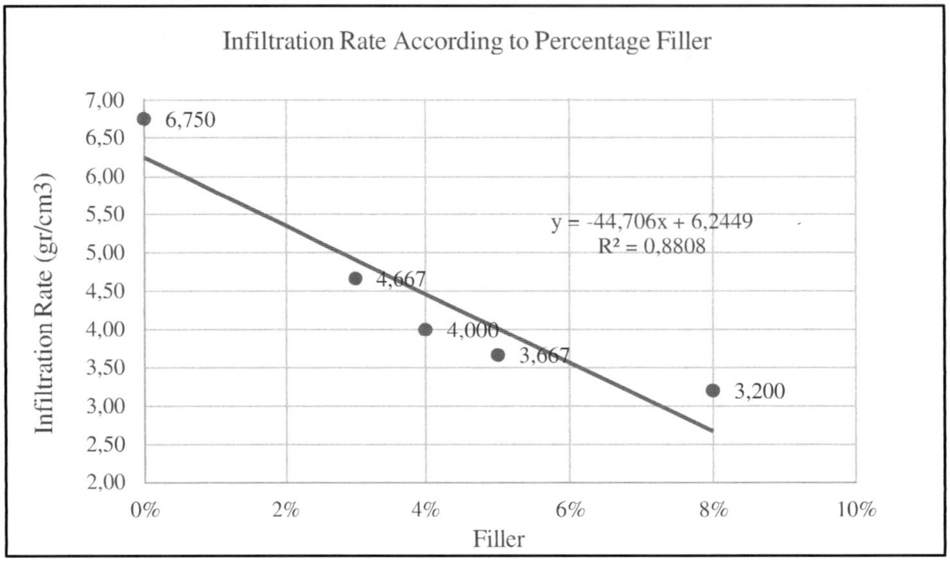

Figure 4. Infiltration Testing Graph

From figure 4, it can be seen the relationship between the characteristics of the filler reduction gradation, it can be concluded that the more fillers the infiltration rate decreases or in other words it reduces the speed of water flow. This is because the air cavity is filled with filler. Analyzing these results clearly reveals that filler percentage has a close relationship to infiltration rate of the material. In this study the highest infiltration rate was obtained at 0% filler variation which is the smallest aggregate filler gradat ion and has a larger cavity so that the water is easily flowed as high as having a high infiltration rate.

4.4 CBR Relationship with Infiltration Rate

Based on the 5 gradations taken aiming to obtain the maximum CBR value. Five gradations consisting of 8%, 5%, 4%, 3%, 0% filler are gradations within DGH. The purpose of determining this gradation is to get the maximum CBR and have the optimum infiltration rate so that the foundation layer results that function as a subdrain.

Table 4. Relationship between CBR Value and Infiltration Rate in the research

Percentage Filler	CBR value (%)	Infiltration constant mm/minutes
8%	74,654	3,200
5%	65,598	3,667
4%	67,345	4,000
3%	63,763	4,667
0%	62,400	6,750

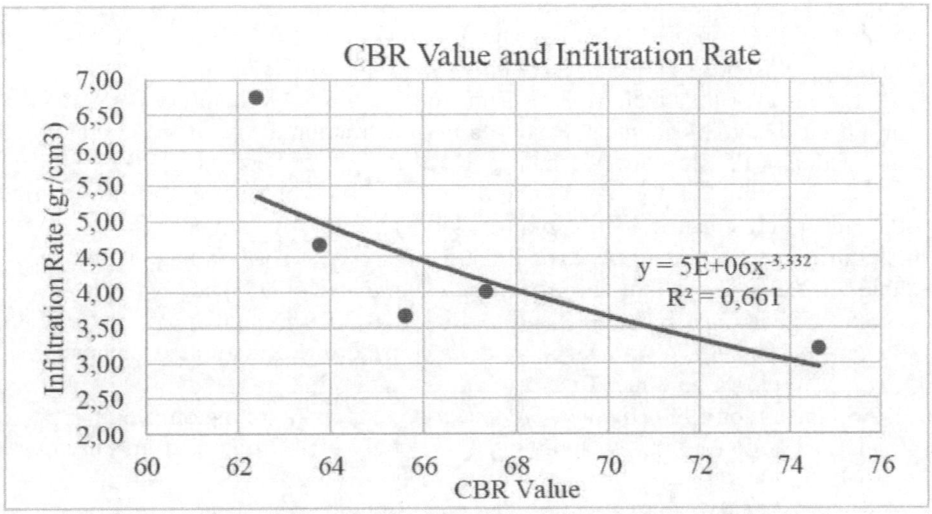

Figure 5. Graph of CBR Value Relationships with Infiltration Rate

In this research using a proofing ring with a maximum load of 10,000 lbs with a calibration factor of 7.492 lbs for 1 reading of the proving ring or equivalent to 3.358 kg.

The calculation method is in accordance with the calculation of laboratory CBR values which generally use the graph method to calculate the CBR value. Analyzing these results clearly reveals that filler percentage has a close relationship to infiltration rate of the material

In table 6 shows the relationship between CBR value and infiltration rate. It can be seen that the greater the CBR value, the lower the infiltration rate is constant. This shows that the maximum CBR has more fine grains so that the ability of water to pass through a sample is slower. Meanwhile, if it has a low CBR, it has fewer fine grains so that the infiltration rate becomes faster.

Besides being influenced by the filler percentage, CBR values are also influenced by the strength of the granules. If the sample mixes with water, the aggregate will decrease due to the flow of water so that the pores in the specimen increase and decrease the CBR value.

Based on the results of the infiltration and CBR tests, the gradation that is considered to be used as the subrain foundation layer is a percentage of 4%. The consideration taken is to have a high CBR and an optimal infiltration rate of 74.654% and 4,000 mm / minute.

5. Conclusion

1. The gradations used are Bina Marga Class A, fillers 8%, 5%, 4%, 3% and 0% are gradation references by reducing the percentage of filler that has been distributed.
2. The more filler percentage in a feeding gradation the infiltration rate becomes more impermeable, the water passing through the specimen becomes slowed. If the percentage of filler is smaller than the infiltration rate is getting bigger.
3. In the relationship graph of the gradation with CBR values shows that the 8% filler is 74.56% is the gradation that has the highest CBR because it has the most fillers. While the lowest CBR is in the 0% filler, which is 62.44%.
4. From the results obtained according to the infiltration rate and CBR values, the gradation that qualify for the subdrain foundation layer is a 4% filler gradation because it has a fast infiltration rate and maximum CBR value.

Reference

[1] Adlinge S S and Gupta A K 2013 Pavement Deterioration and its Causes *IOSR J. Mech. Civ.*

Eng., pp. 2278–1684
[2] Matuszkova R, Heczko M, Cepil j, and Radimsky M 2018 Issues of using Longer Heavier Vehicles on Roads *IOP Conf. Ser. Mater. Sci. Eng.* vol. **317**, no. 1
[3] M. S. Baghini, A. Bin Ismail, M. R. Karim, and F. Shokri, "Laboratory Performance Evaluation of Physical and Mechanical Road Base Construction Properties Laboratory Performance Evaluation of Physical and Mechanical Road Base," no. September, 2015.
[4] Pais J C, Amorim S I R, and Minhoto M J C 2013 Impact of Traffic Overload on Road Pavement Performance *J. Transp. Eng.* vol. **139**, no. 9, pp. 873–879
[5] Sholichin I, and Rumintang A 2018 Relation analysis of road damage with excessive vehicles load on Kalianak road Surabaya *J. Phys. Conf. Ser.*, vol. **953**, no. 1
[6] Kalore S A, Sivakumar Babu G L, and Mallick R B 2019 Design approach for drainage layer in pavement subsurface drainage system considering unsaturated characteristics *Transp. Geotech.*, vol. **18**, no. June 2018, pp. 57–71
[7] F. H. Administration 1990 *Highway Subdrainage Design* (Washington, D. C)
[8] J. Zhang and T. Portfolio, "A laboratory scale study of infiltration from Pervious Pavements," no. May, 2006.
[9] Y. Zaika and L. Djakfar, "Gradation Band of Some Types Material for," vol. 11, no. 25, pp. 2486–2492, 2016.
[10] S. Patra and A. K. Bera, "California Bearing Ratio of Fine Grained Soil and its Correlation California Bearing Ratio of Fine Grained Soil and its Correlation," no. June 2017, 2018.
[11] V. References, "Methods of Sieve Analysis With Particular Reference," 1949.

Evaluation of Side Friction In IHCM For Highway 6 Lanes 2 Ways Divided

Najid[1]
Lecturer, Civil Engineering Department, Tarumanagara University

Fara Yuniarti[2]
Student, Civil Engineering Department, Tarumanagara University

E-mail: najid@ft.untar.ac.id

Abstract. IHCM (Indonesia Highway Capacity Manual) was issued and came into force in 1997. After 20 years there has been a change in traffic from the number and composition, as well as the traffic regulation policy. As a result of this, the determination of IHCM's road capacity is often incorrect. Therefore it is necessary to evaluate IHCM. This study tries to evaluate IHCM through side obstacle factors, as one of the factors that determine the value of road capacity. Research is limited to evaluating the weight of each side obstacle factor and the effect of land use from the survey results using statistical analysis. From the analysis, it is found that the conditions for the use of shops, markets, hotels, gas stations are in the category of moderate side barriers and the ranking of the relative weights of the side barriers at IHCM is still appropriate.

1. Introduction

1.1. Background

Transportation system is a road network system that physically connects an activity space with other activities space, which influences the development of a region (space of activity) and the development of a space of activity will require an increase in the transportation service system (Tamin, 2000).

The development of activity space (activity) mainly occurs on the side of the road. Activities on the side of the road have an impact on traffic performance which can hamper traffic flow, which is then called side friction. Because it influences the flow of traffic it means it also affects the capacity of the road.

Side friction in the IHCM (Indonesian Highway Capacity Manual) are classified consisting of non-motorized vehicles (weight 0.4), pedestrians (0.5), left-right access roads (weight 0.7) and parked / stopped vehicles (weight 1, 0) (IHCM, 1997). The effect of side friction on capacity is the accumulation of the number of each occurrence of the side resistance factor multiplied by its weight.

In addition to IHCM, 1997 has been too long, the calculation of road capacity based on IHCM is often lower than the volume of traffic observed. Therefore IHCM needs to be evaluated. This research is

[1] To whom any correspondence should be addressed.

IOP Conf. Series: Materials Science and Engineering **650** (2019) 012055 doi:10.1088/1757-899X/650/1/012055

expected to contribute to the research to revise IHCM in terms of the side effect factors on road capacity.

1.2. Identification of Problems

a. Relation of each side friction factor to traffic volume.

b. The relationship of each side resistance factor to the speed of traffic.

1.3 Problem Formulation

From the identification of the above problems can be formulated:

a. What is the relationship between the conditions of land use and the magnitude of each factor in the side friction?

b. How does the traffic condition affected by the value of each side friction factor?

c. How is the difference in the side friction based on IHCM with the survey results?

1.4 Research Objectives

The purpose of this study:

a. Analyze the value of side friction in different land uses.

b. Arrange the correlation between land use with the value of the side friction factor.

c. Evaluate each side friction factor IHCM version

1.5 Limitation of Problems

This research is limited by the following scope:

1. Highway configuration 6/2 D (six lane two ways divided)
2. Study in the city of Jakarta
3. Other factors and capacity calculation formulas follow IHCM.

2. Literature Review

2.1. Side Friction Factors Against Capacity Based on IHCM

Side Friction are grouped in five classes, from very low to very high class as an accumulation of the four side obstacle factors. Classification of side Friction can be seen in Table 2.1 below:

IOP Conf. Series: Materials Science and Engineering **650** (2019) 012055 doi:10.1088/1757-899X/650/1/012055

Table 2.1. Classification of Side Fricton (source: IHCM 1997)

Side Friction Factor Level	Code	Amout of All Number of Side Friction per hour	Land Use Condition
Very Low	VL	<100	Residential area; almost no activity
Low	L	100-299	Residential area; in the form of public transportation and so on.
Medium	M	300-499	Commercial area; very high road side activity
High	H	500-899	Industrial area; several shops on the side of the road
Very High	VH	>900	Commercial area; market activity beside the road

Table 2.2. **Side Friction Factors for Adjustment Higway Capacity**

Highway Type	Side Friction Factor Level	Adjustment Factors for Side Barriers and Kerb - Barrier Distance (FCsf)			
		Distance Kerb-Barrier			
		≤ 0,5	1,0	1,5	≥2,0
4/2 D	VL	0,95	0,97	0,99	1,01
	L	0,94	0,96	0,98	1,00
	M	0,91	0,93	0,95	0,98
	H	0,86	0,89	0,92	0,95
	VH	0,81	0,85	0,88	0,92

Based on IHCM to calculate the capacity of the highway use the formula:

$$\mathbf{C = Co \times FCw \times FCsp \times FCsf \times FCcs} \quad \text{.....................................(1)}$$

Where:
C = Capacity (pcu/hour)

Co = Base capacity (pcu/jam)

FCw = Adjusment Factor for width oh lane

FCsp = Adjusment Factor for direction split (non divided highway)

FCsf = Adjusment Factor for side friction

FCcs = Adjusment Factor for city size

3. Data Collection

3.1. Survey Traffic Volume

Survey method with field observations carried out on the side of the road, the surveyor takes data on traffic volume, traffic speed and to the four side obstacle factors. Time for morning data collection (07.00-09.00), daytime (12.00-14.00) and afternoon (16.00-18.00 hours). Survey location at Cideng Highway that has 6 lanes and divided by median as canal.

Survey was conducted in two days by distribution, on Tuesday for direction Tomang and on Thursday for direction Tanah Abang. Data traffic volume on Cideng highway can be seen in Table 3.1 and 3.2 below:

Tabel 3.1 Traffic Volume of Cideng Highway on Tuesday

Time	Traffic Volume (vehicle/hour)			Q(Traffic Volume) (Pcu/hour)
	Motor Cycle	Car	Truk/ Bus	
07.15-08.15	1782	1611	8	2066.1
07.30-08.30	1859	1505	8	1979.35
07.45-08.45	1959	1406	10	1907.75
08.00-09.00	1882	1452	9	1933.3
08.15-09.15	1716	1461	9	1900.8
11.15-12.15	**2484**	**1454**	**51**	**2136.2**
11.30-12.30	2380	1441	56	2103.2
11.45-12.45	2252	1423	57	2054.4
12.00-13.00	2251	1458	49	2079.55
12.15-13.15	2202	1464	42	2064.9
17.15-18.15	3096	1104	36	1921.2
17.30-18.30	2943	1002	26	1768.95
17.45-18.45	2677	890	21	1584.45
18.00-19.00	2791	910	18	1629.35
18.15-19.15	2765	869	19	1583.05

Table 3.1. show peak hour volume occurs on the hour 11.15-12.15 that has traafic volume 2136,2 pcu/hour (pcu value for each vehicle based on IHCM).

Tabel 3.2 Traffic Volume of Cideng Highway on Thursday

Time	Traffic Volume (vehicle/hour)			Q(Traffic Volume) (Pcu/hour)
	Motor Cycle	Car	Truk/ Bus	
07.15-08.15	6679	2025	28	3728.35
07.30-08.30	**6987**	**2030**	**31**	**3813.95**
07.45-08.45	6699	2038	35	3754.75
08.00-09.00	6375	1902	38	3541.35
08.15-09.15	6204	1914	46	3520.2
11.15-12.15	3968	1772	52	2826.4
11.30-12.30	4034	1759	52	2829.9
11.45-12.45	4048	1746	48	2815.6
12.00-13.00	4051	1703	41	2764.95
12.15-13.15	4097	1759	37	2827.65
17.15-18.15	6214	1793	11	3359.7
17.30-18.30	6200	1851	16	3420.2
17.45-18.45	5877	1829	26	3329.45
18.00-19.00	5636	1836	26	3276.2
18.15-19.15	5994	1884	25	3412.5

Table 3.2. show peak hour volume occurs on the hour 07.30-08.30 that has traafic volume 3813,95 pcu/hour (pcu value for each vehicle based on IHCM).

To know the condition of land use at the side of Cideng highway, we can see at the table 3.3 below:

Table 3.3. Land Use and Side Friction Survey

Survey Day	Direction	Survey	Based on IHCM
Tuesday	Tomang	Shop, Office	Commercial area; very high road side activity
Thursday	Tn.Abang	Market, Hotel, Shop, Gas Station	Commercial area; very high road side activity

Table 3.3 show that categorized survey location close to Commercial area; very high road side activity in IHCM.

To know the relationship land use at the side of Cideng highway with land use category in IHCM, we can see at the table 3.4 below:

Table 3.4: Percentage Appropriate Survey and IHCM

Direction	Time	Total Number Side Friction	Level Classification	FCsf	Percentage M Level
Tomang	08.15-09.15	300-499	M	0,98	
	Others	100-299	L	1	6,67%
Tanah Abang	07.15-12.15	300-499	M	0,98	
	Others	100-299	L	1	46,67%

Table 3.4 show that on direction Tanah Abang categorized land use more appropriate with IHCM as level Medium.

4. Data Analysis

Data analysis based on Comparing the weight of side friction factor in IHCM with the value of Pearson Correlation the traffic volume and total amount of side friction value, and the comparisson we can see at Table 4.1:

Table 4.1 Relationship of Traffic Volume with Side Friction

Side Friction	**IHCM**	**Pearson Correlation (R)**					
		Tuesday			**Thursday**		
		Morning	**Midday**	**Afternoon**	**Morning**	**Midday**	**Afternoon**
Non Motorized Vehicle	**0,4**	0.53	**0.31**	**0.37**	0.67	**0.36**	**0.55**
Pedestrian	**0,5**	0.75	**0.50**	**0.50**	0.68	**0.61**	**0.44**
Left-Right Acces	**0,7**	0.55	**0.57**	**0.51**	0.48	**0.63**	**0.44**
Parked/ Stopped Vehicle	**1**	0.42	**0.65**	**0.1**	0.55	**0.63**	**0.1**

Table 4.1 show the relationship between weight of each side friction factor with correlation each side friction factor and traffic volume look appropriate. In the afternoon on both direction there is not any vehicle parked at the side of the road.

5. Conclusion

- In land use of Market, Hotel, Shop, Gas Station is quite similer with Commercial area; very high road side activity that categorized in IHCM.
- Weight Sequence in IHCM for side friction more close with Correlation analysis for midday and afternoon of two direction.
- The value of weight side friction in IHCM and data analysis is quite difference.

IOP Conf. Series: Materials Science and Engineering **650** (2019) 012055 doi:10.1088/1757-899X/650/1/012055

References

[1] Tamin, O Z, (1997). Planning and transportation modeling, Publisher ITB, Bandung.
[2} IHCM (1997), Indonesia Highway Capacity Manual, Directorate General of Highways, Ministry of Public Works Indonesia.
[3] Blunden, W.R. (1971), The land Use - Transport System: Analysis and Synthesis, Oxford Pergamon Press)

IOP Conf. Series: Materials Science and Engineering **650** (2019) 012056 doi:10.1088/1757-899X/650/1/012056

Integrated multimodal disaster mitigation management for urban areas: a preliminary study for 2-d flood modeling

R M S Prastica[1*] and D Wicaksono[2]
[1]Department of Civil Engineering, Vocational College, Universitas Gadjah Mada, Jl. Yacaranda, Yogyakarta, 55281
[2]Department of Geo-technology, Vocational College, Universitas Gadjah Mada, Yogyakarta, 55281

E-mail: *rian.mantasa.s.p@ugm.ac.id

Abstract. Flooding in urban areas is a common phenomenon in megacities due to the high population and settlement growth. Lack of knowledge from urban societies triggers them to intervene in the capacity of the river for housing. This impact affects various disasters, both quality and quantity of water. Kali Belik experiences flooding every year, although several solutions have been installed to mitigate, for example, Embung. Integrated multimodal disaster mitigation management is proposed to solve the problems and answer the research gaps. The preliminary research only focuses on flood analysis. The analysis uses 2-D modeling with HEC-RAS. The study concludes that the main factor of flooding in the research area is decreasing river capacity due to social behavior. Further research would be conducted to obtain a comprehensive understanding and solution by involving the government, academics, and society

1. Introduction

Flood occurs most worldwide [1]. Most cities in South East Asia are situated in the flood-prone areas [2, 3], for example, Semarang, Jakarta, and Yogyakarta. Climate change triggers more variability of rainfall [2] and directly impacts the environment. The cities would experience more flood events, including Jakarta [2, 4]. The primary victims of the flood disaster is a low-income population with a lack of disaster preparedness system [5]. Management of water bodies, both rivers, reservoirs, and drainage, is needed to preserve the environment. The large number of natural disasters that occur today are primarily due to physical factors that have decreased their quality function in maintaining the natural balance. This problem is due to the rapid development of infrastructure along with the increasing number of people. In addition to settlements, the infrastructure built is intended to mitigate natural disasters [6, 7]. For example, normalizing rivers to overcome flooding [8].

However, today's infrastructure development that departs from the problem of natural disasters has not heeded the ecological factors of the environment. This case causes the survival of living things such as fish, macro-invertebrates, etc. are not stable. Declining quality of water bodies as a habitat for living things indicates that humans live in unhealthy environments. This problem will take root in the failure of the United Nation's objectives, as stated in the 17 Sustainable Development Goals (SDGs). One of the areas that were reviewed by this study was urban areas, precisely in the Klitren area, which was passed by Kali Belik and emptied into Embung Langensari. Problems in the area have been flooded over the past decade that has not been overcome, poor sanitation, and poor quality of river and pond water. In general, the main study aims to carry out physical, chemical, biotic, and social analyzes in urban areas to conserve environments with various multidisciplinary research methods. Meanwhile, this paper only aims to identify the physical aspects of the inundated area. Further analysis would take place in future research.

2. Urban Flooding in *Kali* Belik: Research Gap

The flood in Belik River is nothing new because there are many studies on the analysis of floods in this region. Praja and Suyono [9] conducted an evaluation of the capacity of the Kali Belik channel in Yogyakarta and concluded that the Belik River channel at the UGM Valley and on Jalan Batikan was unable to accommodate the existing runoff. The same thing was done by Pratama and Sujono [10] who evaluated the Kali Belik flood by analyzing the application of the UGM detention pool and infiltration wells. This study concludes that the detention pond can reduce flooding by 0.6 m^3/s or 4.6 m^3/s. Flood reduction is also said to be done by building 87,113 infiltration wells for the 10-year discharge rainfall. Alternative solutions to flooding were also carried out by Fitri and Ulfa [11]. The solutions offered are the same as Pratama and Sujono [10] by making shallow infiltration wells.

Based on existing research [9 – 11], researchers only focused on physical problems. Solutions offered have not been through comprehensive analysis with Penta-helix stakeholder management. The Office of Public Works, Housing, and Settlement Areas of the Yogyakarta City Government is one of the policymakers in the arrangement of this area. Also, discussions or focus group discussions are needed for affected people. This study will conduct a study through the stakeholder management Penta-helix model. In addition to the physical aspects, environmental resilience also needs to pay attention to the chemical and biological aspects. Chemical characters can be assessed through the analysis of water quality. Iskandar, Sudjono, and Supraba [12] reviewed the quality of Kali Belik water and its effect on raw water sources at the Umbul Pace UGM Campus. However, it has not carried out further analysis of applied applications to improve water quality in Belik River. The last aspect is the ecology. Many infrastructure developments in handling floods do not heed the ecological issues so that many aquatic biotas that should be able to live could not adapt to an environment that is not supportive. In addition to water biota such as macroinvertebrates, the presence of fish has the potential to decrease due to the absence of a fish-way to swim upstream. This entire study will discuss the three aspects as a gap from the research that already exists in Belik River. Meanwhile, this preliminary research focuses on the scope of the physical.

3. Methodology

3.1. Flood analysis

In conducting flood analysis, many previous studies used conventional methods that have been widely developed, such as the Nakayasu Method [8], Gama I Method, SCS Method, and other methods. In addition to these methods, researchers began to develop flood analysis using 2-D and 3-D methods. Two-dimensional modeling uses software based on finite element methods such as HEC-RAS [8, 13 - 15] and HEC-HMS. While 3-D modeling also began to be developed to conduct flood analysis. Based on previous research in the analysis of floods in Belik River, the study used 2-D HEC-RAS modeling with channel view data as secondary data. This study will use primary data on the condition of the existing channel and modeled on a 3-D basis using photogrammetric methods with the help of Geographic Information System (GIS) software [14].

3.2. Environmental and macro-invertebrates analysis

One of the audit activities carried out is assessing the condition of the waters. To audit, the decline in water quality can be known physically, chemically, and biologically [16]. Physically it can be seen in plain views, such as changes in the color of the water. Chemically chemical parameters can be tested, such as Biological Oxygen Demand (BOD), Chemical Oxygen Demand (COD), and Dissolved Oxygen (DO). Biologically we can see from the diversity of biota in these waters.

In determining the biotic index there are several methods including Simpson's Diversity Index (D), Shannon-Wiener Diversity Index (H), Family Biotic Index (FBI, Metric 2- RBP), Biological Monitoring Working Party (BMWP), Average Score Per Taxon (ASPT), and Taxa Richness (TR, Metric 1- RBP II).

3.3. Data acquisition with Unmanned Aerial Vehicle

Data acquisition with UAV consists of five main activities, namely determination of work area, flyway planning, GCP-ICP determination, GCP-ICP measurement, and aerial photo processing. The activity flow can be seen in Figure 1. The DEM extraction example from an aerial photo could be shown in Figure 2.

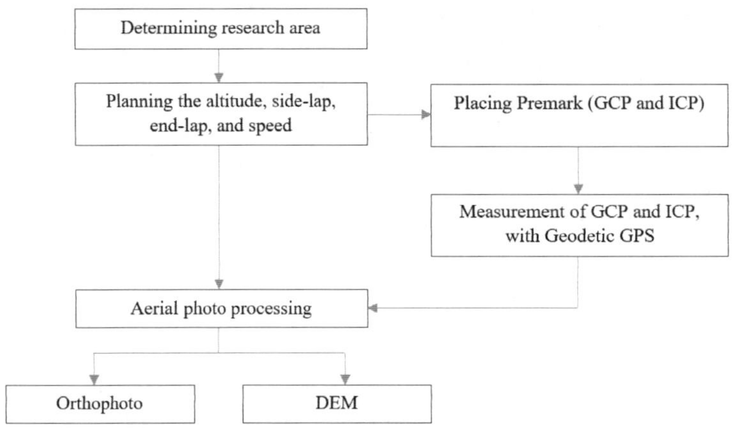

Figure 1. Acquisition data with a serial photo

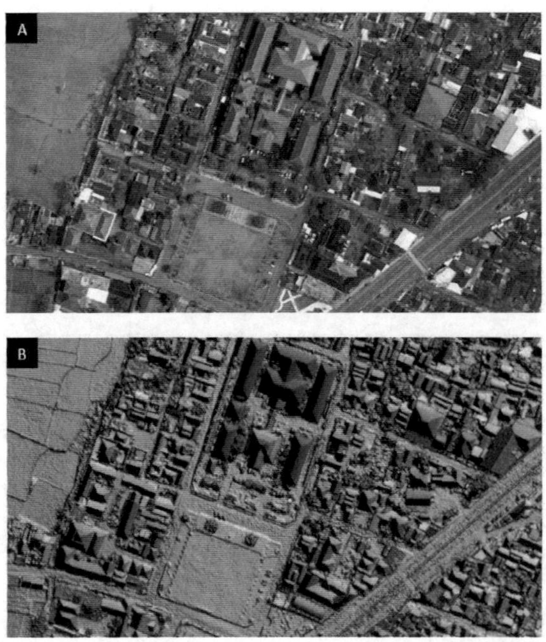

Figure 2. Extraction of DEM from Aerial Photography. A) Photo of Upright Aerial (Orthophoto) and B) DEM resulting from Aerial Photo Processing
(These photos are taken by Angga Dwi Laksono (2018))

4. Results and Discussion

4.1. Flood in Kali Belik

Flood in Kali Belik is a complicated phenomenon for an urban area. Figure 3 depicts the condition of Kali Belik surrounding area in the upper stream part. The population today reaches a critical point because green areas could not be found for rainfall infiltration. This situation triggers a flood disaster every year. People realize that developing settlements near the river is a wrong decision, and the government encourages them to move their settlement. Flood in Kali Belik occurs every year without any further solution. Figure 5 figures the research area.

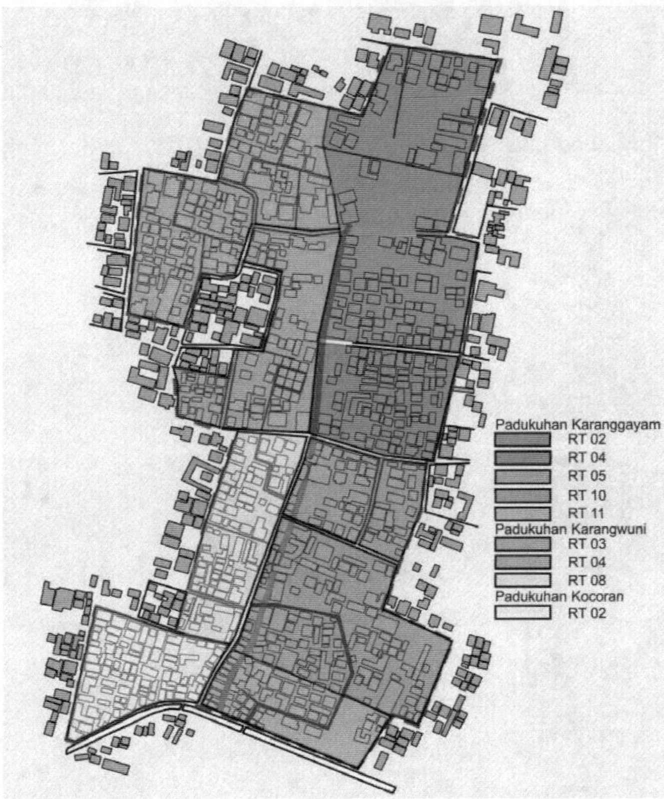

Figure 3. The upper stream part of Kali Belik catchment [17]. The blue line in the middle of the settlement is Kali Belik. The location of the nearest settlement from the river is approximately 0.5 meters. When a flood occurs, the houses are directly impacted. Water enters the settlements with another problem: bad odor, a water-borne disease, etc.

4.2. Rainfall analysis

Rainfall data for flood analysis in the research area are driven from Gemawang rainfall station in Sleman, Yogyakarta, from the year 2004 – 2017. Figure 4 depicts the result of rainfall intensity for 2-year, 5-year, 10-year, 25-year, 50-year, and 100-year rainfall.

This preliminary research only focused on flood analysis. Further analysis is still undertaken and does not get any full result yet. The study discusses the flood in the downstream part of Kali Belik. According to the preliminary study of discharge for 10-year period flooding, the discharge for seven segments of the downstream part of Kali Belik is 4.7; 2.2; 2.8; 6.1; 2.6; 3.2, and 4.7 m^3/s, respectively, until the last segment in Embung Langensari. The chosen return period is the best representative of rainfall characteristics in the research area [9]. According to Figure 6, we could see the actual event in Figure 7. Figure 7 is the flood event in 2014.

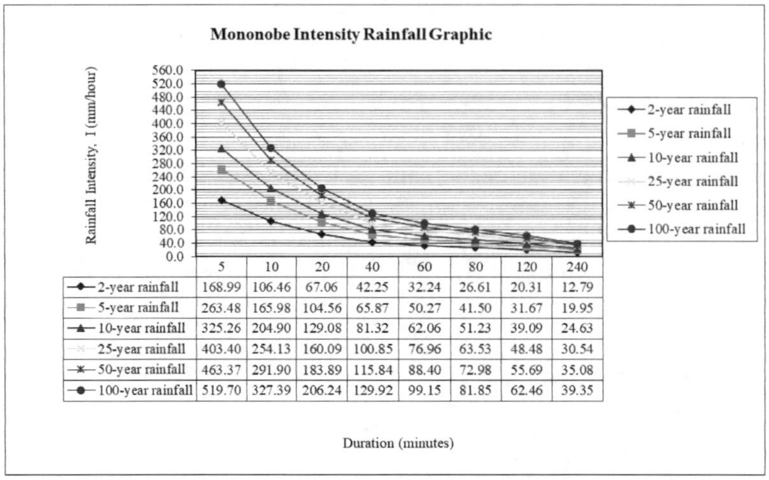

	5	10	20	40	60	80	120	240
2-year rainfall	168.99	106.46	67.06	42.25	32.24	26.61	20.31	12.79
5-year rainfall	263.48	165.98	104.56	65.87	50.27	41.50	31.67	19.95
10-year rainfall	325.26	204.90	129.08	81.32	62.06	51.23	39.09	24.63
25-year rainfall	403.40	254.13	160.09	100.85	76.96	63.53	48.48	30.54
50-year rainfall	463.37	291.90	183.89	115.84	88.40	72.98	55.69	35.08
100-year rainfall	519.70	327.39	206.24	129.92	99.15	81.85	62.46	39.35

Figure 4. Mononobe rainfall intensity result

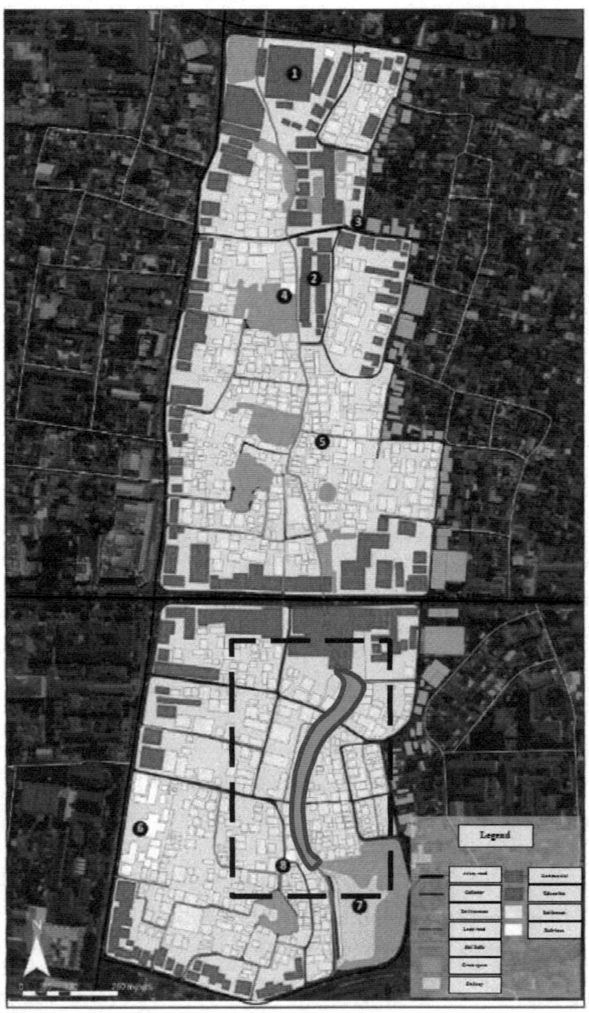

Figure 5. The research area of Kali Belik [18]. According to the picture, there are several notable buildings: 1. Syantikara boarding house, 2. Stella Duce Senior High School, 3. Al-Ikhlas Samirono Mosque, 4. An-Nahr Mosque, 5. Al Husna Iromejan Mosque, 6. GKJ Gondokusuman, 7. Embung Langensari, 8. Al-Muthmainnah Mosque. The dashed picture of Figure 6 describes the flooded area.

IOP Conf. Series: Materials Science and Engineering **650** (2019) 012056 doi:10.1088/1757-899X/650/1/012056

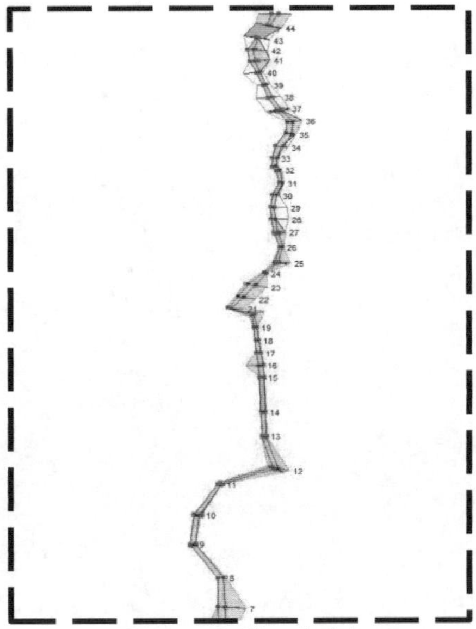

Figure 6. Flooded Area

(a) Before flood event (b) After flood event

Figure 7. Flooding event in Kali Belik

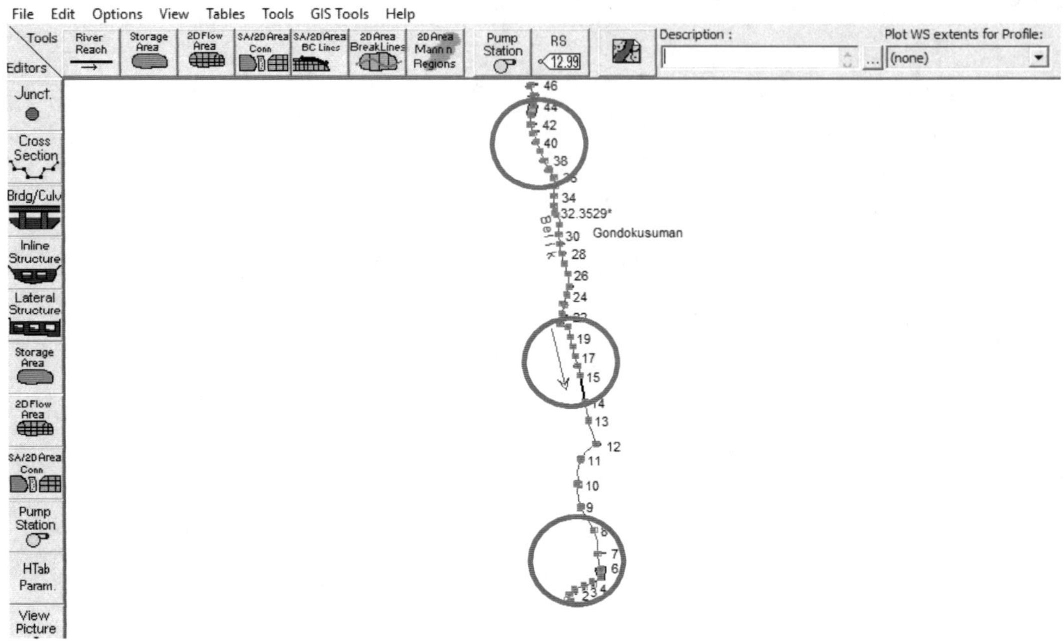

Figure 8. The user interface of 2-D modelling HEC-RAS of Kali Belik

Kali Belik in the downstream part is designed for 2-D flood modeling using HEC-RAS. Figure 8 depicts the HEC-RAS interface for Kali Belik. The initial survey for this research takes 3 location samples to analyze the flood disaster. According to the figure of the downstream part of Kali Belik, the investigation takes three samples in the upper, middle, and downstream. Figure 9 – 11 depict the condition of the river with 10-year flood analysis. The water overflows. The river capacity could not accommodate the discharge.

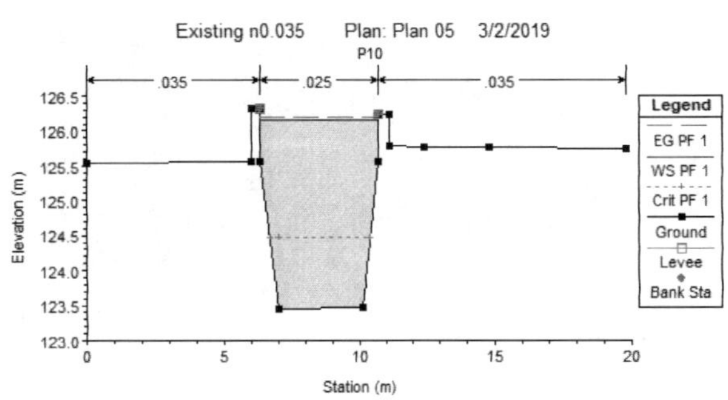

Figure 9. Upper part condition of river

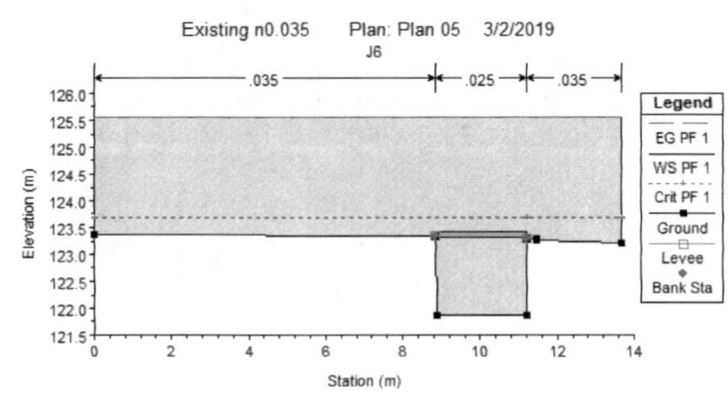

Figure 10. Middle part condition of river

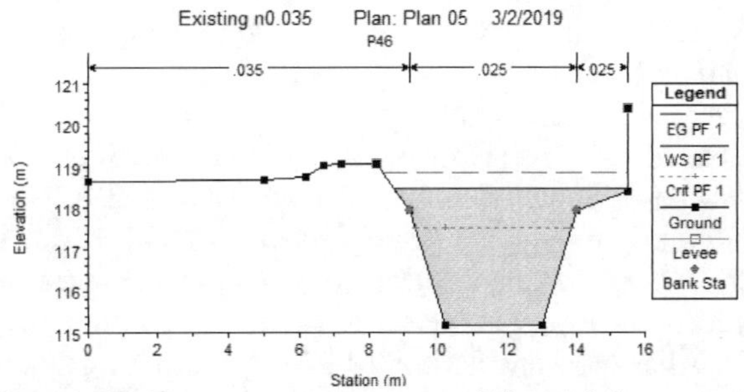

Figure 11. Downstream part condition of river

5. Conclusion and Suggestion

In conclusion, the capacity Of Kali Belik could not accommodate the incoming rainfall. As a result, the inundated area along the river occurs and affects the lack of social life. This preliminary research results that the capacity of Kali Belik reaches critical conditions and needs comprehensive attention to mitigation. According to this initial result, the research suggests conducting a further comprehensive analysis, especially for flood analysis both in 2D and 3D modeling by using UAV. Further research would also conduct integrated multimodal disaster management in the research area. The research recommends involving the government as a policymaker to take parts in disaster mitigation.

Acknowledgments

The authors wish to thank research grant No. 3943/UN1/DITLIT/DIT-LIT/LT/2019 from Universitas Gadjah Mada as financial support to this research. The present work was performed as a part of leading research entitled Integrated Multimodal Disaster Mitigation Management for Water Sensitive City. The authors wish to thank the Hydraulic and Water Infrastructure Laboratory of Vocational College, Universitas Gadjah Mada.

References

[1] Centre for Research on the Epidemiology of Disasters CRED 2016 *2015 disasters in numbers: International disaster database*

[2] The World Bank 2010 *Climate risks and adaptation in Asian coastal megacities* A synthesis report (Washington DC: The World Bank)

[3] Pelling M 2003 Natural disaster and development in a globalizing world *Routledge*

[4] Nicholls R J, S Hanson, C Herweijer, N Patmore, S Hallegatte, J Corfee-Morlot, J Chateau, R Muir-Wood 2008 *Ranking port cities with high exposure and vulnerability to climate extremes: Exposure estimates OECD Environment Working Papers No. 1* (Paris: OECD Publishing)

[5] Berquist M, Daniere A, and Drummond L 2015 Planning for global environmental change in Bangkok's informal settlements *Journal of Environmental Planning and Management* **58(10)** pp 1711–1730

[6] Bangkok Metropolitan Administration 2015a *Bangkok statistics 2015* (Bangkok: Strategy and Evaluation Department, Bangkok Metropolitan Administration)

[7] Bangkok Metropolitan Administration 2015b *Annual Bangkok water management report 2015* (Bangkok: Bangkok Metropolitan Administration)

[8] R M S Prastica, C Maitri, A Hermawan, P C Nugroho, D Sutjiningsih and E Anggraheni 2018 Estimating design flood and HEC-RAS modelling approach for flood analysis in Bojonegoro city *IOP Conf. Ser.: Mater. Sci. Eng.* **316** 012042

[9] Praja S A and Suyono 2013 Evaluation of the Capacity of the Belik River in Yogyakarta *Jurnal Bumi Indonesia* **2** No. 3 pp. 55 – 62

[10] Pratama Y A and Sujono J 2015 Hydrological Analysis of Kali Belik in UGM with Detention Pool

by Using HEC-HMS *Thesis* (Civil and Environmental Engineering Faculty of Engineering Universitas Gadjah Mada)

[11] Fitri A and Ulfa A 2015 Planning for the Implementation of the concept of Zero Run-off and Agroforestry based on River Debit, Study in Belik Sub-catchment, Sleman, Yogyakarta *Jurnal Perencanaan Wilayah dan Kota* **26** No. 3 pp. 192 – 207

[12] Iskandar, Sudjono, and Supraba 2017 *Study of Kali Belik's Water Quality and Its Effect on Raw Water Source (Umbul Pace) of Kampus Universitas Gadjah Mada* Thesis Civil and Environmental Engineering Faculty of Engineering Universitas Gadjah Mada

[13] Ramli, Adli M W, Alias N E, and Taib S M 2018 Evaluating Transportation Modes and Routes for Disaster Relief in Kelantan Using Geographical Information System, in Yusop Z, Aris A, Alias N E, Annammala K V, William L and Waugh Jr (ed.) Improving Flood Management, Prediction and Monitoring *Community, Environment and Disaster Risk Management* **20** Emerald Publishing Limited, pp. 63 - 71

[14] Ezz and Hesham 2018 Integrating GIS and HEC-RAS to model Assiut plateau runoff *The Egyptian Journal of Remote Sensing and Space Science* **21** Issue 3 pp. 219-227

[15] Quiroga V M, Kure S, Udo K, and Mano A 2016 Application of 2D numerical simulation for the analysis of the February 2014 Bolivian Amazonia flood: Application of the new HEC-RAS version 5 *RIBAGUA - Revista Iberoamericana del Agua* **3** Issue 1 pp. 25-33

[16] Yang, Nan, Li Y, Zhang W, Wang L, and Gao Y 2019 Reduction of bacterial integrity associated with dam construction: A quantitative assessment using an index of biotic integrity improved by stability analysis *Journal of Environmental Management* **230** pp. 75-83

[17] Ma'dan, A 2015 *Study of Kali Belik Slum Settlement Area Project Report Universitas Gadjah Mada*

[18] Sulaiman, Muhammad, Inez 2018 *Identification of Segment 3 in Kali Belik, Sleman, Yogyakarta*, Hydraulic and Water Infrastructure Laboratory, Civil Engineering Department, Vocational College, Universitas Gadjah Mada

IOP Conf. Series: Materials Science and Engineering **650** (2019) 012057 doi:10.1088/1757-899X/650/1/012057

Assessment of the quality of well water in the spatial terraces of river code in Yogyakarta Indonesia

E Sriyono*, Sardi, and N C Kresnanto

Civil Engineering Department, Faculty of Engineering, Janabadra University

E-mail: edysriyono@janabadra.ac.id; seiacons_cv@yahoo.co.id; nindyo_ck@janabadra.ac.id

Abstract. Many residents live on the banks of the Yogyakarta Code River. Some of them are still there who dispose of liquid waste directly into the river so that it has the potential to pollute the environment. Pollution also occurs in population wells. In general the condition of the riverbank is terraced, so that liquid waste originating from terrace 1 (the top) will flow to terrace 2 (the middle part), then flow to terrace 3 (bottom), and finally enter the river. The purpose of this study was to assess the quality of well water on the banks of the River Code, based on DO parameters, pH, DHL, Temperature, TDS, and E-Coli bacteria and to ascertain whether there was a direct correlation of pollution accumulation from high elevation to low elevation. The methods used include: identification of wells, measurement of the position of the well with GPS, plotting the location of wells to maps, sampling of well water, analyzing the quality of well water in the laboratory, and assessing the spatial water pollution of the well. Based on the results of the study, it can be concluded that there is a decrease in the quality of well water from high elevation to low elevation so that the quality of well water at low elevations is worse than the quality of well water at higher elevations. So that there is a direct correlation of accumulated pollution from well water at high elevations to lower elevations in the terraces of the Code River banks.

1. Introduction

The city of Yogyakarta is passed by the Code River, Winongo River, and Gajahwong River. The most pollution was the Code River because the central part of the Code River drainage area crossed a densely populated urban area. Code River pollution mainly comes from household waste, because around the river is the most densely populated [1].

Urbanization accelerates population growth in Yogyakarta. The more population increases, the land resource needs for settlements increase, one of the locations considered strategic is the riverbank [2]. Some residents still dispose of liquid waste without processing into the river, so that the potential to pollute the environment [3]. Slum settlements around the Code River also trigger environmental health problems, such as garbage accumulation, increased groundwater pollution mainly by Nitrogen and coliform bacteria.

Pollution besides occurring in rivers also occurs in population wells, as a result of poor sanitation systems. These problems are exacerbated by terrace driver conditions. Terrace 1 is at the top of the cross section of the river. Terrace 2 is located below terrace 1, and terrace 3 is located at the bottom. Waste originating from septictank terrace 1 and terrace 2 will flow downward, and pollute the wells of residents living on terrace 3 as accumulated septic tanks from settlements above them. The purpose of

IOP Conf. Series: Materials Science and Engineering **650** (2019) 012057 doi:10.1088/1757-899X/650/1/012057

this study was to assess the condition of the well water quality of residents, based on DO parameters, pH, DHL, Temperature, TDS, and E-Coli bacteria and to ensure there is a direct correlation of accumulated well water pollution at high elevations to low elevations in terraced residential areas Code River banks.

The coordinate for the study area is along the Code river which entered the urban area of Yogyakarta consisting of Gowongan sub-district, Jetis sub-district, Kotabaru sub-district, Gondokusuman sub-district, Tegalpanggung and Suryatmajan sub-districts of Danurejan sub-district.

Coliform bacteria can be a signal to determine whether the water source has been contaminated by pathogens or not [4]. These bacteria are decomposing bacteria that produce various kinds of poisons which can cause disease if the amount is excessive in the body [5]. These bacteria also have a higher resistance than pathogens and are more easily isolated and grown [6].

The characteristics of these bacteria include aerobic or facultative anaerobes, including gram negative bacteria, do not form spores, and can ferment lactose to produce acids and gases at temperatures of 35° C-37° C [7]. The presence of coliform bacteria in food or drinks shows the possibility of the presence of enteropathogenic and/or toxigenic microorganisms that are harmful to health. E coli is a bacterium that comes from animal and human feces [8]. EPA recommendations for household water supply, for treatment, the number of fecal coliform is less than 2000 colonies/100 mL, and for drinking water standards less than 1 colony/100 ml [9], [10].

Oxygen is a very important parameter in water. Most living things in water need oxygen to sustain life, both plants and aquatic animals, depending on dissolved oxygen. Fish are aquatic creatures with the highest oxygen demand, then invertebrates, and the smallest oxygen needs are bacteria [11]. If at one time organic material in water becomes excessive as a result of the entry of human activity waste, causing the growth rate of microorganisms to multiply, which also means increasing oxygen demand, while supply oxygen from the air is fixed. In these conditions, the equilibrium between oxygen entering the water and those used by aquatic biota is not balanced, resulting in a deficit of oxygen dissolved in water. As with aerobic microbes, anaerobic microbes will also utilize carbon from organic matter. From this anaerobic respiration methane gas (CH4) is formed in addition to forming foul-smelling sulfide acid gas (H2S) [12].

pH is an acid base level of a solution measured on a scale of 0 to 14. The high and low pH of the water is strongly influenced by the content of other minerals contained in water. A pH value that is good for health is not smaller than 6.5 and not greater than 9. Larger values can change some chemical compounds into poisons that can interfere with health [11].

Electric conductivity (DHL) in water is a numerical expression that shows the ability of a solution to deliver an electric current. Measurements made based on the ability of cations and anions to deliver the electric current flowing in water samples can be used as an indicator, where the greater the value of electrical conductivity directed at the conductivity meter means the greater the ability of cations and anions to deliver electric current [11].

Total Dissolved Solids (TDS) is a parameter of the amount of minerals dissolved in this material, which can include carbonates, bicarbonates, chlorides, sulfates, phosphates, calcium nitrates, magnesium, sodium, organic ions and other ions. High concentration TDS can also reduce water clarity, provide a significant decrease in photosynthesis, and a combination with toxic compounds and heavy metals can increase the water temperature [11].

The temperature in water also greatly affects water quality, because the temperature can affect the reaction of the performance of microorganisms in water.

2. Methods

2.1. Study Area

The research location is along the Code river which entered the urban area of Yogyakarta consisting of Gowongan sub-district Jetis sub- district, Kotabaru sub-district Gondokusuman sub-district, Tegalpanggung and Suryatmajan sub-districts of Danurejan sub-districtas shown in Figure 1. The basis for consideration of the selection of research sites is the surrounding or Code Riverbanks. The urban area is a dense residential area, there are industrial activities, shops, offices, restaurants, hotels, laundry businesses and others.

Figure 1. Map of research location

2.2. Research Stages

The stages used in this study consist of:

- Survey preparation.
- Identification of population wells in the research location.
- Measuring the position of the population well with GPS.
- Plotting the location of residents'wells into the map of the research location.
- Taking water samples from residents'wells.
- Analysis of the quality of clean water from samples of water from wells in the laboratory.
- Assessment of water quality of community wells in the terraced area of the Code River by using a Geographic Information System (GIS).

IOP Conf. Series: Materials Science and Engineering **650** (2019) 012057 doi:10.1088/1757-899X/650/1/012057

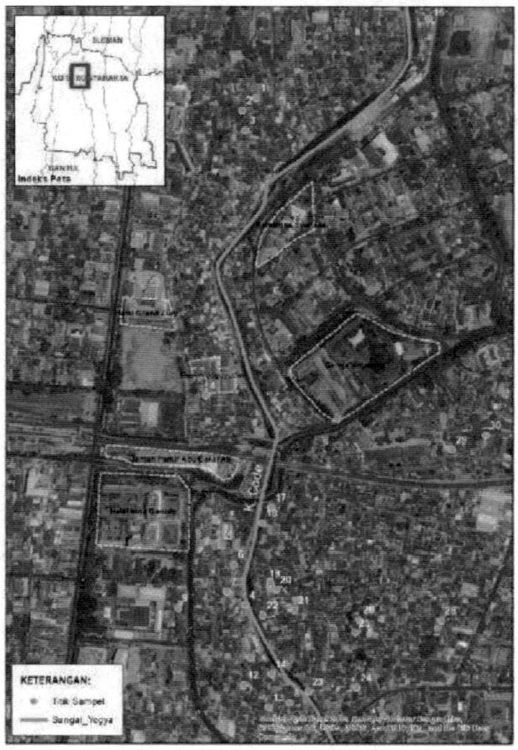

IOP Conf. Series: Materials Science and Engineering **650** (2019) 012057 doi:10.1088/1757-899X/650/1/012057

3.1.2. Results of examination of well water quality

Quality checks of well water samples were carried out on parameters: DO, pH, DHL, Temperature, TDS, and E-Coli. Table 1 shows the results of research on sampling water quality.

Table 1. Results of the quality of well water assessment

No	Location Kelurahan	Coordinate X	Y	Z	Kode sample	DO (ppm)	pH	DHL (µmhos /cm)	Suhu (OC)	TDS	E-Coli (MPN/100 ml)
1	Suryatmajan	430253	9138622	130	9	6.6	6.2	433	29	194	1100
2	Suryatmajan	430336	9138650	120	7	6.3	6.4	458	28	203	2400
3	Suryatmajan	430366	9138627	116	4	5.7	6.7	481	29	207	2400
4	Suryatmajan	430375	9138538	124	5	5.9	6.8	318	30	144	1100
5	Suryatmajan	430409	9138511	118	14	5.1	7.4	418	28	180	2400
6	Suryatmajan	430363	9138493	121	12	7.2	7	365	30	160	1100
7	Suryatmajan	430328	9138726	127	5	6.2	6.9	601	31	253	2400
8	Suryatmajan	430347	9138695	126	6	6	7	449	30	193	2400
9	Suryatmajan	430420	9138420	108	11	5.1	7	370	28	160	75
10	Suryatmajan	430430	9138418	108	15	5.4	6.9	304	28	135	460
11	Suryatmajan	430405	9138456	111	13	5.6	7.4	567	29	240	210
12	Suryatmajan	430440	9138463	111	8	5.7	7.4	384	28	167	2400
13	Tegalpanggung	430472	9138482	114	23	6	7.6	383	29	166	2400
14	Tegalpanggung	430553	9138489	116	24	6.1	7.2	448	29	196	1100
15	Tegalpanggung	430550	9138581	128	25	5.3	7.3	426	28	192	2400
16	Tegalpanggung	430557	9138603	131	26	5.3	7.1	360	28	158	75
17	Tegalpanggung	430693	9138602	121	28	5.4	7.3	371	28	164	1100
18	Tegalpanggung	430710	9138888	130	29	5.2	7.5	333	30	151	240
19	Tegalpanggung	430767	9138912	133	30	4	7.3	350	32	155	9
20	Tegalpanggung	430409	9138790	121	17	6	7.5	381	30	168	2400
21	Tegalpanggung	430394	9138764	123	18	3.7	7.5	381	30	168	1100
22	Tegalpanggung	430408	9138664	120	19	7.4	6.9	450	28	192	1100
23	Tegalpanggung	430418	9138653	121	20	5.5	7.5	431	29	186	93
24	Tegalpanggung	430395	9138610	121	22	5.6	7.2	407	28	184	2400
25	Tegalpanggung	430448	9138619	123	21	4.9	7.1	450	29	185	1100
26	Kotabaru	430672	9139610	137	16	4.3	7.5	381	30	164	1100
27	Gowongan	430361	9139455	133	2	4.2	6.9	366	29	160	1100
28	Gowongan	430366	9139422	130	3	5.3	7	371	29	162	1100
29	Gowongan	430384	9139476	132	1	6.1	7.5	375	29	162	3
30	Gowongan	430300	9138978	131	4	5.4	7.6	491	30	209	460

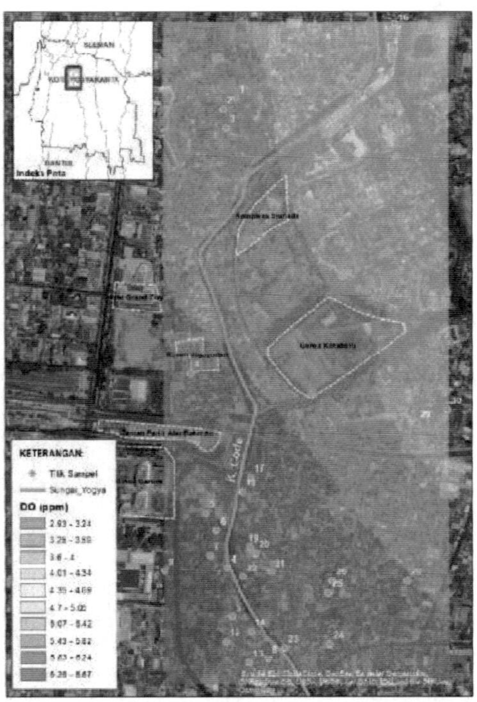

Figure 3. Results of DO sampling

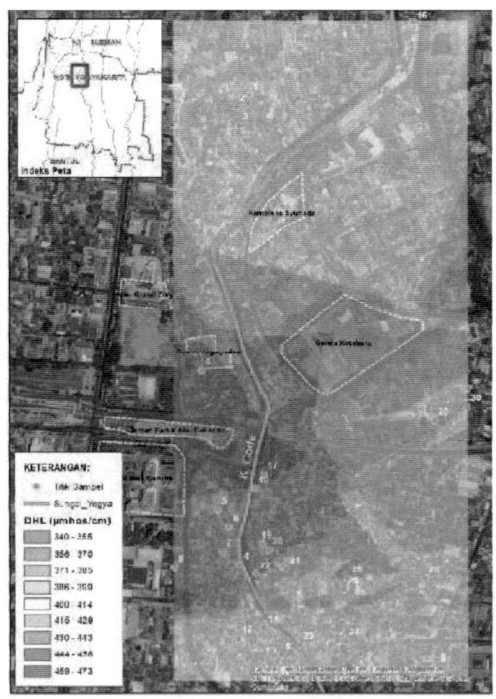

Figure 4. Results of DHL sampling

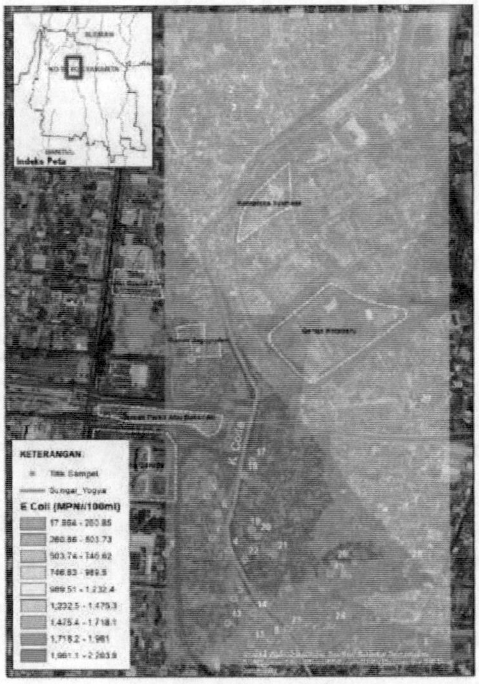

Figure 5. Results of E-Coli sampling

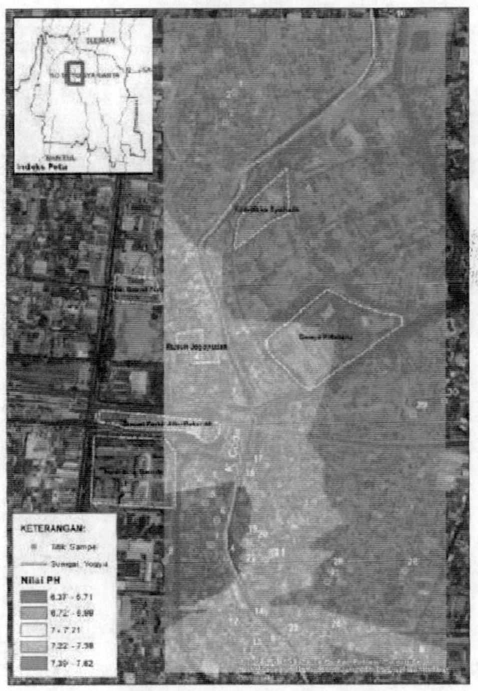

Figure 6. Results of pH sampling

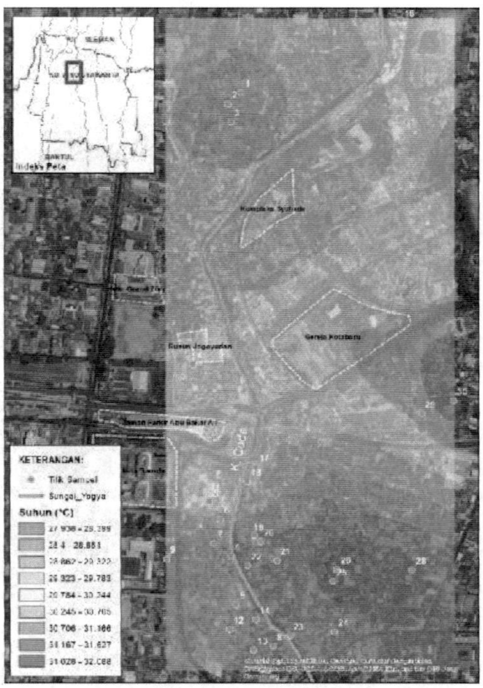

Figure 7. Temperature sampling results

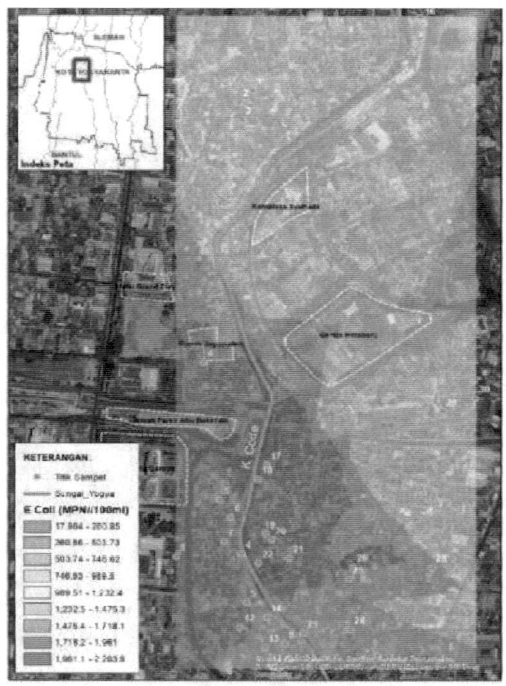

Figure 8. TDS sampling results

4. Conclusion

- Based on the data from the research results mapping the well water quality sampling consisting of parameters: Dissolved Oxygen (DO), DHL (Electrical Conductivity), Coliform Bacteria (EColi), pH,Temperature, and TDS (Total Dissolved Solids) as shown in Fig. 3, 4, 5, 6, 7, and 8 can be seen that there is a decrease in the quality of well water from high elevation to low elevation so that the quality of well water at low elevations is worse than the quality of well water at higher elevations, both in the Gowongan village, Kotabaru village, Suryatmajan village, and Tegalpanggung village. So that it can be concluded that there is a direct correlation of the accumulation of well water pollution at high elevations to low elevations in the residential area terraced by the Code River.
- The decrease in the quality of well water occurs because of waste originating from septic tank terrace1and terrace 2 will flow downward, and pollute the wells of residents living on terrace 3 as accumulated septic tanks from settlements above them.
- In more densely populated residential areas the quality of well water is getting worse compared to other regions. So it can be concluded also that there is also a decline in the quality of well water in densely populated residential areas.

Acknowledgments
Thank you to the Institute for Research and Community Service (LP3M) of Janabadra University and the Faculty of Engineering, for the funding of this research.

References

[1] Eko 2008 *Code Paling Tercemar* (Yogyakarta: Harian Jogja Edisi: Sabtu, 27 Desember 2008)
[2] Laily A R 2011 *Menggali Akar Permasalahan di Balik Fenomena Permukiman Kumuh di Bantaran Sungai Code* (Yogyakarta: Program Studi Ilmu Lingkungan Sekolah Pasca Sarjana Universitas Gadjah Mada Yogyakarta)
[3] Astuty A T, Arum H R T, and Hasanbasri M 2008 *Pengelolaan Kawasan Sehat. Study Kasus di Kampung Jetisharjo dan Cokrokusuman Kelurahan Cokrodiningratan Jetis Kota Yogyakarta* (Yogyakarta: Magister Kebijakan dan Manajemen Pelayanan Kesehatan Universitas Gadjah Mada Yogyakarta)
[4] Burhansyah 2011 *Bakteri Coliform dan Colitinja* http://burhan-syah.blogspot.com/2011/12/bakteri-coliform-dan-colitinja.html (Accessed Juni 2, 2014)
[5] Pracoyo N E 2006 *Penelitian Bakteriologik Air Minum Isi Ulang di Daerah Jabotabek Cermin Dunia Kedokteran* 152 pp. 37-40
[6] Doyle M P and Erickson M C 2006 *Closing The Door On The Fecal Coliform Assay* pp.162-163
[7] Hajna A A and Perry C A 1943 Comparative Study Of Presumptive And Confirmative Media For Bacteria Of The Coliform Group And For Fecal Streptococci Am *J Publ Hlth* **33** pp. 550-556
[8] Nengsih 2010 *Bakteri Koliform* http://nengsiha.blogspot.com/2010/01/bakteri-koliform.html (Accessed May 31, 2011)
[9] Anonim 2010 Bakteri Fekal (Fecal Coliform) *http://www.bangkoyoy.com/2010/10/bakteri-koliform-fekal-fecal-coliform.html* (Accessed May 31, 2011)
[10] Anonim 2010 *Bakteri Koliform yang Bersifat Anaerob* http://1sthumanwinner.wordpress.com/2010/12/16/bakteri-koliform-yang-bersifat-anaerob.html (Accessed May 31, 2011)
[11] Anonim 2012 *Standar Kualitas dan Kuantitas Air Bersih untuk Pemenuhan Kebutuhan Manusia* http://bhybhaeg.blogspot.com/2012/03/standar-kualitas-air-bersih.html (Accessed June 1, 2014)
[12] Anonim 2011 *Kimia Dahsyat Blog* http://kualitasair-minum.blogspot.com/2011/12/parameter-kualitas-air-bersih-fresh.html (Accessed June 5, 2014)

IOP Conf. Series: Materials Science and Engineering **650** (2019) 012058 doi:10.1088/1757-899X/650/1/012058

Study of suspended sediment transport discharge of Serayu River, Central Java, Indonesia in laboratory

W A Pranoto[1], J Johan[1] and Reynaldo[1]

[1]Civil Engineering Department, Faculty of Engineering, Universitas Tarumanagara

[*] watip@ft.untar.ac.id; johnyjohan@gmail.com; reynaldokristianjayakusuma@gmail.com

Abstract. Sedimentation is a natural phenomena that occurs in rivers in Indonesia, including Serayu River in Central of Java, Indonesia. The sedimentation will result in siltation on the riverbed that reducing river function. Suspended load is one of the main factor that cause of sedimentation, therefore research is needed to to determine the amount of sedimentation. knowledge about suspended sediment transport has been developed by several experts in the water sector. This research is done by creating a simulation of the flow of water so that made a susoended load drift. This research was done by using the sediment that pass sieve No.200. The result from the research is knowing the correlation between water discharge and suspended sediment discharge. The result from laboratory then will be compared with Enstein's; Chang, Simon and Richardson's; and Lane and Kalinske's approach.

1. Background

River is one part of public waters and is a natural channel that carries water from upstream to downstream. Indonesia itself has 5,590 main rivers, one of which is Serayu River. Serayu River is one of the largest rivers in Java Island which is located in Central Java Province which crosses several districts, such as Wonosobo, Banjarnegara, Purbalingga, Banyumas and Cilacap Regencies. The length of the main river reaches 180km with 11 tributaries.

Because the flow rate is large, of course, the Serayu River not only flows water, but the flow also carries various particles with water flow. The particles carried by the flow of water are the result of eroded soil and then undergo sedimentation in several parts of the Serayu River. Most of the results of erosion will settle in the estuary area in the form of mud and sand.

In the process of transporting sediment due to the flow of the river will be divided into two types of sediment transport, basic sediment transport (bed load transport) and transport of floating sediments (suspended load transport). The amount of sediment transported due to river water flow is determined by flow velocity. Apart from being influenced by flow velocity, the transport of floating sediments (Suspended Load Transport) is also influenced by the length of the river, the slope of the riverbed, the cross-sectional area, the depth of the river and several other parameters that influence the transport process of sedimentation.

Suspended load is a particle that moves in a vortex of flow that tends to continue to float along the flow. Loads of flying sediments move along with river water flow, consisting of fine sand which is

IOP Conf. Series: Materials Science and Engineering **650** (2019) 012058 doi:10.1088/1757-899X/650/1/012058

always supported by water, and has very little interaction with the river bed because it has been pushed up by flow turbulence.

The material used in the study was taken from the Serayu River estuary :

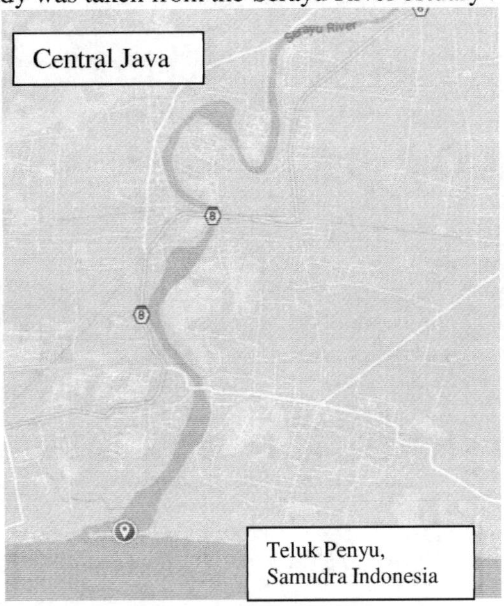

Figure 1. Sample Taken Location

2. Theory

There are several factors that can influence the sedimentation process, namely, the size and shape of sediment-forming materials, sediment loads, and changes in the speed of water flow. These things will greatly affect the sedimentation process itself. If there is a change from several factors above, it is normal to change from the channel channel shape which will adjust from slope, depth, thickness, pattern, layer composition and vegetation around the channel.

There are several general techniques for determining grain size gradations, namely (I) direct measurement with macros or microscopes, (II) using sieve analysis, and (III) with Hydrometer. Filter analysis is used to determine the size of the grain from a rather large to fine sand. To determine the diameter of granules such as silt and clay, the Hydrometer method must be used to determine the diameter size of the dominating sample.

The result of the research will be compared with three expert approach that develop studies in suspended load transport, as follow :

- Einstein's (1950)

  Einstein is assumed that β = 1 and k = 0.4. Einstein approach is based by shear velocitu and grain roughness or grain zise of the suspended load. The equation of Einstein approach :

$$q_{sw} = 11.6\, U^{*} C_{a} a \left\{ \left[2.303 \log \frac{3.02D}{\Delta} \right] I_1 + I_2 \right\} \quad (1)$$

where:

q_{sw} = Suspended load transport discharge
U^* = U'^* = Shear velocity
Ca = Concentration by dry weight
a = Thickness of bedload in the riverbed
D = Depth of water
I_1, I_2= Numeric Intregrity from **Figure 3.** and **Figure 4.**

IOP Conf. Series: Materials Science and Engineering **650** (2019) 012058 doi:10.1088/1757-899X/650/1/012058

Δ = Sediment size x (correction factor on **Figure 2.**)

$$Z = \frac{\omega}{0.4\, U'^{*}}$$

ω = Fall velocity

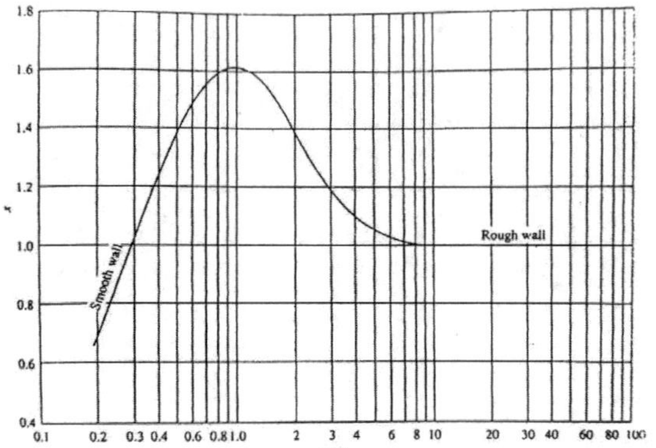

Figure 2. Correction Factor (Yang, 1996)

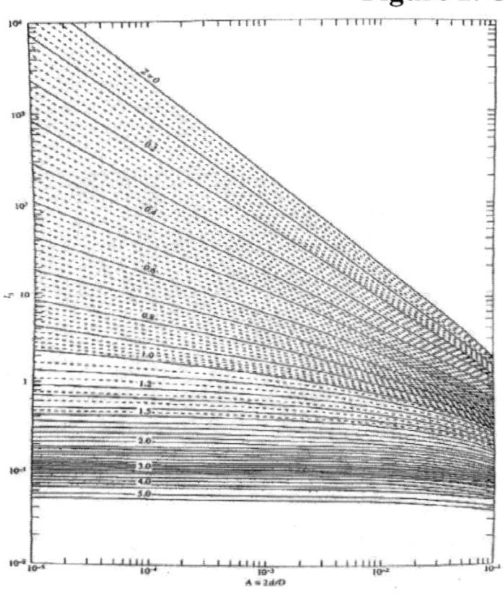

Figure 3. I_1 (Yang, 1996)

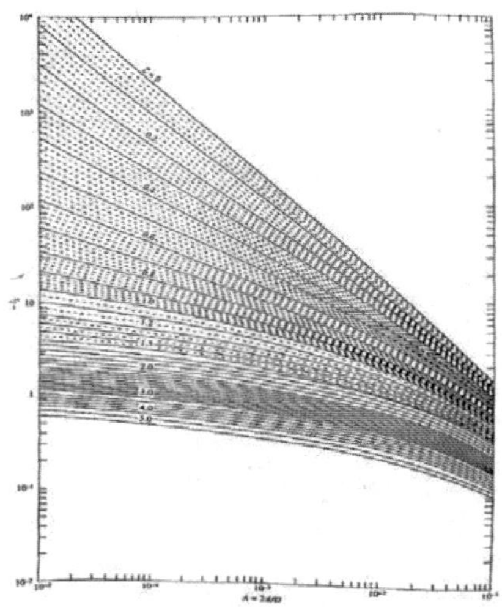

Figure 4. I_2 (Yang, 1996)

- Lane and Kalinske's (1941)
  Lane and Kalinske assumed that $\beta = 1$ and $\varepsilon_s = \varepsilon_m$, and the Lane and Kalinske approach is as follow as :

$$q_{sw}=q\ C_a\ P_L \exp\left(\frac{15\ \omega a}{U^*D}\right) \quad (2)$$

Where :

qsw	= Suspended load transport discharge
Ca	= Concentration by dry weight
PL	= Coefficient Relative With ω and U_* from **Figure 5.**
	= Fall velocity according by dominant sediment size
U_*	= Shear velocity
D	= Depth of water

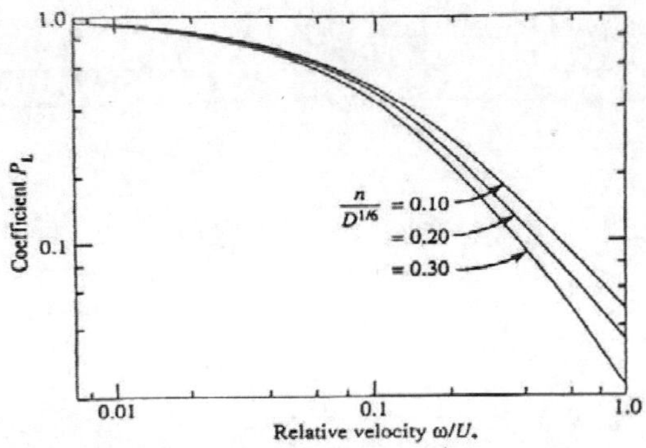

Figure 5. PL Coefficient Relative With ω and U_* (Yang, 1996)

- Chang, Simon and Richardson's (1965)
  The Chang, Simon and Richardson approach is based by measured in weight per unit volume of water-sedimen mixture. The Chang, Simon, and Richardson approach is follow as:

$$q_{sw}=DC_a\left(V\,I_1-\frac{2\,U^*}{k}I_2\right) \quad (3)$$

Where :

D	= Depth of water
Ca	= Concentration by dry weight
V	= Water velocity
I_1 & I_2	= From **Figure 6.** And **Figure 7.**
U_*	= Shear velocity
k	= konstanta Von Karman (0,4)
Z_2	$= \frac{2\omega}{\beta U^* k}$
ξa	$= \frac{a}{D}$

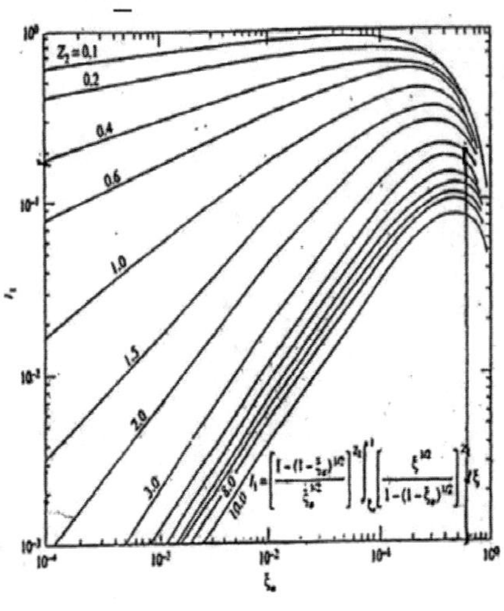

Figure 6. I_1 Base on ξa and Z_2 (Yang, 1996)

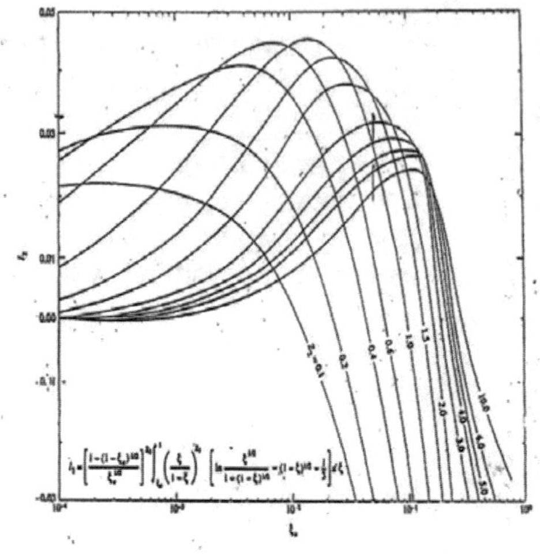

Figure 7. I_2 Base on ξa and Z_2 (Yang, 1996)

3. Research Methodology

The step taken in this research is as follow :

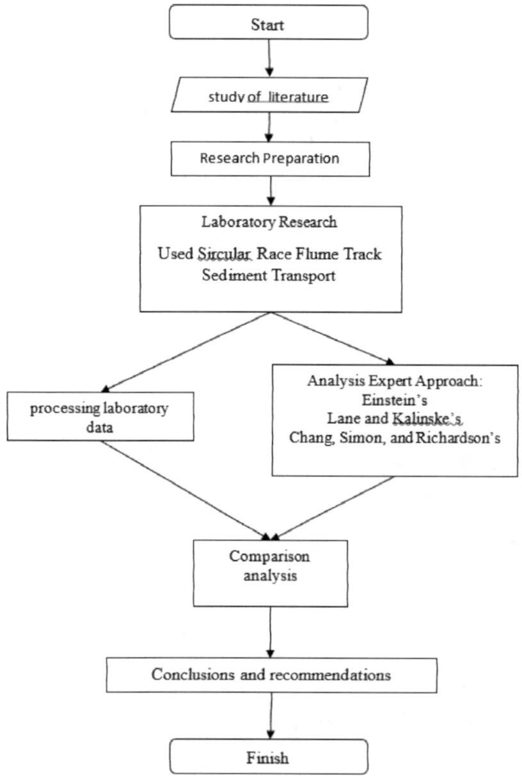

Figure 8. Research Flowchart

IOP Conf. Series: Materials Science and Engineering **650** (2019) 012058 doi:10.1088/1757-899X/650/1/012058

4. Results

This research was done by using the sediment that pass sieve No.200. The specific sediment size is used determine with Hydrometer test. From hydrometer test ,the most particle size is the most dominance is ⌀ = 0,0321 mm with percent retained is 17,66%. Hydrometer test result is as follow :

Table 1. Hydrometer Test

Particle Diameter (mm)	Percent Retained (%)
0.0321	17.66
0.0254	10.88
0.0221	7.25
0.0201	7.25
0.0155	12.69
0.0118	7.25
0.0087	7.25
0.0062	3.63
0.0050	1.26
0.0045	0.55
0.0032	1.81
0.0023	1.81
0.0013	1.81

After determine the particle size, the research will Compared the water discharge with suspended sediment transport discharge between Laboratory result and expert approach (Enstein's; Chang, Simon and Richardson's; and Lane and Kalinske's approach).

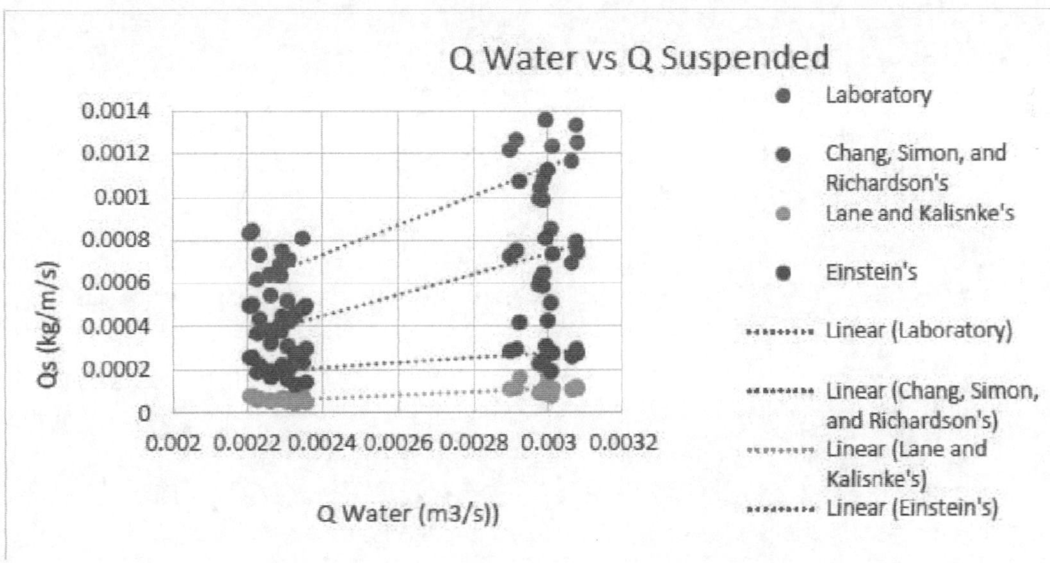

Figure 9. Q Water Against Q Suspended Load Transport

From the Figure 9. the bigger Q water, the bigger too Q suspended load transport from the result of Laboratory research and from the equation approach. the biggest results are the results of laboratory research and the lowest is from Lane and Kalinske's equation. the results closest to the lab results are the results from Chang, Simon and Richardson's approach. The main factor of all result is the water velocity, because the bigger water velocity it will bring more the particle of suspended sediment. The result of Chang, Simon and Richardson's equation approach is the closest to Laboratory research result, so it more suitable to used compared with another equation approach.

To get the equation that is more suitable for the condition of the Serayu river, more research is needed to obtain the specific equations for the river Serayu and also requires deep learning of sedimentary material, both from soil type and chemical properties.

References

[1] Diansari R 2014 *Analisis Perhitungan Muatan Sedimen (Suspended Load) pada Muara Sungai Lilin Kabupaten Musi – Banyuasin* Jakarta.

[2] Laboratorium Hidrolika Program Studi Teknik Sipil Universitas Tarumanagara *Penuntun Praktikum Hidrolika* Jakarta

[3] Laboratorium Mekanika Tanah Program Studi Teknik Sipil Universitas Tarumanagara. *Penuntun Praktikum Mekanika Tanah* Jakarta

[4] Oktavia K *Analisis Sedimentasi Pada Muara Sungai Komering Kota Palembang*

[5] Pranoto W A , and Sekarningrum A 2014 *Pengaruh Debit, Diameter Sedimen, Dan Kemiringan Saluran Pada Debit Sedimen Dasar Di Sungai Serayu* Jakarta.

[6] Pranoto W A, and Sumanton, L 2017 *Studi Angkutan Sedimen Dasar Sungai Serayu Di Laboratorium* Jakarta

[7] Sengupta S M *Introduction of Sedimentology.*

[8] Yang C T 1996 *Sediment Transport Theory and Practice.* The McGraw-Hill Companies, Inc, United States Of America.

IOP Conf. Series: Materials Science and Engineering **650** (2019) 012059 doi:10.1088/1757-899X/650/1/012059

STUDY OF SEDIMENT TRANSPORT DISCHARGES OF CITARUM RIVER, WEST JAVA IN LABORATORY

Wati A. Pranoto[1] dan Marvin Subari[2]

[1]Department of Civil Engineering Universitas Tarumanagara
[2]Department of Civil Engineering Universitas Tarumanagara

[1]watip@ft.untar.ac.id
[2]marvinsubari21@gmail.com

Abstract. Citarum River is one of the longest and largest rivers in Indonesia located in West Java with a length of 300 km. The Citarum River carries the sediment in the water stream from the erosion. This study focused on bed load sediment transport. The study of sediment transport is still a few, but some experts have proposed the formula to calculate the amount of sediment transport such as Einstein, Shields, and Schoklitsch. A research in Hydraulics Laboratory and Soil Mechanics Laboratory is conducted to examine the bed load transport from the Citarum River. The purpose of this study is to analyze and compare the result of discharge obtained from the laboratory with the Einstein, Shields, and Schoklitsch approach.

Keywords : bedload sediment transport, Citarum River, discharge

1. BACKGROUND

Citarum River is one of the rivers in Indonesia. With a length of 300 km, the Citarum river is the largest and longest river in West Java Province. This river crosses the Regency / City, starting from a spring located in Mount Wayang (Bandung Regency) and then ending in the Muara Gembong area by passing Bandung / West Bandung District, Cianjur Regency, Purwakarta Regency, and Karawang / Bekasi Regency.

This Citarum River has been around since the Kingdom of Taruma. The Citarum river has long been navigated by small boats. Many people use the Citarum River for fisheries (fishing, fishing, fish farming) and for drinking water supplies for a portion of Jakarta's population.

The water that flows in the Citarum River besides carrying water, this river also carries various kinds of particles. The particles carried by the water flow are caused by erosion that occurs in this river. After that these particles can settle in several parts of the river because of the sedimentation process.

In the process of sedimentation, there are many factors that affect sediment transport. Besides being influenced by flow velocity, bed load transport is also influenced by the length of the river, the slope of the river bed, river cross section, and various other parameters that affect the sedimentation process.

IOP Conf. Series: Materials Science and Engineering **650** (2019) 012059 doi:10.1088/1757-899X/650/1/012059

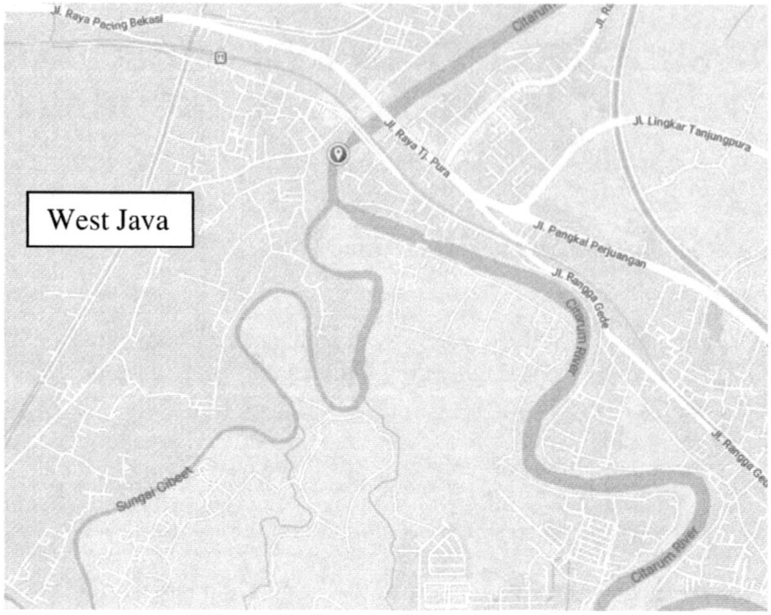

Figure 1. Sample Taken Location

2. THEORY

Many approaches are used by experts to research and develop studies on bed load transport. The study was conducted to compare the results in the laboratory with the results of the approach proposed by several experts, as follow :

- Einstein

  Einstein was the first expert to proposed the bed load sediment transport equations with the theoretical approach, namely statistical theory.

$$\Phi = \frac{q_b}{\gamma_s}\left(\frac{\gamma}{\gamma_s-\gamma} x \frac{1}{g\,D^3}\right)^{1/2} \quad (1)$$

Where :
Φ = instensity parameter bed load
Q_b = bed load sediment transport (N/m.s)
γ dan γs = specific weight of water and specific weight of sediment (kg/m^2)
g = gravity (m/s^2)
D = diameter of sediment (mm)

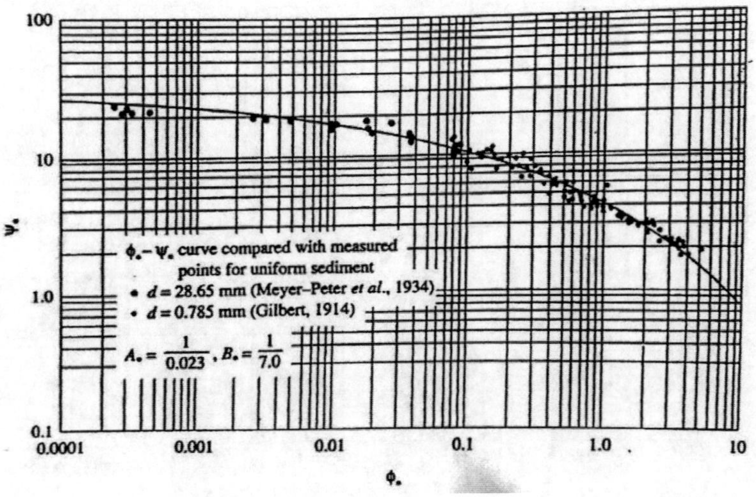

Figure 2. Relationship between Ψ and Φ for Einstein's bed load function (Einstein, 1950)

- Shields

  Shields (1936) measured flow conditions with sediment transport greater than zero, and then extended the relationship to obtain the flow condition corresponding to incipient motion. Thus, a semiempirical equation for bed load is obtained, namely

$$\frac{q_b\,\gamma_s}{q\,S\,\gamma} = \frac{10\,(\tau-\tau_c)}{(\gamma_s-\gamma)d_{50}} \tag{2}$$

$$\tau_c = c\,(\gamma_s - \gamma)d \tag{3}$$

$$U_* = (gDS)^{\frac{1}{2}} \tag{4}$$

$$R_e = \frac{U_* d}{v} \tag{5}$$

Where :
Q_b = bed load sediment transport discharge ((ft^3/s)/ft)
q = water discharge ((ft^3/s)/ft)
d_{50} = diameter of sediment (mm)
$\tau = \gamma DS$ (lb/ft^2)
D = depth of water (ft)
γ = specific weight of water (lb/ft^3)
S = channel slope (assumption 0.5%)
τ_c = critical tractive force along the bed (lb/ft^2)
U_* = shear velocity (ft/s)
v = kinematic viscosiry of water
R_e = Reynold constants
C = dimensionless shear stress (Figure 3)

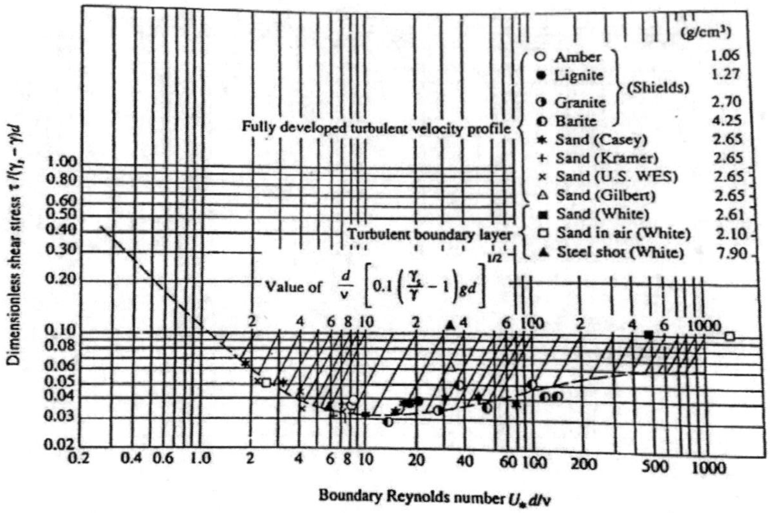

Figure 3. Shields Diagram (Dake,M. K Jones, 1985)

- Schoklitsch

  Schoklitsch (1934) is a expert who suggests the influence of water discharge that occurs to determine the amount of bed load sediment transport, as follow :

$$Q_b = 7000 \frac{s^{3/2}}{d^{1/2}} (q - q_c) \quad (6)$$

$$q_c = \frac{0.00001944\, d}{s^{4/3}} \quad (7)$$

Where :
Q_b = bed load sediment transport ((kg/s)/m)
d = diameter of sediment (mm)
q = water discharge ((m^3/s)/m)
q_c = crititcal water discharge ((m^3/s)/m)

3. RESEARCH METODHOLOGY

the steps taken in this study are as follows :

IOP Conf. Series: Materials Science and Engineering **650** (2019) 012059 doi:10.1088/1757-899X/650/1/012059

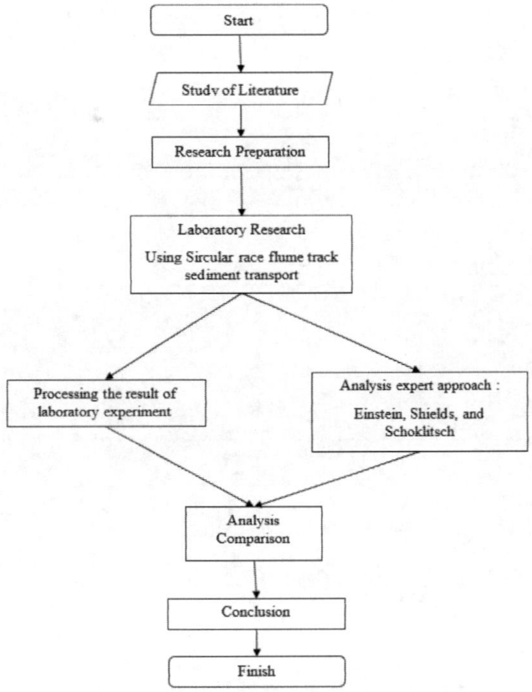

Figure 4. Research Flowchart

4. RESULT

The sediment that is use for the research must be analyzed for the dominant sand size gradations in the soil using sieve analysis. The result is as follow :

Table 1. Sieve Analysis Data

Sieve No 1	Diam. (mm)	Wt. retained	% reatained	% passsing
4	4.75	0.4	0.08	99.92
10	2	6.7	1.34	98.58
20	0.85	21.6	4.32	94.26
40	0.425	110.7	22.14	72.12
60	0.25	229.7	45.94	26.18
140	0.106	40.5	8.1	18.08
200	0.075	0.2	0.04	18.04
pan	-	0.3	0.06	
sum		410.1		

After determine the diameter particle sediment, compare the water discharge and the bed load sediment transport discharge between the research results and the 3 approach (Einstein, Shields, and Schoklitsch).

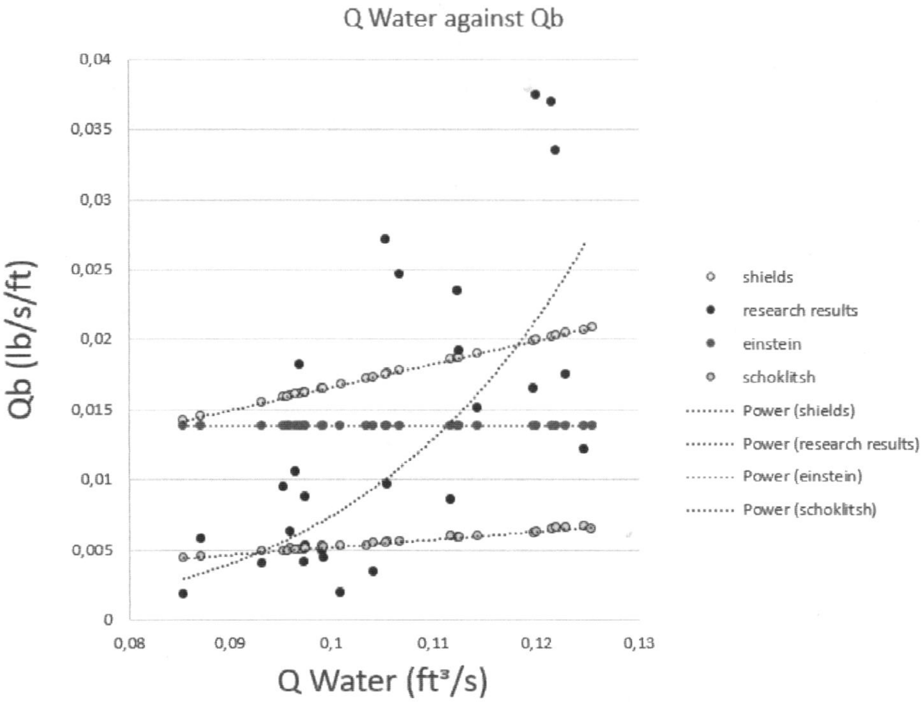

Figure 5. Q water against Qb from 3 approach

From **Figure 5.** we can conclude that the sediment discharge in Citarum river obtained in the research is not in accordance with the discharge obtained from 3 approaches (Einstein, Shields, and Schoklitsch). This can happen because the Einstein approach did not pay attention to the speed and only focused with the diameter of the sediment. Whilst the Schoklitsch approach is only applicable from the velocity range of 0.004 to 0.007 and Shields approach is more variegated, hence is more suitable is for multifarious elements.

To get the equation that is more suitable for the condition of the Citarum River, more research and deep learning is required to obtain specific equations for the Citarum River. Also for the soil type and chemical properties.

5. REFERENCES

[1] Laboratorium Hidrolika Program Studi Teknik Sipil Universitas Tarumanagara., (2019). "Penuntun Praktikum Hidrolika", Jakarta.

[2] Laboratorium Mekanika Tanah Program Studi Teknik Sipil Universitas Tarumanagara., (2019). "Penuntun Praktikum Mekanika Tanah", Jakarta.

[3] Oktavia, K., "Analisis Sedimentasi pada Muara Sungai Komering Kota Palembang".

[4] Pranoto, W.A. and Sekarningrum, A., (2014). "Pengaruh Debit, Diameter Sedimen, dan Kemiringan Saluran pada Debit Sedimen Dasar di Sungai Serayu", Jakarta.

[5] Pranoto, W.A. and Sumanton, L., (2017). "Studi Angkutan Sedimen Dasar Sungai Serayu di Laboratorium". Jakarta.

[6] Sengupta, S.M., *Introduction of Sedimentology*

[7] Yang, C.T., (1996). *Sediment Transport: Theory and Practice*. The McGraw-Hill Companies, Inc, United Stated of America.

IOP Conf. Series: Materials Science and Engineering **650** (2019) 012060 doi:10.1088/1757-899X/650/1/012060

Distribution fitting on rainfall data in Jakarta

V Kurniawan[1]

[1]Department of Civil Engineering, Universitas Tarumanagara, Indonesia

*vkurniawan@ft.untar.ac.id

Abstract. The estimation of design flood or rainfall for various return periods is not easy due to the difficulty in finding the fitting distribution for the corresponding data. This study focuses on selecting the best distribution in representing rainfall data from Kemayoran rain gauge (Jakarta, Indonesia) for the last 20 years. The objective is to probe the most fitting distribution for Jakarta. This study applies goodness of fit tests to Kemayoran rainfall data. The result is quite unexpected as the best fitting distributions are not commonly used in flood frequency analysis (FFA) subject and most of them are distributions with three parameters. The referred distributions in Indonesia national standard for flood estimation (SNI 2415:2016) produce mixed results. The classic distributions in FFA e.g. Lognormal 3P and Log-Pearson 3 show good compatibility to the corresponding data while the less prevalent distributions such as Lognormal and Normal are not as compatible.

1. Introduction

Flood frequency analysis (FFA) is an important practice in hydraulic engineering and water resources engineering as the determination of design flood is one of the key factors in designing hydraulic structures e.g. dam, flood control structure, urban drainage system etc [1] [2]. Underestimating design flood could inflict heavy financial damage and at worst, it could claim lifes. On the other hand, overestimating design flood means excessive resources being wasted needlessly. Therefore, it is crucial to determine design flood as accurate as possible in order to achieve effective and efficient design.

Design flood determination is a perennial issue. The process starts from determining rainfall for specific return period or extreme rainfall which later will translate into design flood. However, the determination of extreme rainfall hitherto remains a difficult practice although it has been researched for decades. The studies can be traced back to the work of Das in 1952 [3] or Buishand in 1977 [4]. Despite of such long period of research on stochastic aspect on rainfall, the estimation of extreme rainfall is still an ongoing subject especially in distribution fitting aspect. The accurate distribution fitting means more accurate design flood estimation hence it compels many researchers to perfect it.

This paper will focus on finding the most representative distribution for a particular rain gauge in Jakarta, Indonesia. The nation has a national standard (will be referred as SNI 2415:2016 henceforth) to estimate the design flood [5]. The standard mentions 8 (eight) distributions which are likely to describe the behaviour of the series of maximum annual rainfall in Indonesia. The distributions are :

1. Gumbel;
2. Pearson 3;

IOP Conf. Series: Materials Science and Engineering **650** (2019) 012060 doi:10.1088/1757-899X/650/1/012060

3. Log-Pearson 3;
4. Gamma;
5. Log-Gamma;
6. Normal;
7. Log-Normal (2 parameters); and
8. Log-Normal 3 parameters.

Although there are eight distributions referred in SNI 2415:2016, there are only four distributions which are customarily used in Indonesia i.e. Normal, Lognormal, Gumbel, and Log- Pearson 3. Gumbel and Log-Pearson 3 are the most frequently selected distributions to model rainfall data while Normal and Lognormal often serve as the benchmarks for other distributions in goodness of fit test.

The main objective of this study is to investigate the most fitting distributions in representing the corresponding rainfall data. It will assess if the referred distributions in SNI 2415:2016 fit the assessed rainfall data. Furthermore, it is plausible if the pattern of rainfall data in Indonesia follows other distribution apart from the distributions mentioned in SNI 2415:2016. Therefore, the study also assesses the possibility if there are feasible alternative distributions in modelling Jakarta rainfall.

2. Case Study

The source of the rainfall data is from Kemayoran meteorology station which is located in Jakarta, Indonesia. Its geographical position is at -6.16 latitude and 106.84 longitude and 4 meter above mean sea level. It is registered in World Meteorological Organization with the ID number 96745. The station is positioned in the midst of the city with the prevailing tropical climate. There are two seasons i.e. dry season and wet season. The latter roughly lasts 4 months in a year.

Kemayoran rain gauge is chosen because it is located in the headquarter of Badan Meteorologi, Klimatologi, dan Geofisika (the national body of meteorology, climatology, and geophysics; abbreviated as BMKG). Therefore, it grants the accuracy, the reliability, and the availability of the retrieved data.

The assessed rainfall data comes from the observation of Kemayoran rain gauge in the last 20 years (1999-2018). The data is retrieved online from BMKG official website. The maximum rainfall in a day for each year are displayed in table 1 (abbreviated as R)

3. Methodology

This study utilises MathWave software to perform the goodness of fit analysis. It examines the data sample against three sorts of goodness of fit test i.e. Kolmogorov-Smirnov, Anderson- Darling, and Chi-Squared. The test will determine which distribution is the most fitting for the given data sample from the available 61 distributions.

Kolmogorov-Smirnov is a test which calculates the maximum difference between an empirical and a hypothetical cumulative distribution [6]. Anderson-Darling examines if the data sample conforms to a population from particular distribution [7]. Lastly, Chi-Square divides the sample data into several intervals and compare the factual number of data to the expected number of data in each interval [8]. These test of goodness of fit methods are very commonly employed in statistics calculation including rainfall analysis. The result of implementing these tests to the Kemayoran rainfall data will be displayed in the next section.

IOP Conf. Series: Materials Science and Engineering **650** (2019) 012060 doi:10.1088/1757-899X/650/1/012060

Table 1. The annual maximum rainfall in Jakarta in the last 20 years

Year	R (mm/day)
1999	147.2
2000	94.8
2001	82.2
2002	168.5
2003	199.7
2004	129.3
2005	124.1
2006	72.0
2007	234.7
2008	192.7
2009	122.5
2010	93.0
2011	119.2
2012	105.2
2013	193.4
2014	147.9
2015	277.5
2016	124.5
2017	179.7
2018	104.6

4. Result and Discussion

The brief descriptive statistics of the data in table 1 is shown in table 2. Meanwhile, the result of the implementation of the aforementioned goodness of fit tests on Kemayoran rainfall data is displayed in table 3.

Table 2. The annual maximum rainfall in Jakarta in the last 20 years

Statistic	Value
Sample size	20.0
Mean	147.2
Standard deviation	94.8
Variance	82.2
Skewness	168.5
Median	199.7

The meaning of each column in table 3 is as follows :

- column (1) : the rank of the most fitting distribution, sorted from the value of column (8);
- column (2) : distribution type;
- column (3) : the result of Kolmogorov-Smirnov test;
- column (4) : the distribution's rank in Kolmogorov-Smirnov test;
- column (5) : the result of Anderson-Darling test;
- column (6) : the distribution's rank in Anderson-Darling test;

IOP Conf. Series: Materials Science and Engineering **650** (2019) 012060 doi:10.1088/1757-899X/650/1/012060

Table 3. The rank of the most fitting distribution for Jakarta rainfall 1999-2018

Rank	Distribution	KS		AD		CS		Rank Sum
		Stat.	Rank	Stat.	Rank	Stat.	Rank	
(1)	(2)	(3)	(4	(5)	(6)	(7)	(8)	(9)
1	Gamma (3P)	0.083	1	0.165	2	0.189	1	4
2	Weibull (3P)	0.089	6	0.169	5	0.206	2	13
3	Johnson SB	0.092	8	0.145	1	0.513	7	16
4	Fatigue Life	0.085	2	0.167	3	0.745	12	17
5	Inv. Gaussian	0.086	4	0.169	4	0.753	11	26
6	**Lognormal (3P)**	0.089	5	0.174	10	0.735	11	26
7	Pearson 6 (4P)	0.086	3	0.172	9	0.791	15	27
8	Log-Logistic (3P)	0.089	7	0.198	18	0.5	6	31
9	Pearson 5	0.093	9	0.183	12	0.901	18	39
10	Pearson 5 (3P)	0.093	10	0.183	13	0.905	19	42
...	...	...	...	...	...	...	...	...
13	Gen. Extreme Value	0.097	13	0.171	8	1.265	34	55
...	...	...	...	...	...	...	...	...
15	**Log Pearson 3**	0.103	17	0.170	6	1.364	40	63
16	**Log Gamma**	0.104	18	0.171	7	1.362	39	64
26	**Gumbel**	0.113	22	0.193	16	1.441	42	80
30	**Gamma**	0.126	28	0.223	20	1.270	36	84
32	**Lognormal**	0.115	23	0.200	19	1.476	44	86
35	**Normal**	0.169	38	0.466	30	0.973	26	94

- column (7) : the result of Chi-Square test;
- column (8) : the distribution's rank in Chi-Square test; and
- column (9) : the sum of the rank in each goodness of fit test i.e. (4) + (6) + (8).

The top ten most fitting distribution as well as the rank of the distributions referred in SNI 2415:2016 are revealed in table 3. The top ten most fitting distributions are dominated by the distributions which are not prevalent in FFA subject and most of them feature three parameters.

The discussion of the top ten most fitting distribution is as follow :

1 Gamma (3P)
Three parameter Gamma or Generalized Gamma distribution fits the rainfall data better than other distributions in this case. It resonates with the finding of Chen [10] which concluded that Generalized Gamma distribution fits the flood data for various return period reasonably well. It also agrees with USGS [11] which referred this distribution to model rainfall data. However, Generalized Gamma distribution is still less popular than its counterpart i.e. Gamma distribution (with 2 parameter) as the latter is widely implemented in FFA [12].

2 Weibull (3P)
Three parameter Weibull distribution closely shares the same account with Gamma (3P) distribution. So far there is no previous study which supports the reliability of Weibull (3P) distribution in representing extreme rainfall. However, Weibull distribution (2 parameter) is a commonly used in hydrology field of study.

3 Johnson SB
This distribution is commonly used in biology and nature field of study. There are only few rainfall-related studies which utilizes this distribution such as D'Adderio et al who estimated rainfall rate, reflectivity, and diameter [9]. However, so far there is no study which connects Johnson SB distribution with extreme rainfall or FFA.

IOP Conf. Series: Materials Science and Engineering **650** (2019) 012060 doi:10.1088/1757-899X/650/1/012060

4 Fatigue Life (3P)
It is usually employed to model the lifetime of material before it suffers failure. Its applicability in FFA or in hydrology field of study is unheard. It is doubtful that this distribution can be universally used to represent extreme rainfall.

5 Inv. Gaussian (3P)
Inverse Gaussian or Wald distribution is widely used in analysing lifetime data [13] or in bioscience [14]. Judging from its characteristics and its past usage, it is unlikely that this distribution is able to extensively fit rainfall data.

6 Lognormal (3P)
Three parameter Lognormal distribution shares almost similar characteristics with Lognormal distribution (2P). The difference is Lognormal (3P) introduces an additional parameter on which the value of data is reduced. Lognormal (3P) distribution is less popular than Lognormal distribution. But, Sangal and Biswas [15] proved that the former produced reasonable success when projected on the data of flood flow (note that the data derives from the flood flow observation, not from the rainfall data). In this case, Lognormal (3P) displays better fitting than other distributions referred in SNI 2415:2016.

7 Pearson distribution family
Four parameter Pearson 6, Pearson 5, and three parameter Pearson 5 appear in the top ten list while Log-Pearson 3, which is very prevalent in FFA, surprisingly fares marginally worse. Although Pearson 6 4P, Pearson 5, and Pearson 5 (3P) outperfrom Log-Pearson 3 in this case, there is no proof which supports that the former would be accurate for other FFA-related cases.

8 Log-Logistic
Log-Logistic distribution is employed for various purposes such as medic [16], economy [17], and FFA as conducted by Ahmad [18]. In that particular case, Ahmad proved that Log- Logistic distribution performed well and even better than the more traditionally prevalent distribution such as Generalized Extreme Value (GEV) distribution. Here, in this study, Log-Logistic distribution again performs better than GEV although more extensive studies are required to prove if the former can be universally used in FFA.

9 Gen. Extreme Value
Although Generalized Extreme Value (GEV) sits outside of top ten list, but the distribution worths mentioning as it is one of the most generally used distribution in modelling any extreme events including natural extreme events [19] and FFA [11] [20] [21] [22] [23]. There are plenty of studies which prove its accuracy in modelling the extreme flood or rainfall [24] [25]. Its extensive use in the world raises an expectation for GEV to perform better in modelling rainfall data in Jakarta, however the result here still shows satisfactory reliability.

The following points are the elaboration about the distributions referred in SNI 2415:2016 which yield mixed result in the goodness of fit tests :

1 Log-Pearson 3
Log-Pearson 3 has a strong history in representing the distribution of flood discharge in

USA [11] [26] and it also shows good fit at other parts of the world such as Nigeria [2] or Australia [27]. From the conducted test above, the distribution is also compatible with Jakarta rainfall data as well.

2 Log-Gamma
While Gamma or Generalized Gamma distribution are used quite frequently in FFA, there is no study which supports the implementation of Log-Gamma in this field. Here in Jakarta, the distribution fits the rainfall data well, however, the question is raised if it can be universally applied to other regions.

3 Gumbel
Gumbel or Extreme Value (EV) Type 1 or Fisher-Tippett distribution is also adopted in FFA [28] [29] although it is less popular than Generalized Extreme Value (GEV) distribution. Nevertheless the former is referred in SNI 2415:2016 while the latter is not. EV Type 1 is inferior compared to GEV in fitting Jakarta rainfall data which is quite expected as GEV is more prevalent in FFA.

4 Lognormal
Lognormal distribution is quite often employed in goodness of fit test in FFA although it often loses to other distribution (e.g. GEV, Log-Pearson 3) in fitting the rainfall or flood data. Although there are two cases from Malaysia which resulted in Lognormal distribution's prevailing over GEV or Log-Pearson 3 [30] [31] in fitting the rainfall data, GEV and Log-Pearson 3 remain the more favourable distributions in FFA. It agrees with the finding of this study as Lognormal distribution is only positioned at rank 32nd while GEV and Log-Pearson 3 are at 13rd and 15th.

5 Gamma
Gamma distribution is one of the most prevalent distribution in FFA [11] [21], but the outcome of the goodness of fit test here does reflect its prevalence. It is placed at rank 30th and even losing to Log-Gamma which are much less popular. However, similar to the discussions of the other distributions, perhaps the distribution might be incompatible for Jakarta rainfall while it works well in other data sets.

6 Normal
Normal distribution is probably the most popular stochastic distribution, but its applicability in FFA is not as popular. Normal distribution is often used in FFA, but it is merely as a comparison benchmark with other distributions. It rarely works as compatible as GEV or Log-Pearson 3 in modelling rainfall or runoff data except for rare occasion [32] [33]. The goodness of fit tests on Jakarta rainfall data again reflects the incompatibility of Normal distribution in FFA as it is only positioned at rank 35th, placed below all the distributions referred in SNI 2415:2016.

5. Conclusions

The list of the most fitting distributions for Jakarta in table 3 produces quite an unexpected result as it is dominated by the distributions which are uncommon in FFA. The appearance of Johnson SB, Fatigue Life, Inverse Gaussian etc in fitting rainfall data is very rare if not unheard of. Further studies on different data sets on different locations are required to figure out whether those distributions are generally applicable for FFA.

The top ten list is dominated by distribution with 3 (three) parameters. It is uncertain if distributions with three (or more) parameters could fit rainfall or flood data accurately or it only applies to Jakarta's rainfall only. However, distribution with two parameters are more practical to work with while the additional parameters tend to be relatively difficult to determine. Practicality is also as important as accuracy especially when it serves as a code or standard which should hinder the usage of tedious mathematical processes. As finding the balance between accuracy and practicality is important in code, it would be better to employ distributions with two parameters instead of distributions with more than two parameters although the latter might be more accurate.

The distributions referred in SNI 2415:2016 generate mixed outcome. Log-Pearson 3 is one of the most common distribution for FFA in Indonesia and it produces satisfactory result. Lognormal (3P) and Log-Gamma also show positive outcome although they are rarely used. Gumbel distribution is employed as frequently as Log-Pearson 3 in modelling Indonesia rainfall data, but the former is less compatible than the latter in fitting Jakarta rainfall data. Gamma distribution is quite popular in FFA subject, but it is not as popular in Indonesia. Here it does not show excellent result either. Lognormal and Normal distributions does not produce exceptional outcome and it sort of justifies its practice in FFA as they often serve as benchmarks for other popular distributions (e.g. GEV, Log-Pearson 3), but not as main distributions which represent the corresponding data.

References

[1] Tao D Q, Nguyen V T V, Bourque A 2002 *Proc. Ann. Conf. of the Canadian Soc. for Civil Eng.*
[2] Olofintoye O O, Sule B F, Salami A W 2009 *New York Science J.* **2(3)** 1-12
[3] Das S C 1952 *Australian J. Physics* **8** 298
[4] Buishand T A 1977 *Stochastic Modelling of Daily Rainfall Sequences* (Wageningen : Com. Agric. Univ.Wageningen)
[5] Badan Standardisasi Nasional 2016 *SNI 2415:2016 Tata cara perhitungan debit banjir rencana* (Jakarta)
[6] Massey F J M 1951 *J. Amer. Stat. Assn* **46** 68-78
[7] Anderson T W and Darling D A 1954 *J. Amer. Stat. Assn* **49(268)** 765-769
[8] Cochran W G 1952 *Ann. Math. Stat.* **23(3)** 315-345
[9] D'Adderio L P, Cugerone K, Porcu` F, Michele C D, Tokay A 2016 *Adv. Water Resources* **23(3)** 315-345
[10] Chen L, Singh V P, and Xiong F 2017 *Entropy* **19(6)** 239
[11] USGS 1982 *Guidelines For Determining Flood Flow Frequency* (Virginia)
[12] Yue S 2001 *Hydrological Processes* **15(6)** 1033-1045
[13] Basak P and Balakrishnan 2010 *J. Statistical. Computation and Simulation* **82(7)** 1055-1072
[14] Wise M E, Osborn S B, Anderson J, Tomlinson R W S 1968 *Math. Biosci* **2** 199-224
[15] Sangal B P and Biswas A K 1970 *Water Resources Research* **6(2)** 505-515
[16] Bennett S 1983 *Royal Stats. Soc.* **32(2)** 165-171
[17] Kantam R R L, Rao G S, and Sriram B 2007 *J. Appl. Stats.* **33(3)** 291-296
[18] Ahmad M I, Sinclair C D, and Werritty 1988 *J. Hydrology.* **98(3-4)** 205-224
[19] Smith R L 1986 *J. Hydrology.* **86** 27-43
[20] Smith J A 1987 *Water Resour. Res.* **23(8)** 1657-1666
[21] Onoz B and Bayazit M 1995 *J. Hydrology* **167** 195-208
[22] Morisson J E and Smith J A 2002 *Water Resour. Res.* **38(12)**
[23] National Engineering Research Council 1975 *Flood Studies Report Vols. 1-5* (London)
[24] Zalina M D, Desa M N M, Nguyen V T V, and Kassim A H M 2002 *Water Sci. Tech.* **45(2)** 63-8
[25] Alam M A, Emura K, Farnham C, and Yuan J H 2018 *Climate* **6(9)**
[26] Chow V T, Maidment D R, and Mays L W 1988 *Applied Hydrology* (New York : McGraw Hill)
[27] McMahon T A and Srikanthan R 1981 *J. Hydrology* **52** 139-147
[28] Todorovic P 1978 *Water Resour. Res.* **14(2)** 345-356
[29] Onen F and Bagatur T 2017 *Arab J Sci Eng* **42** 3895-3906
[30] Suhaila J and Jemain A A 2007 *J. Appl. Phys* **7(14)** 1880-1886
[31] Deraman W H A W, Mutalib N J A, and Mukhtar N Z 2017 *J. Phys.: Conf. Series* **890**
[32] Markovic R D 1965 *Probability Functions of Best Fit to Distributions of Annual Precipitation*

and Runoff (Colorado : Hydr. Papers Colorado State Univ.) [33] Yue S 1999 *Water Int.* **24(3)** 248-254

Identification of flood risk zones in the region of Yogyakarta

Nursetiawan[1]**, and R Faizah**[1]

[1]Department of Civil Engineering, Faculty of Engineering, Universitas Muhammadiyah Yogyakarta, Bantul, Daerah Istimewa Yogyakarta, Indonesia

*nursetiawan@umy.ac.id ; restu.faizah@umy.ac.id

Abstract. There are three rivers that cross the center of the Yogyakarta region and through Sleman Regency, Yogyakarta City and Bantul District, namely Code River, Gajah Wong River and Winongo River. The contribution of these three rivers to flood events in Yogyakarta is considered significantly high because the river passes through urban areas, settlements, agricultural land, and densely populated areas. In this study a hazard level map and the level of vulnerability of floods are created in those three watersheds. Map making starts with a study of hazards and vulnerabilities using the Analytical Hierarchy Process (AHP) method, weighting, scoring, and continued with a map depiction using Geographic Information Systems (GIS). Map of research results in the form of hazard level maps and flood vulnerability levels for Code watershed, Gajah Wong and Winongo. From the map, it can be concluded that most watershed areas of the Code are included in the level of moderate danger and low level of vulnerability. The entire Gajah Wong watershed is included in a moderate level of danger and most of them are at a high level of vulnerability, while most of the Winongo watershed has a high level of danger and vulnerability.

Keyword: flood, flood risk, flood hazard, flood vulnerability, GIS

1. Introduction

Yogyakarta's natural conditions are very varied and contoured, because there is Mount Merapi on the north side and bordering the sea on the south side. It provides the natural beauty of Yogyakarta which is one of the tourist destinations in Indonesia. In addition to many valleys and hills, there are also rivers that are mostly upstream in the Merapi area and eventually empty on the south coast of Java. Among the major rivers that divide the Daerah Istimewa Yogyakarta (DIY) region are the Progo River, Opak River, Oya River, Winongo River, Boyong-Code River, Gajah Wong River, Bedog River, and Serang River.

At present flooding is one of the serious problems that occur in the DIY region. The BPBD DIY Pusdalop (need an explanation of the abbreviation) reported that the peak flood event in 2017 in the DIY area occurred in January and November. The highest percentage of occurrences in the Bantul area (30%), following the Sleman region (22%), Kulon Progo (22%), Yogyakarta City (13%) and GunungKidul (13%). 2 Floods are often followed by landslides, especially in mountain areas Kidul (Gunung Kidul) and Kulon Progo, and a small portion in Bantul and Sleman. Some flood events have caused casualties, damaged infrastructure and agricultural land.

Flood disaster management activities in DIY are very necessary, especially in the pre-disaster period as a prevention effort so that disasters do not recur when the rainy season comes. One effort to prevent or reduce flood risk is mitigation, both structural and non-structural mitigation, especially in areas that are considered vulnerable. Regional vulnerability to flooding can be expressed by high levels of hazard and / or vulnerability

In this study, 3 major rivers were taken through the DIY region, namely the Code river, Gajah Wong and Winongo river as the research sites. The three rivers go through the center of the DIY region, so that it seems to divide the city of Yogyakarta by the distance between rivers 1 - 2 km. Based on the observations of researchers in the field, the condition of the rivers in 2017 looks dirty, cloudy and full of garbage.

Gajah Wong River, Code River and Winongo River cross Sleman Regency in the upstream, Yogyakarta City in the middle, and upstream in Bantul Regency. Overall the Gajah Wong watershed covers eight sub-districts with an area of 40.97 km², the DAS Code River watershed covers 18 sub-districts with an area of 40,254 km², and the Winongo River watershed of ± 47.83 km². The three rivers are used as sources of drinking water, irrigation, fisheries and as a means of livelihood for the residents of the river banks.

The study conducted on the 3 watersheds (the locations have been mentioned above) included a study of the level of hazard (hazard) and the level of vulnerability (vulnerability), with the assumption that the three watersheds were significant enough to assess the level of risk of flood disaster in the DIY region. The results of this research are expected to provide an overview of the potential for flood disasters in the DIY region and disaster management activities can be carried out immediately in areas that have high levels of hazard and vulnerability. The results of the study are also presented in the form of maps that are processed using the Geography Information System (GIS)

Disaster is an event concentrated in space and time, where the community or one of the subdivisions experiences physical damage and social disturbances, so that all or some important functions of society or subdivisions are disrupted (Fritz, quoted from Lindell, 2013). One of the disasters that often hit (need the proper word) the Indonesian region is flooding. Flooding is the flow of river water whose height exceeds the normal water level so that it overflows from the riverbed causing inundation on low land on the river side (BAKORNAS PB, 2007). The negative impact is given due to the occurrence of floods in areas where there is human activity (activities), which can cause fatalities, material losses, and psychological effects or trauma (Seniarwan etc., 2013).

Floods can be caused by 2 (two) types of factors, namely natural and human factors. Natural factors that can cause flooding such as river topography and geophysics, high rainfall, land subsidence, damage to flood control buildings, erosion and sedimentation of river capacity and inadequate drainage, and so on. Whereas in human factors including careless garbage disposal, planning of flood control systems is not appropriate, changes in land use, slum areas along the river, etc. (Razikin et al., 2017).

In this study the area that will be reviewed the level of risk of flooding is Yogyakarta, especially in the Gajah Wong watershed. Gajah Wong River is one of the three largest rivers in the Special Region of Yogyakarta in addition to the Code River and Winongo River. A watershed is defined as an area of land where all rainfall enters the same place in the direction of the same part of the stream or an area with a low topography. Watersheds are simple enough to be identified in mountainous or hilly areas because their boundaries are determined by mountains (Edwards et al., 2015). (mentioned aboved)

To find out the level of risk of flooding in the river, research or analysis must be done. The level of flood risk is determined by three factors, namely, the level of risk of flooding, flood vulnerability, and flood capacity. Each has parameters that need to be assessed and analyzed. The method that can be done to get an assessment of the level of flood risk is by scoring parameters and map modeling using Geographic Information Systems (GIS). According to Aronoff (1989) in Rosdania et al., (2015) Geographic Information Systems (GIS) or Geographic Information System (GIS) is a computer-based information system used to process and store geographic data or information. GIS has the ability to connect, analyze and map the results of various data at a certain point on earth.

IOP Conf. Series: Materials Science and Engineering **650** (2019) 012061 doi:10.1088/1757-899X/650/1/012061

2. Research Methodology

According to Hooijer et al. (2004) (in Idris and Darmashiri, 2015) (please find the original paper) flood risk is defined as a function of possible flood hazards and potential damage, most of the flood reduction measures aim to reduce the possibility of flooding and minimize potential damage. The level of flood risk in the Gajah Wong watershed is influenced by the level of danger, level of vulnerability, and level of capacity. The three relationships between the level of danger, level of vulnerability, and level of capacity can be written in the following formula:

(question number) $$Risk = Hazard\ x\ \frac{Vulnerability}{Capacity}$$

The parameters in each category for the hazard level, level of vulnerability and capacity level are as shown in the diagram in Figure 1.

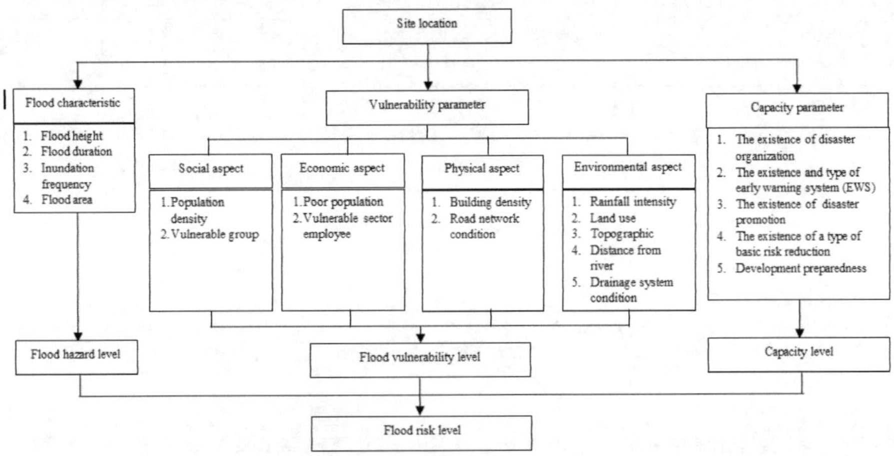

Figure 1. Diagram of parameters to determine risk level

The level of flood risk is obtained through three approaches, namely the hazard level, level of vulnerability, and level of capacity. Determination of the level of flood hazard in this study is based on a questionnaire to several government agencies and experts in the field of disaster and interviews with the community at the research site. The data needed to determine the level of flood vulnerability in this study was obtained from several agencies by searching directly or limited to downloading data from the official website of the relevant agencies. In addition, updates are needed for hazard level data and flood vulnerability. Then for capacity level data was obtained by conducting questionnaires/interviews with local agencies and local communities. Data that has been obtained, then given a value and scoring, weighting and classifying classifies it into low, medium, or high class. The final step is drawing maps using GIS software with map overlay techniques.

3. Result and Discussion

3.1. Flood hazard assessment

In the initial stage, the determination of flood hazard parameters was carried out by using the Analytical Hierarchy Process (AHP) method from the results of practitioners' querial results and expert interviews. The AHP results are then supplemented by tracing related references and previous research, so that the results as shown in Table 1 are obtained.

The assessment of weights and scores are then applied to data obtained by each village in the Code watershed, Gajah Wong and Winongo. Because of the large amount of data tabulation, it cannot be included as a whole in this paper. Weighting of the watershed code is carried out in 32 villages, Gajah Wong watershed 24 villages, and 16 villages in Winongo.

Table 1. Parameter and indicator to evaluate flood hazard level

Parameter and weight	Class	Indicator	Value	Score
Flood depth	Low	< 20 cm	1	0,4
Weight: 40%	Medium	20 - 50 cm	2	0,8
	High	> 50 cm	3	1,2
Flood duration	Low	< 12 hours	1	0,2
Weight: 20%	Medium	12 - 24 hours	2	0,4
	High	> 24 hours	3	0,6
Inundation frequency	Low	< 5 times	1	0,2
Weight: 20%	Medium	6 - 20 times	2	0,4
	High	> 20 times	3	0,6
Inundation area	Low	< 100 m^2	1	0,2
Weight: 20%	Medium	100 - 300 m^2	2	0,4
	High	> 300 m^2	3	0,6

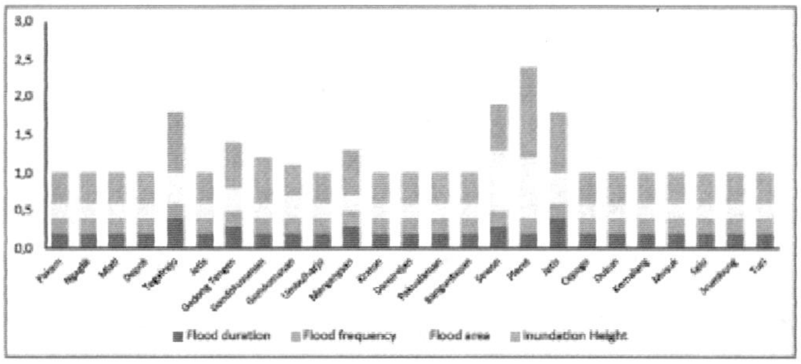

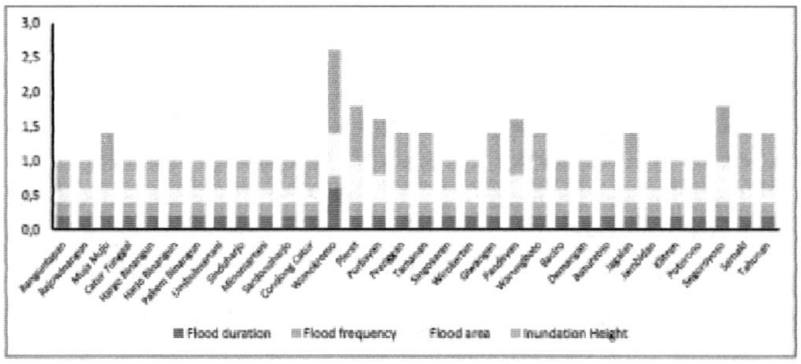

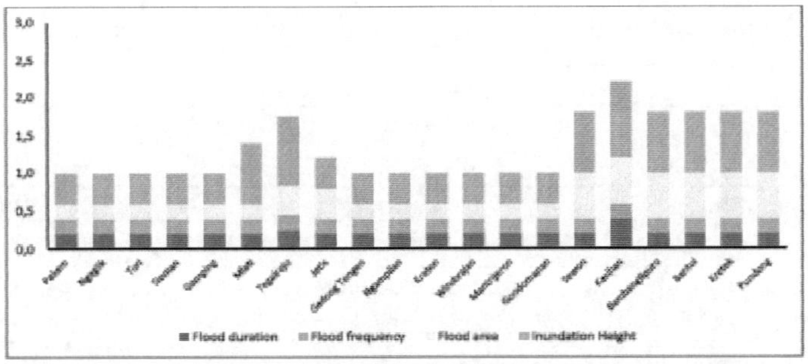

(the letter too small)

Figure 2. Graph of flood hazard level in a) Ccode, b) Gajah Wong and c) Winongo watershed

Based on the flood hazard map in Figure 2, it is known that the Code watershed is mostly in the area with a moderate hazard level which is shown in yellow (Figure 2), which is 69%. The rest is included in the low hazard level of 22% and the high hazard level is 10%. Some villages that are included in the high danger level of the Watershed Code watershed are Sinduadi, Sorosutan, Brontokusuman and Terban villages (better the name of village replace or added an explanation by a general word, upstream, middle or downstream). Based on direct observations in the field and interviews with communities around the watershed, some of these villages often experience flooding with a large enough pool (more than 300m2) with a duration of more than 24 hours, so it is included in the high hazard level.

The Gajah Wong watershed is on a map (Figure 3). It looks overall yellow, or is included in the moderate danger level. This is in accordance with the information of the people who live around the Gajah Wong watershed, that some villages have experienced flooding but with a maximum inundation height of 30 cm and the inundation length does not reach 12 hours. Floods in several villages averaged 1 time in one year, except for Prenggan village 6 times, Muja-muju 5 times, Caturtunggal 3 times, and Banguntapan and Pandean 2 times.

Meanwhile, most of the Winongo watershed has a low hazard level (45%), followed by 37.5% high hazard level and 17.5% moderate hazard level. Areas that have a high level of danger include the villages of Mlati, Tegalrejo, Sleman, Jetis, Kasihan and Sewon. The causes of the high level of danger of flooding in these 6 villages are on average due to high inundation when floods can reach 100-150 cm and some villages experience frequency of flooding more than 6 times in one year.

This hazard level map can be used as a basis for making risk maps, where the risk of flooding will be high when the danger level is high while vulnerability is also high. By knowing the level of danger of a region against a disaster, mitigation can be done immediately, so that disasters can be prevented or the impact minimized.

Scoring results for each parameter are mapped using the GIS program to obtain the hazard parameter map, then overlaying all parameters is done so that the flood hazard level map is obtained. The flood hazard map for Code watershed, Gajah Wong and Winongo based on the tabulation of the data obtained is presented in the appendix.

3.2. Flood vulnerability assessment

Vulnerability describes the inability of a region to solve problems that arise due to the threat of danger. A high level of vulnerability when met with high hazards will also have the potential for disasters. The vulnerability study in this study uses the National Guidelines for National Disaster Management Agency No.2 of 2012 concerning General Guidelines for Disaster Risk Assessment.

There are four vulnerability parameters studied in this study, namely the parameters of social, economic, physical and environmental vulnerability. Furthermore, a study of each of these parameters is carried out using the weighting and scoring method, based on the assessment indicators set out in Perka BNPB No.2 Year 2012, summarized in Table 2 and Table 3.

IOP Conf. Series: Materials Science and Engineering **650** (2019) 012061 doi:10.1088/1757-899X/650/1/012061

Table 2. Parameter and indicator to calculate flood vulnerability

Parameter		Indicator	Index
Social vulnerability	-	Population Density	0.6
Weight 0.4	-	Vulnerable group	0.4
Economic vulnerability	-	Poor family	0.6
Weight 0.25	-	Vulnerable sector employee	0.4
Physical vulnerability	-	Building density	0.6
Bobot 0.25	-	Road network condition	0.4
Environmental vulnerability	-	Rainfall intensity	0.25
Weight 0.10	-	Land use	0.25
	-	Topographic condition	0.20
	-	Distance from river	0.20
	-	Drainage system condition	0.10

Table 3. Measurement Criteria for each flood vulnerability indicator

Parameter	Class	indicator	Value	Score
Population density	Low	< 500 people/km^2	1	0,6
Weight 0,6	Medium	500-1000 people /km^2	2	1,2
	High	>1000 people /km^2	3	1,8
Vulnerable group	Low	< 20%	1	0,4
Weight 0,4	Medium	20 – 40%	2	0,8
	High	>40%	3	1,2
Poor family	Low	< 20%	1	0,6
Weight 0,6	Medium	20 – 40%	2	1,2
	High	>40%	3	1,8
Vulnerable sector	Low	< 20%	1	0,4
employee	Medium	20 – 40%	2	0,8
Weight 0,4	High	>40%	3	1,2
Building density	Low	< 18 unit/ha	1	0,6
Weight 0,6	Medium	18 – 34 unit/ha	2	1,2
	High	> 34 unit/ha	3	1,8
Road network	Low	> 70%	1	0,4
condition	Medium	30 – 70%	2	0,8
Weight 0,4	High	< 30%	3	1,2
Rainfall intensity	Low	< 1000 mm	1	0,25
Weight 0,25	Medium	1000 - 2500 mm	2	0,5
	High	> 2500 mm	3	0,75
Land use	Low	Unoccupied land dll > 50%	1	0,25
Weight 0,25	Medium	Agriculture and services > 50%	2	0,5
	High	Settlement and industry > 50%	3	0,75
Topographic condition	Low	> 300 Masl	1	0,2
Weight: 20%	Medium	20 – 300 Masl	2	0,4
	High	< 20 Masl	3	0,6
Distance from river	Low	> 1000 m	1	0,2
Weight: 20%	Medium	500 – 1000 m	2	0,4
	High	< 500 m	3	0,6

Drainage system condition Weight: 10%	Low	> 70%	1	0,1
	Medium	30 – 70%	2	0,2
	High	< 30%	3	0,3

After obtaining a vulnerability rating score, then the map is drawn using the GIS method. The result is a graph of the level of vulnerability for Code Watershed, Gajah Wong and Winongo, as shown in Figure 3.

Figure 3 also shows that most of the Code Watershed area is green, which means it has a low level of vulnerability, covering 70% of the entire Code Watershed area. The remaining 29% is included in the high level of vulnerability and only 1% is at the level of moderate vulnerability. Areas that are at a high level of vulnerability consist of 18 villages in 13 sub-districts. The high level of vulnerability is caused by the distance of the village which is very close to the river (<500 m) and the amount of productive land in the village such as agricultural and building / residential land. This area which has a high level of vulnerability has the potential to experience large losses if it is hit by a flood.

The highest level of flood hazard is in Wonokromo Village and the level is in Pleret Village. While the other 24 villages are classified as low hazard levels. The highest flood inundation score is in Wonokromo Village. The height of the inundation can reach more than 1 meter, inundation that occurs around 1-2 days, and the extent of floods that can flood even up to 2 broad hamlets. The frequency of flood events in the Gajah Wong watershed area is relatively low due to flooding that occurs only 1 - 5 times a year. Figure 1 shows the distribution of flood hazard levels in the form of maps.

The vulnerability level of the Gajah Wong watershed (Figure 3) is mostly included at a high level, which is 40.7% of the Gajah Wong watershed. The next sequence is 33.4% classified as moderate level of vulnerability, and the remaining 25.9% is at a low level of vulnerability. Areas with high levels of vulnerability are scattered in 16 villages, most of which are in the downstream or southern regions. This high vulnerability is dominated by parameters of social vulnerability and physical vulnerability, where conditions of high population density and many residents belonging to vulnerable groups according to the criteria in Law No. 24 of 2007, namely pregnant / lactating women, toddlers, the elderly, and people with disabilities. Observing this phenomenon, then to reduce the level of flood vulnerability in the Gajah Wong watershed can be focused on handling social problems.

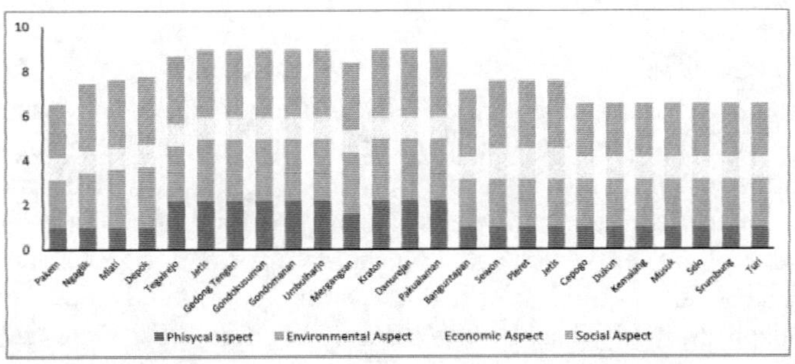

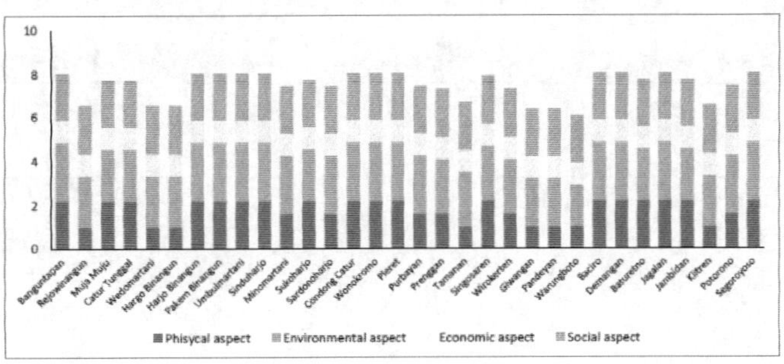

IOP Conf. Series: Materials Science and Engineering **650** (2019) 012061 doi:10.1088/1757-899X/650/1/012061

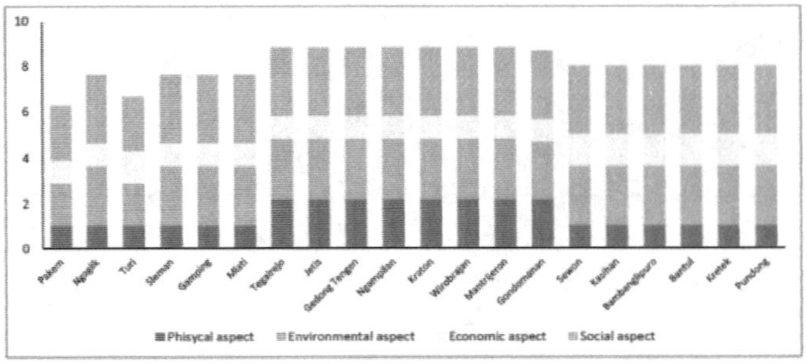

Figure 3. Graph of flood vulnerability level in a) code, b) Gajah Wong and c) Winongo

There are several villages in the Winongo watershed area which are included in the high level of vulnerability, especially in the Mlati, Jetis, Gedongtengen, Ngampilan, Wirobrajan and Gondomanan sub-districts. Meanwhile, the rest is categorized as low and medium. The high level of vulnerability in several villages in the Winongo watershed is largely due to the distance of buildings too close to the river lip, high population density and the function of land as a settlement. This condition has a high exposure potential if it is flooded, because the possibility of losses will be high. In addition, high levels of density often make it difficult for disaster management actions, especially disaster emergency conditions. With this map information, disaster risk reduction activities are expected to be determined. By observing the hazard level map and the level of flood vulnerability of each watershed, it can be seen that the Code watershed has a large portion of the area with moderate hazard levels and a low level of vulnerability, so the Code watershed area has a low to moderate risk of flooding. If this condition is equipped with good community capacity, the risk level will be smaller.

The entire Gajah Wong watershed is classified as being at a moderate level of danger, but most of them have a high level of vulnerability, which needs to be considered in some of these villages. Risk reduction efforts can be carried out by mitigating and increasing community capacity.

As for the Winongo watershed, most of the area is included in the level of danger and high level of vulnerability. A high level of danger when meeting with a high level of vulnerability will have a very high risk. So it is recommended that in several villages in the Winongo watershed disaster risk reduction is immediately carried out.

3.3. Capacity level assessment

The definition of capacity according to Perka BNPB No. 02/2012 is the ability of regions and communities to take action to reduce the level of threats and the level of losses due to disasters. According to Nugraha et al., (2015) community capacity is divided into two aspects, namely individual capacity (influenced by knowledge, local wisdom, and action plans) and institutional capacity (influenced by the existence of programs, information, leadership, local wisdom and facilities). The capacity level consists of five parameters namely, the existence of a Disaster Management Organization (OPB), an Early Warning System (EWS) or an early warning system, disaster education, the existence of a type of reduction in basic risk factors, and the development of preparedness on all lines. The following is a capacity level assessment table with each parameter and equation.

Table 4. Weighting and Classification of Capacity Level Parameters

Parameter	Value	Weight (%)	Score	Class
Rules and Institutions for Disaster Management	1	21	0,21	Low
	2		0,42	Medium
	3		0,63	High

IOP Conf. Series: Materials Science and Engineering **650** (2019) 012061 doi:10.1088/1757-899X/650/1/012061

Early Warning and Disaster Risk Assessment	1	19	0,19	Low
	2		0,38	Medium
	3		0,57	High
Disaster Education	1	23	0,23	Low
	2		0,46	Medium
	3		0,69	High
Reduction of Basic Risk Factors	1	16	0,16	Low
	2		0,32	Medium
	3		0,48	High
Development of Preparedness at All aspects	1	21	0,21	Low
	2		0,42	Medium
	3		0,63	High

(Modified form Perka BNPB No.02/2012 and AHP result)

Calculation of capacity level scores is (0.21 × disaster management rules and institutional values) + (0.19 × value of early warning and disaster risk assessment) + (0.23 × disaster education value) + (0.16 × factor reduction value basic risk) + (0.21 × value of preparedness development on aspects).

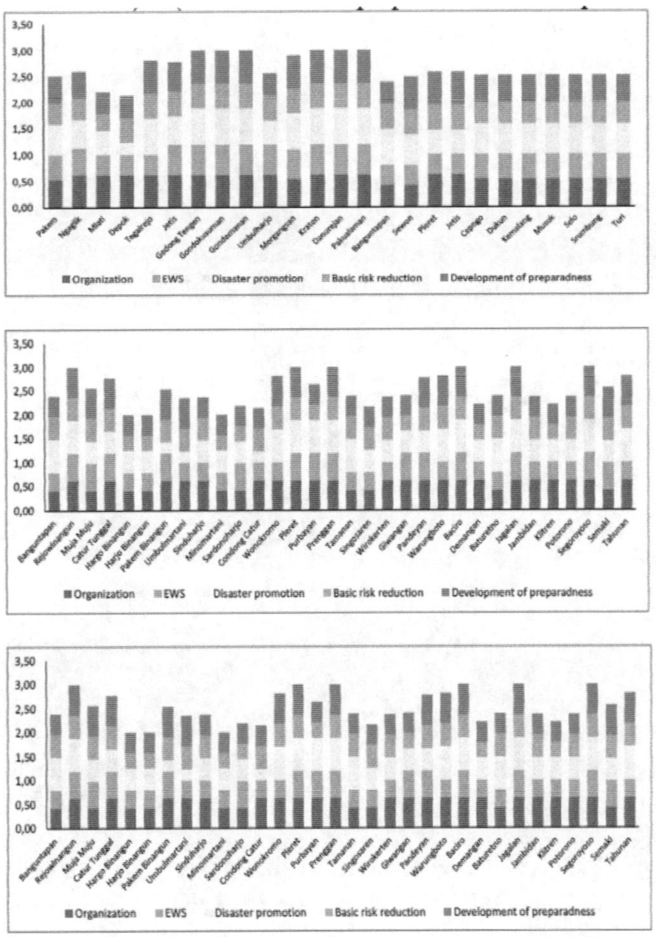

Figure 4. Graph of capacity level in a) code, b) Gajah Wong and c) Winongo

IOP Conf. Series: Materials Science and Engineering **650** (2019) 012061 doi:10.1088/1757-899X/650/1/012061

Based on the results of the capacity level analysis of 26 Gajah Wong watershed villages, it is quite good. There are 8 villages classified as medium capacity levels and 18 villages with high capacity. This shows that the community and the government have made good disaster prevention efforts. In general, the conditions and availability of the five parameters above are good enough. Communities already know and play a role in increasing capacity to cope with the adverse effects of disasters. Figure 4 shows a graph of the distribution of capacity levels in 3 watershed.

3.4. *Analysis of Flood Risk Level*

The level of flood risk in an area is influenced by the level of flood hazard, the level of flood vulnerability, and the level of flood capacity. To determine the level of flood risk class, it is necessary to use the interval score that has been made

The results of the assessment of the parameters obtained are then mapped using a geographic information system in the form of ArcGIS software to obtain a map of each parameter. After the map of each parameter is completed, then overlaying or overlapping the map is done so that it becomes a map of the hazard level, level of vulnerability, capacity level, and level of risk of flood disaster. Overlay is a technique that intends to assist in the need to respond to and outline the recently developed global scientific maps so as to produce a visually interesting comparison, very readable, and potentially useful for policy making (Rafols et al., 2010).

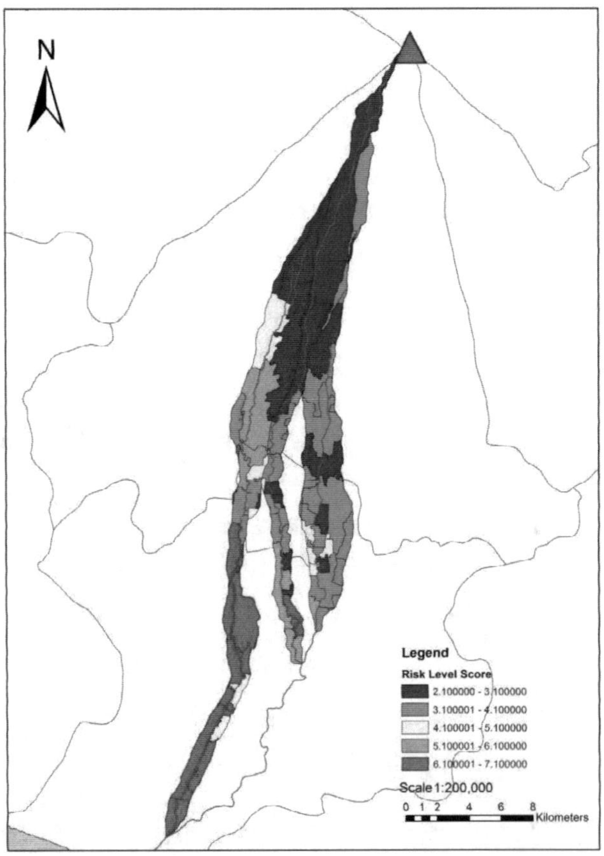

Figure 5. Map of Flood Risk Level a) Code, b) Gajah Wong, c) Winongo (pointed the name of watershed, which one the Code River so on),

IOP Conf. Series: Materials Science and Engineering **650** (2019) 012061 doi:10.1088/1757-899X/650/1/012061

The results of the analysis obtained indicate that there are 26 villages in the Gajah Wong watershed that have a low risk of flooding. Regions with a low level of risk have a relatively low level of flood hazard, a low - moderate level of vulnerability, and a medium - high level of capacity. Figure 5 shows a map of the distribution of flood risk levels in the Gajah Wong watershed.

4. Conclusion

From the analysis of the flood hazard level and flood vulnerability level in the Code watershed, Gajah Wong and Winongo, the following conclusions can be obtained:

1. Most watershed areas of the Code are included in the level of moderate danger and low level of vulnerability.
2. The entire Gajah Wong watershed is included in the moderate hazard level and most of it is included in the high level of vulnerability
3. Most of the Winongo watershed has a high level of danger and vulnerability.
4. The level of flood disaster capacity in the Gajah Wong watershed shows that there are 18 villages including medium capacity, and 8 high capacity villages. A good level of capacity indicates that the community has been able to anticipate disasters.
5. The level of flood risk in the Gajah Wong watershed is low. This is because the level of flood hazard which is mostly classified as low class, the level of vulnerability is classified as low - moderate, and the capacity level is classified as medium - high. In addition, the ability of the community to deal with disasters is quite good, so that the risk of disasters caused can also be reduced.

References

[1] Bahn, S 2012 Workplace hazard identification: What do people know and how is it done *Annual Conference of the Association of Industrial Relations Academics Australia and New Zealand* 1-9.

[2] Darmawan, K, Hani'ah, dan Suprayogi A 2017 Analisis Tingkat Kerawanan Banjir Di Kabupaten Sampang Menggunakan Metode Overlay Dengan Scoring Berbasis Sistem Informasi Geografis *Jurnal Geodesi Undip* **6(1)** 31-40.

[3] Edwards P J, Williard K W J, dan Schoonover J E 2015 Fundamentals of Watershed Hydrology *Journal of Contemporary Water Research and Education* Issue **154** 3-20

[4] Fristyananda M A, dan Idajati H 2017 Tingkat Bahaya Bencana Banjir di Kali Lamong Kabupaten Gresik *Jurnal Teknik ITS* **6(1)** 56-59

[5] Hagelsteen M, dan Becker P 2014 Fowarding a Challenging Task: Seven Elements for Capacity Development for Disaster Risk Reduction *Global Risk Forum Davos* **2(2)** 94-97.

[6] Idris S, dan Dharmasiri L M 2015 Flood Risk Inevitability and Flood Risk Management in Urban Areas: A Review *Journal of Geography and Regional Planning* **8(8)** 205-209.

[7] Jefferson T L, dan Johannes T W 2016 Using Geographic Information System to Support Decision Making in Disaster Response *Intelligent Decision Technologies* 193-207.

[8] Lindell M K 2013 Disaster Studies *Current Sociology* **61(5/6)** 797-825.

[9] Nasiri H, Yusof M H M, dan Ali T A M 2016 An Overview To flood Vulnerability Assessment Methods *Sustainable Water Resources Management* **2** 331–336.

[10] National Board for Disaster Management 2008 *Peraturan Kepala Badan Nasional Penanggulangan Bencana Nomor 4 Tahun 2008 Tentang Pedoman Penyusunan Rencana Penanggulangan Bencana*

Acknowledgment

The author would like to thank the Institute for Research, Publication and Community Service of the University of Muhammadiyah Yogyakarta (LP3M UMY) for providing funding for the implementation

IOP Conf. Series: Materials Science and Engineering **650** (2019) 012061 doi:10.1088/1757-899X/650/1/012061

of this research.

Sediment Characteristics of Bed Load Transport in Downstream of Progo River, Indonesia

J Ikhsan[1]*, C Wardhana[1], and D K Widiyarto[1]

[1]Department of Civil Engineering, Faculty of Engineering, Universitas Muhammadiyah Yogyakarta, Indonesia

* jazaul.ikhsan@umy.ac.id

Abstract. Progo River is one of the rivers that has a sediment supply from volcanic and non-volcanic areas. Sediment transport, especially the bed load transport, it is very influential on the phenomenon of river degradation and aggradation. Therefore, a study to determine the characteristics of bed load transport is needed and important. The research was carried out in the downstream part of Progo River. Sampling locations were taking in four locations, namely Ancol Bridge, Kebon Agung II Bridge, Kebon Agung I Bridge and Bantar Bridge. The bed load transport measurement tool was used Helley Smith (WMO, 1980) equipment. Every river cross section is taken three times. The results showed that the sediment type of bed load transports is silty sand, with specific gravity values of 2.67 to 2.70. The average diameter values (D50) of the bed load transports vary greatly from 0.129 mm to 0.454 mm. The values of bed load transports vary greatly and are influenced by the values of the discharges. The correlation between bed load transport and discharge is linear, with R values are 0.85 to 0.99.

Keywords: Bed load, Helley Smith, Progo River, Volcanic Area.

1. Introduction

A knowledge of sediment transport will make it possible to estimate suspended sediment carried by the stream or sediment moving on the riverbed. An understanding of the phenomenon of sediment transport, especially bedload transport, is needed to be able to predict more valid sedimentation and erosion processes in a river. Some experts have issued a formula for bedload transport based on laboratory experiments with predetermined conditions. As it is generally known, the calculation result of the of sediment transport using one formula comparing another formula, will give different results, and sometimes the difference is quite large. Therefore, a research on the characteristics of bedload transport still needs to be done, especially on rivers in Indonesia which have their own characteristics.

The problem related sediment that occurs in the Progo River is the formation of sediment deposits in the lower reaches of the river which causes changes on river morphology in a relatively short time. These sediment deposits are caused by excessive sediment supply from the 2010 Merapi eruption (Harsanto, et al., 2015). Changes in river morphology will change the flow conditions such as water level, flow velocity, and shear stress. Flow hydraulic plays an important role in the process of aggradation/ sedimentation and degradation/erosion of riverbed (Manonama 2003). The process of erosion and sedimentation greatly affects the balance of the riverbed configuration. The riverbed

IOP Conf. Series: Materials Science and Engineering **650** (2019) 012062 doi:10.1088/1757-899X/650/1/012062

configuration factors are strongly influenced by speed, flow time and flow depth (Suwartha, 2001). In addition, people tend to use as much sediment as possible to support regional development. However, sediment mining can also have a negative impact on ecosystems and reduce river regulatory safety (Ikhsan, et al., 2010).

The eruption of Mount Merapi in 2010 caused the Progo River to experience changes in river morphology, changes in physical sediments and riverbed material porosity values. To know how much sediment transport occurs on the River Progo will help prevent undesirable things from happening. Based on this reason, it is important to analyze the sediment transport in Progo River.

2. Eruption of Mount Merapi

It is well known that Mount Merapi is one of the most active volcanoes in the world. It is 2,968 m high and is located in the island of Java at 7°32'26.99" S 110°26'41.34" E on the border between the Central Java and the Yogyakarta Special Provinces. It has erupted 41 times in the last 200 years including 15 major eruptions. It generally erupts every 3 years with a major eruption every 9 years. The history of its eruptions is shown in Figure 1.

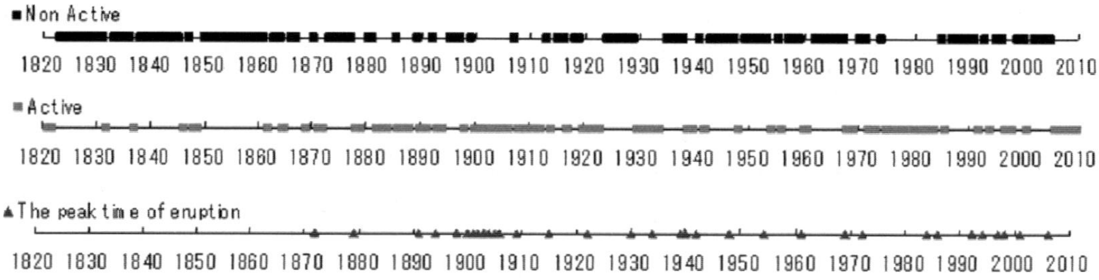

Figure 1. The Historical Activities of Mount Merapi (Ikhsan, 2010)

Volcanic activity occurred on the western slope during 1830–1870 and on the north western slope at the end of the 1800s. During 1903 and 1904, the volcanic activities moved to the eastern slope, then to the south eastern slope during 1905 and 1906, and then to the north western slope during 1909–1913.

In the period from 1940–1994, volcanic activities were confined mainly to the southwestern slope, except for a short period on the northern slope during 1954–1956. During 1999–2001, volcanic activities moved from the southwestern slope to the western slope. The eruption in June 2006 occurred on the south eastern slope. And the latest eruption took place on the north western and the south eastern slopes in October 2010.

Its eruptions have produced large amounts of volcanic material as ash falls, lava, and pyroclastic flows. Produced sediment is deposited on the slopes of Mount Merapi (as shown in Figure 2) and is partly transported by water flow to the downstream areas through the tributaries that originate in the volcano. Some tributaries of Progo River are originated in the north western slope of Mount Merapi. The morphology and transport sediment in Progo River are induced by the sediment from Mount Merapi.

2.1 The Progo River

The Progo River is a river that flows in the Central Java and the Yogyakarta Special Provinces. The river is originated on the slopes of Mount Sumbing, which cross to the southeast and its estuary is the Indian Ocean, at Trisik Beach, Bantul Regency. The main river length is ± 138 km and the watershed area is 2830 km2 (Mananoma et al., 2003). On the Progo River, there are many tributaries that originated on the Mount Merapi, namely the Krasak River, Batang River, Putih River, Blongkeng River and Pabelan River.

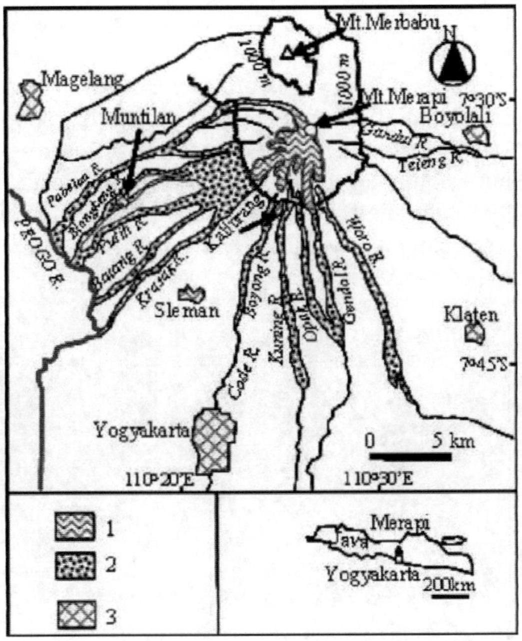

Figure 2. Location of Sediment Deposits in the Mount Merapi. (1: Summit lava dome and andesitic lava flow, 2: deposits from Mount Merapi, 3: main cities) (Modified from Lavigne and Thouret 2002; Ikhsan, 2010)

3. Study areas and Methods

3.1 Study areas

The research locations of bedload transport in the Progo River were at Ancol Bridge in the Bligo village (Ngluwar Sub Distrik, Magelang Regency). Kebon Agung II Bridge in Sendang Minggir village (Sleman-Kulon Progo Regency), Kebon Agung I Bridge (Kulon Progo-Sleman Regency) and Bantar Bridge (Bantul-Kulon Progo Regency). The distance between Ancol Bridge-Kebon Agung II Bridge, Kebon Agung II Bridge-Kebon Agung I Bridge, and Kebon Agung I Bridge- Bantar Bridge are 7.65 km, 4.64 km, and 8.74 km, respectively. The study area is shown in Figure 3.

Figure 3. Location of Study Area in Progo River, Yogyakarta, Indonesia

IOP Conf. Series: Materials Science and Engineering **650** (2019) 012062 doi:10.1088/1757-899X/650/1/012062

3.2 Hydrometry survey

3.2.1. Equipment. The data needed in this study are sediment transport, flow velocity, river depth, slope of bank river, and cross section of the river. The sediment transport data is taken every 2 hours. The equipment needed in the research in the field includes floats, stopwatches, odometer, depth measurement and Helley Smith equipment. The equipment used in the laboratory is a tool used to determine the specific gravity of the material from sediment transport and gradation of the material. Figure 4 shows the Helley Smith equipment.

Figure 4. Helley Smith Equipment

Methods

Flow velocity measurement was conducted by using floats. The floats are used as a flow rate measurement tool if flow velocity is required with a relatively small degree of accuracy, because what is measured is surface velocity. To get a velocity that represents a cross section of the river, then the results of surface velocity measurements need correction with a constant value of 0.85 - 0.90. Measurements of water level were carried out manually by carrying out the elevation of the water surface indicated on the normal water gauge. Measurements of cross-sectional area require water level; and measurement of water level were done using a stick with a measuring sign. Measurement of flow width is also used to find out the width of the riverbed which will be used to get river cross section area. Measurement of flow width was carried out using a width measuring instrument and measuring channel width was taken using odometer.

The method for measuring bedload and equipment are still in the developing stage, so there is no single method or equipment suitable for all field conditions. For this study, sediment transport data collection was carried out in the rainy season between March and April. Data collection techniques were from the upper part of bridges using the Helley Smith tool (WMO, 1980) with the help of pulleys. Then, the results of measurements of bedload are taken to the laboratory to be analyzed for specific gravity and gradation. Furthermore, an analysis of the characteristics of sediment transport at the points used as testing/survey is carried out.

This bedload transport study is a direct research in the field, taking samples using Helley Smith equipment. This research took in four locations, namely Ancol Bridge, Kebon Agung II Bridge, Kebon Agung I Bridge and Bantar Bridge. To get the bedload value that is close to its original condition, the Helley Smith equipment must be modified.

Efficiency Calculation of Helley Smith Equipment

$$e = \frac{Ka}{Kr} \tag{1}$$

where,

e = Efficiency of the bedload measuring equipment (%)
Ka = Quantity of sediment captured by the bedload equipment
Kr = Quantity of sediment transported if the bedload equipment is not placed on

$$e = \frac{0.261}{0.35} \text{ X } 100\,\%$$

e = 74.57% > 70%, take a middle value (the value of efficiency of the equipment between 40% - 100%)

Sediment Transport Analysis After Modification

The amount of sediment transport captured by Helley Smith equipment can be calculated using the equipment below.

$$qb = \frac{100W}{ebt} \qquad (2)$$

where,

Qb = Bedload transport per unit width after being modified based on the equipment efficiency
W = The weight of the sample caught by the equipment during a period
e = Efficiency of the equipment (%)
b = Width of the equipment
t = Time of measurement

Bedload Transport of Full Cross Section

$$Y2 = \frac{X2.Y1}{X2} \qquad (3)$$

where,
X1 = Mouth area of the Helley Smith equipment (m^2)
X2 = Bedload transport by the Helley Smith equipment (kg/hour)
X3 = Area of flow cross section (m^2)
Y2 = Bedload transport carried by river flow (kg/hour)

Bedload Transport of 1/8 h Cross Section

Related sediment transport in a river, there are three layers of sediment transport, namely bedload, suspended load and surface load Then, it can be assumed by reviewing the cross section which is at 1/8 of the depth of the river. In analyzing the calculation of this paper taken 1/8 from the depth of the river, for example, to a depth of 6 meters, the calculation of the sediment is 0.75 meters. This value is compared to the empirical method calculation.

$$Y2 = \frac{X2.Y1}{X1} \qquad (4)$$

where,

X1 = Mouth area of the Helley Smith equipment (m^2)
Y1 = Bedload transport by the Helley Smith equipment (kg/hour)
X2 = River cross section area (m^2)
Y2 = Bedload transport carried by river flow (kg/hour)

4. Results and Discussion

4.1 Hydrometry Data

The measurement results of river hydrometry at the selected locations are shown in the Table 1 as follows:

Table 1. Hydrometry Parameters at the Survey Locations

Location	**Date**	**Velocity (m/s)**	**Cross Section Area (m^2)**	**Discharge (m^3/s)**
Ancol	13 March 2017	1.02	110.87	113.58
Ancol	14 March 2017	1.09	110.90	120.61
Ancol	15 March 2017	1.13	128.58	144.87
Kebon Agung 2	26 March 2017	1.10	119.70	131.67
Kebon Agung 2	26 April 2017	1.16	120.20	139.43
Kebon Agung 2'	27 April 2017	1.20	123.10	147.72
Kebon Agung 1	22 March 2017	1.161	104.23	121,01
Kebon Agung 1	23 March 2017	1.286	113.10	145.44
Kebon Agung 1	26 March 2017	1.574	150.62	213.28
Bantar	16 March 2017	0.832	190.83	158.72
Bantar	17 March 2017	0.612	162.76	99.64
Bantar	19 March 2017	0.644	170.87	110.11

Table 1 shows that the velocity at each survey location is vary between 0.612 m/s to 1.574 m/s. The lowest velocity was measured at the Bantar Bridge on March 17, 2017, and the largest velocity at the Kebon Agung 1 Bridge was measured on March 26, 2017. The lowest velocity occurred in Bantar Bridge because this location is downstream of the Progo River, and as well as at this location, there is a groundsill in the lower part of the bridge, causing the slope of the river to be mild which causes the velocity at this location to be the lowest compared to other locations. The largest velocity value occurred at Kebon Agung 1 Bridge, because this location has the largest slope as well as the location is a sediment transport area. In addition, the biggest velocity occurred on March 26, 2017, because on that date there was rain in the upper reaches of the Progo River. Based on **Table 1** also shows that the measured discharge varies from 99.64 m^3 s to 213.28 m^3/s. The lowest discharge was measured on March 17, 2017 at the Bantar Bridge and the largest discharge was measured on March 26, 2017 at the Kebun Agung 1 Bridge.

4.2 Sediment Transport

The results of the sediment transport values at each point are shown in **Table 2** as below.

Table 2. Results of Sediment Transport Measurements at the Survey Locations

Location	**Date**	**1/8 H (m^3/s)**	**Full Cross Section (ton/day)**
Ancol	13 March 2017	0.19	1.54
Ancol	14 March 2017	0.21	1.69
Ancol	15 March 2017	0.35	2.85
Kebon Agung 2	26 March 2017	0.53	4.25
Kebon Agung 2	26 April 2017	0.51	4.10
Kebon Agung 2	27 April 2017	0.65	5.26
Kebon Agung 1	22 March 2017	2.88	22.54
Kebon Agung 1	23 March 2017	8.27	66.50
Kebon Agung 1	26 March 2017	19.22	153.76
Bantar	16 March 2017	3.37	27.68
Bantar	17 March 2017	2.16	17.31
Bantar	19 March 2017	3.04	23.82

Based on **Table 2**, it is shown that the results of measurements of sediment transport are vary between 1.54 ton/day up to 153.76 ton/day. The smallest sediment transport was measured at the Ancol Bridge and the largest one was measured at the Kebun Agung 1 Bridge. From the measurement data, it shows that there is a tendency for sediment transport to occur from the upstream (Ancol Bridge) to the downstream (Kebun Agung 1 Bridge), getting bigger, even though the value of the water discharge is relatively balanced. This shows that between the Ancol Bridge to Kebun Agung I Bridge is a sediment transport area, and a large erosion occurs on this area. In the river part between the Agung Kebun 1 Bridge and Bantar Bridge, based on the measurement results, there is a tendency for the deposition process to occur. The variation of sediment transport that occurs is very large among the location being reviewed, especially in Ancol Bridge and the Kebun Agung 2 Bridge, which they are small because the distance between the bridge and riverbed at Ancol Bridge and Kebun Agung 2 Bridge is very large, causing difficulties in lowering and raising the Helley Smith equipment. This condition causes the possibility of the sediment caught in the Helley Smith equipment lost during the lifting process of the equipment. In addition, with a distance is too large, to put the mouth of Helley Smith's equipment in the direction of the flow that occurs is more difficult comparing a short distance.

When the mouth of Helley Smith equipment does not parade perfectly, the sediment caught by the equipment is not optimal.

4.3. Characteristics of Sediment Transport Material

To identify the transported material, in this paper author used two parameters, namely a specific gravity and average diameter (d50) of sediment transport material. The results are shown in **Table 3**.

Table 3. Results of Measurement of Specific Gravity and Gradation of Sediment Transport Material

Location	Specific Gravity (ton/m^3)	D50 (mm)	Note
Ancol	2.68	0.129	*Silty Sand*
Kebon.Agung 2	2.70	0.408	*Silty Sand*
Kebon Agung 1	2.67	0.454	*Silty Sand*
Bantar	2.69	0.169	*Silty Sand*

Table 3 shows that the transported material is relatively uniform, and its type is silty sand, with the values of density varying between 2.67 to 2.70. The gradation of sediment transport material is also varying, with the D50 averaging between 0.129 mm to 0.454 mm. From this data, sediment transport materials at Ancol Bridge and Bantar Bridge are softer than in the Kebun Agung 2 Bridge and the Kebun Agung 1 Bridge. This is due to the greater velocity and steeper slope of the riverbed in Kebon Agung Bridge compared to the same parameters at Ancol Bridge and the Bantar Bridge.

4.4. Correlation Between Discharge and Sediment Transport

By using data of sediment transport discharge calculation, their correlation can be found, whether positive, negative or uncorrelated.

$$r = \frac{\sum (X-X)(Y-Y)}{\sqrt{\sum (X-X)^2 . \sum (Y-Y)^2}} \tag{5}$$

$$r = \frac{\sum xy - \frac{\sum X . \sum Y}{n}}{\sqrt{\left(\sum x^2 - \frac{(\sum X^2)}{n}\right) . \left(\sum Y^2 - \frac{(\sum Y^2)}{n}\right)}} \tag{6}$$

The value of r is always between -1 and +1 ($-1 < r < +1$)

where,
r = +1, it means there is a perfect positive correlation between X and Y.
r = -1, it means there is a perfect negative correlation between X and Y.
r = 0, it means there is no correlation between X and Y.

IOP Conf. Series: Materials Science and Engineering **650** (2019) 012062 doi:10.1088/1757-899X/650/1/012062

Table 4. Correlation Value between Water Discharge and Bedload Transport in the Survey Locations

No	Location	Discharge m^3/s (X)	Bedload (ton/day) (Y)	r value
1	Ancol Bridge	113.58	1.54	0.99
2	Ancol Bridge	120.61	1.69	
3	Ancol Bridge	144.87	2.85	
4	Kebon Agung 2 Bridge	131.67	4.25	0.85
5	Kebon Agung 2 Bridge	139.43	4.10	
6	Kebon Agung 2 Bridge	147.72	5.26	
7	Kebon Agung 1 Bridge	121.01	22.54	0.88
8	Kebon Agung 1 Bridge	145.44	66.50	
9	Kebon Agung 1 Bridge	213.28	153.76	
10	Bantar Bridge	158.72	27.68	0.99
11	Bantar Bridge	99.64	17.31	
12	Bantar Bridge	110.11	23.82	

Table 4 shows that there is a close relationship between water discharge and bedload transport. The correlation between the two parameters is indicated by the value of r which varies between 0.85 to 0.99. The smallest r value obtained in Kebon Agung 2 Bridge is 0.85. This is due to the very high condition of the Kebun Agung 2 Bridge, so that to measure bedload transport faces more problems compared to other locations. The largest r value occurs in the Bantar Bridge and Ancol Bridge, with r value of 0.99. Based on Table 4, it can be concluded that the greater water discharge, it means the greater bedload transport that occurs.

5.Conclusions and Recommendations

The conclusions from the research on the characteristics of bedload transport in the Progo River using the Helley Smith equipment are as follows:

1. Measured water discharges are varying from 99.64 m^3/s to 213.28 m^3/s.
2. The variation of bedload transport is very large among the survey locations, which varies between 1.54 ton/day to 153.76 ton/day.
3. Specific gravity of sediment transport materials varies between 2.67 to 2.70 and the type is silty sand type. The mean diameter (D50) is between 0.129 mm to 0.454 mm.
4. The relationship between bedload transport and water discharge is directly proportional, with the value of varying between 0.85 to 0.99.

Based on experience during measuring basic sediment transport with Helley Smith tools, several recommendations can be proposed by the author for future researchers if they will conduct research with Helley Smith tools as follows:

1. If the distance between a river water surface and a bridge is too high, an alternative location is needed, because it will cause serious difficulties, especially when lowering and raising the equipment, which can give impact in measurement of sediment transport.
2. The Helley Smith equipment is better when equipped with a camera to guide/assist in directing the mouth of the equipment so that it can be in the same direction with the water flow, finally the sediment can be captured by the equipment optimally.

Acknowledgements
The author would like to express his deepest gratitude to Dany Kurniawan Widiyarto and Chandra Wardhana for their assistance in conducting research and the Muhammadiyah University of Yogyakarta which has provided funding for this research.

References

[1] Iqbal Z, and Mirakhor A 2012 Financial inclusion: Islamic finance perspective *Journal of Islamic Business and Management* **2(1)** 35-64.
[2] Harsanto P, Ikhsan J, Pujianto A, Hartono E, Fitriadin A A, Kuncoro A H B 2015 Sediment Disaster Characteristics on the Volcanic River. *Proceedings of the 5th National Seminar on Civil Engineering Seminar in 2015, ISSN: 2459-9727* pp. H 200-H 206. (in Indonesia)
[3] Mananoma T, Legono D, and Rahardjo A P, 2003 Natural Phenomena of Erosion and Sedimentation of Lower Progo River *Journal and Development of Water (Jurnal dan Pengembangan Keairan)* **10 (1)** pp. 1-70 (in Indonesia)
[4] Suwartha. N 2001 *Hydraulics Study of Sediment Transport Pattern in the Lower Progo River* (Thesis) Yogyakarta:Universitas Gadjah Mada Indonesia (in Indonesia)
[5] Ikhsan J, Fujita M, and Takebayashi H 2010 Sediment Disaster and Resource Management in the Mount Merapi Area, Indonesia *International Journal of Erosion Control Engineering* **3(1)** pp. 43-52
[6] Ikhsan J 2010 Study on Integrated Sediment Management in an Active Volcanic Basin (Dissertation) Kyoto University Japan (unpublished)
[7] Lavigne F, and Thouret J C 2002 Sediment transportation and deposition by rain triggered lahars at Merapi volcano, Central Java, Indonesia *Journal of Volcanology and Geothermal Research* Vol. **49** pp. 45-69.

Institute of Physics
Dirac House, Temple Back
Bristol BS1 6BE UK

ISSN: 1757-8981
ISBN 978-1-5108-9847-9

9 781510 898479

CURRAN ASSOCIATES INC.
proceedings.com